The Structural Basis of Muscular Contraction

The Structural Basis of Muscular Contraction

John Squire
Imperial College of Science and Technology
London, England

PLENUM PRESS • NEW YORK AND LONDON

Library of Congress Cataloging in Publication Data

Main entry under title:

The Structural basis of muscular contraction.

Bibliography: p.
Includes index.
1. Muscle contraction. 2. Ultrastructure (Biology) I. Squire, John, 1945-
[DNLM: 1. Muscle contraction. WE 500 S927]

QP321.S912 591.1'852 81-2321
ISBN 0-306-40582-2 AACR2

A Division of Plenum Publishing Corporation
233 Spring Street, New York, N.Y. 10013

Printed in the United States of America

Preface

Muscular contraction provides one of the most fascinating topics for a biophysicist to study. Although muscle comprises a molecular machine whereby chemical energy is converted to mechanical work, its action in producing force is something that is readily observable in everyday life, a feature that does not apply to most other structures of biophysical interest. In addition, muscle is so beautifully organized at the microscopic level that those important structural probes, electron microscopy (with the associated image analysis methods) and X-ray diffraction, have provided a wealth of information about the arrangements of the constituent proteins in a variety of muscle types. But, despite all this, the answer to the question "How does muscle work?" is still uncertain, especially with regard to the molecular events by which force is actually generated, and the question remains one of the major unsolved problems in biology.

With this problem in mind, this book has been written to collect together the available evidence on the structures of the muscle filaments and on their arrangements in different muscle cells, to extract the common structural features of these cells, and thus to attempt to define a possible series of mechanical steps that will describe at molecular resolution the process by which force is generated.

The book cannot be considered to be an introductory text; in fact, it presents a very detailed account of muscle structure as gleaned mainly from electron microscopy and X-ray diffraction. But neither is it written on the assumption that the reader is an expert on the structural methods used. An attempt has been made throughout to present the often complicated structural arguments in such a way that those not familiar with diffraction methods need not feel out of their depth. To promote this, Chapters 2 and 3 on techniques have not been written for the experts (for whom there are already several excellent mathematical presentations, e.g., Fraser and MacRae, 1973); rather they are presented as a liberally illustrated qualitative treatment that, it is hoped, those readers without any background in diffraction methods will find valuable.

One of the advantages of a structure such as muscle is that it is amenable to a great variety of experimental approaches (e.g., biochemical, physiological, and structural). But, in order for real progress to be made, researchers with quite different training need to meet on common ground. Each needs to understand at least the basic ideas behind the techniques and results of his colleagues. Chapters 2 and 3 are an attempt to allow physiologists and biochemists with no formal training in diffraction to find common ground with the structuralists. A little study and application of these chapters by such people and by students new to muscle research should make the rest of the book palatable and meaningful.

But the book is not written just for nonstructuralists. It is, indeed, written very much with structuralists in mind. Each chapter is a review of what is known about particular aspects of muscle structure at the time of writing (early 1980). But, as other structuralists will be well aware, there is a great deal of controversy in this subject. Where this is so, both sides of the argument are stated, but if there seems to be good reason for preferring one interpretation to another, then the reasons for the preference are given. So, in many cases where controversies exist, a definite point of view is clearly stated in the belief that the conclusions are the most reasonable deductions from the data at the present time. However, there are one or two areas (e.g., the contractile machinery of vertebrate smooth muscle) where there really is an enormous amount of uncertainty but in which there seems to be a common preference for a particular view without there being strong enough evidence on which to form an opinion. In such cases, a "devil's advocate" approach has been adopted in order to provoke discussion. Even so, there are bound to be areas in which the informed reader disagrees with what is said. Such things are healthy and help to improve our understanding. However, a sincere effort has been made to make sure that nowhere are the clear results of a particular piece of research in any way misrepresented.

The main object of the book is to provide stimulus for thought and so to further the progress of muscle research. An attempt has been made to separate what has actually been proved from what is commonly assumed to occur during muscular contraction. For this reason, it will be found that many new ideas are presented as are many new interpretations of data that have been available for some time. Although the reference list is very extensive, no attempt has been made to provide encyclopedic coverage of every paper that has ever been written on muscle structure. Rather, the book provides an account of the present state of the art, and it draws on those papers that have been most instrumental in establishing particular features of muscle. It is hoped that the book will therefore provide a useful handbook on muscle structure for the 1980s.

Finally, there are several people, in addition to those listed in the Acknowledgments section, to whom the Author owes a great debt of gratitude that will be very hard to repay. Expert opinions on various chapters have been given by Drs. Arthur Elliott, Gerald Offer, Pauline Bennett, John Kendrick-Jones, and Ed O'Brien, and their very valuable comments have much improved the text. Needless to say any errors that remain are the fault of the Author. I also acknowledge gladly the help of the members of my research group (the Biopolymer Group) here at Imperial College. These are Drs. Pradeep Luther and Alan Freundlich, Mr. Peter Munro, and Mrs. Elizabeth Jelinek. Apart from providing some of the figures used in the book, they have all helped in its preparation and thus eased the burden on the Author. But I would like to put in writing here my indebtedness to my own family: to my parents, who made all things possible and who have never failed in their encouragement; and to my wife, Melanie, and children, Deborah, Emily, Katherine, and Elizabeth, who have sustained me throughout and who have given up nearly four years of normal family life so that this book could be written. It is to them that the book is dedicated.

John M. Squire

London

Acknowledgments

I am happy to acknowledge here the contribution that many individuals and publishers have made to this book; without their help it could not have been produced.

Many individuals very readily sent copies of their original plates and diagrams for inclusion in the book. These include: Dr. Doreen Ashhurst, Dr. Carolyn Cohen, Dr. Roger Craig, Dr. Arthur Elliott, Dr. Marshall Elzinga, Dr. Bruce Fraser, Dr. Alan Freundlich, Dr. Margaret Ann Goldstein, Dr. John Haselgrove, Professor Sir Andrew Huxley, Dr. Aaron Klug, Dr. Jack Lowy, Dr. Pradeep Luther, Dr. Peter Munro, Dr. Gerald Offer, Dr. David Parry, Professor Frank Pepe, Professor Mike Reedy, Dr. Michael Sjöström, Dr. Vic Small, Dr. Apolinary Sobieszek, Professor Andrew Somlyo, Dr. Peter Vibert, and Dr. John Wray. My thanks are due to them for their generosity.

I am also indebted to the following publishers for their permission to reproduce copyrighted material here: Academic Press, Annual Reviews, Inc., Cold Spring Harbor Laboratory, Journal of Physiology, Longmans, Macmillan Journals, Ltd., North-Holland Press, Palo Alto Medical Research Foundation, Pergamon Press, Prentice Hall, Inc., Rockefeller University Press, Science, and The Royal Society.

All figures in the book are original work unless stated otherwise in the caption. Much of the original work reported here was part of the research of the Biopolymer Group at Imperial College which has been generously supported by Project Grants from the Science Research Council, the Medical Research Council, and also from the Muscular Dystrophy Association of America. I am pleased to be able to acknowledge the help of these grant-giving bodies in supporting the Biopolymer Group.

Finally, there are a few people who have helped me enormously by carrying out many of the time-consuming clerical aspects of producing a book such as this. In particular, I acknowledge with thanks the two typists, Mrs. Eileen Williams and Mrs. Shelagh Vaughan-Davies, who transformed my original hand-written notes into a presentable manuscript, and also Mrs. Elizabeth Jelinek, Mr. Jeff Harford, and my wife, Melanie, for their invaluable help in compiling reference lists and other related material.

JMS

Contents

1. Introduction

2. X-Ray Diffraction Methods in Muscle Research

4. Protein Conformation and Characterization

5. Thin Filament Structure and Regulation

6. Structure, Components, and Interactions of the Myosin Molecule

7. Vertebrate Skeletal Muscle

8. Comparative Ultrastructures of Diverse Muscle Types

9. Molecular Packing in Myosin-Containing Filaments

10. Structural Evidence on the Contractile Event

11. Discussion: Modeling the Contractile Event

1

Introduction

1.1. Introduction: Muscles and Movement

When you picked up this book, did it occur to you that it was those obedient servants, your muscles, that actually made your action possible? Now, as you read this page, your eyes are made to focus and scan carefully along each line of print by means of delicately balanced muscular forces. All day, every day, the muscles in our bodies are continually contracting and relaxing as we carry out those everyday actions that are normally taken for granted.

It is one of the basic characteristics of animals that they can move, and this movement can often be a thing of great beauty. Think for a moment of the flight of birds and insects, of the graceful swimming of fish, and of man's delight in music making, dancing, and sport. All of these are possible because nature has evolved a sophisticated type of machine, the muscle, which can exert controlled levels of force and which can cause movement at controlled speeds. The motile machinery in man and other animals is therefore a fascinating topic for study.

Muscles are molecular machines that convert chemical energy, initially derived from food, into mechanical force. The molecular mechanism by which this conversion process occurs is the subject of a large part of the current research into the molecular biophysics of muscle, since it remains one of the major unsolved problems in biology. It is also the principal topic covered by this book, which is an attempt to gather together the available structural evidence on different muscles so that a possible contractile mechanism can be identified.

1.2. Classification of Muscle Types

The muscles in different animals are by no means all identical in structure, even though they now appear to generate force by the same basic mechanism. Variations on the same theme have evolved in such a way that the muscles in different animals or in different parts of the same animal are specialized for particular functions ranging from the rapid driving of the wings of insects to the prolonged closure of the shells of clams. The common features and the specializations of different muscles will be described in Chapters 5 to 9 of this book, but here the method of classifying these different muscles is briefly outlined.

Vertebrates, of which humans, rabbits, rats, chickens, and frogs are the principal examples used in muscle research, contain two main classes of muscle (Fig. 1.1) which were originally distinguished by their appearance in the light microscope. Those muscles that act directly on the bones and therefore cause movement of the skeleton (the skeletal muscles) appear cross-striated in the light microscope and are under voluntary control. The muscles in the vertebrate intestines, veins, and arteries do not show such a striated appearance; they are termed smooth muscles and tend to be involuntary muscles in the sense that their contractile behavior is not regulated primarily by the nervous input from the brain. Heart (cardiac) muscle is rather a special case since, like the smooth muscles, it is not under voluntary control, but in the light microscope it appears cross-striated like the skeletal muscles.

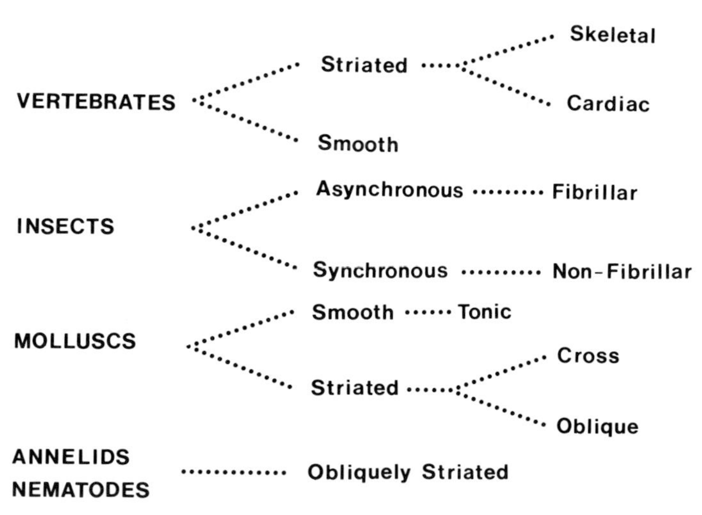

FIGURE 1.1. Classification of the major muscle types. The list is not exhaustive but includes the main types used in muscle research and hence, the main types described in this book.

It is common knowledge that if a muscle from, say, the leg of a frog is stimulated electrically it will shorten or contract (Fig. 1.2). In fact, this was known as long ago as the 18th Century. Luigi Galvani (1737–1798) was the first to demonstrate the phenomenon, and Allesandro Volta (1745–1827) later explained it in terms of electrical stimulation. Muscles that exhibit this kind of single twitch response to a single electrical stimulus (i.e., a rapid generation of tension followed by complete relaxation; Fig. 1.2) are commonly called synchronous muscles. All vertebrate skeletal muscles are of this kind.

Although all of the muscles in insects are striated, some of them do not show the synchronous kind of response found in vertebrate skeletal muscles. The asynchronous insect muscles, mostly the insect flight muscles, when stimulated electrically and linked to a suitable resonant mechanical oscillator (such as the wing/thorax assembly in the insect), go through an alternating (oscillating) cycle of contraction and relaxation. Even when stimulated repetitively, the frequency of oscillation is generally different from and faster than the frequency of stimulation. In the insect flight muscles, this clearly permits the insect's wings to beat extremely rapidly, sometimes up to 1000 times per second, and this can be a clear advantage in flight.

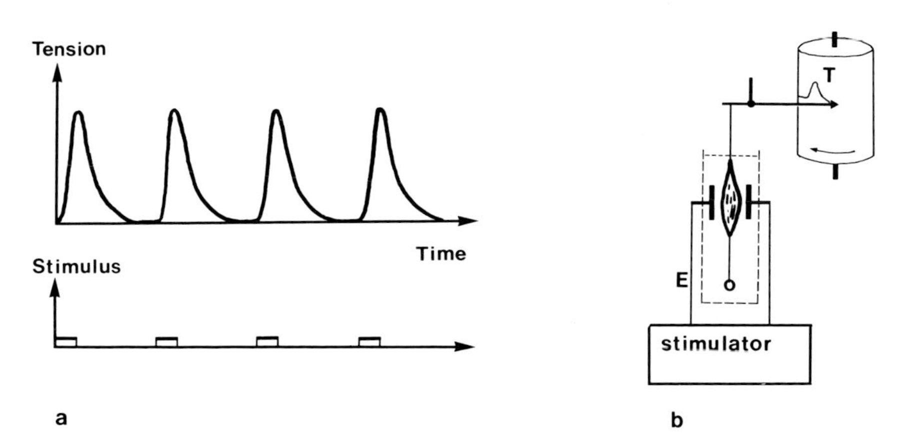

FIGURE 1.2. (a) A twitch response in a fast vertebrate skeletal muscle. Each time the muscle is stimulated by an electric shock, the tension in the muscle increases rapidly to a peak and then relaxes. One such twitch will last about 0.1 to 1 sec from beginning to end depending on conditions. (b) Simple experimental setup for stimulating a muscle and recording its tension. The muscle is in a chamber containing a suitable bathing medium (see Chapter 3). The chamber contains electrodes for stimulating the muscle and a connection from the muscle to a tension recording device (in this case a lever supported on a torsional wire mounting). The tension measurement is recorded on a suitable device, in this case a pointer on the lever tip that traces the lever movements on a moving chart.

However, not all insect muscles oscillate when activated. Some, such as the leg muscles and some flight muscles, have a twitch response and are therefore called synchronous muscles. These two muscle types can also be distinguished by their internal structure, and asynchronous muscles are known as fibrillar muscles, whereas synchronous muscles are known as nonfibrillar.

The muscles that occur in molluscs (e.g., oysters, mussels) are also of two main kinds. A few molluscan muscles are striated, but most of them are smooth muscles, rather different from vertebrate smooth muscles in their structure and properties, but smooth in appearance nonetheless. A unique property of many of the smooth muscles in these bivalves is the ability to exhibit the "catch" or tonic response. When this occurs, the muscle generates and maintains tension over a considerable period of time (sometimes many hours) with comparatively little expenditure of energy. It is this kind of response that enables the animals to hold their shells closed for long periods. A variation on this theme occurs in the scallop (*Pecten*) where part of one muscle (the adductor) holds the shells together while another part is used for swimming.

The last main distinct class of muscles to be considered in this book is obliquely striated muscle. Such muscles commonly occur in annelids and nematodes (earthworms, leeches, etc.), but some of the striated molluscan muscles are also of this kind. As the name implies, the striations that occur in this muscle type are not perpendicular to the length of the muscle (as in the other striated muscles) but are angled. In fact, the angle of obliquity actually changes with muscle length as described in Chapter 8.

The classification of these different muscles is summarized in Fig. 1.1. The structure and molecular arrangements in each muscle type will be described in Chapters 5 to 9.

1.3. Vertebrate Skeletal Muscle

1.3.1. Introduction

Much of the initial ultrastructural work on muscle was carried out using vertebrate skeletal muscle. In particular, one of the leg muscles (sartorius) of the frog and the back muscle (psoas) of the rabbit were commonly used (Fig. 1.3). Muscles of this type possess a considerable degree of structural regularity, and some consist of a parallel bundle of cylindrical muscle cells or fibers that run the whole length of the muscle. This relatively simple organization makes them ideally suited to the main types of structural technique available to the biophysicist: X-ray diffrac-

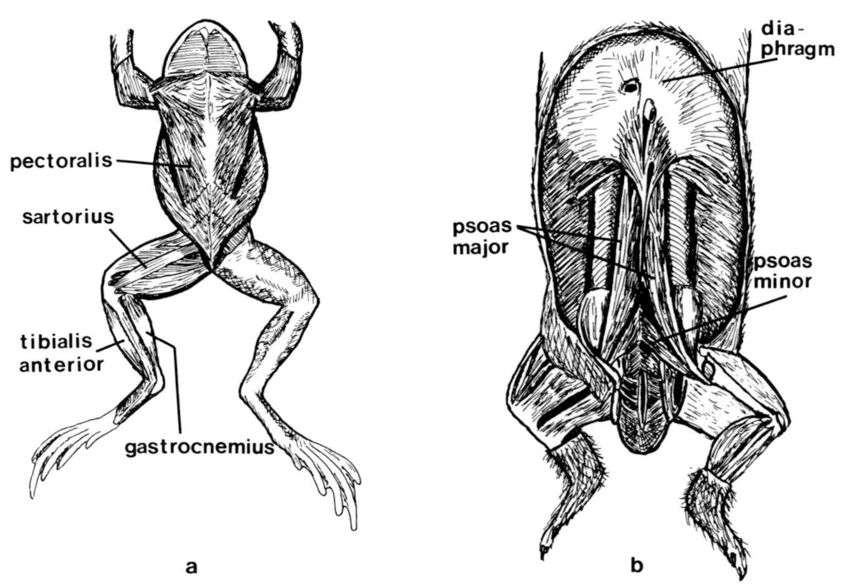

FIGURE 1.3. (a) Diagram showing the anatomical locations of the sartorius and gastrocnemius muscles of the frog. (b) Position of the psoas muscle in the guinea pig (ventral view with the gut removed). The rabbit is similar.

tion and electron microscopy. In addition, their almost uniform cross section permits calculation of such things as force per unit area to be carried out relatively easily.

In this book, vertebrate skeletal muscle will be used as a model to illustrate some of the basic features of muscle structure and of muscular contraction and will serve as a standard against which the structures and properties of other myscle types can be measured. The present section gives a brief introductory description of the structure of vertebrate skeletal muscle together with an account of some of the current ideas on the mechanism of force generation. This will provide a framework within which the structures of specific muscle components can be considered in detail.

1.3.2. The Sarcomere

Figure 1.4 illustrates the structure of a whole vertebrate skeletal muscle. It is made up of many cells or fibers, often about 20 to 100 μm across, each of which comprises a bundle of cylindrical myofibrils about 1 to 2 μm in diameter. The fibers in the muscle are linked together by collagenous connective tissue, and at the muscle ends this forms the tendons between the muscle and the skeleton on which it acts. Within

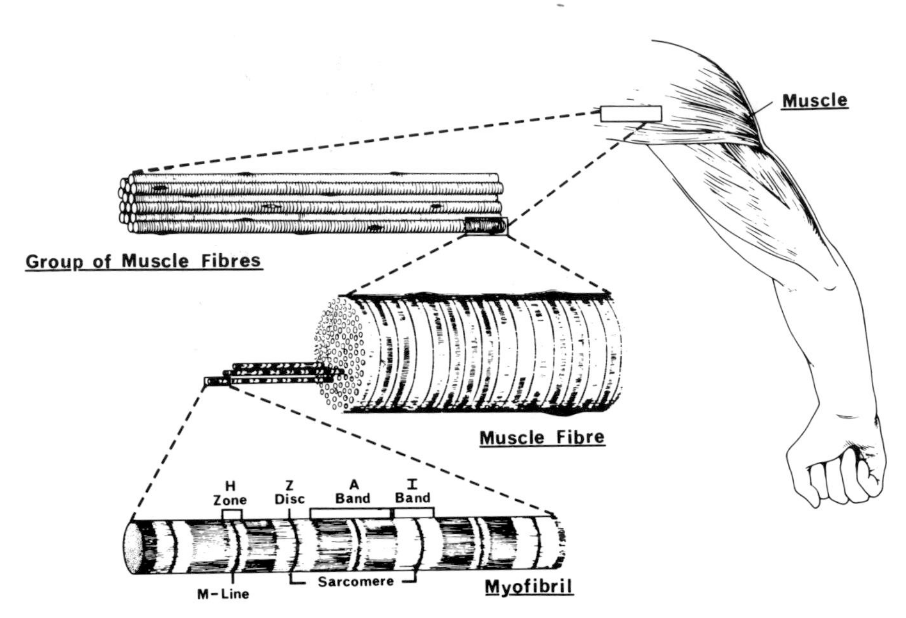

FIGURE 1.4. Levels of structural organization in a typical vertebrate skeletal muscle. The muscle comprises long, cylindrical muscle fibers (about 100 μm across), each of which is a single multinucleate muscle cell. Each fiber contains a bundle of parallel cylindrical myofibrils (about 2 to 3 μm across) mainly composed of the contractile material. This material is organized into repeating units, the sarcomeres, which are commonly about 2 to 3 μm long. This gives the myofibrils, the fibers, and the whole muscle their characteristic cross-striated appearance. The positions of the A-band, I-band, Z-band, and H-zone are also shown (see text). (From W. Bloom and D. W. Fawcett, *A Textbook of Histology*, 10th ed., Philadelphia, W. B. Saunders Co., 1975.)

each myofibril, the contractile material is organized into repeating units called sarcomeres. These can be seen in light micrographs of myofibrils as an alternating pattern of light and dark bands (Fig. 1.5). The dark band (when viewed in the light microscope as shown in Fig. 1.5) is termed the A-band (for Anisotropic band), and the light band is the I-band (for Isotropic band). Since these bands tend to be in good axial register in adjacent myofibrils, they give the fibers, and indeed the whole muscle, their characteristic cross-striated appearance.

It is also possible in the light microscope to see a dark line in the center of the I-band (Fig. 1.5). This was originally called the Z-line (from the name Zwischensheibe), but the terms Z-disk or Z-band are now often preferred because of the band's finite thickness. The sarcomere is defined as the repeating unit between one Z-band and the next Z-band along the myofibril. In vertebrate muscles the separation between adjacent Z-bands (the sarcomere length) is commonly about 2.0 to 3.0 μm.

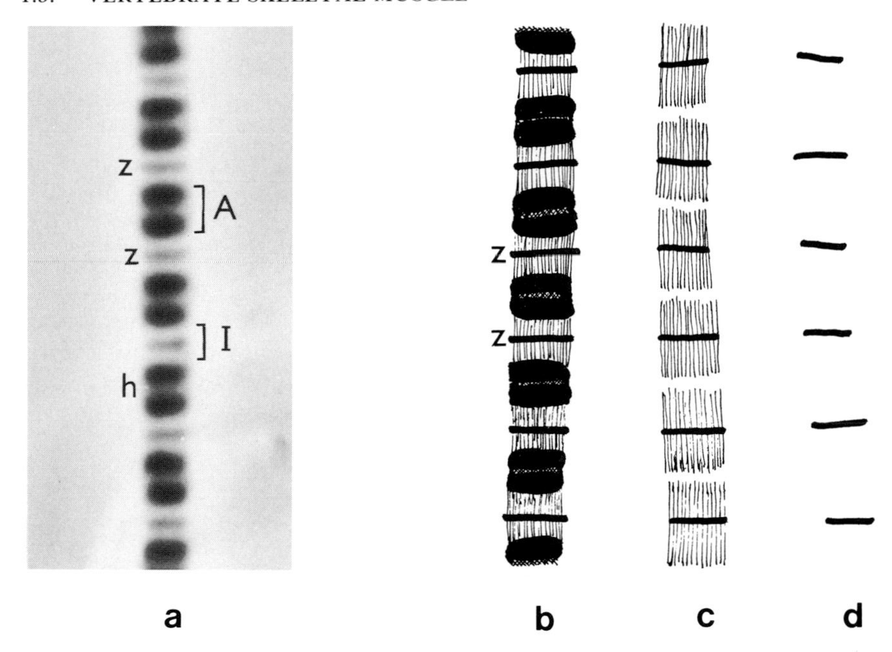

FIGURE 1.5. (a) Light micrograph of a myofibril showing the alternating dark (A-) and light (I-) bands. In the middle of the I-band is a dense line called the Z-band (z) and in the middle of the A-band is the H-zone (h). (Courtesy of R. Craig.) Figures (b), (c), and (d) illustrate diagrammatically the appearance in the phase contrast microscope of a stretched myofibril from glycerol-extraced rabbit psoas muscle after various treatments. (b) The original myofibril as in (a); (c) has had the myosin extracted; (d) has had actin extracted as well. (After Hanson and Huxley, 1955.)

When a muscle shortens during a contraction, the sarcomere length reduces by a proportionate amount, indicating that the sarcomere is the unit in the muscle where active shortening is produced. Another important feature of the myofibril shown in Fig. 1.5 is that half way along each A-band, there is a less dark region known as the H-zone. Its origin will become clear in the next section.

Apart from a large amount of water and a small amount of soluble protein, the myofibril is composed largely of the proteins myosin and actin as shown in a series of extraction experiments by Hanson and Huxley (1953) and Hasselbach (1953). They were able to demonstrate that the protein in the A-band consists mainly of myosin together with some actin and that the I-band consists mainly of the protein actin with no myosin. To show this, they made use of a preparation of myofibrils separated from muscle that had previously been soaked in glycerol. Glycerol tends to disrupt the muscle membrane (sarcolemma) and then dissolves out most of the soluble proteins in the cytoplasm of the muscle cell (sarcoplasm), leaving only the contractile apparatus itself. It had previously been possible to extract myosin from homogenized (minced) muscle using a solution of 0.6 M KCl and a little pyrophosphate and magnesium chloride which relax a muscle (see Chapter 3, Section 3.2.3 or Table 3.2). When such a solution was applied to the isolated myofibrils, the dark A-band was found to be extracted when viewed in the light microscope (Fig. 1.5c). A similar treatment with 0.6 M KI, known to remove actin from minced muscle, removed all of the I-band except the Z-band (Fig. 1.5d). An intriguing additional observation by Hanson and Huxley was that myofibrils extracted so that only Z-bands appeared to be left still showed mechanical continuity; it appears that some remaining structure mechanically links the Z-bands, an idea that we shall return to in Chapter 7.

1.3.3. The Sliding Filament Model

Before about 1953, a multitude of contraction theories had been proposed (see Chapter 11), but it was widely supposed that shortening of the sarcomere was brought about by some kind of folding up or coiling of a continuous structure containing both actin and myosin (actomyosin) and extending from one Z-band to the next. One could imagine, for example, a helical structure containing actin and myosin molecules that shortened on activation by reducing the pitch of the helix, just as successive turns in a coiled spring come closer when the spring is compressed.

However, this type of mechanism was disproved by a series of experiments reported in 1953 to 1956. These are summarized briefly here.

1. Myosin is confined to the A-bands (as described above; Hasselbach, 1953; Hanson and Huxley, 1953).

2. The length of the A-band, about 1.6 μm, does not alter when a muscle is stretched (A. F. Huxley and Niedergerke, 1954; H. E. Huxley and Hanson, 1954), but the I-band and the H-zone both lengthen. These two sets of authors studied, respectively, whole fibers (immersed in a medium of refractive index that matched the average refractive index of the fibers to eliminate artifacts) and isolated myofibrils. Both types of preparation were studied at different sarcomere lengths and during contraction in the interference or phase contrast microscope. The results were taken to imply that contraction involved the relative sliding of two overlapping sets of protein filaments.

3. Thin sections of striated muscle viewed in the electron microscope (H. E. Huxley, 1956) subsequently showed directly that the sarcomere does consist of two interdigitating sets of filaments (Fig. 1.6). Thick filaments of myosin occurred in the A-bands, and thin filaments of actin occurred in the I-bands, and these two sets of filaments overlapped at

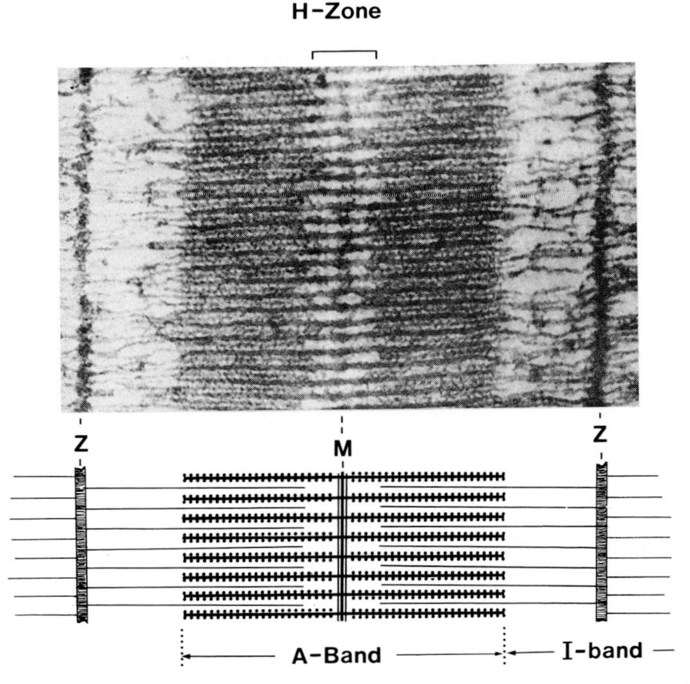

FIGURE 1.6. High-power electron micrograph of a longitudinal section of rabbit psoas muscle showing that the A- and I-bands consist of separate arrays of interdigitating filaments as illustrated in the lower diagram. The A-band comprises thick myosin filaments covered with projections together with interdigitating thin filaments, and the I-band is formed by thin actin filaments alone. The overlap region between these two sets of filaments can be seen clearly as can the H-zone which is the part of the A-band not overlapped by actin. At the center of the A-band is a structure called the M-band which cross-links the myosin filaments. (Micrograph courtesy of H. E. Huxley.)

the ends of the A-band by an amount that varied linearly with sarcomere length.

Taken together with a result of Astbury (1947) who had found that the high-angle X-ray diffraction pattern from active muscle was little different from the pattern from relaxed muscle, these results suggested strongly that the two interdigitating filament arrays must slide past each other when a muscle shortens (Fig. 1.7) and that the filaments themselves do not change significantly in length. This fundamentally important concept (A. F. Huxley and Niedergerke, 1954; H. E. Huxley and Hanson, 1954), on which current models about contraction are largely based, was termed "the sliding filament model of contraction." It was clearly not a theory of contraction in the sense that it described at the molecular level the mechanism of force generation. What it did show was

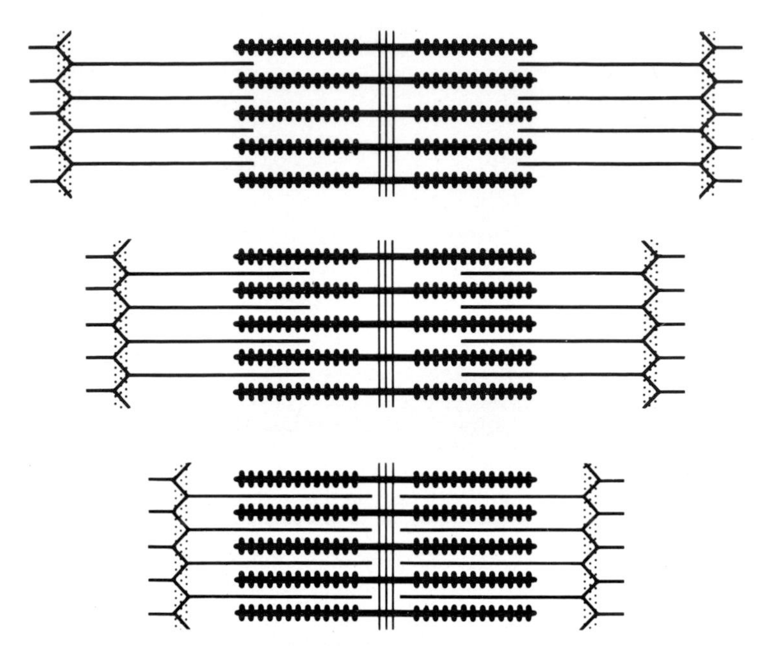

FIGURE 1.7. Diagram to show the change in overlap of the thick and thin filaments with changing sarcomere length. The A-band length remains constant, but the H-zone and I-band lengths reduce as the sarcomere becomes shorter. (After Offer, 1974.)

the way in which the relative sliding of the actin and myosin filaments in a sarcomere could give rise to an overall shortening of the whole muscle.

The crucial question facing the muscle biologist therefore changed in 1953–1954 from the macroscopic "How does a muscle shorten?" to the microscopic "How is the sliding force between the two sets of filaments generated?" This central problem in muscle research will be considered in Chapters 10 and 11. However, some of the current views are aired briefly in the next section to provide a background for the structural descriptions of different muscles given in Chapters 5 to 9. A point of nomenclature should be mentioned here. This is that the term "contraction" is used in a special sense by the muscle biologist. A muscle is said to contract if it is stimulated and produces tension. A muscle can therefore contract in this sense without changing its length, in which case the contraction is said to be *isometric.* A contraction in which a muscle shortens against a fixed load is said to be *isotonic.* These two types of contraction just happen to be convenient in experimental work, but many other contraction conditions are possible, and *in vivo,* the load will often change continuously.

1.3.4. Force Generation

A number of alternative theories of force generation have been proposed (see Chapter 11 and A. F. Huxley, 1974), but one theory in particular has gained wide acceptance. This present-day view of the nature of the force-generating mechanism was originally based on a variety of observations by H. E. Huxley, A. F. Huxley, and others, and these are summarized briefly below. They will be described in detail in Chapters 7, 10, and 11.

1. Electron micrographs of longitudinal sections (i.e., sections cut in a direction parallel to the muscle length) of striated muscle (Fig. 1.6) show that there are projections or knobs on the surface of the thick myosin-containing filaments in the A-band. These occur everywhere except in a short region (the bare zone) in the middle of the filament (H. E. Huxley, 1957).

2. The myosin projections tend to form cross-links (or cross bridges) to the thinner, actin-containing filaments in certain types of muscle preparations (H. E. Huxley, 1957).

3. Theories that assume that the actin and myosin filaments interact through a series of independently acting force generators can readily explain the different observations on the mechanical performance of vertebrate skeletal muscles (A. F. Huxley, 1957). In particular, the existence of independent force generators would explain very well the form of the length–tension curve of striated muscle. This is the graph of the isometric tension (T) which the muscle can actively generate (when stimulated to give a maximal response) as a function of the sarcomere length (S) at which the tension is measured. The form of this curve is shown in Fig. 1.8 and was determined initially by Ramsey and Street (1940). It was repeated with considerably greater accuracy by Gordon *et al.* (1966). Gordon *et al.* showed that the tension could be correlated satisfactorily with the amount of overlap that occurred between the actin and myosin filaments. For example, very little tension occurred at a sarcomere length (S) of about 3.6 μm where no overlap occurred (S is A-band length, about 1.6 μm, plus twice the thin filament length of about 1 μm plus the Z-band thickness). The tension then increased linearly to a maximum value as the sarcomere length was reduced down to about $S = 2.1$ μm. At this point, the actin filaments had been drawn into the A-band right up to the edge of the bare zone of the myosin filaments. The tension developed in each contraction then remained the same as the bare zone was traversed by the actin filaments and then decreased again below a sarcomere length of about $S = 2.0$ μm where the actin filaments in the two halves of the sarcomere began to

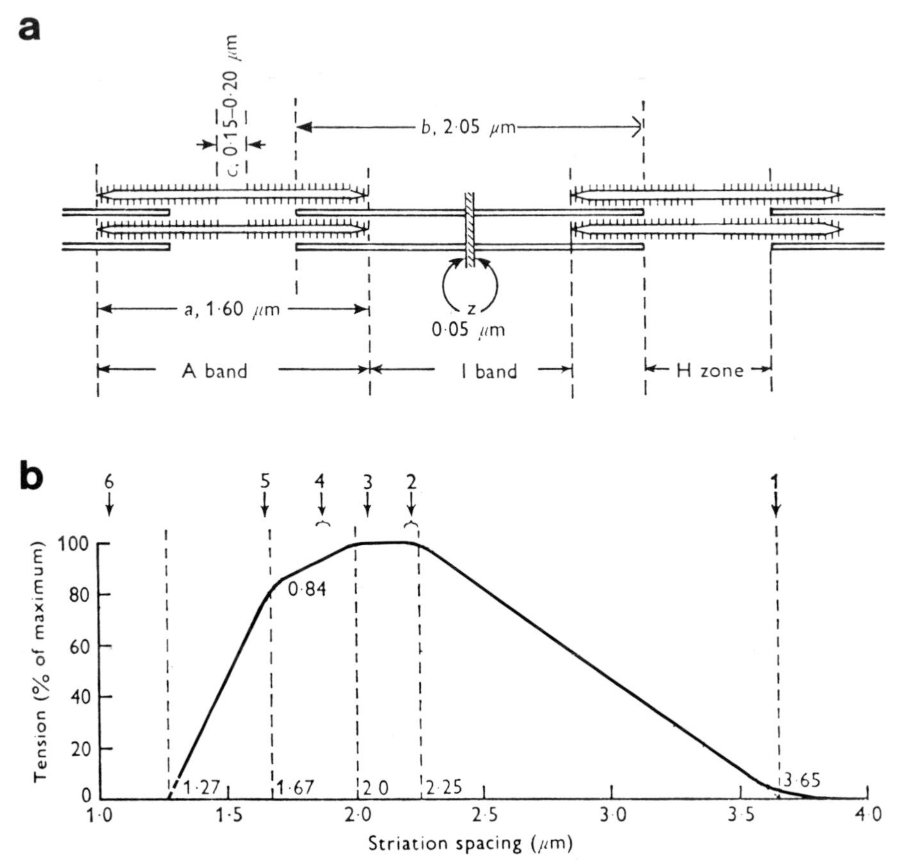

FIGURE 1.8. The alteration of tension actively generated by the contractile apparatus as a function of sarcomere length (b) and its interpretation in terms of the changing overlap of the thick and thin filaments (a and c). For details see text. (From Gordon *et al.*, 1966.)

overlap and interfere with each other. A further reduction in tension occurred when the myosin filaments came up against the Z-band at about $S = 1.6\ \mu m$. All of these stages in the length–tension curve can be seen clearly in Fig. 1.8 which is from Gordon *et al.* (1966). The important part of this curve *in vivo* is between sarcomere lengths of just less than 2 μm and about 3 μm, and clearly, in this range, the maximum isometric tension is related almost linearly to the amount of filament overlap.

All of these results were clearly consistent with the idea that the projections on the myosin filaments are involved in the generation of force and that they act independently of each other. An increasing amount of filament overlap would then correspond to the interaction with actin of an increasing number of cross bridges and hence to a greater tension.

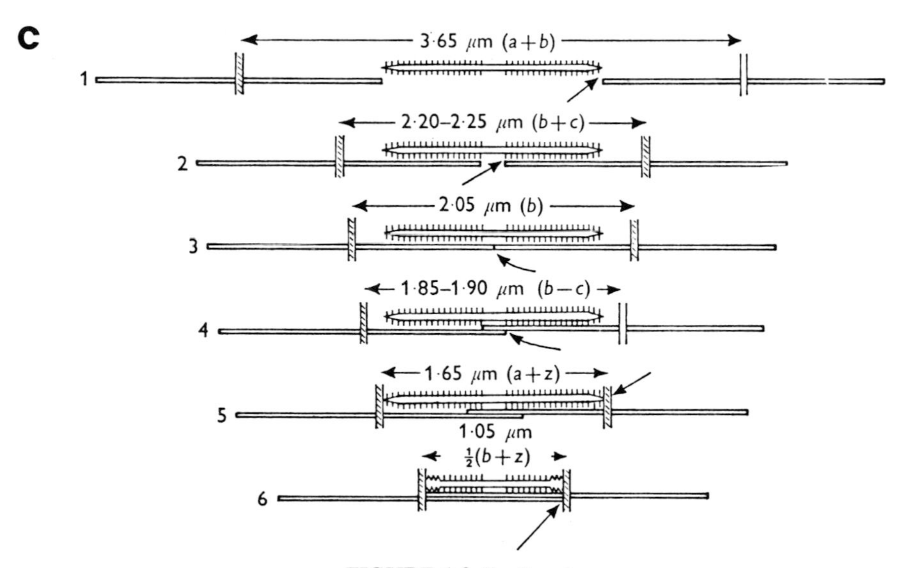

FIGURE 1.8 *Continued*

It was shown by Rice (1961a,b, 1964) and later by Zobel and Carlson (1963) and H. E. Huxley (1963) that the myosin molecules that aggregate to form the myosin filaments are match-shaped and consist of a long cylindrical tail about 1500 Å long on the end of which is a large globular region (Fig. 1.9). The form and properties of these molecules will be considered fully in Chapter 6. Suffice it to say here that the globular region is now known to comprise two large globular components (heads) together with some smaller protein chains. Myosin molecules possess the ability to split (hydrolyze) adenosine triphosphate (ATP) (Fig. 1.10) into adenosine diphosphate (ADP) and inorganic phosphate (P). We can write

$$\mathrm{ATP} \leftrightharpoons \mathrm{ADP} + \mathrm{P}$$

a reaction that is illustrated in Fig. 1.10. The ATP molecule, which is synthesised from ADP and P using the free energy of oxidation of glucose and fatty acids, is the immediate source of chemical energy in muscle. When ATP is split, a considerable amount of chemical energy is released, and this is converted into mechanical energy by the force generators mentioned earlier. The ATP-splitting (or ATPase) activity of myosin actually resides in the two globular heads of the myosin molecule. This part of the myosin molecule is also soluble under physiological conditions, whereas the bulk of the tail part of the myosin molecule is insoluble. It was therefore suggested that the myosin tails

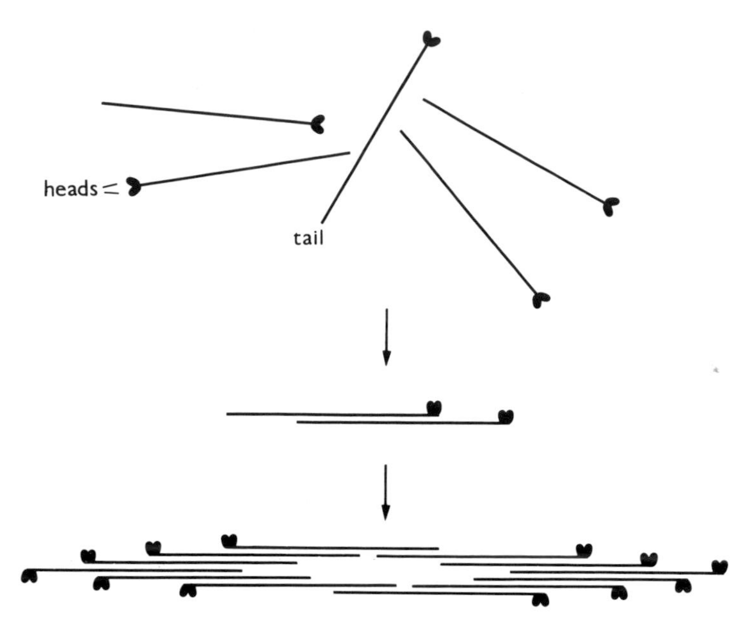

FIGURE 1.9. Schematic diagrams showing the form of one myosin molecule (a tail on the end of which are two globular heads) and the way such molecules are thought to aggregate to form bipolar myosin filaments with a central bare zone. (From Offer, 1974, after Huxley, 1963.)

aggregate to form the backbones of the myosin filaments and that the globular heads form the projections or cross bridges (H. E. Huxley, 1963). It was also suggested that the molecules are packed in an antiparallel fashion near the middle of the filament (Fig. 1.9), thus giving rise to the filament bare zone, and that elsewhere the molecules are packed parallel to form the two ends of the filaments.

Since the myosin crossbridges possess ATPase activity, and since this activity is enhanced by the presence of actin (Eisenberg and Moos, 1968, 1970), it appeared to be very likely that these heads must be intimately involved in the machinery for converting chemical energy into mechanical work. But the mechanism involved is only poorly understood and is the subject of current work by many muscle biophysicists. However, a number of working models have been suggested of which the most notable is probably that of H. E. Huxley (1969). He suggested that a cross bridge cycle could occur in which a cross bridge would first attach to an actin filament in a particular orientation. The release of chemical energy that accompanies the release of the products ADP and P would then result in a change in orientation of the actin-attached cross bridge, and this would result directly in the relative sliding of the actin and myosin filaments. The cross bridge would then detach from the actin filament and prepare itself for another attachment–detachment cycle. Such a

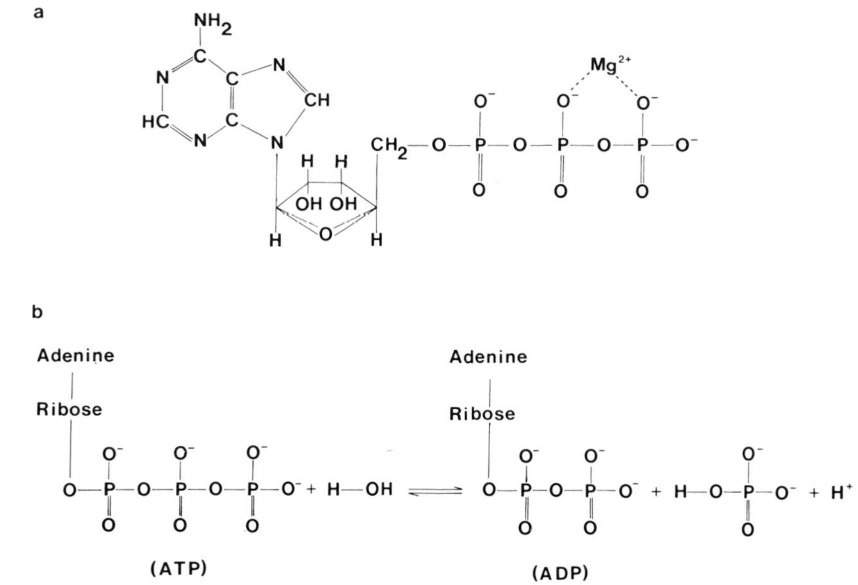

FIGURE 1.10. (a) Structure of the ATP molecule showing the adenine group, the ribose ring, and the three terminal phosphate groups. Also indicated is the position of the chelated magnesium ion on the last two phosphates. This is the normal Mg-ATP form of ATP found in muscle. (b) Structural representation of the ATP hydrolysis reaction in which ADP and inorganic phosphate are the products.

scheme is illustrated in Fig. 1.11 which also shows a simplified cycle of the biochemical steps that could correspond to the mechanical steps mentioned above (Lymn and Taylor, 1970,1971). It has now been shown (Eisenberg and Moos, 1968,1970) that a myosin head attached to actin will remain attached (in state AM) until it binds a molecule of ATP (Step 1 in Fig. 1.11). As mentioned earlier, when a muscle is soaked in glycerol, the soluble components of the sarcoplasm including the ATP are extracted. This means that in glycerinated muscle, the cross bridges attached to actin cannot bind ATP and must, therefore, remain attached to actin. For obvious reasons, this permanent attachment renders the muscle stiff. Such a glycerol-extracted muscle is said to be *in rigor.* This is by analogy with rigor mortis which is itself a rigid muscle state induced by the depletion of the ATP supply in the body. It has been found that in glycerinated muscle the myosin cross bridges are indeed angled to the actin filaments as in the state AM in Fig. 1.11 (see Chapters 5 and 10). Unfortunately, although this rigor configuration probably is one of the stages in the cross bridge cycle in active muscle, little is known conclusively about the other steps in the cycle, especially step 4 in which force is believed to be generated. The available evidence on the various steps in the cross bridge cycle will be assessed in Chapters 10 and 11.

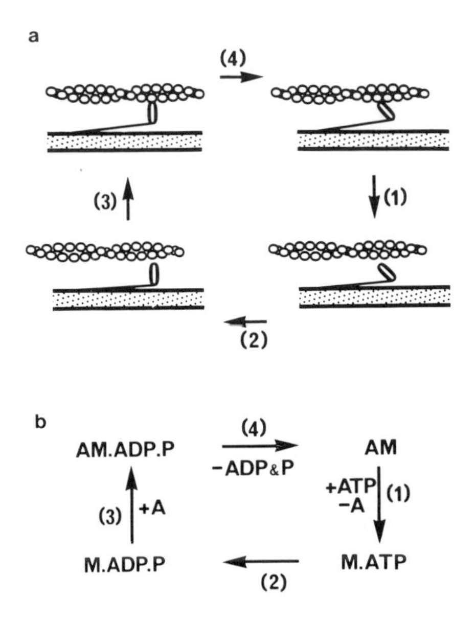

FIGURE 1.11. Comparison of the mechanical steps thought to be involved in the force-generating cross bridge cycle in muscle (a) and the corresponding biochemical steps (b). In step 3, a cross bridge attaches to the adjacent actin filament; in step 4, it changes its angle of attachment, causing relative sliding of the actin and myosin filaments (ADP and P are released); in step 1, the cross bridge detaches after binding ATP; and in step 2, it reverts to the configuration in which it can restart its attachment cycle. (From Offer, 1974, after E. W. Taylor.)

1.4. Introduction to Muscle Physiology

1.4.1. Contractile Response in Vertebrate Skeletal Muscles

It was shown in Fig. 1.2 that the effect of a brief electrical stimulus on a rapidly contracting vertebrate skeletal muscle (such as the frog sartorius) is that the muscle contracts and produces tension. In fact, the form of the contractile response depends to a large extent on the form of the applied stimulus. If a muscle is held at constant length and is stimulated by short, well-separated electrical pulses, then a series of single twitches is obtained in which the tension increases over a period of up to 200 msec (the exact time scale depending on the particular muscle involved and on its temperature) and then decays again back to the resting level (Fig. 1.12a). But if the time interval between pulses is reduced, a situation is reached in which the muscle is restimulated before it has completely relaxed after the previous stimulus. In this case, the tension gradually builds up (Fig. 1.12b) to give a bumpy tension trace at a tension level that is generally higher than the peak tension in a single twitch. Eventually, when the stimulation frequency is sufficiently high, the individual twitches fuse together to give a *tetanus* (Fig. 1.12b) in which tension rises steadily to a level plateau and then remains at that level for a relatively long time. The frequency at which this occurs is termed the fusion frequency (it is sometimes rather uncertain). The peak tetanic tension of a previously well-rested muscle is the maximum tension that the muscle can generate.

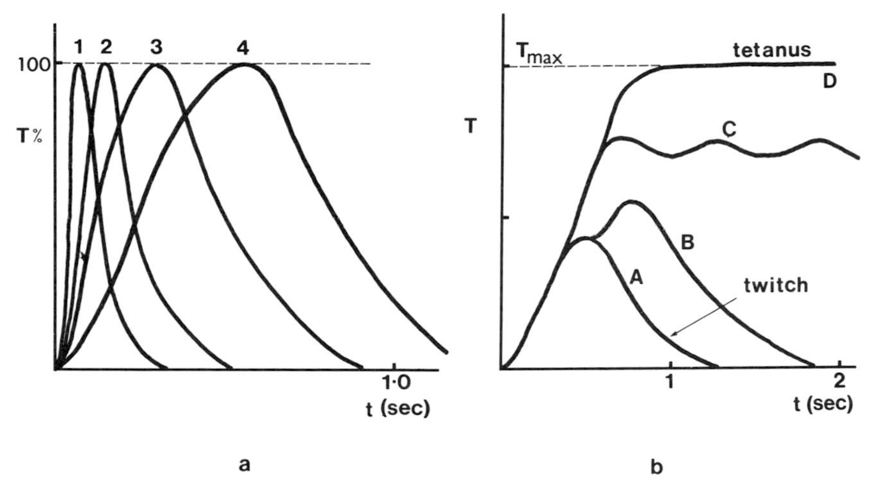

FIGURE 1.12. (a) Different time scales of the twitch responses in (1) a mammal gastrocnemius muscle, (2) a mammal soleus muscle, and a frog sartorius muscle at 10°C (3) and at 0°C (4). (b) Variation of tension response in a twitch muscle with different stimuli: (A) response to a single shock; (B) response to two closely spaced shocks; (C) unfused tetanus produced by repetitive stimulation at moderate frequencies; (D) tetanus induced by repetitive shocks arriving faster than the fusion frequency. (After Wilkie, 1976.)

Note that in practice the measurement of isometric tension can be achieved in two main ways. One is to use an isometric lever (Fig. 1.13a), and the other is to use a mechanoelectrical tension transducer (strain gauge; Fig. 1.13b). In either case, the tension can only manifest itself as a slight movement (of the isometric lever or of the transducer arm), and for this reason, the system can never be truly isometric. But stiff transducers allow quite large tensions to be measured without at the same time permitting appreciable shortening of the muscle, and they are therefore commonly used now. Such transducers usually involve the alteration, by slight movements of the transducer arm, of the electrical properties either of a triode, in which case the arm is linked directly to the anode, or of a silicon strain gauge that forms parts of a resistive circuit. The resulting electrical signals are then recorded on an oscilloscope or chart recorder. Since these transducers have relatively little inertia, they can follow very rapid changes in tension.

If the maximum tetanic tension of a muscle is measured as a function of muscle length, then a tension variation of the form shown in Fig. 1.14a is obtained. Carrying out a similar series of tension measurements on an unstimulated muscle gives the passive tension curve shown dotted in Fig. 1.14a. The difference between these two curves represents the tension actively generated by the contractile apparatus and is the type of curve shown in Fig. 1.8 as a function of sarcomere length. It is known as

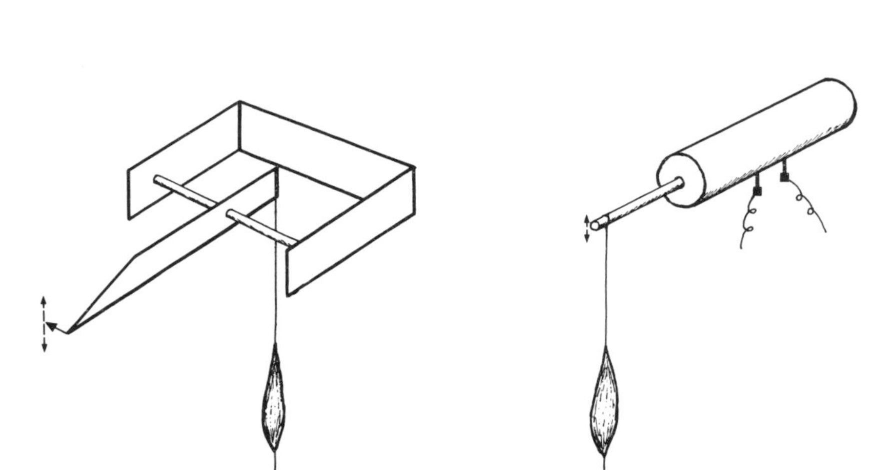

FIGURE 1.13. Two methods of recording tension: (a) an isometric lever supported on a torsion wire (see Fig. 1.2); (b) a stiff mechanoelectrical force transducer. See text for details.

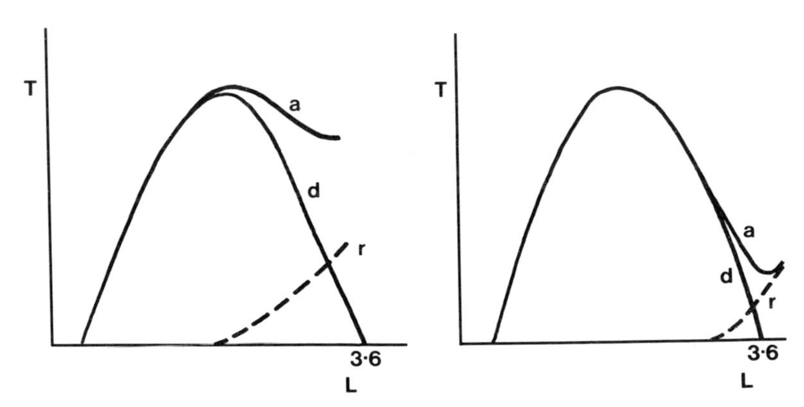

FIGURE 1.14. The comparative mechanical properties of frog sartorius muscle (left) and frog semitendinosus muscle (right) as a function of muscle length. In each case are shown the resting tension(r), the total active tension curve (a) and the difference, the active increment curve (d), which represents the tension actively generated by the contractile apparatus as in Fig. 1.8. The position marked 3.6 represents the sarcomere length in micrometers at which the actin and myosin filaments are just not overlapping. (After Wilkie, 1976.)

the active increment curve or active length–tension curve. [Note that Fig. 1.8 is more detailed than Fig. 1.14, since Gordon *et al.* (1966) used a very refined experimental protocol involving single fibers of which the sarcomere length was monitored accurately using a device termed a spot-follower.] The resting (passive) tension curve results from some form of elastic component in the muscle, partly in the form of connective tissue which is largely made up of collagen and partly from the sarcolemma. The passive tension curve can be very different in different muscle types. For example, frog sartorius muscle cannot be stretched to sarcomere lengths larger than about 3.0 μm without producing serious structural damage (Fig. 1.14a), but frog semitendinosus muscle can be stretched to very long sarcomere lengths (Fig. 1.14b), even to the point where the myosin and actin filaments no longer overlap each other, without any apparent permanent ill effects.

1.4.2. Comparative Innervation and Response in Different Muscles

The Action Potential. One of the most fundamental requirements of a muscle system *in vivo* is that its contractile behavior can be controlled according to the particular role of the muscle in the body. Different muscles are clearly required to respond in different ways. For example, heart muscles are required to contract rhythmically and continuously in order to maintain life, whereas skeletal muscles are required to respond rapidly and by a controlled amount only when the appropriate signal from the brain is received. We have already seen that electrical stimulation will cause a skeletal muscle to contract. In the body, the corresponding effect is usually produced via the nervous system. When the brain initiates the signal for a particular muscle to contract, this signal is passed down the appropriate motor nerves in the form of a propagated electrical depolarization of the nerve membrane (see Katz, 1966; Aidley, 1971). When this electrical impulse (action potential) passes down a small branch of the motor nerve and arrives at the neuromuscular junction (Fig. 1.15), the relatively small signal carried by the nerve is "amplified" by the motor *end plate* and this in turn triggers the release of acetylcholine. In a muscle fiber at rest, a particular distribution of ions exists on each side of the fiber membrane as governed by the Donnan equilibrium and the Nernst equation (see Katz, 1966; Aidley, 1971). In brief, the equilibrium situation is one in which the bulk of the K^+ ions are on the inside of the fiber membrane, while there is an excess of Cl^- and Na^+ ions on the outside. The net effect of this distribution of ions is that an electrical potential difference of about -90 mV exists between the inside and the outside of the cell. When acetylcholine is released at the motor end plate, the membrane is in some way (as yet undiscovered) rendered

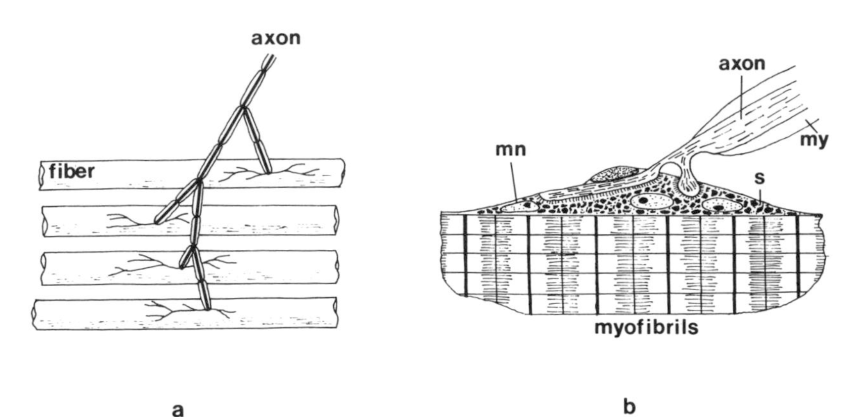

FIGURE 1.15. (a) Diagram showing the innervation of a very small motor unit of muscle fibers by a single axon. In fact, there may be many more fibers in a single motor unit, and they need not be adjacent to each other. (After Aidley, 1971.) (b) Schematic diagram of a motor end plate showing the axon surrounded by a membranous envelope (the myelin sheath; my; see Aidley, 1971) the sacroplasm (S), the cell nuclei (mn) and the myofibrillar material. (After Couteaux, 1960.)

more permeable to all ions, and a rapid redistribution of ions results. The effect is that the membrane potential increases rapidly to about +30 to 40 mV and then returns to its former value; this is the action potential that leads to the activation of the contractile apparatus. An action potential at one spot on the membrane also triggers a similar redistribution of ions at neighboring points on the membrane so that the depolarization is propagated along the muscle. This kind of depolarization is frequently an all-or-none effect in which total depolarization results provided the stimulus is above a certain threshold value. The action potential can be mimicked *in vitro* by direct electrical stimulation of the muscle membrane (as described in Section 1.4.1; Fig. 1.12), by increasing the K^+ concentration on the outside of the membrane (to give the so-called potassium contracture), and by altering the properties of the membrane either by mechanical disruption or by the direct application of acetylcholine and some other drugs.

Vertebrate Skeletal Muscles. Different types of fibers in vertebrate skeletal muscles are distinguished by the kind of innervation that they receive. In some fibers, the fiber membrane is excitable as is the nerve membrane, and the all-or-none action potential generated at the neuromuscular junction is propagated along the length of the fiber so that the whole of it contracts. Fibers of this kind normally have a single motor end plate (very occasionally two) and are described as having *uniterminal innervation* (Fig. 1.16a). Included in this category are most of the twitch fibers in the skeletal muscles (such as the sartorius) of frogs and toads. Other fibers have membranes that are electrically inexcitable, and in these, the action potential at the neuromuscular junction only causes a localized stimulation of the contractile apparatus. For this reason, such fibers (called *tonic* fibers) tend to have multiterminal innervation (Fig. 1.16b). The response of tonic fibers to a stimulus is also different. A single stimulus has relatively little effect, but repetitive stimulation causes the gradual buildup of tension. In fact, the amount of depolarization of the fiber membrane depends directly on the frequency of the nervous input and is, therefore, quite different from the all-or-none effect described earlier for twitch fibers. Note that twitch and tonic fibers are sometimes referred to as fast and slow fibers respectively, but this terminology can lead to confusion since there are also fast and slow twitch fibers (see Section 1.4.5).

Arthropod Muscles. Many variations in the method of innervation occur throughout the animal kingdom. For example, the muscles in arthropods are usually multiterminally innervated, but they also commonly have excitable membranes. However, they do not show all-or-none propagated action potentials; the response is graded in proportion to the initial depolarization. As well as being multiterminal, arthropod muscles are often polyneuronal and can be innervated by two or three independent axons (Fig. 1.16c). For example, the hind leg of the locust is innervated by a fast excitor, a slow excitor, and an inhibitor axon. The fast excitor serves over three-quarters of the fibers, and the slow excitor serves about one-quarter, some of which are also stimulated by the fast axon. In addition, some of the fibers are also innervated by the inhibitor axon which, when stimulated, tends to relax fibers that have already been stimulated by the slow excitor axon.

Graded Responses in Muscles. Since muscles can only pull and cannot push, skeletal movements require the opposed action of at least two antagonistic muscles (Fig. 1.16d). But in order to produce finely controlled movements, it is clearly necessary that the level of tension generated by each muscle be under close control. In principle, there are two ways in which this can be achieved. One is to control the number of fibers that are active at any instant in a particular muscle; the other is to control the amount of tension generated by each fiber. In the case of twitch fibers which show an all-or-none response, it is clearly necessary to control the number of active fibers. In practice, this can be achieved, since each muscle is innervated by a large number of motor axons each of which divides and serves only a small fraction (perhaps several hundred) of the fibers in the muscle (Fig. 1.15a). These fibers are said to form a *motor unit.* Different motor units are activated at different times to provide a given average tension level, and such asynchronous activation also helps to smoothe out the tension in the muscle. A minor control in addition to this is the frequency of the nervous input from a particular axon, since if this is sufficiently fast, it will give rise to a tetanic response in which the mean tension level is higher and of longer duration than in

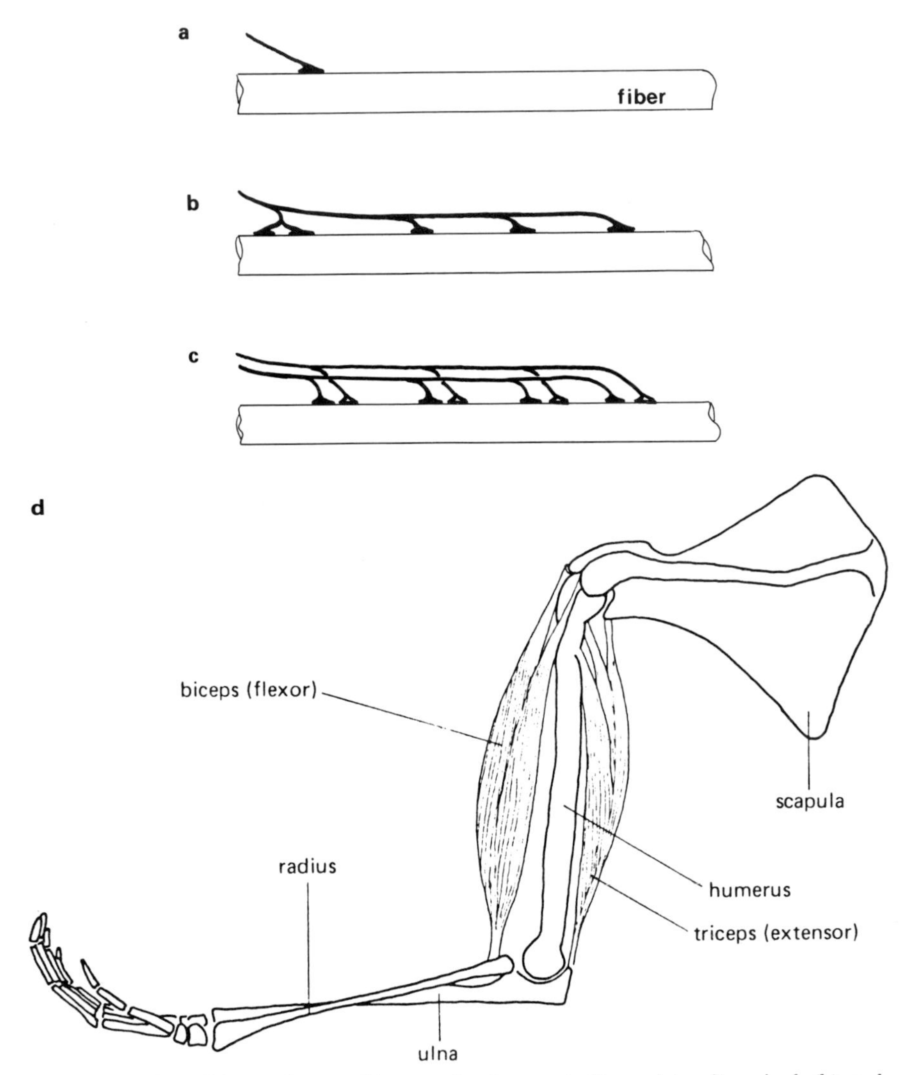

FIGURE 1.16. Different forms of innervation in muscle fibers: (a) uniterminal; (b) multiterminal; (c) multiterminal, polyneuronal. (After Aidley, 1971.) (d) Antagonistic action of muscles at the elbow joint. (From Mackean, 1962.)

a twitch. The response of the muscle is detected by a kind of feedback mechanism to the nervous system which involves a specialized structure known as the muscle spindle.

In muscles that have multiterminal and polyneuronal innervation, such as vertebrate tonic muscles and arthropod muscles, the level of tension can be controlled directly. For this reason, many arthropod muscles only have a small number of axons serving the whole muscle. But vertebrate tonic muscles are also served by a large number of different axons, and control of the number of active fibers can also be achieved.

Vertebrate Heart Muscles. Heart muscles can be divided into two main categories: those that contract in response to external nervous stimuli (the neurogenic hearts) and those in which the contraction is initiated spontaneously in the muscles themselves (the myogenic hearts). The heart muscles of vertebrates are of the myogenic type, and these will be considered briefly here.

Figure 1.17 shows a schematic drawing of the typical structure of a vertebrate heart muscle. Unlike the fibers in skeletal muscles (Fig. 1.4), the cardiac fibers are relatively short and branched, and they are linked together at their ends by structures called intercalated disks. These

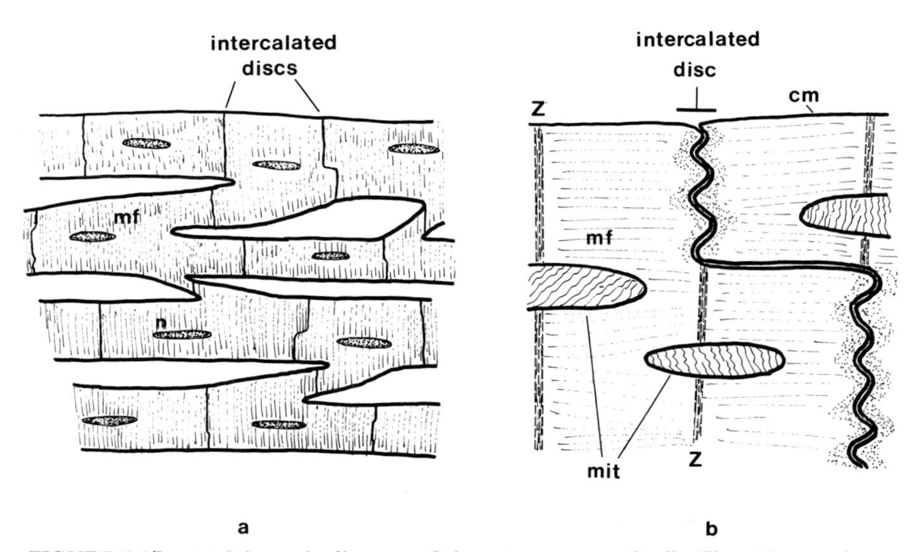

FIGURE 1.17. (a) Schematic diagram of the arrangement of cells (fibers) in vertebrate cardiac muscle, showing the cell branching and the intercalated disks between cells. (b) Enlargement of the intercalated disk region of a cardiac muscle showing the dense granular structure on each side of the intercellular boundary. (cm, cell membrane; mit, mitochondria; Z, Z-band; mf, myofilament material).

structures are thought to provide low-resistance electrical connections between fibers so that an action potential generated in one cell can easily be propagated to its neighbors. In addition, mammalian hearts also contain a unique conducting system comprising special fibers known as Purkinje fibers. These and the intercalated disks between fibers help to insure that all of the fibers in the heart contract in synchrony with the fastest spontaneously active cells—the pacemaker cells. Uncoordinated contractions would clearly be ineffectual.

Figure 1.18 shows the form of the potential changes generated in a typical cardiac fiber and the corresponding tension response. A number of features are worth noting. One is that the potential changes are periodic so that rhythmical contractions are produced. The second is that there is a delay in the repolarization of the membrane so that the duration of one depolarization is very much longer than in a twitch fiber (Fig. 1.12). The third is that the duration of each contraction is similar to the duration of the change in potential, and this makes it impossible for a fused tetanus to be produced (cf. Fig. 1.12b).

Finally, it should be mentioned that although the contractile response in these muscles is myogenic in origin, it can be modified by the action of the sympathetic and parasympathetic nerve fibers. Excitation

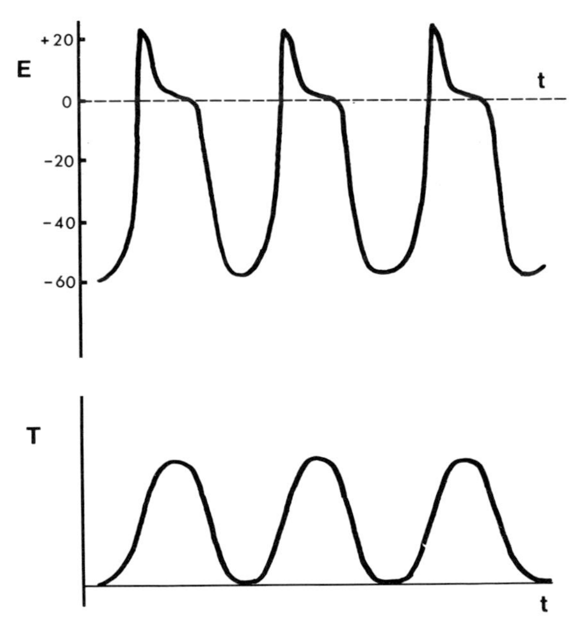

FIGURE 1.18. Spontaneous electrical membrane activity (E) of a typical vertebrate cardiac muscle and the corresponding rhythmical tension response (T). For details see text.

of these nerves respectively produces an increase in the excitation frequency or a diminution of spontaneous activity, and they can therefore have a controlling effect on the contractile activity of the heart according to the immediate needs of the body.

Vertebrate Smooth Muscles. Like heart muscles, most vertabrate smooth muscles also exhibit spontaneous activity. But the conduction between cells is relatively poor, and the result is that waves of contraction can be propagated through the muscle. Since these muscles are used either to move the contents of certain hollow organs of the body (apart from the heart) or to hold their shape, such contractile behavior is well suited to its purpose. The typical structure of a vertebrate smooth muscle is illustrated in Fig. 1.19. It consists of spindle-shaped cells that are linked together by an extensive network of connective tissue consisting largely of collagen fibers. Typical smooth muscle cells are about 50 to 200 μm long and about 2 to 5 μm in diameter in their extended state, but these values clearly change as the muscle contracts. The activity of the cell membranes consists of slow waves of variable amplitude together with intermittent all-or-none action potentials (Fig. 1.20). As in heart muscle, smooth muscle activity can be modified to some extent by the action of extrinsic nerves, by adrenaline, and in some cases (e.g., uterine muscle), by the action of particular hormones. Indeed, the smooth muscle in the vas deferens does not exhibit spontaneous activity at all and is totally controlled by the extrinsic nerves. *In vitro,* spontaneous activity in smooth muscles can be suppressed by the action of certain drugs or by reduction of the temperature of the muscle from its *in vivo* level (usually 37°C) to about 10–15°C. As detailed in Chapter 8, this property has important uses in the study of the smooth muscle contractile apparatus.

Molluscan Muscles. The molluscan muscles that exhibit the catch response are remarkable in that the relaxation rate following tension generation depends on how the muscle is stimulated (Lowy and Millman, 1963). Phasic responses in which the relaxation rate is relatively

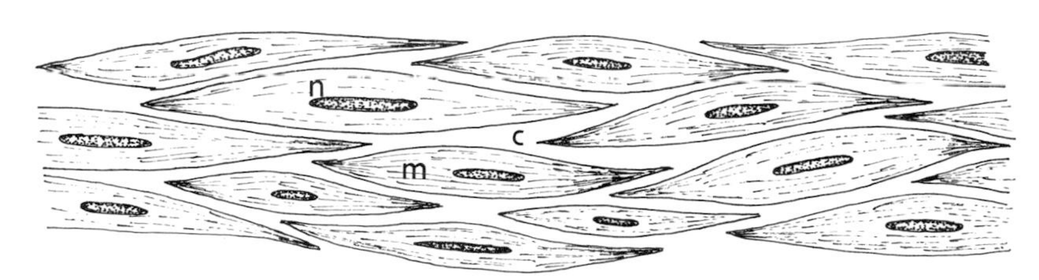

FIGURE 1.19. Schematic diagram showing the typical structure of a vertebrate smooth muscle in longitudinal section. The cells (m) are spindle-shaped, contain a single nucleus (n), and are linked by collagenous connective tissue (c) in the intercellular space.

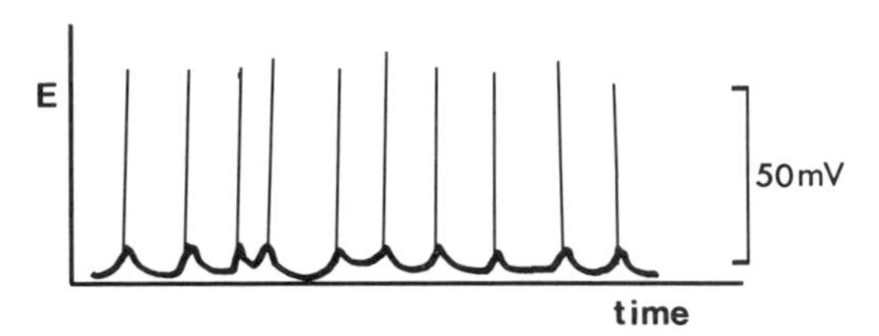

FIGURE 1.20. Typical record of the spontaneous membrane activity (i.e., membrane potential, E) of a vertebrate smooth muscle. (After Bulbring *et al.*, 1958.)

rapid after stimulation has stopped (Fig. 1.21) can be produced by repetitive brief electrical pulses or by alternating-current stimulation. On the other hand, tonic catch responses can be produced by direct-current stimulation or by the action of acetylcholine, and in this case, relaxation is so slow that a high tension level can be maintained for up to several hours. However, addition of the relaxant 5-hydroxytryptamine causes the abolition of the tonic response, and the muscle relaxes at the phasic rate. With 5-hydroxytryptamine present, only phasic responses can be produced whatever the method of stimulation.

Insect Asynchronous Muscles. Possibly the most unusual type of muscle response is exhibited by the asynchronous insect flight muscles. Here, measurement of the resting tension as a function of length gives graphs such as the lower curve in Fig. 1.22. It can be seen that resting tension rises very rapidly with small length increases, indicating that the muscle is very stiff. Stimulation of a muscle held isometrically then causes an increase in the tension level (Fig. 1.22, upper curve) as in other muscles. But, if the muscle is suitably loaded, for example, with the wing–thorax assembly in the insect, then stimulation causes an oscillatory re-

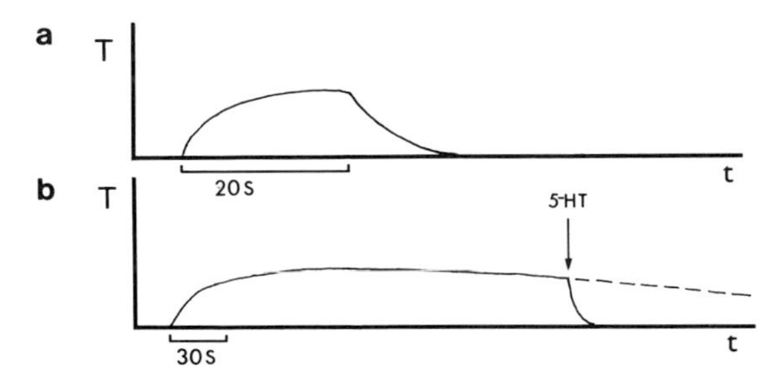

FIGURE 1.21. Tension responses in molluscan "catch" muscles to two different forms of stimulus. (Tension, T; time, t.) (a) A phasic response to alternating-current stimulation. The muscle relaxes when the stimulus stops (here after 20 sec). (b) The same muscle treated with acetylcholine for 30 sec. The tension remains high even after the acetylcholine is removed [note different time scale from (a)] but will drop at the phasic rate if 5-hydroxytryptamine (5-HT) is added. (After Lowy and Millman, 1963.)

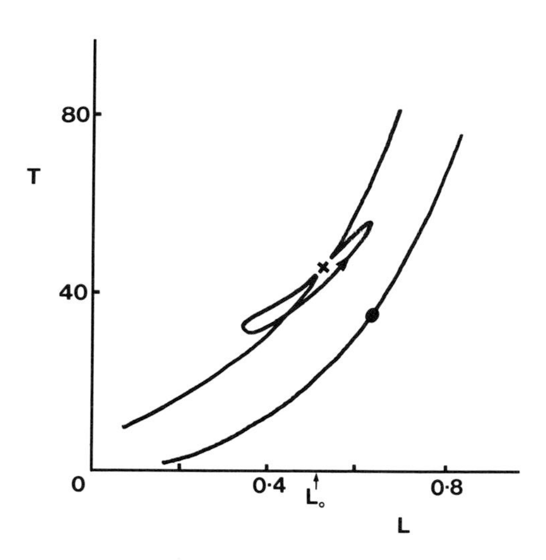

FIGURE 1.22. Various forms of tension response (T) in an asynchronous insect muscle (rhinocerus beetle basalar muscle, Machin and Pringle, 1959) as a function of muscle length (L). The most interesting feature is the oscillatory response (here at 25 Hz) denoted by the closed loop. The loop is traversed counterclockwise, and the muscle is doing oscillatory work, measured in terms of the area of the loop.

sponse (Fig. 1.22, closed loop) in which the tension is higher as the muscle shortens following a stretch than it is when it lengthens. In this way, oscillatory work is done by the muscle on the external system and is measured in terms of the area of the closed loop in Fig. 1.22. As mentioned in Section 1.2, the frequency of the oscillation is dependent on the nature of the load and on the properties of the muscle and is not directly governed by the frequency of repetitive stimulation if this occurs. The delayed increase of tension following a slight stretch of the muscle is known as *stretch activation,* and the origins of this and also of the tonic response in molluscan muscles provide two of the outstanding puzzles in current muscle research.

1.4.3. Excitation–Contraction Coupling

So far it has been stated that contraction occurs in a fiber as a result of depolarization of the fiber membrane, and in fact this is generally true whatever the cause of the depolarization. But the link between this depolarization and the contractile apparatus itself has not been mentioned. It is now known that the mediator of the process is the Ca^{2+} ion (see Ebashi *et al.,* 1969). Depolarization of the muscle membrane causes the release of Ca^{2+} ions within the cell, and these diffuse into the myofibrillar structure, bind to certain myofibrillar proteins (as described in Chapters 5 and 6), and thus activate the actin and myosin filaments so that contraction can occur. The effect of the Ca^{2+} ion can be demonstrated on a glycerinated muscle containing Mg-ATP, since an increase in the concentration of Ca^{2+} ions from 10^{-7} M to 10^{-5} M causes complete contraction of the fiber (Fig. 1.23).

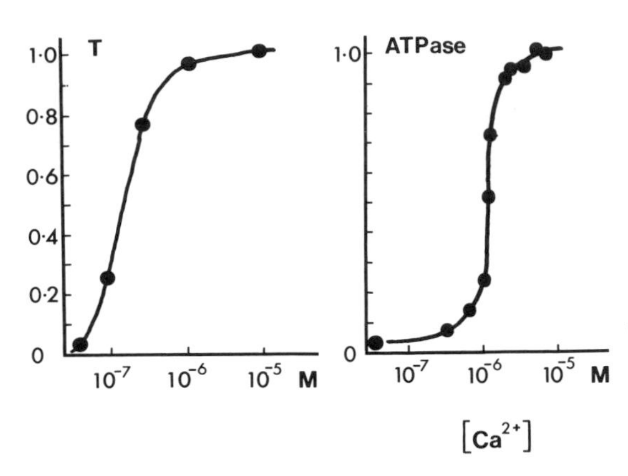

FIGURE 1.23. Variation of muscular tension (*T*) and myofibrillar ATPase as functions of the concentration of calcium ions. Both follow a sigmoid curve centered between 10^{-7} and 10^{-6} M Ca^{2+}. (After Hellam and Podolsky, 1969, and Weber and Herz, 1963.)

In an elegant experiment by Ashley and Ridgeway (1968), the increase in the calcium concentration following stimulation of a fiber was directly demonstrated. They used the novel technique of injecting a protein, aequorin, into the large muscle fibers of the barnacle, *Balanus nubilis*. Aequorin, which can be isolated from a bioluminescent jellyfish, has the property of emitting light as a burst the duration of which depends on the concentration of calcium ions. Electrical stimulation of the barnacle fibers containing aequorin caused the emission of a small but detectable amount of light, as illustrated in Fig. 1.24. As discussed in Chapters 5 and 6, it has been found that Ca^{2+} binding to certain regulatory proteins in muscle as a function of Ca^{2+} concentration follows the

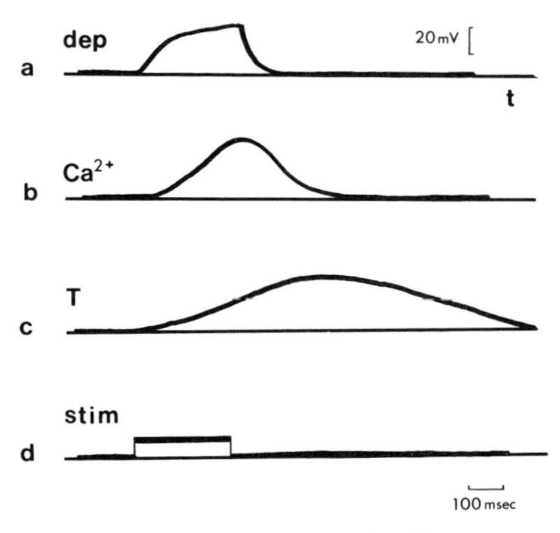

FIGURE 1.24. Calcium transients in *Balanus* muscle fibers measured by the aequorin technique: (a) membrane potential; (b) variation in calcium level; (c) tension trace; (d) duration of stimulus. (After Ashley and Ridgeway, 1968.)

same kind of sigmoid curve as the rise in tension and ATPase (Fig. 1.23), a result that implicates the calcium ion directly in the activation process of the myofibrillar material.

In certain kinds of cells, such as those in vertebrate smooth muscles, the diameter of the cell is sufficiently small (2 to 5 μm), and the contractile response is sufficiently slow, that it is reasonable to account for the activation effect in terms of the diffusion of calcium ions from the cell surface following activation. But in the case of vertebrate skeletal muscles (such as the frog sartorius) and certain other muscles, the fiber diameters are so large (greater than 100 μm), and the contractile response is so rapid, that the activation effect could not be achieved by a simple diffusion process. This would be much too slow. Apart from speed, it has also been shown that the amount of calcium entering the cell from the fiber membrane would be much too small to cause significant activation of the whole fiber.

A. F. Huxley and Taylor (1955, 1958) were the first to demonstrate that there is, in fact, a special conducting system within each fiber which transmits the membrane action potential into the fiber interior. They showed this by placing a microelectrode onto different parts of an isolated single frog fiber under the light microscope and found that depolarizing currents produced local contractions, but only when the electrode was placed in line with the Z-bands in the fiber. Stimulation elsewhere had no effect. The Z-band itself was found not to be the conducting system, since in other types of fiber (e.g., in crab muscles), the conducting system is located at the boundary between the A- and I-bands.

The conducting structure itself was discovered by Porter and Palade (1957). It was found to be a membranous structure in close proximity to, but not part of, the sarcoplasmic reticulum (SR) of the fibers. The sarcoplasmic reticulum is a system of membrane-bound vesicular elements arranged around the myofibrils in the manner shown in Fig. 1.25 for vertebrate skeletal muscle. The conducting structures are in the form of transverse tubules which occur alongside the Z-bands and which are referred to as the T-system in the muscle. On each side of such a transverse tubule, but not continuous with it, are structures called *terminal cisternae* (or outer vesicles) which form part of the SR and act as a calcium store. A transverse tubule (T-tubule) and the two adjacent cisternae are collectively known as a *triad*. The terminal cisternae are connected through longitudinal elements or tubules to a fenestrated "collar" structure in line with the M-band.

It is clear that the transverse tubules are in the right position (near to the Z-band) to be the conducting elements that were suggested by the results of Huxley and Taylor. They were later shown to be continuous

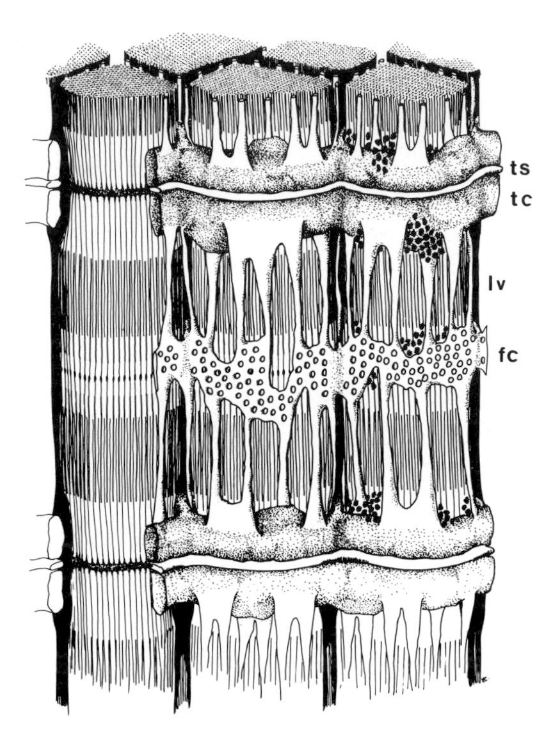

FIGURE 1.25. Schematic diagram showing the arrangement of the sarcoplasmic reticulum around myofibrils in a vertebrate skeletal muscle. This shows the transverse tubules (ts) forming the T-system, the terminal cisternae (tc), the longitudinal vesicles (lv), and the fenestrated collar (fc). (From Peachey, 1965.)

with the fiber membrane by conventional electron microscopy (Franzini-Armstrong and Porter, 1964) and by electron microscopy of fibers soaked in ferritin (Page, 1964; H. E. Huxley, 1964). In this latter experiment, the ferritin granules were found to be present in the T-tubules but not in the SR, a result that clearly demonstrated that the T-tubules must be continuous with the exterior of the fiber. Since then, evidence has been obtained that suggests that the outer membrane of the T-system may be linked directly to the terminal cisternae by membranous bridging structures (Somlyo, 1979).

It is now thought that the effect of an action potential propagating along the fiber membrane is to produce some kind of action potential in the T-system and that this in turn causes the release of calcium ions from the terminal cisternae into the sarcoplasm. If no further stimulus arrives, these calcium ions are then pumped back into the sarcoplasmic reticulum, probably by the longitudinal vesicles, and the fiber relaxes (see review by Ebashi, 1976). This relaxing system actually comprises an ATP-driven calcium pump (Hasselbach, 1964). Finally, it should be noted that caffeine, which activates a muscle but does not alter the membrane potential, is thought to act directly on the sarcoplasmic reticulum and thus to cause the release of calcium. It is also appropriate to mention here that a ubiquitous protein called desmin (or skeletin) has been found to form a cross-linking structure at the level of the Z-band between adjacent myofibrils. Skeletin has also been found in smooth muscle cells (see review by Lazarides, 1980).

1.4.4. The Energy Supply

It has already been mentioned that the local energy source in a myofibril is the molecule adenosine triphosphate (Figs. 1.10 and 1.23). But the primary source of energy in all animals is food. Since the level of ATP in a muscle is usually quite small (only enough for about eight twitches), it is clear that the energy supply must involve two separate processes. One process involves the steady synthesis of high-energy phosphate from foodstuffs, and the other involves the rapid replenishment of the ATP that has been hydrolyzed to ADP by the activity of the contractile machinery.

The energy reserve closest to ATP is phosphocreatine (PCr). In the presence of this molecule and the enzyme creatine phosphotransferase (CPT), regeneration of ATP occurs.

$$\mathrm{ADP} + \mathrm{PCr} \underset{\mathrm{CPT}}{\leftrightarrows} \mathrm{ATP} + \mathrm{Cr} \tag{1}$$

This restores the ATP hydrolysed by the reaction

$$\mathrm{ATP} \leftrightarrows \mathrm{ADP} + \mathrm{P} \tag{2}$$

Although reaction (1) is reversible, the equilibrium constant is such that the forward reaction is favored, and the net result during contraction is, therefore, that the ATP level is maintained at the expense of phosphocreatine which is gradually used up. Obviously, this depleted level of PCr in the muscle needs to be replenished during and after a contraction, and this is achieved by a complex recovery process in which foodstuffs are broken down in a series of energy-conserving reactions during which ADP is rephosphorylated. The increased level of ATP then reverses reaction (1), and phosphocreatine is resynthesised.

Figure 1.26 illustrates an important observation by Wilkie (1968) in which this resynthesis of PCr was prevented. He found that, under those

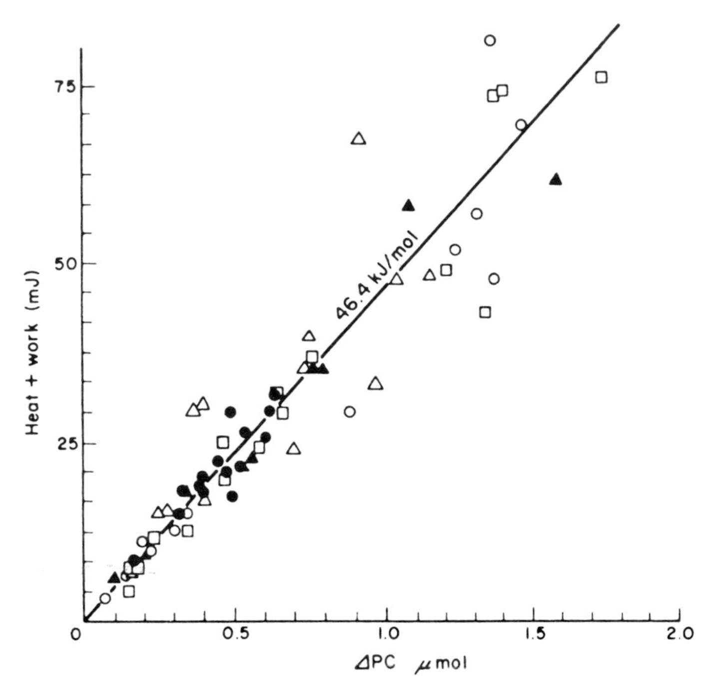

FIGURE 1.26. Experimental results showing the linear relationship between the heat plus work output of a muscle and phosphocreatine (PC) breakdown no matter what the type or duration of the stimulus. Phosphocreatine recovery was prevented by excluding oxygen and by preventing lactic acid formation with iodoacetate. (From Wilkie, 1968.)

conditions, the heat plus work produced by a contracting muscle is directly proportional to phosphocreatine breakdown. This shows the direct link that exists between ATP usage and the output of the muscle.

The form in which foodstuffs are used in the energy supply is either as glycogen, a polymer of glucose $(C_6H_{12}O_6)_n$, or as fatty acids. The breakdown of glycogen into simpler molecules is carried out in such a way that the free energy release during many of the steps is conserved by the resynthesis of ATP from ADP. The particular process involved depends on the muscle and on whether few or many mitochondria are present. In the absence of oxygen, the process involves the division of glycogen into six-carbon units $(C_6H_{12}O_6)$ that are then broken down into three-carbon units ending with pyruvic acid ($CH_3COCOOH$) and finally with lactic acid ($CH_3CHOHCOOH$). This process is known as the glycolytic pathway and is illustrated schematically in Fig. 1.27a. It gives a net yield of three ATP per glucose residue in glycogen, and since the hydrogens used in the final stage of lactic acid formation can be the hydrogens generated during an earlier step in the pathway (after step 7

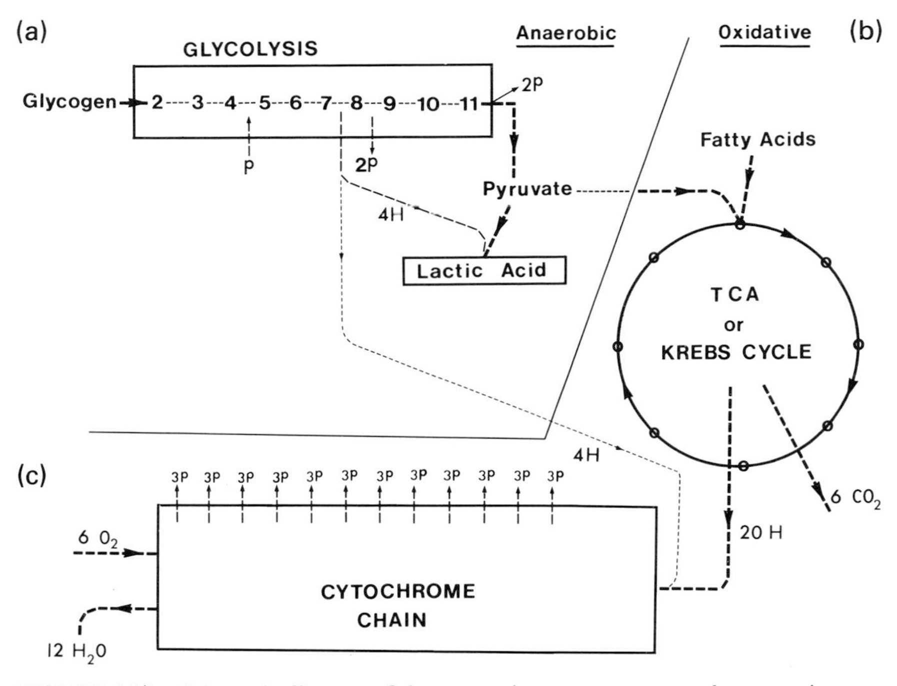

FIGURE 1.27. Schematic diagram of the two main energy processes that occur in muscles, the glycolytic pathway (a) which can continue in the absence of oxygen and which leads to lactic acid formation and the oxidative phosphorylation system involving (b) the tricarboxylic (TCA) or Krebs cycle and (c) the cytochrome chain. For details see text. (After Starling and Lovatt-Evans, 1962; for comprehensive account see Lehninger, 1975.)

in Fig. 1.27a), the glycolytic pathway is a self-contained process that can continue in the complete absence of oxygen.

If many mitochondria are present, then quite a different process can occur. This time, highly productive phosphorylation can occur either at the end of the glycolytic pathway via pyruvate or at the end of another common process in animals, the fatty acid oxidation cycle (Lehninger, 1975). The products of either of these pathways take part in the tricarboxylic acid (TCA) or Krebs cycle (Fig. 1.27b), in which a further 20 usable hydrogens are obtained. These hydrogens, together with the hydrogens generated after step 7 of the glycolytic pathway, and all of which are present in combination with the hydrogen carrier nicotinamide–adenine dinucleotide (NAD) as NADH, then take part in a highly productive phosphorylating system coupled to what is known as the cytochrome chain (Fig. 1.27). Here considerable phosphorylation is possible provided that oxygen is present and there is a good supply of NADH.

Whichever of these processes occurs, the net effect during exercise or lack of oxygen is to increase the level of NADH which reacts with pyruvic acid to form lactic acid.

$$CH_3COCOOH + NADH + H^+ \leftrightharpoons CH_3CHOHCOOH + NAD^+$$

Lactate then appears in large quantities in the blood, and respiration and other regenerative processes must continue for some time after the active period in order to repay the oxygen debt. Some of the lactate is oxidized by other organs including the heart, and some is converted back to glycogen.

1.4.5. Classification of Vertebrate Fiber Types

We have seen that vertebrate muscles can be divided into tonic and phasic (twitch) types depending on their innervation and response. However, in mammals, almost all of the muscles are phasic and yet show a considerable range of contractile behavior. One of the main reasons for this is that different muscles contain varying proportions of individual fiber types that have distinct properties. Not all fibers in a particular muscle are identical in structure or in contractile behavior, and certain characteristic fiber types can be identified. As described in Section 1.4.4., fibers can be distinguished by whether their energy supply is largely glycolytic or is dominated by the mitochondria. Generally the glycolytic fibers are relatively fast in contractile response, and they appear white, whereas fibers that have an abundance of mitochondria are said to be oxidative, they are relatively slow in contractile response, and they appear reddish in color because of the presence of the oxygen carrier myoglobin. But this simple division is by no means the whole story, since there is considerable overlap between the two energy sources, especially in the so-called intermediate fibers which contain much glycogen and many mitochondria. These are generally termed fast oxidative/glycolytic fibers, and they are reddish in color.

The slow and fast fibers are often referred to as Type I and Type II fibers, respectively, and the Type II (fast) fibers are then further subdivided into Type IIA (fast oxidative/glycolytic) and Type IIB (fast glycolytic). These fiber types were originally distinguished most reproducibly by histochemical methods (Dubowitz and Brooke, 1973). Such methods included testing thick sections of frozen muscle for the ATPase properties of the fibers (Fig. 1.28) after incubation under either neutral, acid, or alkaline conditions, for succinic dehydrogenase activity (which is involved in the respiratory chain and occurs largely in mitochondria), and for phosphorylase activity (which is associated with glycogen breakdown). As a general rule (Table 1.1), Type I fibers have acid-resistant,

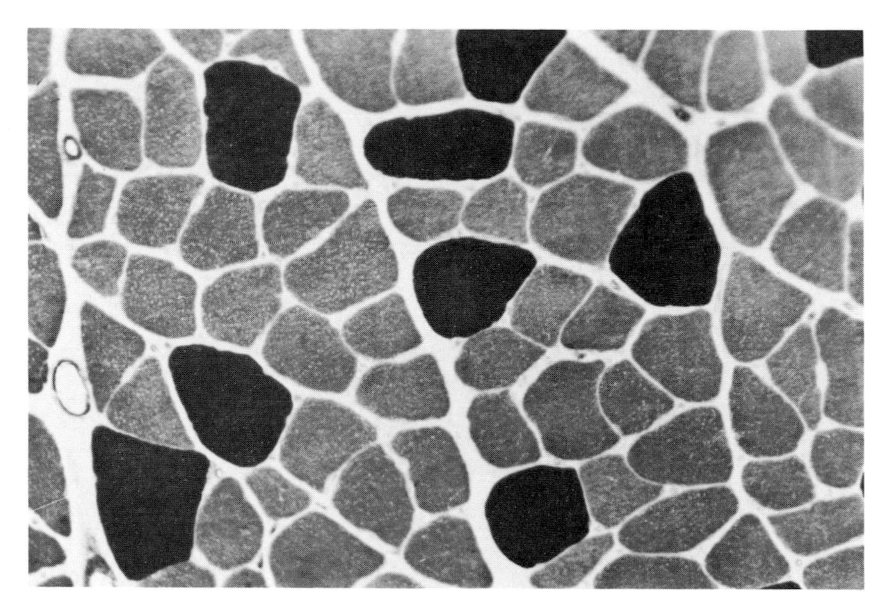

FIGURE 1.28. Histochemical ATPase tests applied to vertebrate muscle. Transverse section from fresh frozen human m. tibialis anterior treated for visualization of myofibrillar ATPase at pH 9.4. Lightly stained fibers have an acid stable ATPase and are Type I, and heavily stained fibers have an alkali stable ATPase and are Type II. (Micrograph courtesy of Dr. M. Sjöström, Umea University.) For details of method, see Dubowitz and Brooke (1973). Note that within each clear group of fibers (the fascicles) there is a random arrangement of fiber types.

alkali-sensitive myofibrillar ATPase properties, relatively high mitochondrial contents (and hence high succinic dehydrogenase activity), low glycogen content and phosphorylase activity, a relatively poorly developed SR, low contraction speeds, and high resistance to fatigue, whereas Type IIB fibers have the converse properties. The intermediate (Type IIA) fibers have intermediate properties, and this helps to illustrate the diversity of properties that occurs and the difficulty that sometimes arises in classifying a particular fiber. In fact, there is often such a gradation of properties and a disparity among the physiological, histochemical, and structural natures of fibers in different muscles and animals that a simple systematic fiber type classification scheme is extremely hard to define. However, as will be shown in Chapter 7, it is now becoming apparent that different histochemically defined fiber types can have characteristically different ultrastructures, and this should help to define a useful classification system, at least for a given species.

Apart from the problem of classification, an extremely crucial question concerns the origins of these fiber type differences. It was shown by

TABLE 1.1. Method of Classification of Common Fiber Types in Vertebrate Skeletal Muscle[a]

Fiber type and name	ATPase[b]			SDH[c]	Phosphor.[d]			Resistance	
	pH 9.4	pH 4.3	pH 4.6	(Mit)	(Glyc.)	SR[e]	Speed	to fatigue	Color
Type I (slow oxidative)	Low	High	High	High	Low	Poor	Slow	High	Red
Type IIA[f] (fast oxidative/glycolytic)	High	Very low	Very low	Moderate	High	Rich	Fast	Moderate	White
Type IIB[f] (fast glycolytic)	High	Very low	High	Low	High	Rich	Fast	Low	White

[a] Note that this table is generalized, and exceptions to these rules may be found.
[b] ATPase tested histochemically after incubation of tissue section under either acid or alkaline conditions as indicated.
[c] Succinic dehydrogenase (SDH) tested histochemically; it correlates with mitochondrial content.
[d] Phosphorylase activity (Phosphor.) correlates with glycogen content (Glyc.).
[e] SR is sarcoplasmic reticulum.
[f] Note that a third Type II fiber type (Type IIC) can also be distinguished by these tests but, according to Brooke and Kaiser (1970), it is more of an undifferentiated or precursor fiber to Types IIA and IIB. For details of all histochemical methods, see Dubowitz and Brooke (1973).

Buller *et al.* (1960) that cross-innervation experiments in which the motor nerves of a fast twitch muscle and a slow twitch muscle (each of nearly uniform fiber type) were cut, exchanged, and then allowed to innervate their new host muscles, caused the slow muscle to become faster and the fast muscle to become slower. It therefore appeared that the subtle differences in contractile behavior among different muscles and different fiber types must be associated in some way with the particular type of innervation they receive.

This kind of approach was extended by Salmons and Sreter (1976) to identify which particular property of the motor nerves caused this characteristic response. They considered the possibility that it might be the pattern of impulse activity reaching a muscle that determines its properties. The motor nerves to a slow muscle generate a sustained low-frequency pattern of activity, whereas those to a fast muscle provide intermittent bursts of more intense activity (Eccles *et al.*, 1968). By an ingenious series of experiments and controls, Salmons and Sreter were able to show that it is indeed the pattern of activity that controls the fiber typing, and they ruled out the possibility that some chemical difference (in the form of a trophic factor) between fast and slow motor nerves is the controlling factor. In another type of experiment, it was shown that different muscles contain myosins with different ATPase properties and that the proportions of these myosins in a muscle change as a result of cross-innervation (Buller *et al.*, 1969; Barany and Close, 1971; Weeds *et al.*, 1974; Streter *et al.*, 1974; Hoh, 1975). It has also been shown that the relative proportions of polymorphic forms of some of the proteins that regulate contractile activity (see Chapter 5) also change as a result of cross innervation (Amphlett *et al.*, 1975). It was therefore suggested by Salmons and Sreter (1976) that the impulse activity in the motor nerve must in some way have a regulating effect on protein synthesis in the muscle (or motor unit) so that changes in innervation cause some genes to be switched on and others to be suppressed. Such a process could clearly have a crucial role in the differentiation processes that occur during muscle growth (myogenesis). Also, since different muscles can contain different proportions of the different fiber types, those of each type being part of a motor unit of that particular type of nerve, it is clear that the observed gradation of properties in different muscles is just what might be expected if fiber type were controlled directly by the specific innervation pattern.

1.5. The Molecular Biophysicist's Approach to Muscle

In the preceding pages a brief sketch has been given of some of the current views on the contractile process and on muscle physiology. The question, "How is chemical energy (from ATP) converted into mechanical work?" can be thought of in the context of the cross bridge cycle in terms of the question, "Why and in what way does the myosin cross bridge change its attachment configuration on actin when ADP and phosphate are released?" A second crucial question concerns the means by which the actin–myosin interaction is controlled at the molecular level by calcium. The way in which the molecular biophysicist has approached these questions is based on his assertion that the more knowledge there is about the structure of a muscle, the easier it will be to discover how a muscle works. In fact, the available data on the structures of a number of different muscles are considerable, and much of this data has been obtained in the last few years. The purpose of this volume is, therefore, to describe and evaluate the data on the molecular arrangements in the main muscle types that have been studied (Chapters 5 to 9) and then to consider in detail the nature of the cross bridge cycle itself (Chapters 10 and 11). These last two chapters will include a brief consideration of the biochemical and mechanical evidence on the cross bridge cycle along with a discussion of various published contraction theories so that, when taken together with the structural evidence, some limits can be put on the type of mechanism that might be involved in force generation.

But first, as an introduction to the description of the molecular architectures in different muscles, it is necessary to describe briefly some of the structural techniques that are involved in current ultrastructural research. One of the most important of these, since it can be applied to intact living muscle, is the technique of X-ray diffraction. Unfortunately, to the nonspecialist, this technique often appears to be unintelligible and forbidding. But in the next chapter, the basic ideas of X-ray diffraction, as required for an understanding of the remainder of the book, are presented in a descriptive nonmathematical way that, it is hoped, will provide for the newcomer a palatable introduction to the subject. For this reason it is believed that a little patient study of Chapter 2 by those new to the X-ray diffraction technique will have its reward.

2

X-Ray Diffraction Methods in Muscle Research

2.1. Introduction

The rapid advances that have taken place in biology in the last 50 years have largely resulted from the advent of sophisticated new ultrastructural techniques. Molecular biology is itself the child of these new techniques and is motivated by a completely new approach to biology which asks how things work at the molecular level. The main techniques that made this new approach possible were X-ray diffraction, advanced light microscopy, and later electron microscopy, and all of these techniques have been applied with great effect in muscle ultrastructural research. At the same time, parallel studies involving the application of newly developed biochemical and physiological methods have provided considerable information on the biochemistry of the muscle proteins involved in contraction and on the mechanical properties of the contractile machinery.

It is not within the scope of a book of this kind to give detailed, rigorous accounts of all of the various structural techniques involved in present-day muscle research. The aim of Chapters 2, 3, and 4 is rather to provide simplified descriptions of these techniques sufficient for the nonspecialist reader to be able to follow intelligently the arguments used in later chapters of the book. They also aim to provide him with a feel for the current structural approaches to the problem of muscular contraction. References to rigorous treatments of each technique will be given for those who wish to read further.

2.2. Principles of Diffraction

2.2.1. Interference of Waves

To many researchers into muscle structure and function (even some of the most eminent), the technique of X-ray diffraction is often considered to be extremely difficult to comprehend, and for this reason little effort is made by them to come to terms with it. It is true that rigorous mathematical formulations of the theory of X-ray diffraction are often forbidding; but it is also true that the basic ideas behind that theory are quite easy to understand and can be described in relatively simple nonmathematical terms. The aim of this section is to provide just such a simple description of diffraction theory as it applies to the muscle ultrastructural research outlined in this book.

The single central principle behind all diffraction theory lies in the interference properties of two or more waves. In simple terms, the result of adding two wave motions together (Fig. 2.1) depends on whether they are in step or out of step. Each wave can be thought of as an alternating

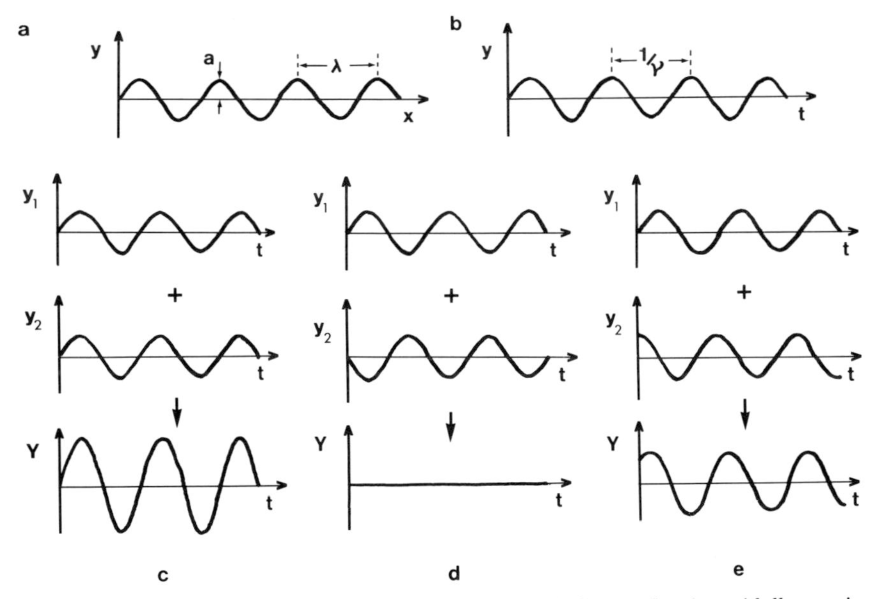

FIGURE 2.1. (a) Representation of the amplitude variation y of a sinusoidally varying wave as a function of position x. Note the definitions of wavelength (λ) and the maximum amplitude a. (b) As for (a) but plotted for a single point in (a) as a function of time (t). Note the definition of the frequency (ν). (c, d, and e) Addition of two similar sinusoidal waves y_1 and y_2 when they are oscillating in step or in phase (c), completely out of step (d), or somewhere between (e). The resulting wave Y has an aplitude $2a$ in (c), zero in (d), and of intermediate value in (e).

series of positive and negative peaks and can be represented diagrammatically (Fig. 2.1a) by a sinusoidal variation of amplitude (y) as a function of position in space (x) or (Fig. 2.1b) as a function of time (t). The maximum amplitude of the wave is a, and the wave can be described in terms of a wavelength λ and a frequency (cycles per second). Figure 2.1c shows that if two such waves are added together when they are in step (i.e., the positive peaks in both occur at the same time), then when added together they produce a resultant wave with a much larger amplitude (constructive interference). On the other hand, if the waves are out of step (Fig. 2.1d), then the negative troughs in one wave will coincide with the positive peaks in the other, and the combined amplitude of the waves will be reduced—in this case to zero (destructive interference).

Visible light and X-rays are both forms of electromagnetic radiation, and this radiation is associated with alternating electric and magnetic fields. For many purposes we need only consider the oscillations of the amplitude of the electric field. Two electromagnetic waves will interfere with each other either destructively or constructively (as in Fig. 2.1) depending on whether their electric fields are in step or out of step.

In practice, instead of describing two wave motions as being "out of step," it is normal to say they are out of phase. The term phase is used frequently in diffraction theory and has a special meaning. In Fig. 2.2a, the amplitude variation (y) of the wave of Fig. 2.1a at one point in space (say A) is plotted as a function of time. At this point, the amplitude of the electric vector varies sinusoidally with time. The amplitude variation can therefore be thought of as a cyclic phenomenon as indicated on the right of Fig. 2.2a. Here a vector (**OQ** or **a**) of length a (the maximum aplitude of the wave) is rotating steadily about O with the same frequency (v) as that of the electric field. The amplitude (y) of the electric field at any instant of time (t_α) can be thought of as the length (a_α) of the vector (**a**) when projected in one particular direction, e.g., along **OP.** The angle which the vector **a** makes with some defined direction (in this case **OP**) is called the *phase angle,* and this can clearly be anything from 0 to 2π (360°). Two wave motions can be represented by two such diagrams (Fig. 2.2b) or by two vectors on the same diagram (Fig. 2.2c), and the wave motions are said to be *in phase* if the two vectors have the same phase angle at the same instant of time (Fig. 2.2d). Similarly they are *out of phase* if the phase angles are π (180°) apart (Fig. 2.2e). In each case, the amplitude of the resultant **OQ**′ is obtained by adding the vectors together. In fact the term "phase" for a single wave is not really meaningful since the origin that defines the phase angle is arbitrary. But the *phase difference* (ϕ) or *relative phase* between two waves (Fig. 2.2c) is a significant quantity. If the phase difference is 0 or 2π, then constructive interference will occur (Fig. 2.2d), but if it is π, the

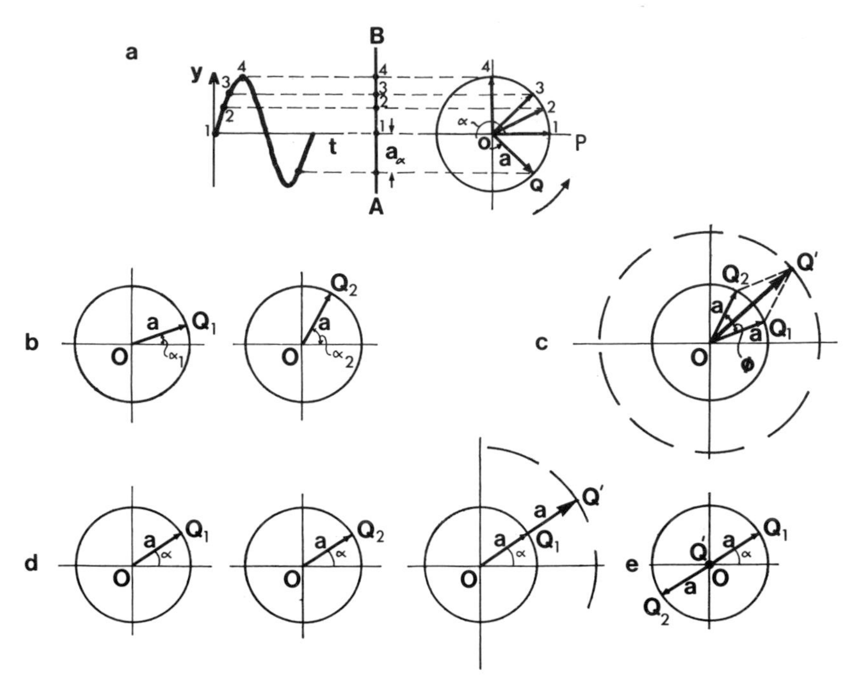

FIGURE 2.2. (a) Cyclic representation of the wave in Fig. 1b. The vector **OQ** of length a rotates steadily about O, and at any instant of time t_α makes an angle α, the phase angle, with some defined direction **OP**. The projection of **OQ** in the direction of **OP** gives the amplitude a_α of the wave at time t_α. (b) The equivalent cyclic vector representations of two waves of equal amplitude a but different phases defined by angles α_1 and α_2. (c) The resultant of adding these two waves together. The phase difference (ϕ) between the waves is $\phi = \alpha_2 - \alpha_1$, and the phase ($\alpha_R$) of the resultant is $\alpha_R = \alpha_1 + \phi/2$. (d) A similar scheme to (b) except that the two waves have the same amplitude a and phase $\boldsymbol{\alpha}$. The resultant **OQ′** has amplitude $2a$ and phase α. But if the two waves had been exactly out of phase (i.e., $\alpha_2 = \alpha_1 + 180°$), then the resultant would have been zero (e).

waves will interfere destructively (Fig. 2.2e). At angles other than 0, π, 2π, etc. intermediate effects occur (Figs. 2.2c and 2.1e), and the combined wave (**OQ′**) has a different phase from its two compontnts.

When X-rays interact with an object, the associated electric field causes the electrons in the object to oscillate. Such oscillating charged particles are known to radiate X rays with a frequency identical to the oscillation frequency. The radiation is emitted in all directions, and the electron therefore acts as a point source. An atom in a typical X-ray beam (of wavelength 1 to 2 Å) can also be considered to be a point source of secondary X rays, since the wavelength is sufficiently long that all of the electrons will be stimulated more or less in phase. The amplitude of the scattered wave therefore depends on the number of electrons in the atom. When an X-ray beam is incident on a group of atoms, each atom in the group will act as a source of secondary X rays, and, in a particular direction in space, the secondary X rays from each atom will add up, just as in Figs. 2.1d and 2.2c to give a resultant wave that will have a particular amplitude and relative phase. In some directions the amplitude (a) of the resultant will be large, and its intensity ($\propto a^2$) can then easily be recorded on a photographic film. In other directions, destructive interference may occur so that nothing is recorded on the film. As a whole, the film will show a characteristic pattern of varying intensity—the *diffraction pattern* from the object. Any object will give rise to a diffraction pattern, but only regular periodic objects will give rise to regular diffraction patterns. It is such patterns that are of direct use in ultrastructural research. X-ray diffraction patterns therefore tend to provide information about regularly periodic arrangements of atoms.

It should be clear by now that the basic ideas behind X-ray diffraction are relatively simple. The complicated mathematical formulations of X-ray diffraction theory only appear when it is required to describe exactly, for a particular diffracting object, how the waves scattered by the atoms in the object interfere with each other in a particular direction in space. This type of formulation is obviously necessary for practitioners of X-ray diffraction when quantitative deductions are required. However, underlying these mathematical formulations are one or two "rules of diffraction" which can be demonstrated in nonmathematical terms and which explain qualitatively the appearance of the diffraction patterns of relatively complicated structures such as those that occur in muscle.

In the next section, this sort of nonmathematical approach to X-ray diffraction phenomena is developed, and the technique of optical diffraction, which can visually demonstrate the form of particular diffraction patterns, is introduced.

2.2.2. Diffraction from Periodic Arrays

Thoughout this section (and indeed the remainder of the book), liberal use will be made of the technique of optical diffraction to illustrate the nature of particular diffraction patterns. This technique is one in which a simple optical bench is used to record the diffraction pattern of a two-dimensional object (e.g., an opaque card in which holes have been punched). The arrangement of the optical bench, called an optical diffractometer, is illustrated in Fig. 2.3. A laser beam is focused by a short-focal-length lens (L_1) onto a small aperture A (50 to 100 μm diameter) which acts as a point source of coherent light. The expanding beam is made parallel by a second lens (L_2), and the mask is placed in this

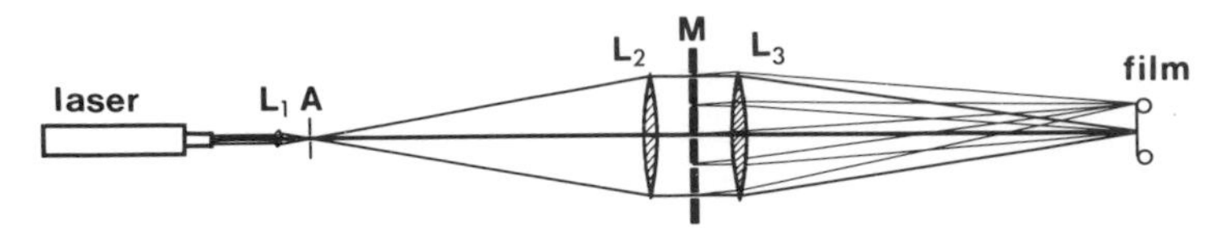

FIGURE 2.3. Layout of the optical diffractometer. L_1, L_2, and L_3 are lenses, A is a small defining aperture (point source), M is the diffracting mask (object), and the diffraction pattern can be viewed or recorded at the film position. For further details see text.

parallel beam. The holes in the mask diffract the light, and the diffraction pattern is collected and focused by a third lens (L_3) of long focal length. At the focal plane of this lens, the diffraction pattern can be viewed on a ground glass screen or recorded on film.

The optical diffraction pattern is in many ways analogous to an X-ray diffraction pattern. If the mask used represents a two-dimensional projection of an arrangement of atoms, then the optical diffraction pattern of that mask is closely related to the X-ray diffraction pattern. This means that the forms of the diffraction patterns of rather complicated objects can easily be visualized by taking optical diffraction patterns of two-dimensional representations of the objects. In this chapter, such an approach will be used to demonstrate some of the important forms and properties of diffraction patterns.

A Row of Pinhole Scatterers. Just as atoms in an X-ray beam can be considered to be small sources of secondary X rays, so can very small pinholes in an opaque mask placed in the optical diffractometer be considered as individual point sources of secondary light waves. If we look at Fig. 2.4, we can see how the waves from a row of pinholes with equal separation d will interfere with each other. The way we do this is to find the phase difference between the waves as a function of the angle of diffraction (ϕ) and then find the angles at which this is equal to 0, 2π, 4π, etc. In fact, this process is equivalent to finding the optical path difference between the waves and recognizing that constructive interference will occur if this path difference is equal to an integral number (n) of wavelengths (Fig. 2.1c). The angles of diffraction for which this is so will be the angles in which interference peaks will be seen. It can be seen in the ray diagram in Fig. 2.4 that in going to point F, the wave from O has to travel a distance of OP more than the wave from A. The length of OP is equal to $d \sin \phi$. Hence the required condition for constructive interference to occur is $d \sin \phi = n\lambda$. (Note that the lens preserves this optical path difference so that, at F, the waves have the same optical path difference as at the wavefront AP)

This rule tells us several things about diffraction patterns in general.

1. If λ is fixed and we consider a periodic array of spacing d, then there are several angles ϕ at which constructive interference will occur,

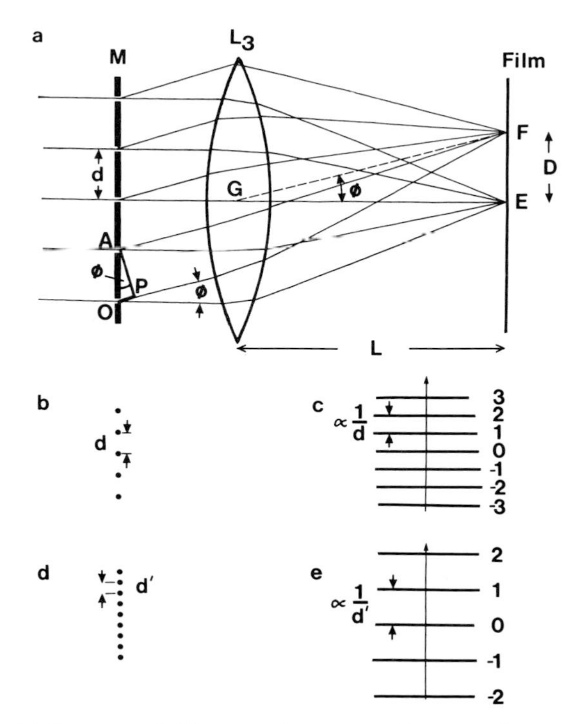

FIGURE 2.4. (a) Geometry of diffraction from a one-dimensional array of spacing d. For scattering at angle ϕ to the incident direction (along GE), the wave from O travels OP further than the wave from A. OP is equal to $d \sin \phi$. For constructive interference to occur, OP must be a whole number of wavelengths, so $d \sin \phi = n\lambda$. Experimentally, ϕ can be determined by measuring the separation D of a diffraction spot from the center E of the pattern, and if L is the focal length of the lens then $\tan \phi = D/L$. (b and d) Two one-dimensional arrays of spacing d and d', respectively. (c and e) Their diffraction patterns. Since d' is less than d, the peaks in the diffraction patterns are more widely separated in (e) than in (c) (since $\sin \phi \propto 1/d$). Note the numbering of the diffraction peaks from O (the undiffracted beam) to 1 (the fundamental), 2, 3, and so on (the second, third, etc. orders).

depending on the value of n. If $n = 1$, we have what is called a first-order diffraction peak (or the fundamental). When $n = 2(3)$, we have second (third) orders. The zero order peak occurs if $n = 0$ and corresponds to the undiffracted beam ($\phi = 0$).

2. Considering only the fundamental (first order) for a particular wavelength (λ), then an increase in the value of d will produce a decrease in the value of ϕ (since $\sin \phi \propto 1/d$). In Fig. 2.4, we have two arrays (b and d) with different d spacings. The interference (diffraction) pattern, c, from b has peaks closer together than those in the pattern e from d, since the d spacing in b is larger than that in d.

3. If we have a diffraction pattern recorded under known conditions (Fig. 2.4a), we can work out the value of ϕ from the distance D of a

reflection from the center of the pattern and the focal length (L) of the lens L_3. The value of ϕ is given by $\tan \phi = D/L$ (from triangle EFG). We also know that $n\lambda = d \sin \phi$. This means that we can work out the value of d using a combination of these expressions.

$$d = n\lambda/\sin [\tan^{-1}(D/L)]$$

(Note that in practice the angles involved are sometimes so small that the expression reduces to $d = n\lambda L/D$ since $\sin \phi \simeq \tan \phi$.)

Extent of the Row. It may be shown that the number of holes in the array has an important effect on the diffraction pattern. If the number is increased, then the diffraction peaks become sharper (or narrower), but their separation must, of course, be the same. As a general rule, it is therefore possible to calculate from the width of a diffraction peak the approximate extent of the periodic array from which the diffraction pattern comes. An example of the relation between the number of holes and the sharpness of the peaks is shown in Fig. 2.5.

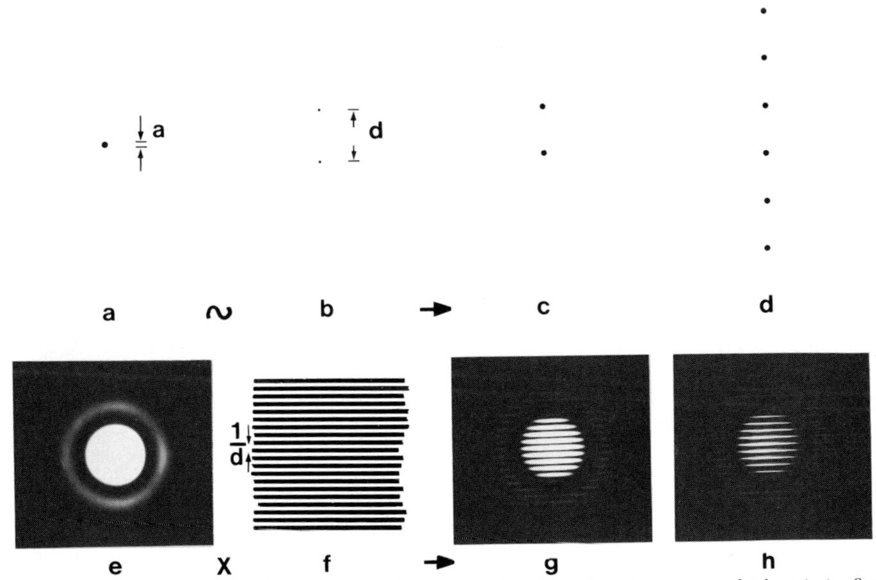

FIGURE 2.5. An array of two large holes (c) can be thought of as a convolution (~) of a single large hole (a) with an array of two small points (b). From the convolution theorem, the diffraction pattern (g) of the array (c) can be generated by forming the product (×) of the diffraction pattern (e) of (a) and the diffraction pattern (f) of (b). (d) An array of similar spacing to that in (c) but containing more points. Since the spacing in the array is unchanged, the diffraction peaks in (h) must be in the same positions as those in (g). The effect of extending the array is to make each of the diffraction peaks narrower or sharper (h).

Large Circular Apertures. So far we have considered diffraction from very small holes in the mask. What happens if we make the holes bigger? Figure 2.6a shows that if the diffracting object is a single large aperture, then its diffraction pattern can be calculated if the aperture is considered to consist of many very closely spaced pinholes. We therefore divide up the aperture into small areas, each of which is considered to be a point scatterer of light, and we add up all the waves from these point scatterers to work out the values of ϕ for which constructive interference occurs. In fact, the mathematics involved is complex, but for our purpose here it is sufficient to let the optical diffractometer do the work for us. Figure 2.6b shows a circular diffracting object (of diameter a) and its diffraction pattern Fig. 2.6(d). The latter consists of a central circular peak surrounded by a series of concentric rings of intensity. This pattern is commonly known as the Airy disk pattern. It can be shown that the middle of the gap (the first minimum) between the central peak and the first ring is at an angular separation (ϕ) from the center given by $\sin \phi = 0.61\ \lambda/a$. The form of the scattering function from an atom (taken to be spherical) is somewhat similar to this Airy disk pattern. Figures 2.6b,c show apertures of different diameters a_1 and a_2, and the aperture with the larger diameter gives the smaller diffraction pattern as expected

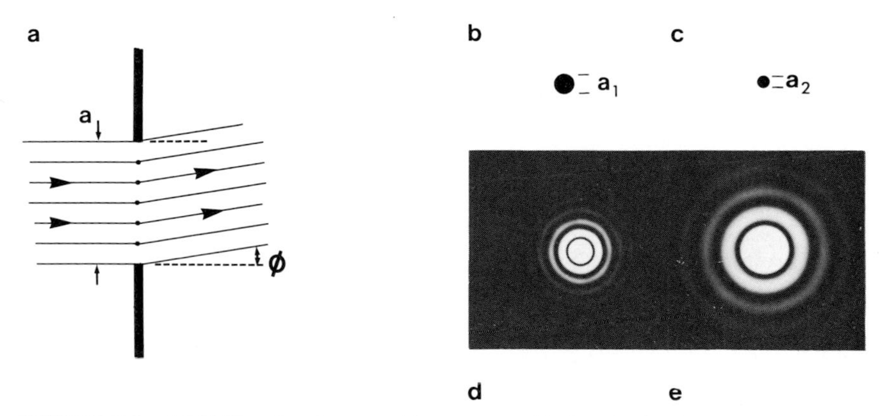

FIGURE 2.6. (a) Diffraction from a wide aperture. Each point in the aperture is considered to act as a point scatterer, and the waves from each point are summed to find where constructive interference occurs. (d and e) Optical diffraction patterns from circular holes (b) and (c). They consist of Airy disk patterns containing a central circular peak surrounded by concentric rings of gradually decreasing intensity. Since the hole in (b) is larger than the hole in (c), the whole Airy disk pattern in (d) is smaller (i.e., the rings are more closely spaced) than that in (e). Note that here and elsewhere in the book black in diagrams represents either high density or high intensity, whereas all photographs of optical diffraction patterns have intense regions shown as white on a black background.

from the reciprocal relation between the scale of a diffracting object and that of its diffraction pattern.

A Row of Large Apertures. We now consider the scattering from a row of scattering objects as in Fig. 2.4 but with each object a large circular aperture as in Fig. 2.6b. We can start by considering just two apertures, each of diameter a, separated by a distance d (Fig. 2.5c). Obviously, if we were to take just one aperture or just the other, the diffraction pattern of each would be as in Fig. 2.6d. But since we have two objects, the two diffraction patterns will interfere with each other, and we have already worked out the rule for this interference. If $n\lambda = d \sin \phi$, then constructive interference will occur; but if not, the waves will interfere destructively. The total effect is demonstrated in Fig. 2.5. Figure 2.5a shows a single large aperture, and Fig. 2.5e its diffraction pattern (an Airy disk). Figures 2.5b and f show an array of two very small points of separation d and its simulated diffraction pattern; Fig. 2.5c shows the combined effect of two objects as in a separated by a distance d as in b. The result is, in fact, that the amplitudes of the two diffraction patterns have been multiplied together (Fig. 2.5g) so that the underlying Airy disk pattern remains, but it is now crossed (sampled) by the interference lines that represent the diffraction pattern of two point objects separated by d. As described earlier, the effect of adding more large apertures to the array is that the individual interference lines become narrower (or sharper), as can be seen comparing Figs. 2.5g and h.

The effect on the diffraction pattern of placing a complicated structure (let us call it A) on a lattice (call it B) follows a general rule in diffraction theory and is an example of what is called the *convolution theorem.* The term convolution needs to be explained. We say that the lattice of objects in Fig. 2.5c is a convolution of one object (Fig. 2.5a) with the two points (Fig. 2.5b). We can think of a convolution of two objects as being produced physically when one object is picked up by some point (say the center of the aperture in Fig. 2.5a), and this point is then placed successively on each point in the second object (it is translated to each point without rotation). The convolution theorem then states that if an object C can be described as the *convolution* of two simpler objects A and B (written $A \sim B$), then the diffraction pattern of C is the product of the diffraction patterns of A and B. (Here the term *product* means that at a particular point in the diffraction pattern of C, the amplitudes of the patterns of A and B at the same point have been multiplied together.)

In Fig. 2.5, the convolution of (a) with (b) gives (c), so the diffraction pattern of (c) is obtained by multiplying the diffraction patterns (e) and (f) of (a) and (b).

In fact, the converse of this is also true, and the convolution theorem also states that if C can be thought of as a product of two simpler objects A and B, then the diffraction pattern of C is the convolution of the diffraction patterns of A and B. An example of this aspect of the theorem will be given a little later on.

2.2.3. Diffraction from Two-Dimensional Arrays

Introduction. The convolution theorem is a valuable simplifying tool in diffraction theory, and liberal use of it will be made in this section. It is hoped that the first few examples given in the following pages will help to demonstrate both the idea of convolution and the application of the theorem. Here we consider the diffraction pattern of a two-dimensional array of points. The sort of mask we are dealing with is shown in Fig. 2.7c. Figures 2.7a and b show that this kind of array can be considered as the convolution of a one-dimensional array of points (Fig. 2.7a) with another one-dimensional array (Fig. 2.7b) running in another direction.

We can think of picking up one array by one of its points (say point A in a) and then translating this point and placing it successively on the points of the other array (b). From the convolution theorem all we need do is multiply together the diffraction patterns (two sets of parallel lines;

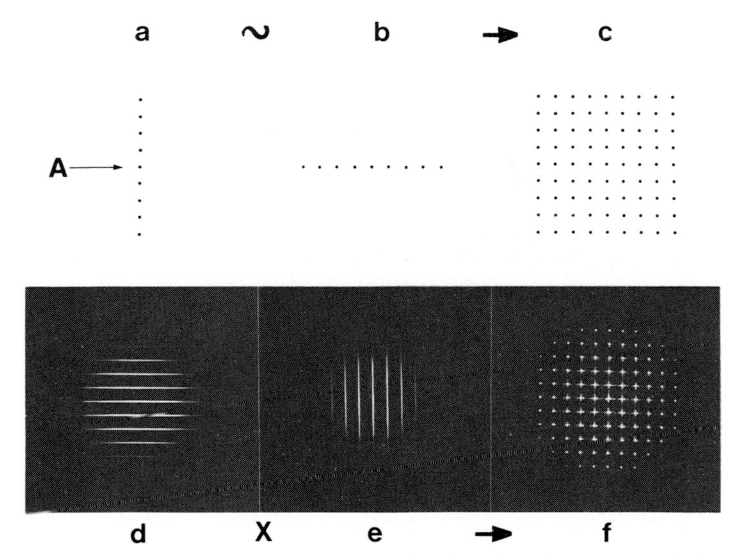

FIGURE 2.7. A two-dimensional array (c) can be thought of as the convolution of a 1-D array (a) with another 1-D array (b) of different orientation. The product of the diffraction patterns (d and e) of the two 1-D arrays therefore generates the diffraction pattern (f) of the 2-D array. The diffraction pattern of a 2-D array is therefore a 2-D array of diffraction peaks.

Figs. 2.7d and e) from the two one-dimensional arrays, and the diffraction pattern of the two-dimensional array is generated. Figure 2.7f shows that the diffraction pattern is just what would be expected. The product of the two intersecting sets of parallel lines is just a two-dimensional array of points. The diffraction pattern of a lattice of points is therefore another lattice of diffraction peaks.

Let us think for a minute what this means. It means that we could have expressed the diffraction pattern itself as a convolution; it is the convolution of one row of diffraction spots (Fig. 2.8a) with another row (Fig. 2.8b). A row of points like this is the diffraction pattern of a set of lines (i.e., a grating) orientated at right angles to the direction of the row of points. This is shown in Figs. 2.8d and e. Two such gratings (as in Figs. 2.8d and e) multiplied together produce an array of points as in Figs. 2.7a and 2.8f. If the lines in the gratings are at spacings d_1 and d_2, then the two sets of points that convolute together to give the diffraction pattern (Fig. 2.8c) will be at regular positions defined by diffraction angles ϕ_1 and ϕ_2 given by $d_1 \sin \phi_1 = n\lambda$ and $d_2 \sin \phi_2 = n\lambda$ as before. This means that if we are presented with a diffraction pattern as in Fig. 2.8c, then we can work out the separations d_1 and d_2 of the two sets of lines that generate the two-dimensional object lattice.

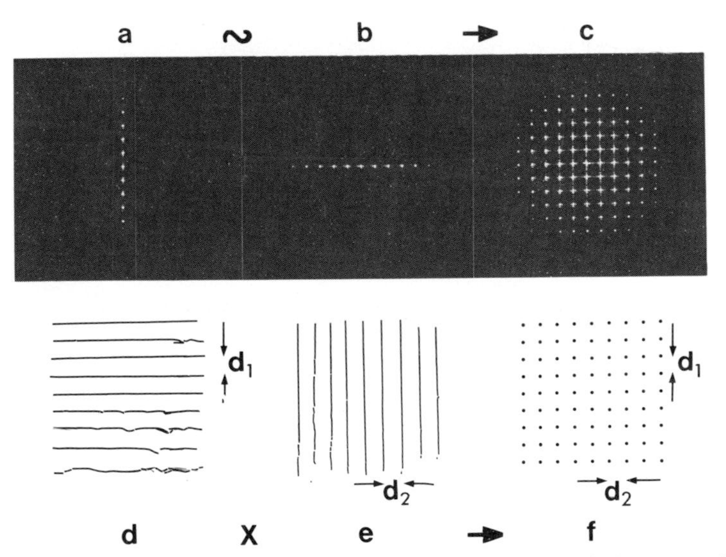

FIGURE 2.8. Following from Fig. 2.7, the diffraction pattern (c) of a 2-D array (f) can be thought of as the convolution of two diffraction patterns (a) and (b), each of which is a row of spots. The object that would give a diffraction pattern such as (a) or (b) is a set of parallel straight lines of the appropriate separation (d_1 or d_2) given by $n\lambda = d_1 \sin \phi_1$ and $n\lambda = d_2 \sin \phi_2$. ϕ_1 and ϕ_2 are the angles of diffraction for the nth order in (a) and (b), respectively. The product of the two sets of parallel lines (d and e) is the required 2-D array (f) as in Fig. 2.7(c).

Clearly Figs. 2.7 and 2.8 represent two alternative ways of thinking about the diffraction pattern from the same two-dimensional lattice. In fact, it is usual to think of the problem as in Fig. 2.7 (in terms of point sources of secondary radiation) but to imagine all of the points to be situated on a lattice defined by sets of imaginary lines (Fig. 2.9). Here it will be seen that there is in fact a very large number of different sets of straight lines that can be drawn through two or more points in the lattice. Alternative sets are shown in Fig. 2.9a. Clearly, the two-dimensional lattice of points could have been generated by choosing any two of these sets of lines. Each set would give a diffraction pattern consisting of a row of points at right angles to the lines (Fig. 2.9c), and these two patterns would convolute together as in Fig. 2.8 to generate the diffraction pattern of the two-dimensional lattice. But whatever sets of lines are chosen,

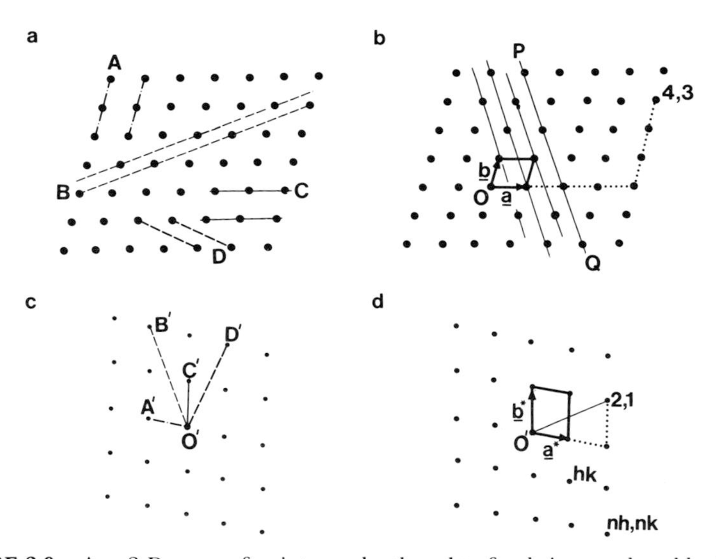

FIGURE 2.9. Any 2-D array of points can be thought of as being produced by forming the product of any two of a large number of alternative sets of parallel lines such as those shown in (a). Each set of lines can be thought of as producing its own set of diffraction peaks in a direction perpendicular to the direction of the lines (c). Thus A′, B′, C′, and D′ in (c) can be thought of as coming from sets of lines parallel to A, B, C, and D in (a). Such a 2-D array (b) can be conveniently described in terms of a repeating unit, the unit cell (bold outline), whose sides are the vectors **a** and **b**. If a set of lines such as those parallel to **PQ** cut **a** into h segments and **b** into k segments, then those lines are called the (h,k) lines. Here **PQ** is a (2,1) line. The diffraction pattern (d) from (b) has peaks in it that can also be defined in terms of h and k. The fundamental (first order) diffraction peak from the (h,k) lines is called the h,k reflection, and the nth order is the nh,nk reflection. The 2,1 reflection from the lines **PQ** in (b) is shown in (d). The diffraction pattern can itself be described in terms of two vectors **a*** and **b*** that define a reciprocal unit cell in the reciprocal lattice. The h,k reflection is reached by going h**a*** along **a*** and then k**b*** parallel to **b***.

rays. Each atom gives a diffraction pattern (Fig. 2.11d) somewhat similar to the Airy disk pattern (Fig. 2.5c). The X rays scattered by each atom interfere with each other to give the characteristic diffraction pattern of the whole crystal. This time the unit cell in the lattice of three-dimensional motifs is defined by three vectors **a, b,** and **c** (which together define a parallelopiped), and the atoms can be thought to lie on different sets of crystal planes defined by three indices h, k, and l. The values of h, k, and l are determined using exactly the same rule as before, but now

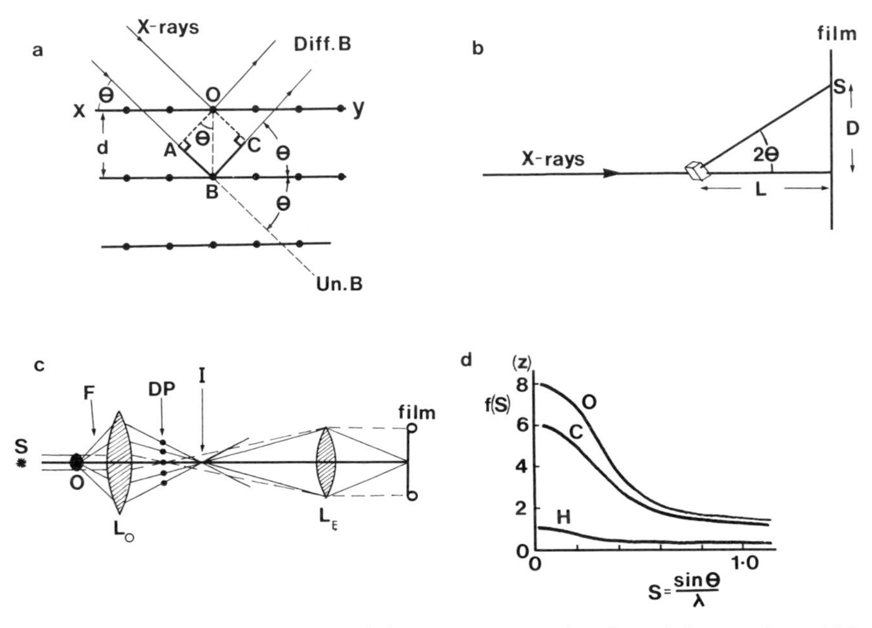

FIGURE 2.11. (a) Bragg suggested that X-ray diffraction from 3-D crystals could be thought of in terms of the reflection of X rays from successive crystal planes and their subsequent interference. The beam through O in (a) travels a distance $AB + BC$ less than the beam through B. AB and BC are both equal to $d \sin \theta$, where θ is the angle between the incident X rays and the crystal planes. Constructive interference therefore occurs when $AB + BC = 2d \sin \theta = n\lambda$ (Bragg's Law). A diffraction pattern (b) recorded at a distance L from the object has a peak at a distance D from the center of the pattern; hence, $2\theta = \tan^{-1}(D/L)$. (c) In the light microscope, the light from the source (S) is diffracted (F) by the object (O); the diffracted beams are collected by the objective lens (L_o) which forms an image at I that is enlarged by the projector lens (L_E) and focused onto the film. The diffraction pattern can be observed in the back focal plane of the objective lens (DP). Unfortunately, X rays cannot be focused in this way, and all that can be done is to record the X-ray diffraction pattern from the object at a position equivalent to F in the light microscope. But in that record, the phase information contained in the various diffracted beams is destroyed. (d) The atomic scattering factors $f(S)$ of three different atoms are shown as a function (S) of scattering angle (2θ), where $S = \sin \theta/\lambda$. The atomic scattering factor is spherically symmetrical about the origin, and at the origin $f(S) = Z$, the atomic number of the atoms.

we imagine the vectors **a, b,** and **c** to be divided into *h, k,* and *l* segments, respectively, by a set of planes of atoms.

A three-dimensional set of points can be thought of as being produced by the convolution of two-dimensional lattice with a row of equidistant points in the third dimension. The result is a diffraction pattern consisting of a three-dimensional array of point diffraction peaks. In other words, possible diffraction directions are described by a three-dimensional (3-D) reciprocal lattice defined by vectors **a*, b*,** and **c*,** any point on the lattice being defined by the indices *h, k,* and *l.* In fact, it is difficult on this basis to imagine what is going on in three dimensions, but the whole problem was made easier to consider because W. L. Bragg suggested a relatively simple way of discussing diffraction from crystals (see Bragg, 1933). He realized that the scattering from every atom in a two-dimensional layer of a crystal (such as the layer *xy* seen on edge in Fig. 2.11a) will be in phase provided the scattered beam makes the same angle (θ) to the atom layer as does the incident beam. It is almost as if the layer reflects the incident X-ray beam. Successive parallel layers of atoms would each reflect X rays, and the reflected beams would interfere with each other as exemplified in Fig. 2.1. The resultant amplitude would depend on the relative phases of the various reflected beams; these, in turn, would depend both on the separation, *d,* of the atom layers and on the angle (θ) that the layers make with the X-ray beam. Figure 2.11a shows that the path difference between the beams reflected from two successive atom layers (i.e., $AB + BC$) is equal to $2d \sin \theta$. The condition for constructive interference to occur must therefore be $n\lambda = 2d \sin \theta$, a condition that is usually called *Bragg's Law.* Note that here the angle of diffraction is 2θ and that this rule is different from the one that applies in optical diffraction ($n\lambda = d \sin \phi$). Bragg's Law enables us to work out the spacings (d) of the relevant layers (or planes) in the crystal if λ is defined and we can determine the value of θ. Experimentally, this is fairly easily done as shown in Fig. 2.11b. The diffraction pattern from the crystal is recorded on a flat film at a distance L from the crystal. A diffraction spot (S) is seen at a distance D from the center of the diffraction pattern. The angle of diffraction, 2θ, is then given by $\tan 2\theta = D/L$, and d is given by $d = n\lambda/2 \sin [\frac{1}{2} \tan^{-1} (D/L)]$.

Finally, one of the most crucial features of diffraction from crystals should be noted. If the crystal planes separated by d are not tilted to the X-ray beam (of wavelength λ) at the angle θ given by Bragg's Law for that d value, then no diffraction peaks from those crystal planes will be observed. As a result of this, diffraction corresponding to only a few reciprocal lattice points will be recorded when a single crystal is placed with a fixed orientation in a monochromatic X-ray beam. More points can be recorded either if the crystal is rotated (rotating crystal method), or if a powdered (polycrystalline) sample is used as the specimen (Debye–Scherrer method), or if the X-ray beam contains many wavelength components (Laue method). For details of these methods see, for example, Woolfson (1970).

The Phase Problem. Clearly, the principal use of X-ray diffraction is to enable one to determine, from the diffraction pattern, the structure of the diffracting object (i.e., the motif). In theory, the diffracted X rays carry all of the information needed to do this. But, unfortunately, as soon as the diffraction pattern is recorded, a crucial part of the required information is lost. As a result, it is not usually possible to work out the structure of the motif directly from analysis of the recorded diffraction pattern. Reference can be made to the principle of image formation in the simple light microscope to illustrate the problem. Figure 2.11c illustrates the optics of image formation in such a microscope according to Abbé's theory. An object is illuminated by parallel light, and this light is scattered by each point in the object (0) into a number of beams (F) corresponding to different orders of diffraction. The diffracted beams are then collected and focused by the objective lens (L_O) onto a plane (DP). In this plane each diffraction peak has a specific amplitude and relative phase. Finally, the eyepiece or projector lens L_E of the microscope adds the diffracted beams together and combines them to form an enlarged image. The image is actually the diffraction pattern of the diffraction pattern—this is another important property of diffraction and image formation. The crucial feature about the lenses involved is that they preserve the relative phases of the various diffracted beams right through the microscope, so that these beams interfere correctly to produce a good final image of the object.

Unfortunately, comparable X-ray lenses cannot be made, and so it is simply not possible to produce an X-ray microscope (there are other problems too). All that can be done is to place a film in a position analogous to F in Fig. 2.11c, thereby recording the diffraction pattern from the object. But when this is done, all of the information about the relative phases of the diffracted beams is destroyed, and without this vital phase information, an image of the object cannot be reconstructed. This problem is referred to as the *phase problem* in X-ray diffraction, and much attention has been directed to finding ways in which the phase problem can be solved or circumvented. The reader is referred to Blundell and Johnson (1976) for a discussion of the various methods now available for solving the phase problem when dealing with proteins. Suffice it to say here that with certain specimens it can be done (although with difficulty) and that the structures of a rapidly increasing number of complex protein molecules (such as myoglobin, hemoglobin, lysozyme, etc.) have now been determined.

2.3. Diffraction from Helical Structures

2.3.1. Importance of Helices

The molecules of which muscle is composed (i.e., proteins) are polymers. Polymers are molecules formed from repeating subunits joined together end to end like beads on a string (Fig. 2.12a). It will be shown in Chapter 4 that the monomers along one protein chain need not all be identical; indeed, their differences are crucial and define the structure and properties of a particular protein. But protein monomers (amino acids) are all variations on a single type of structure. For our purpose here it is therefore sufficient to consider initially the properties of polymers of identical subunits.

A three-dimensional crystal is one of nature's ways of packing together identical motifs so that they are all in exactly equivalent (i.e., identical) environments. What is obtained is a minimum energy configuration and, therefore, one that is highly stable. Exactly the same situation occurs with polymer molecules. In fact, there is only one type of configuration a polymer molecule can adopt such that all of the monomers along the chain have identical situations. As shown in Fig. 2.12b, this configuration is a helix. Two apparently simpler structures, a straight chain molecule (Fig. 2.12a) and a circular molecule (c), can both

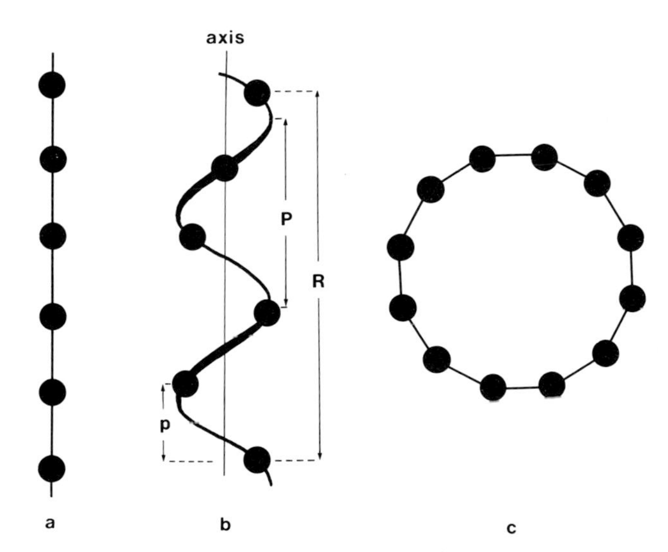

FIGURE 2.12. Conformations of polymer molecules in which all the subunits or monomers (black circles) are equivalent: (a) Straight chain; (b) helix of pitch P, repeat R, and subunit axial translation p; (c) circular molecule. The forms in (a) and (c) can be thought of as special cases of (b).

be considered as special cases of the helix, one with infinite pitch and one with zero pitch. The monomers on the helix can be thought of as analogous to the steps on a very long spiral staircase. Each is exactly equivalent to every other monomer (or step) on the spiral. The amount climbed vertically by each monomer (step) is called the *subunit axial translation* (p); the amount climbed in one turn of the spiral is called the *pitch* (P); and the amount climbed to reach a monomer (step) exactly over the first (lowest) is called the axial repeat distance (R). If there is a whole number of monomers (steps) in one turn of the helix (i.e., $P/p = N$, N integral), then R and P are clearly identical.

Since virtually all of the important structures in the muscle myofibril involve either helical arrangements of atoms or helical arrangements of molecules, it is clearly of fundamental importance to know something about the general form of the X-ray diffraction pattern from such helical structures. Fortunately, we can make use one again of the optical diffractometer to illustrate some of the principles involved.

2.3.2. The Continuous Helix

What we are trying to find is the form of the diffraction pattern from a three-dimensional array of helical molecules in which the monomer contains several atoms. This may seem a formidable task—indeed the mathematical formulation involved in a rigorous derivation is quite sophisticated (see Fraser and MacRae, 1973). But as before, it is our purpose to avoid such mathematics, and we shall rather attempt to express the 3-D crystal of molecules as a series of convolutions and products of much simpler structures of which we already know (or can easily find) the form of the diffraction pattern.

In this approach, we recognize first of all (Fig. 2.13c) that a single continuous helical wire (e.g., a coiled spring) of pitch P can be thought of as a convolution of one turn of the helix (a) with a row of equidistant points of separation P aligned along the axis of the helix (Fig. 2.13). The diffraction pattern from a row of points (P apart) is a set of lines of intensity (or planes in 3-D) orientated perpendicular to the row and of spacing proportional to $1/P$ (Fig. 2.13e). If we can find the diffraction pattern of a single turn of the helical wire, we can therefore use the convolution theorem to find the diffraction pattern of the whole helix. When the diffraction pattern of one turn is calculated, the wire is thought of as being composed of many small diffracting elements along a helical path, and the scattered X-rays from each element are added together to give the form of the diffraction pattern. Once again we can let the optical diffractometer show us what the answer looks like. Figure 2.13h shows that the diffraction pattern of a single turn of a helix has the

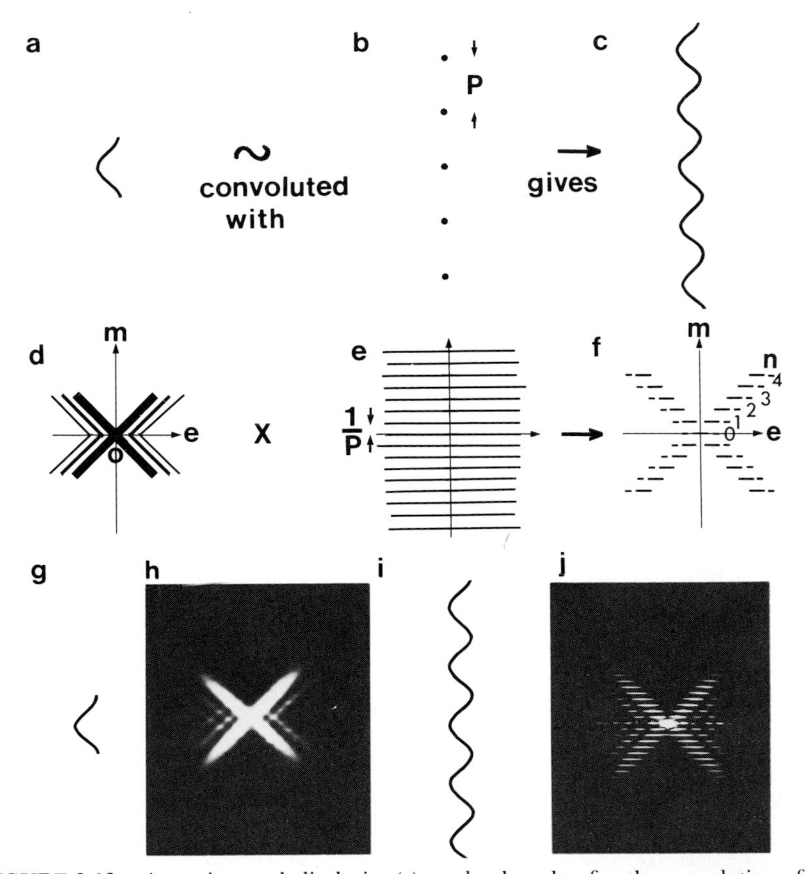

FIGURE 2.13. A continuous helical wire (c) can be thought of as the convolution of one helical turn (a) with a set of points P apart arrayed along the helix axis (b). The diffraction pattern (f) of the continuous helix is therefore the product of the diffraction pattern (d) (the helix cross) of (a) and the diffraction pattern (e) of (b) (a set of parallel lines $1/P$ apart and perpendicular to the helix axis). The result (f) is a helix cross that has been broken up into parallel layers that can be described in terms of a number n starting with $n = 0$ through the origin. This is the equator of the pattern. As confirmation of the theory, (h) is the optical diffraction pattern of one helix turn (g) as in (a), and (j) is the optical diffraction pattern of the extended continuous helix (i). The peaks on the layers in the cross gradually get weaker and move further from the axis (meridian) of the diffraction pattern (marked m in f). The value of n, therefore, gives a measure of the position and strength of the nth layer for a helix of a particular radius. Note that if the helix in (i) had a larger radius, the arms of the helix cross would be closer to the meridian than in (j), and vice versa.

form of a cross (or **X**) of intensity with subsidiary peaks on the outsides of the central **X** shape (simulated in Figure 2.13d). One important feature of this pattern is that there is no intensity [except at the center (*0*) of the pattern] on the axis (meridian) of the diffraction pattern. The meridian is the direction (**m**) in the diffraction pattern (Fig. 2.13d) which passes through *0* and is parallel to the helix axis. This absence of meridional intensity is a characteristic diffraction feature of continuous helical structures. Note also the effect of altering the radius of the helix (not illustrated): if the radius of the helix is larger, then the arms of the cross move nearer to the meridian and vice versa.

To generate the diffraction pattern of the whole helical wire, a convolution (Fig. 2.13c) of one helical turn (Fig. 2.13a) with a row of points of separation P (Fig. 2.13b), we need to multiply the cross in Fig. 2.13d by the diffraction pattern of the row of points (i.e., a set of parallel lines $1/P$ apart in Fig. 2.13e), and the result (Fig. 2.13f and j) is a cross that has been broken up into horizontal layers. Since the separation of the layers is proportional to $1/P$, a diffraction pattern such as that in Fig. 2.13j can be used to work out the helix pitch. The layers are usually numbered such that the nth layer is at a distance $Z = n/P$ from the equator ($n = 0$) of the pattern (**e** in Fig. 2.13f). It will be seen that as n increases, the intensity on the layer moves out to a larger radius from the meridian and also diminishes. The value of n therefore gives a measure of the strength and position of a peak in the helix cross for a given helix radius.

2.3.3. The Discontinuous Helix

Simple Helices. So far we have been talking about a continuous uniform helical wire. But helical polymers are discontinuous in the sense that the chain is broken up into monomers. For simplicity, we can start to approach this modification by considering a helical arrangement of point scatterers (Fig. 2.14c). Such a structure can be thought of as being the product of a continuous helix (Fig. 2.14a) and a set of planes of unit density orientated perpendicular to the helix axis and of separation p, where p is the subunit axial translation (Fig. 2.14b). The planes of unit density would be very thin and would be separated by regions of zero density. Since the discontinuous helix can be thought of as the product of (a) and (b), the convolution of their diffraction patterns will generate the diffraction pattern of the helix. The planes of density of separation (p) will give rise to a pattern of point diffraction peaks along the meridian of the pattern with spacings proportional to $1/p$ (Fig. 2.14e). The pattern from the continuous helix (Fig. 2.14d) then needs to be placed successively on each of these points to generate the whole pattern (Fig. 2.14f). The result is a multiple cross pattern in which the different crosses may (and usually do) overlap each other to give a seemingly very complex result. The peak at b in Fig. 2.14f is called the first meridional reflection, and its spacing tells us the axial separation (p) of the

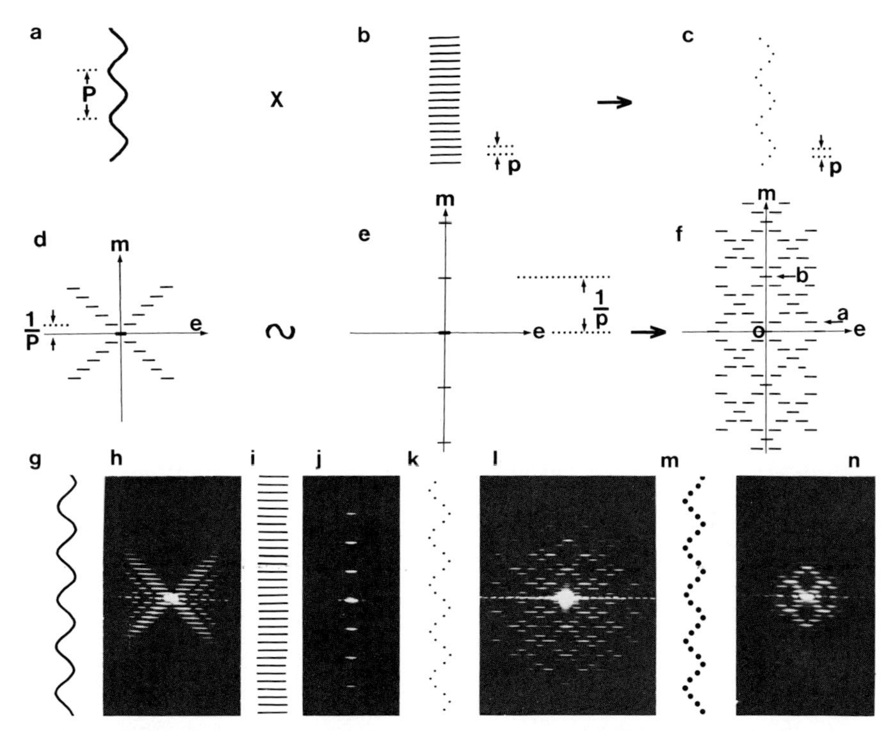

FIGURE 2.14. A discontinuous helix is the product of a continuous helix (a) and a set of planes of spacing p (the subunit axial translation) arrayed perpendicular to the helix axis. The diffraction pattern (f) of (c) is therefore the convolution of the diffraction patterns (d) and (e) from (a) and (b), respectively. It consists of a series of helix crosses centered on a series of meridional positions m/p from the origin. Reflection b in (f) is the first meridional reflection ($m = 1$), and its spacing is therefore related directly to the value of p. Figures (g) to (l) show the optical diffraction equivalents of Figs. (a), (d), (b), (e), (c), and (f), respectively, where all of the predicted features can be seen. A helix of atoms (m) can be thought of as a convolution of one atom with the discontinuous (point) helix in (c) or (k). The diffraction pattern (n) from the helix of atoms is therefore the product of (l) with the diffraction pattern from one atom (Fig. 2.6e). Only the central part of the pattern is strong.

monomers along the helix. Other meridionals occur at spacings from the origin (*0*) proportional to m/p where m is an integer (positive, negative, or zero).

A helix of monomers (Fig. 2.14m) where each monomer is a single atom (taken to be spherically symmetrical) can be thought of as a convolution of one atom (Fig. 2.6a) with a helix of points (Fig. 2.14k). This means that the diffraction pattern from the point helix needs to be multiplied by the pattern from a single atom (see Fig. 2.6e). The result is shown in Fig. 2.14n; only the central part of the pattern is strong. As mentioned earlier, the atomic scattering function is somewhat similar to the Airy disk pattern. In fact, in the two-dimensional model illustrated in Fig. 2.6m, the pattern actually is an Airy disk pattern. In three dimensions, the atomic scattering function is slightly different; in particular, the intensity falls off more slowly at higher angles of diffraction (Fig. 2.11d).

Nonintegral Helices. To summarize the conclusions so far, the form of the diffraction pattern from a discontinous helix of single atoms has been visualized by simple arguments based on the convolution theorem. It has been found that the layer separation in the helix cross from a continuous helix is related to the pitch (P) of the helix and that the additional helix crosses that are generated if the helix is discontinuous are centered on meridional reflections at spacings (m/p) governed by the subunit axial translation (p) in the helix. Unfortunately, the apparent simplicity of this result is complicated to some extent when there is a nonintegral number of subunits per turn (i.e., P/p is nonintegral). This means that the layers in the helix cross centered on the first meridional reflection (for example) do not coincide with the layers in the cross

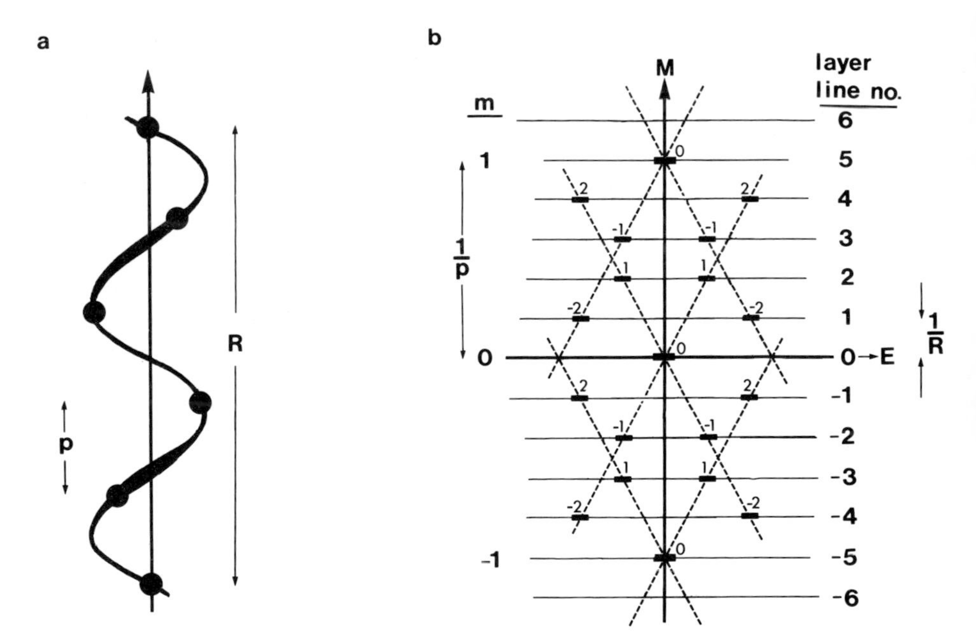

FIGURE 2.15. The diffraction pattern from a nonintegral helix (a) is more complex than Fig. 2.14 but is derived in a similar way. A set of helix crosses with layers separated by n/P from their origins are placed with their origins on meridional positions m/p from the center of the diffraction pattern. Peaks will therefore appear at distances from the equator equal to $m/p + n/P$. These peaks are said to lie on a set of layer lines (layer planes in 3-D) such that the layer line number (l) is given by $l = nt + um$, where there are exactly u monomers in t turns of the helix (for details see text). An example of such a helix will be discussed in Chapter 5.

centered on the origin (Fig. 2.15). In general, a layer of intensity will occur at a distance Z along the meridian from the origin where

$$Z = \frac{n}{P} + \frac{m}{p} \tag{1}$$

and m and n are integers as defined before. The true repeat (R) of the helix will occur after t turns of the helix if Pt/p is integral (say u). u is then the number of monomers in t turns of the helix giving a repeat R. From this,

$$R = Pt = pu \tag{2}$$

The whole of the diffraction pattern is then defined in terms of a series of so-called *layer lines** (layer *planes* in 3-D) that have spacings Z related to the true repeat R of the helix such that

$$Z = \frac{l}{R} \tag{3}$$

and l is called the layer line number or index. Combining (1) and (3):

$$Z = \frac{l}{R} = \frac{m}{p} + \frac{n}{P}$$

which gives

$$\ell = ZR = \frac{mR}{p} + \frac{nR}{P} = \frac{m(pu)}{p} + \frac{n(Pt)}{P}$$

when (3) is incorporated. This results in the general expression $l = um + tn$ which enables one to calculate the values of n that are possible on a particular layer line of number l if the number of subunits u in t turns of the helix is known. Remember that m can have any integral value, positive, negative, or zero. Earlier it was shown (Fig. 2.13f) that the value of n gives the position and intensity of the reflection (for a given helix radius) on a particular layer in the helix cross, and on this basis an easier way to determine the n value on a particular layer line is to count the layers from the origin of the particular helix cross involved (at m/p).

2.3.4. Complex Helical Molecules

Most helical molecules have monomers consisting of a group of several atoms rather than just one. When this is the case, equivalent

*Note that here and elsewhere in the book the term layer refers always to a single helix cross, whereas layer line refers to the whole diffraction pattern from a discontinuous helix.

atoms in successive monomers can be considered to lie on their own helix of pitch P at its own particular helix radius. Different atoms in the monomer will be on coaxial helices of different radii and at different orientations around and translations along the helix axis. But each will have the same pitch P and subunit axial translation p. Each helix of atoms will give rise to a diffraction pattern similar to that in Fig. 2.16b, and if there are N atoms in the monomer, then N diffraction patterns of this kind need to be added together vectorially to give the diffraction pattern of the whole helical molecule. Figure 2.16e shows an example of this in which the monomer contains two atoms. The general effect is to produce rather complex distributions of intensity along each layer line (Fig. 2.16f).

The conclusion from this is that the distribution of intensity along a layer line gives information about the arrangement of atoms in the monomer, but the axial positions of the layer lines give information about the symmetry of the helical arrangement of monomers.

2.3.5. Three-Dimensional Arrays of Helical Molecules

Helical molecules often packed together to form 3-D arrays or crystals. In this case, the array can be considered to be a convolution of one long helical molecule with a 2-D lattice as illustrated in Fig. 2.17a. The repeat in the third dimension is obviously the repeat R of the helix itself. The diffraction pattern of the 2-D lattice consists of a series of lines of intensity orientated parallel to the helix axis and to the meridian of the diffraction pattern. The pattern in Fig. 2.17c from a single helix will therefore be multiplied by this line diffraction pattern so that it is, in effect, "sampled" by many lines (called *row lines*) parallel to the meridian as in Fig. 2.17e. The positions of these row lines therefore tell us about the lateral packing of the helical molecules. A similar effect would be observed if the individual molecules were more complex, as in Fig. 2.16e. Note also that even if there is no axial register between the different helices, layer lines will still appear, but only the equator of the diffraction pattern will be sampled.

2.3.6. Summary

We have now achieved our objective of demonstrating the form of the diffraction pattern from a 3-D crystalline array of polymer molecules. It is therefore appropriate to summarize briefly the conclusions that have been reached.

First, the whole pattern will consist of a series of layer lines (Fig. 2.17f), some or all of which may be sampled by the reciprocal lattice points of the 2-D array on which the molecules are situated. The sam-

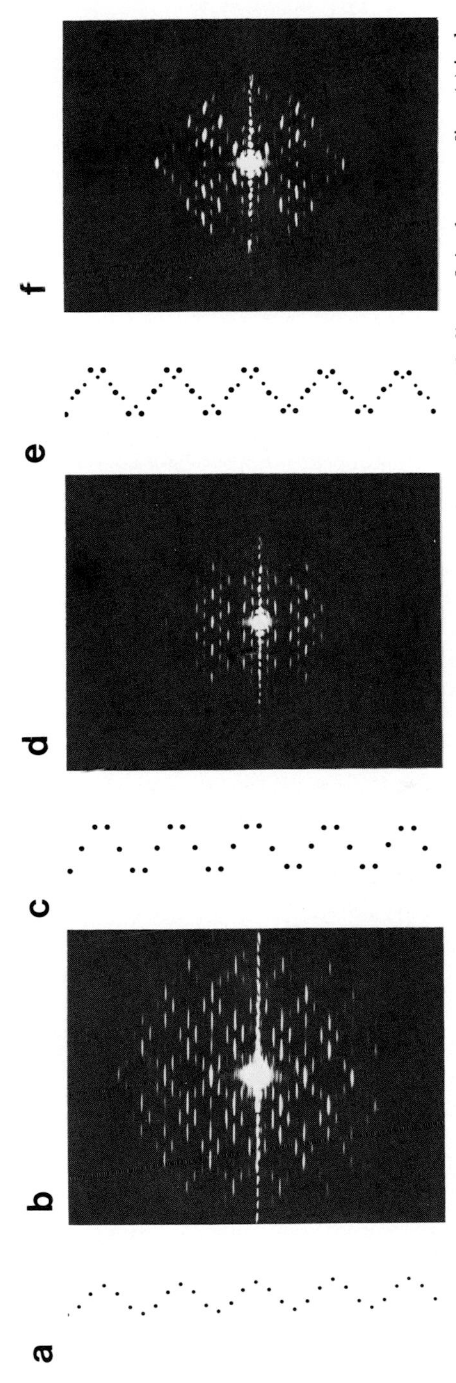

FIGURE 2.16. A helix in which each monomer is two (or more) atoms (e) is the sum of two (or more) helices of single atoms [i.e., (e) is the sum of (a) and (c)]; (b) and (d) are the diffraction patterns of (a) and (c), respectively, and adding these diffraction patterns together (taking account of the relative phases at each point in the diffraction pattern) gives the total diffraction pattern (f) from the complex molecule (e).

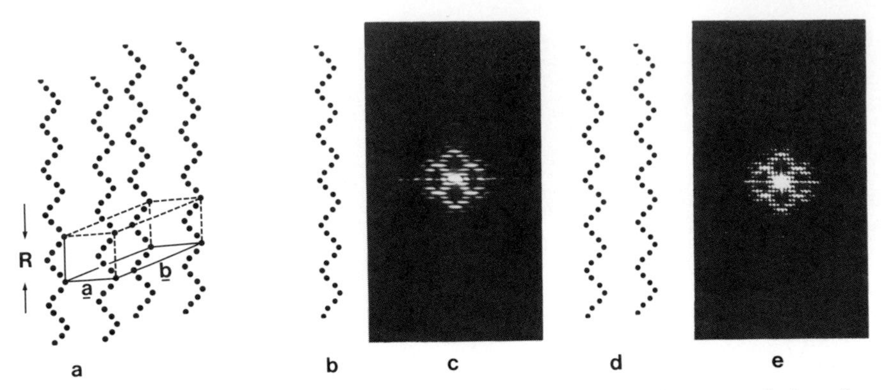

FIGURE 2.17. A crystal of helical molecules (a) can be thought of as a convolution of one molecule with a 2-D lattice defined by lattice vectors **a** and **b.** The repeat R of the helix defines the third lattice vector **c.** The diffraction pattern (c) from one molecule (b) must therefore be multiplied by the appropriate reciprocal lattice to generate the diffraction pattern of the whole crystal. As a simple example of this, (d) shows an array of two molecules (a convolution of one molecule with two points), and (e) shows its diffraction pattern. This is similar to (c) except that it is now crossed by interference lines parallel to the meridian. These lines (called row lines) are just the diffraction pattern of the two-point array on which the helices are situated.

pling peaks lie along lines called *row lines,* and the spacings of these from the meridian (i.e., along the equator of the diffraction pattern) tell us about the size and shape of the 2-D lattice on which the helices are arranged. The relative intensities of the peaks on the different layer lines will depend on the structure of an individual helical molecule. Intensity on the meridian occurs at spacings from the equator related to m/p where p is the subunit axial translation. The layer lines themselves occur at positions (z) equal to l/R where R is the repeat of the helix and l is the layer line number.

It is therefore generally possible to determine directly from the diffraction pattern the values of p and R and the shape and size of the unit cell of the 2-D array on which the helical molecules are arranged. The value of the pitch P is often more difficult to determine unless the helix is integral, in which case P and R are the same. The determination of the structure of a monomer from the overall intensity in the diffraction pattern is also a complicated procedure (requiring solution of the phase problem), but it is often possible to determine the mean radius of the helix from the overall intensity distribution provided that the pitch P is already known.

2.3.7. Multistranded Helices

Helical structures met with in biology are not always helical polymer molecules. Some are helical arrangements of globular molecules. In this case, the helix can be thought of as a helix of monomers in which each monomer is itself a molecule of high molecular weight. The thin and thick filaments in muscle are both examples of this type of structure. The theory of helical diffraction that has already been presented obviously applies equally well to this second type of helical structure, but some additional complexities often occur. For example, it often happens that the repeating subunits are arranged not on a single helix but on a set of identical coaxial helices. The result is a multistranded helix of molecules as illustrated in Fig. 2.18c. This example is a three-stranded continuous helix. The three helices are related by a threefold rotation axis parallel to the helix axis (i.e., they are related by a 120° rotation about this axis). If the helices are continuous, then the pitch (P) of the helix is effectively divided into identical repeats of length $P/3$ (or P/N for an N-stranded helix). The effect of this on the diffraction pattern is that layers in the helix cross of n (Fig. 2.13f) other than integral multiples of 3 (or N) are missing (Fig. 2.18d). If the helices are discontinuous, then a similar result is obtained, but now several such helix crosses occur centered m/p along the meridian. An example that occurs in Chapter 7 is a three-stranded helix with nine subunits per turn on each helix. Only layer lines on which $n = 3 \times$ integer are seen in the diffraction pattern.

This type of effect poses a problem when one is presented with a diffraction pattern and tries to work out the number of strands in the helix giving rise to it. In general, a layer line at a particular spacing ($1/R$) could be given by any N-stranded helix for which N/P is equal to $1/R$. It

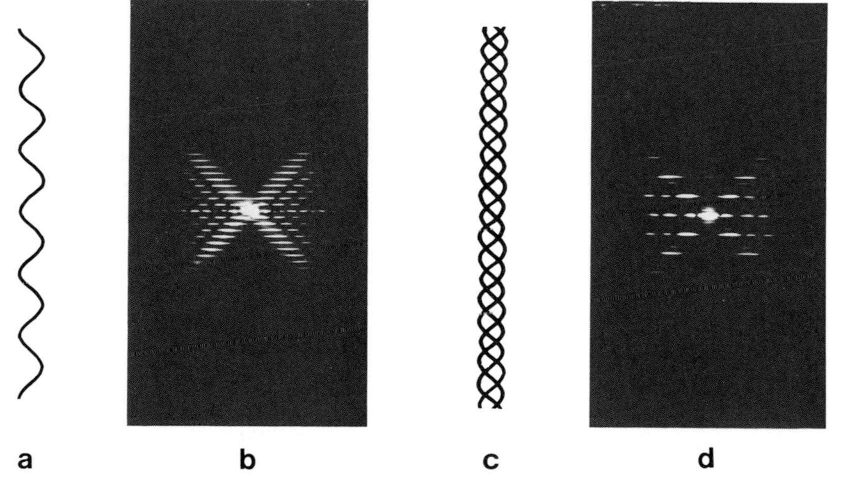

FIGURE 2.18. The effect of increasing the number of strands in a continuous helix from one (a) to three (c) is to cut out all layers in the helix cross except those for which $n = 3 \times$ integer. If there are N strands related by an N-fold rotation axis, layers are only seen when $n = N \times$ integer.

will be remembered that the position of a diffraction peak along a layer line depends both on the radius of the helix and on the n value of the layer of the helix cross which occurs on that particular layer line. In the case of the N-stranded helix of repeat P/N, the layers seen in the diffraction pattern will be associated with values of N such that $n = N \times integer$. Therefore, a particular layer line peak could come from any helix of pitch $P = N \times R$ provided that the radius of the helix is such that it brings the intensity peak to the right position for the particular n value involved. The general result of this is that the value of N or P cannot be determined unless the radius of the helix is known or, alternatively, that the radius cannot be found unless N (or P) is known. This ambiguity is one of the problems involved in determining the structure of the myosin filaments in vertebrate skeletal muscle, so we shall return to it in Chapter 7.

2.4. The Jargon of X-Ray Crystallography

Now that a nonmathematical outline of diffraction theory has been presented, it is appropriate to conclude by explaining some of the jargon that professional X-ray crystallographers use. It has already been mentioned that the diffraction pattern from a crystal of atoms can be considered as the result of adding together the secondary X rays from each atom as exemplified for two waves in Fig. 2.1. Each diffracted beam (where constructive interference occurs) is associated with one of the reciprocal lattice points with indices h, k, and l and can be described in terms of an amplitude and relative phase. These are both determined by the structure of the motif itself, and for this reason they are called the *structure amplitude* and the *phase*. The single mathematical expression that defines both the structure amplitude and the phase of the diffracted beam associated with the reciprocal lattice point (h, k, l), is usually called the *structure factor* and is written $F(h, k, l)$. The description of the diffraction pattern of an object in terms of its structure factors is called the *Fourier transform* of the object (after the scientist who originally formulated the mathematics involved). The process of reconstructing an image of the object from a knowledge of the structure factors for all of the different h,k,l values is called *Fourier synthesis*. An equivalent process is what occurs in the image-forming part of the light microscope. Mathematically, Fourier synthesis is equivalent to forming the Fourier transform of the Fourier transform of the object. Earlier the phase problem in X-ray diffraction was mentioned, and this clearly has to overcome in order to carry out a Fourier synthesis. One method of doing this is to use a technique called the *isomorphous replacement method*. Here heavy atoms are added to the molecules in a crystal without altering the crystal structure, and this produces a characteristic change in its diffraction pattern. By comparing the Fourier transform of the original crystal with the Fourier transform of crystals of two heavy-atom derivatives, it is possible to work out the phases of the structure factors of the original crystal. In this way, the structures of myoglobin, hemoglobin, lysozyme, and many other globular proteins have been worked out. As described later, similar techniques are now being applied to some of the muscle proteins.

2.5. Practical X-Ray Diffraction Methods

2.5.1. Introduction

Muscles are relatively weak diffracters of X rays, and the periodicities (*d* spacings) involved are large relative to atomic dimensions (up to about 1000 Å). This means that both very intense sources of X rays and special cameras for recording the diffraction patterns need to be used. In this section, the design of modern X-ray diffraction equipment and its application to muscle research is briefly discussed.

Figure 2.19a illustrates the basic technique involved in recording an X-ray diffraction pattern. An X-ray generator produces a beam of X rays consisting of many different wavelength components by bombarding a particular target material (usually copper) with electrons. In the simplest X-ray cameras, the X-ray beam is reduced in size by means of a set of collimating pinholes that select out a small pencil of X rays. The collimated beam then reaches the object where scattering occurs, and the resulting diffraction pattern is recorded on film. Because the X-ray beam contains many wavelength components associated with both a continuous and a line spectrum, the observed diffraction pattern would tend to be blurred, since each different wavelength will produce a diffraction pattern of a different size. For this reason, it is usual to try to make the X-ray beam monochromatic. The simplest method when using a Cu target is to place a thin foil of nickel in the beam between the X-ray source and the diffracting object. Figure 2.19b shows that the X-ray spectrum characteristic of Cu for a bombarding electron energy of 50 kV can be modified by preferential absorption by Ni of the Cu Kβ line relative to that of the Cu Kα line (Both lines are actually closely spaced doublets.). The latter ($\lambda = 1.5418$ Å) is the line normally used for diffraction work.

It is clear from Fig. 2.19 that, since the pinholes in the collimating system have a finite diameter, the diffraction spots recorded on the film

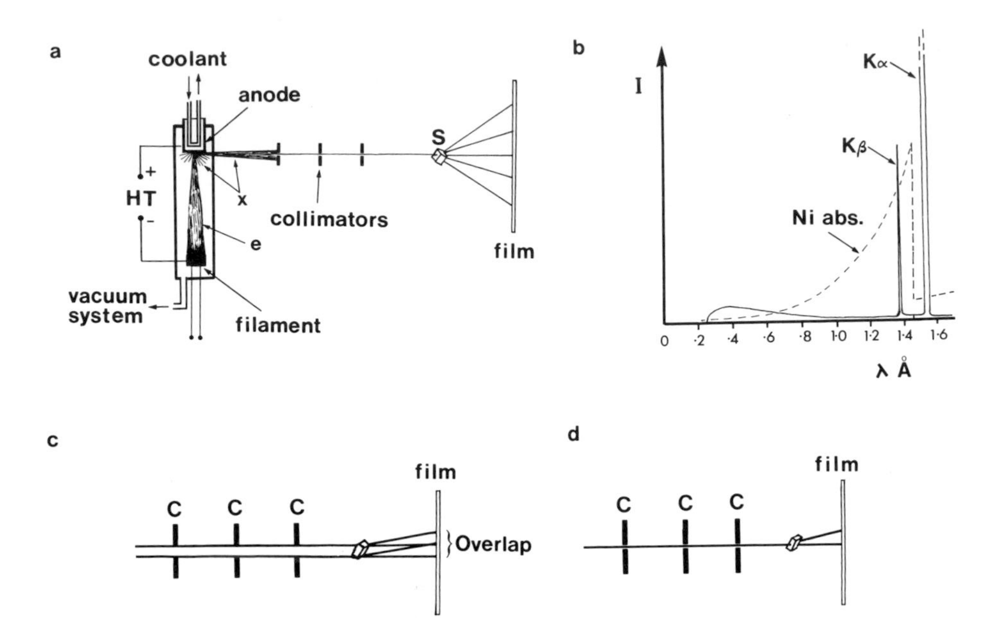

FIGURE 2.19. (a) Schematic representation of the head (tube) of a typical X-ray generator. A heated filament in the evacuated tube emits electrons (e) which are accelerated towards the anode. The electrons incident on the anode cause the emission in all directions of X-rays (×) either by the excitation of orbital electrons in the atoms of the anode material, which then drop down to lower energy levels accompanied by the emission of the characteristic line spectrum, or alternatively, by Bremmstrahlung (the continuum—see text). The process is very inefficient, much anode heating occurs, and the anode therefore needs to be cooled. The X-ray beam emitted through a suitable window in the tube (often a beryllium window) is collimated by a series of pinholes, and it then interacts with the specimen (S) where diffraction occurs. The diffraction pattern can then be recorded on an appropriately placed film. (b) When a monochromatic X-ray beam is required using a copper anode (usually the Cu Kα doublet line), a thin filter of nickel can be placed in the beam. Here, the unfiltered spectrum from copper is shown superimposed on the absorption spectrum for nickel. A nickel absorption edge (Ni abs.) falls between the Cu Kα and Cu Kβ peaks so that the Kβ line is preferentially absorbed when a nickel foil is placed in the beam. (c and d) The effect of altering the pinhole size in the collimation system. With large pinholes (c) the beam intensity is high, but low-angle reflections will overlap. Better low-angle resolution can be achieved if the pinholes are smalled (d), but a considerable loss in intensity is inevitable.

will have a size that depends on the width of the X-ray beam as defined by the collimator as well as on the diffraction properties of the object itself. If the crystal spacings (d) in the object are small (say up to 10 Å), then the beam size does not normally pose a problem. However, if large d spacings are involved, then the values of θ in Bragg's Law will be correspondingly small, and the possibility then arises that the collimated beam width at the film will be of the same order as (or greater than) the separation of the diffraction peaks (Fig. 2.19c). In principle, this problem could be overcome by reducing the size of the pinholes in the collimator (Fig. 2.19d). But a significant loss in beam intensity (for a given X-ray source) would clearly result, and with weakly diffracting objects, prohibitively long exposures (days or weeks) would be needed. For this reason the use of pinhole collimation is usually restricted to the study of specimens such as single crystals of proteins (or of other molecules) where only high-angle diffraction data are required.

2.5.2. Focusing X-Ray Cameras

Methods that have been devised to produce a small X-ray beam size at the film without a corresponding loss of X-ray intensity have all relied on the use of a variety of focusing devices for X rays. It has already been mentioned that it is not possible to produce X-ray lenses. An X-ray focusing device must therefore, of necessity, make use of mirrors. The principle of the focusing X-ray camera is relatively simple (Fig. 2.20). A curved X-ray mirror is placed close to the X-ray source so that it collects a relatively large part of the X-ray beam. The curvature of the mirror is then adjusted so that the beams reflected from different parts of the mirror all converge to a particular point, F, the focus of the camera.

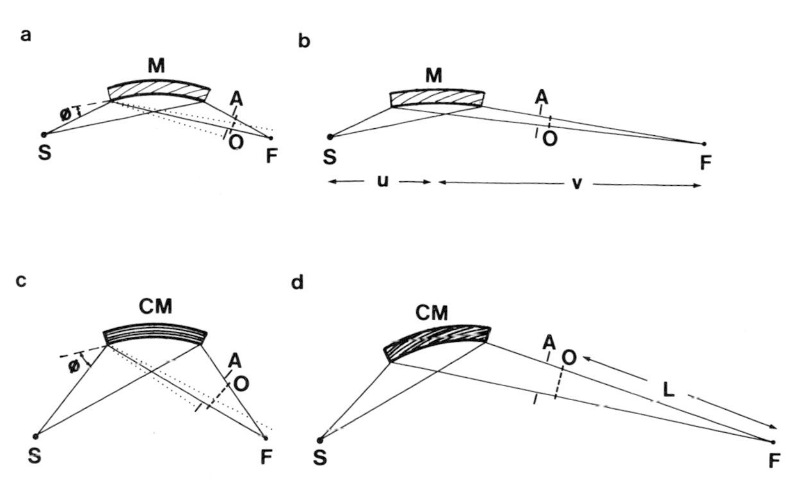

FIGURE 2.20. The use of curved glass mirrors (M) in (a) and (b) or curved crystal monochromators (CM) in (c) and (d) to focus the X-ray beam is illustrated. (a and c) Symmetrical configuration. (b and d) The mirror and crystal are placed asymmetrically so that a relatively long distance (L) from specimen (O) to the film (F) is possible. The apertures (A) limit the parasitic scatter from the mirror or crystal. The angle ϕ for a crystal is about 13°, whereas for a mirror it is only about ½°. Therefore, the crystal collects X rays over a wider angle than an equivalently placed mirror.

Ideally the mirror curvature would follow an elliptical path and the X-ray source (S) and focus (F) would then be at the two foci of the ellipse. The diffracting object (0) is placed between the mirror and the focus. If the mirror is placed in the symmetrical position (Fig. 2.20a), then the focus and source would be of the same size. If the mirror is placed asymmetrically (Fig. 2.20b), the focus and source size would tend to be in the ratio v/u. It is therefore clear that the source size itself must not be too large or the advantages of the focusing system will be lost. It is also clear that the beam width at the specimen is relatively large, and specimens of a comparable size need to be used to take full advantage of the extra intensity available.

The main problems in such devices are, first, that no appreciable reflection of X rays by a glass or gold surface occurs at angles of incidence (ϕ in Fig. 2.20a) greater than about ¼° or ½° and, second, that appreciable parasitic scatter occurs for which the angle of reflection is not equal to ϕ. The first of these means that very flat reflecting surfaces need to be used (preferably flat to within about 500 Å) and that the jig used to bend the glass has to be very accurately made, since very careful adjustments of the mirror curvature will be needed. The second of these, the general effect of the parasitic scatter, is to surround the central direct beam by a halo of diffusely reflected radiation. This increase in the effective size of the undiffracted beam must be limited by placing defining apertures (A) between the mirror and the specimen (Fig. 2.20), especially since this beam is usually many times more intense than any of the diffracted beams.

An alternative means of focusing the X-ray beam is to make use of the diffraction properties of crystals. X rays are "reflected" from the atomic planes in a crystal, and if the crystal is curved, a focus will be produced (Fig. 2.20c). In fact, the focusing properties of such curved crystals are very good. They also have the added advantage that, since they work by diffraction, they can be set up to reflect only one wavelength of X rays (usually the Cu $K\alpha_1$ line; $\lambda = 1.54051$ Å). They are therefore known as "curved crystal monochromators." Quartz crystals (often about 6 cm by 2 cm) are normally used (graphite is also used), and they produce very clean diffraction patterns. The crystal planes normally used for reflection are the (101) planes which have a d spacing of 3.33 Å. The corresponding value of θ in Bragg's Law for Cu $K\alpha_1$ radiation is about 13° which is much larger than the angles involved in reflection from glass- or gold-coated mirrors and allows a much greater part of the X-ray beam from the generator to be collected and focused. The beam width at the specimen is also very large (~1 cm), and only extended specimens, such as muscles of a suitable size, can therefore take advantage of the full size of the beam. Unfortunately, like mirrors, crystals also give rise to parasitic scatter, and extra collimators (slits) need to be included between the crystal and the specimen to keep the effect of this to a minimum.

The adjustment of the curvature of the crystals and their orientation relative to the X-ray beam direction is extremely critical. Crystal holders with very sensitive adjustments are therefore an important part of the design of a crystal monochromator camera.

Figure 2.20c illustrates the optics of a curved crystal camera in which the crystal has been cut so that its faces are parallel to the diffracting planes. If the faces are cut at an angle to these planes, then an asymmetrical configuration is possible (Fig. 2.20d) in which the crystal-to-source distance can be kept small, thus increasing the effective area over which X rays from the source are collected, while the crystal-to-focus distance can be made relatively large. The size of the diffraction pattern recorded on the film clearly depends on the specimen-to-film distance (L). Since the focus from a crystal can be kept small even when the value of L is increased, it is clear that the larger L is, the higher will be the resolution of the camera. In muscle research, L values between about 12 cm and 2 m have been used.

A number of different mirror/crystal camera configurations have been found to be useful in research on muscle (and on some other biological materials). The common configurations are illustrated in Fig. 2.21. The simplest of these, the Franks camera, consists of two curved glass mirrors in a crossed configuration so that they focus the beam in two mutually perpendicular directions to give a "point" focus (Fig. 2.21a). Cameras of this sort were used in early low-angle diffraction studies of muscle (G. F. Elliott, 1963, 1964), and they still have a useful application in present-day studies of biological materials (Figure 2.22f) especially for very small specimens such as single muscle fibers (e.g., Matsubara and Elliott, 1972). A focusing camera giving a truly point focus has been designed by A. Elliott (1965). Here (Fig. 2.21b), the reflecting surface is toroidal (barrel-shaped) and approximates a section from an ellipsoidal surface. The source and focus are at the foci of the ellipsoid. The "toroid camera" is useful in that it collects a relatively large part of the X rays generated by the source, thereby giving a high beam intensity. On the other hand, it cannot be used for very-low-angle diffraction work since it gives virtually a 1 to 1 magnification between the focus and the source without allowing a very long specimen-to-film distance. This type of camera therefore finds its widest application in muscle research in recording the moderate- and high-angle diffraction patterns (Fig. 2.22a).

Very-low-angle diffraction studies of muscle have all made use of curved crystal monochromators often combined with a single Franks

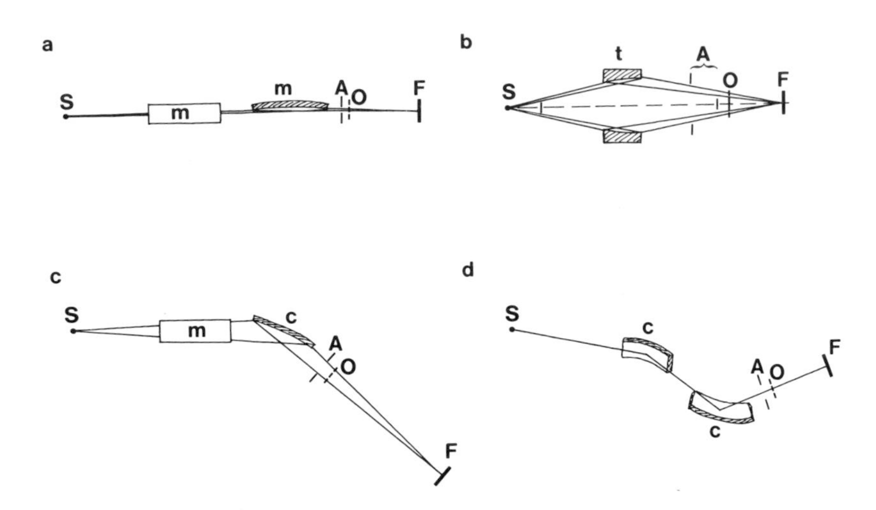

FIGURE 2.21. Schematic representations of different forms and combinations of mirrors and crystals in conventional focusing X-ray cameras. (a) The Franks camera (Franks, 1955, 1958) has two crossed glass mirrors (m). Crossed means that they focus in two mutually perpendicular planes. (b) The Elliott toroid camera (A. Elliott, 1965) has a barrel-shaped (toroidal) gold-coated reflector (t) and produces a point focus from a point source. (c) A Huxley–Holmes mirror monochromator camera combining a glass mirror (m) with a curved crystal (c) in a crossed configuration (H. E. Huxley and Brown, 1967). (d) A double-crystal monochromator combining two crystals in a crossed configuration. In all diagrams S is the X-ray source, A is the scatter-limiting aperture (with a stop in b), O is the diffracting object, and F is the film.

mirror (Fig. 2.21c) (or sometimes with two parallel mirrors; H. E. Huxley and Brown, 1967). In conjunction with the best fine-focus X-ray generators (see Section 2.5.3.) which give an effective source size of about 0.1 × 0.1 mm, such a camera will produce an elongated focus of dimensions about 0.1 × 0.5 mm (the second figure depending on the position of the glass mirror). The resolution in one direction can therefore be extremely high if the specimen-to-film distance is long enough. In the other direction, the resolution is not as good. Since in muscle the important axial periodicities tend to be rather closely spaced and some have very large values (up to 1000 Å or more), it is usual to record diffraction patterns from muscles oriented so that the good focusing occurs in the axial (meridional) direction (Fig. 2.22d and e). On those occasions when good resolution in the equatorial direction is required, either the muscle orientation in a mirror/monochromator camera can be changed by 90° or use can be made of a double crystal monochromator camera (Fig. 2.21d). In the latter case, exceptionally good 2-D focusing can be achieved, but the camera suffers from the disadvantage of being extremely cumbersome and difficult to set up. Note also that sometimes only 1-D focusing is needed, for which a single monochromator without a mirror is sufficient.

As a final comment, it should be noted that air scatter of the X-ray beam can be considerable, and for this reason evacuation of the X-ray path, at least between the specimen and film, is normally necessary.

2.5.3. Specimen Mounting for X-Ray Diffraction

The general methods of dissecting and treating muscles for ultrastructural work are given in Chapter 3. Although muscles that have been allowed to dry out are occasionally used as X-ray diffraction specimens, it is generally necessary to maintain a muscle specimen in as near to its *in vivo* state as is possible while its diffraction pattern is being recorded. To do this, special specimen chambers need to be incorporated into the X-ray camera in use. To maintain a muscle in a healthy "living" state for hours or days, it is normally desirable to bathe it in cooled (~4°C) oxygenated Ringer's solution which mimics the external environment of the muscle *in vivo* (Chapter 3). Sometimes antibiotics and metabolites are also added when experiments on active muscle are being performed. In this case, the muscle chamber must contain, in addition, both electrodes for stimulation of the muscle and facilities for recording (usually via transducers) the tension in the muscle and its length. A schematic diagram of such a muscle chamber is given in Fig. 2.23. It should be noted here that since the bathing medium will absorb and scatter X rays, satisfactory recording of the X-ray diffraction pattern of the muscle can only be achieved if the X-ray path through this medium is kept to a minimum.

2.5.4. Methods of Recording the Diffraction Pattern

The most generally useful method of recording X-ray diffraction patterns is to make use of photographic film. In no other way is it possible at present to obtain an overall impression of the general distribution of intensity in the pattern.

In most low-angle X-ray diffraction cameras, the films are flat, and d spacings in the specimen can be determined from spacings D on the film from the expression $d = n\lambda/2 \sin[\frac{1}{2}\tan^{-1}(D/L)]$ given earlier (Fig. 2.11b). However, when focusing cameras are used at fairly short specimen-to-film distances, it is sometimes desirable to record the X-ray diffraction pattern on a cylindrical film because the beam reaching the specimen is convergent, and the diffracted beams from the specimen do not focus onto a plane but onto a spherical (sometimes cylindrical) surface. Short of using a spherical film, the cylindrical arrangement indi-

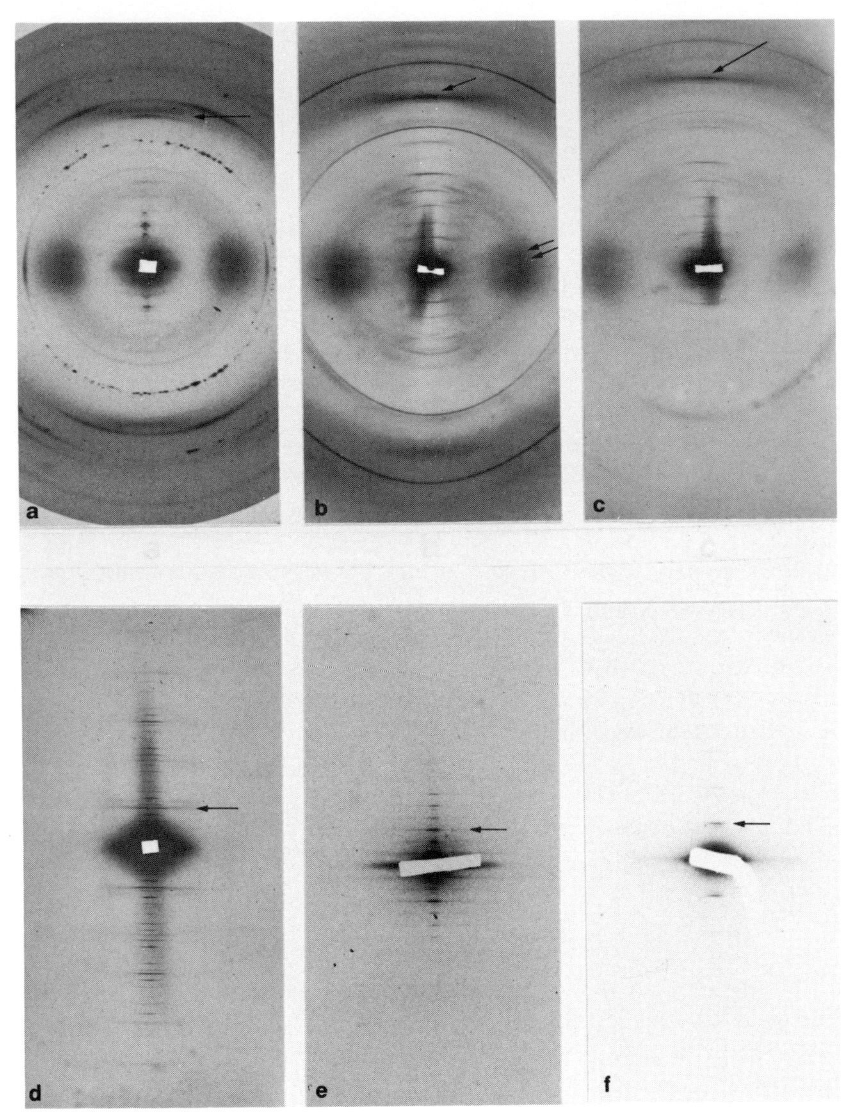

FIGURE 2.22. Diffraction patterns recorded using different types of X-ray camera. (a) Pattern from a dried molluscan muscle taken on a toroid camera. The arrowed reflection is at a spacing of about 5.1 Å and is typical of diffraction from an α-helix (see Chapter 4). Specimen-to-film distance about 4 cm. (b) Similar specimen to (a) but taken on a mirror monochromator camera with 12 cm specimen-to-film distance. The horseshoe-shaped reflections around the equator (double arrow) are typical of diffraction from a coiled coil α-helical structure (Chapter 4). The specimen and cylindrical film holder were tilted to the X-ray beam (see Section 2.5.4) for the 5.1 Å meridional reflection (single arrow). A weaker exposure of a similar pattern (c) shows that the resolution in the 5.1 Å region is extremely good (arrow). (d) Similar specimen to (a), (b), and (c), but pattern recorded on a mirror

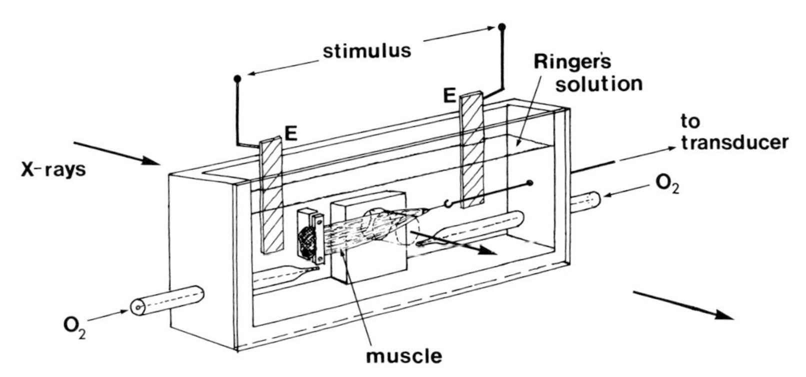

FIGURE 2.23. Schematic diagram of a muscle cell suitable for use on an X-ray camera to record the diffraction pattern from an active muscle. It has been drawn as though the front of the cell is transparent. The muscle is bathed in Ringer's solution (see Chapter 3) that is continuously oxygenated. It can be stimulated via the electrodes E, and its tension can be recorded via a transducer. The X rays (arrows) enter the middle back of the cell through a transparent window and leave through a similar front window. The X-ray path through the Ringer's solution is kept to a minimum because of scatter. The whole cell would normally be cooled (to about 4°C) either by an external cooling jacket or electrically through a thermoelectric cooler.

cated in Fig. 2.24 is used. Here the specimen and the film lie along the circumference of the cylinder of which the axis is perpendicular to the direction of the undeviated X-ray beam.

Worth noting here is one of the important features of diffraction from fibrous structures such as muscle. This is simply the fact that if the specimen is oriented perpendicular to the X-ray beam, then strictly, from Bragg's Law, no meridional reflections should be observed in the diffraction pattern. These should only be seen if the specimen is tilted by the angle θ to the X-ray beam so that the atom planes giving rise to a particular meridional reflection are in the correct orientation for Bragg reflection to occur (Fig. 2.24b). To record high-angle meridional reflections accurately, it is therefore necessary to tilt the fiber by the angle θ, but at the same time the focal surface also moves. In fact, it moves such that the specimen must be kept tangential to the cylindrical film.

---

monochromator camera with a specimen-to-film distance of about 40 cm. Because of the increased specimen-to-film distance, the low-angle resolution is much improved. The arrowed reflection corresponds to a *d* spacing of about 59 Å and comes from the actin filaments in the muscle (see Chapter 5). (e) Similar camera to (d), but specimen is a live frog sartorius muscle. Pattern is less exposed and more enlarged than (d). The arrowed reflection (from the myosin filaments, Chapter 7) corresponds to a *d* spacing of 143 Å and is the third in a series of layer lines based on a repeat *R* of 429 Å. (f) Weakly exposed pattern from a molluscan muscle showing the 143-Å reflection typical of myosin-containing filaments (as in e) but taken on a Franks camera. Specimen-to-film distance 12 cm.

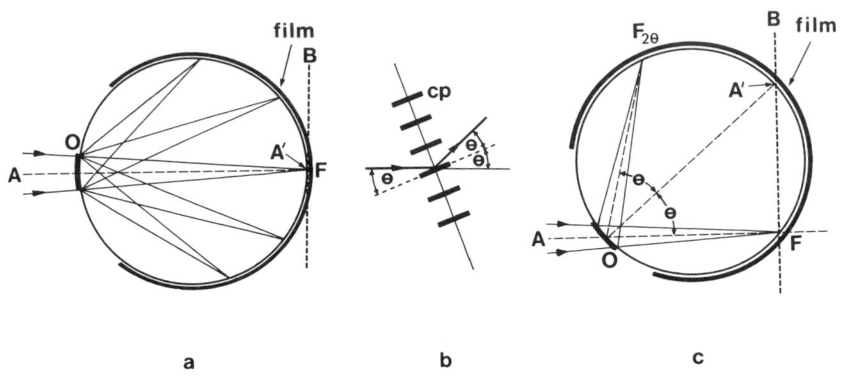

FIGURE 2.24. (a) A convergent beam incident on the diffracting object will be diffracted as shown so that the various foci lie on a circular (spherical in 3-D) surface. Strictly, the specimen should be curved too. Cylindrical films follow this circle at least in one direction. (b) Meridional reflections should not strictly be observed according to the Bragg condition unless the crystal planes (cp) giving rise to them are tilted at the appropriate angle θ to the X-ray beam. In such a case, the diffraction peak from a convergent beam can still be in focus on a cylindrical film [as in (a)] provided that the film diameter is also tilted by θ to the X-ray beam direction (c). This can be achieved using a Rowland mounting (A. Elliott, 1968) in which the specimen O is constrained to move along AF in (a) and A′ is constrained to move along FB which is perpendicular to AF. The result is shown in (c).

For this reason, focusing cameras used to record moderately-high-angle diffraction patterns (e.g., $d = 5$ Å or less) make use of cylindrical film holders on which the specimen itself is also fixed, and the film holders are placed on what is called a Rowland mounting (Fig. 2.24c). This mounting is such that the film holder and specimen can be rotated together by a particular angle θ while keeping the undeviated beam and the required diffracted beam in focus on the cylindrical surface. Figures 2.22b and c show typical patterns obtained using such a camera. The resolution in the high-angle region is very good. Note, however, that for low angles of diffraction, the meridional reflections from muscle can usually be recorded without the need for tilting because of the imperfect alignment of the fibers inherent in the normal muscle structure.

The measurement of intensities on film can be carried out using a microdensitometer which gives optical density as a function of position on the film (in digital form if required). But it is important to use X-ray film only over the range for which it is linear. An alternative method of recording X-ray diffraction information is to use an X-ray detector such as a proportional counter. Such a device clearly measures intensities directly. Unfortunately, until recently, it has only been possible to record one diffraction peak at a time using a single counter. Even so, such a method has been used successfully (as described in Chapter 10) to help to follow the changes that occur in the diffraction pattern of a muscle as it contracts.

This type of experiment has now been rendered easier by the introduction of position-sensitive X-ray detectors. Such detectors consist essentially of a hollow chamber containing a gas (perhaps xenon or argon) along the middle of which runs a fine uniform wire. The wire is the anode and the chamber the cathode in a high-tension circuit. When an X-ray photon arrives at a particular point in the chamber, ionization occurs, and a small pulse either travels in both directions along the central wire or is detected by a part of the chamber; by timing electronically the arrival of the pulses at the end of the counter, one can estimate the position of the ionization event in the counter. A multichannel analyzer can then be used to build up a plot of ionization events (of which the number is proportional to the intensity of the X-ray beam at that point) as a function of position along the counter. A number of variations in design have been used (Figure 2.25; Faruqi, 1975; Cork *et al.*, 1973; see Barrington-Leigh and Rosenbaum, 1976), and such counters have provided some fascinating results on muscle, as described in Chapter 10. More recently, two-dimensional detectors have been developed (see

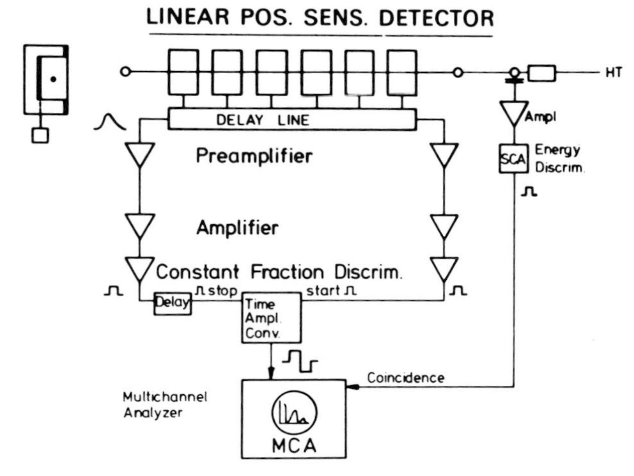

FIGURE 2.25. Schematic diagram of one type of linear position-sensitive detector. Upper left is a cross section of the detector cell showing the central metal wire of diameter 25 μm. The X rays enter from the right through an aluminized mylar window. The cell is filled with xenon gas. The back cathode plane is divided into sections 2.5 mm wide that pick up an induced charge signal from the local avalanche of ionizations produced by each absorbed X-ray photon. The signals from an electrode travel along an external delay line to each end of the cell. From the times of arrival of the two signals from each event, the location of the event can be determined. The resulting information is then fed to a multichannel analyzer (MCA) where the events are counted for each location along the detector. (From Barrington-Leigh and Rosenbaum, 1976.)

Faruqi and Huxley, 1978) as have image-intensifying screens coupled to television monitors (Arndt *et al.,* 1972), so it may soon be possible to record diffraction patterns in two-dimensions with very good time resolution.

2.5.5. X-Ray Generators

It has been shown in the section on X-ray camera design that what is required from an X-ray generator is a small intense source of X rays. As mentioned earlier, X rays are produced when electrons, accelerated through a high voltage, impinge on a metal target such as copper. The X rays are generated either when electron jumps occur between different energy levels in an atom following excitation by the incident electron beam (line spectrum) or when the incident electrons are deflected from their path by electrostatic attraction to the nucleus and, as a result, emit Bremmstrahlung radiation (the continuous spectrum). Unfortunately, the combined efficiency of these processes is less than 1%, and the remaining 99% or more of the energy of the electron beam appears as heat. Cooling of the anode (by circulating water or oil) is therefore crucial if it is not to melt. Clearly, when the electron beam is made smaller and more intense (as required for a small, bright X-ray source), a point will be reached at which local melting of the anode will occur. It has been found that an electron beam power of 150 W concentrated into a target area of about 1 mm $\times$ 0.1 mm is about the maximum that can be achieved with conventional X-ray generators. This corresponds to a specific loading of 1500 W/cm^2. Loadings of up to ten times this value have been achieved in what is called a rotating anode generator (e.g., Elliott GX series) in which the anode is a rotating drum faced with copper that continually presents a new area of target to the electron beam. The electron beam energy is therefore dissipated over a much larger target area, and melting is avoided. Anodes of diameters 3 inches and 18 inches rotating at speeds of 3000 or 6000 r.p.m. have been used.

In the last few years, muscle research has also moved into the realms of "Big Science," since the most recent innovation in X-ray diffraction work has been the use of the X radiation emitted in synchrotrons. The intensity of the radiation is extremely high, and gains in speed, using crystal monochromator cameras, of up to 100 times are expected to be achieved when this application of synchrotron radiation has been fully developed.

Recent reviews by Barrington-Leigh and Rosenbaum (1976) and by Mitsui (1978) summarize the application of synchrotron radiation in biological research. The main drawbacks of its use at the moment are the enormous expense involved and the fact that the synchrotrons are normally run by nuclear physicists for their own use. In the past, peripheral users such as X-ray diffractionists have had to make use of the radiation as and when they could. It is hoped that the advent of particle storage rings [for example at DESY (Hamburg) and at Daresbury (England)] will reduce these problems. The enormous potential of this technique in the study of the kinetics of biological events will then be realizable. An important aspect of this is clearly that it will be possible to follow, with very good time resolution, the structural events involved in muscular contraction.

3

Muscle Preparation, Electron Microscopy, and Image Analysis

3.1. Introduction

Following the early development of electron microscopes to a stage where a better resolution was achieved than was possible with light microscopes, attempts were made to find methods of preparing biological material so that it could be viewed in the new instruments. The initial problem was simply that the columns of electron microscopes had to be under high vacuum and the specimen to be viewed therefore had to be stable under such conditions. With biological tissue, this meant primarily that all of the water in the tissue had to be removed prior to viewing. A second problem was that at the electron-accelerating voltages used (about 100 kV), considerable absorption of the electron beam occurred, and the specimen therefore needed to be both thin enough to allow sufficient penetration of the electron beam and stable enough to withstand electron bombardment. Because of these requirements, it was soon realized that bulk biological tissue, such as muscle, would need to be preserved in some way and then sliced into thin sections. The preparation of thin sections itself demanded that the tissue should be rendered rigid enough to cut easily.

One of the original ideas was to stiffen the tissue by freezing it and then to cut sections while the tissue was still frozen. But this approach was still being developed when a different technique was introduced: the tissue was made rigid by embedding it in plastic. It was then found that satisfactory sections could be obtained using one of the newly developed ultramicrotomes. It turned out that the plastic embedding method gave very good and reasonably consistent results and rapidly increased the available data on muscle structure. For this reason, the alternative freezing methods were not developed further.

In muscle research, the plastic embedding method proved especially satisfactory, and a great deal of new ultrastructural information on muscle was obtained in a relatively short period of time. Electron microscopy of plastic sections was particularly useful as a complementary technique to X-ray diffraction. It has already been shown in Chapter 2 that X-ray diffraction can yield a great deal of information about the structures of regular periodic objects. Muscle is such an object—it is very regular indeed—and X-ray diffraction patterns from muscle are probably more informative than those from any other intact biological tissue. One advantage of the X-ray technique is that diffraction patterns can be recorded from muscle that are known to be alive in the sense that they will still contract when stimulated. X-ray diffraction can therefore give information on the *in vivo* molecular architecture of a muscle. Its main disadvantage is that it does not lead to a direct image of the diffracting object, and interpretation of the X-ray diffraction pattern can often be ambiguous (as an example, see Chapter 2, Section 2.3.7). The ability to visualize muscle structure in the electron microscope is therefore of crucial importance. However, electron microscopy is not sufficient in itself because in processing a muscle, considerable disruption of the *in vivo* structure is bound to be produced. In fact, what is seen in the electron microscope can only be a relic of the original structure. But together, X-ray diffraction and electron microscopy have proven to be powerful allies. Electron micrographs help to interpret the X-ray diffraction evidence; X-ray diffraction provides information about the structures seen in micrographs, but as they are *in vivo* rather than after the harsh preparative procedures have taken their toll. Because of their usefulness, much of the structural work described in this book is based on these two complementary techniques.

Recently, with the improved X-ray methods now available, an increasing number of very detailed diffraction patterns have been obtained from muscle. But on their own these can seldom be interpreted unambiguously, and there is little complementary evidence about them from electron micrographs of plastic-embedded muscle. This is because of the limited preservation of structure obtainable with present plastic-embedding methods. In an attempt to improve the preservation in sections, the rapid freezing method of rendering the tissue rigid enough for sectioning has been revived and developed. It has been found that if a muscle is treated with a cryoprotective agent to reduce ice crystal formation, then this new freezing method can indeed provide some of the required improvement in structural preservation.

Whether plastic-embedded or frozen tissue is used for sectioning, a further problem is encountered when sections are viewed in the electron microscope. This problem results from the generally low electron-scattering power of the atoms that make up most biological materials. These are mainly carbon, hydrogen, oxygen, and nitrogen, all of which have atomic weights, and therefore scattering powers, that are relatively low. The effect of the low scattering power is that micrographs of biological tissue are of low inherent contrast. This problem has been overcome reasonably well by making use of stains that contain atoms of high atomic weight and therefore high scattering power. By labeling the principal structures in a section with such stains, adequate contrast in the section is obtained. But, unfortunately, this happens to some extent at the expense of structural preservation and resolution.

Once a micrograph has been obtained, whether it be of muscle, of an isolated muscle protein, or of an assembly of proteins, it is often found that an underlying regular arrangement of repeating units is apparent, but that disorder in the arrangement, produced either by the preparation procedure or by nonuniform staining, tends to obscure the detailed information that is required from the micrograph. For this reason, a variety of methods have been developed for extracting such information from micrographs. These techniques of image analysis are becoming increasingly important as the search continues for more and more detailed ultrastructural information.

The purposes of this chapter are to give brief accounts of the procedures (e.g., plastic embedding and cryosectioning) that are commonly used in the preparation of muscle and muscle components for the electron microscope and then to describe some of the methods of image analysis that have been found to be useful in muscle research. First, however, the important initial steps in processing—dissection and mounting of the muscles—are briefly described. These initial preparative methods are applicable both to the X-ray diffraction method described in Chapter 2 and to electron microscopy.

3.2. Muscle Dissection and Initial Treatment

3.2.1. Dissection

One of the most critical steps in the whole of the preparative procedure for the electron microscopy or X-ray diffraction of muscle is the initial step of dissection. If this step is done poorly, then poor results will be obtained however excellent the treatment of the muscle may be after dissection. For this reason, dissection must be carried out with great care. Clearly what is required is that the muscle should suffer as little mechanical damage as possible. When whole individual muscles are required (such as the sartorious or semitendinosus of the frog), then it is normal to use dissection scissors to cut the connective tissue between the muscle and its neighbors. During this procedure, it is important not to cut into the muscle itself. If the muscle has tendons at its ends or is connected directly onto bone, then it is simplest to tie cotton onto the tendon or bone at each end of the muscle and then to excise the muscle so that it is held solely by the cotton. Before dissection, it is sometimes useful to note the length of the muscle and then to tie the dissected muscle onto a holder or former after stretching it back to exactly the same length. When required for X-ray diffraction experiments, the muscle would be tied into a cell such as that shown in Fig. 2.23.

In the case of very large muscles (such as the chicken breast or rabbit psoas), only small pieces of the muscle need to be dissected. Large muscles of this kind are often very fibrous, and it is a fairly easy matter to separate out a fiber bundle from the bulk of the muscle using a needle or scalpel. Two pieces of cotton or two clamps are then fixed to the ends of the fiber bundle before final excision. The muscle can then be tied onto a former. This technique is employed when human muscle is being used because it is clear that during muscle biopsy only a small piece of muscle can be dissected from the patient and that the dissection needs to be carried out under local anesthesia. An alternative method of mounting fiber bundles that is frequently used with insect flight muscle is to glue each end of the fiber bundle to glass or wire rods mounted on a suitable holder. A satisfactory glue is made from cellulose acetate dissolved in acetone.

During and following the dissection, it is normal to bathe the muscle in a Ringer's solution that mimics the external environment of the muscle *in vivo*. Ringer's solution is a balanced mixture of ions (e.g., Na^+, Mg^{2+}, K^+, Cl^-, etc.) buffered to the appropriate physiological pH (usually 7.0 to 7.2); it may also contain metabolites and antibiotics. The exact recipe for the Ringer's solution depends on the type of muscle involved. Details of these recipes for the muscles discussed in this book are given in Table 3.1. In each case, the recipes are by no means fixed, and many variations on them are possible.

3.2.2. Adjustment of Sarcomere Length

If the muscle specimen is to be set at a particular sarcomere length when, for example, the myosin and actin filaments are required to over-

TABLE 3.1. Muscle Bathing Media (Ringer's Solutions)[a]

Content	Mammalian muscle		Frog muscle		Insect muscle[f]	Vertebrate smooth muscle[g]	Frog iodoacetic acid rigor[e]
	Skeletal[b]	Cardiac[c]	Resting[d]	Active[e]			
Solutes (mM)							
$CaCl_2$		3	1.8[b]	1.8		1.2	1.8
KCl	2	7.5	2.5	2.5	10	4.7	2.5
$MgCl_2$	2	1			40	1.05	
$MgSO_4$				1.25		1.2	
NaCl	100	130	115	66.5	100	118.9	115
$NaHCO_3$						24.9	
NaOH				11.5			
Na/Pi (buffer)	5	14	3	15	10		3
KH_2PO_4						1.2	
Sodium glutamate				20			
Sodium pyruvate				5			
Fumaric acid				6			
D-Glucose	5.6			168	100	5.6	
Penicillin				0.01			
Streptomycin				0.01			
Iodoacetic acid							1
Gases (%)							
Oxygen	95	95	95	95	100	95	
CO_2	5	5	5	5	0	5	
pH	7.0	7.2–7.3	7.0	7.0	6.8	7.0–7.4	

[a] For molluscan muscle, use aerated sea water. From Lowy and Millman (1963).
[b] Calcium may be omitted and EGTA (see Table 3.2) added for resting muscles. From Craig (1977).
[c] From Matsubara *et al.* (1977b).
[d] From Haselgrove and Huxley (1973).
[e] From H. E. Huxley and Brown (1967).
[f] From Miller and Tregear (1972).
[g] From Devine and Somlyo (1971).

lap by a set amount (or not at all), then it is possible to define this length in a living muscle by making use of laser diffraction. When a laser beam (e.g., from a 1-mW He/Ne laser, $\lambda = 6328$ Å) is directed through a fresh striated muscle, it will be found that a clear diffraction pattern can be seen on a white screen placed 5 to 10 cm behind the muscle. A very simple practical arrangement for obtaining such a pattern is shown in Fig. 3.1a. The muscle on its former is placed in a small petri dish containing Ringer's solution, and this is held on top of a small laboratory tripod. The laser beam is reflected by a mirror down through the muscle, and the diffraction pattern from the muscle sarcomeres can be seen on a white card placed on the table top. An alternative method is to have the muscle in a small cell, such as the cell described earlier (Fig. 2.23) for use

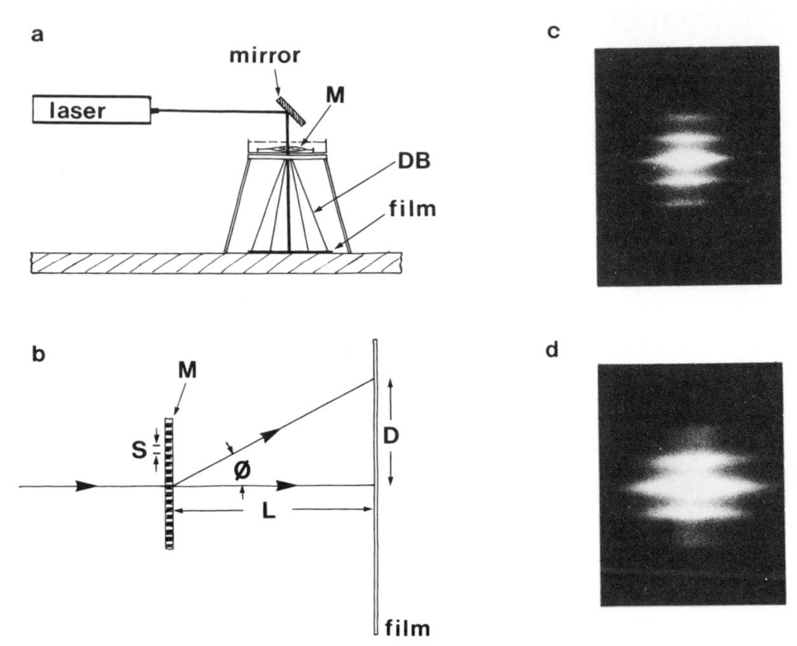

FIGURE 3.1. The determination of sarcomere length by laser diffraction. (a) The experimental arrangement (M, muscle; DB, diffracted beams). (b) Geometry of the diffraction pattern (M, muscle; *S*, sarcomere length). (c and d) Two laser diffraction patterns from the same muscle at different sarcomere lengths (muscle-to-film distance, 3 cm). Sarcomere length is 3.4 μm (c) and 2.8 μm (d).

on X-ray cameras, and to shine the laser beam through the cell windows. In fact, this method is sometimes used to determine sarcomere length before or after the X-ray diffraction pattern of a muscle is recorded.

The geometry of the diffraction pattern is indicated in Fig. 3.1b. Each sarcomere acts as a repeating unit in a one-dimensional array of spacing, *S*. Since the muscle is at right angles to the laser beam, the geometry of diffraction is just as in Fig. 2.4 for light diffraction in the optical diffractometer, and in this case, diffraction peaks will occur at angles ϕ given by $n\lambda = S \sin \phi$. In Fig. 3.1, it will be seen that ϕ is given by $\phi = \tan^{-1}(D/L)$, and hence $S = n\lambda/\sin(\tan^{-1}D/L)$. Since λ is fixed for a particular laser (often 6328 Å), the angles ϕ and lengths D for a particular sarcomere length (S) in the muscle can easily be determined. The screen can therefore be calibrated in terms of sarcomere length (for $n = 1$), and the length of a muscle altered until its first-order diffraction peak lies in the required position on the screen. Figures 3.1c,d show examples of the kind of pattern that is obtained. It should be noted that this procedure is best done on a thin, freshly dissected muscle. In other kinds of preparation (e.g., glycerinated muscle—see Section 3.2.3., the muscle is often too opaque for the laser diffraction pattern to be seen clearly.

3.2.3. Glycerol-Extracted Muscle

It was discovered in 1949 by Albert Szent-Gyorgi (1949, 1951) that the effect of bathing a muscle in a solution containing glycerol (~50%) is to disrupt the muscle membrane (sarcolemma), thus allowing the extraction of soluble proteins and other soluble cell constituents such as ATP without at the same time destroying the contractile machinery itself. The effect of the removal of ATP, mentioned briefly in Chapter 1, is to form permanent attachments (rigor complexes) between the heads of the myosin molecules and actin. This renders the muscle stiff and produces a state (rigor) akin to rigor mortis. Subsequently, the glycerol can be washed out of the muscle using a standard salt solution that mimics the internal environment of the cell *in vivo* (i.e., in ionic strength, pH, etc). If the muscle is then placed in standard salt solution containing ATP and a calcium-chelating agent (e.g., EGTA; Table 3.2), the rigor complexes are broken, and an artificial relaxed state is produced. Contraction of the muscle can then be produced by the addition of Ca^{2+} ions which activate the myofibrillar components. Since a glycerinated muscle is viable, in this sense, and since the membranes are disrupted, such a muscle is a valuable tool; it can be permeated with a solution of any desired composition. Such a solution might have a chosen ionic content or pH, it might contain an antibody label specific for one of the myofibrillar proteins, or it might be capable of extracting a specific part (e.g., the M-band) of the myofibrillar material. In addition to this, glycerol acts as a good antifreeze, and this means that glycerol-extracted muscle can be kept in the deep freeze for a long period (at least several months) without a serious loss of viability. This storage possibility means that a single animal can be used for many experiments, a considerable help in the case of specimens (e.g., tropical insects) that are in short supply.

Once again, the glycerol and standard salt solutions that are commonly used depend on the particular muscle type involved and often also on the laboratory concerned. Table 3.2 summarizes some basic recipes that are commonly used for the main muscle types.

3.2.4. Single Fibers

By careful dissection under the binocular microscope of a fibrous muscle, preferably with large fibers, it is possible to separate a single fiber attached at each end to a tendon. Such a fiber can then be mounted

TABLE 3.2. Glycerol and Standard Salt Solutions

	Glycerol solutions[a]		Standard salt and relaxing solutions[d]					
			Frog muscle solutions[c]			Insect muscle solutions[d]		
				Relaxing				
Contents	Frog, rabbit, insect[b] (1)	Vertebrate smooth muscle[b] (2)	Standard (3)	I (4)	II (5)	Std. (6)	Relax. (7)	Active (8)
Na/Pi (buffer)	20 mM	20 mM	6.67 mM			5.0 mM		
Histidine/Cl (buffer)				10 mM	10 mM			
KCl	100 mM		100 mM	100 mM	100 mM	40 mM	65 mM	65 mM
Phenyl methyl sulfonyl fluoride	10 μM	10 μM						
Glycerol	50%	50%						
$MgCl_2$			1.0 mM	5.0 mM	5.0 mM	5.0 mM	5.0 mM	5.0 mM
EGTA[e]			4.0 mM	4.0 mM	4.0 mM		2.0 mM	
Ca^{2+}								≤2.0 mM
ATP					5.0 mM		5.0 mM	5.0 mM
pH	7.0	7.0	7.0	7.0	7.0	7.0	7.0	7.0

[a] The method for frog muscle glycerination and use is as follows. Place muscle in 50% glycerol (solution 1) at 4°C for 24 hr. Change solution and keep at 4°C for an additional 24 hr. Store at −20°C for at least 3 weeks. For use, wash in standard salt solution (solution 3 or 4) for 1 hr. Change solution or, for relaxed muscle, transfer to solution 5. If desired, Ca^{2+} ions can then be added to produce activity.
[b] Adapted from Szent-Gyorgyi (1949).
[c] From Rome (1972).
[d] From Pringle (1967).
[e] Ethyleneglycol-bis-(β-aminoethyl ether)-*N,N'*-tetraacetic acid.

in a suitable cell for X-ray diffraction experiments or tied onto a holder for microscopic analysis.

In addition, it is also possible to remove the fiber membrane (sarcolemma) using a very sharp knife. Such a "skinned" fiber can then be used as a model structure rather similar to a glycerinated fiber in that it can be permeated with any desired solution (e.g., April *et al.*, 1971,1972; Matsubara and Elliott, 1972). But it has the advantage that it will have suffered less disruptive treatment than a glycerinated muscle. Because of the small size of the fibers, it is necessary to record X-ray diffraction patterns using a Franks camera (Section 2.5.2) which has a narrow beam.

3.3. Preparative Methods in Biological Electron Microscopy

3.3.1. Embedding Methods

Fixation. The initial procedure in any embedding method is to place the mounted muscle into a chemical fixative such as glutaraldelyde so that the muscle is stabilized prior to the dehydration and embedding procedures. The mode of action of such fixatives is not completely understood, but basically they polymerize to form chemical crosslinks between particular groups (e.g., $-NH_2$ groups) in the molecules that form the various cell organelles. Apart from glutaraldehyde, other commonly used fixatives are osmium tetroxide, achrolein dichromate, and uranyl acetate. The last of these also serves as a stain because of the heavy uranium atoms in it, and its use at this stage is frequently referred to as block staining or staining *en bloc.* Recipes for these various fixatives are given in Table 3.3. The most commonly used combination of fixatives is primary fixation with glutaraldehyde (2 to 5% in buffer) followed by postfixation is osmium tetroxide (~ 1% in buffer); this is despite the fact that osmium has a highly disruptive effect on muscle ultrastructure (Reedy and Barkas, 1974).

In an attempt to improve fixation of muscle prior to embedding, Reedy (private communication) has used methods that supplement primary glutaraldehyde fixation by converting protein side groups such as carboxyl groups (see –COOH, Table 4.1) so that they form links with short chain diamino groups. These can subsequently be crosslinked by further glutaraldehyde fixation. Some early successes with this approach have already been reported (Reedy, 1976). Note also that much improved preservation can be achieved by fixing tissue prior to dissection by immersion, but only the superficial parts of the tissue should then be used for ultrastructural studies.

TABLE 3.3. Standard Fixation Procedures

Procedure	Concentration (%)	Additives	Buffer	pH	Temp. (°C)	Time (hr)
Primary fixation						
Glutaraldehyde[a]	2.5		0.1 M sodium cacodylate	7.0	22 or 4	1–2
Acrolein + potassium dichromate[b]	1.5 1.0	6% sucrose		7.35[c]	0–4	3–4
Postfixation						
Osmium tetroxide[a]	1.0	7% sucrose	0.1 M sodium cacodylate	7.0	4	2
Osmium tetroxide + potassium dichromate[a]	1.0 1.0			7.0[c]	4	2

[a] Glauert (1975).
[b] Robison and Lipton (1969).
[c] pH adjusted with 5N KOH.

Dehydration. Few embedding media are water soluble, so a fixed muscle needs to be dehydrated before it is embedded. Dehydration is a very drastic procedure, and without previous fixation the structure of muscle would rapidly be destroyed. The usual methods of dehydration are to pass the muscle (or other tissue) through a graded series of alcohol/water or acetone/water solutions from about 10% to 50% in water through to 95% in water. The tissue is then placed in pure alcohol or pure acetone. The procedure is summarized in Table 3.4.

During the process of dehydration, some lipids and proteins are also extracted. Fixation of the tissue reduces but does not eliminate this effect. Dehydration after osmium tetroxide fixation is reported to extract about 4% of the tissue proteins (Luft and Wood, 1963).

Embedding in Epoxy Resins. The first stage in the embedding procedure is usually to transfer the dehydrated muscle from 100% ethanol (or acetone) into a mixture of 50% epoxy resin and a suitable solvent (e.g., propylene oxide) for ½ to 1 hr. This step aids penetration of the resin into the tissue, but ethanol or acetone may also be used because of the unpleasant properties of propylene oxide. The muscle is then placed in pure resin mixture and left in an open container to allow complete evaporation of any solvents that may have been used. Specimens are then transferred to small capsules that are filled with fresh resin mixture. Finally, the resin is hardened by warming to about 60°C in an oven for 1 or more days. The quality of the block obtained in this way depends, among other things, on the recipe that has been used to make the resin.

A number of different resin mixtures have been used successfully in

TABLE 3.4. Embedding Procedures Using Epoxy Resins

Step	Alcohol process	Acetone process	Temp.	Time
1	50% Alcohol	10% Acetone[a]	22°C	15 min
2	70% Alcohol	25% Acetone	22°C	15 min
3	95% Alcohol	60% Acetone	22°C	15 min
4	100% Alcohol	95% Acetone	22°C	15 min
5	100% Alcohol (Dried)	100% Acetone	22°C	15 min
6	100% Propylene oxide (PO)[b]	—	22°C	15 min
7	PO + Embedding medium (EM)[c] (1 : 1)	Acetone + Embedding medium[c] (1 : 1)	22°C	30 min
8	PO + EM (1 : 2)	Acetone + EM (1 : 2)	22°C	30 min
9	PO + EM (1 : 3)	Acetone + EM (1 : 3)	22°C	30 min
10	Embedding medium	Embedding medium	4°C	Overnight
11	Embedding medium	Embedding medium	60°C	36 hr

[a] Same concentrations as alcohol series to Step 5 can be used as well.
[b] Ethanol can be used throughout instead of PO.
[c] Resin mixtures (see Luft, 1961; Hayat, 1972; Glauert, 1975) consist of the resin, an accelerator, a hardener, and a plasticizer. The most frequently used resins are Araldite[R] (Ciba) and Epon[R] (Shell). Araldite is prepared by combining 8.2 g of CY 212 (resin), either 0.5 ml of 2,4,6-tridimethylaminomethylphenol (DMP-30) or 0.4 ml of benzyl dimethylamine (BDMA) (accelerator), 10 g of dodecenyl succinic anhydride (DDSA) (hardener), and 0.5 g of dibutyl phthalate (DBPth) (plasticizer). Epon is prepared by combining 20 g or Epon 812, 1 g of DMP-30, 12.5 g of DDSA, and 16.5 g of methyl nadic anhydride (MNA).

muscle research (see review by Glauert, 1975). Table 3.4 summarizes two convenient methods used in the author's laboratory. Most resin mixtures consist of three basic ingredients: the resin itself, a hardener, and an accelerator. A satisfactory mixture (see Glauert, 1975) is required, among other things, to have (1) a low initial viscosity so that tissue infiltration is reasonably fast, (2) uniform polymerization characteristics so that the final block has an even hardness, (3) good sectioning characteristics, and (4) good stability in the electron beam. Araldite and Epon (Table 3.4) have very reasonable properties in this respect, and they are very commonly used.

The hardness of the final block is very critical if good sectioning is to be achieved, and use is sometimes made of plasticizers or flexibilizers in the resin mixture to reduce the final hardness of the block. Blocks that are too brittle tend to chip easily, and the sections obtained often break up. On the other hand, sectioning of very soft blocks is difficult, and excessive compression of the sections can occur.

Sectioning Procedures. The basis of modern sectioning procedures (Fig. 3.2) is that the block to be sectioned is moved repetitively past the edge of a very sharp knife, and at the same time it is gradually moved forward by a small controlled amount so that a thin section is cut at each stroke (see review by Reid, 1975). The specimen advance (which can usually be set between about 50 Å and 1 or 2 μm) can be produced either

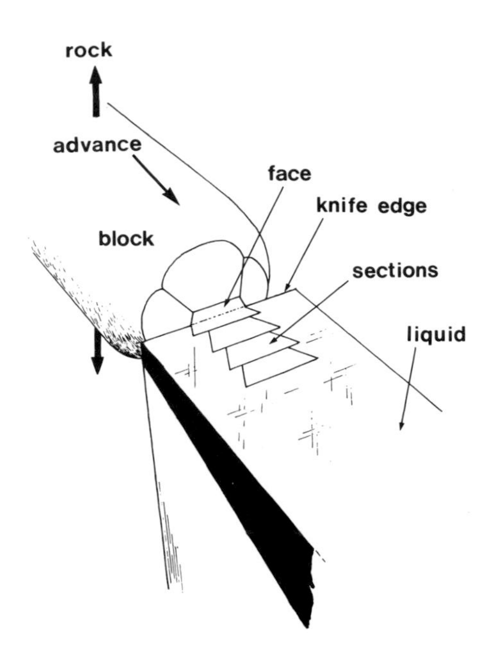

FIGURE 3.2. Schematic diagram to show the basic procedure for obtaining thin sections of embedded muscle. The block containing the muscle is moved up and down and advances towards the cutting edge after each stroke by a small controlled amount. Sections float off from the knife edge onto a liquid surface.

thermally (e.g., as in the LKB ultramicrotomes) or mechanically (as in the Cambridge "Huxley" microtome). The knives used to cut the block are either the fracture edges produced during a controlled break of a glass strip (glass knife) or the edge of a highly polished diamond. Glass knives are relatively cheap and are disposable, and they can give very good results. Diamond knives are expensive (about $750 per millimeter edge) and require careful maintenance. But they consistently give very good results and can often section "difficult" blocks more easily than can glass knives.

In muscle research, two different types of section are normally cut, depending on whether the cut is parallel or perpendicular to the long axis of the muscle. For obvious reasons these are referred to as longitudinal and transverse sections, respectively. Only rarely are oblique sections required, since the arrangement of the myofilaments makes such sections difficult to interpret unless they are very thin.

Biological sections are usually cut onto a liquid surface behind the knife edge. The knives are fitted with troughs (which can either be permanent fixtures as on diamond knives or temporary arrangements made with sticking plaster or tape). These troughs are filled with water to give a very flat surface as judged by reflection from the water surface. When the water surface can be seen to be uniformly reflecting, thin sections floating on it will show different interference colors according to their thickness. As a general guide, these colors are summarized in Table

3.5 as a function of section thickness. More accurate estimates of section thickness can be obtained either by reembedding a section and cutting it transversely or by measuring section folds in the electron microscope (Small, 1968; Reedy, 1968).

Since micrographs of muscle are often required to provide spatial measurements of various structural features, it is important that as little distortion and shrinkage as possible be produced in the sections. Muscles embedded in Araldite or Epon tend to show little shrinkage compared with other embedding media, and fairly hard blocks will suffer relatively little compression. Wrinkles or folds in the section can often be reduced when the section is floating on the trough liquid by holding a small brush soaked in chloroform over the top of the section (without touching it). But some distortion is almost bound to occur, and care should be taken to cut the section in such a way that the distortion will interfere as little as possible with the particular features of the muscles that are being investigated.

Page and Huxley (1963) and Page (1964) carried out a systematic study of the reliability of spatial measurements on electron micrographs of muscle, and they found that different processing procedures produced some changes in filament lengths in striated muscles but that conditions and procedures could be found that minimized these changes. Glutaraldehyde fixation, araldite embedding, and sectioning with the knife edge parallel to the direction in the muscle in which measurements are to be made helps to minimize spatial distortions.

Often, when sectioning a muscle for electron microscopy, it is useful to monitor the sections by taking occasional thick sections and viewing these on the light microscope. Light microscopy can also be a valuable aid in checking and controlling the orientation of the block in the microtome so that truly longitudinal or transverse sections are cut (Luther and Squire, 1978).

TABLE 3.5. Interference Colors of Thin Sections[a]

Color	Section thickness (Å)
Gray	600
Silver	600 to 900
Gold	900 to 1500
Purple	1500 to 1900
Blue	1900 to 2400
Green	2400 to 2800
Yellow	2800 to 3200

[a] From Peachey (1958).

Sections are collected from the trough either by placing electron microscope grids (e.g., copper grids with 200 or 400 mesh lines per inch) directly on top of them or by picking them up in a wire or plastic loop and using this to place them onto the grid. Grids are often coated with a thin film of Formvar or collodion coated with carbon prior to use. This provides a mechanical support for very thin sections, tends to conduct heat away, and helps to stabilize sections in the electron beam.

Staining Methods. It has already been mentioned earlier in this section that uranyl acetate can serve both as a fixative and as a block stain. But even if this staining *en bloc* is used, adequate contrast is not normally obtained unless the sections themselves are stained. A variety of different staining solutions has been used for this in the past, but the combination most commonly used is that of uranyl acetate (e.g., 2% in water) followed by lead citrate (4% in water). Some of the recipes involved are given in Table 3.6. Other stains that can be used include phosphotungstic acid, potassium permanganate, and uranyl magnesium acetate.

Each of these stains has a different effect and binds in a specific way to different cell components. However the uranyl acetate/lead citrate combination has been found to give good overall staining. In particular, the principal myofibrillar components are stained well, and for this reason such a combination is very commonly used in muscle work.

Evaluation of Structural Preservation in Plastic Sections of Muscle. Because of the inherent structural regularity in striated muscle, it is possible to evaluate the various steps in the plastic-embedding process using the X-ray diffraction technique. For example, it has been found (Reedy and Barkas, 1974 and personal communication) that relatively little change from the diffraction pattern of a live muscle (Fig. 3.3a) occurs following brief glutaraldehyde fixation. On the other hand, postfixation, dehydration, and plastic embedding cause a gradual deterioration of the diffraction pattern. Figure 3.3b shows that even prolonged glutaraldehyde fixation can have a marked effect on muscle structure. This pattern is characteristically different from the "relaxed" muscle diffraction pattern (Fig. 3.3a), showing that the glutaraldehyde treatment has caused a specific structural change in the muscle. The changes involved will be discussed in Chapters 10 and 11.

Despite this, the immediate result of importance is that brief glutaraldehyde treatment causes relatively little structural disruption. It is used during the cryosectioning procedure for preparing muscle for the electron microscope, and, as will be described in Section 3.3.3., this freezing method preserves certain aspects of muscle ultrastructure better than does conventional plastic embedding.

Finally, it should be noted that the structural preservation in sec-

TABLE 3.6. Useful Staining Solutions

Stain	Application	Recipe	Temp.	Time
Uranyl acetate[a]	En bloc (i.e., prior to embedding	2% in 10% acetone or ethanol	22°C	15 min during dehydration
	Sections	2% aqueous or in 10% ethanol	22°C	60 min
	Negative staining[b]	1% aqueous (pH 4.6)		
Lead citrate[c]	Sections	1.33 g lead nitrate, 1.76 g sodium citrate, 30 ml distilled water; shake for 20 min, then add 8.0 ml 1 N NaOH. Dilute to 50 ml with dist. H_2O and shake until clear.	22°C	1 to 15 min
Uranyl magnesium acetate[d]	Sections	2% aqueous	60°C	30 min
Potassium permanganate[e]	Sections	1% aqueous	22°C	5 to 15 min
Ammonium molybdate[f]	Negative staining[b] cryosections, A-segments)	2% aqueous (pH 7.3)		
Potassium phosphotungstate[f]	Negative staining[b]	1% aqueous (pH 7.0)		

[a] H. E. Huxley and Zubay (1960).
[b] Specimens for negative staining are either sections or isolated particles (molecules or molecular aggregates) on a coated grid.
[c] Reynolds (1963).
[d] Frasca and Parks (1965).
[e] Should not be used with embedding medium (e.g., Epon) containing MNA (see Table 3.4) (Reedy, 1965; Robertson *et al.*, 1963).
[f] Craig (1977) and Sjöström and Squire (1977a,b).

tions can be tested objectively by obtaining optical diffraction patterns from the electron micrographs and comparing these with the X-ray diffraction pattern from living muscle.

3.3.2. Negative Staining and Shadowing of Isolated Particles

When the properties and characteristics of proteins are described in the next chapter, it will be shown that there are various combinations of mechanical and chemical preparative methods by which individual components of the myofibril (e.g., whole filaments or single molecules) can be isolated and sometimes purified. Such components are obviously

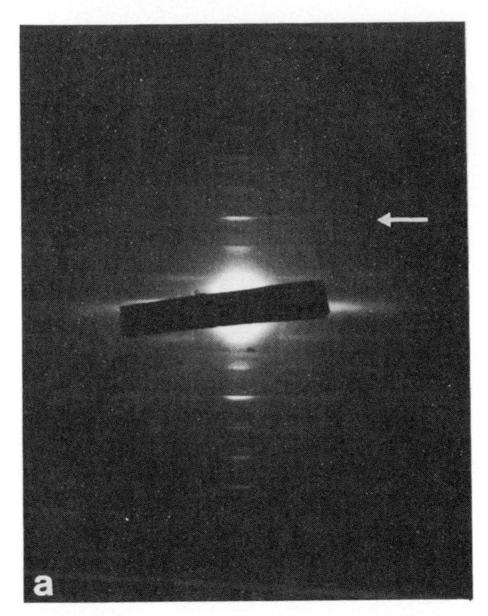

FIGURE 3.3. Examples of the X-ray diffraction monitoring of muscles taken through the preparative procedure for electron microscopy. (a) From a live resting frog sartorius muscle (4°C, 40 cm mirror/monochromator camera). (b) From the same muscle after prolonged glutaraldehyde fixation (overnight). Considerable changes have occurred in the diffraction pattern as described in Chapter 10. In both pictures, the 143-Å meridional reflection is arrowed.

small and can be allowed to settle from suspension directly onto a coated electron microscope grid. After this, they need only be contrasted in some way before being viewed in the electron microscope. There are two principal methods of producing contrast. One involves the use of stains such as uranyl acetate, and the other involves the deposition of a thin layer of heavy metal onto the surface of the specimen grid.

A simple staining procedure is to take a drop of the suspension of particles and place this onto a carbon-coated grid. The particles are allowed to settle, and the liquid drop is removed by touching a filter paper to the edge of the grid. A drop of stain is then added. At this point, two alternative methods can be used. In one, after excess stain is removed, the remainder is allowed to dry down onto the grid, and in so doing, it piles up against the edges and depressions or gaps on the protein particles. Since the bulk of the stain is then situated between and around regions of high protein density, this technique is termed the negative staining method. The alternative method is to remove the stain using a filter paper as before and then to rinse the grid thoroughly with distilled water. In this way, most of the surplus stain is removed, and the bulk of the stain that remains is thought to be that which is bound firmly to specific sites on the particles themselves. This method is termed the positive staining method. It should be noted, however, that the division between these two methods is not clear-cut, since negatively stained particles may also be positively stained as well, and some of the "positive stain" effect may in fact be caused by residual negative staining. Figure 3.4 summarizes diagrammatically the ideal effects of these techniques and shows some examples of their use.

Heavy metal shadowing of particles is carried out in the vacuum coating units that are used routinely for carbon coating of electron microscope grids. A wire of the metal to be used (e.g., platinum, gold/palladium, or tungsten) is coiled around a filament through which a suitable current can be passed. When the filament heats up, evaporation from the metal wire occurs, and a thin layer of metal is deposited on the

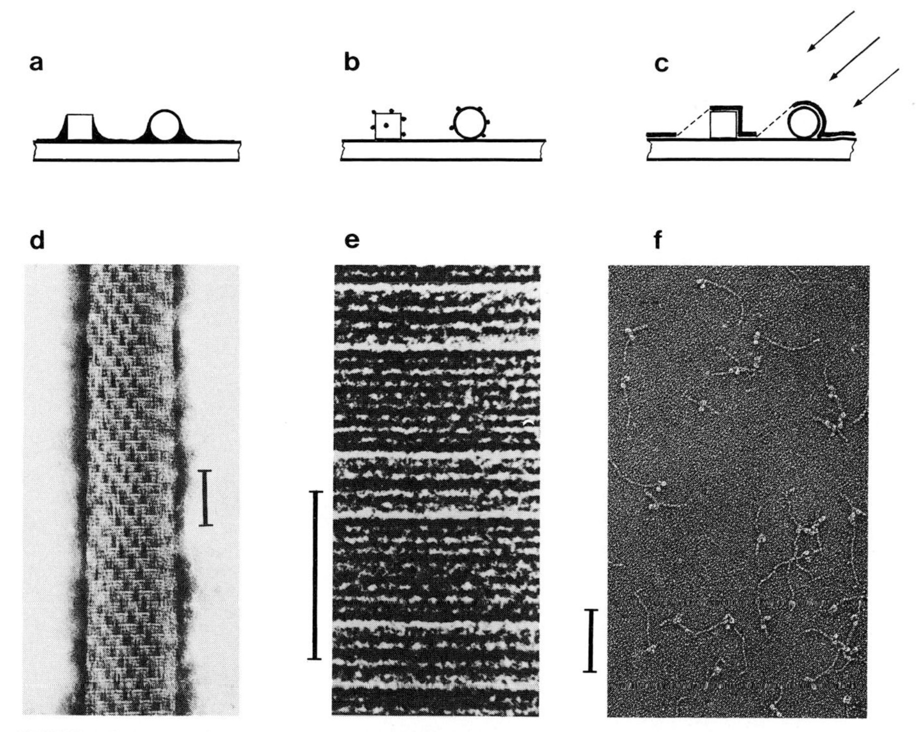

FIGURE 3.4. Methods of contrasting isolated particles for electron microscopy. (a) The negative staining process. (d) An example of a negatively stained myosin-containing paramyosin filament from a molluscan muscle (bar about 720 Å; from A. Elliott and Lowy, 1970). (b) The effect of positive staining. (c) A positively stained paracrystal of tropomyosin (bar about 400 Å). (d and e) Stained with uranyl acetate. [(e) From Stewart and McLachlan (1976).] Heavy metal shadowing (c) and shadowed isolated myosin molecules (f) prepared using the freeze-drying method by A. Elliott *et al.* (1976). Bar is about 1000 Å. For further details see text.

specimen grid. Metal evaporation onto the grid from the side (Fig. 3.4c) causes nonuniform coating of the particles and, in particular, they cast metal-free shadows on the grid. The ultimate resolution obtainable in micrographs of such preparations depends critically on the size of the metal granules that are deposited on the grid. Operating conditions must therefore be chosen to reduce this size to a minimum. At present, the technique is capable of revealing the presence of rod-shaped particles only 20 Å thick (e.g., the tails of myosin molecules), but no internal detail of these or indeed of larger globular proteins (e.g., the myosin heads) has yet been obtained.

To date, some of the most successful results on shadowed muscle proteins have been obtained using a freeze-drying method (A. Elliott *et al.*, 1976). Droplets of the suspension of particles (in this case myosin molecules) were sprayed onto a cold mica surface (liquid nitrogen temperature) using an aerosol. While cold, the mica sheet was transferred to the vacuum chamber, and the particles were allowed to dry for about 4 hr while held at about −50°C. The freeze-dried particles were then shadowed using platinum, and the whole sheet coated with a layer of carbon. The platinum–carbon replica was then floated off onto water and collected on copper grids. Figure 3.4f shows an example of the type of preparation. Detailed results from this work will be discussed in Chapter 6.

As a final comment on the staining and shadowing methods, it should be noted that the latter can only reveal the surface structure of a particle, whereas stain can penetrate deeply into a particle and may show up some of its internal structure. In this sense, the two methods are not alternative techniques, but, rather, they can provide complementary information about a particular type of specimen.

3.3.3. Cryosectioning and Other Freezing Methods

The Freezing Process. The fundamental problem in using freezing methods on biological tissue is that ice crystal formation, if it occurs, is liable to have a severe effect. Both mechanical distruption of the tissue and an effective reduction of the free water content (thus altering such things as ion concentration) would result. Care must therefore be taken to freeze the tissue in such a way that ice crystal formation is minimized.

As a general rule, the faster the tissue is cooled, the smaller will be the resulting ice crystals. For this reason, it is usual to freeze very small pieces of tissue by sudden immersion in inert fluids at liquid nitrogen temperatures. But even this type of treatment is insufficient to reduce ice crystal formation to a point at which it is no longer troublesome. It is usual, therefore, when freezing tissue for ultrastructural work, to bathe the tissue prior to freezing in a solution containing an antifreeze or cryoprotective agent. Glycerol is an efficient antifreeze, and since it does not destroy the tension-generating properties of muscle, it is the cryoprotective agent of choice in muscle research.

Before antifreeze treatment, it is often helpful to fix a muscle briefly using glutaraldehyde since this fixative is known to disrupt muscle structure very little, and it clearly helps to preserve the muscle structure through the freezing and sectioning procedures.

Once a tissue has been frozen, there are two principal methods that can be used to reveal ultrastructural information. In one, the freeze-etching method, the tissue is cleaved while frozen, and a replica is made of the fractured surface. This replica is then viewed in the electron microscope. In the other, the cryosectioning method, an ultramicrotome fitted with a cooled knife and specimen holder is used to cut ultrathin sections directly from the frozen tissue. These two techniques are described briefly below.

The Freeze-Etching Technique. The technique of freeze-etching has been largely developed for muscle studies by Rayns and his collaborators (Rayns, 1972; Rayns *et al.*, 1975). The principal steps in this technique are as listed below:

1. Dissection and slight stretching of the muscle.
2. Pretreatment with glutaraldehyde for fixation and glycerol for cryoprotection.
3. Rapid freezing of the tissue to liquid nitrogen temperatures (−196°C).
4. Placing the tissue on a cold table (~−150°C) in the bell jar of a vacuum evaporator.
5. Evacuation of the bell jar to about 10^{-6} torr.
6. Fracturing (or cleaving) of the specimen using a cooled razor blade. (The specimen temperature is sometimes raised to about −100°C before cleaving is carried out.)
7. Etching of the fracture surface by leaving the specimen in the bell jar at about −50°C to allow sublimation from the specimen surface. Sublimation occurs more slowly from protein structures than it does from the surrounding water/ice matrix, and etching therefore causes the protein to protrude from the specimen surface relative to its surroundings.
8. The etched surface is then shadowed from the side using, for example, gold/palladium or platinum/carbon, and the shadowed surface is coated with a relatively thick layer of carbon to give a replica of the etched surface.
9. The carbon replica, after being cleaned (e.g., in 40% chromium dioxide) and rinsed, can then be viewed in the electron microscope.

FIGURE 3.5. Freeze-etched preparations of vertebrate skeletal muscle (a) showing ice crystal formation in unprotected muscle and (b) the lack of ice crystals in muscle cryoprotected with glycerol. (A, A-band; m, mitochondria; Z, Z-band; sr, sarcoplasmic reticulum). (From Sjöström and Thornell, 1975.)

Figure 3.5 shows some examples of typical freeze–etch replicas of muscle. The cleavage surface often tends to follow directions of weakness in the tissue such as between the two halves of a membrane or between adjacent rows of myofilaments. But sometimes the surface goes through the middle of a structure such as a myosin filament, thus revealing its internal structure. In addition, it provides a good means of studying ice crystal formation in rapidly frozen biological tissue. Sublimation from ice is faster than from protein, and the positions and size of any ice crystals can therefore be determined very simply. Freeze–etching is therefore a useful adjunct to the cryosectioning method described next.

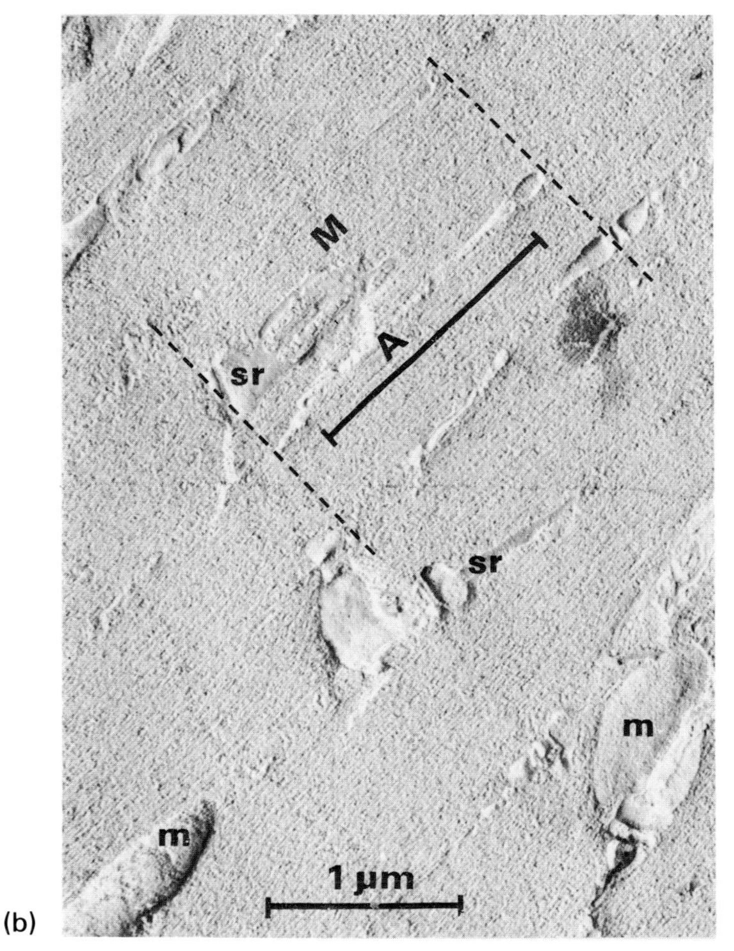

FIGURE 3.5. *Continued*

Cryosectioning. The cryosectioning technique as applied to muscle has been developed largely by Sjostrom and his collaborators at Umea University, Sweden (see Sjöström and Squire, 1977a,b). In their work, they have made use of the L.K.B. Ultrotome III® ultramicrotome fitted with the L.K.B. CryoKit®. (Alternative equipment is available from Reichert or DuPont.)

In the CryoKit, the temperatures of both the specimen head and the knife are independently controlled using a balance between liquid nitrogen cooling and localized warming by an electrical heating element. Specimen and knife temperatures between 0°C and −170°C can be obtained in this way.

Sectioning can either be carried out on a dry knife (the dry cutting variant), sometimes fitted with an antiroll plate to prevent curling of the sections, or on a knife with an antifreeze (e.g., DMSO or glycerol) as trough liquid. This wet sectioning method is sometimes easier and is more generally useful for ultrastructural work. The dry sectioning method is useful when the muscle has to be treated with as few chemical solutions as possible as, for example, when X-ray microanalysis is to be carried out (see Sjöström and Thornell, 1975).

To date, the most useful procedure for ultrastructural studies of cryosectioned muscle involves the following steps (Sjöström and Squire, 1977a,b):

1. Dissection and slight stretching of a muscle.
2. Prefixation using 2.5% glutaraldehyde in Ringer's solution for 5 to 15 min.
3. Cryoprotection using 30% glycerol in buffer.
4. Rapid cooling in Freon-12 cooled by liquid nitrogen.
5. Wet sectioning on the L.K.B. CryoKit with 50% DMSO as trough liquid and with specimen and knife temperatures lower than −80°C (preferably −100°C) and about −50°C, respectively.
6. Collection of sections using a plastic loop with which the sections are transferred to a formvar-coated grid.
7. Negative staining of the sections while still wet and at room temperature using a 2% solution of ammonium molybdate (pH 7.3) or 0.5% solution of uranyl acetate (pH 4.6).
8. Air drying followed by storage in a desiccator prior to viewing in the electron microscope.

The principal advantages of this procedure over conventional plastic-embedding methods are that it is very fast (it takes only hours rather than the days normally involved), that the processes of chemical dehydration and plastic embedding are avoided, and that use can be made of negative staining methods which are capable of yielding good resolution. But an additional very important advantage of cryosectioning is that the sections themselves can be treated with special solutions. It is therefore not necessary to make use of glycerinated muscle or of skinned fibers if the internal milieu of the cell is to be altered. This means that such procedures as extraction of specific myofibrillar components, enzyme cytochemistry, or labeling of particular molecules with antibodies can be carried out directly on the sections. The advantages of this include the possibility of carrying out a series of different treatments on serial sections from the same muscle and the great saving that results when, for example, treatment of a tissue with antibody solutions is carried out. A much smaller quantity of antibody is needed to label a section than is needed to label a whole muscle, and so less wastage of the precious antibody occurs. Penetration of antibody into intact fibers is also rather poor.

As with most procedures involved in electron microscopy, a large number of variations are possible in the cryosectioning method. Some of these are shown in Fig. 3.6 where a few possible omissions or extra steps in the procedure (e.g., block staining or encapsulation in gelatin) are indicated.

It will be shown in Chapter 7 that, despite its infancy, this type of cryosectioning method of studying muscle has revealed several ultrastructural features of vertebrate skeletal muscle that had not been seen before.

Two examples of cryosections of muscle are shown in Fig. 3.7. Here the effects of the negative stain can be seen. The protein appears white, and the stain has piled up against prominent features such as the Z-band and M-band and has accumulated in the A-band where there is much material.

Objective tests using optical diffraction methods (see Section 3.5.4. and Chapter 7) have indicated that the structural preservation of the myofibrillar material is better in cryosections than it is in comparable plastic sections. The comparison has not yet been made between the preservation in cryosections and in plastic sections prepared using Reedy's new fixation method. However, it is known that prefixation using Reedy's method prior to cryosectioning does not significantly alter the appearance of the resulting sections (Sjostrom, personal communication). This confirms that the preservation in cryosections is already inherently good. Another important feature of cryosections is that less shrinkage seems to occur than in comparable plastic sections, and therefore direct spatial measurements on cryosections are more reliable.

As a final comment, and to redress the balance a little, it is worth noting that if only low-magnification survey sections of a muscle are required, then plastic sectioning is still the method of choice, since uniform negative staining of large areas of a cryosection is difficult to

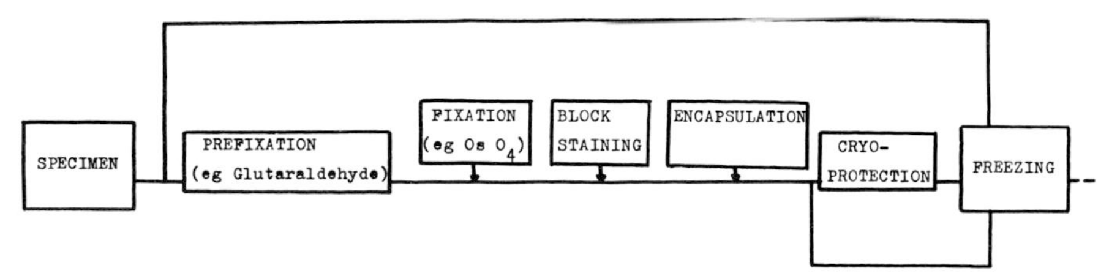

FIGURE 3.6. Schematic flow diagram to show possible variations in the preparative procedure prior to cryosectioning (From Sjóstróm and Squire, 1977b.)

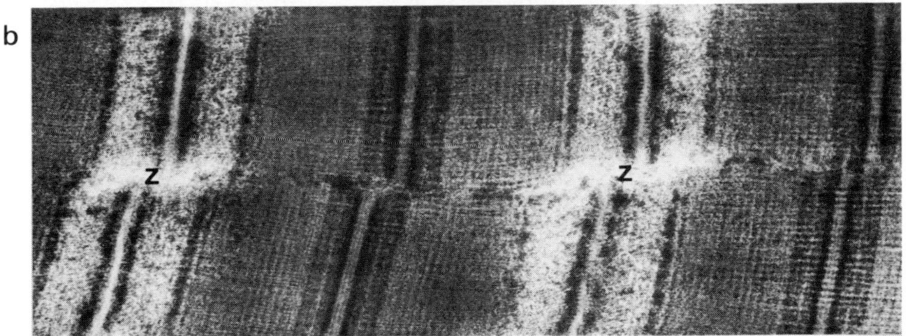

FIGURE 3.7. Examples of negatively stained longitudinal cryosections of frog sartorius muscle. (a) Survey section showing the accumulation of stain (ammonium molybdate) against the Z-bands (z) and M-bands (m) and generally throughout the A-band (A). (b) Higher magnification micrograph showing considerable detail in the A-band and a very clearly defined H-zone. (a) × 13,000; (b) × 37,000 (reduced 30% for reproduction). (Unpublished micrographs of Dr. A. Freundlich.)

achieve, whereas uniform staining of a plastic section is relatively simple. In addition, visualization of glycogen granules and some other structures of morphological interest is also relatively poor in cryosections.

3.4. Biological Electron Microscopy

3.4.1. Introduction

The study of biological material makes a number of special demands on an electron microscope. Among these are the necessity to preserve rather fragile specimens for as long as possible while in the microscope and the need to enhance the contrast in the specimens by instrumental means without, at the same time, lowering the resolution too much. In this section, these problems of specimen preservation, contrasting, and resolution will be discussed after a brief introductory description has been given of the basic design principles of a typical electron microscope.

3.4.2. Basic Electron Microscope Design

Although the principles of image formation in an electron microscope are rather different from those in a light microscope, the two types of microscope are analogous in the overall layout of their components. One form of the light microscope (Fig. 3.8a) consists essentially of a light source, a condenser system that focuses light from the source onto the specimen, an objective system that produces a first image of the object, and then a projector lens that produces a highly magnified image on a ground glass screen. The corresponding features in an electron micro-

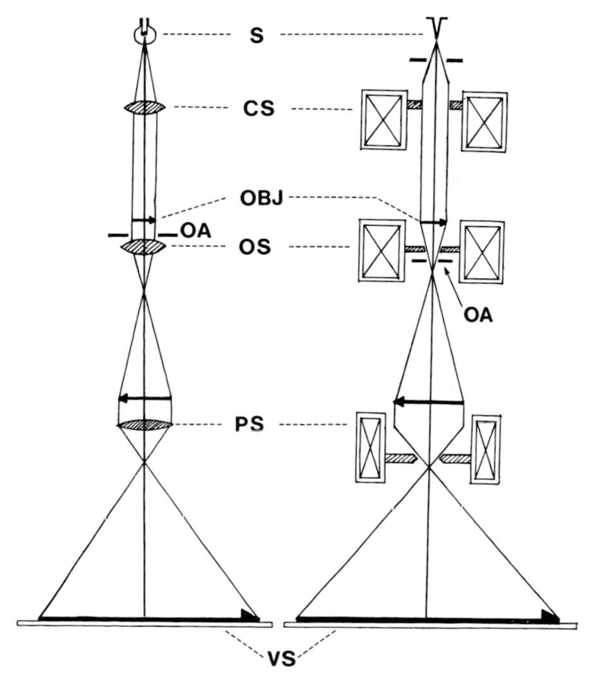

FIGURE 3.8. Schematic diagrams showing the layout of components in a light microscope (left) and in an electron microscope (right). S, source; CS, condenser System; OBJ, object; OA, objective aperture; OS, objective system; PS, projector system; VS, viewing screen.

scope are indicated in Fig. 3.8b. Electrons from a suitable source (usually a heated filament) at the top of the evacuated microscope column are focused by a condenser system of electromagnetic lenses onto the specimen. Electrons from the specimen are then focused by an electromagnetic objective lens system to give an intermediate image. After further enlargement, using an electromagnetic projector lens system, a final image of the object is produced on a fluorescent screen where it can be viewed.

It has already been mentioned in Chapter 2 that according to Abbé's theory of the light microscope, the object diffracts the incident light, and the diffracted beams are collected by the objective lens which focuses them to produce an image of the object. But because the objective lens has a finite diameter, some of the diffraction orders will fall outside the objective, and the information carried by those orders will be lost when the image is reconstructed. The image will therefore be imperfect. In fact, in this case, the imperfect image of a point object is an Airy disk pattern, and the width D of the central peak (Fig. 3.9) can be shown to be given by $D = 1.22\ \lambda/n \sin \alpha$, where λ is the wavelength of light, α is the semi-angular aperture of the lens, and n is the refractive index of the medium in the which the object is situated. This can be thought of in terms of the convolution theorem, since the lens effectively multiplies the diffraction pattern from the object by a function that is unity over the area of the lens but zero outside the lens. The near perfect image of the object that would be produced by a perfect lens of very large diameter must therefore be convoluted with the diffraction pattern (the Airy disk) of a circular object of diameter equal to that of the lens.

In the light microscope, the ultimate resolving power is defined as the closest separation of two points in the object that will give rise to Airy disk patterns such that the central maximum of one falls on the first

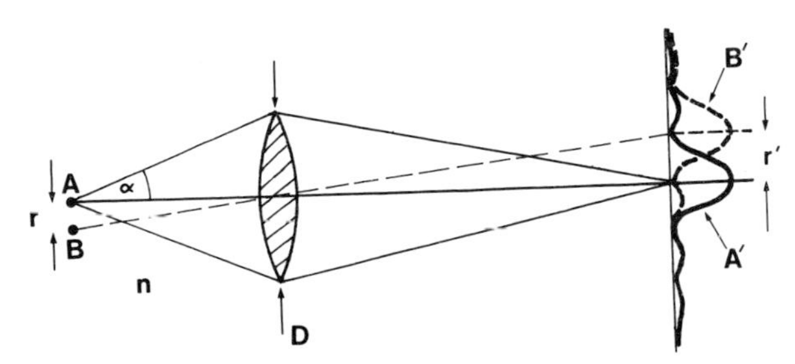

FIGURE 3.9. Illustration of the definition of resolution in the light microscope. D, lens diameter; α, angular aperture; objects A and B produce images A′ and B′ in the form of Airy disk patterns. The resolution r is defined as the object separation for which the central peak in A′ falls on the first minimum in B′. n is the refractive index of the medium in which the objects A and B are situated.

minimum of the other (Fig. 3.9). From the expression for D above, this limit is clearly $r = 0.61\lambda/n \sin \alpha$. For light of wavelength about 5000 Å, the ultimate resolution is therefore about 2000 Å. This resolution can be achieved in practice, since corrected glass lenses for which $n \sin \alpha$ approaches 1.4 can be made.

Electrons accelerated through a voltage of, say, 100 kV would have a wavelength of only about 0.04 Å, and electron microscopes should therefore, in theory, be well able to resolve atomic spacings of about 1 Å. In fact, the ultimate resolution so far obtained in an electron microscope is only about 1.5 Å under very favorable conditions. The large difference between the theoretical and actual resolving powers of electron microscopes stems from the very severe defects (or aberrations) of even the best of the available electromagnetic lenses. As with a glass lens, the aberrations increase with increasing aperture, so they are less of a problem if only the central region of a lens is used. But the effect of this is to reduce the effective angular aperture of the lens, and the resolution, r, will be poorer (i.e., r will increase). In electron microscope design, the angular aperture of the system is defined by an objective aperture (OA in Fig. 3.8). The diameter of this aperture is chosen to give a compromise between the effects of spherical aberration of the electromagnetic lenses (better with a smaller aperture) and the diffraction limit on resolution (which requires a larger aperture).

3.4.3. Image Formation in the Electron Microscope

Two main processes of electron scattering by the object, *elastic* and *inelastic* scattering, are involved in image formation in the electron microscope.

When elastic scattering occurs, an electron is deflected from its original path without losing energy by its interaction with the charges in the relatively massive nuclei of the atoms in the object. Clearly, the number of electrons scattered in this way and the amount by which they are scattered depend on the number of nuclei in a given area of the specimen and on the charge (atomic number) of the nuclei with which they interact. If an aperture (the objective aperture) is placed behind the specimen, then many of the scattered electrons will fall outside this aperture, and these have effectively been "absorbed" (i.e., removed from the electron beam) by the specimen. Regions of high mass density in the object therefore "absorb" more electrons than do regions of low mass density, and amplitude contrast in the image is produced. Since the effect of the objective aperture is to remove some of the elastically scattered electrons from the beam, its diameter will determine what proportion of electrons are removed. Alteration of the amplitude contrast in the image can therefore be produced by altering the diameter of the objective aperture. At the same time, it is clear that fast electrons will be scattered less than slow electrons, and therefore amplitude contrast can also be altered by altering the accelerating voltage in the electron gun of the microscope. Note that the above description could equally well have been given in terms of Abbé's theory, since the loss of high orders of diffraction is what gives rise to an imperfect image and therefore to contrast.

Inelastic scattering of electrons occurs when an incident electron encounters an orbital electron in an atom in the object. Since the interacting objects are of the same mass, the incident electron loses a significant amount of energy during the collision, and its wavelength will change. The angle of scatter involved is very small, and most inelastically scattered electrons will pass through the objective aperture and contribute to the image. However, since their wavelength has changed, they will gradually move out of phase with the unscattered beam and will interfere with it during image formation to give what is called *phase contrast.* The phase contrast effect can be increased if the microscope is moved slightly out of focus, since the relative phases of the scattered and unscattered beams are thereby altered.

3.4.4. Contrast Enhancement in Biological Specimens

One of the basic problems in the microscopy of biological material is that since it contains atoms of low atomic number, it tends to be transparent. A perfect image of the object would, therefore, reveal nothing. Put another way, the inherent amplitude contrast in biological tissue is very low. The electron microscopy of unstained biological material therefore depends on the ability of the microscope to produce a slightly imperfect image of an object.

It has already been mentioned that stains are commonly used to produce "synthetic" amplitude contrast in the specimen. But in addition, the electron microscope itself can be used to produce intrumental contrast enhancement. From the previous section, it is clear that this can be done either by reducing the diameter of the objective aperture, by reducing the accelerating voltage of the electrons (e.g., to 60 kV), or by slightly defocusing the image of the specimen. In practice, all of these methods are used. As it happens, the molecules used to stain biological materials are so large (say, 5 Å, at least) that there is no point in trying to push the electron microscope to give an intrumental resolution greater than 3 Å. This means that contrast enhancement using small objective apertures is quite acceptable despite the loss in resolution that results.

In sections of tissue, it is usually reckoned that the resolution limit caused by aberrations is about 20 Å.

Contrast can also be enhanced by what are termed darkfield methods. These methods involve the blocking of the unscattered beam so that only scattered electrons contribute to the final image. In the case of a periodic object, it is possible to select only a single diffraction order and to allow that order alone to form the image. This is normally carried out by deflecting the main beam so that it is brought to a position outside the objective aperture while the required diffraction peak is brought onto the optic axis of the microscope.

Finally, it is worth noting that additional contrast in an electron micrograph can be achieved by giving the micrograph a relatively long exposure so that the optical density on the developed micrograph is high, and this may be further increased during printing.

3.4.5. Specimen Deterioration in the Electron Microscope

Deterioration of biological specimens in the electron microscope principally results from two causes associated with radiation damage and the deposition of organic vapors. Radiation damage apparently has the effect of completely destroying the native conformation of a protein under the normal conditions for transmission electron microscopy, and at the same time it is very likely that the stain molecules in a section are disturbed as a result. The extent of damage depends on the wavelength of the electrons and hence on the accelerating voltage. For this reason, the high-voltage (1 MeV) microscopes now available cause less radiation damage, but only at the expense of a reduction in contrast.

As a general rule, speed is all important when taking micrographs, and for the highest resolution, very low beam intensities can be used (Henderson and Unwin, 1975).

Contamination caused by organic materials results from the imperfect vacuum in the microscope column. When organic molecules come to rest on the specimen being viewed, the electron beam can decompose them into a form of carbon. A layer of carbon can therefore build up on the specimen at a considerable rate (about 5 Å thickness per sec). There are two principal ways of reducing this effect. The most commonly used method is to surround as much as possible of the specimen with a surface cooled by liquid nitrogen (Heide, 1958, 1960, 1963, 1964). The organic molecules condense preferentially on the anticontaminator and leave the specimen itself in a relatively clean state. The second method, which has only recently been introduced, is simply to provide a sufficiently high vacuum near the specimen. For example, the Philips E. M. 400® electron microscope has a column vacuum of 10^{-7} torr which is claimed to be sufficiently good to render the use of an anticontaminator device largely unnecessary.

3.5. Methods of Image Analysis

3.5.1. Introduction

It has already been mentioned that with the ever increasing demands for finer and finer structural details from micrographs, the techniques of image analysis and processing are rapidly becoming more important. These techniques range from simple photographic methods to sophisticated optical diffraction methods and to the use of computers. What they all do is to provide some means of helping the researcher to extract, from the somewhat jumbled array of information contained in the distribution of optical density in the electron micrograph, the underlying form of the structure that gave rise to that image. In many cases, the image can be seen to be periodic, but the details may be partly obscured because of poor structural preservation of the specimen or of uneven staining effects. In these cases, a clearer idea of the form of the repeating motif in the regular array can often be determined from an average image. Such an average is produced by superimposing all of the different repeating units and adding them up. In this way, genuine structural features of the motif will tend to be enhanced, but artifacts, which will generally be different in each repeat, will tend to be smeared out. Averaged images of this kind have, in fact, been found to be very useful in muscle research. They can be produced in a variety of ways, either manually, using direct multiple exposure photographic methods, or automatically, using particular optical instruments. On the other hand, if the optical density information is available in digital form, then image averaging can be carried out using computer methods.

Another way of revealing structural features in an image is to make use of optical diffraction. This time, instead of using a mask of holes as the diffracting object (see Chapter 2), a copy of the electron micrograph itself can be used. The optical diffraction pattern from the micrograph will reveal the presence and relative spacings of periodic structures. But once again, if the information in the micrograph is put into digital form, a computer can be used instead of the optical diffractometer to calculate the diffraction pattern of the micrograph.

In this section, these alternative methods are briefly evaluated and demonstrated, since later in the book their applications to muscle research will be seen to be numerous.

3.5.2. Photographic Methods

The simplest photographic method of image averaging is to measure the periodicity on a micrograph and then to use a photographic enlarger to print the micrograph, using a multiple exposure technique. Here, the photographic paper is shifted between each exposure by a distance equal to the periodicity (Markham *et al.*, 1964). Such a procedure is effective, even if somewhat laborious, and has been used in the analysis of Z-band structure (R. W. D. Rowe, 1973). A variation of the technique is possible when a single object has a certain rotational symmetry (Markham *et al.*, 1963). This time, between each exposure, the micrograph is rotated by $1/n$th of a revolution for an object with n-fold symmetry, and n exposures are made. Used in this way, the rotational averaging technique can be very informative. But attempts to reverse this method as a way of finding the unknown rotational symmetry of an object are fraught with dangers and are seldom convincing (Friedman, 1970). A more reliable method using computation has been developed by Crowther and Amos (1971).

3.5.3. Automatic Image-Averaging Methods

Introduction. A variety of automatic image-averaging methods has been devised including the use of multiple pinhole cameras (Warren and Hicks, 1971) and a machine called a periodograph (Foster, 1930). The latter can pick out the periodicity of any micrograph by, in effect, altering the translation between several superimposed identical images of the micrograph until reinforcement of the periodicity occurs. Figure 3.10 demonstrates that what all of these methods do is produce a convolution of the micrograph with a one-dimensional array of points or lines of separation equal to the periodicity in the micrograph. The latter is usually called the averaging function. All one- or two-dimensional image-averaging methods produce some form of convolution of a micrograph with an averaging function.

Convolution Camera. In order to illustrate the principles of image averaging, use will be made here of a single very simple optical system called a convolution camera (A. Elliott *et al.*, 1968b) which is capable of producing convolutions of many different kinds. Although it will be shown in Section 3.5.5. that some analogous image-averaging processes can be carried out on the optical diffractometer (Fraser and Millward, 1970), the convolution camera provides a means of directly visualizing the processes involved.

All that is needed to make a convolution camera (Fig. 3.11) is a

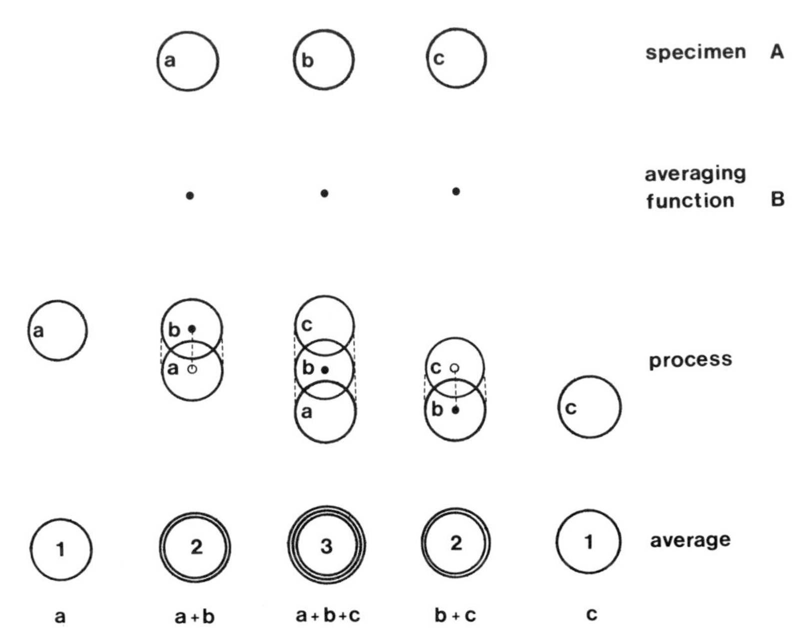

FIGURE 3.10. The basis of convolution averaging. An object (A) in the form of an array of similar but not identical motifs a, b, and c is convoluted with a similar array of small points (B) by the process shown. The final convolution is such that the central feature is an average image produced by the superposition of a, b, and c.

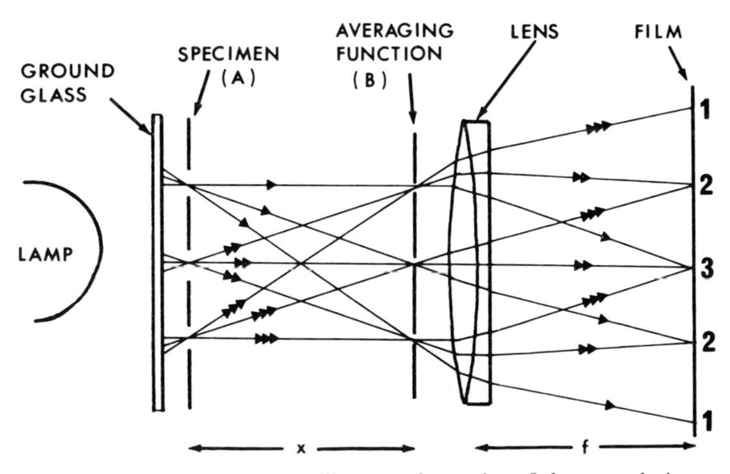

FIGURE 3.11. Schematic diagram to illustrate the optics of the convolution camera in this case for an object and averaging function as in Fig. 3.10.

wide-aperture lens (e.g., *f*/2.5 anastigmat with 5- to 20-cm focal length), two holders on which to mount transparencies of the micrograph (A) and the averaging function (B) (one of which can be rotated about the optic axis of the lens), and a film holder with which to record the convolution. The film (or a ground-glass screen for direct viewing) is set in the focal plane of the lens. The mask A is illuminated evenly by a ground-glass screen in front of a tungsten lamp. The lens and film holder are conveniently part of the same photographic camera, in which case the lens must be set at ∞ and the aperture opened out fully. Such a camera also provides a convenient shutter mechanism.

As will be seen from the optics of the camera (Fig. 3.11), if mask B is an array of holes in an opaque card, then the camera acts as a multiple pinhole camera. Each pinhole produces an image of the object A, but these images are laterally separated by the periodicity in the averaging function. In this way, convolution of mask B with mask A inverted through the optic axis of the camera is obtained on the film.

As an example of the type of process involved in image averaging, Fig. 3.10 shows a limited one-dimensional array (A) consisting of three large circular objects and an averaging function (B) consisting of three small points on the same array. The convolution of A with B consists of an array of five images of which the central feature is an average of all three objects in A. The ray diagram in Fig. 3.11 illustrates how the average is produced in this particular case on the convolution camera.

Unlike many other image-averaging methods (except optical diffraction), a similar process on the convolution camera can be carried out very easily in two dimensions using a 2-D object array and a 2-D averaging function (Fig. 3.12). In this example, the central image is an average of all seven motifs in the object.

Two alternative kinds of convolution can also be produced using the convolution camera. The kind originally described by A. Elliott *et al.*

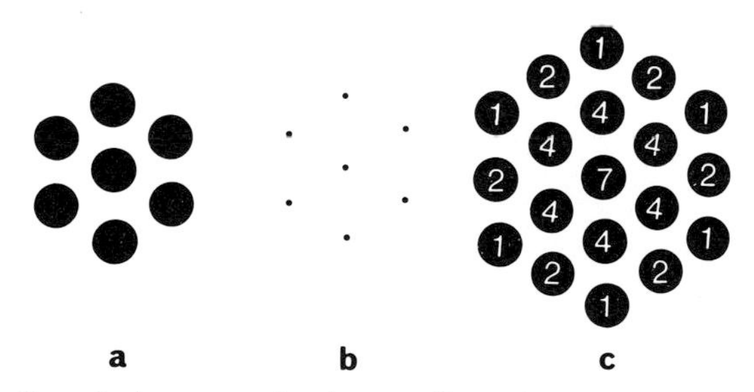

FIGURE 3.12. Convolution averaging in two dimensions. A 2-D array of motifs (a) is convoluted with an averaging function (b), and the convolution (c) has a central feature that is an average of all seven objects in (a).

(1968b) is a self-convolution produced by convoluting a micrograph (A) with an identical micrograph (B) inverted through the origin (Q-convolution). Paradoxically, this is achieved in practice by putting two micrographs with identical orientations in positions A and B on the convolution camera—since the inversion of one mask is carried out automatically by the camera. A. Elliott *et al.* found that the self-convolution of a micrograph (otherwise known as its convolution square or autocorrelation function) could reveal structural detail that was not revealed readily by optical diffraction.

The other kind of convolution is a convolution of a micrograph with itself without inversion (produced by placing two micrographs in positions A and B in the camera but with one rotated by 180° about the optic axis). This type of convolution, called P-convolution by Hosemann and Bagchi (1962), is actually capable of revealing more than a self-convolution, since the latter is always a function with twofold rotational symmetry (i.e., centrosymmetry) even though the micrograph itself may not possess such symmetry. The P-convolution, on the other hand, preserves the intrinsic symmetry of the micrograph while at the same time producing a useful average image. These two methods of convoluting two identical masks are illustrated diagrammatically in Fig. 3.13, where the origin of the centrosymmetry of the self-convolution is apparent.

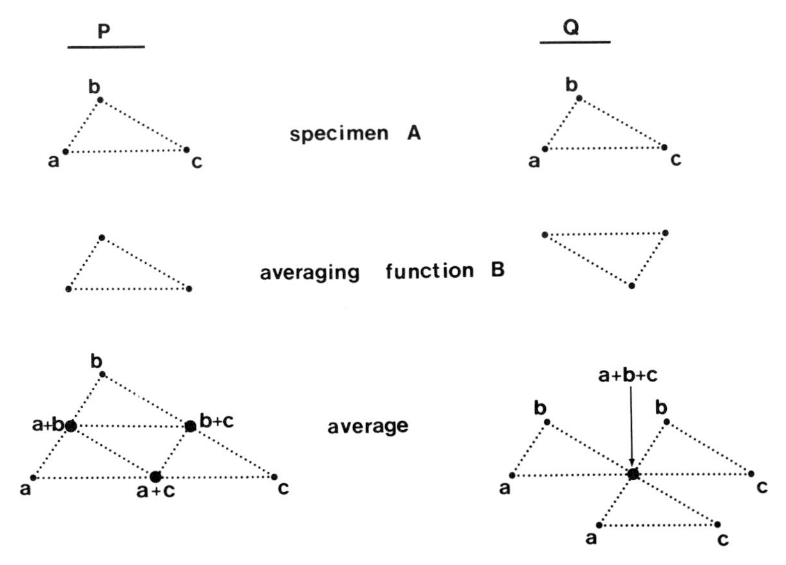

FIGURE 3.13. Self-convolution (Q-) and P-convolution of model structures as an illustration of the processes involved. The self-convolution is automatically centrosymmetric whatever the symmetry of the original specimen (A), but P-convolution preserves more of the symmetry of the specimen.

Practical examples of self-convolution, P-convolution, and convolution averaging are illustrated in Fig. 3.14. The micrograph in this case shows the surface of negatively stained thick myosin-containing filament from the Oyster *Crassostrea angulata* (Fig. 3.14a). The convolution of the

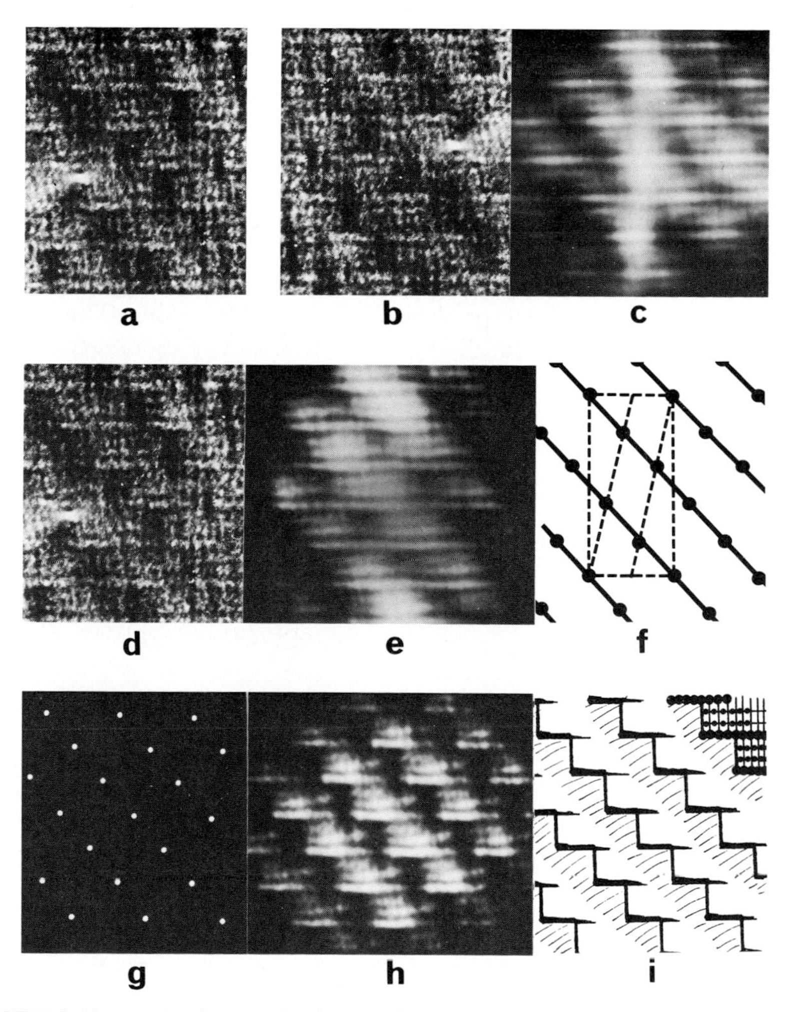

FIGURE 3.14. (a) A micrograph of a negatively stained paramyosin filament (as in Fig. 3.4d), (b) its inversion, and (c) the self-convolution of (a) produced by convoluting (a) with (b). (e) The P-convolution produced by convoluting (a) with itself without inversion (d). (f) The average lattice deduced from (c) and (e). (g) An averaging function derived from (f). The convolution average (h) produced by convoluting (a) with (g). The interpretation of what can be seen (i) will be discussed in Chapter 8; suffice it to say here that marked diagonal ridges of stain can be seen (each ridge having a zig-zag shape) and that a small lateral periodicity can be seen as indicated in the top right of (i).

filament (a) with its own inversion (b) (i.e., the self-convolution) is shown in (c). The P-convolution produced by convoluting (a) with itself without inversion (d) is shown in (e). From both the self-convolution (c) and P-convolution (e), it is possible to define the average lattice (f) on which the repeating units in the micrograph (a) are arranged. A mask of holes (g) prepared from this average lattice (f) can then be used as the averaging function to produce the convolution average shown in (h). The details that can be seen in this average, summarized in (i), will be described in Chapter 8.

Special Averaging Functions. Fraser and Millward (1970), in their account of image averaging using the optical diffractometer, described several different forms of averaging function in addition to those representing just the simple underlying lattice on which the motifs are arranged. For example, many synthetic aggregates of fibrous proteins and also some muscle sections show only clear 1-D order. This usually has the form of periodic banding across the width of the image. Figure 3.15 a shows an example of this type of structure, which is a synthetic aggregate of tropomyosin molecules (located in muscle in the thin filaments), together with some averaging functions and the resulting convolution averages. A useful average image from this structure is a laterally smeared image (Fig. 3.15c) which can be produced by convolution of the micrograph with a single very narrow slit (Fig. 3.15b) lined up parallel to the cross-striations in the micrograph. If desired, the repeating unit in this laterally smeared image can then be further reinforced by convolution with an array of points of the appropriate periodicity lined

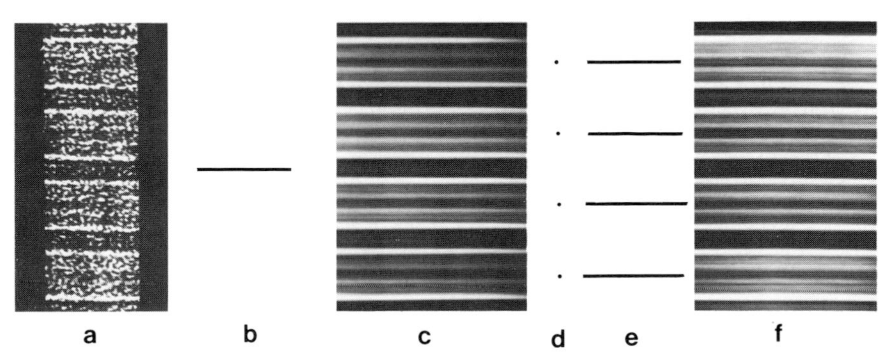

FIGURE 3.15. Lateral averaging by convolution. (a) A negatively stained (uronyl acetate) paracrystal of tropomyosin (compare Fig. 3.4e which is positively stained). (b) A single slit averaging function. (c) The laterally smeared image produced by convoluting (a) with (b). Further image enhancement could be produced by convoluting (c) with a set of points of the same periodicity lined parallel to the long axis of the paracrystal (d), but the process can be carried out in one step by convoluting (a) with an averaging function (e) that is itself a convolution of (b) and (d). The result is shown in (f).

up parallel to the long axis of the aggregate (Fig. 3.15d). But it is clear that both of these processes can be carried out simultaneously using an averaging function of the form shown in Figure 3.15e (which is just Figs. 3.15b and d convoluted together).

It should be clear from this that the convolution camera can carry out virtually any kind of averaging procedure, and since it is such a simple and inexpensive device, its use can be recommended.

Practical Hints. One or two practical features of the convolution camera are worthy of mention. Referring to Fig. 3.14 which shows a ray diagram through the camera, it should be noted that with the camera lens set at ∞, two masks prepared on the same scale will yeild a convolution function also on the same scale, provided the separation (x) of the masks A and B is equal to f, the focal length of the lens. However, movement of mask A, keeping mask B close to the lens to make full use of its large aperture, will alter the magnification but not the form of convolution. In this way, different levels of detail in the micrograph can be studied visually using either the self-convolution, P-convolution, or convolution-averaging methods. The magnification is equal to f/x.

In preparing a convolution, it is essential that the two masks A and B should be strictly parallel in their orientation about the camera axis. It is therefore necessary to have at least one of the masks mounted so that it can be rotated about this axis or one parallel to it. The masks can be lined up by viewing the convolution using a screen and pocket lens or by using the viewfinder of a reflex camera. One of the masks is rotated until the sharpest convolution pattern is obtained. Fortunately, in practice, it is usually very easy to judge this position.

Finally, it should be noted that the convolution camera can be used to pick out periodicities in an image (i.e., its dominant spatial frequencies) in a way analogous to the periodograph of Foster (1930) or Warren and Hicks (1971). Here a periodic averaging function of arbitrarily chosen periodicity is placed in position B on the convolution camera (Fig. 3.11). The micrograph being analyzed is placed at A. The separation of the masks A and B is x. The average image is then viewed using the camera, and the separation (v) of the viewing screen and the lens is gradually altered until optimum reinforcement of the periodicity in the image is obtained. In practice there may be a number of positions where such reinforcement occurs. By noting these positions, the periodicity in the micrograph can be determined by simple geometrical optics if the geometry of the camera is known (i.e., the values of x,v, and f).

In essence, this analysis of the principal spatial frequency in the object relies on the simple premise that convolution of A and B is produced, even if they are not on the same scale, provided the focus of the lens is adjusted accordingly.

In summary, it can be seen that the convolution camera, despite its simplicity, is a very versatile instrument. In fact, it can carry out procedures in real space using incoherent light that are entirely analogous to the Fourier transform manipulations that can be carried out using the optical diffractometer as described in the next section.

3.5.4. Optical Diffraction

Analysis of Periodicities. The optical diffractometer, described in Chapter 2, was originally designed as an analogue of X-ray diffraction following a suggestion by Bragg (1939). The instrument was developed largely by Lipson and Taylor (Lipson and Taylor, 1958; Taylor and Lipson, 1964; Lipson, 1972) for just this kind of application, and much use of the technique has been made in Chapter 2 of this book to illustrate general diffraction principles. However, it was realized by Klug and Berger (1964) that optical diffraction would also have its application to the analysis of electron micrographs. A micrograph placed in the object position of the diffractometer will give rise to a diffraction pattern in the focal plane of the objective lens. If the micrograph carries periodic information, the reflections caused by this periodicity will be seen in the optical diffraction pattern. From the geometry of diffraction (which gives the diffraction angle), it will then be possible to determine the periodicity in the micrograph. This method of analysis of electron micrographs has found a wide application and has been applied successfully to a variety of tissues including muscle (e.g., Reedy, 1967; H. E. Huxley, 1968; Moore *et al.*, 1970, O'Brien *et al.*, 1971, Small and Squire, 1972 and many others as described in later chapters). All of the diffraction theory outlined in Chapter 2 for one- and two-dimensional diffracting objects is directly applicable to the analysis of electron micrographs, and the diffraction condition $n\lambda = d \sin \phi$ will give the d spacing in the micrograph for a particular reflection at angle ϕ.

Several practical points about optical diffraction analysis of electron micrographs are listed below.

1. It is not desirable or necessary to use the original micrograph itself as the diffracting object. It is often convenient to produce on film a very-high-contrast copy of the micrograph in which the detail being studied is clear against a dark background. This also permits the magnification of the micrograph to be altered to suit the performance of the particular diffractometer being used.

2. The surface of most photographic films are far from flat, and to avoid the introduction of phase artifacts in the optical diffraction pattern, it is desirable to sandwich the film being analyzed between two glass

optical flats (preferably flat to at least $\lambda/10$) with a layer of immersion oil between each flat and the film (Fig. 3.16c).

3. As illustrated in Fig. 3.16, two alternative layouts of a diffractometer are possible. In one (a), the diffractometer components are mounted on a rigid vertical steel girder, and the micrograph is therefore situated in a horizontal plane. This is the original design of Lipson and Taylor (1958). In the other arrangement (b), all of the components are mounted on a conventional horizontal optical bench, and the micrograph lies in a vertical plane. In this case, it is convenient to mount the specimen in a liquid gate of the kind illustrated in Fig. 3.16d. Here the micrograph can simply be slotted into a holder in an oil bath bounded by two glass flats. The advantage of this method is that very rapid changing of micrograph negatives is possible without having to move the glass flats as is necessary in the vertical system.

4. It is useful to be able to select a particular small area of the micrograph for analysis. This is conveniently achieved by using an adjustable rectangular aperture or by using an iris diaphragm. Either of these is placed as close as is practicable to the micrograph. Since each

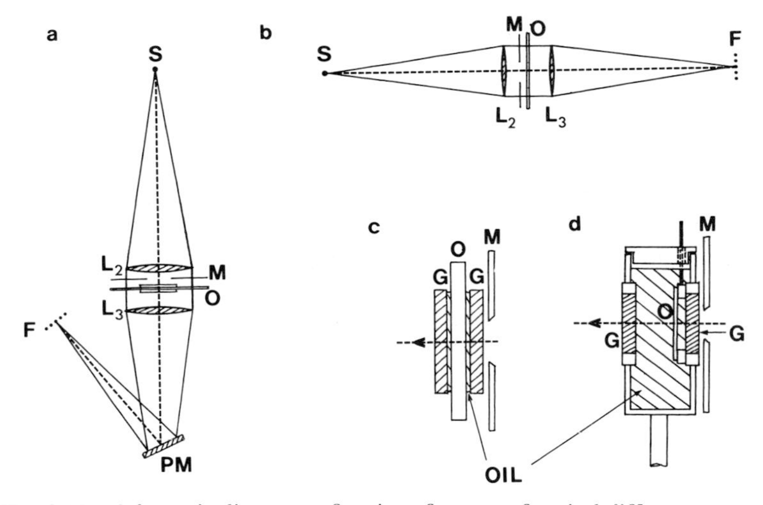

FIGURE 3.16. Schematic diagrams of various features of optical diffractometers. (a, b) Vertical and horizontal arrangements, respectively, of the optical components. PM is a plane mirror (a), L_2 and L_3 are lenses, O is the diffracting object, M is the defining aperture or mask, S is the source, and F is the film position where the diffraction pattern can be observed or recorded. (c) The method of sandwiching an object film (O) in oil between glass optical flats (G). A similar method can be adapted (d) for use on a horizontal diffractometer (b). Here the object is placed in an oil bath bounded by glass optical flats. This arrangement allows very rapid changing of objects and is a very satisfactory method of mounting the specimen.

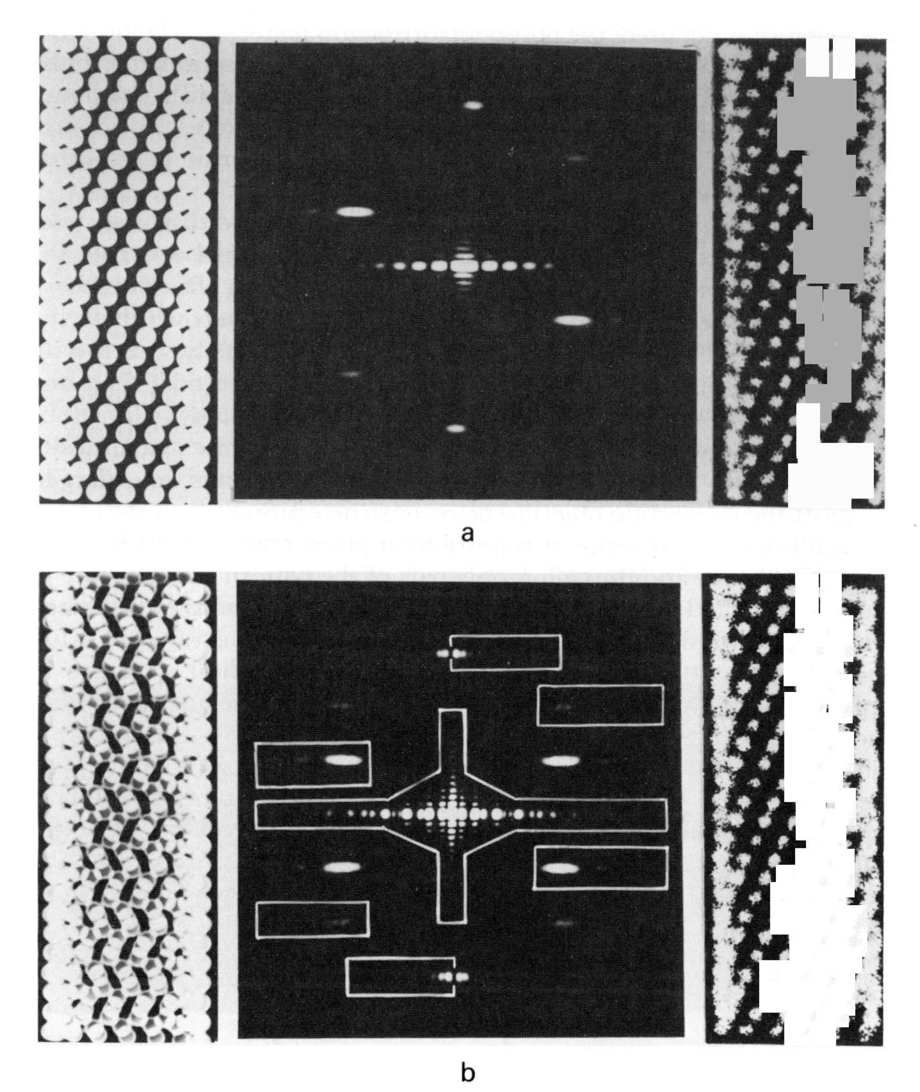

FIGURE 3.19. The use of optical filtering to separate the front and back images of a helix. (a) One side of a helical object (left), its optical diffraction pattern (center), and its reconstructed image (right). Note that the diffraction pattern is asymmetrical about the meridian. (b) A similar object to that in (a) except that both sides (front and back) of the helix can be seen. Its diffraction pattern is shown in the center, and right is the reconstruction produced by allowing only those reflections outlined in the diffraction pattern to contribute. Since these are the peaks in the diffraction pattern in (a), the reconstruction is one side of the model helix only. (From Klug and DeRosier, 1966.)

3.5.5. Image Averaging by Optical Diffraction

As mentioned earlier in the discussion of image averaging with the convolution camera, analogous processes can be carried out using the optical diffraction method. Fraser and Millward (1970) presented a detailed description of the processes involved and the kinds of average that can be obtained. We have already seen (Fig. 3.10) that averaged images are produced when an object, A is convoluted with an averaging function, B. In Chapter 2, the convolution theorem was discussed, and it was found that the diffraction pattern of the convolution of two objects is obtained by forming the product of their diffraction patterns. This means that an averaged image can be reconstructed in the optical diffractometer if the diffraction pattern of the micrograph (A) is multiplied by the diffraction pattern of the averaging function (B). The averaging process therefore demands that the diffraction pattern of the averaging function must be calculated and then represented approximately by cutting (or etching) apertures of the appropriate shape and position in an opaque mask. This mask is then placed in the diffraction plane of the diffractometer, and the resulting reconstructed image is then the required average. Some illustrations of this approach to image averaging are shown in Fig. 3.20 (from Fraser and Millward, 1970). It is clear that the averaging procedure on the diffractometer can be very successful. But researchers without such an instrument should bear in mind that virtually equivalent image averaging can be carried out on the relatively simple and inexpensive convolution camera.

3.5.6. Computer Methods and Three-Dimensional Reconstruction

With the advent of two-dimentionsal recording microdensitometers giving a digital output, modern computer methods involving fast Fourier transform programs can be used either to calculate directly the form of the diffraction pattern of a micrograph or to produce any of the kinds of averaged image that can be produced using a convolution camera or optical diffractometer. But the most important recent application of computer methods is in the 3-D reconstruction of an object from one or more electron micrograph images. This has been developed by DeRosier and Klug (1968) and is based on the premise that information from two or more different views of the same object can be used to give a 3-D reconstruction of the object. A simple everyday illustration of this premise is that with two eyes that see two slightly different views of the same scene, it is possible for us to visualize our surroundings in three dimensions. The main difference between this everyday example of three-dimensional viewing and the study of electron micrographs of

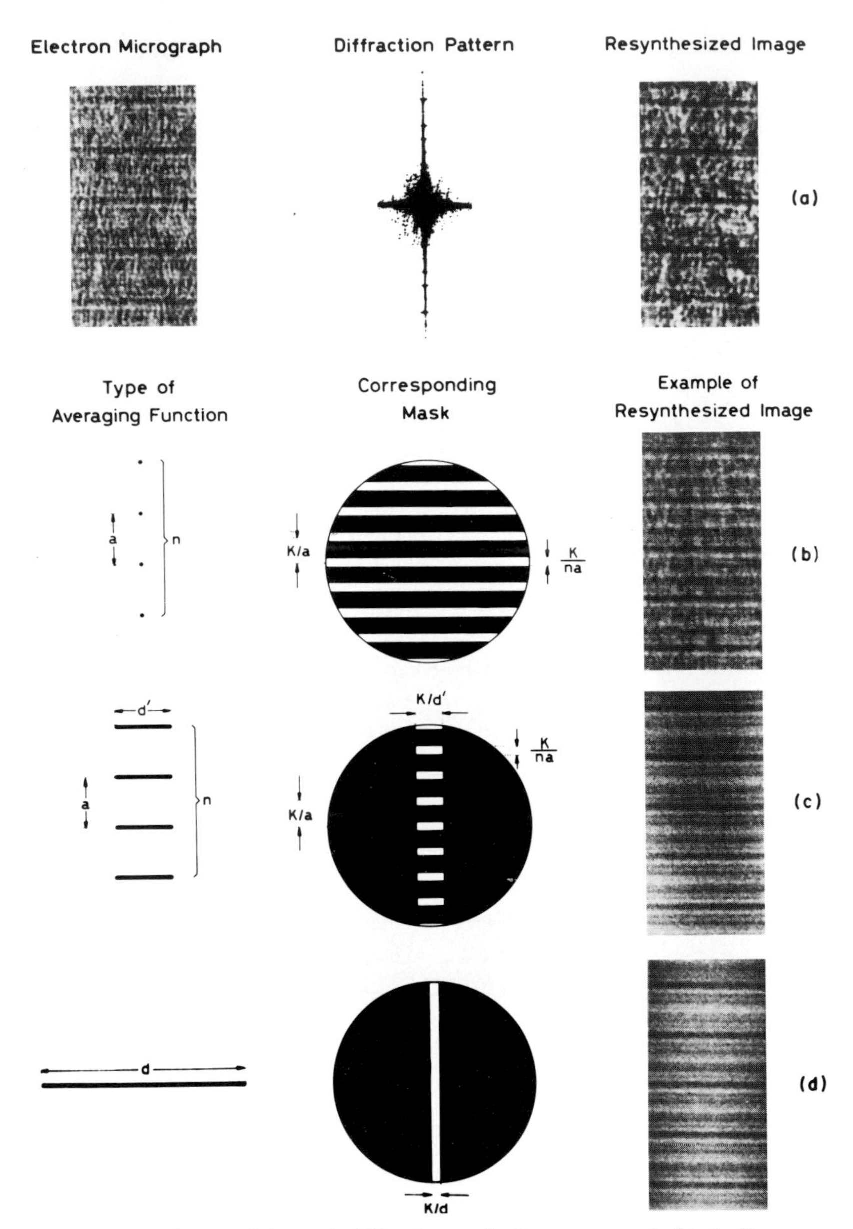

FIGURE 3.20. The use of the optical filtering method to carry out the kind of image averaging shown in Fig. 3.15. As before, the object is a negatively stained paracrystal of tropomyosin. (a) Resynthesised image without filtering; (b) produced using a point lattice as an averaging function; (c) slit smearing included as well (compare Fig. 3.15e); (d) a single slit averaging function (compare Fig. 3.15b). In each case, the diffraction pattern from the micrograph has been filtered by a mask representing the diffraction pattern of the appropriate averaging function. (From Fraser and Millward, 1970.)

stained objects is that in the latter, the image represents the projected density through the object, whereas normally our eyes only see surface features in an object.

In a projected view of a 3-D object, some overlapping of detail at different levels in the object will occur. But some idea of the distribution of features through the object can be obtained if the object is viewed at two different angles. This may be done by taking two electron micrographs with the object at two different tilts relative to the electron beam. These two micrographs can then be viewed simultaneously, using a special viewer, to give a three-dimensional effect. The use of such stereo pairs is now common in electron microscope investigations. They are especially useful with shadowed preparations where only surfaces are visualized. But whether applied to negatively stained, sectioned, or shadowed material, this approach usually gives only a subjective idea of the structure even if many stereo paris are studied.

A more objective approach is to calculate the Fourier transforms (diffraction patterns), in terms of amplitude and phase, of a series of different projections of the same object. It is known from diffraction theory that the Fourier transform of a projected structure represents a particular part (called a central section) of the whole Fourier transform of the object. By carrying out similar computations on different projected views of the structure, it is possible to fill in enough of the Fourier transform of the object to make a complete Fourier synthesis possible. (Remember that a Fourier synthesis is effectively a way of producing a 3-D image of an object from the known amplitudes and phases of the various structure factors.)

The problem is much simplified if the object itself has rotational or helical symmetry, since a single view of it will consist of images of identical subunits in different orientations. In fact, many biological structures do consist of identical subunits. In particular, many structures are helical, and if the helices have a sufficient number of subunits per repeat, then many different views of the subunit will be seen in a single micrograph of the helix. Even when this is so, it is a good idea to average the information from a number of micrographs to even out any artifacts that may be present.

This type of 3-D reconstruction has been applied to the study of actin filaments, and further reference will therefore be made to the technique in Chapter 5.

4

Protein Conformation and Characterization

4.1. Amino Acids, Polypeptides, and Proteins

The macromolecules that form the bulk of the myofibrillar material and are intimately involved in the force-generating process itself are proteins. The purposes of this chapter are to describe the basic types of protein structure that occur in muscle and then to summarize briefly the principal features which determine the conformations which these proteins adopt.

The monomeric repeating units in proteins are derived from amino acids. These have a general formula NH_2-CHR-COOH, where R can be any one of a number of different side groups. In solution at physiological pH, they occur in the ionized form NH_1^+-CHR-COO^-. Amino acids join together to form linear polymers by what is essentially a condensation (water-eliminating) reaction between the carboxyl group (-COOH) in one amino acid and the amino group (-NH_2) of the next.

$$\mathrm{NH_2\text{-}CH}R\mathrm{\text{-}C}\begin{matrix}\nearrow \mathrm{O} \\ \searrow \mathrm{OH}\end{matrix} \quad + \quad \begin{matrix}\mathrm{H} \searrow \\ \mathrm{H} \nearrow\end{matrix}\mathrm{N\text{-}CH}R\mathrm{\text{-}COOH}$$

$$\longrightarrow \mathrm{NH_2\text{-}CH}R\mathrm{\text{-}CO\text{-}NH\text{-}CH}R\mathrm{\text{-}COOH} \quad + \quad \mathrm{H_2O}$$

The -CO-NH- link is called the amide or peptide bond, and for this reason, long polymers derived from amino acid monomers are called polypeptides. All proteins are polypeptides. Note that each polypeptide chain ends with $-NH_2$(N-terminal) and $-COOH$ (C-terminal) groups.

Many kinds of amino acid occur in proteins, and their side groups (R) can have markedly different properties. In fact, it is the sequence of these different amino acids along a particular protein chain that determines its three-dimensional structure, its interactions, and its chemical properties. Table 4.1 lists the common R groups of the amino acids found in proteins in terms of their structural formulae, their names, and their commonly used abbreviations. Two amino acids are very different in character from the others and are worthy of special mention. One is proline (Pro) which is an amino acid in which the R group links back to the main chain, thus restricting rotation about the C–C bond (Fig. 4.1). The other is cystine (Cys) which is produced by the covalent bonding between two cysteine (CysH) residues (R group is $-CH_2SH$) to give a disulfide bridge ($-CH_2-S-S-CH_2-$). Such a covalent cystine link can occur either between different parts of the same protein chain (intramolecular disulfide) or between different chains (intermolecular disulfide). Apart from this and one or two other less common covalent interactions, a number of weaker interactions are very important in defining protein structure. As indicated in Table 4.1, the different amino acids can be divided into three main groups according to the character of their side-chains (R) at physiological pH.

1. Glycine, alanine, valine, leucine, isoleucine, proline, phenylalanine, tryptophan, and methionine are all amino acids with nonpolar side chains. These hydrophobic residues are the least soluble of the amino acids found in proteins, and they are often found clustered in the interior of proteins to the exclusion of water. However, some hydrophobic residues do occur on the protein surface where they can play an important role in the function of the protein. Hydrophobic forces appear to be involved in the initial folding of a polypeptide chain and subsequently in helping to stabilize the folded chain structure.

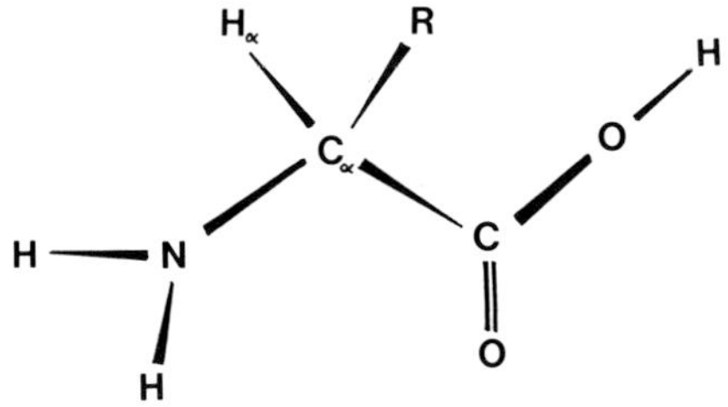

FIGURE 4.1. Outline structure of an L-amino acid. The common R groups are listed in Table 4.1. A D-amino acid is the mirror image of this.

TABLE 4.1. The *R* Groups of the Amino Acids Commonly Found in Proteins

Name	Code	Symbol	Amino acid	Side group (*R*)
Nonpolar amino acids (apolar, hydrophobic)				
Glycine[a]	Gly	G		-H
Alanine	Ala	A		$-CH_3$
Valine	Val	V	$^{-}OOC-C(H)(NH_3^+)-R$	$-CH(CH_3)_2$
Leucine	Leu	L		$-CH_2-CH(CH_3)_2$
Isoleucine	Ile	I		$-CH(CH_3)(CH_2-CH_3)$
Phenylalanine	Phe	F		$-CH_2-C_6H_5$
Tryptophan	Try (Trp)	W		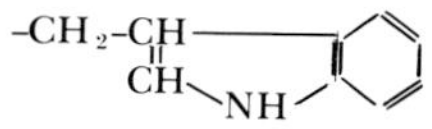 $-CH_2-C$=CH–NH– (indole)
Methionine	Met	M		$-CH_2CH_2-S-CH_3$
Proline	Pro	P	$^{-}OOC-C(H)-CH_2\!>\!CH_2$, $NH-CH_2$ (ring)	
Uncharged polar amino acids				
Serine	Ser	S		$-CH_2-OH$
Threonine	Thr	T		$-CHOH-CH_3$
Tyrosine	Tyr	Y		$-CH_2-C_6H_4-OH$
Asparagine	Asn	N		$-CH_2-CONH_2$
Glutamine	Gln	Q		$-CH_2-CH_2-CONH_2$
Cysteine	CysH	C		$-CH_2-SH$
Cystine	Cys			$-CH_2-S-S-CH_2-$
Charged polar amino acids				
Lysine	Lys	K		$-CH_2-CH_2-CH_2-CH_2-NH_3^+$
Arginine	Arg	R		$-CH_2-CH_2-CH_2-NH-C(=NH_2^+)NH_2$
Histidine	His	H		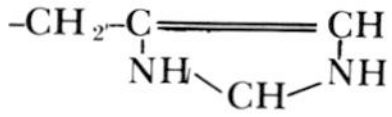 $-CH_2-C$=CH, NH–CH–NH (imidazole ring)
Aspartic acid	Asp	D		$-CH_2-COO^-$
Glutamic acid	Glu	E		$-CH_2-CH_2-COO^-$

[a]Sometimes classified as uncharged polar.

2. Uncharged polar side chains occur in serine, threonine, tyrosine, asparagine, glutamine, and cysteine. These are more soluble than the apolar amino acids and can hydrogen bond to water.

3. Charged polar side chains occur in lysine, arginine, and histidine (positive charge) and in aspartic and glutamic acids (negative charge). Such charged polar groups usually occur on the outside surfaces of proteins, and they are important in stabilizing the structure of individual molecules and the interactions among molecules.

A weak bond that has fundamental importance in defining protein structure is the hydrogen bond. This occurs between two electronegative atoms, one of which (A in Fig. 4.2) is bonded to hydrogen. In this situation, the hydrogen carries an induced partial positive charge and is attracted towards another electronegative atom (B). The result is that the two electronegative atoms (A and B) approach closer to each other than normal Van der Waals contact would allow. In proteins, the most important electronegative atoms for hydrogen bond formation are oxygen and nitrogen in descending order of electronegativity. Hydrogen bond lengths lie in the range 2.7 to 3.1 Å and tend to occur with a bond angle (Fig. 4.2) of less than about 30°. Such hydrogen bonds can occur between several different pairs of amino acid side groups. But they also occur between the carbonyl oxygen (>C = 0) and the peptide nitrogen (>N–H) in the amide groups of different residues. Such interpeptide hydrogen bonds are very important in stabilizing the various regular backbone conformations (e.g., the α-helix) that polypeptides can adopt. Some of these conformations are described in the next section.

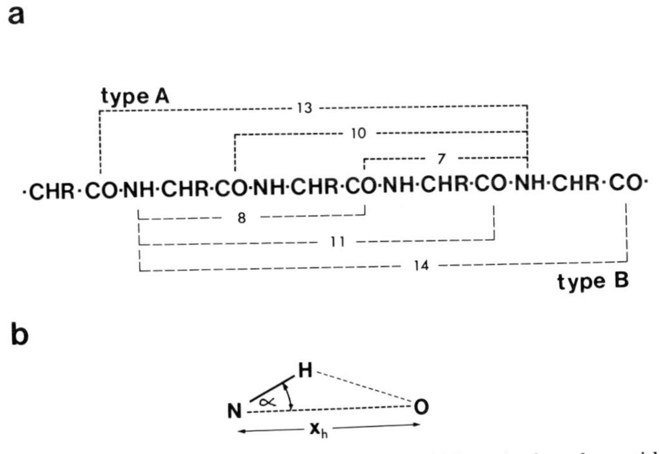

FIGURE 4.2. (a) Possible hydrogen-bonding schemes within a single polypeptide chain. In each case, the numbers indicate how many main-chain atoms there are in a hydrogen-bonded loop (see text). (b) Indication of the linearity and length of a typical hydrogen bond. The bond distance (x_h) is normally between 2.7 and 3.1 Å, and α is usually less than 30°.

4.2. Regular Protein Conformations

4.2.1. Basic Ideas

Following the original ideas of Huggins (1943) and of Bragg *et al.* (1950) about regular polypeptide folding, Pauling and his colleagues (1951, 1953) considered several possible polypeptide conformations in terms of their potential energy. They based their arguments on a number of different criteria.

They recognized that the amide group would be planar (i.e., the H, N, C, and O atoms lie in a plane) with rotation about the C–N bond restricted by its partial double-bond character. This can be thought of either as resulting from resonance between H–N–C = 0 and H–N$^+$ = C–0$^-$ structures or from the overlapping of the π orbital electrons in the >C = 0 group with the p orbital electrons in the nitrogen of the >N–H group (Fig. 4.3). So in Fig. 4.4, the rotation angle ω about the C–N bond would be 180°.

It was assumed that the bond lengths and angles would be similar to those found in small amide molecules (Fig. 4.4).

Rotation about the Cα–C and N–Cα bonds was taken to be possible (Fig. 4.4), but it was recognized that the most satisfactory conformations would be those in which the rotational potential energy was minimal.

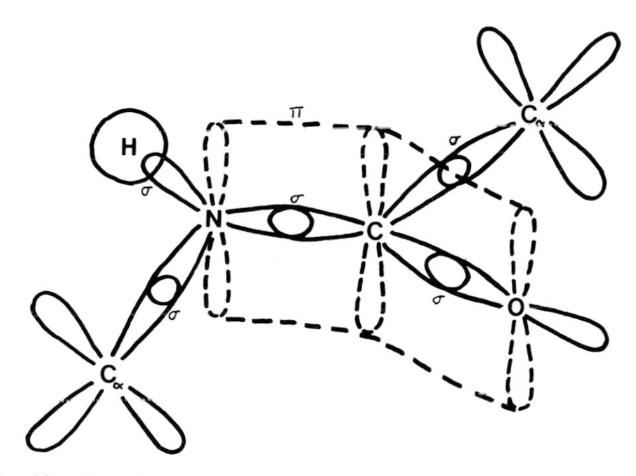

FIGURE 4.3. Sketch to show the origin of the planarity of the amide group. In addition to the σ-bonds between the N, C, and O atoms, there is also a delocalized π-bond (dashed line) formed from the p_z electron orbitals (dashed orbitals) of N, C, and O (assuming that the N–C–O plane is the x,y plane). Cα is an sp^3 hybrid, and N and C are both sp^2 hybrids (see Coulson, 1961). The extra π-bond gives the N–C bond partial double-bond character, and this restricts rotation and makes the group planar.

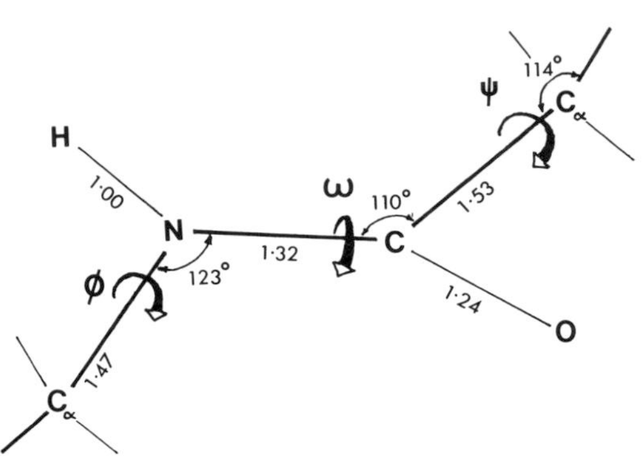

FIGURE 4.4. Diagram to show the main bond lengths and bond angles in the peptide repeating unit and the definitions of the angles ϕ, ω, and ψ. Note that $\phi = \omega = \psi = 180°$ is a fully extended chain and that looking down a bond, clockwise rotation of the far end is counted as positive.

Rotation angles about these two bonds are usually denoted by ψ and ϕ respectively (Fig. 4.4).

As suggested by Bragg *et al.* (1950), it was assumed that the conformations would be stabilized by hydrogen bonding between different amide groups and that in the most energetically favorable conformations such bonding would be linear ($\alpha < \pm\ 30°$ in Fig. 4.2), and as many bonds as possible would be made. By combining these special criteria with normal steric considerations (e.g., permissible Van der Waals contacts), two preferred types of structure were discovered: the α-helix and the β conformation.

4.2.2. The β Conformation

Figure 4.5 illustrates the β conformation in which the protein chains are essentially extended and linear and adjacent chains are linked through hydrogen bonds. In principle, two main types of β conformation are possible depending on whether the interacting extended chains are parallel or antiparallel (Fig. 4.5). In practice, it has been found that the antiparallel form is by far the more common and that the chains are not completely extended but are "pleated" to give the so-called antiparallel β-pleated sheet. The pleating (Fig. 4.5b) allows the *R* groups to project out from the surfaces of the hydrogen-bonded sheet and thus allows some of the more bulky side groups to be accommodated. A special type of antiparallel β conformation occurs when a single chain folds back on itself to give what is termed the cross-β structure.

β-pleated sheets have been found to occur as the principal conformation in a number of structures including certain silks (see Fraser and MacRae, 1973). But significant stretches of parallel and antiparallel

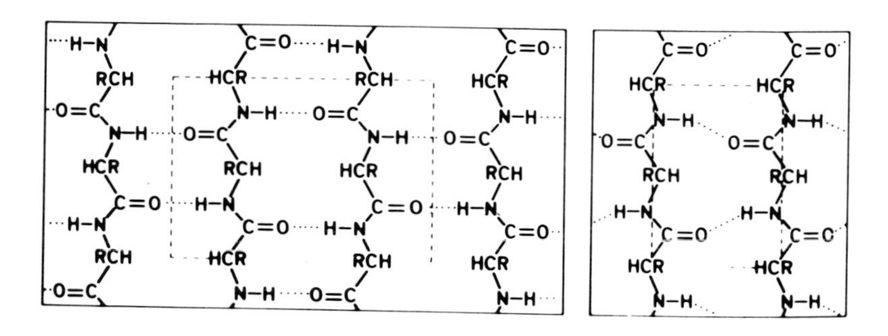

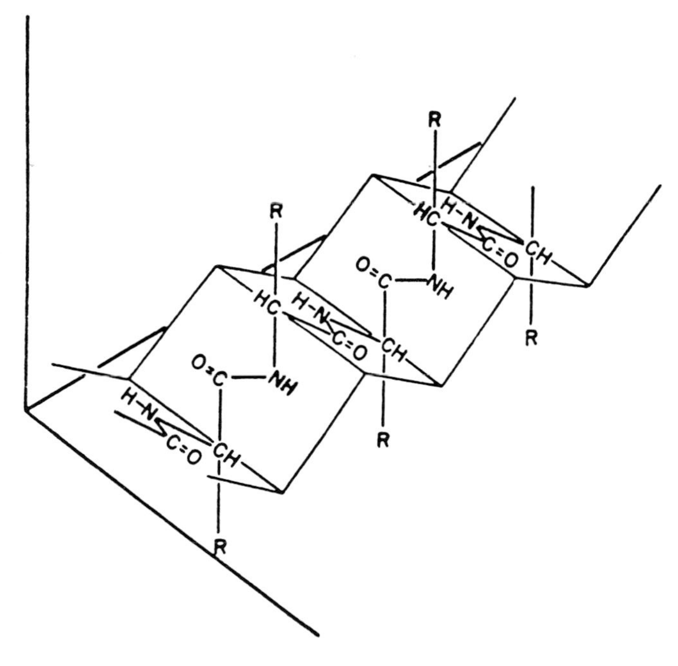

FIGURE 4.5. (a) Schematic diagrams showing the hydrogen-bonding schemes in an antiparallel β structure (left) and a parallel β structure (right). In the cross-β structure, the chains in the antiparallel arrangement are parts of a single molecule that folds back on itself. (From Fraser and MacRae, 1973.) (b) Illustration of the β-pleated-sheet structure in which the *R* groups point out of the plane of the sheet. (After Barrow, 1966.)

β structure have also been found in the majority of globular proteins so far studied (Section 4.4).

4.2.3. The α-Helix

Possible hydrogen-bonding schemes involving amide groups along a single polypeptide chain (intramolecular H-bonds) were discussed originally by Huggins (1943) and then more systematically by Bragg *et al.* (1950). Figure 4.2a illustrates schematically the many possible N–H to C = 0 interactions that could occur along such a chain (although some of these are not sterically feasible) and shows that these can be divided into two different types of interaction. Type A interactions lead to the general scheme for one hydrogen-bonded loop

$$\text{—C}\begin{smallmatrix}\nearrow \text{O}\text{-------------}\text{H}\searrow \\ \searrow (\text{NH–CH}R\text{–CO})n' \nearrow\end{smallmatrix}\text{N—}$$

in which there are $N_a = 3n' + 4$ main chain atoms in one loop. Type B interactions follow the general scheme

$$\text{—C}\begin{smallmatrix}\nearrow \text{O}\text{-------------}\text{H}\searrow \\ \searrow (\text{NH–CHR–CO})n'\text{–CH}R \nearrow\end{smallmatrix}\text{N—}$$

and this time there are $N_a = 3n' + 5$ main chain atoms in a hydrogen-bonded loop. Out of all of these many helical structures, it was found by Pauling *et al.* (1951) and Pauling and Corey (1953), on potential energy considerations that the Type A structure with $n' = 3$ seemed to be the most favorable. This helix, called the α-helix, has 3.6 residues per turn with $N = 13$ main chain atoms in the hydrogen-bonded ring. For this reason, it was originally termed a 3.6_{13} helix. But this nomenclature has largely been superseded by one with the general form u/t, where u is the number of residues in t turns of the helix (see Chapter 2, Section 2.3.1). Using this designation, the simplest α-helix is an 18/5 helix.

Although a number of other helical forms are sterically possible (notably the so-called ω-helix with $N_a = 13$, $u = 4$, $t = 1$ and the π-helix with $N_a = 10$, $u = 3$, $t = 1$), the α-helix is the one that proliferates in muscle proteins, and it alone will therefore be described in detail here. In theory, four forms of the α-helix are possible, since amino acid residues can be either right-handed (D) or left-handed (L) about the tetrahedral *alpha* carbon atom (C_α) (Fig. 4.1), and in each case, the helix that is formed can itself follow either a right-handed or a left-handed screw. Figure 4.6 indicates the forms of the right-handed and left-handed α-helices (respectively) of a polypeptide consisting of L amino acids. In fact, it has been found that only L amino acids occur in proteins, and Huggins (1952) showed on theoretical grounds that the right-handed α-helical arrangement of these L-amino acids is likely to be more stable than the left-handed form. In subsequent studies of α-helical structures,

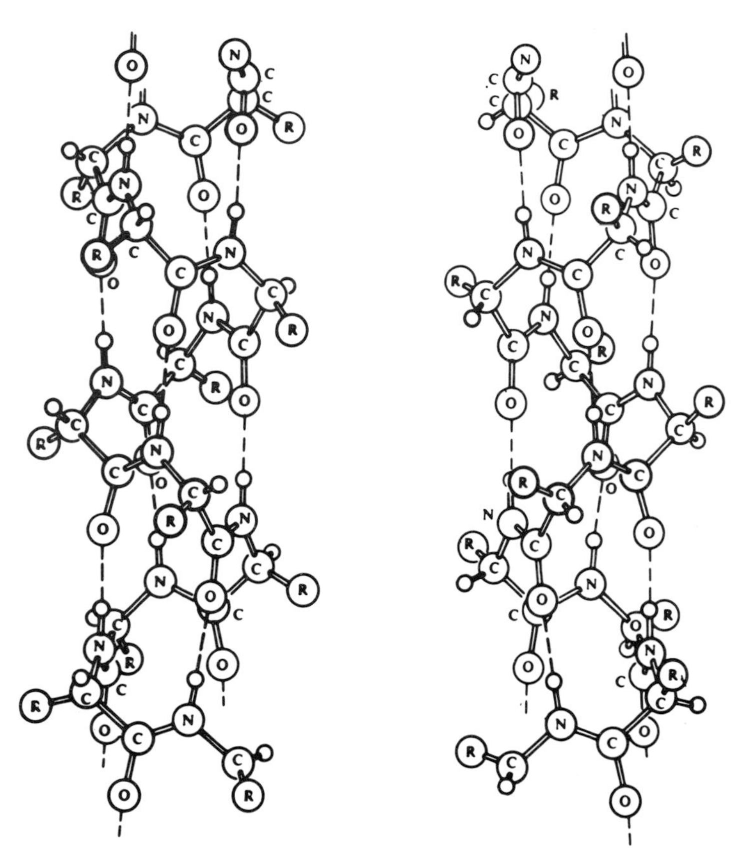

FIGURE 4.6. Drawings of the left-handed (left) and right-handed (right) α-helical forms of a polypeptide containing L-amino acids. (From Low and Edsall, 1968.)

this suggestion has been largely substantiated (A. Elliott and Malcolm, 1958).

The simple 18/5 α-helix has a pitch (P) of 5.4 Å, a subunit axial translation (p) of 1.5 Å, and a true repeat (R) equal to 27 Å. Table 4.2 lists the atomic coordinates of the main chain atoms of an L residue in a right-handed α-helix (Fig. 4.6) as determined by Parry and Suzuki (1969). Other helical parameters are $\phi = -51.45°$, $\psi = -52.74°$, and $\omega = 180°$ (i.e., the amide group is planar). Note here that ϕ, ψ, and ω are measured so that $\phi = \psi = \omega = 180°$ in a fully extended chain, and right-handed rotation is positive. This means that looking along the direction of a bond, the group at the far end rotates clockwise relative to that at the near end (Fig. 4.4).

TABLE 4.2. Atomic Coordinates of an L-Residue in a Right-Handed α-Helix[a]

Atom	x (Å)	y (Å)	z (Å)	r (Å)	ϕ (deg)
N	1.360	−0.744	−0.873	1.550	−28.69
H	1.485	−0.559	−1.847	1.586	−20.63
C^{α}	2.280	0	0	2.280	0
H^{α}	2.989	−0.699	0.467	3.069	−13.17
C^{β}	3.049	1.029	−0.830	3.218	18.65
C′	1.480	0.718	1.089	1.645	25.88
O	1.765	0.579	2.288	1.858	18.15

[a] 18/5 α-Helix, $P = 5.4$ Å, $\phi = -51.45°$, $\psi = 52.74$ Å, $\omega = 180°$; **z** is along helix axis. (From Parry and Suzuki, 1969.)

4.2.4. Diffraction from an α-Helix

From the helical diffraction theory outlined in Chapter 2 and knowledge of the helical parameters involved, it is a fairly easy matter to visualize the form of the diffraction pattern of an α-helix. The α-helix can first be thought of as a continuous helix (Fig. 4.7a) of pitch 5.4 Å and mean radius about 2 Å (Table 4.2). By analogy with Fig. 2.13, the diffraction pattern of such a continuous helix (Fig. 4.7d) will consist of a single helix cross centered on the origin of the diffraction pattern and in which the layers are $n/5.4$ Å^{-1} from the equator. The position of the peak along a particular layer is governed both by n and by the radius of the helix.

If the helix is now considered as a discontinuous helix of point scatterers with a subunit axial translation (p) equal to 1.5 Å (Fig. 4.7d), additional helix crosses will occur centered in the diffraction pattern at positions given by $m/1.5$ Å^{-1} (Fig. 4.7f). Since we are not dealing with an integral helix (i.e., $P/p \neq$ integer), the layers in the cross centered on, say, the first meridional reflection ($m = 1$) will not coincide with those in the cross centered on the origin ($m = 0$). In this case, the repeat of the helix is 27 Å (i.e., 18 × 1.5 Å). The pattern can therefore be thought of in terms of layer lines with spacings z along the meridian given by $Z = l/27$ Å^{-1} where l is the layer line number. On this basis, the layers at $n/5.4$ Å^{-1} in the helix cross centered on the origin become layer lines of number l equal to $5n$, and the meridional reflections are on layer lines with l equal to 18 m. In general, the layer line number of the nth layer centered on the mth meridional reflection is given by $l = 18m + 5n$ (where m and n can have positive, zero, or negative values).

The form of the diffraction pattern of an 18/5 helix of point atoms is shown in Fig. 4.7f. Two main features should be noted. These are that

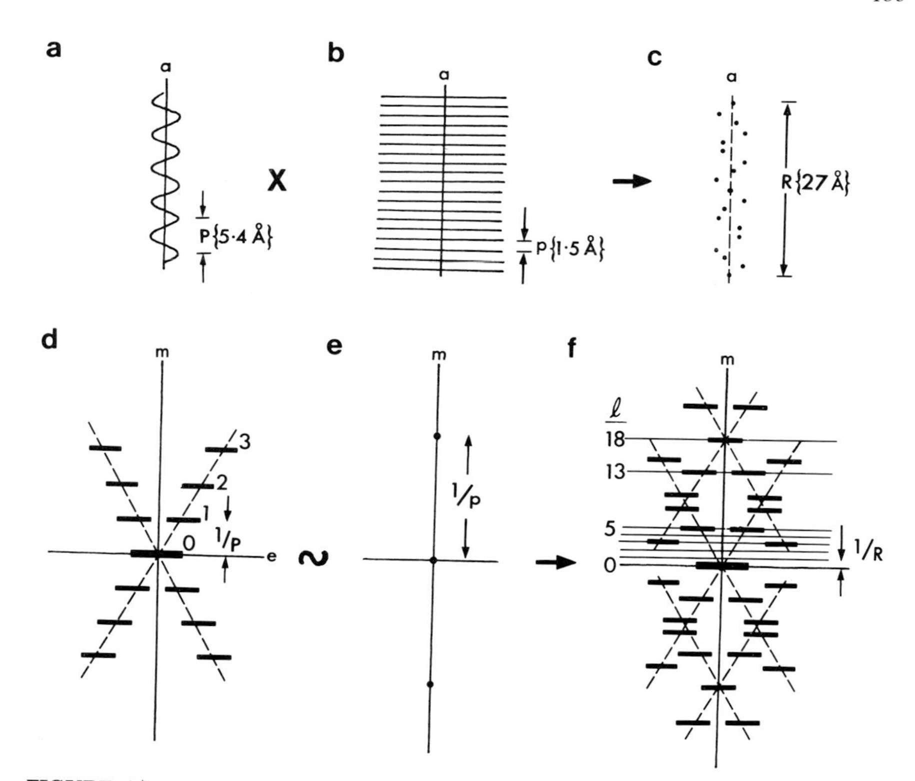

FIGURE 4.7. Generation of the form of the diffraction pattern from an 18/5 α-helix. Such a helix (c) can be considered as the product of a continuous helix of pitch 5.4 Å (a) and a set of density planes spaced 1.5 Å apart (b). The diffraction pattern (d; a helix cross) from (a) must therefore be convoluted with the diffraction pattern (e; a set of meridional points) from (b) to give the diffraction pattern (f) from the α-helix. Strong reflections occur on layer lines for which l is 5, 13, and 18.

strong intensity occurs on the fifth layer line at $Z = 1/5.4$ Å^{-1} ($m = 0$, $n = 1$), on the 13th layer line ($m = 1$, $n = 1$), and on the meridian (first meridional; 18th layer line) at a spacing $Z = 1/1.5$ Å^{-1}. Optical diffraction patterns (Fig. 4.8) confirm this conclusion. The diffraction pattern from a full α-helix is generated by considering equivalent atoms in successive amino acid residues to be on their own helix and to be giving their own diffraction pattern. Each helix will give a diffraction pattern similar to that in Fig. 4.8 but at a different cross angle. The contributions from all the helices of atoms must then be added vectorially to give the diffraction pattern of the complete structure. The final pattern is clearly more complex than in Fig. 4.8, but the strong intensity on layer lines 5, 13, and 18 remains. In fact, these reflections are a diagnostic characteristic of α-helical structures.

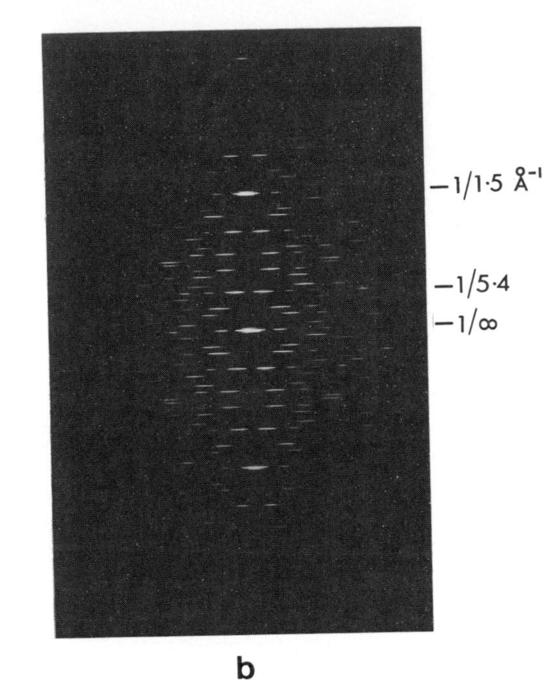

FIGURE 4.8. A model (a) and its diffraction pattern (b) of a 43/12 helix ($P/p = 3.58$). Although the helical parameters are different from those of the 18/5 α-helix in Fig. 4.7c, the intensity distribution in (b) is very similar to that in Fig. 4.7f and, in particular, the 5.4- and 1.5-Å layer lines are strong.

Up to this point, the discovery of the energetically favorable α-helix conformation and the derivation of the form of its diffraction pattern were both the result of theoretical analyses. It was Perutz (1951) who was the first to demonstrate experimentally that the diffraction pattern from a particular synthetic polypeptide was similar to that expected from an α-helical structure according to the theory of Cochran *et al.* (1952). Synthetic polypeptides are usually chemically synthesized polymers of identical amino acid monomers. In this case, the amino acid involved was called benzyl glutamate—a derivative of glutamic acid with the general formula

$$\begin{array}{l} \quad\;\; \mathrm{COOH} \\ \qquad\;\; | \\ \mathrm{H-C-CH_2-CH_2-CO-O-CH_2-C_6H_5} \\ \qquad\;\; | \\ \quad\;\;\; \mathrm{NH_2} \end{array}$$

The diffraction pattern of poly-γ-benzyl glutamate contained, in particular, a strong meridional reflection at a spacing of 1/1.5 $Å^{-1}$ and strong intensity on a layer line at 1/5.4 $Å^{-1}$ as expected.

Since that time, many other synthetic polypeptides have been found to occur in the α-helix configuration. But in some of these polypeptides it was found that although in each case the ratio P/p was very nearly equal to 3.6, there were not exactly 18 residues in 5 turns. Only slight twisting on an 18/5 α-helix will result in a helix with quite different helical parameters but very similar structure. For example, in poly-L-alanine (R group is $-CH_3$), a 47/13 α-helix was found ($P/p = 3.62$), and in poly-γ-methyl glutamate a 69/19 α-helix was found in some preparations and an 18/5 helix in others (see Fraser and MacRae, 1973). Despite these variations in the helical parameters which result in very slightly altered layer line spacings, the distribution of intensity in the respective diffraction patterns remained quite comparable. The conclusion were reached that α-helical structures are common in polypeptides, and the ratio P/p is always close to 3.6, but there is nothing special about the numbers $u = 18$ and $t = 5$. Note that the helix in Fig. 4.8 is actually a 43/12 α-helix but that layer lines at spacings corresponding to 5.4 and 1.5 Å are strong as in Fig. 4.7f.

4.2.5. Structures of Synthetic α-Helical Polypeptides

In addition to slight differences in the u and t values, several other significant features about real α-helical structures are worth noting. The first of these became apparent when an attempt was made to compute the intensity distribution in the diffraction pattern of α-helical poly-L-alanine (Brown and Trotter, 1956). It was found that serious differences existed between the computed and the observed diffraction patterns. Later it was found by A. Elliott and Malcolm (1958) and confirmed by Arnott and his co-workers (1967) that the discrepancies could be largely removed if it were recognized that an α-helical molecule has a sense or polarity and that there is no particular reason why in a synthetic polypeptide fiber any particular lattice point should be occupied by an "up" rather than a "down" molecule. Elliott and Malcolm used optical diffraction to evaluate the effect of putting "up" and "down" molecules randomly into the known lateral packing arrangement for poly-L-alanine so that on each lattice point there was a statistically averaged structure consisting of half an "up" molecule and half a "down" molecule. By this means, reasonably good agreement with the observed X-ray diffraction patterns of poly-L-alanine was obtained provided that the α-helices were right-handed. This statistical poly-L-alanine structure was later refined by Arnott and Wonacott (1966) who obtained very satisfactory agreement between observed and calculated intensity values.

The second feature about real α-helical structures was that meridional reflections were sometimes seen at positions where the Fourier transform of an α-helix should be zero. For example, in the observed diffraction patterns from poly-L-alanine, meridional reflections occurred at *d* spacings of 4.4 and 2.2 Å, whereas a simple α-helix would give no meridional reflections at *d* spacings larger than 1.5 Å. This was explained by the fact that the poly-L-alanine molecules were packed into a hexagonal unit cell. The steric interaction between the methyl (*R*) groups on the outsides of adjacent molecules would therefore cause some local distortion of the α-helical structure (which clearly does not possess sixfold symmetry), and this would occur periodically along its length. The "forbidden" meridional reflections were thought to be caused by the periodicity of this distortion.

The third feature of a number of synthetic polypeptides was that the *R* group side chains were sometimes sufficiently long and flexible that they could interact with each other to form regular side-chain structures that had a different symmetry and repeat from the α-helical backbone of the molecules. Since they were regular, these side-chain configurations contributed to the X-ray diffraction pattern to give diffraction features that were not directly related to a structure with α-helical symmetry. Indeed, in one case at least (poly-γ-benzyl aspartate: $R = -CH_2-CO-C_6H_5$), the side-chain interactions were so strong that they distorted the backbone structure from α-helical to ω-helical symmetry (i.e., from 3.6 to 4 residues per turn) (Bradbury *et al.* 1962).

For these (and other) reasons, studies of synthetic polypeptides have given an insight into the perturbations that can occur in α-helical molecules both from their environment and their side-chain characteristics. In the next section it will be shown that in biological material as well, α-helical molecules tend to be distorted in a characteristic way because of specific side-group interactions.

4.3. Structure of Fibrous α-Proteins

4.3.1. Introduction: The Coiled Coil

As early as 1947, Astbury had demonstrated marked similarities among the X-ray diffraction patterns of a variety of fibrous biological materials [i.e., keratin, myosin (in muscle), epidermin, and fibrinogen].

Later it was found that certain silks gave a characteristically different kind of pattern. These two classes of diffraction pattern were termed respectively α and β patterns. In some cases (e.g., keratin), it was then found that a structure giving the α pattern could be converted into a structure giving the β pattern simply by stretching it under suitable conditions.

After the α-helix and β conformations for polypeptides had been discovered by Pauling and Corey (1951), it was supposed that the α pattern (which included a meridional reflection at about $d = 1.5$ Å) was the result of structures in the α-helical configuration. Similarly, the β pattern was thought to be caused by molecules with a β conformation. On this basis, one would naturally expect one structure to convert to the other by stretching.

But there were certain features of the α pattern that were puzzling. One was that the α-helix reflection at 1/5.4 $Å^{-1}$, which should be near but not on the meridian, was absent from the α pattern. Instead, a meridional reflection was observed close by at a spacing of about 1/5.1 $Å^{-1}$. Another feature was that near to the equator, some unexpected but very strong layer lines were observed with spacings in the meridional direction somewhere between 1/50 $Å^{-1}$ and 1/100 $Å^{-1}$. These near-equatorial layer lines together with a nearby equatorial reflection gave rise to a characteristic horseshoe-shaped intensity distribution along each half of the equator (see Fig. 2.22b).

Crick (1952, 1953a,b) and Pauling and Corey (1953) both attempted to explain these features of the α-pattern in terms of some kind of periodic distortion (supercoiling) of an underlying α-helical structure. Pauling and Corey thought that they might be caused by periodic distortions of intramolecular origin resulting from a specific repeating amino acid sequence. But Crick had a different idea. He considered what kind of distortion might be produced when two or more α-helical molecules are packed together. To do this, he envisaged the molecular surface as consisting of projections or "knobs" formed by the amino acid side chains alternating with "holes" in the regions between side chains. On this basis he suggested that an energetically favorable conformation would probably be one in which some of the knobs on one molecule could fit regularly into some of the holes on the surface of a neighboring molecule, just as the teeth of two cog wheels might mesh together.

It is possible to investigate this kind of possibility by making use of what is called a radial projection of a helix. This can be thought of as being produced by wrapping a piece of paper into a cylinder around the surface of the helix in question. The positions of all of the monomers on the helix are then traced onto the paper, and the paper is opened out flat again. The result for an α-helix is shown in Fig. 4.9b. Here the extreme edges of the radial projection correspond to the same line along the surface of the helix. To see how two α-helices could pack together with knob-into-hole packing, one must imagine that one is on the inside (0 in Fig. 4.9a) of one molecule (B) and looking out towards the outside of an adjacent molecule (A) having the same screw sense. In this case, the two molecular surfaces in contact would be represented by radial projections that have opposite handedness (Fig. 4.9b). The radial projection of the

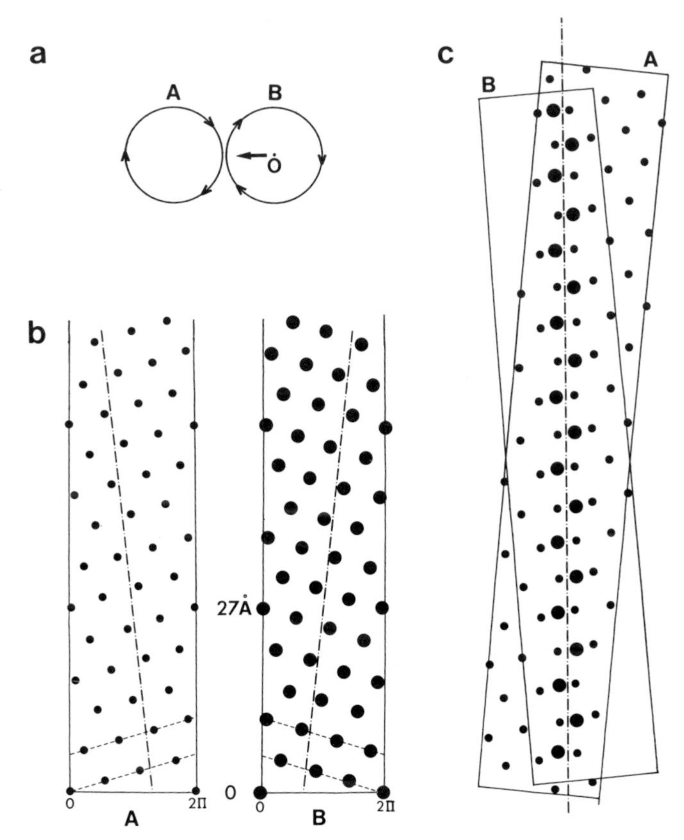

FIGURE 4.9. Demonstration of the knob-into-hole α-helix packing scheme suggested by Crick (1952). (a) Two adjacent identical α-helical molecules, A and B, in cross section. O is the position of an imaginary observer looking in the direction of the arrow towards the line of contact between the two molecules. (b) The radial projections (see text) of the surfaces of the molecules A and B in (a) according to what the observer in (a) can see. Although the two molecules have the same screw sense, the fact that the observer sees one helix from the inside and the other from the outside means that the radial projections have the opposite hand. Crick suggested that if the line of contact between the two molecules were along the dotted lines in (b), then good knob-into-hole packing of the amino acid side groups could be achieved. This can be seen in (c) where the two radial projections in (b) have been superimposed so that the two dotted lines coincide. Along this line in (c), the subunits in the two molecules intermesh in a very regular and systematic way. Note that for clarity only those residues in molecule B that lie adjacent to the line of contact have been included. Further along the chains from the region shown in (c), the two molecules would cease to overlap assuming that they are both straight. They would, however, continue to touch along their entire length if the two molecules could twist around each other (as in Fig. 4.10 or Fig. 4.12c) so that their axes followed helical paths.

molecule from which the observation is being made can then be moved around and tilted relative to that of the other molecule to see how the two surfaces will fit together.

Crick (1952, 1953a,b) found that, in the case of an α-helix with 3.6 residues per turn, knob-into-hole packing could be achieved if the axes of the two interacting helices were tilted to each other by about 20° in the manner shown in Fig. 4.9c. But this figure shows that such packing only occurs over a very small region of each molecule if the molecules themselves are straight. However, the same type of knob-into-hole packing could occur along the entire length of each molecule if the two molecules could twist around each other so that their axes followed helical paths of very long pitch (Fig. 4.10). The resulting structure was called a coiled coil, and since the arrangement demonstrated in Figure 4.10a contains two molecules, it was termed a two-stranded rope of coiled-coil molecules. Note also that in theory the two α-helical molecules could be parallel or antiparallel. The pitch of the coiled coil (the major helix) depends on the radius of the α-helical molecules (the α-helix is the minor helix) and the exact number of residues per turn in the undistorted helices. From known α-helix dimensions, the pitch of the two-stranded coiled coil was estimated to be between 140 and 200 Å. Note that such a coiled-coil structure does need to be stabilized by additional forces between the chains as will be described in Section 4.3.2.

An alternative kind of coiled coil, which was also considered by Crick, was one in which three α-helical molecules pack together with a knob-into-hole arrangement to produce a three-stranded rope of parallel coiled-coil molecules (Fig. 4.10b). The three chains would be related by a threefold rotation axis coincident with the axis of the coiled coil.

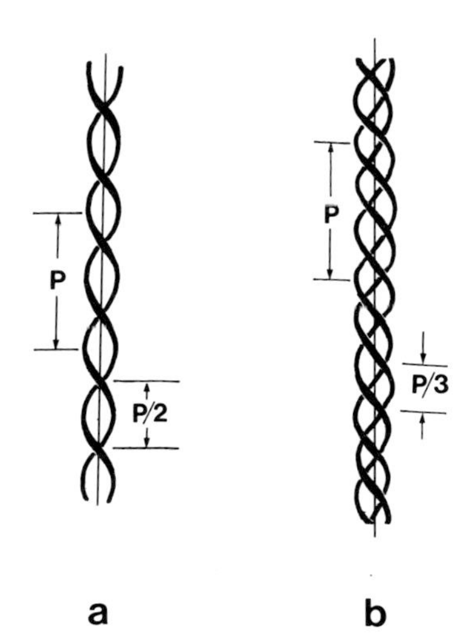

FIGURE 4.10. Sketches of the forms of two-chain and three-chain ropes of molecules as in α-helix coiled-coil molecules. The repeats are respectively ½ and ⅓ of the pitch *P* of a single strand.

4.3.2 Diffraction from a Coiled-Coil Structure

In order to test the existence of such coiled-coil structures in α-proteins, it was clearly necessary to deduce the form of their X-ray diffraction patterns. It was Crick himself (1952, 1953a,b) who originally suggested a way in which this could be done (although another, more direct, method has been used since then; see Cohen and Holmes, 1963). As with the presentation in Chapter 2, the mathematical formulations involved in these approaches will be avoided here. Instead, the form of the diffraction pattern of a coiled coil will be derived here in simple qualitative terms with the aid of optical diffraction patterns. First, if we forget for a moment that the two or three strands are actually α-helical molecules, we can find the form of the diffraction pattern of a two- or three-stranded arrangement of continuous helices. One strand would give rise to a simple helix cross (as in Fig. 2.14), but since the pitch is very long and the helix radius rather small, a very shallow helix cross will be produced (Fig. 4.11e). The effect of adding a second strand is to reduce the axial repeat to $P/2$, and so only layers with spacings (z) along the meridian equal to $0/P$, $2/P$, $4/P$, $6/P$, etc. (i.e., every second layer) will be observed. For a three-stranded rope, the rule would be that layers are seen when $z = 0/P$, $3/P$, $6/P$, etc. This shallow cross pattern could be the origin of the near-equatorial layer line in the α-pattern, and since this has a spacing (z) equal to 1/70 to 1/100 Å^{-1}, the pitch of a two-strand rope would need to be between 2×70 and 2×100 Å and that of a three-strand rope would need to be between 3×70 and 3×100 Å.

If it is now recognized that each strand is in fact an α-helix, then clearly the structure in each strand must be discontinuous. It will be seen in Fig. 4.9 that the line of contact between the two α-helical molecules in a two-strand coiled coil runs through every seventh residue on each of the α-helices. This clearly must mean that (to a first approximation) every seventh residue is in an equivalent environment and, therefore, that a repeating unit of seven amino acid residues occurs along each coiled-coil molecule. If the α-helices had been straight, then the axial translation between groups of seven residues would have been 7×1.5 Å $= 10.5$ Å. But since the helices are super coiled, this repeat distance reduces slightly to about 10.2 Å (depending on the radius and pitch of the supercoil).

The next step in the analysis is, therefore, to find the diffraction pattern of a two-stranded discontinuous helix of pitch P equal to between 140 and 200 Å and subunit axial translation (p) equal to 10.2 Å. Clearly, this must be a pattern in which a shallow helix cross (as Fig. 4.11a) is convoluted with a series of meridional reflections of spacing (z) equal to $m/10.2$ Å^{-1} (m is integral). The result is shown in Fig. 4.11h for a

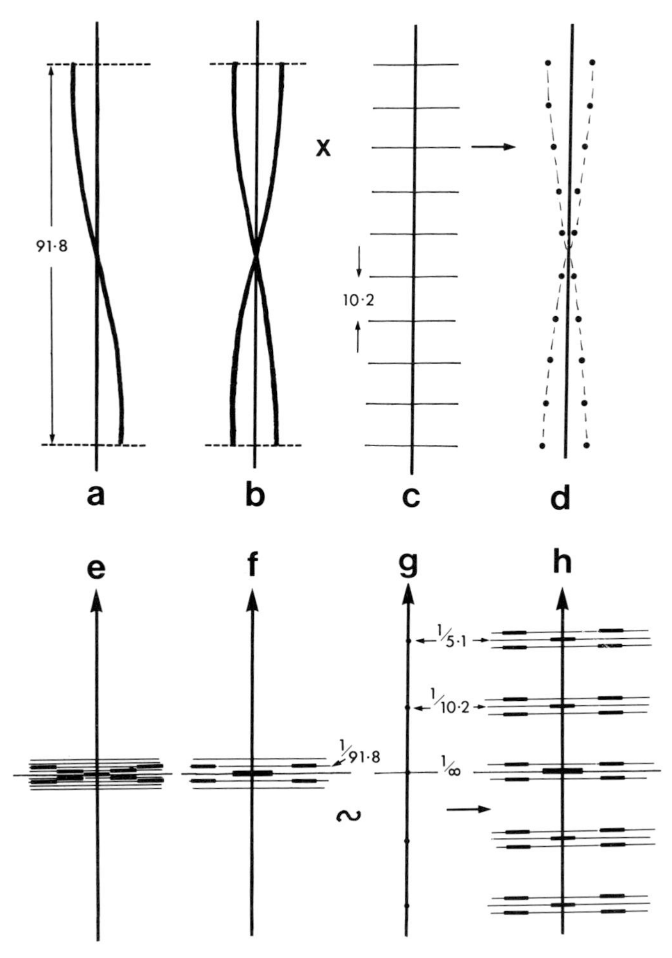

FIGURE 4.11. Generation of the form of the diffraction pattern from a two-chain coiled-coil α-helical molecule. The pitch of the coiled coil is here taken to be 183.6 Å, and the subunit repeat along each strand is 183.6/18 = 10.2 Å. The diffraction pattern (e) of a single continuous helical strand of pitch 183.6 Å and relatively small radius (about 5 Å) is a very shallow helix cross with layers at positions n/183.6 Å from the equator. The introduction of a second strand coaxial with the first (b) halves the repeat to 91.8 Å (as in Fig. 4.10a), and only layers for which n is even in (e) will be seen in the diffraction pattern (f). A discontinuous helix (d) is obtained by multiplying (b) by a set of planes of density 10.2 Å apart (c). The diffraction pattern (h) of (d) is therefore the convolution of (f) with (g) which is the diffraction pattern of the set of planes (c). In a coiled-coil α-helix, the 10.2-Å repeats contain substructure (the α-helical chain) with dominant 5.4- and 1.5-Å periodicities, and only these regions of the diffraction pattern (h) are therefore strong in the final pattern (Fig. 4.12).

two-start helix of pitch about 183 Å (18 × 10.2 Å). This pattern clearly includes a meridional reflection at a d spacing of 5.1 Å as observed in the α-pattern and another reflection (not shown) at a d spacing close to 1.5 Å ($m = 7$). But, in addition, several other meridionals are apparent (especially one at a d spacing of 10.2 Å) that are not present in the α patterns of proteins. The reason for this is simply that each repeating unit along one of the coiled-coil strands is not just a point scatterer as we have assumed but constitutes two turns (slightly distorted) of an α-helical molecule (Fig. 4.12). The two dominant repeats in the α-helix are, of course, the axial translation of 1.5 Å between amino acid monomers and the 5.4-Å pitch of the helix.

In one form of the rigorous derivation of the Fourier transform of a coiled coil, each atom on each of the seven residues on one repeating unit is imagined to lie on its own helix, and the Fourier transforms of all these helices of atoms are added vectorially to give the total diffraction pattern. When this is done, it is found that, for obvious reasons, the overall intensity distribution in the pattern is strongest on the regions

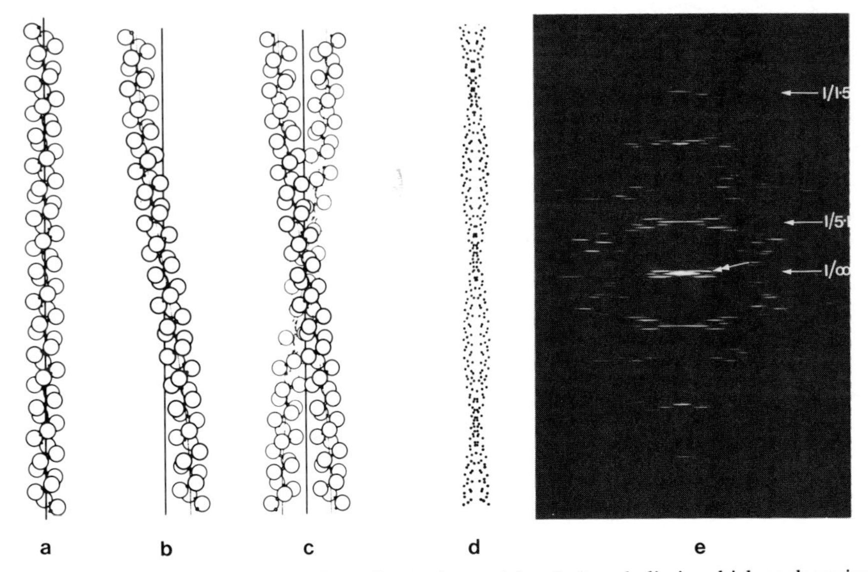

FIGURE 4.12. (a) Representation of a single straight-chain α-helix in which each amino acid residue is represented by a single circle. (b) The same structure as in (a) except that the axis of the helix itself follows a helical path as in a coiled coil. (c) Two coiled coils as in (b) put together to form a two-chain rope. (d) Diffraction mask derived from several repeats of the structure shown in (c); (e) its optical diffraction pattern. This has the features typical of α-patterns from fibrous proteins (see Figs. 2.22b and 4.13a) and shows the 1.5-Å and the 5.1-Å meridional reflections (arrow) and the strong near-equatorial layer lines (double-headed arrows). [Figures (a),(b), and (c) are from Fraser and MacRae, 1973.]

that correspond to the repeats of 5.4 Å and 1.5 Å in a single α-helical chain. The result is that the 10.2- Å meridional reflection is weak, but the 5.1-Å reflection is strong. This effect is synthetically simulated in Fig. 4.12 where the optical diffraction pattern is shown of a projection of a coiled-coil structure in which each repeat consists of seven point scatterers in two turns of the minor helix. In the case of the three-stranded coiled coil, very similar arguments apply, and the type of pattern produced is similar but not identical to the pattern from a two-stranded structure.

4.3.3. Evaluation of the Coiled-Coil Model

From the presentation given above, the impression may have been gained that the coiled-coil helical model has explained the α pattern from fibrous proteins. In fact, the situation is by no means as simple as that.

Cohen and Holmes (1963) and Fraser *et al.* (1965) attempted to test the various coiled-coil models by computing the layer line intensities in the diffraction patterns of each and comparing the results with the observed α patterns from a number of muscles and other tissues. The general conclusion was that a coiled-coil structure could explain reasonably well the overall form of the observed diffraction patterns. It was also found that the agreement between observed and calculated layer line intensities was better for a two-stranded coiled coil than it was for a three-stranded arrangement. On the other hand, the agreement in the region of the 5.1 Å meridional reflection was in no case particularly good but was different for a two-stranded rope depending on whether the two strands were parallel or antiparallel.

In addition, it would be expected that the spacings of the meridional reflections near to 5.1 Å and 1.5 Å should be exactly in the ratio of 7 : 2. But A. Elliott *et al.* (1968a) using molluscan muscle, obtained measurements in the range 5.084 to 5.157 Å and 1.485 (±0.005) Å that were clearly not in the expected ratio. Subsequently Parry and Elliott (1967), Squire (1969), and Squire and Elliott (1972) obtained diffraction patterns from a particular form of the synthetic polypeptide poly-γ-benzyl-L-glutamate ($R = -CH_2CH_2COOCH_2C_6H_5$) that indicated that the structure involved comprised straight-chain α-helices. But, in addition, the diffraction pattern contained near-equatorial layer lines and a meridional reflection at a spacing of about 5.1 Å (Fig. 4.13b). These unusual features were explained in terms of a regular arrangement of the long side chains in this particular polypeptide. Since the computation of the Fourier transform of a coiled coil carried out by Cohen and Holmes (1963) did not include any contribution from the side chains, it

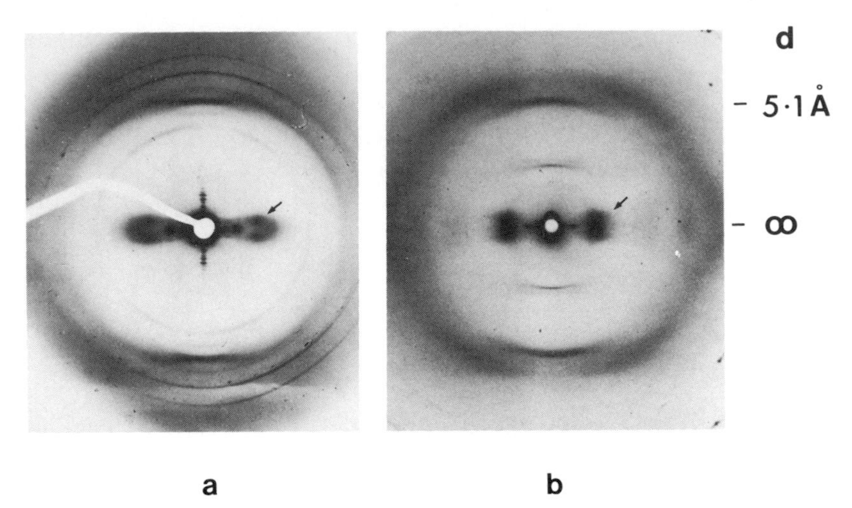

FIGURE 4.13. (a) The high-angle diffraction pattern from a molluscan muscle (*Mytilus edulis* adductor muscle) showing the meridional 5.1-Å reflection and the near-equatorial reflections (arrows) at a spacing from the equator of about 1/70 $Å^{-1}$. (Courtesy Dr. A. Elliott.) (b) A comparable high-angle diffraction pattern from an oriented fiber of the synthetic polypeptide poly-γ-benzyl glutamate swollen with benzene. Like (a), this also shows a meridional 5.1-Å reflection and a near-equatorial reflection. But the structure here contains straight α-helical molecules (Squire and Elliott, 1969; Squire, 1969), and the near-equatorial reflections are thought to be caused by a regular arrangement of the benzyl glutamate side chains.

seemed to A. Elliott *et al.* (1968a) to be quite possible that regular side-chain arrangements could make a significant contribution to the observed α patterns. This idea is supported by the fact that structures such as side chains at a relatively large radius will affect the intensity in the 5.1-Å region markedly more than that of the 1.5-Å region. In this way, both the observed intensity in the 5.1-Å region and the relative spacings of the 5.1-Å and 1.5-Å reflections might be explained. Another possibility is that the two chains might not be symmetrically related (Crick, 1953a).

Fraser *et al.* (1965) carried out Fourier transform computations of coiled-coil models in which some account was taken of the presence (but not the structure) of side chains and of the water environment of the molecules. They found that the inclusion of these features, combined with a slight change in the radius of the coiled coil, could give a marginally better fit than that obtained by Cohen and Holmes (1963).

Finally, it should be noted that Parry (1969, 1970) has shown that an intensity distribution similar to that from a coiled coil could also be produced in the diffraction pattern of a segmented rope of straight α-helices provided that these contain a repeating sequence of seven residues (e.g., Lys–X–Glu–X–Asp–X–X)$_n$.

To summarize, the coiled-coil or segmented-rope α-helix structures will both account reasonably well for the overall appearance of the α-pattern. But clearly the amino acid side chains must contribute significantly to the diffraction pattern. The interpretation of the α pattern is therefore, as yet by no means complete. Fortunately, in the last year or two, evidence has been obtained that provides considerable support for Crick's basic ideas about knob-into-hole packing and the coiled-coil structure. One of his original ideas was that the interaction between the two intertwining molecular chains might well be stabilized by hydrophobic forces so that the residues along the line of contact between the two molecules would very likely be hydrophobic residues. Otherwise coiled coils might be expected to occur in homopolypeptides such as poly-L-alanine.

Recently, Smillie and his co-workers (Hodges *et al.,* 1972; Sodek *et al.,* 1972; Stone *et al.,* 1974) have determined the entire amino acid sequence of the muscle α-protein tropomyosin (see Chapter 5). They found that, in every group of seven amino acids, hydrophobic residues always occurred in the same two positions at intervals of three and then four amino acid residues. This sequence is just what would be expected to give a coiled-coil structure according to Crick's ideas, and it strongly supports his suggestions. Tropomyosin structure will be discussed fully in Chapter 5. Finally, it should be noted that biochemical results have confirmed that the α proteins in muscle contain only two chains, and these results, together with electron micrographs of molecular aggregates which sometimes have a polar structure, have indicated that the two chains are parallel.

4.3.4. Three-Dimensional Packing of Coiled-Coil Molecules

One of the features of the α pattern from muscle proteins that has not yet been discussed is the presence of interference maxima (i.e., Bragg reflections) along the equator of the diffraction pattern. These maxima occur at *d* spacings of about 20 Å and 10 Å, and it is thought that they are caused by the lateral packing of coiled-coil molecules that would be expected to have an outer diameter of about 20 Å. Further details of coiled-coil packing were provided by A. Elliott *et al.* (1968a) who obtained improved α patterns from molluscan muscles by soaking the muscles in various acetone/water mixtures. The kind of diffraction pattern they obtained is shown in Fig. 4.14a. The patterns showed very clear equatorial reflections at spacings of about 17 to 20 Å and 9 to 10 Å,

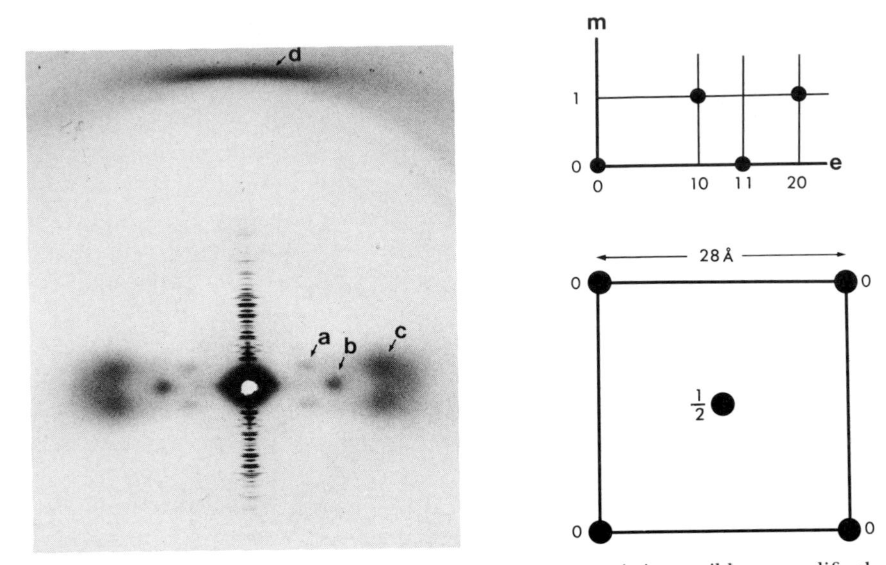

FIGURE 4.14. By soaking a molluscan muscle in acetone, it is possible to modify the structure of the myosin-containing paramyosin filaments in the muscle so that the equatorial and near-equatorial regions of the X-ray diffraction pattern (left) are clearly sampled by an interference function. Reflection (d) is the 5.1-Å meridional reflection, (b) an equatorial reflection at a spacing from the meridian of about 1/20 $Å^{-1}$, and (a) and (c) sampling peaks (cf. Fig. 2.17) on the 70-Å near-equatorial layer line. They have spacings from the meridian of about 1/28 and 1/14 $Å^{-1}$. Reflections (a),(b), and (c) can be indexed respectively as the 110, 101, and 201 reflections (top right) from a tetragonal unit cell of base 28 Å square (lower right) containing two molecules. Since the sampling on layer line 1 (top right) is different from that on the equator and follows the selection rule $h + k + l = 2N$ (where N is an integer), it can be concluded that the molecule in the center of the unit cell (lower right) must be shifted axially by one-half of the c-axis repeat (70 Å) in the unit cell relative to the molecules at the corners. This is consistent with the kind of molecular packing shown in Fig. 4.15. The intermolecular distance is about 20 Å. (After Elliott *et al.*, 1968a; diffraction pattern courtesy of Dr. A. Elliott.)

depending on the solutions used. But in addition, interference maxima also occurred on the near-equatorial layer line. It was found that the row lines on which these equatorial and near-equatorial maxima occurred had spacings of about 28, 20, 14, and 10 Å, which could be indexed as the 10, 11, 20, and 21 reflections of a tetragonal unit cell of side 28 Å. This is also shown in Fig. 4.14. The fact that the 28-Å and 14-Å reflections did not appear on the equator suggested that the unit cell contained two two-stranded ropes of molecules positioned at $x = 0, y = 0$ and at $x = ½, y = ½$, and that the z coordinates of the two molecules were different ($z = 0$ and ½ respectively). Here x, y, and z are fractional unit cell coordinates.

In trying to explain why adjacent coiled-coil molecules should be axially shifted in this way, it was suggested by Rudall (1956) and Caspar *et al.* (1969) that the important feature would be the closeness of packing of adjacent ropes. Since two-chain helical ropes have profiles that are periodically undulating (Fig. 4.15), it is clear that close packing can be achieved if the adjacent ropes are staggered axially by $P/4$ where P is the pitch of the two component chains in each rope. The reason for this is that the true repeat along a two-chain rope of continuous helices is $P/2$, and in a view from the side, the wide parts and narrow parts of a single rope are separated axially by $P/4$. If the wide part of one rope is to fit into the narrow part of its neighbor, then they must clearly be shifted by $P/4$. Such an arrangement can be followed systematically in three dimensions through a tetragonal array of molecules as in Fig. 4.14, but, as discussed by Longley (1975), this does not give the closest possible molecular packing (see Chapter 9, Section 9.2).

4.4. Globular Proteins

4.4.1. General Description

Although very little is known about the detailed structure of the globular proteins that are intimately involved in the contractile mechanism in muscle, one or two general features of such molecules are worth noting. It has already been mentioned in Chapter 2 (Section 2.2.4) that if a globular protein can be crystallized, then it is reasonably certain that its structure can be determined by the methods of X-ray crystallog-

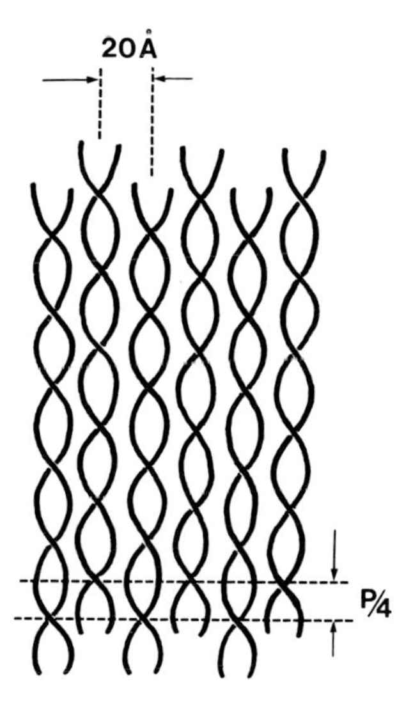

FIGURE 4.15. Sketch to show that two-chain coiled-coil molecules will tend to close pack with an axial stagger of $P/4$ between adjacent molecules, where P is the pitch of a single helical chain. The intermolecular packing distance is shown to be 20 Å as in Fig. 4.14.

raphy. The structures of a rapidly growing number of globular proteins have been determined in this way, and it is now possible to list some of their common features.

Clearly the fundamental feature of globular proteins is that they fold up in a somewhat irregular fashion rather than forming extended regular arrays of α-helix or β structure. The specific way in which the molecules fold is determined by the amino acid sequence along the chain and, as mentioned earlier, the middle of the globular protein tends to be populated by hydrophobic (apolar) residues, whereas the polar residues tend to be near the molecular surface. The whole three-dimensional molecular arrangement is stabilized by hydrogen bonding, ionic bonding, and sometimes also by disulfide bridges. Although in globular proteins the arrangement of the amino acid residues does not follow a regular configuration, globular proteins do possess significant stretches of α-helix and of both parallel and antiparallel β conformations. For example, myoglobin and lysozyme are 70% and 25% α-helical, respectively. Similarly, lysozyme contains about 14 residues in an antiparallel β conformation, and carboxypeptidase A contains a large sheet of mixed parallel and antiparallel β structure involving eight extended chains and 55 residues (see review by Blundell and Johnson, 1976).

4.4.2. Levels of Structure

Some globular muscle proteins, such as actin, form aggregates (in this case actin filaments) that have a structural as well as an enzymatic function. In protein aggregates of this kind, four levels of structure are recognized. The *primary structure* of a protein is the sequence of amino acid residues along the protein chain. The *secondary structure* of the molecule (or region of a molecule) defines the local folding of the protein chain: whether it forms as an α-helix, a β structure, or another irregularly folded arrangement. The *tertiary structure* of a molecule describes how the α-helix, β-sheet or irregular parts of the chain are folded together on a larger scale to give the specific three-dimensional structure of the molecule. Finally, the *quaternary structure* describes how the protein molecules aggregate to form larger structures, whether dimers, trimers, or whole filaments comprising many molecules.

4.4.3. Structural Influence of Specific Amino Acids

One of the present ambitions in molecular biology is to be able to predict the complete three-dimensional structure of a globular protein simply from a knowledge of its primary structure. As yet, this is still a dream, although a certain amount of progress has been made.

A useful initial step is to assess different amino acid residues, on the basis of known protein and polypeptide structures, in terms of their ability or lack of it to form, for example, an α-helix. Table 4.3 summarizes the assessment of amino acids by Chou and Fasman (1974) in terms of factors related to the likelihood that a particular residue will form an α-helix, β-sheet, or random chain. It can be seen here that residues such as Ala, Glu, Leu, and His are significantly above average in their likelihood of forming an α-helix, whereas Arg, Asn, Gly, Pro, Ser, and Tyr are very likely to form an irregular structure. It is not surprising that the former tend to be abundant in the α proteins, whereas the latter are generally much less so. To date the prediction of α-helix formation in globular proteins of known structure has achieved an average success rate of about 70 to 80%. Although in itself this is quite an achievement, it is clear that if protein structures are ever to be satisfactorily predicted (and it is by no means certain that this is going to be possible), then much better evaluation of the many factors involved will need to be carried out.

TABLE 4.3. Conformational Parameters for Amino Acid Residues[a]

	α-Factor	β-Factor	Coil factor
Ala	1·45	0·97	0·66
Arg	0·79	0·90	1·20
Asn	0·73	0·65	1·33
Asp	0·98	0·80	1·09
Cys	0·77	1·30	1·07
Gln	1·17	1·23	0·79
Glu	1·53	0·26	0·87
Gly	0·53	0·81	1·42
His	1·24	0·71	0·92
Ile	1·00	1·60	0·78
Leu	1·34	1·22	0·66
Lys	1·07	0·74	1·05
Met	1·20	1·67	0·61
Phe	1·12	1·28	0·81
Pro	0·59	0·62	1·45
Ser	0·79	0·72	1·27
Thr	0·82	1·20	1·05
Trp	1·14	1·19	0·82
Tyr	0·61	1·29	1·19
Val	1·14	1·65	0·66

[a] The factors give the likelihood of a particular residue forming an alpha helix, beta sheet, or random coil. A high factor means that that particular structure is very likely to occur. (Values from Chou and Fasman, 1974.)

5

Thin Filament Structure and Regulation

5.1. Introduction

Actin must rank as one of the most ubiquitous proteins in nature. Thin filaments consisting largely of actin molecules have been found to occur in all of the different muscle types that have so far been studied. In addition, actin molecules with virtually identical properties to those in muscle have been found in a variety of nonmuscle cells, ranging from slime molds and amoeba to the vertebrate brain, blood platelets, and spermatozoa.

Actin was first discovered in 1942 by Straub, and in the same year it was shown that synthetic fibers containing both actin and myosin (actomyosin) would contract when immersed in a suitable salt solution containing ATP (Szent-Gyorgyi, 1942). As described in Chapter 1, Section 1.3.3, actin was shown to be present in the I-band of the vertebrate sarcomere in a series of extraction experiments carried out in the 1950s (Hanson and Huxley, 1953; Hasselbach, 1953; Szent-Gyorgyi *et al.*, 1955; Perry and Corsi, 1958). These results were confirmed by comparing negatively stained thin filaments mechanically separated from relaxed muscle with synthetic filaments reconstituted from preparations of purified actin molecules (H. E. Huxley, 1961, 1963; Hanson and Lowy, 1963). Structurally similar filaments were obtained by both methods.

Extraction studies of myofibrils showed that the I-band contained, in addition to actin, a significant amount of the protein tropomyosin. The proportion of actin to tropomyosin in different extracts was found to be more or less constant, suggesting that thin filaments might be composed of fixed proportions of the two molecules.

The first significant advances in the understanding of thin filament structure were reported by Bear (1945), Cannan (1950), and Selby and Bear (1956). These authors carried out a detailed analysis of the excellent X-ray diffraction patterns from dried molluscan muscle which they had obtained using the then newly developed low-angle X-ray diffraction cameras (Cannan, 1950; H. E. Huxley, 1953a). Together, the results of Bear (1945), Astbury and Spark (1947), and Astbury (1947, 1949) had shown that part of the characteristic diffraction pattern from molluscan muscle was similar to that from synthetic fibers of actin. The pattern consisted of a series of meridionally oriented X-ray reflections which, at low resolution, appeared to be successive orders of an axial repeat of about 55 Å. The high-angle pattern was characteristically different from the α-pattern described by Astbury (1947) and now known to correlate with α-helical structures. In fact. the observed X-ray diffraction patterns showed that the axial repeat of the actin filaments in molluscan muscle was in the range 350–410 Å with particularly strong orders close to 55 Å, 27 Å, etc. (Bear, 1945).

These results were extended by Selby and Bear (1956) who interpreted their observations in terms of a number of alternative arrangements of diffracting centers within the thin filaments. Table 5.1 lists the axial spacings of the actin reflections observed by these authors together with their tentative indexing on repeats of either about 410 Å or about 355 Å. Also shown are the spacings of reflections in the X-ray diffraction patterns from actin films and fibers. The alternative structural models suggested by Selby and Bear are illustrated in Fig. 5.1. They are either helical nets of scattering centers with helical repeats of 410 or 355 Å or the related planar nets of similar repeats. Figure 5.1 also shows the form of the diffraction patterns of these alternative structures. It will be seen that they are, in fact, so similar that the inability of Selby and Bear to interpret their data unambiguously is not surprising. One thing they were able to show, however, was that the helical structures of which Figs. 5.1a and 5.1c would be the radial projections would have their scattering centers at a helix radius of about 25 Å, giving a maximum outer thin filament diameter of about 80–100 Å. This was in reasonable agreement with electron microscope observations of thin filaments in sections of muscle (H. E. Huxley, 1953a).

A dramatic visual demonstration of the basic structure of the thin filament was provided by Hanson and Lowy (1962, 1963). They used the then relatively new method of negative staining to study the structure of isolated actin filaments in the electron microscope. Figure 5.2 illustrates the kind of results they obtained. In their micrographs, actin filaments

TABLE 5.1. X-Ray Reflections from F-Actin Preparations and Dried Muscle

Alternative indexing[a]		Tinted *Venus* adductor muscle[b] (dry)		Actin film[c]		Dry actin fiber[d]
(l_a	l_b)	*d* (Å)	I[e]	*d* (Å)	I[e]	*d* (Å)
1		400	s			
2		195	m			
7	6	59	vs	54	vs	53
8	7	51	s			
14	12	28.2	w			
15	13	26.9	w	26.8	m	26.7
21	18	19.0	w			
22	19	18.8	m	18.1	m	18.3
30	26	13.6	m	13.5	s	13.4
31	27	12.9	w			
37	32	10.9	m	10.9	s	10.9
45	39	9.0	m	9.1	m	9.0
52	45	7.9	w	7.9	s	7.7
60	52	6.8	m	6.8	vw	6.8

[a] Indexing is on a repeat of about 406 Å (l_a) or 351 Å (l_b).
[b] From Selby and Bear (1956).
[c] From Astbury (1949).
[d] From Cohen and Hanson (1956).
[e] Abbreviations: s, strong; m, medium; w, weak; v, very.

appeared as two intertwining helical strands of globular repeating units. Crossover points between the two strands were observed to be at a measured separation of about 350 Å, and between crossover points there were very nearly 13 globular units with an axial separation of about 55 Å. These observations clearly suggested that the Selby and Bear model illustrated in Fig. 5.1a is correct and that the actin filament is helical with about 13 residues in six turns of the genetic helix (the genetic helix is represented by the lines in Fig 5.1a). A model of this structure is shown at the side of Fig. 5.2 and can be compared with Hanson and Lowy's micrograph.

Following this work, attempts were made to determine precisely the axial repeat in the thin filament using improved X-ray diffraction and electron microscope methods. Although Worthington (1959) had claimed that the axial repeat of actin filaments is close to 410 Å and contains 15 repeating units, more recent work by Hanson (1967), H. E. Huxley and Brown (1967), Lowy and Vibert (1967), Miller and Tregear (1972), and Haselgrove (1975a) has tended to confirm that, although slightly variable, the repeat of the crossover of the two actin strands (Fig. 5.2) is generally in the range 360 to 385 Å. At the same time it was

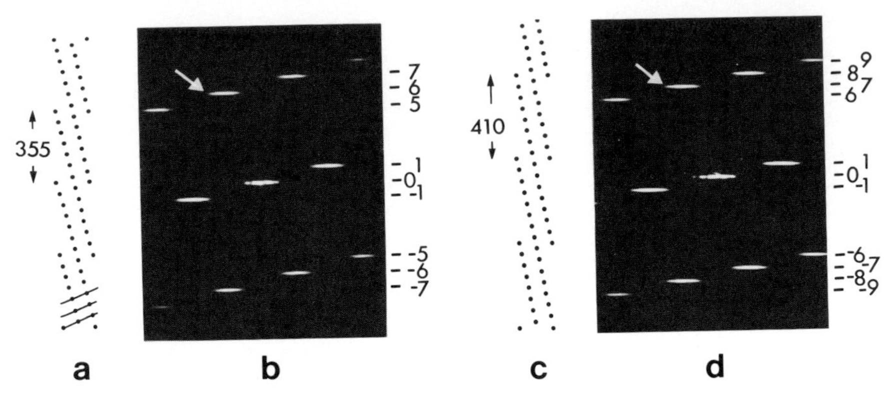

FIGURE 5.1. The two alternative actin nets suggested by Selby and Bear (1956): (a) and (c) can be taken as true planar nets or as the radial projections of the related helical structures. The net in (a) has a subunit axial repeat of 355/13 or 27.3 Å and repeats after 13 subunit translations (or six turns of the helix); that in (c) has a subunit axial repeat of 410/15 or 27.33 Å and repeats after 15 subunit translations (or seven turns of the helix). The optical diffraction patterns of these two structures are shown in (b) and (d), respectively, and these are clearly very similar. In particular, a strong reflection (arrow) occurs at a spacing from the equator (0) of about 1/59 $Å^{-1}$ in both patterns. The diffraction patterns from the helices related to (a) and (c) would have reflections at the same axial positions as in (b) and (d), but the pattern would be symmetrical about the meridian (vertical). The bold lines indicate one of the two genetic helices that can be drawn through the structure when (a) is considered as a radial projection of a helix. This genetic helix has a pitch of 59 Å. A second genetic helix of opposite hand and pitch 51 Å can also be drawn.

realized that in the thin filament there need not necessarily be an integral number of actin subunits in the crossover repeat. In fact, it is now generally agreed that there are probably just over 13 subunits in six helical turns in living muscle, although the repeat probably varies slightly in different muscle types, in different synthetic actin preparations, and possibly also in different muscle states.

Earlier it was mentioned that in addition to actin the I-band also contains the protein tropomyosin. Much of the initial work on tropomyosin, including its isolation, was carried out by Bailey (1946, 1948), Bailey *et al.* (1948), and Tsao *et al.* (1951). Since that time, it has been shown to be a rodlike, two-chain α-helical coiled-coil molecule of length about 400 Å and diameter about 20 Å. When Hanson and Lowy had demonstrated the arrangement of actin molecules in the thin filament (Fig. 5.2), it seemed possible to them that the long thin tropomyosin molecules might well lie end to end along each of the two long pitched "grooves" between the two helical strands of actin molecules. Since the tropomyosin molecules are about 400 Å long, it was thought that this arrangement might also help to account for the periodicity of about 400 Å that had often been seen along the I-bands in

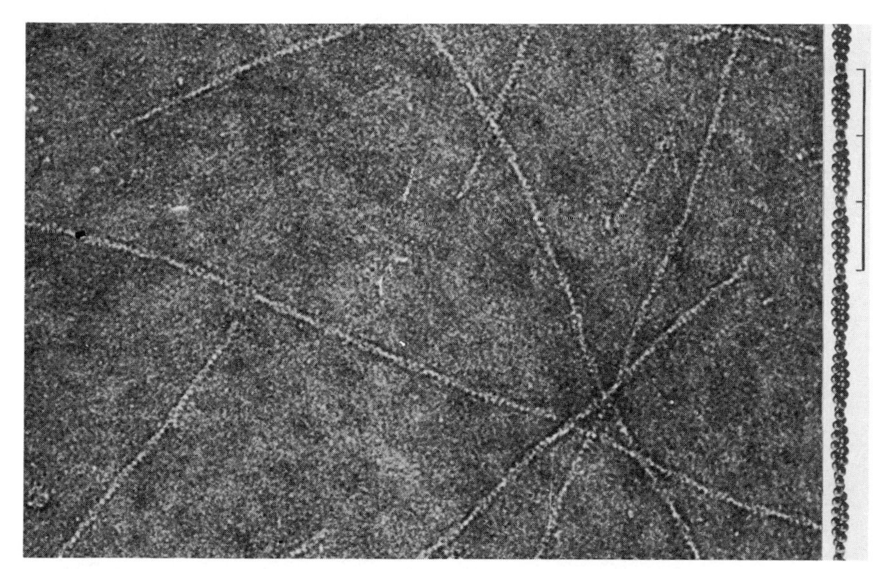

FIGURE 5.2. Electron micrograph of negatively stained F-actin filaments showing the characteristic appearance of two twisting strands of globular units as indicated by the model structure on the right. The measured distance between crossover points of the two strands is about 350 to 380 Å. (From Hanson and Lowy, 1963, and Hanson, 1973.)

sections of striated muscles (Fig. 5.3; Page and Huxley, 1963; Draper and Hodge, 1949). But thin filaments containing actin and tropomyosin alone might not be expected to give such a marked 400-Å periodicity. Hanson and Lowy (1964) therefore suggested that the actin–tropomyosin assembly could be responsible for the generation of the 400-Å repeat, but that some form of additional material might be labeling this repeat to give rise to the very obvious stripes of density.

In fact, this suggestion has now been substantiated, and the "additional material" is a protein known as troponin. But troponin itself was discovered by quite a different approach. One of the characteristics of muscle action is clearly that, as well as having the ability to contract, it must also be able to relax again. A muscle that is unable to relax would be of little use. However, in the early work of Szent-Gyorgyi (1942), synthetic preparations of actin and myosin (actomyosin) were found to superprecipitate (analogous to contraction) by the addition of ATP; but subsequent removal of ATP did not cause the actomyosin to swell again (i.e., no relaxation occurred). Later it was found in the case of myofibrils that the ATPase activity depended on the presence of a very small amount of Ca^{2+} ion (Weber, 1959) and that relaxation could be achieved by the removal of Ca^{2+}ions from the contractile system (Ebashi, 1960).

A similar effect was later demonstrated in actomyosin. Previous to

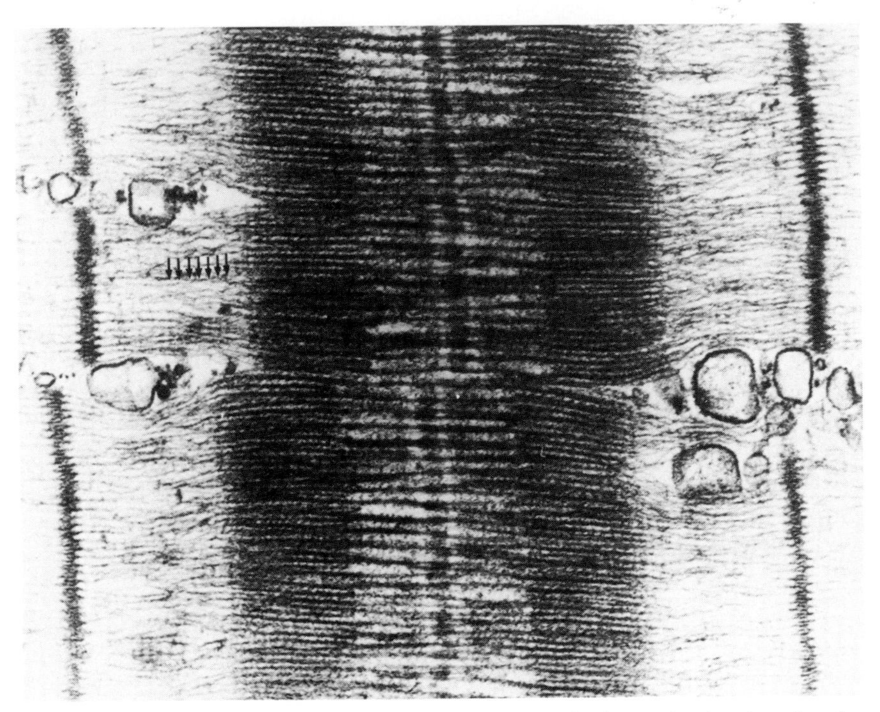

FIGURE 5.3. Longitudinal section of a vertebrate skeletal muscle showing the clear transverse stripes (arrow) across the I-band. The spacing between the stripes is about 400 Å. (× 44,000; reduced 18% for reproduction.)

this, Perry and Grey (1956) had found that synthetic actomyosin (prepared from pure actin and myosin) was different from natural actomyosin (which was largely unpurified) in the sense that its ATPase activity did not depend on the presence of the chelating agent ethylene diamine tetraacetic acid (EDTA). The ATPase of natural actomyosin, on the other hand, was considerably reduced when EDTA was present.

The significance of this result was not fully realized at the time, but later when it was shown the Ca^{2+} ions have a regulatory role, the effect of EDTA as a calcium-chelating agent became obvious. When this work was followed up, a correlation was found between the purity of the actin preparation in different actomyosins and the Ca^{2+} sensitivity of their ATPase activity (Weber and Winicur, 1961). The "impurity" in the actin preparations was finally separated out by Ebashi and Ebashi (1964), and they found that by adding it to pure synthetic actomyosin, full Ca^{2+} sensitivity was developed.

Further investigation showed that the new protein preparation (the impurity) was in some ways similar to the tropomyosin prepared by Bailey (1946, 1948; Bailey *et al.*, 1948), and, indeed, by treating the preparation with methods used for tropomyosin extractions, a considerable amount of new protein was obtained. The new protein preparation was therefore called "native tropomyosin." It was later shown by Ebashi and Kodama (1966) that native tropomyosin comprised tropomyosin molecules bound to an additional globular protein which was called troponin. Addition of troponin to purified tropomyosin generated fully the properties of native tropomyosin. Troponin was therefore identified as the third thin filament component. Since it was found to possess the ability to bind or release Ca^{2+}ions depending on the Ca^{2+}ion concentration, it became clear that troponin must be intimately involved in the regulation of the thin filament and hence of the actomyosin interaction.

The picture of the thin filament that emerged was one in which its basic components are actin, tropomyosin, and troponin and in which its activity is regulated by alterations in the Ca^{2+}ion concentration in the muscle cell. Estimates of the relative proportions of these three proteins and their molecular weights indicated that in vertebrate skeletal muscle the thin filament probably contains one tropomyosin molecule and one troponin molecule for every seven actin molecules in the thin filament. The tropomyosin molecule is just about the right length (400 Å) to interact with seven actin molecules ($7 \times 55 = 385$ Å) if it lies along one of the two grooves in the actin helix (as suggested by Hanson and Lowy, 1963). As a result, the general structural scheme illustrated in Fig. 5.4 for the thin filament was suggested by Ebashi *et al.* (1969). Here, the troponin/tropomyosin repeat is 385 Å (for reasons to be given later), and this is taken to be different from the crossover repeat of the actin helix which is shown here as 360 to 370 Å.

For our purposes, this model for the thin filament serves as a good basis for a deeper discussion of thin filament structure and regulation. In the remainder of this chapter, accounts will be given of the structures, properties, and interactions of actin, tropomyosin, and troponin and also of the recent wealth of evidence that has revealed the likely structural role of tropomyosin and troponin in thin filament regulation.

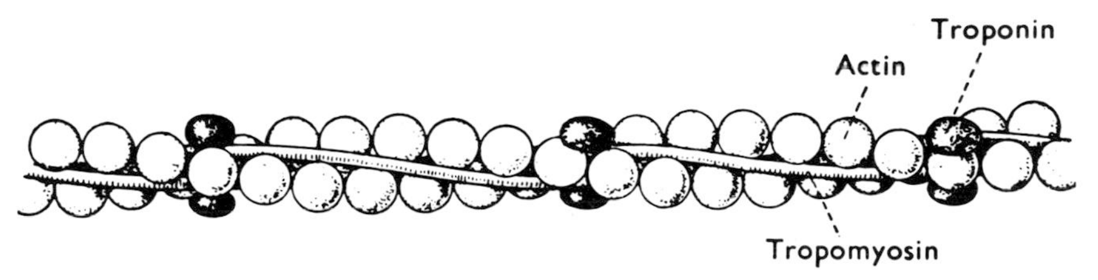

FIGURE 5.4. Schematic model of the complete thin filament comprising actin, tropomyosin, and troponin according to Ebashi *et al.* (1969). The tropomyosin molecules run along the grooves of the actin helix.

5.2. Actin

5.2.1. Characterization of G-Actin

The original method used by Straub (1943) to prepare actin from muscle involved the initial treatment of the muscle with acetone after which actin molecules could be extracted with low-ionic-strength solutions. The water-soluble extract was found to contain globular actin molecules (G-actin) together with a considerable amount of impurity. Purified actin preparations were obtained by taking advantage of the fact that under suitable conditions (e.g., increased ionic strength; 0.1 M KCl and 0.1 mM $MgCl_2$), the G-actin molecules polymerize into filaments (known as F-actin), thus increasing rapidly the viscosity of the solution. The polymerized actin was then separated from the impurities by centrifugation, after which the pellet could be redissolved and the F-actin depolymerized in a suitable low-ionic-strength medium. Two or three repetitions of such a polymerization–depolymerization cycle were found to yield relatively pure actin preparations (Szent-Gyorgyi, 1951), especially if the initial extraction was carried out at 0°C. An alternative depolymerization method using KI solutions was also used by Szent-Gyorgyi (1951). This method is now known to yield actin together with tropomyosin and troponin.

Both of these methods are still used to extract actin, and preliminary purification by repeated polymerization cycles is usually carried out. If necessary, gel filtration and other more specific purification methods are then used to remove any residual impurity. Table 5.2 summarizes the contents of various thin-filament preparations according to Hanson *et al.* (1972).

In low-ionic-strength solutions, actin has the properties of a single-chain globular protein (hence the term G-Actin) with a molecular weight of about 42,000. It binds to Ca^{2+} ions and (noncovalently) to ATP in a mole-to-mole ratio (Rees and Young, 1967; Tsuboi, 1968). Analysis by a number of different methods of the content of α-helix and β conformation in G-actin (Nagy and Jencks, 1962; Nagy, 1966; Murphy, 1971; Greenfeld and Fasman, 1969; Wu and Yang, 1970) has suggested that only about 26 to 30% of the molecule is α-helical, about 26% is in the β-pleated sheet form, and that the remainder of the chain follows a random-coil configuration. The low α-helix content is typical of a globular protein and fits in well with the known amino acid composition of the molecule (Table 5.3; Johnson and Perry, 1968), since the content of the α-helix-breaking amino acids proline and glycine is relatively high compared with their abundance in highly α-helical molecules such as tropomyosin (Table 5.3).

TABLE 5.2. Composition of Various Thin Filament Preparations According to SDS Electrophoresis and the Correlation between Troponin Content and the Appearance of Stripes on Synthetic Thin Filament Paracrystals[a]

Preparation	Composition (SDS electrophoresis)	Ca^{2+}-sensitizing activity	Cross-striation
Purified actin[b]	A	Absent (Greaser and Gergely, 1971)	Absent
KI actin[c]	A, TM, TP	Present (J. Spudich, pers. commun.)	Present
20°-extracted actin[d]	—	Present (Katz, 1966)	Present
0°-extracted actin[d]	A, trace TM	Very low (Katz, 1966)	Absent
Purified actin + Mueller TM[e]	—	Present (Mueller, 1966)	Present
Purified actin + TM, TP[f]	A, TM, TP[f]	Present (Spudich and Watt, 1971)	Present
Purified actin + purified TM[g]	A, TM	Absent (Schaub and Perry, 1969)	Absent
Purified actin + TM + TP[f]	A, TM, TP[f]	Present (Spudich and Watt, 1971)	Present

[a] From Hanson *et al.* (1972).
[b] Most preparations were purified by a method based on the work of Martonosi (1962); others were purified according to Rees and Young (1967).
[c] Prepared by the method Szent-Györgyi (1951). The polymerized actin was collected by centrifugation and resuspended in 0.1 M KCl before the crystals were formed. SDS gel electrophoretograms show that α-actinin is absent from the washed polymers and that the proportions of A, TM, and TP in the polymers are about the same as myofibrils.
[d] Prepared by the Straub method (Katz and Hall, 1963). The G-actin was extracted from the acetone-dried residue either at room temperature (about 20°) or at 0°. In each case the actin was polymerized in 0.1 M KCl, collected by centrifugation, and resuspended in 0.1 M KCl before the crystals were formed.
[e] The TM was prepared in the presence of dithiothreitol by the method of Mueller (1966) except that it was precipitated only once at pH 4.3 and only once in 70% saturated $(NH_4)_2SO_4$.
[f] Preparations generously donated by Dr. J. Spudich (see Spudich and Watt, 1971).
[g] Two preparations were used, one generously donated by Professor S. V. Perry, the other prepared according to Noelken (1962). SDS gel electrophoretograms of the latter material show the usual two bands at positions corresponding to chain weights of about 35,000 and 37,000 (see Cohen *et al.*, 1971a,b).

TABLE 5.3. Amino Acid Compositions of the Thin Filament Components

	Actin[a]	Tropomyosin[b]	TN-T[c]	TN-I[d]	TN-C[e]
Alanine	8.0[f]	12.1[f]	26.0[g]	14[g]	13.8[g]
Arginine	4.8	4.8	24.0	16	7.3
Aspartic acid/Asn	9.0	10.1	20.7	15	23.0
Glutamic acid/Gln	11.3	24.7	55.7	33	33.5
Glycine	7.5	1.3	9.0	8	11.7
Half cystine	1.4	1.0	—	3	0.9
Histidine	2.0	0.7	5.8	4	1.2
Leucine	7.6	3.8	8.1	22	9.8
Lysine	7.0	10.9	19.8	24	10.1
Methionine	4.1	4.4	4.6	9	8.4
Phenylalanine	3.1	0.5	5.3	3	9.4
Proline	4.9	0.2	9.6	5	0.9
Serine	5.9	4.4	8.7	10	8.4
Threonine	6.9	2.7	6.4	3	7.5
Tryptophan	1.2	—	2.0	1	—
Tyrosine	4.0	1.9	4.3	2	2.3
Valine	5.4	3.4	11.7	7	6.9

[a] Rabbit skeletal muscle (Johnson and Perry, 1968). Also has 3-methylhistidine.
[b] Rabbit skeletal muscle (Hodges and Smillie, 1970).
[c] Rabbit skeletal muscle (Pearlstone *et al.*, 1976).
[d] Rabbit skeletal muscle (Wilkinson and Grand, 1975).
[e] Rabbit white skeletal muscle (Head and Perry, 1974).
[f] All values given as residues per 100 residues.
[g] Given as residues per molecule.

The entire primary structure (amino acid sequence) of actin has recently been determined by Elzinga and his collaborators (Elzinga and Collins, 1972; Collins and Elzinga, 1975). G-actin comprises 374 amino acids in the sequence shown in Fig. 5.5, thus defining the molecular weight accurately at 41,700. Unfortunately, the three-dimensional structure of G-actin molecules has not yet been determined. For this reason it is not possible to assign functions to the various parts of the G-actin chain. However, for a variety of reasons (Elzinga and Collins, 1972), it is likely that the region including Tyr-69 and His(Me)-73 lies near the molecular surface, as does Lys-113. It is also likely for obvious reasons that the three long sequences of mostly nonpolar residues (120 to 153; 292 to 309; 336 to 357) lie in the interior of the molecule.

Comparative studies on actin from different sources (Carsten and Katz, 1964; Totsuka and Hatano, 1970; Bridgen 1971; Elzinga and Collins, 1972; Bray, 1972) have indicated that whatever the source, the actin obtained is virtually identical. The highly critical test in which amino acid sequences are compared (Bridgen 1971; Elzinga and Collins 1972) has revealed that substitutions do occasionally occur (in rabbit, bovine heart, and trout actins) but that when they do, they tend to be conservative in

CB-13→
Ac-Asp-Glu-Thr-Glu-Asp-Thr-Ala-Leu-Val-Cys- 10
Asp-Asp-Gly-Ser-Gly-Leu-Val-Lys-Ala-Gly- 20
Phe-Ala-Gly-Asp-Asp-Ala-Pro-Arg-Ala-Val- 30
Phe-Pro-Ser-Ile-Val-Gly-Arg-Pro-Arg-His- 40
CB-1→ CB-10→
Gln-Gly-Val-Met-Val-Ser-Met-Gly-Gln-Lys- 50
Asp-Ser-Tyr-Val-Gly-Asp-Gly-Ala-Gln-Ser- 60
Lys-Arg-Gly-Ile-Leu-Thr-Leu-Lys-Tyr-Pro- 70
Ile-Glu-His(ᵀMe)-Trp-Gly-Ile-Ile-Thr-Asn-Asp- 80
CB-11→
Asp-Met-Glu-Lys-Ile-Trp-His-His-Thr-Phe- 90
Tyr-Asn-Glu-Leu-Arg-Val-Ala-Pro-Glu-Glu- 100
His-Pro-Thr-Leu-Leu-Thr-Glu-Ala-Pro-Leu- 110
Asn-Pro-Lys-Ala-Asn-Arg-Glu-Lys-Met/Thr- (120)
CB-4 →
Gln-Ile-Met-Phe-Glu-Thr-Phe-Asn-Val-Pro- (130)
CB-15 →
Ala-Met-Tyr-Val-Ala-Ile-Gln-Ala-Val-Leu- (140)
Ser-Leu-Tyr-Ala-Ser-Gly-Arg-Thr-Thr-Gly- (150)
Ile-Val-Leu-Asp-Ser-Gly-Asp-Gly-Val-Thr- (160)
His-Asn-Val-Pro-Ile-Tyr-Glu-Gly-Tyr-Ala- (170)
CB-7 →
Leu-Pro-His-Ala-Ile-Met-Arg-Leu-Asp-Leu- (180)
Ala-Gly-Arg-Asp-Leu-Thr-Asp-Tyr-Leu-Met- (190)
CB-17 →
Lys-Ile-Lys-Thr-Glu-Arg-Gly-Tyr-Ser-Phe- (200)
Val-Thr-Thr-Ala-Glu-Arg-Glu-Ile-Val-Arg- (210)
Asp-Ile-Lys-Gln-Lys-Leu-Cys-Tyr-Val-Ala- (220)
CB-12 →
Leu-Asp-Phe-Glu-Asn-Glu-Met-Ala-Thr-Ala- (230)
Ala-Ser-Ser-Ser-Leu-Glu-Lys-Ser-Tyr-Glu- (240)
Leu-(Pro, Asx, Glx, Gly, Ile, Val)-Thr-Ile-Gly- (250)
Asn-Glu-Arg-Phe-Arg-Cys-Pro-Glu-Thr-(Phe, (260)
CB-6 → (270)
Leu, Phe, Gln, Pro, Ser, Ile, Gly)-Met-Glu-Ser-
Ala-Gly-Ile-His-Glu-Thr-Thr-Tyr-Asn-Ser- (280)
CB-8 →
Ile-Met-Lys-Cys-Asp-Ile-Asp-Ile-Arg-Lys- (290)
CB-2 →
Asp-Leu-Tyr-Ala-Asn-Asn-Val-Met-Ser-Gly- (300)
CB-3 →
Gly-Thr-Thr-Met-Tyr-Pro-Gly-Ile-Ala-Asp- (310)
CB-5 →
Arg-Met-Gln-Lys-Glu-Ile-Thr-Ala-Leu-Ala- (320)
CB-16 →
Pro-Ser-Thr-Met-Lys-Ile-Lys-Ile-Ile-Ala- (330)
Pro-Pro-Glu-Arg-Lys-Tyr-Ser-Val-Trp-Ile- (340)
Gly-Gly-Ser-Ile-Leu-Ala-Ser-Leu-Ser-Thr- (350)
CB-9 →
Phe-Gln-Gln-Met-Trp-Ile-Thr-Lys-Gln-Glu- (360)
Tyr-Asp-Glu-Ala-Gly-Pro-Ser-Ile-Val-His- (370)
Arg-Lys-Cys-Phe. (374)

FIGURE 5.5. The complete sequence of actin from rabbit skeletal muscle as determined by Elzinga and Collins (1972).

nature. Even a preliminary comparison between actins from sources as different as rabbit and amoeba suggests that the sequences are similar (Weihing and Korn, 1972; Elzinga and Collins, 1972). It will be seen later in this chapter that actin is involved in so many different interactions that it is unlikely that significant sequence mutations could occur without destroying the function of the molecule. The fact that actins from different sources are so similar may not, therefore, be very surprising.

One useful consequence of a knowledge of the primary sequence of actin is that it might be possible to prepare a well-characterized but modified form of the actin molecule that has lost the ability to polymerize and thus might be more likely to crystallize. It is already known that polymerization is selectively inhibited in two derivatives: photooxidized actin (Martonosi, 1968) and partially nitrated G-actin (Múhlrad *et al.*, 1968). But attempts to crystallize these two modified actins have not, as yet, been successful. On the other hand, recent reports describing the crystallization of aggregates containing actin and profilin (from calf spleen) (Carlsson *et al.,* 1976) or actin and DNase (Kabsch *et al.,* 1980) and tubes of actin formed in the presence of gadolinium (Dickens, 1978) are very interesting and may augur well for the detailed analysis of actin structure in the foreseeable future.

5.2.2. F-Actin Formation and Structure

As described earlier, F-actin is produced from solutions of G-actin by increasing the ionic strength (e.g., by addition of 0.1 M KCl). The polymerization process can be summarized by the reaction

$$n(\mathrm{G\cdot ATP}) \rightarrow (\mathrm{G\cdot ADP})_n + n\mathrm{P}_1$$

(although the presence of ATP is not essential). The viscosity of the solution increases very rapidly when G-to-F polymerization occurs, and if the protein concentration is above about 5 mg/ml, a thick gel is produced.

We have already seen that the work of Hanson and Lowy (1963) and Selby and Bear (1956) has given rise to a picture of the F-actin structure (Figs. 5.1 and 5.2) in which G-actin molecules appear to be arranged on two helical strands of pitch about 700 to 760 Å, giving a near repeat in the F-actin filament of about 350 to 380 Å. Depue and Rice (1965) showed that these long-pitch helices are also right-handed. Since much of the present detailed evidence on thin filament structure has come from diffraction studies, it is appropriate here to consider the form of the diffraction pattern from this F-actin structure.

In terms of one of the two genetic helices (i.e., the two alternative single helices that can be drawn through all of the subunits in the helix), the F-actin structure can be thought of (Fig. 5.6) as a helix of actin subunits that has a pitch of about 59 Å, a subunit axial translation of about 27.5 Å, and a repeat after about six turns in a distance of 350 to 380 Å. For simplicity, it will be assumed initially that the structure repeats exactly after six turns in which there are 13 G-actin monomers. Considering this structure first as a continuous helix of pitch 59 Å (Fig. 5.6a) the diffraction pattern will consist of a single helix cross (Fig. 5.6d) with layers of intensity at distances (n/59) Å^{-1} from the equator of the pattern. But the helix is really discontinuous with a subunit axial translation of 27.5 Å, and this discontinuous structure (Fig. 5.6c) can be thought of (Chapter 2, Section 2.3.3) as the product of a continuous helix (Fig. 5.6a) and a set of parallel planes of density separated axially by 27.5 Å (Fig. 5.6b).

From the convolution theorem (Chapter 2), the diffraction pattern

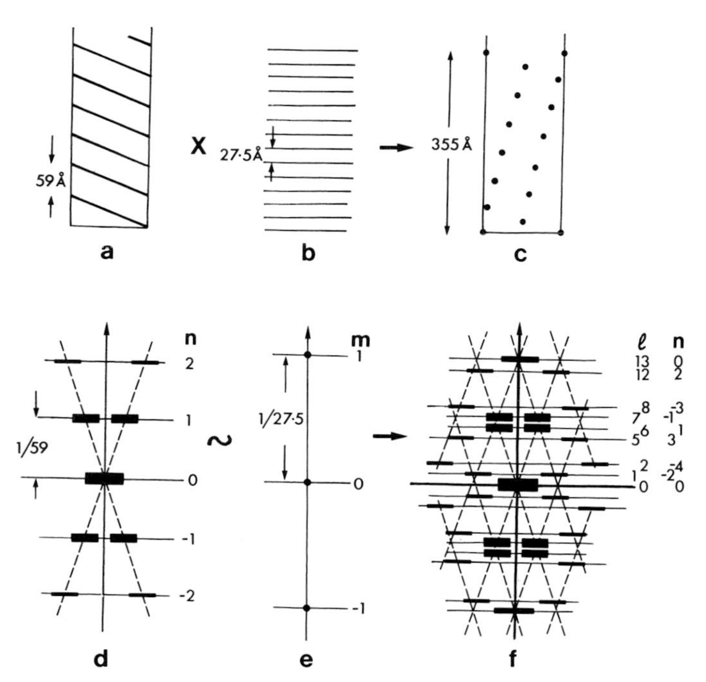

FIGURE 5.6. Generation of the Fourier transform of the F-actin helix. (a) Radial projection of a continuous helix of pitch 59 Å. When this is multiplied by a set of density planes 27.5 Å apart (b), a discontinuous helix is generated (c) which has about 13 subunits in six turns. The diffraction pattern of (a) is a single helix cross with layers at spacings (Z) equal to $n/59$ Å^{-1}. (e) Diffraction pattern from (b), a set of meridional spots spaced $m/27.5$ Å^{-1} from the equator. The convolution of (d) with (e) is shown in (f) and is the diffraction pattern of the 13/6 helix shown in (c). Layer lines occur in (f) at $Z = n/59 + m/27.5$ Å^{-1}, and the most prominent are $l = 6$ and 7 ($n = 1$ and -1) and $l = 13$ ($n = 0$). The n values on the other most prominent layer lines are listed on the right of (f).

of a helix with the symmetry of the F-actin helix must therefore consist of a set of helix crosses of the kind shown in Fig. 5.6d spaced along the meridian at distances Z from the equator equal to $(m/27.5)$ Å^{-1}. The resulting pattern is illustrated in Fig. 5.6f. The most prominent features of this pattern are the sixth and seventh layer lines at spacings from the equator of (1/59) Å^{-1} ($n = 1$, $m = 0$) and (1/51) Å^{-1} ($n = -1$, $m = 1$), respectively, and the first meridional reflection at a spacing of (1/27.5) Å^{-1} ($m = 1$, $n = 0$). But these are only part of a series of layer lines that have spacings (Z) from the equator given by the general expression $Z = l/R = (n/59 + m/27.5)$ Å^{-1}. The first layer line occurs at $Z = 1/R = 1/354$ Å^{-1} ($m = 1$, $n = -2$), the second layer line at $Z = 2/R = 1/177$ Å^{-1} ($m = 2$, $n = -4$), and so on. The effect of altering slightly the helical parameters from exact 13/6 helical symmetry is merely to alter slightly these layer line positions without significantly altering the form of the F-actin diffraction pattern.

The structure so far considered is a helix of points arranged with the symmetry of the F-actin helix. In fact, a G-actin molecule will be located on each point, and this will modify the intensity distributuion in the diffraction pattern. The resulting distribution of intensity can be determined to a first approximation by computing the Fourier transform of a model structure in which each G-actin monomer is taken to be a spherical molecule of radius about 24 Å and in which the centers of these monomers are at a helix radius of about 24 Å. The transform on the first few layer lines can then be computed using a standard expression for the Fourier transform of a helix. The result has the form shown in Fig. 5.7 (Parry and Squire, 1973). It is clear from this that layer lines 1, 6, and 7 are all expected to be strong.

This kind of theoretical diffraction pattern can be compared directly with a number of experimentally observed diffraction patterns (Fig. 5.8). These are (a) the low-angle X-ray diffraction pattern from intact molluscan muscle (Vibert *et al.*, 1972), (e) the optical diffraction pattern from electron micrographs (d) of F-actin paracrystals (O'Brien *et al.*, 1971; Moore *et al.*, 1970), and (c) optical diffraction patterns from a model structure (b). All of these patterns include the characteristic actin layer lines at about 360 Å, 59 Å, 51 Å, and 27 Å (Table 5.4), and the simple conclusion can be reached that both in intact muscle and when isolated, the thin filaments have a symmetry that is similar to that of the model in Fig. 5.8b.

However, one of several characteristic differences between the observed diffraction patterns and the computed pattern is the marked difference in the relative intensities of the peaks on the 59-Å and 51-Å layer lines. In the computed diffraction pattern (Fig. 5.7), these peaks have very comparable intensities, whereas in the observed patterns (Fig. 5.8a,e) the 51-Å reflection is relatively much weaker. Clearly the model structure, although possessing the correct symmetry, is inadequate in that it represents the G-actin molecules simply as spheres, whereas the true shape of these molecules may be significantly different. It is the shape of the actin monomers that will largely determine the relative intensities of the 59-Å and 51-Å layer lines (Vibert, 1968; O'Brien *et al.*, 1971). Unfortunately, it is not possible as yet to determine the G-actin shape directly from X-ray diffraction patterns. But Moore *et al.* (1970) have applied the novel technique of three-dimensional reconstruction from electron micrograph images to various actin-containing structures (including F-actin), and this work, together with more recent studies (see Section 5.2.3), has given the first glimpse of the likely shape of G-actin.

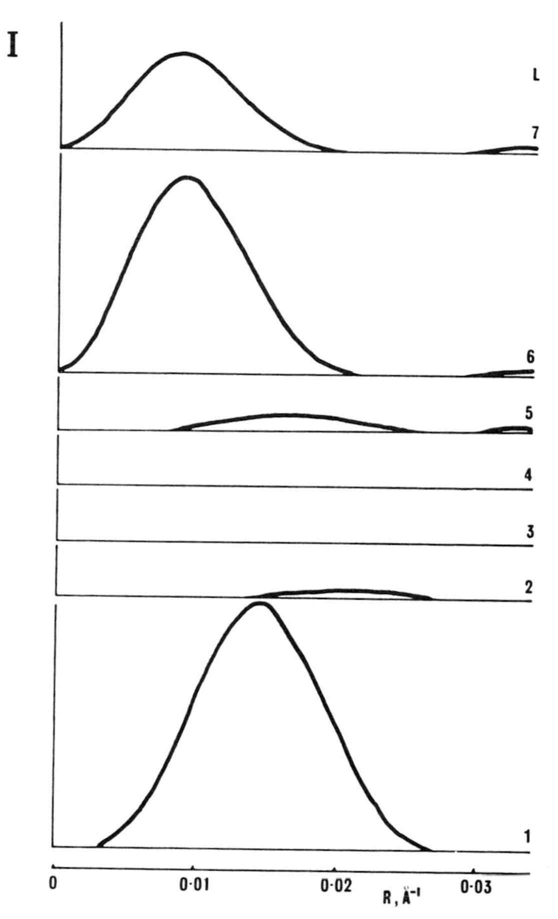

FIGURE 5.7. The computed intensity distribution on the first seven layer lines in the diffraction pattern from a 13/6 helix of spherical subunits each of radius 24 Å and centered at a distance of 24 Å from the helix axis. Note that the second layer line is relatively weak.

5.2.3. Three-Dimensional Reconstruction from Paracrystals of F-Actin

Following the original work of Hanson (1967), paracrystals of F-actin in a number of polymorphic forms have been prepared (Hanson 1967, 1968; Kawamura and Maruyama, 1970; O'Brien *et al.*, 1971, Yamamoto *et al.*, 1975). Figure 5.8d illustrates one of these forms. Here (Hanson, 1967, 1968), the F-actin filaments are parallel and packed side by side in axial register so that adjacent filaments are related by a simple

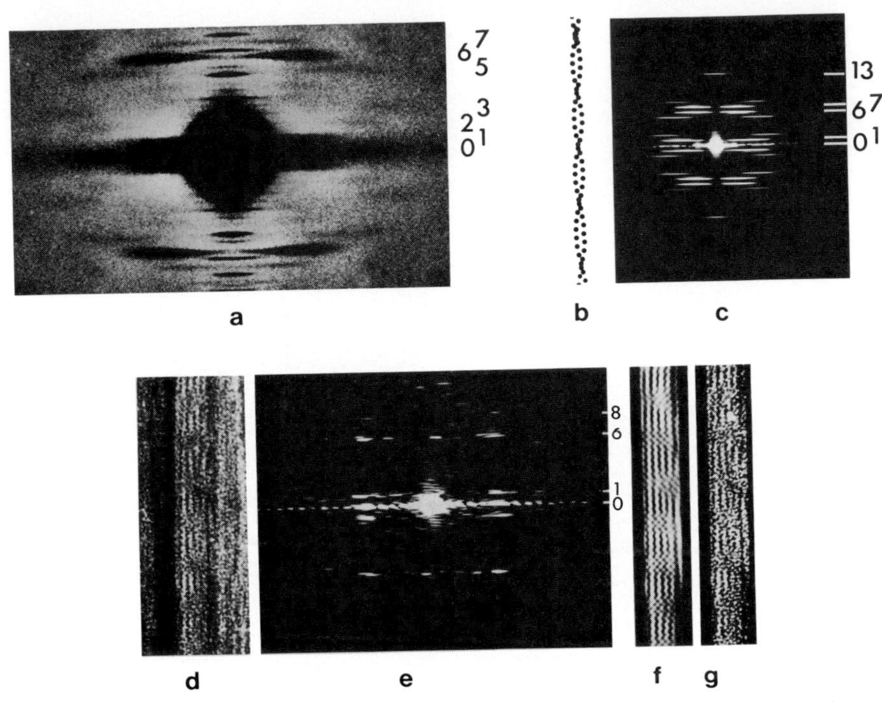

FIGURE 5.8. (a) The low-angle X-ray diffraction pattern from a living anterior byssus retractor muscle of *Mytilus* showing a very well-developed actin pattern. The figures to the right indicate the layer line numbering in terms of a 13/6 helix. From Vibert *et al.* (1972). (b) Model of a 13/6 helix and its optical diffraction pattern (c). (d) A paracrystal of actin plus impure tropomyosin negatively stained with 1% uranyl acetate after fixation with glutaraldehyde. (e) The optical diffraction pattern from (d) which is indexed as in (a). (f and g) Reconstructions of the paracrystal in (d), (f) with and (g) without optical filtering. (d–g) From O'Brien *et al.* (1971). Note that (a) and (e) both show the strong first and sixth layer lines as in (c). But, unlike the simple model in (c), the seventh layer line is weak in (a) and (e). See text for details.

translation perpendicular to their length. In the others, the filaments form a network in which the projection of the repeating unit is diamond shaped and has a side of length about 340 Å (Kawamura and Maruyama, 1970; Yamamoto *et al.*, 1975).

Paracrystals of the type shown in Fig. 5.8d have been used extensively for optical diffraction studies of F-actin (e.g., O'Brien *et al.*, 1975) and for three-dimensional reconstruction work (e.g., Moore *et al.*, 1970). They were used primarily because F-actin filaments tend to be rather flexible, and good micrographs of individual straight F-actin filaments, which would be suitable for image analysis, are rarely obtained.

Optical diffraction analysis of these paracrystals by O'Brien *et al.*

TABLE 5.4. Actin X-Ray Layer Line Spacings from Live Muscle[a]

Calculated layer line spacings[b]			Observed spacings				
			Abrm	Guinea pig	Frog sartorius muscle		
$d(Z^{-1})$	m	n	d^c	d^c	d^c	d^d	d^e
384.6	1	−2	381	372	400	—	366
192.3	2	−4	196	184	180	—	186
128.2	3	−6	132	124	125	—	120.2
84.74	−2	5	—	—	—	—	87.8
69.44	−1	3	70.0	70.9	70.0	—	69.6
58.95	0	1	59.2	59.4	59.1	59.1	59.06
50.84	1	−1	51.4	51.1	51.1	50.97	50.96
						(39.0)	
29.47	0	2	—	—	—	29.8	—
27.30	1	0	—	—	—	27.3	
19.65	0	3				(20.4)	
18.66	1	1				18.75	
13.65	2	0				13.66	
13.16	3	−2				—	
11.08	2	1				11.06	
9.10	3	0				9.08	
7.88	3	−1				7.85	
6.83	4	0				6.87	

[a] Dashes indicate that the appropriate spacing was not quoted, not that the reflection could not be observed. Values of m and n are quoted for a right-handed genetic helix. In fact, this helix is left-handed.
[b] Calculated for a helix with $P = 58.95$ Å and $p = 27.30$ Å. This is very close to a 28/13 helix. All quoted spacings given in Å.
[c] From Vibert *et al.* (1972); all values represent summary from resting, active, and rigor states.
[d] From H. F. Huxley and Brown (1967); living resting muscle.
[e] From Haselgrove (1975a); rigor muscle.

(1975) and Moore *et al.* (1970) has shown that in this form, if not in muscle, the F-actin helix has exactly 13 monomers in six turns (within experimental error) and an axial repeat of about 350 to 360 Å. In addition, an interesting feature of the paracrystals is that the first row line spacing (giving the lateral separation of adjacent filaments) is only about 57 to 60 Å (O'Brien *et al.*, 1971; Moore *et al.*, 1970). O'Brien *et al.* took this close approach of filaments, nominally about 80 Å in diameter, to imply that some interdigitation of adjacent filaments must occur. Optically reconstructed "filtered" images of F-actin paracrystals (Fig. 5.8f) tended to confirm this conclusion. They also lent support to the idea that the G-actin subunits are probably elongated in shape rather than spherical.

The 3-D reconstruction technique [formulated by DeRosier and

Klug (1968) and DeRosier and Moore (1970) and used by Moore *et al.* (1970) on actin filaments] applies directly to an image of a single helical structure. The necessity of using paracrystals of F-actin in order to obtain images of straight, regular filaments was therefore an added complication, and an additional procedure known as deconvolution (Moore and DeRosier, 1970) was required to separate out from the image of the paracrystal the form of a single "average" filament. The three-dimensional structure of this filament was then reconstructed from a knowledge of the Fourier transform of the projected image of the filament using a procedure known as inverse Fourier–Bessel transformation (Klug *et al.,* 1958). Ideally, this procedure requires information to be included from all of the different layer lines in the diffraction pattern including the equator. In addition, the technique is based on the assumption that the image is of a truly helical structure. No account is taken of possible flattening of the structure onto the electron microscope grid or of other distortions of the structure. In fact, any departure of the structure from cylindrical symmetry would have a marked effect on the equator. For this reason, Moore *et al.* (1970) omitted the equatorial information from their reconstruction. The effect of this was to render uncertain the zero-density cut-off level in their reconstructions, which can be thought of as three-dimensional contour maps of density. A subjective assignment of this cut-off therefore had to be made. Information was included in the reconstruction out to a distance from the center of the transform of about 1/23 $Å^{-1}$ in a radial direction and 1/45 $Å^{-1}$ in an axial direction, thus defining the likely resolution of the final image.

Figure 5.9 shows the kind of result obtained by this method. It represents a structure with about average features in a series of reconstructions from different micrographs. The characteristics of this model are as follows. It consists of the expected helical arrangement of nonspherical subunits with the whole structure having an outer diameter of about 80 Å. Each subunit measures approximately 55 Å axially, 35 Å radially, and 50 Å tangentially and has an approximate volume of 45,000 $Å^3$. This accounts for about 90% of the volume expected for a G-actin molecule of molecular weight 41,700 and partial specific volume 0.74 cm^3/g. Such agreement is fairly good if it is remembered that the resolution in the electron microscope is relatively poor and that there is considerable uncertainty in the assignment of the cutoff level in the computed electron-density map.

Apart from showing up the approximate shape of the F-actin subunit, the form of the reconstructed image tended to confirm that the observed subunit was equivalent to a single G-actin molecule. In addition, the reconstruction seemed to highlight the four interactions that a

FIGURE 5.9. Side view (top) and end view of the three-dimensional reconstruction of F-actin carried out by Moore *et al.* (1970). They used paracrystals similar to that in Fig. 5.8d. The dotted outlines indicate the probable extent of a single G-actin molecule.

single actin molecule makes with its four nearest neighbors in the F-actin structure. In Fig. 5.9, these interactions are actually visualized as bridges of density between adjacent subunits, but the true significance of these remains to be determined.

The results of Moore *et al.* (1970) have been followed up by Wakabayashi *et al.* (1975) who analyzed the structure of F-actin filaments with tropomyosin and troponin added. In their 3-D reconstruction work, the G-actin monomer appeared to be rather triangular (or boot-shaped) and thus gave the whole filament a definite structural polarity. Their results will be discussed more fully in Section 5.5.4.

5.2.4. Actin Interactions

The actin interactions that have been discussed so far are those which a G-actin molecule can make with its four immediate G-actin neighbors in the thin filament. In addition, actin can make interactions with tropomyosin and troponin (as will be described in Section 5.4.3), with a protein called α-actinin which is involved in the Z-band structure (as described in Chapter 7), and of course with myosin. The nature of the actin–myosin interaction will be considered in detail in Chapters 10 and 11. But it is appropriate here to consider the structure of F-actin filaments that have been labeled with certain subfragments of myosin molecules. It was mentioned in Chapter 1 (Section 1.3.4) that myosin molecules consist of a long thin rod-shaped tail on one end of which occur two globular subunits. The myosin molecule can be fragmented by proteolytic enzymes (Fig. 5.10a) to give a rod [light meromyosin (LMM)] and the head assembly called heavy meromyosin (HMM) containing the two globular subunits. Heavy meromyosin can be further digested to separate the two globular subunits (heads), each of which is called heavy meromyosin subfragment 1 (HMM S-1). These S-1 subunits retain the ATPase and actin-binding properties of the myosin molecule.

In a classic experiment, H. E. Huxley (1963) prepared isolated thin filaments or synthetic F-actin filaments and allowed these to interact with a solution of HMM moieties in the absence of ATP, thus producing

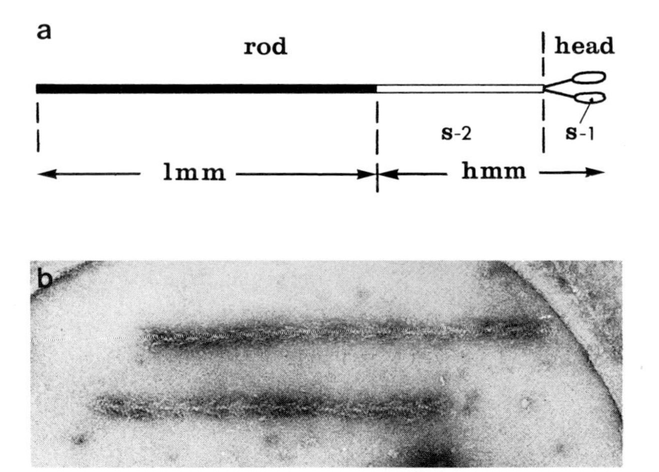

FIGURE 5.10. (a) Schematic diagram showing the structure of a myosin molecule and the definition of its main proteolytic fragments LMM, HMM, and HMM S-1 and S-2. (b) Electron micrograph of a negatively stained preparation of actin filaments labeled with HMM showing the characteristic arrowhead appearance. From H. E. Huxley (1963). Thin filaments decorated with HMM S-1 show a similar appearance.

rigorlike actin–HMM complexes. He then found that the HMM-labeled thin filaments (commonly termed "decorated" thin filaments) had a very characteristic appearance in negatively stained preparations. Each repeat in the structure looked rather like an arrowhead in profile (Fig. 5.10b), and it therefore possessed a definite polarity. This indication of thin filament polarity was an important result in itself. Huxley also showed, by decorating thin filaments still attached to the Z-band, that on each side of the Z-band the thin filaments have opposite polarity. This clearly supported the fact that the thick myosin filaments in striated muscles have a bipolar structure and that they would therefore be expected to interact with thin filaments of opposite polarity at opposite ends of the sarcomere (Fig. 1.6). Similar results are obtained if actin is labeled with the HMM S-1 moieties.

The arrowhead appearance of decorated thin filaments is such a characteristic structure that S-1-labeling experiments are now commonly used as a means of showing up the actin-like properties of "actins" from a variety of nonmuscle cells such as slime molds, blood platelets, and the vertebrate brain. But in addition, this structure has also been subjected to further three-dimensional reconstruction work by Moore *et al.* (1970), and the details of the reconstruction have given the first indication of the nature of the actin–S-1 interaction and the position of the myosin binding site on G-actin.

It was found that the individual decorated thin filaments were straight enough for the reconstruction technique to be applied directly to them. This avoided the extra step of deconvolution that was necessary in the analysis of the F-actin paracrystals.

The results of Moore *et al.* are shown in Fig. 5.11. The actual reconstruction is shown in Fig. 5.11a, and Figs. 5.11b,c show simplified models which Moore *et al.* suggested might account for the reconstruction and the arrowhead appearance. The conclusions that they reached can be summarized as follows.

1. The HMM S-1 subunit is approximately 150 Å long, 45 Å wide (in an axial direction), and 30 Å wide radially.
2. The long axis of the S-1 is slightly curved, making it rather banana shaped.
3. The angle between the long axis of the S-1 unit and the filament axis is about 50° (it corresponds to the rigor interaction; the AM state in Fig. 1.11).
4. The angle between a radius through a G-actin molecule and the inner end of the S-1 unit attached to it is about 60°, increasing to about 120° at the outer end of the S-1 unit because of its curvature. (The angles are measured counterclockwise when the filament is viewed down its axis towards the point of an arrowhead.)

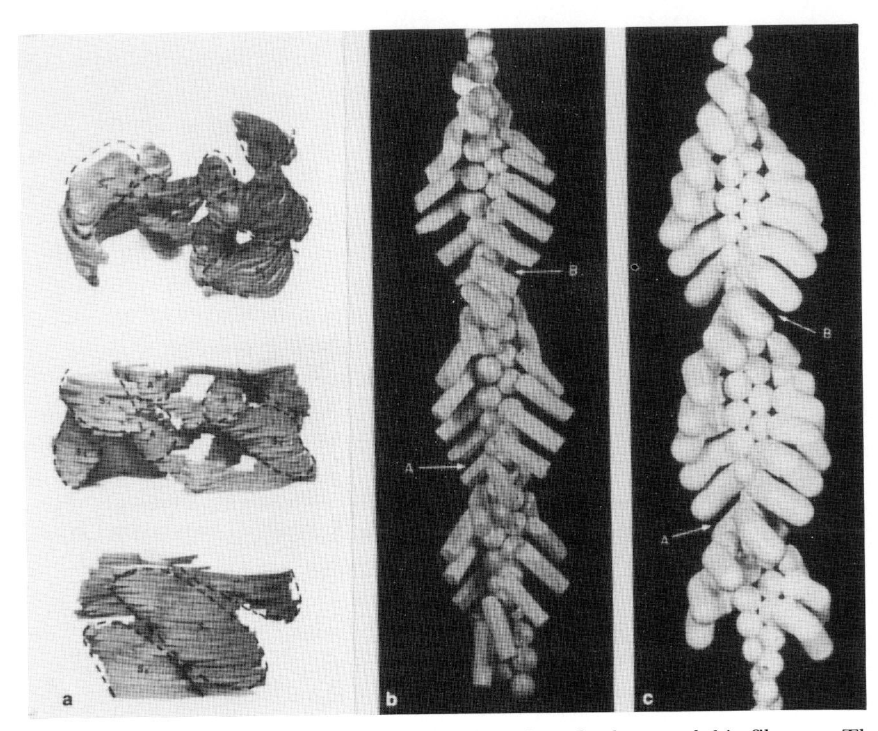

FIGURE 5.11. (a) Three views of a model of a portion of a decorated thin filament. The probable positions of the HMM S-1 units and of the G-actin subunits are indicated by dotted lines. The S-1 units are slightly curved and are applied somewhat tangentially to the actin monomers. The long axis of the S-1 units is both tilted and slewed relative to the axis of the thin filament. Top, nearly end-on view; middle, side view with S-1 units on each side of the centrally placed actin subunits; bottom, a second side view showing the entire length of one S-1 subunit (the model has also been inverted). (b and c) Models illustrating the appearance of a thin filament that has been decorated with S-1 units. As in (a), these units are tilted and slewed relative to the thin filament axis. The S-1 units are represented by straight rods in (b) and by curved models in (c). The latter represents more closely the appearance in (a). Points A and B represent successive crossovers in the actin helix. Between these points the polar unsymmetrical variation in the projected view of the S-1 units is what is believed to give rise to the characteristic arrowhead appearance in decorated thin filaments (Fig. 5.10b). (From Moore *et al.,* 1970.)

5. The S-1 units interact somewhat tangentially with the actin monomers as shown in Figure 5.11a, at an angle ϕ (see Fig. 5.27) of about 30 to 40°.

The information obtained from this reconstruction was therefore quite considerable. However, Fig. 5.11b shows that the arrowhead appearance could be produced by S-1 units with the correct orientation to the thin filament without the need for them to be curved as in (c). In view of other conflicting evidence on the shape of the S-1 unit (to be described in Section 6.2.4), it is reasonable to take this reconstruction as a general indication of the nature of the S-1–actin interaction without putting too much weight on the details of the shape of the S-1 unit which the reconstruction gives. The shape and properties of the myosin molecule and S-1 will be considered in detail in Chapter 6 (Section 6.2.4), and the geometry of the rigor-complex labeling of thin filaments will also be considered again in Chapters 10 and 11 in the discussions of the contractile event.

5.3. Tropomyosin

5.3.1. Preliminary Characterization of Tropomyosin

Soon after the discovery of tropomyosin by Bailey in 1948, it was shown by Astbury (1949) and Perutz (1951) that films of tropomyosin molecules give the characteristic α X-ray diffraction patterns now known to be caused by an α-helical structure. It was later shown by A. G. Szent-Gyorgyi and Cohen (1957) that the α-helix content, as indicated by the value of b_0 in the Moffitt–Yang equation for optical rotatory dispersion (Moffit and Yang, 1956), was very close to 100%. Analysis of the amino acid content of tropomyosin (Hodges and Smillie, 1970) showed it to be rich in α-helix-favoring residues (Table 5.3) and low in those residues that do not favor α-helix formation. In 1953, Crick suggested that the form of the α-pattern from tropomyosin might result from supercoiling of the α-helical chains giving rise to the coiled-coil structure as described in Chapter 4. Detailed X-ray analysis of the analogous protein paramyosin by Cohen and Holmes (1963) indicated that paramyosin probably consisted of a two-chain α-helical coiled coil, and the inference was made that tropomyosin would probably have a similar structure. More direct support for this conclusion was given by Fraser *et al.* (1965).

Tropomyosin is normally extracted by the method of Bailey (1948) which relies on its solubility at neutral pH. However this gives a tropomyosin contaminated to some extent by troponin. Purified tropomyosin can be prepared by the methods of Hartshorne and Mueller (1967), Greaser and Gergely (1971), and Cummins and Perry (1973).

The molecular weight of tropomyosin as originally determined by hydrodynamic studies (Holtzer *et al.*, 1965; Woods, 1967) was about 68,000, and Woods also showed that by using denaturing and reducing agents the molecule could be dissociated into two similar chains of molecular weight 33,500 ± 2,000 (Woods, 1967), thus tending to confirm the structure as a two-chain coiled coil.

On the basis of this molecular weight, the mean residue weight (Table 5.3), and the known mean residue axial translation in a coiled coil (1.485 Å; Fraser *et al.*, 1965; A. Elliott *et al.*, 1968a), a molecular length in the range 405 to 456 A would be predicted. This compares well with early measurements made on electron micrographs of individual molecules (Rowe, 1964; Ooi and Fujime-Higashi, 1971) which gave a value of about 400 Å. Studies of synthetic aggregates of tropomyosin (to be described in Section 5.3.2) have confirmed that the molecular length is close to 400 Å.

Purified tropomyosin, when run on SDS polyacrylamide gels, shows at least two peaks of slightly different chain weights, the two components being known as the α and β chains. The α chain in skeletal muscle is the more abundant and is distinguished by having only one cysteine residue, compared with the two cysteines that occur in the less abundant β chain. Cummins and Perry (1972, 1973, 1974) have correlated the proportions of the two chain types in a muscle with its physiological properties and have demonstrated by immunologic methods that differences occur in the proportions of the two chains in different muscles. Their results and those of Cohen *et al.* (1972) indicate that individual fiber types may contain only a single kind of chain. They also mention that traces of tropomyosin species other than the α and β chains may be present in some preparations, and this is supported by the work of Hodges and Smillie (1972a,b).

Smillie and his collaborators have also been instrumental in determining the complete amino acid sequence of the α chain. Details of this sequence will be given Section 5.3.3

5.3.2. Analysis of Tropomyosin Crystals and Tactoids

Introduction. Extracted tropomyosin will form a number of synthetic aggregates under suitable conditions. These include regularly striped tactoids of various kinds (Cohen and Longley, 1966) and extended true crystals (Bailey, 1948). Studies of these aggregates have yielded much useful information on the length, structure, and interaction properties of tropomyosin. Tactoids are produced when divalent cations are added to neutral or slightly alkaline solutions of tropomyosin (Cohen and Longley, 1966; Caspar *et al.*, 1969), whereas true crystals are produced in solutions having a pH close to the tropomyosin isoelectric point (pH 5.1) and an ionic strength of about 0.3 to 0.4 (Bailey, 1948, 1954).

Tactoids. Tsao *et al.* (1965), Cohen and Longley (1966), and Caspar *et al.* (1969) have described the various polymorphic forms that tropomyosin molecules can produce when precipitated with divalent cations. Long-period fibrous tactoids are produced by precipitation with Mg^{2+} (or Ca^{2+}), Ba^{2+}, and Pb^{2+}. Although similar tactoids can sometimes be produced using different cations, in each case the aggregates produced are characteristic of the use of a particular ion. Mg^{2+} tactoids are dihedral in structure with an axial repeat close to 395 Å (Fig. 5.12a), although slight variations of this repeat have been observed along a single tactoid. Substriations across the negatively stained tactoids divide this 395 Å repeat into two bands about 150 and 250 Å wide and edged by narrow, lightly stained stripes about 30 Å in width.

The Pb^{2+} tactoids (Fig. 5.12b,c) are distinguished by the clearly polar band patterns. Once again, the repeat is very close to 395 Å. Since it is thought that the two chains in one tropomyosin molecule are virtually identical (Woods, 1967), the occurrence of a tactoid with a polar band pattern implies that the molecule itself is polar and, therefore, that the two component chains in the coiled-coil structure must be parallel (i.e., their C-terminals are both at one end and their N-terminals both at the other).

The tactoids produced by Ba^{2+} precipitation occur in two forms: one is similar to the dihedral Mg^{2+} tactoid with a repeat of 395 Å, and the other, characteristic of Ba^{2+} precipitation, has a repeat about 10 Å longer (~405 Å). The difference is significant (Fig. 5.12d), since the mismatch between two such tactoids, which happen to lie side by side in the same micrograph, can easily be detected (Caspar *et al.*, 1969). Like the Mg^{2+} tactoid, this characteristic Ba^{2+} tactoid has dihedral symmetry, but the two bands within each repeat are more nearly equal in width (about 200 Å) even though they stain differently. Some of these tactoids have been observed to terminate in a fringe about 200 Å wide (Caspar *et al.*, 1969).

Tsao *et al.* (1965) and Cohen and Longley (1966) reported the existence of tactoids of repeat 800 Å in preparations of chicken gizzard tropomyosin. These also had a dihedral structure and were thought to be cases of alternate head-to-head and tail-to-tail interactions between successive tropomyosin molecules.

The Mg^{2+} tactoids of tropomyosin have been the subject of numerous structural studies (see Fig. 3.15). In particular, the spacings of the subbands with the 395 Å repeat have been analyzed (Caspar *et al.*, 1969; Parry and Squire, 1973), and an approximate repeat of about 28 Å, which is about 1/14th of the long repeat of 395 Å, was clearly detected. This repeat was evident in optical diffraction patterns of micrographs of the tactoids and in X-ray diffraction patterns of fibers of Mg^{2+} tactoids, both of which give a very strong 14th order meridional reflection (Caspar *et al.*, 1969) of spacing 28 Å. Recently this repeat has been investigated thoroughly by analysis of the amino acid sequence of tropomyosin, and it will therefore be discussed further in Section 5.3.3.

FIGURE 5.12. Tropomyosin tactoids prepared by precipitation in the presence of different divalent cations (× 100,000; reduced 30% for reproduction). All are negatively stained. (a) Tactoid prepared using magnesium and showing a nonpolar structure with a 395-Å repeat and two major subbands of width about 150 and 250 Å. Details of this kind of paracrystal are given in Fig. 5.19. (b) Tactoids prepared with lead (right) and magnesium (left). The lead tactoid is polar and indicates that the two chains in the tropomyosin are parallel (if they are identical). (c) A lead tactoid as in (b). (d) Two tactoids formed using barium but having quite distinct structures. The tactoid on the left is similar to the magnesium tactoids (a), but that on the right has a period that is slightly larger (by 2.5%) and contains two subbands about 200 Å wide. (From Caspar *et al.*, 1969.)

Self-convolution analysis of the Mg^{2+} tactoids (Parry and Squire, 1973) has shown that a lateral periodicity of about 38 Å occurs and is ordered for at least 500 Å across the tactoid. This repeat was taken to indicate a pairing of the antiparallel tropomyosin molecules in the tactoid. A similar repeat was previously reported to occur in paramyosin aggregates (A. Elliott *et al.*, 1968b) in which a two-molecule unit cell was known to occur (A. Elliott *et al.*, 1968a; Fig. 4.14). Separate analysis of the 250-Å and 150-Å bands in the 395-Å tactoid repeat showed that the lateral order was quite good in the shorter band and rather poor in the 250-Å band (Parry and Squire, 1973).

Crystals. Crystals of tropomyosin have been studied extensively in the electron microscope (Caspar *et al.*, 1969; Fujime-Higashi and Ooi, 1969; Cohen *et al.*, 1971b, 1972) and by X-ray diffraction (Caspar *et al.*, 1969; Cohen *et al.*, 1971b, 1972; Phillips *et al.*, 1979). In the electron microscope many different netlike tropomyosin aggregates have been reported (Fig. 5.13). One form was a simple square net of side 400 Å (Fig. 5.13b, right; Caspar *et al.*, 1969), but more commonly nets are formed in which successive crossover points are separated alternately by about 220 Å and 175 Å (Figs. 5.13c,d,e and 5.13a center). Many different net forms show these two intercrossover separations (Caspar *et al.*, 1969), but one structure in particular (Fig. 5.13e) is similar to the structure of the large single crystals of tropomyosin that have been studied by X-ray diffraction. A combined study of such crystals by these two methods has in fact helped to determine the crystal structure. In the electron microscope, each mesh in the crystal net (Fig. 5.13e and see also Fig. 5.23a) has the shape of a kite (Fig. 5.14a). The rectangular unit cell in the structure is very variable and has dimensions of about $b = 235$ Å, and $c = 292$ Å, with the c/b ratio varying between about 1.1 and 1.3 in different regions. In each case the diagonals (along which the molecules run in this projection) always have the same length of about 380 Å. Figure 5.14a also shows the interpretation by Caspar *et al.* (1969) of this net pattern in terms of pairs of strands of tropomyosin molecules which interact head to tail to produce each strand. The strands of molecules are also curved, and, since the molecules are themselves coiled coils, the net structure is an assembly of slightly coiled coiled-coil molecules.

X-ray diffraction analysis of crystals of tropomyosin has shown that the structure can be described in terms of an orthorhombic unit cell (like a rectangular box: Fig. 5.14b) with unit cell sides of about $a = 122$ to 127 Å, $b = 230$ to 250 Å, and $c = 286$ to 300 Å (Caspar *et al.*, 1969; Cohen *et al.*, 1971b).* The variability is a genuine reflection of different unit cell parameters in different crystals. In each case, the unit cell parameters

*For crystallographers, the space group is $P2_12_12$; see Phillips *et al.* (1979).

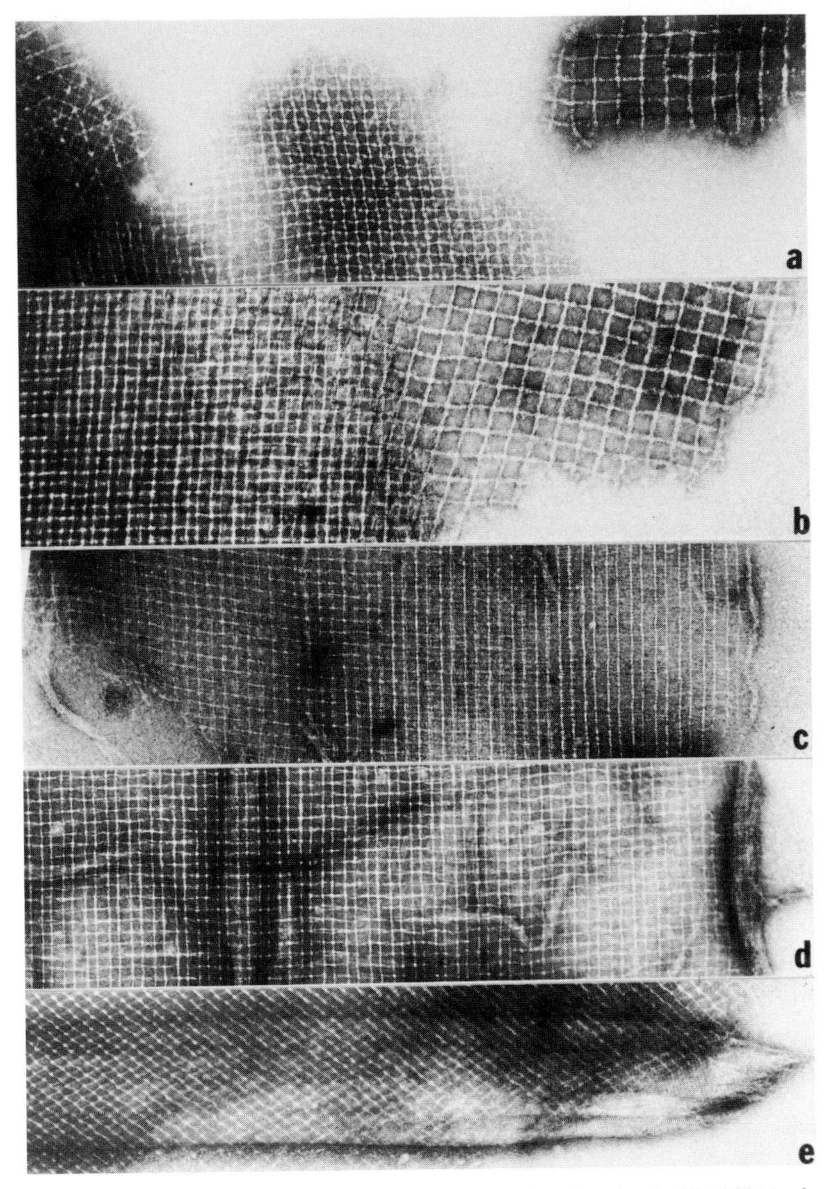

FIGURE 5.13. Tropomyosin nets formed near the isoelectric point (× 100,000) (reduced 25% for reproduction). (a) Upper right, 400-Å square net; center, crystal net with kite-shaped repeating unit; left, 400-Å square net superimposed on quartered square net. (b) 400-Å square net and quartered square nets as in (a). (c) Transition from a square mesh to an array of thick strands with long–short separations connected by thin wavy strands. (d) Long–short spacing rectangular net. (e) Short spacing tactoid together with the crystal net as in (a). (All from Caspar *et al.*, 1969.)

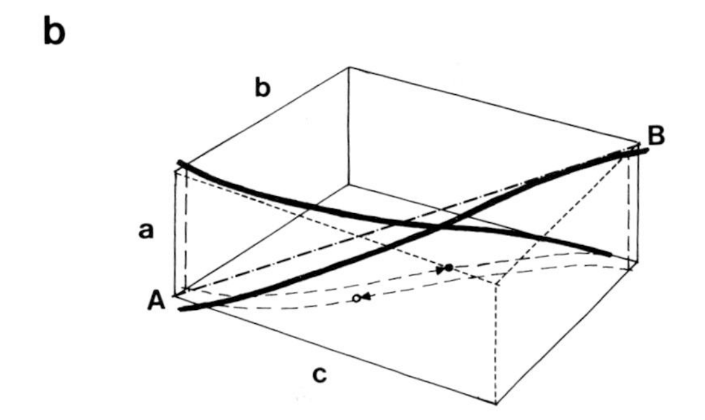

FIGURE 5.14. (a) The unit cell and tropomyosin locations in the kite-shaped crystal nets shown in Fig. 5.13a,e according to Cohen (1975). (b) The unit cell in (a) drawn in three dimensions to indicate how the tropomyosin strands follow approximately the body diagonals of the unit cell (the line AB). The dashed lines on the **bc** plane in (b) show the projections of the tropomyosin strands as seen in (a). Only two of the several tropomyosin strands in the unit cell are shown. (After Cohen, 1975.)

were defined within 0.4%, but as a and c increased, b was found to decrease in such a way that the body diagonal (AB in Fig, 5.14b) remained constant at almost exacaly 402 Å (Cohen *et al.,* 1971b). This indicated that the molecular strands of tropomyosin run along the body diagonals as indicated in Fig. 5.14, and it gives a very precise minimum value of 402 Å for the length of a tropomyosin molecule. If allowance is made for the bending of the strands in the unit cell, then the length of an undistorted molecule as measured in three-dimensional electron density maps (calculated from the X-ray diffraction data from crystals) is 410.3 ± 1.4 Å (Phillips *et al.,* 1979).

5.3.3. Amino Acid Sequence and Structure of Tropomyosin

Introduction—The Coiled-Coil Structure. Recent work by Smillie and his collaborators has culminated in the determination of the entire amino acid sequence of the α chain of the tropomyosin in rabbit skeletal muscle (Stone *et al.,* 1974). The sequence shown in Fig. 5.15 consists of 284 residues, giving a calculated chain molecular weight of 32,758.

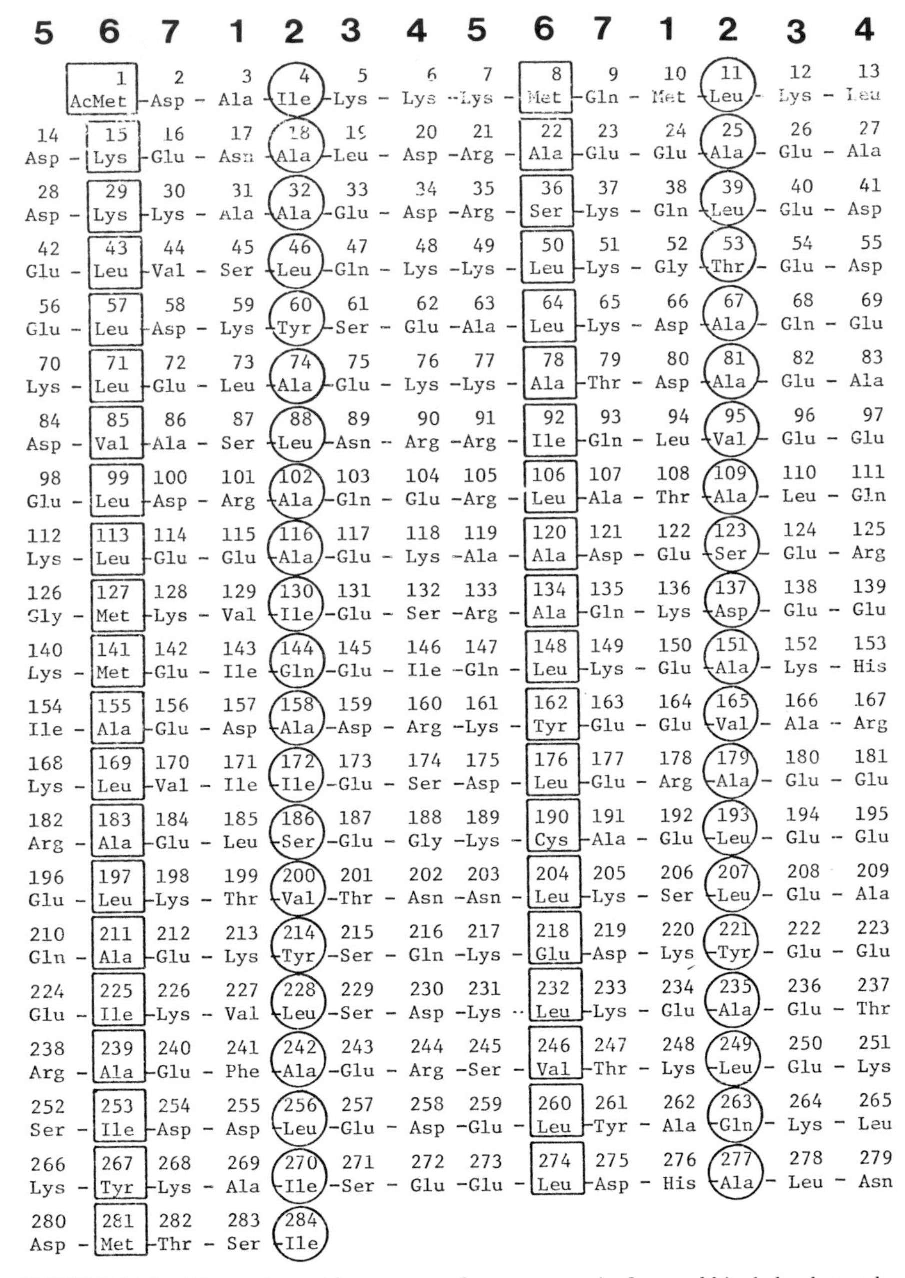

FIGURE 5.15. The amino acid sequence of α-tropomyosin from rabbit skeletal muscle according to Stone *et al.* (1974). The residues have been grouped in sevens (see numbering at top of figure) to show the repeating heptad structure in which hydrophobic residues occur systematically in positions 2 and 6.

It was mentioned in Chapter 4 (Section 4.3) that one of Crick's predictions about a protein with the two-chain coiled-coil structure (Crick, 1953a,b) was that the interaction between the two chains might well be stabilized by hydrophobic forces. A periodic arrangement of hydrophobic residues might then be expected along the line of contact between the two chains (Fig. 4.9). The amino acid sequence of tropomyosin provides striking evidence for the validity of this idea. In the sequence given in Fig. 5.15, the residues have been grouped in two heptads (sevens) to represent the seven-residue repeating unit in a coiled coil. It can be seen that, almost without exception, positions 2 and 6 in each heptad are consistently occupied by nonpolar residues (i.e., Ala, Val, Leu, Ile, Pro, Phe, Met). This remarkable result clearly provides very strong support for the ideas behind the coiled-coil structure. The way in which two tropomyosin chains would fit together over a short region to give knob-into-hole packing (Figure 4.9) and good hydrophobic interactions is illustrated in Fig. 5.16 (Hodges *et al.,* 1972; Stone *et al.,* 1974).

In addition to the predominance of apolar residues in positions 2 and 6, it was also found that many of the other positions in the heptad were largely filled by residues of a particular type (Stone *et al.,* 1974; Parry, 1974, 1975a). For example, position 3 is more acidic and position 5 is more basic than would be expected on average. It was pointed out by Miller (quoted in Parry, 1975a,b) that these two positions will act as acidic and basic stripes alongside the two apolar stripes. This might be the cause of the parallel aggregation of the two α-helical tropomyosin chains, since an antiparallel arrangement would lead to positive–positive and negative–negative interactions between the stripes on the two chains, and this would tend to destabilize the structure.

Functional Periodicities in the Tropomyosin Sequence. Because of the possibility mentioned earlier that tropomyosin molecules might lie along the grooves of the F-actin helix and interact with seven (or possibly 14) actin molecules, the published amino acid sequence of tropomyosin was rapidly analyzed by several researchers to see if a corresponding repeat could be detected in the amino acid sequence. Parry and Squire (1973) had already suggested that the observed 28-Å repeat in the Mg^{2+} tactoid could represent some form of pseudorepeat of the sequence along the molecule and that this repeat might coincide with the axial repeat between actin molecules along the thin filament.

Early inspection of the sequence showed that there was certainly no exact sequence repeat in tropomyosin. However, detailed sequence analyses by Miller (1975), Parry (1974, 1975a), Stone *et al.* (1974), Stewart and McLachlan (1975), and McLachlan *et al.* (1975) showed that a very significant pseudorepeat does exist (as anticipated by Cohen *et al.,*

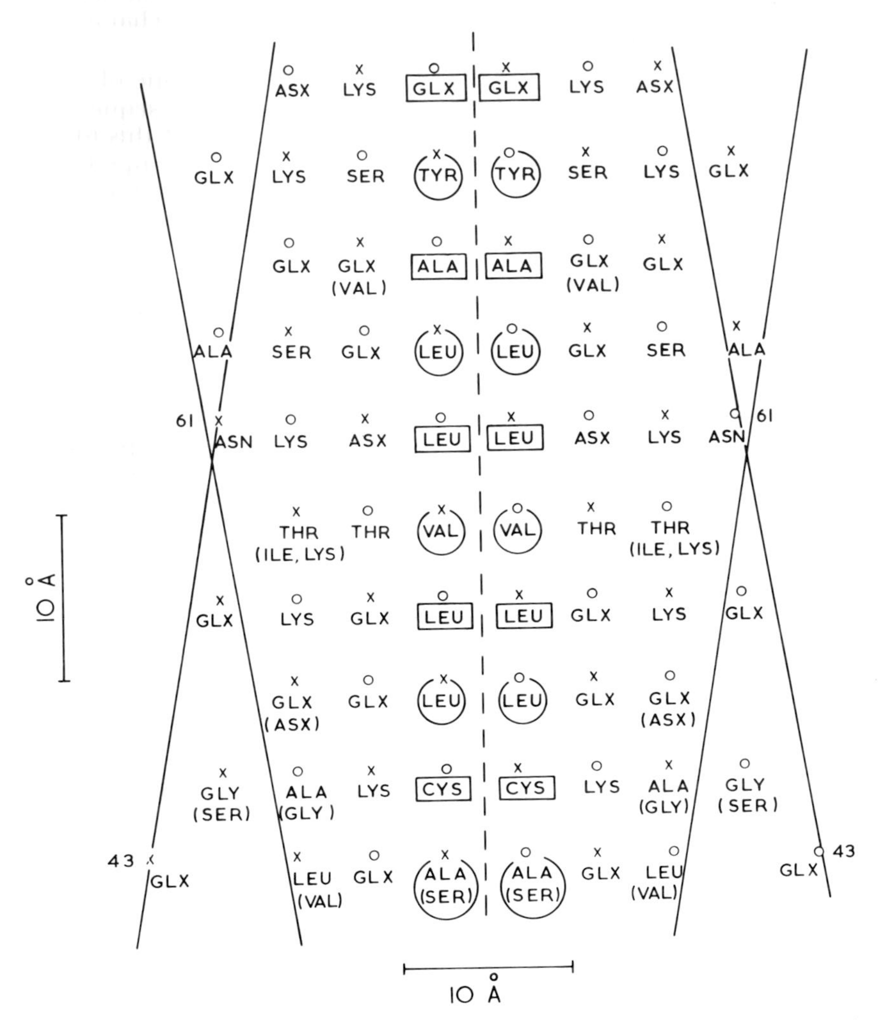

FIGURE 5.16. The fit of two α-helical tropomyosin chains so that knob-into-hole packing occurs between the hydrophobic residues in positions 2 and 6 on the two chains (see Fig. 5.15). This can be compared with the theoretical coiled-coil structure in Fig. 4.9. (From Hodges *et al.*, 1972.) Amino acids marked ○ belong to one chain; those marked × to the other.

1971b and Parry and Squire, 1973), and it has a period of about 19.5 residues. This could be detected in the distribution of acidic residues, basic residues, and apolar residues (Fig. 5.17). Parry (1975a) also showed that a significant repeat about every 40 residues occurred both in the distribution of glutamic acid residues and in the distribution of the α-helix factors (see Table 4.4) of all of the amino acids along the chain.

Taking a broader view of the sequence, one can describe the

tropomyosin molecule (Stone *et al.*, 1974) as a set of bands about eight residues wide, which are predominantly basic and apolar, alternating with bands about 12 residues wide which (apart from positions 2 and 6) are more acidic than average. The molecule, therefore, consists of about 14 pseudoperiods of alternate acidic and basic–apolar bands superimposed on which is a subtler periodicity of twice the spacing dividing the molecule into seven pseudorepeats. It was suggested by McLachlan and Stewart (1976) that this latter repeat may be a result of gene duplication during the evolution of tropomyosin.

Stagger between Chains. The amino acid sequence has also been analyzed to find out if the two chains in the tropomyosin molecule are in register or are staggered axially by an integral multiple of the seven-residue coiled-coil repeat. Early analysis suggested that a shift of seven or 14 residues might well occur (Stone *et al.*, 1974; Parry, 1974). However, a more direct approach in which it was shown that disulfide cross-links could be produced between the two chains (Johnson and Smillie, 1975; Stewart, 1975) clearly indicated that the chains must be in axial register. Further analysis of the tropomyosin sequence has tended to support this conclusion (McLachlan and Stewart, 1975).

One important feature of the unstaggered (symmetrical) model of tropomyosin is that the alternating basic and acidic bands in each component chain will lie side by side and will therefore give the whole molecule a similar banded structure. This aspect of the structure of tropomyosin is clearly related to its interaction with actin and will be discussed further in Section 5.5.4.

End-to-End Interaction between Molecules. One of the original reasons for suggesting that the two chains in a tropomyosin molecule might not be in axial register was that it was thought likely that the observed end-to-end interaction between molecules could be stabilized by the interaction between the overlapping single-chain ends of successive tropomyosin molecules. However, since then it has been shown both that the chains are in register and that very strong interactions could occur if the last nine residues in both chains in one tropomyosin molecule overlap the first nine residues in both chains of the next molecule (McLachlan and Stewart, 1975). This arrangement is illustrated schematically in Fig. 5.18. On this basis, the overlap zone would be about 13.5 Å long. Support for this conclusion is the fact that an α-helical molecule containing 284 residues would be expected to be about 423 Å long (i.e., 284 $\times$ 1.49 Å), but the results from tropomyosin crystals (Cohen *et al.*, 1971a) gave a molecular length of 410 ($\pm$4) Å. An overlap between molecular ends of 13.5 Å would account for this discrepancy.

Analysis of Magnesium Paracrystals. Stewart and McLachlan (1976) have used the published sequence of α-tropomyosin to analyze

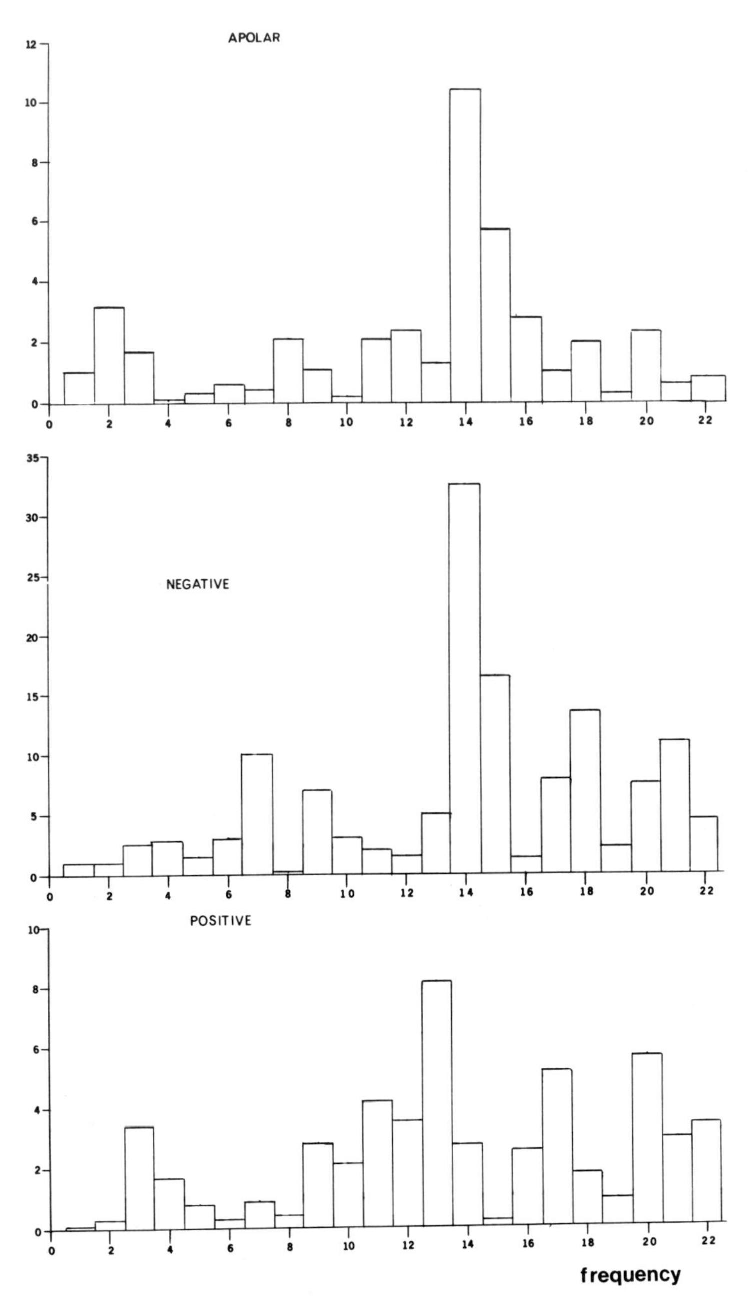

FIGURE 5.17. Analysis of repeats in the tropomyosin sequence shown in Fig. 5.15 (Parry, 1975a). The graphs are Fourier transforms of the positions of amino acids of different character. The top graph shows the positions of apolar residues not in positions 2

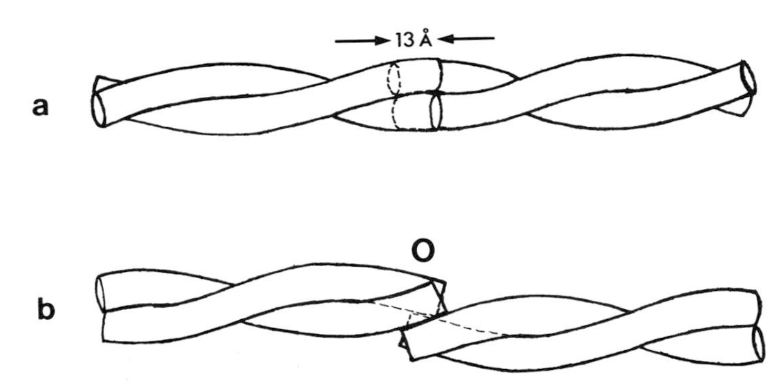

FIGURE 5.18. Illustrations of the possible form of the end-to-end overlap between successive tropomyosin molecules along a single tropomyosin strand. Note that Phillips *et al.* (1979) have suggested from their electron density maps of tropomyosin crystals that the overlap may be much less regularly α-helical, perhaps being much more globular in nature.

the structure of positively stained magnesium paracrystals of tropomyosin. In their analysis they assumed that sain would only bind to the acidic residues along the tropomyosin chains and that it should, therefore, be possible to relate the appearance of the positively stained paracrystals to the known distribution of acidic residues in each molecule. Figure 5.19 shows the form of the dihedral paracrystals (cf. Fig. 5.12a) together with the probable molecular arrangement and an average density profile across the paracrystal. By altering the amount of molecular overlap in the C-C overlap region, McLachlan and Stewart were able to show that the best agreement between the observed density profile and the profile computed from the distribution of acidic residues occurred with an overlap of about 173 to 177 residues. Similar analysis of the N-N overlap region was less conclusive but seemed to be consistent with end-to-end overlaps between tropomyosin molecules from −2 to +13 residues. However, in no instance could the two deep troughs in the trace in Fig. 5.19 at the ends of the N-N overlap region be reproduced. Clearly, some additional feature, as yet unidentified, must be involved in the staining pattern.

---

and 6. The middle graph shows the positions of negatively charged residues (aspartic and glutamic acids) and the lower graph shows the distribution of positively charged residues (lysine and arginine). The top two transforms show prominent peaks at about the 14th or 15th order of 284 residues, corresponding to a linear periodicity of 19.6 (±0.7) residues. The distribution of negative residues also shows a small peak corresponding to a repeat of 40.6 (±2.7) residues or about one-seventh of the tropomyosin length. A similar peak (not shown) occurs in the distribution of α-helix factors (see Table 4.3). There is no repeat of the same significance in the distribution of positive residues. The vertical scale is in arbitrary units.

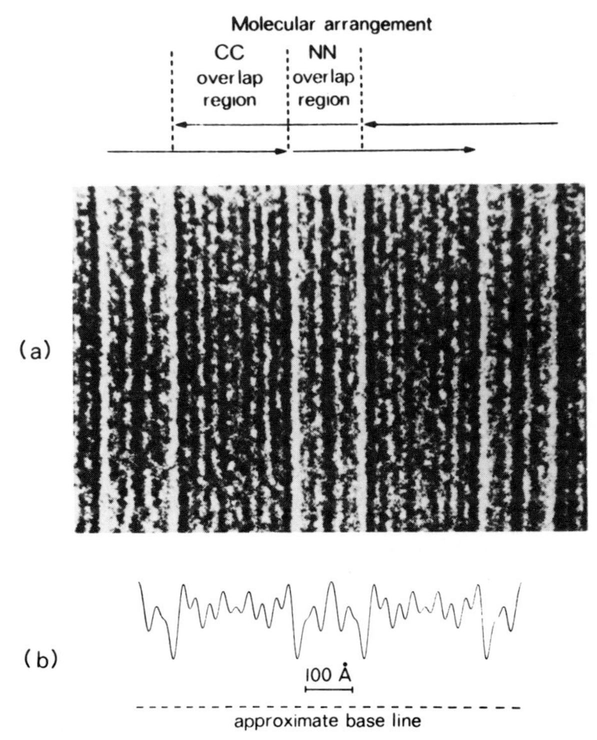

FIGURE 5.19. (a) Segment of a magnesium paracrystal of α-tropomyosin positively stained with uranyl acetate. (b) Average filtered axially projected density trace of the structure in (a). The averaging process is analogous to that in Fig. 3.15. Note the characteristic pattern of 14 bands in each repeat, 9 in the C–C overlap region and 5 in the N–N overlap region. The probable arrangement of molecules in the paracrystal is shown above (a) with the arrowhead at the C terminus. As shown in Fig. 5.18, there may be a small overlap between the C terminus of one molecule and the N terminus of the next. (From Stewart and McLachlan, 1976.)

5.3.4. Structure of Actin Filaments Containing Tropomyosin

For many years the working model of thin filament structure was one in which the tropomyosin molecules were located in the grooves of the F-actin helix as indicated in Fig. 5.4. However, no direct evidence for this was obtained until the early 1970s when it was found that new X-ray diffraction evidence from relaxed and contracting muscles (H. E. Huxley, 1971a,b; Vibert *et al.*, 1972) could only be explained by models with the tropomyosin located in this position (H. E. Huxley, 1971b; Haselgrove, 1972; Parry and Squire, 1973). O'Brien *et al.* (1971) came to the same conclusion using electron microscopy. Since that time, the 3-D reconstruction work of Spudich *et al.* (1972), Wakabayashi *et al.* (1975), and O'Brien and Couch (1976) on F-actin filaments containing tropomyosin has confirmed the model. But the full significance of the results obtained is not apparent until the role of troponin in the thin filament is taken into account. These results will therefore be described in detail in Section 5.4.3 of this chapter after a discussion has been given of the structure and properties of troponin.

5.4. Troponin

5.4.1. Components of the Troponin Complex

As mentioned earlier, troponin was originally isolated by Ebashi and Kodama (1966) from "native tropomyosin." Troponin and tropomyosin can be separated by using isoelectric precipitation or hydroxyapatite columns (Ebashi and Endo, 1968; Ebashi *et al.*, 1968; Eisenberg and Kielley, 1973). The troponin molecule so obtained (of molecular weight about 80,000) was shown by Ebashi and Kodama (1966) to confer Ca^{2+} sensitivity to the actomyosin ATPase and to have a high affinity for Ca^{2+}ions above a certain threshold Ca^{2+} concentration (about 10^{-6}M). Troponin was therefore strongly implicated as an important part of the muscle regulatory process.

Not long after its discovery, it was shown by Hartshorne and Mueller (1968) that the 80,000-dalton troponin molecule contained subunits with quite different properties. Since that time, work in many laboratories has involved the preparation, purification, and characterization of these troponin subunits (see review by Weber and Murray, 1973). Initially there was uncertainty about the number of subunits involved, but by about 1972, it was generally agreed that there are three, and their molecular weights in rabbit muscle were thought at the time to be about 18,000, 23,000 and 37,000 to 39,000 (Greaser and Gergely, 1971; Spudich and Watt, 1971; Wilkinson *et al.*, 1972). These could be separated from troponin by chromatography on Sephadex gels in the presence of urea (Greaser and Gergely, 1971; and see detailed references in Weber and Murray, 1973).

The three troponin subunits were found to have quite distinct properties: the 18,000-dalton protein was found to bind Ca^{2+}ions, the 23,000-dalton protein was found to have the ability to inhibit on its own the actomyosin interaction, and the 37,000- to 39,000-dalton protein was found to bind strongly to tropomyosin. Because of these properties, it is now normal to refer to these subunits as Troponin-C (TN-C), Troponin-I (TN-I), and Troponin-T (TN-T), respectively. This terminology was especially useful since it was found that equivalent proteins from different muscles could have significantly different molecular weights. For example, TN-T from chicken muscle is reported to have a molecular weight of about 45,000 (Greaser *et al.,* 1972; Hartshorne and Dreizen 1972; Perry *et al.,* 1972).

A considerable amount of work has been carried out on the properties and interactions of these subunits, and these have been admirably reviewed by Weber and Murray (1973). The general conclusions of this work are summarized briefly below.

Troponin-I has a sequence molecular weight of 20,900 (Wilkinson and Grand, 1975), and its amino acid composition and sequence are as shown in Table 5.3 and Fig. 5.20, respectively. It binds to F-actin and to F-actin plus tropomyosin (Potter and Gergely, 1974; Hitchcock, 1975b), and it affects actin in the absence of the other subunits by inhibiting the actin–myosin interaction. Despite the fact that TN-I and tropomyosin do not interact strongly, when tropomyosin is present the TN-I molecule will bind more strongly to actin and will inhibit between four and seven actin monomers. But when it is absent, the same inhibition can only be produced if considerably more TN-I is present. Troponin-I will bind strongly to TN-C and possibly also to TN-T (see below).

Troponin-C, which reversibly binds Ca^{2+} ions as a function of Ca^{2+} concentration, can remove the inhibitory effect of TN-I in the full thin filament assembly [actin (A) + tropomyosin (TM) + troponin (TN)] provided that the [Ca^{2+}] is high. But strangely, it can also reverse the inhibitory effect of TN-I on pure actin even at low concentrations of Ca^{2+} ions. Troponin-C has a high proportion of acidic residues (Table 5.3), and its amino acid sequence has been determined (Fig. 5.20; Collins *et al.,* 1973; Collins, 1974), giving it a calculated (sequence) molecular weight of 17,800. As mentioned above, the evidence that TN-C binds to actin and to TN-I is that it reverses the inhibitory effect of TN-I. But the actual mechanism of this reversal is unknown. TN-C-to-TN-I binding is strongly implicated; binding of TN-C to actin is less certain. Apart from this there is good evidence that TN-C can bind to TN-T (Cohen *et al.,* 1972; see below) but little evidence for binding of TN-C to tropomyosin. Ca^{2+}-binding studies on isolated TN-C have suggested that one TN-C molecule can bind two Ca^{2+} ions (Greaser *et al.,* 1972; Hartshorne and Pyun, 1971).

Troponin-T, of sequence molecular weight 30,500 in rabbit (Pearlstone *et al.,* 1976), is the component of the troponin complex that binds strongly to tropomyosin. It also binds to TN-C. It contains a relatively high proportion of basic residues (Table 5.3) and is rather insoluble under physiological conditions. Its amino acid sequence has now been determined (Fig. 5.20). On its own, TN-T has no effect on the actomyosin ATPase, and its precise role, apart from providing

TROPONIN-T

```
      1                                         10                                        20
Ac-  Ser-Asn-Glu-Glu-Val-Glu-His-Val-Glu-Glu-Glu-Ala-Glu-Glu-Glu-Ala-Pro-Ser-Pro-Ala-
     Glu-Val-His-Glu-Pro-Ala-Pro-Glu-His-Val-Val-Pro-Glu-Glu-Val-His-Glu-Glu-Glu-Lys-
 41  Pro-Arg-Lys-Leu-Thr-Ala-Pro-Lys-Ile-Pro-Glu-Gly-Glu-Lys-Val-Asp-Phe-Asp-Asp-Ile-
 61  Gln-Lys-Lys-Arg-Gln-Asn-Lys-Asp-Leu-Met-Glu-Leu-Gln-Ala-Leu-Ile-Asp-Ser-His-Phe-
 81  Glu-Ala-Arg-Lys-Lys-Glu-Glu-Glu-Glu-Leu-Val-Ala-Leu-Lys-Glu-Arg-Ile-Glu-Lys-Arg-
101  Arg-Ala-Glu-Arg-Ala-Glu-Gln-Gln-Arg-Ile-Arg-Ala-Glu-Lys-Glu-Arg-Glu-Arg-Gln-Asn-
121  Arg-Leu-Ala-Glu-Glu-Lys-Ala-Arg-Arg-Glu-Glu-Glu-Asp-Ala-Lys-Arg-Arg-Ala-Glu-Glu-
141  Asp-Leu-Lys-Lys-Lys-Lys-Ala-Leu-Ser-Ser-Met-Gly-Ala-Asn-Tyr-Ser-Ser-Tyr-Leu-Ala-
161  Lys-Ala-Asp-Gln-Lys-Arg-Gly-Lys-Lys-Gln-Thr-Ala-Arg-Glu-Met-Lys-Lys-Lys-Ile-Leu-
181  Ala-Glu-Arg-Arg-Lys-Pro-Leu-Asn-Ile-Asp-His-Leu-Ser-Asp-Glu-Lys-Leu-Arg-Asp-Lys-
201  Ala-Lys-Glu-Leu-Trp-Asp-Thr-Leu-Tyr-Gln-Leu-Glu-Thr-Asp-Lys-Phe-Glu-Phe-Gly-Glu-
221  Lys-Leu-Lys-Arg-Gln-Lys-Tyr-Asp-Ile-Met-Asn-Val-Arg-Ala-Arg-Val-Glu-Met-Leu-Ala-
241  Lys-Phe-Ser-Lys-Lys-Ala-Gly-Thr-Thr-Ala-Lys-Gly-Lys-Val-Gly-Gly-Arg-Trp-Lys
                                                                                  (259)
```

TROPONIN-I

```
      1                                         10                                        20
Ac-  Gly-Asp-Glu-Glu-Lys-Arg-Asn-Arg-Ala-Ile-Thr-Ala-Arg-Arg-Gln-His-Leu-Lys-Ser-Val-
     Met-Leu-Gln-Ile-Ala-Ala-Thr-Glu-Leu-Glu-Lys-Glu-Glu-Gly-Arg-Arg-Glu-Ala-Glu-Lys-
 41  Gln-Asn-Tyr-Leu-Ala-Glu-His-Cys-Pro-Pro-Leu-Ser-Leu-Pro-Gly-Ser-Met-Ala-Glu-Val-
 61  Gln-Glu-Leu-Cys-Lys-Gln-Leu-His-Ala-Lys-Ile-Asp-Ala-Ala-Glu-Glu-Glu-Lys-Tyr-Asp-
 81  Met-Glu-Ile-Lys-Val-Gln-Lys-Ser-Ser-Lys-Glu-Leu-Glu-Asp-Met-Asn-Gln-Lys-Leu-Phe-
101  Asp-Leu-Arg-Gly-Lys-Phe-Lys-Arg-Pro-Pro-Leu-Arg-Arg-Arg-Val-Arg-Met-Ser-Ala-Asp-
121  Ala-Met-Leu-Lys-Ala-Leu-Leu-Gly-Ser-Lys-His-Lys-Val-Cys-Met-Asp-Leu-Arg-Ala-Asn-
141  Leu-Lys-Gln-Val-Lys-Lys-Glu-Asp-Thr-Glu-Lys-Glu-Arg-Asp-Val-Gly-Asp-Trp-Arg-Lys-
161  Asn-Ile-Glu-Glu-Lys-Ser-Gly-Met-Glu-Gly-Arg-Lys-Lys-Met-Phe-Glu-Ser-Glu-Ser
                                                                                  (179)
```

TROPONIN-C

```
      1                                         10                                        20
     Asp-Thr-Gln-Gln-Ala-Glu-Ala-Arg-Ser-Tyr-Leu-Ser-Glu-Glu-Met-Ile-Ala-Glu-Phe-Lys-
 21  Ala-Ala-Phe-Asp-Met-Phe-Asp-Ala-Asp-Gly-Gly-Gly-Asp-Ile-Ser-Val-Lys-Glu-Leu-Gly-
 41  Thr-Val-Met-Arg-Met-Leu-Gly-Gln-Thr-Pro-Thr-Lys-Glu-Glu-Leu-Asp-Ala-Ile-Ile-Glu-
 61  Glu-Val-Asp-Glu-Asp-Gly-Ser-Gly-Thr-Ile-Asp-Phe-Glu-Glu-Phe-Leu-Val-Met-Met-Val-
 81  Arg-Gln-Met-Lys-Glu-Asp-Ala-Lys-Gly-Lys-Ser-Glu-Glu-Glu-Leu-Ala-Glu-Cys-Phe-Arg-
101  Ile-Phe-Asp-Arg-Asn-Ala-Asp-Gly-Tyr-Ile-Asp-Ala-Glu-Glu-Leu-Ala-Glu-Ile-Phe-Arg-
121  Ala-Ser-Gly-Glu-His-Val-Thr-Asp-Glu-Glu-Ile-Glu-Ser-Leu-Met-Lys-Asp-Gly-Asp-Lys-
141  Asn-Asn-Asp-Gly-Arg-Ile-Asp-Phe-Asp-Glu-Phe-Leu-Lys-Met-Met-Glu-Gly-Val-Gln
                                                                                  (159)
```

FIGURE 5.20. The amino acid sequences in the three troponin subunits of rabbit skeletal muscle. (a) The sequence in troponin-T as determined by Pearlstone *et al.* (1976). There are 259 amino acid residues, giving a total molecular weight of 30,503. The regions from residues 80 to 102 and from 122 to 146 are very likely α-helical, and they are implicated in binding to tropomyosin. (b) The sequence of troponin-I as determined by Wilkinson and Grand (1975). Two highly basic regions occur. In the sequence from residues 102 to 135, there is only a single acidic residue, and a second basic region is from 5 to 27. No sequence homology has been reported between this and the light chains of the myosin molecule. (c) The sequence of troponin-C as determined by Collins (1974). This sequence is reported to have close homology with the sequences of two of the light chains of myosin (the alkali light chains—see Chapter 6) and also with the calcium-binding protein parvalbumin from fish. This clearly suggests that they have a common ancestor.

additional linking to the thin filament of TN-C and indirectly of TN-I, is not completely clear, although there is evidence that it has a role in increasing the Ca^{2+} sensitivity of the TN complex.

5.4.2. Properties of the Whole Troponin Complex

These various properties of the TN components can be summarized diagrammatically as in Fig. 5.21 (after Weber and Murray, 1973) on the assumption that there is one of each subunit in the troponin complex. The molecular weight of the complex (80,000; Hartshorne and Dreizen, 1972) is consistent with this assumption (17,800 + 20,900 + 30,500 = 69,200). The whole troponin complex is rather globular, and ORD measurements (and others) suggest that the α-helix content is fairly low (45%; Murray and Kay, 1972). From the properties of its components, the complex would be expected to bind both to actin and tropomyosin. Clearly, it will also bind Ca^{2+} ions; actually, it appears to bind four Ca^{2+} ions per molecule (Potter and Gergely, 1975), there being two high-affinity sites which also bind Mg^{2+} and two low-affinity sites which do not bind Mg^{2+} but which do appear to be the sites involved in regulation.

The most uncertain part of the interaction scheme in Fig. 5.21 is the TN-I-to-TN-T interaction. Only recently has binding between these two components been demonstrated (Horowitz *et al.*, 1979). However, by carrying out cross-linking experiments, Hitchcock (1975a) found that these two components are rather close to each other (6 Å or less), suggesting that the troponin complex does have the compact form indi-

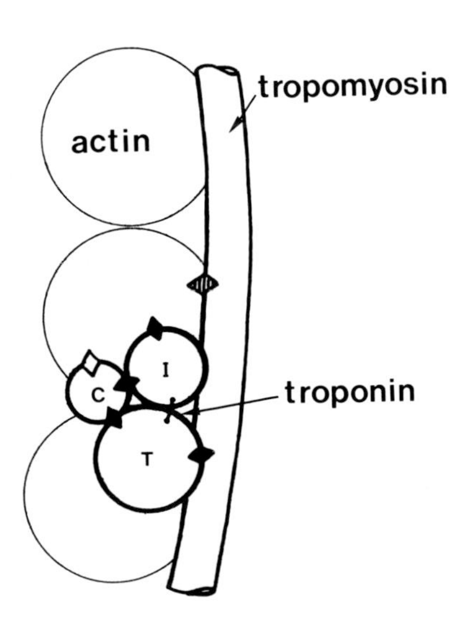

FIGURE 5.21. Schematic diagram showing the interactions between the troponin subunits, tropomyosin, and actin. I, C, and T are the TN-I, TN-C, and TN-T subunits of troponin. The solid diamonds indicate strong interactions, the open diamond indicates a possible interaction, and the link between T and I denotes close proximity but not necessarily a significant interaction. (For details see text.)

cated in Fig. 5.21 rather than the extended form TN-I : TN-C : TN-T : TM.

With regard to the Ca^{2+} sensitivity of the whole TN complex, it has been shown that the affinity of TN for actin is less in the presence of Ca^{2+} (Hitchcock, 1973). A further result of interest is that troponin-I + C will not bind to tropomyosin or actin alone but will bind to actin and tropomyosin together in the absence of Ca^{2+} (Hitchcock, 1975b). This suggests that tropomyosin in some way modifies the structure of actin so that TN-I + C can bind to it.

5.4.3. Location of Troponin on Tropomyosin and in the Thin Filament

It has already been mentioned that longitudinal sections of vertebrate skeletal muscle show striations across the I-band with an axial sep-separation of roughly 400 Å. This is thought to be caused by troponin molecules labeling the tropomyosin repeat along the thin filaments. A meridional reflection at 1/385 $Å^{-1}$ is also seen in X-ray diffraction patterns from this muscle type. The original suggestion by H. E. Huxley and Brown (1967) that this might be caused by the same thin filament repeat, has been confirmed by Rome *et al.* (1973b) who labeled glycerinated muscle with antibody to troponin and found that the intensity of the 385-Å meridional reflection was enhanced. The X-ray diffraction results therefore give a relatively accurate value for the thin filament repeat at 385 Å. This spacing is almost exactly equal to seven times the axial repeat (55 Å) between actin monomers in each of the two long-pitch helices of actin. Since the stoichiometry of actin to tropomyosin to troponin in the thin filament is in the ratio 7 : 1 : 1, it is possible that the thin filament may comprise functional units consisting of seven actin monomers combined with one tropomyosin molecule and one troponin (Parry and Squire, 1973; Cohen *et al.*, 1972). This unit is illustrated in Fig. 5.22a; every seventh actin monomer is shown gray.

With this scheme in mind, two questions remain to be answered: (1) where on a tropomyosin molecule does troponin bind? and (2) are the functional units in opposite grooves of the thin filament in axial register or are they axially staggered?

The first of these questions has been answered to some extent by work on tropomyosin crystals and paracrystals that have been labeled with troponin (Nonomura *et al.*, 1968; Cohen *et al.*, 1971a, 1972; Hitchcock *et al.*, 1973; Yamaguchi *et al.*, 1974; Ohtsuki, 1974). Extra material can be seen in the middle of the wide band (i.e., the C–C overlap region in Fig. 5.19) in Mg^{2+} paracrystals. Stewart and McLachlan (1976), from their analysis of the structure of these Mg^{2+} paracrystals, suggested that the troponin binding site on tropomyosin might be near residue 197 in

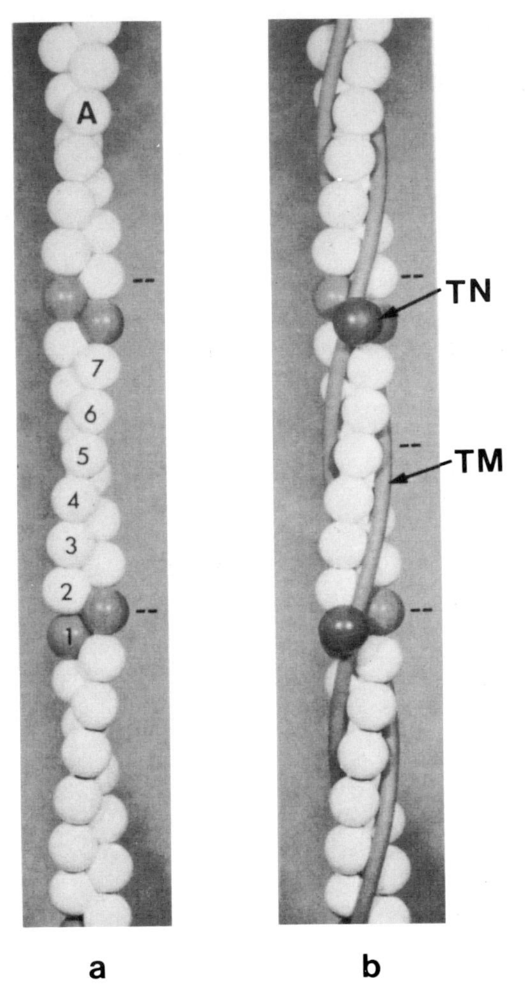

FIGURE 5.22. (a) Model of F-actin showing the seven-actin repeat along each thin filament strand (marked by gray spheres). (b) The complete structure showing the tropomyosin (TM) and troponin (TN) which actually generate the seven-actin functional repeat in the thin filament. Note that the presence of tropomyosin along the grooves of the actin structure in (b) tends to halve the thin filament repeat (see text).

the amino acid sequence (Fig. 5.15), and they noted that in this region the sequence has some unusual features that might be related to TN binding. In a similar way tropomyosin crystals labeled with troponin (Fig. 5.23b) show clear positions of extra density halfway along the long arm of the kite-shaped repeating unit in the crystal (Fig. 5.23a and see Fig. 5.14).

Although the exact axial relationship between the two

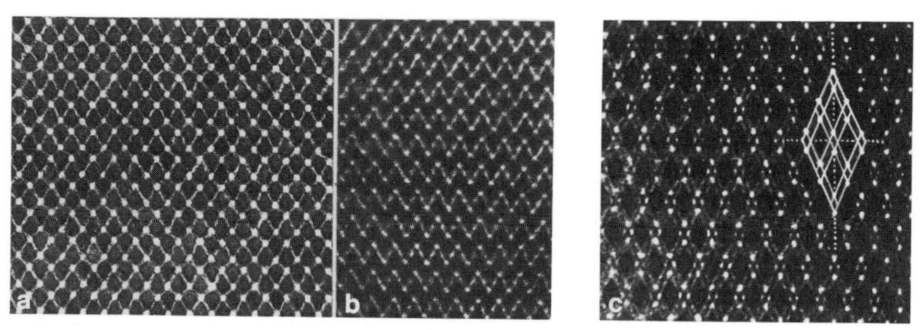

FIGURE 5.23. Electron micrographs of net structures formed by (a) tropomyosin, (b) tropomyosin plus troponin, and (c) tropomyosin plus troponin-T. All are negatively stained and × 200,000 (reduced 45% for reproduction). (a) The crystal has a kite-shaped repeating unit as illustrated in Fig. 5.14. (b) The same type of crystal net, but with troponin labeling halfway along the long edges of the kite. (Both are from Cohen *et al.*, 1971b.) (c) The so-called double diamond net in which dense white blobs at the acute vertices of the diamonds probably represent the location of TN-T dimers linking two tropomyosin molecules. (From Cohen *et al.*, 1972.)

tropomyosin–troponin strands in opposite grooves of a thin filament has not been determined exactly, it is commonly assumed that they are almost in register. A relative axial stagger of about 27 Å would allow each strand to make equivalent interactions with actin and would keep the two strands very nearly in register. A large axial stagger seems to be ruled out by the fact that the 400-Å-spaced stripes in the I-band are very narrow, as are the stripes seen in actin paracrystals containing troponin. Very recent X-ray diffraction evidence (Wray *et al.*, 1978; Maeda, 1978, 1979; Namba *et al.*, 1978) on the intensities of those parts of the thin filament diffraction patterns from crustacean muscles thought to be caused by troponin has confirmed this suggestion. In thin filaments with a 385-Å crossover repeat, the troponin molecules will lie on one of two coaxial 2/1 helices of pitch 2 × 385 Å. The intensities in the diffraction pattern from such a structure clearly depend on the relative axial stagger between the two helices, and it was found that the best agreement with experiment was obtained with the 27-Å axial stagger.

Note, finally, that Cohen *et al.* (1972) and Margossian and Cohen (1973) have studied the interaction of some of the separated troponin subunits with tropomyosin paracrystals. One particular result of considerable interest is that coprecipitation of tropomyosin and TN-T caused the formation of a new type of paracrystal which was termed the double diamond net (Fig. 5.23c). The repeating unit in this net was quite different from the normal kite-shaped unit cell in crystals of TM alone (Fig. 5.23a). Instead of crossover points spaced at 230 Å and 170 Å, the crossovers occurred at intervals of about 100 Å and 300 Å. The relative brightness of the spots at the crossover points located at the acute vertices of the diamond indicated that this must be the location of TN-T and that TN-T itself might be forming a dimer linkage between the crossing tropomyosin strands. It was suggested that such TN-T-to-TN-T dimer formation might normally be blocked by the presence of the other subunits in the complete TN complex.

Margossian and Cohen (1973) used the formation of the double diamond net as a tool by which to evaluate the interactions between TN-T and either TN-C or TN-I. In this way, they confirmed that TN-C binds directly to TN-T and that Ca^{2+} binding to TN-C strengthens this linkage (Ebashi *et al.*, 1972). Similarly, they showed that under certain conditions TN-T and TN-I may bind, although such interaction was difficult to demonstrate directly. In addition, they showed that TN-I + C binding to tropomyosin was weak, since tropomyosin selectively precipitates from a solution containing TM and the TN-I + C complex.

5.5. Thin Filament Structure and Regulation

5.5.1. X-Ray Diffraction Evidence for Changes in Thin Filament Structure during Regulation

Once tropomyosin and troponin had been identified as the regulating components of the thin filament, it was the turn of X-ray diffractionists to throw further light on the regulation process. X-ray diffraction experiments were carried out in the hope that the structural changes in the thin filament that are associated with regulation might be sufficient to give rise to characteristic changes in the observed diffraction patterns. As it turned out, these hopes were fulfilled, and significant results were obtained both from vertebrate skeletal muscle (H. E. Huxley, 1971b, 1972a) and from smooth muscles (Vibert *et al.*, 1972). These observations can be summarized as follows. In the X-ray diffraction patterns from relaxed muscle (Fig. 5.24a) actin layer lines could be seen at axial spacings of 1/380 $Å^{-1}$ (first layer line), 1/127 $Å^{-1}$ (third layer line) 1/77 $Å^{-1}$ (fifth layer line), 1/59 $Å^{-1}$ (sixth layer line), 1/51 $Å^{-1}$ (seventh layer line), etc. The layer line numbering is given for the situation in which the actin helix is integral with 13 residues in six turns. The layer line at 1/190 $Å^{-1}$ (second layer line) was either very weak or absent in these patterns. Similar diffraction patterns taken from muscles that were actively generating tension showed clear differences in the relative intensities of the second and third layer lines (Fig. 5.24b). The second layer line now became visible with an intensity equal to or slightly greater than that of the third. The other layer lines were largely unaltered

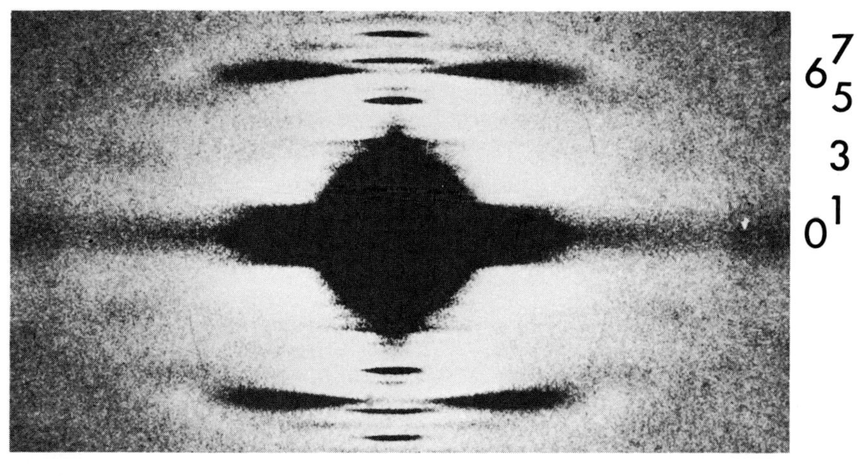

FIGURE 5.24. Low-angle X-ray diffraction patterns from (a) relaxed and (b) active molluscan smooth muscle (anterior byssus retractor muscle from *Mytilus edulis*). The actin pattern in (a) shows a clear third layer line but a missing second layer line, whereas (b) shows a second layer line almost as strong as the third.

(except for a possible change in the intensity of the 59-Å reflection, which will be discussed in Chapter 10). The observed diffraction pattern from rigor vertebrate skeletal muscle showed the same characteristic increase in the intensity of the second layer line, as did patterns from muscle that was rigorized after being stretched to a sarcomere length at which no overlap between the A- and I-bands occurred. The latter result suggested that the change in structure of the thin filaments was not a result of the actomyosin interaction but was a Ca^{2+}-induced change in the thin filaments resulting directly from activation.

5.5.2. Analysis of the Observed Changes

Despite the considerable technical difficulties involved in carrying out these experiments, the results themselves were relatively simple. But the problem remained to interpret the observed changes, especially of the intensity of the second layer line, in terms of structural changes in the thin filament. Two groups worked on this problem independently and arrived at the same conclusions at about the same time in early 1972 (H. E. Huxley, 1972a; Haselgrove, 1972; Parry and Squire, 1973).

The first step in the logic of the analysis was to consider what types of change in thin filament structure might be involved. As it happened, the results of O'Brien *et al.* (1971) and Hanson *et al.* (1972) had already given a clue to the possible answer. They had obtained optical diffraction patterns from paracrystals of F-actin filaments with or without added tropomyosin (and also X-ray diffraction patterns from gels of these filaments), and these patterns clearly showed that the presence of tropomyosin caused a marked increase in the intensity of the second layer line. Such an increase was consistent with the thin filament model in which tropomyosin molecules are located in the grooves of the F-actin helix. This would tend to make the filaments seem rather like four-stranded rather than two-stranded helices (Fig. 5.25a), and the repeat would appear to be halved from 355 Å to 178 Å, thus enhancing the intensity of the second layer line at 1/178 $Å^{-1}$.

It was therefore reasonable to conclude that tropomyosin makes a significant contribution to the second layer line. But presumably, in muscle, tropomyosin must be associated with the thin filaments the whole time, so the observed changes in layer line intensity associated with thin filament regulation could not be explained merely in terms of the presence or absence of tropomyosin. They might, however, be ex-

→

FIGURE 5.25. (a) Radial projection of a 13/6 helix of actin monomers (large solid circles) with two tropomyosin strands located centrally in the grooves of the actin helix (small circles). Note that as in Fig. 5.22, this diagram shows that the tropomyosin in this position tends to halve the thin filament repeat by making it more four-stranded than two-stranded. (b) Calculated intensity distribution in the F-actin diffraction pattern. (c) The corresponding distribution with tropomyosin located centrally in the F-actin grooves and modeled as in (a). Since the tropomyosin tends to halve the thin filament repeat, the second layer line in (c) is strengthened relative to that in (b). The fifth, sixth, and seventh layer lines are not altered at all in the regions shown. (From Parry and Squire, 1973.)

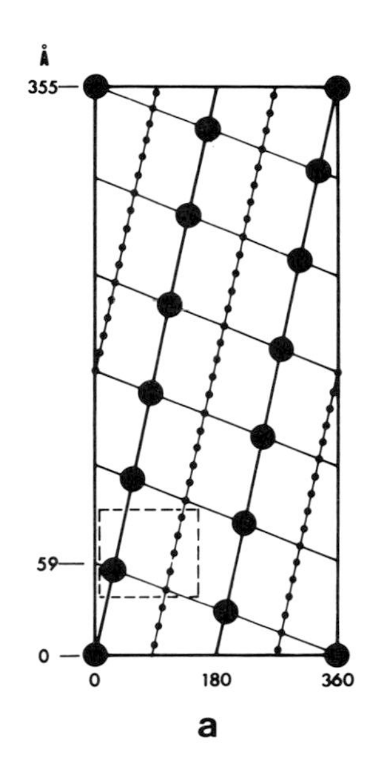

Å
355
59
0
0
180
360
a

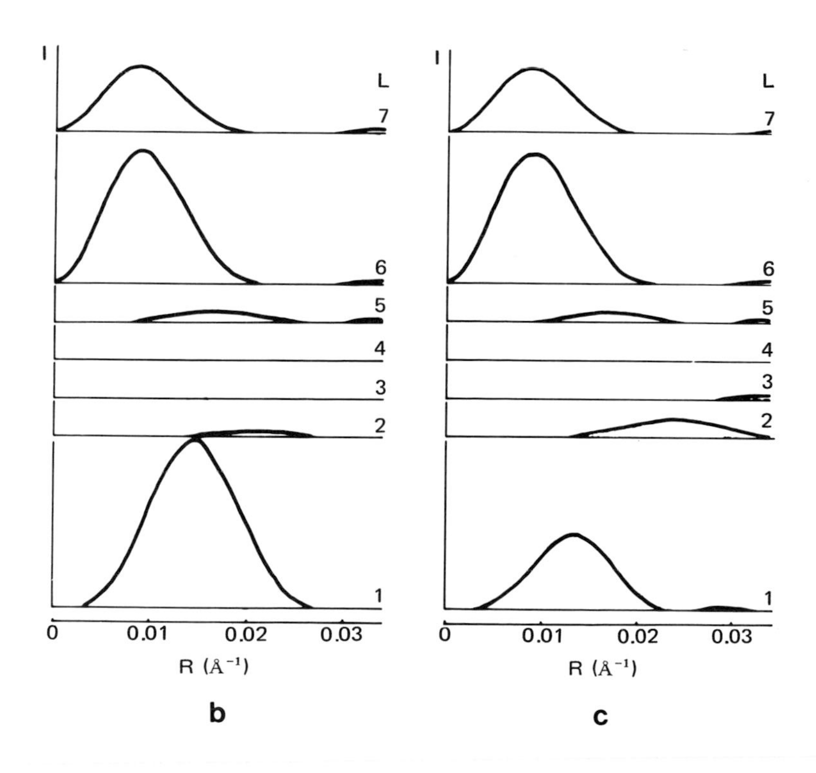

I
L
7
6
5
4
3
2
1
0
0.01
0.02
0.03
R (Å⁻¹)
b
c

pected to result from changes in the structure or location of the tropomyosin molecules in the thin filament grooves.

Two possibilities were considered by Parry and Squire (1973). In one, the effect of a transition between an ordered and disordered tropomyosin structure was investigated and found not to be capable of explaining the observed intensity changes. In the other, the effect of altering the location of tropomyosin in the thin filament grooves was considered, and this time it was found that the observed intensity changes could be explained. H. E. Huxley (1972a) and Haselgrove (1972) also considered this second possibility and came to similar conclusions.

In each case the analysis involved the following steps. First, the thin filament structure was modeled in a simplified way so that its Fourier transform could be calculated reasonably easily. Then, the intensity profiles on the relevant layer lines were computed for each of the different structures it was thought the thin filament might adopt. Finally, the computed layer line intensities were related to the published X-ray diffraction observations. Fortunately, it was then found that the results could be interpreted in terms of a simple mechanism for thin filament regulation, although this has been considerably complicated since then. Some details of the analysis by Parry and Squire (1973) are given below. These also represent the arguments of Huxley (1972) and Haselgrove(1972) unless stated otherwise.

The modeling of F-actin by a set of spheres of radius 24 Å and at a helix radius of 24 Å has already been described (Fig. 5.7). Figure 5.25a shows a radial projection of the F-actin helix, taking it to be a 13/6 helix of repeat exactly 355 Å (13 × 27.3 Å), together with tropomyosin molecules placed centrally in the grooves of the actin helix and modeled as a set of overlapping spheres of axial separation about 11 Å (54.6/5 Å) and radius 8.3 Å. This radius was chosen to give the correct ratio by volume between actin and tropomyosin, assuming that they have the same density. Figure 5.25b,c show the computed intensities of layer lines 1 to 7 for F-actin alone and for F-actin with tropomyosin located centrally in the actin grooves. It will be seen here that the effect of the presence of tropomyosin is to enhance the intensity of the second and third layer lines as observed by O'Brien *et al.* (1971) and Hanson *et al.* (1972). Layer lines 4 to 7 are not altered at all, since the tropomyosin appears as two continuous helical strands, thus giving rise to a single helix cross centred on the origin of the diffraction pattern (Fig. 5.26d). The layers of intensity are only important at spacings 1/355, 1/178, and 1/120 $Å^{-1}$, respectively. On the other layers (where n is higher), the intensity is too far from the meridian to significantly alter the thin filament diffraction patterns. Figure 5.26 demonstrates this effect by means

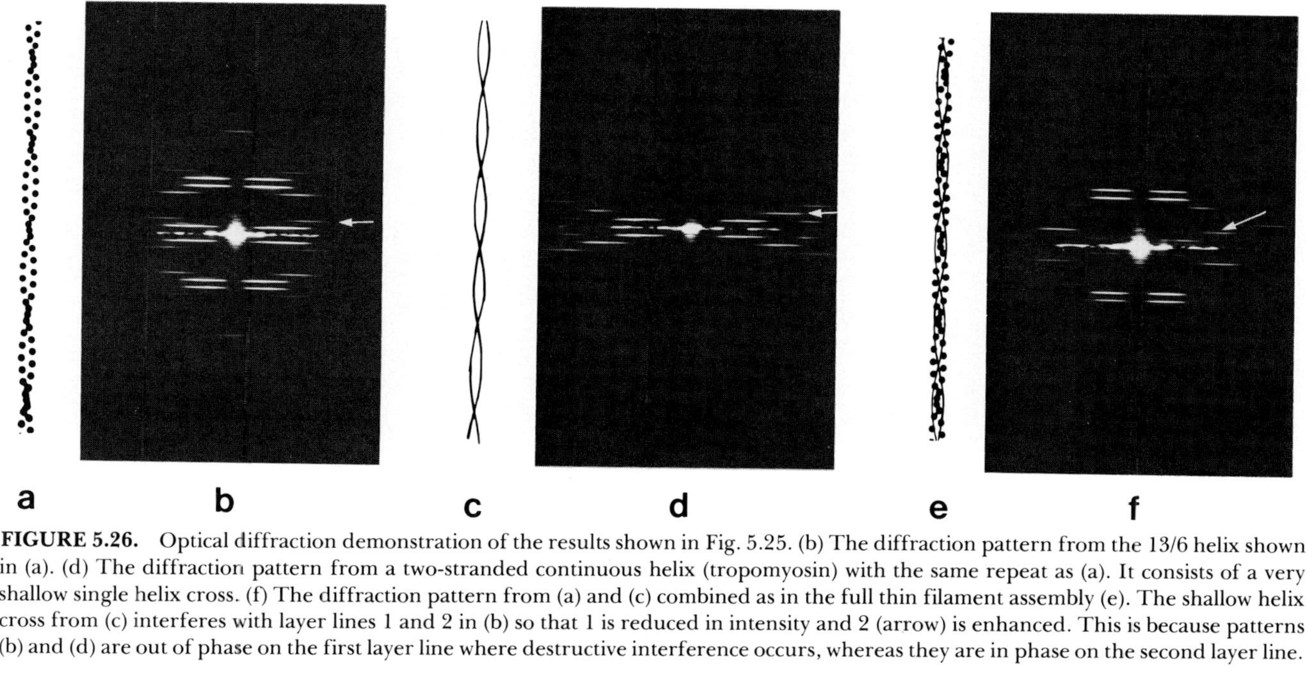

FIGURE 5.26. Optical diffraction demonstration of the results shown in Fig. 5.25. (b) The diffraction pattern from the 13/6 helix shown in (a). (d) The diffraction pattern from a two-stranded continuous helix (tropomyosin) with the same repeat as (a). It consists of a very shallow single helix cross. (f) The diffraction pattern from (a) and (c) combined as in the full thin filament assembly (e). The shallow helix cross from (c) interferes with layer lines 1 and 2 in (b) so that 1 is reduced in intensity and 2 (arrow) is enhanced. This is because patterns (b) and (d) are out of phase on the first layer line where destructive interference occurs, whereas they are in phase on the second layer line.

of optical diffraction. Figure 5.26b is the diffraction pattern from a simple actin helix (a), (d) is the diffraction pattern from the two continuous tropomyosin strands (c), and (f) is the diffraction pattern of the complete structure (e). The second layer line is clearly stronger in (f) than it is in (b).

It has already been mentioned that troponin is probably located at axial intervals of 385 Å along the thin filament and that this undoubtedly corresponds to the axial repeat along the tropomyosin strands. Since the tropomyosin molecules are about 410 Å long (Cohen *et al.*, 1971b, 1972; Phillips *et al.*, 1979), it is clear that they could only be accommodated tightly in the center of the grooves as in Fig. 5.22b if they are slightly supercoiled. A reduction in molecular length to about 395 Å would give a projected axial repeat of about 385 Å if the molecules follow these helical grooves. On the other hand, a fully extended molecule could be located at a helix radius of about 45 Å.

The possibility was therefore considered that the tropomyosin strands might not be centrally placed in the grooves of the actin helix. Fourier transforms were calculated of models in which the tropomyosin was in different positions. These positions were described in terms of the angle ϕ as defined in Fig. 5.27. In this figure, ϕ can vary from 45° to 90° (the center of the groove). In all cases, it was assumed that the tropomyosin and actin molecules were in close contact, and the coordinates given in Table 5.5 were generated. Figure 5.28a shows the intensities of layer lines 2 and 3 as a function of ϕ for these models. It can be seen that very considerable changes in relative intensities occur such that at $\phi = 45°$, I_3 is greater than I_2; whereas at $\phi = 70°$, I_2 is very much greater than I_3. The dotted line in this figure models the effect of allowing those parts of the tropomyosin molecules located between the actin–tropomyosin contact points to vibrate from temperature effects. The general form of the curves is little altered.

The results from vertebrate skeletal muscle (H. E. Huxley, 1971b, 1972a), in which the second layer line is weaker than the third in patterns from relaxed muscle but becomes slightly stronger than the third in patterns from active muscle, could therefore be explained directly in terms of a movement of the tropomyosin strands from $\phi = 45$ to 50° to ϕ

TABLE 5.5. Polar Coordinates of Various Possible Positions of the Tropomyosin Molecules in the Thin Filaments[a]

ϕ (°)	45	50	55	60	65	70	75	80	85	90
r (Å)	45	42.5	39	37	35	33	31	29	27	25

[a] For definition of r and ϕ see Fig. 5.27. From Parry and Squire (1973).

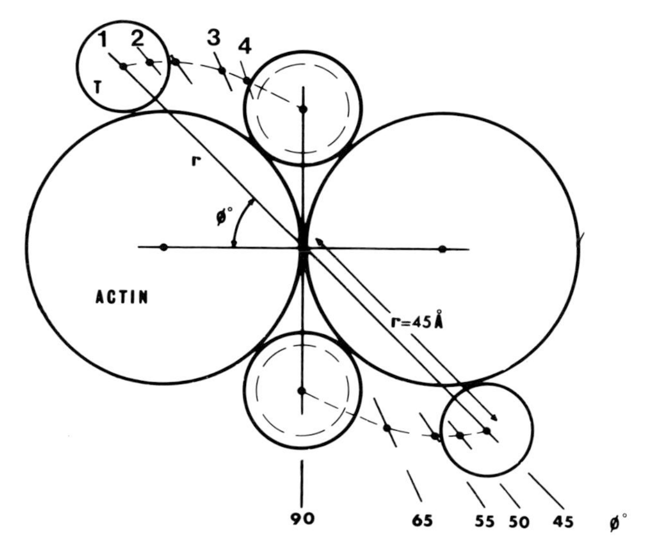

FIGURE 5.27. End-on view of a thin filament (in helical projection) showing possible positions of a tropomyosin strand in the groove of the actin helix. These positions are defined in terms of the tropomyosin radius (r) from the thin filament axis and the angular position ($\phi°$) from the line joining the centers of the two actin strands. At the top, the numbers 1 to 4 indicate four possible physiologically significant tropomyosin positions as detailed in the text. (After Parry and Squire, 1973.) The increased tropomyosin diameter at the 90° position indicates the likely presence of some supercoiling of the tropomyosin, since the helix radius (r) is smaller.

= 65 to 70°, respectively. Note that as it turns out, the helix radius of the tropomyosin strands cannot be defined uniquely, since it has been shown by Haselgrove (1972) that the results summarized in Fig. 5.28, although strongly dependent of the value of ϕ, are little effected by r (provided of course that r is less than the physical maximum of 45 Å).

Since the X-ray diffraction observations could not be explained any other way, these modeling studies were taken to mean that the tropomyosin molecules are definitely located along the grooves of the thin filament and that activation results in a change in position of the tropomyosin within these grooves. The change involved is illustrated in Fig. 5.28c,d.

Since that time a considerable amount of evidence has confirmed these conditions. Hanson *et al.* (1972) and Gillis and O'Brien (1975) have used X-ray diffraction to study synthetic gels of actin with or without tropomyosin present (Fig. 5.29) and also of thin filaments in the presence or absence of Ca^{2+} ions and have observed intensity changes on the second and third layer lines (as in Figs. 5.25 and 5.26) similar to those that were observed by H. E. Huxley (1972a) and Vibert *et al.* (1972) in

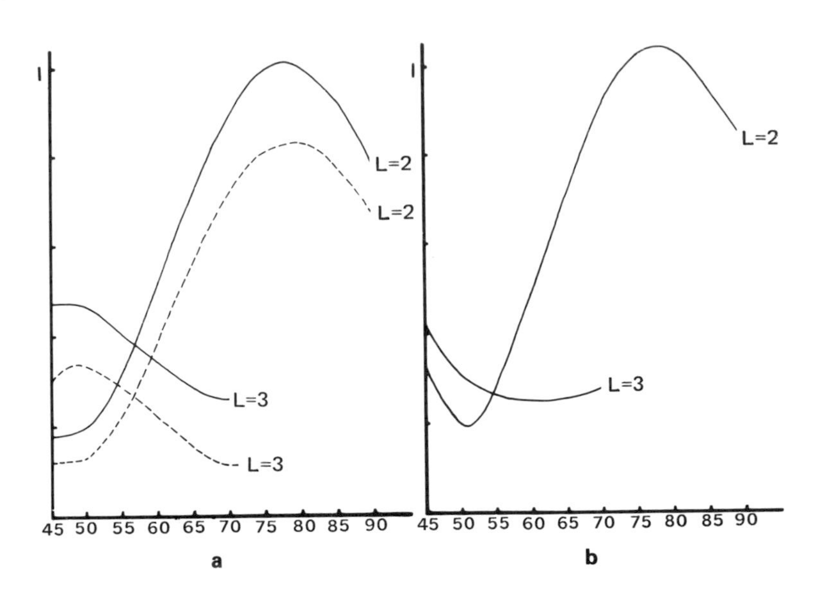

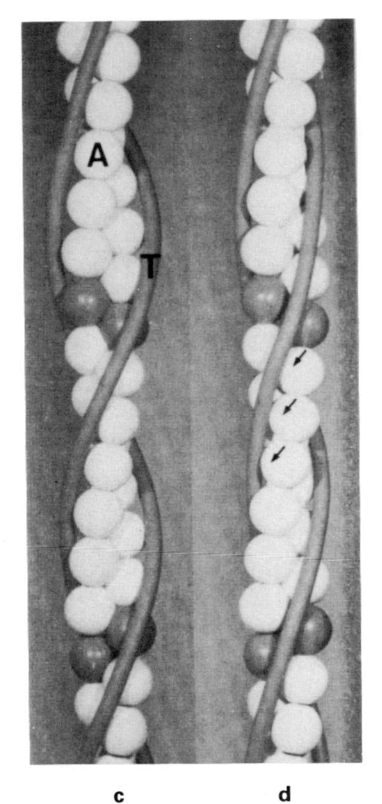

FIGURE 5.28. (a) Solid line: variation of peak intensity as a function of tropomyosin position (defined by ϕ in Fig. 5.27 and Table 5.5) on layer lines 2 and 3 in the thin filament transform. Dashed line, the corresponding variation in intensity if the tropomyosin strands are subject to significant thermal oscillations between successive actin attachment sites. (b) As in (a) (solid line) with the effect of the labeling of the thin filament by myosin cross bridges included (see discussion in Chapter 10). (c and d) Models of the thin filament with tropomyosin at $\phi = 45°$ and $\phi = 65°$, respectively. These correspond to the likely positions of tropomyosin in relaxed and active troponin-regulated thin filaments. Note that the myosin binding site regions (arrows) are exposed in (d) but covered in (c). (From Parry and Squire, 1973.)

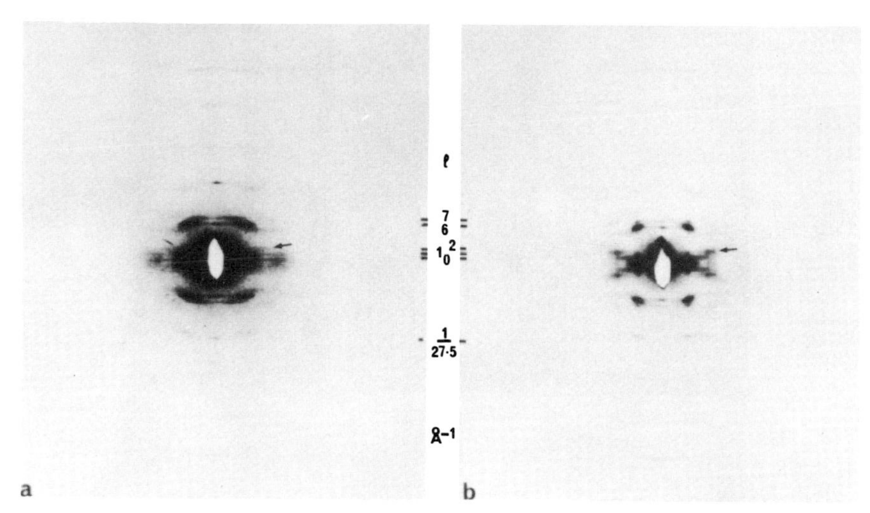

FIGURE 5.29. X-ray diffraction patterns from oriented gels of (a) purified actin and (b) actin containing tropomyosin and troponin. There is an intensification of the second layer line in (b) (arrowed) relative to (a). (From Hanson *et al.*, 1972.)

patterns from intact muscle. They also noted similar changes in the optical diffraction patterns of actin paracrystals prepared in the presence or absence of Ca^{2+} ions. Thin filament structure has been further elucidated by Spudich *et al.* (1972) and Wakabayashi *et al.* (1975) who applied the 3-D reconstruction technique to electron micrographs of actin filaments in various states. Spudich *et al.* (1972) applied the technique to F-actin filaments containing tropomyosin and troponin, and they showed that in the presence of Ca^{2+} the tropomyosin was located towards the center of the groove in the F-actin structure.

Wakabayashi *et al.* (1975) extended this work by applying the technique to paracrystals of F-actin either with tropomyosin alone or with tropomyosin plus TN-I and TN-T, thus mimicking the *on* and *off* states. They demonstrated that the tropomyosin had a different location in these two preparations. Figure 5.30 summarizes these results in terms of radial projections of the computed three-dimensional density maps of the two types of structure. The change in position of the tropomyosin is very clear in these projections. Also shown in Fig. 5.30 are views of the cross section of these structures when helically projected. The larger peaks in (c) correspond to the two strands of actin molecules, and the small peaks to the tropomyosin strands. It will be seen that here the tropomyosin is located at a relatively large radius from the filament axis although being at an angle ϕ (as in Fig. 5.27) of about 70°. In Fig. 5.30d, the two peaks have merged (possibly indicating relatively tight actin–

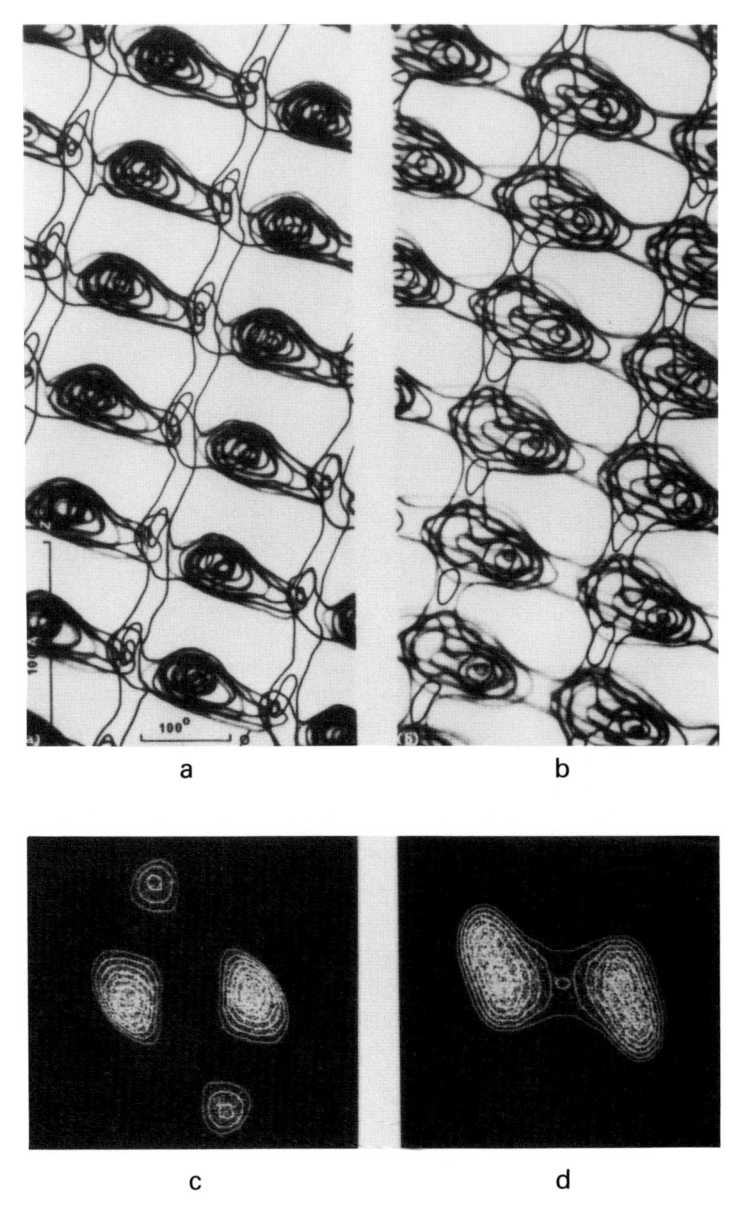

FIGURE 5.30. Cylindrical sections of the three-dimensional reconstructions of (a) actin plus tropomyosin and (b) actin plus tropomyosin plus troponin-T + I by Wakabayashi *et al.* (1975) are shown superimposed in radial projection. The sections are at radii 10, 15, 20, and 25 Å, the contour lines of those at small radii being fainter and less sharp than those at larger radii. The difference in the tropomyosin position can be clearly seen in these projections. The position in (a) is analogous to that in the "on" state, and that in (b) to the

tropomyosin binding), but tropomyosin is now clearly positioned at a much larger value of ϕ.

One feature of the reconstructions shown in Fig. 5.30a,b is that the tropomyosin strands appear to have a regular zigzag structure. Such an appearance could easily be an artifact, since every repeat in the reconstruction will be an "average" structure, and any error in the reconstruction will be reproduced in each repeat to give such a periodic distortion. Although some support for this sinusoidal undulation has been considered by McLachlan and Stewart (1976), it is not supported by recent work of O'Brien and Couch (1976) who have carried out a 3-D reconstruction of the thin filament using structure amplitudes and phases determined directly by optical diffraction methods. In their reconstruction, the tropomyosin strands are more or less linear.

5.5.3. A Model for Thin Filament Regulation

All of these results in themselves would mean little if they could not be related in some way to the regulation of the actin–myosin interaction. But fortunately, as soon as the results were known, a possible regulation scheme became apparent (H. E. Huxley, 1972a; Haselgrove, 1972; Parry and Squire, 1973,) although this has since had to be modified (see Section 5.6). It had already been shown by Moore *et al.* (1970) that the myosin head or subfragment-1 (HMM S-1) attaches to actin in a position which in terms of the angle ϕ in Fig. 5.27 was about 30 to 40° (see Fig. 5.11). This is very close to the position of tropomyosin (ϕ = 45°) in the "switched-off" (Ca^{2+}-free) thin filament, whereas when the thin filament is "switched on" at high Ca^{2+} concentrations, the tropomyosin position at ϕ = 65 to 70° is well clear of this attachment site. It was therefore suggested that tropomyosin in the "off" position might physically block the attachment of myosin to actin (or at least be close enough to the attachment site to modify this attachment). Movement of tropomyosin to the "on" position would then unblock this site and allow attachment to take place. These two situations are illustrated in Fig. 5.31a,b. Direct evidence that the presence of tropomyosin could modify the attachment of myosin to actin was obtained by Spudich *et al.* (1972) who studied preparations of actin with tropomyosin and troponin present which had been labeled with HMM S-1. They found that S-1 binding was less ordered in decorated filaments prepared at low levels of Ca^{2+}.

"off" state. (c and d) Helical projections of the structures shown in (a) and (b), respectively, down the long-pitched two-stranded helix. Density and contour maps are superimposed in each case. Note the large apparent separation of the actin strands (large peaks) and the tropomyosin (small peaks) in (c). (From Wakabayashi *et al.*, 1975.)

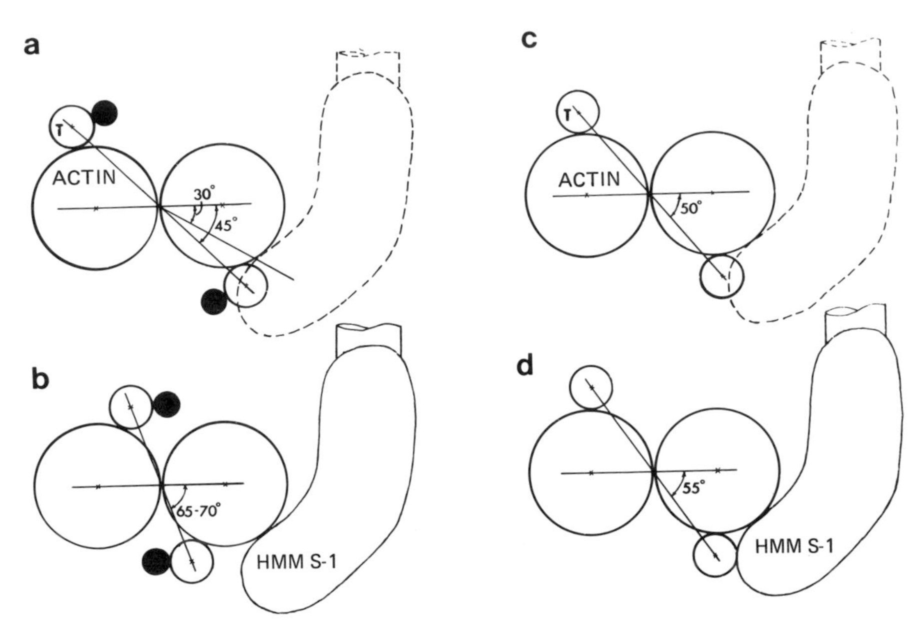

FIGURE 5.31. Tropomyosin movement in thin filaments containing troponin (a and b) and without troponin (c and d). (a) The tropomyosin (T) is in the "off" position (1 in Fig. 5.27) and blocks the attachment of a myosin cross bridge (attachment position outlined). (b) The same filament after calcium has bound to troponin (which is represented here by the filled circles). Tropomyosin has moved to the "on" position (4 in Fig. 5.27), thus allowing cross-bridge attachment. With troponin absent (c), the tropomyosin in position 2 in Fig. 5.27 is moved to position 3 by the attaching myosin head. (From Parry and Squire, 1973.)

The simple regulation scheme outlined above was complicated to some extent by the results from molluscan muscle (Vibert *et al.*, 1972). In this case, changes in the observed layer line intensities indicated that movement of the tropomyosin molecules in the thin filament must take place. But this was despite the fact that in this muscle there is no troponin to mediate the process (Kendrick-Jones *et al.*, 1970). It was concluded by Vibert *et al.* (1972) and Parry and Squire (1973) that this movement must be a consequence of actin–myosin interaction rather than a precursor to it and that attaching myosin heads must be pushing the tropomyosin away from the attachment site and towards the center of the groove (Fig. 5.31c,d). The problem then was to explain how one tropomyosin position (the "off" position) could block the actin–myosin interaction in vertebrate skeletal muscle but allow it to occur in molluscan muscle (and also possibly in vertebrate smooth muscle).

A possible way around this problem was thought to be that in the absence of troponin the tropomyosin position could be at $\phi = 50°$ where myosin attachment is not blocked, and that the effect of troponin is to move the tropomyosin to $\phi = 45°$ where full blocking could occur. The difference in position from 50° to 45° would not be detectable by X-ray diffraction as shown in the calculations summarized in Fig. 5.28. (But note that an alternative possibility is mentioned later.) A further complication was the fact that in the observed diffraction patterns from activated molluscan muscle, the second and third layer lines had comparable intensities (Vibert *et al.*, 1972), whereas in patterns from active vertebrate skeletal muscle, the second layer line became stronger than the third. These results could mean that in the latter the tropomyosin molecules move further into the groove ($\phi = 70°$) than they do when pushed towards the groove by an attaching myosin molecule ($\phi = 65°$). However, this result could also be explained by the fact that the thin filaments in molluscan muscle are very long (see Chapter Section 8.3.3) and that relatively little of a particular actin filament will be in a position for myosin attachment to occur. Only part of the tropomyosin will then be moved to the "on" position by attaching myosin heads, and relatively little change in the intensity of the second layer line might therefore be expected. Unfortunately, there is as yet no way of knowing which of these possibilities is correct, and for this reason it must be allowed that at least three and possibly four different tropomyosin positions might be involved. These are summarized in Figure 5.31.

5.5.4. Structural Details of the Regulation Scheme

With the basic form of the thin filament regulation scheme in mind, it is appropriate to consider in more detail the actin–tropomyosin–troponin interactions that are involved.

Tropomyosin Structure. It is reasonable to start from the assumption that a functional unit occurs in the thin filament [as described in Section 5.4.3. (Fig. 5.22)] and contains seven actin monomers, one tropomyosin molecule, and one troponin complex. Since there is a pseudorepeat in the tropomyosin sequence that is about one-seventh of the repeat along the tropomyosin strands (allowing for a 13.5-Å overlap between successive molecules), then it is also reasonable to postulate that the tropomyosin molecules make reasonably equivalent interactions with successive actin monomers along one long-pitched actin strand. Since the tropomyosin molecule is a two-strand coiled-coil structure, this would require the twist of the coiled coil to be related to the axial stagger between actin molecules. Parry (1975a,b) and Stewart and McLachlan (1975) both suggested that the tropomyosin molecules might twist one-half turn relative to an actin strand in one actin subunit repeat. Such a scheme is illustrated in Fig. 5.32a. This would mean that when the

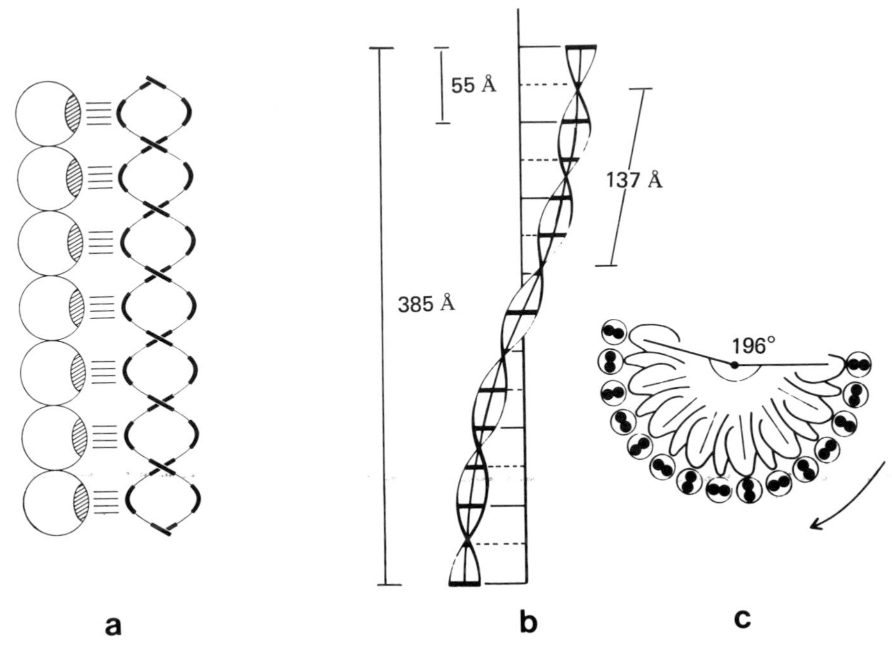

FIGURE 5.32. The fit of the seven-actin repeating unit in the thin filament to seven half-turns of the two-stranded tropomyosin molecule (right). The short dark lines along the tropomyosin chains indicate the 28 pseudoequivalent bands in the tropomyosin molecule. (From Stewart and McLachlan, 1975.) (b) The two-stranded tropomyosin molecule as in (a) twisted around to follow the long-pitch grooves in the thin filament. The effect is that the molecule is untwisted by one half-turn, making six half-turns per repeat while maintaining equivalent interactions with all seven actin monomers in the repeat. This is indicated in axial projection in (c). Note that this diagram is highly schematic, and the actin shape should not be taken literally. (From McLachlan and Stewart, 1976.)

tropomyosin follows the helical path of the actin strand which itself rotates by about half a turn every seven residues, the six half turns of the tropomyosin molecule will appear to the thin filament to present seven equivalent sites to the seven actin monomers. To the outside observer, the tropomyosin still has six half turns in an axial distance of about 385 Å (Fig. 5.32b,c). The pitch in the coiled-coil molecule would then be 1/3 of 410 Å (the repeat distance along the tropomyosin strand) or 137 Å. This is in reasonable agreement with the estimated pitch lengths of other two-strand coiled-coil molecules such as paramyosin (136–140 Å or 178 Å, Chapter 8), α-keratin (140 to 170 Å), and honeybee silk (140 Å; see Chapter 4 and Fraser and MacRae, 1973).

Figure 5.32a from Stewart and McLachlan (1975) also shows how seven of the 28 acidic regions along each tropomyosin molecule will interact with the actin monomers along one strand in the thin filament.

So far, the division of the tropomyosin molecule into seven half twists has been considered; but in fact there is a significant pseudoperiodicity that divides the tropomyosin repeat into 14 subrepeats. Each of these 14 subrepeats contains acidic bands, and McLachlan and Stewart have classed these as seven α bands and seven β bands. Each of the seven major subrepeats then contains one α and one β band (Fig. 5.33b). With the tropomyosin molecule twisting one-half turn in each of the major subrepeats, the α and β bands are separated by about one-quarter of a turn. It was therefore suggested by Parry (1976) and McLachlan and Stewart (1976) that the "on" and "off" configurations of tropomyosin could involve binding between tropomyosin and

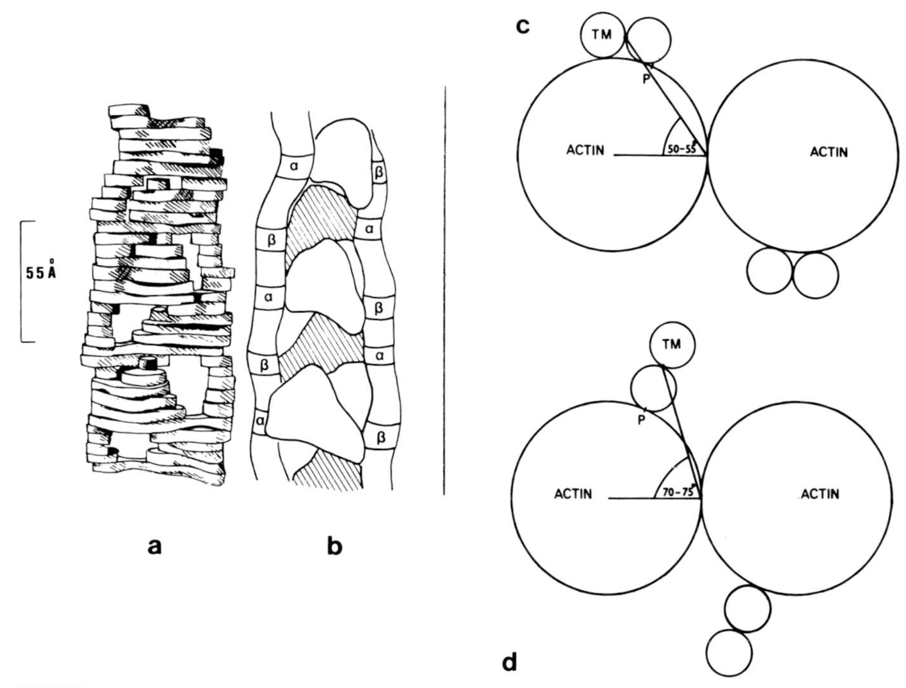

FIGURE 5.33. (a) Sketch of the three-dimensional reconstruction of actin–tropomyosin in the "on" state (as in Fig. 5.30a and c; Wakabayashi *et al.*, 1975) showing boot-shaped polar actin monomers in contact with tropomyosin at the heel and toe positions. (b) Outline of (a) with the positions of the α and bands of the tropomyosin sequence tentatively located. Note that the direction of the tropomyosin strands relative to the polar actin backbone is not known. (From McLachlan and Stewart, 1976.) (c and d) Two states of a simple rolling tropomyosin model of regulation. (c) The "off" state; (d) the "on" state. A roll of 90° changes the structure in (c) to that in (d) and at the same time moves the center of gravity of the tropomyosin strands away from the actin surface (see Fig. 5.30c). The binding to actin could be through the tropomyosin α band in (c) and through the β band in (d). (From Parry, 1976.)

actin through the α bands in one case and the β bands in the other. In any event, a simple 90° roll of the whole tropomyosin strand (Fig. 5.33c,d) would change the interaction site of tropomyosin from an α to a β site and at the same time would involve a movement of the tropomyosin molecule towards the groove of the actin helix. In molluscan muscle, attaching myosin heads would roll the tropomyosin away from the "off" position but not far enough to optimize the interaction between the β sites and actin. The tropomyosin would therefore roll back to the "off" position when the myosin detaches. This model involving rolling tropomyosin molecules (Parry, 1976) would also account for the large apparent separation between tropomyosin and actin in the reconstructed images of thin filaments in the "on" state (see Fig. 5.30a), since the axis of the tropomyosin strand would move away from the actin surface during the roll. Parry has also noted that a local torque produced by rolling of the tropomyosin molecules near to troponin would tend to be more easily propagated along the whole tropomyosin strand than would a simple lateral shift of the molecules. The latter might be expected to be relatively easily accommodated by local bending of the molecules.

Troponin Structure. The next main question to ask is, "How does troponin bring about this change in structure?" With the model for the tropomyosin movement in mind, it is clear that troponin would be required to cause reversible rolling of the tropomyosin molecules when Ca^{2+} ions are bound. It is known from work on the troponin subunit interactions as a function of Ca^{2+} concentration that the interaction between TN-I and actin is weakened when Ca^{2+} is bound to TN-C, but that the TN-T binding to TN-C and to tropomyosin is strengthened. But, of course, there is little evidence as yet about the three-dimensional structure or subunit arrangement in troponin, and any suggestions about the effect of troponin must by hypothetical. Nevertheless, the purely mechanistic scheme illustrated in Fig. 5.34 seems to be consistent with all of the evidence available at present. In this, TN-I normally binds close to the myosin binding site on actin. Inhibition by TN-I alone can therefore occur. Troponin-C binds to TN-I close to this attachment site so that the TN-I-to-actin interaction can be influenced by TN-C. Troponin-C is also close to the TN-T-to-tropomyosin binding position so that this also can be suitably modified by structural changes in TN-C. In the off state TN-I is firmly bound to actin, and this locks the tropomyosin in the "off" position via the TN-C–TN-T–tropomyosin interaction (or possibly through actin—see below). When Ca^{2+} ions are bound to TN-C, the configuration of the troponin complex alters so that tropomyosin is forced to roll (or allowed to roll) towards the groove of the actin helix. This movement could be caused by the tighter binding among TN-C,

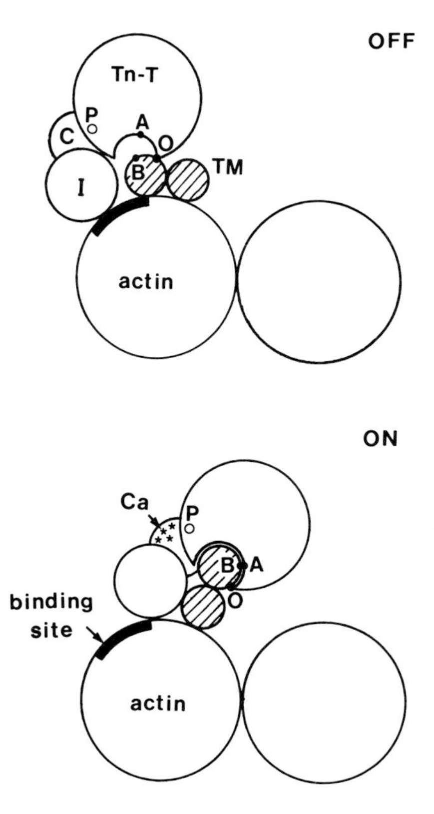

FIGURE 5.34. A regulation model as in Fig. 5.33c, and d but including troponin to show a purely mechanistic model for troponin action. In the "off" state, TN-I is bound to actin near the myosin attachment site. When calcium binds to TN-C to give the "on" state, the binding from tropomyosin to TN-T to TN-C to TN-I is strengthened, but the binding of TN-I to actin is weakened. The tropomyosin is therefore forced to (or allowed to) roll towards the thin filament groove. In this drawing, the rolling is thought of as being forced by a calcium-induced attraction between sites A and B and pivoting of the tropomyosin–TN-T assembly about O. But clearly, this hypothetical scheme should not be taken too literally.

TN-T, and tropomyosin (as modeled here in terms of a transition from single-site binding to double-site binding between each pair of components) and is unhindered by the TN-I-to-actin binding since this is weakened when TN-C binds Ca^{2+} ions.

Such a scheme provides a convenient mental picture of the regulation mechanism but clearly should not be taken too literally. The true nature of the troponin subunit interactions and the effect of Ca^{2+} binding can only be discovered when the full three-dimensional structures are known. Fortunately, TN-C has now been crystallized by a group at Oxford University (Mercola *et al.*, 1975) and it should soon be possible to discuss the properties of at least that part of troponin with more confidence

One extra puzzle to be discussed is that actin–myosin inhibition by TN-I alone can be augmented by the presence of tropomyosin even though TN-I binding to tropomyosin is weak. This could mean that TN-I causes a local structural change in actin which in turn locks the tropomyosin into the "off" position. This lock would then be released in the full thin filament assembly when TN-C binds Ca^{2+}. A mechanism of

concentrations as it was in solutions of S-1 and pure F-actin filaments (Fig. 5.36a).

2. At higher ATP concentrations, the ATPase dropped to the "off" value with calcium absent but remained high in the presence of calcium.

3. At low ATP and calcium concentrations, the calcium affinity of troponin for the last two of the four troponin-bound calcium ions was increased.

4. At high calcium concentrations, when the thin filaments should be fully in the "on" state, the rate of ATP hydrolysis was found to be lower at low myosin concentrations (one myosin active site to 50 actin monomers) than in a comparable situation with tropomyosin absent. The ATPase rate increases smoothly with ATP concentration (Fig. 5.36b).

5. In similar experiments to 4 but with excess myosin, a completely different result is obtained. Here ATPase activity is maximum at low ATP concentrations (where a significant number of rigor complexes must occur), and then it reduces when the concentration of ATP is

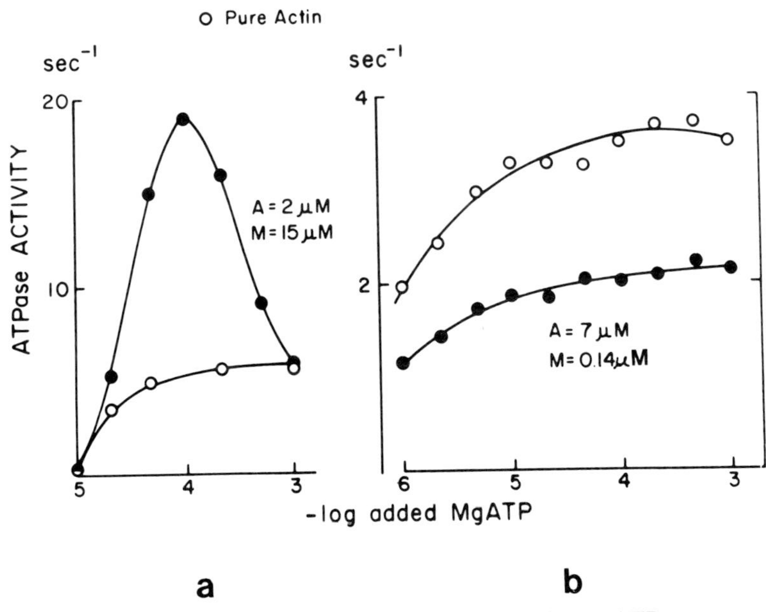

FIGURE 5.36. The rate of ATP hydrolysis as a function of Mg · ATP concentration in the presence of calcium. In each graph a comparison is made between the interaction of S-1 with regulated or with pure actin (i.e., actin with or without tropomyosin and troponin). (a) S-1 is present in sevenfold excess over actin. (b) Actin is present in 50-fold excess over S-1. For explanation, see text. (From Bremel *et al.*, 1972).

increased. The ATPase rate is also considerably higher than in a comparable situation with tropomyosin and troponin absent (Fig. 5.36a).

Weber (1975) has accounted for several of these observations in terms of rigor complexes holding the tropomyosin strands away from the "off" position on the thin filament and thus to some extent mimicking the effect produced when calcium binds to troponin. It was suggested that nucleotide-free myosin can interact with quite high affinity with actin filaments nominally in the "off" state. Once attached, they would then hold the tropomyosin in that region of the thin filament away from its "off" position and thus permit actin binding by nucleotide-bound myosin molecules. The observed increase in ATPase rate would result (1 above). At higher ATP concentrations, the rigor complexes would be too short-lived to allow this kind of regulation to occur to a great extent (2 above).

It was also suggested (see also Squire, 1975) that the movement of tropomyosin caused by the formation of rigor complexes would result in the same conformational change in troponin that would normally be induced by calcium binding. The calcium affinity of troponin would therefore increase as a consequence (3 above) (see also Taylor, 1979).

Observations 4 and 5 are more difficult to account for in this simple way. First, it seems that even in calcium-regulated thin filaments(4), the tropomyosin has a somewhat inhibiting effect on the ATPase compared with the situation with tropomyosin absent. But when an excess of myosin is present at low ATP (5), the tropomyosin then enhances the ATPase rate over the rate observed in the absence of tropomyosin. Weber has accounted for this observation in terms of a cooperative effect in which rigor complex formation induces a conformational change in the associated actin monomer, and the tropomyosin molecule then transmits an appropriate signal to the other actin monomers associated with it so that myosin binding is increased. But whether or not this has to do with tropomyosin position remains to be seen. What is clear, however, is that thin filament regulation is by no means as simple as suggested by the scheme in Fig. 5.31. In reality, it must involve the complex cooperative behavior of a number of proteins as a function of calcium concentration, of ATP concentration, and possibly also during muscular activity of the instantaneous pattern of attachment of the myosin heads on the actin filaments as defined by the geometry of the A-band. Although a great deal has been learned about thin filament regulation, it is obviously necessary to know the detailed structures of the thin filament components in order to understand some of the subtleties of the mechanism.

An additional complexity in the regulation scheme has come to light in a recent paper by Seymour and O'Brien (1980). This concerns the location of the tropomyosin molecules in the thin filament relative to the myosin attachment sites on actin. It was assumed in Fig. 5.31 that the tropomyosin and the myosin heads attach on the ame side of the actin

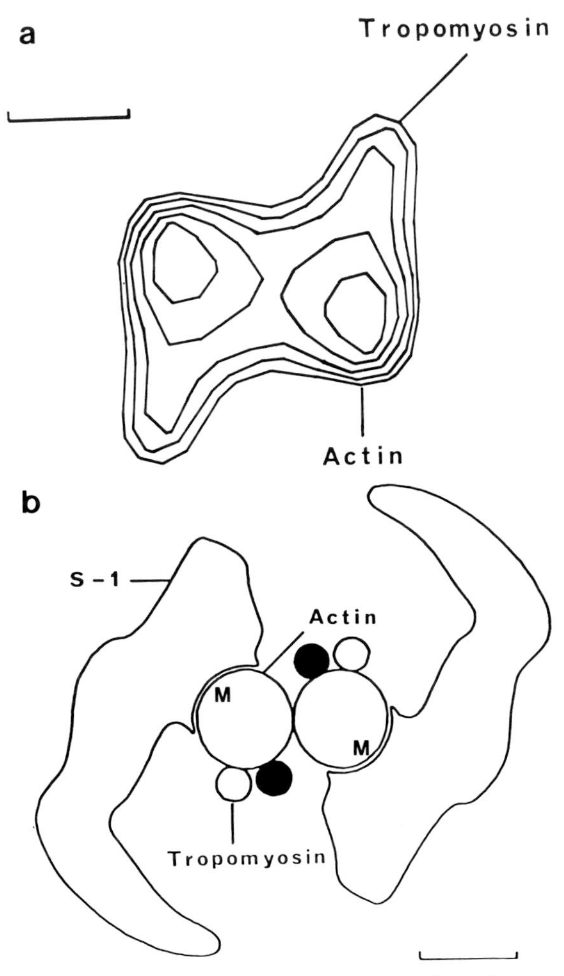

FIGURE 5.37. (a) Helical projection of the structure of thin filaments as determined by Seymour and O'Brien (1980). This corresponds to a view looking towards the Z-band. (Scale bar, 20 Å.) (b) A sketch of the corresponding view of the thin filament labeled with HMM S-1 according to Moore *et al.* (1970). Note that the tropomyosin peaks in (a) (labeled TM) are positioned top right and bottom left of the actin strands, whereas in (b) the myosin (S-1) attachment sites (M) as drawn are at top left and bottom right (i.e., they are on the opposite side of the actin groove). The on and off positions for tropomyosin are shown as filled and open circles, respectively, in (b). Scale bar in (b) is 50 Å. Figure (a) is from Seymour and O'Brien (1980); (b) is based on H. E. Huxley (1972a). Note that the possibility remains that the cross-bridge position in (b) is incorrect and should really be on the opposite side.

groove, since this makes a very sensible arrangement with regard to the possible blocking action of tropomyosin. However, it was an assumption, albeit a very reasonable one at the time, and the people involved in putting forward the steric blocking model of regulation (i.e., H. E. Huxley, 1972a; Haselgrove, 1972; Parry and Squire, 1973) were well aware at the time that their arguments were based on that assumption. The recent paper by Seymour and O'Brien (1980) has provided direct evidence on the position of tropomyosin relative to the myosin attachment site. They studied intact thin filaments still attached to Z-bands (i.e., I-segments) and determined the absolute handedness of the thin filament. In this way they could find the position of tropomyosin relative to the myosin-labeling positions that had already been determined in previous work (H. E. Huxley, 1963).

Figure 5.37 shows a contour map of the helical projection of the thin filament looking towards the Z-band as derived by Seymour and O'Brien from their computer analysis of negatively stained thin filaments in I-segments. When this is compared with the position of the myosin heads in reconstructions of decorated thin filaments (Fig. 5.37b), it is clear that the tropomyosin heads and tropomyosin molecules do not lie on the same side of the actin groove. In trying to relate this remarkable result with possible regulation schemes, one finds it reasonable to suggest two alternatives. One is that the movement of tropomyosin involved in regulation has a cooperative effect on the actin molecules which then regulate myosin head attachment, a suggestion which might tie in well with the results of Weber and her collaborators described above. The second is that the myosin head attachment site indicated by the results of Moore *et al.* (1970) is in some way misleading and that an appreciable part of the myosin head is located near enough to tropomyosin for direct steric blocking to occur. As indicated in Seymour and O'Brien (1980), K. Taylor and L. A. Amos are reexamining the structure of decorated thin filaments in the light of this possibility. The results of this reappraisal are awaited with interest.

6

Structure, Components, and Interactions of the Myosin Molecule

6.1. Introduction

The most abundant protein in many muscles and the functional component common to the thick filaments of all types of muscle is the protein myosin. In fact, muscle thick filaments are often referred to simply as myosin filaments or better as myosin-containing filaments, the latter attempting to signify that thick filaments do contain other proteins as well. The thick filaments in certain molluscan muscles actually contain more of the protein paramyosin than of myosin.

Despite the obvious similarities among the myosin molecules found in different muscles, the thick filaments that they form can have markedly different shapes and sizes ranging from cylinders 150 Å in diameter in vertebrate skeletal muscle to square or ribbon-shaped filaments in vertebrate smooth muscle to large cylinders up to 1500 Å in diameter in some molluscan muscles. Because of this diversity, it is not possible to treat all thick filaments together in a single chapter as was done in the last chapter for actin-containing filaments which appear to have a common basic structure in all muscles. Instead, this chapter will deal in a general way with the structure and the aggregation properties of the component myosin molecules. Detailed discussions of the structures of particular types of myosin-containing filaments will be presented in subsequent chapters that deal with individual muscle types.

Like actin, myosin molecules are found in many cell types apart from muscle cells. These include slime mold, blood platelets, brain, fibroblasts and amoebas (Pollard and Weihing, 1974; Clarke and Spudich, 1974). It therefore seems likely that many kinds of cell movement are associated in some way with the interaction between myosin and actin molecules. The distinctive feature of muscle cells is that these molecules are so abundant and so highly organized that the contractile force is significant at a macroscopic level.

Myosin can be extracted from homogenized muscle at high ionic strength. It was first isolated in an impure state by Kühne in 1859 and studied in detail by Edsall in 1930. It was later shown by Straub and Szent-Gyorgyi (1943) that the "myosin" preparation in fact contained actin in addition to myosin and that the impure "myosin" referred to as actomyosin possessed ATPase activity.

Very pure myosin can now be prepared using a variety of methods. Studies of the properties of the isolated myosin molecule have shown that it possesses three important intrinsic properties: it is an enzyme with ATPase activity, it has the ability to bind to actin, which also activates the myosin ATPase, and it can aggregate with itself to form filaments. Clearly all of these properties are fundamental to the action of myosin in muscle.

The myosin molecule, the largest of the muscle proteins, comprises several distinct polypeptide chains, and much effort has been spent in identifying and characterizing the various subunits of which it is composed. The ability of proteins such as myosin to aggregate can clearly be vital for their function *in vivo*, but it can pose a problem when attempts are made to determine their molecular weights. Largely for this reason, the molecular weight of the whole myosin molecule was found to be very difficult to define. In early studies, the presence of aggregates (dimers, trimers, etc.) in myosin solutions meant that elevated values for the myosin molecular weight were often obtained. Molecular weight values ranging from 420,000 to about 860,000 were, in fact, obtained for the myosin molecules of rabbit skeletal muscle (see Review by Perry, 1967).

Fortunately, more recent work in which care has been taken to avoid aggregation or in which new techniques (such as SDS gel electrophoresis) have been used, has converged on a single agreed model for the structure of the myosin molecule (Lowey, 1971). The molecular weight of the intact myosin molecule is now thought to be close to 500,000. It is made of two similar heavy chains (200,000 daltons each) which interact along part of their length to give a two-chain coiled-coil α-helical structure (the myosin tail) together with a number of so-called "light" chains (about 20,000 daltons each) which combine with the non-α-helical parts of the heavy chains to form two globular regions (heads) at one end of the myosin rod (Fig. 6.1). The myosin molecule is therefore unusual in having a fibrous part (the two-chain coiled coil) directly linked to globular "heads" which have the properties of an enzyme.

In addition to the distinct protein subunits of which myosin is composed, it has also been found that certain proteolytic enzymes (including papain, trypsin, and chymotrypsin) can cleave the myosin molecule at specific sites to yield reasonably homogeneous proteolytic fragments (Szent-Gyorgyi, 1953b). The actions of the enzymes papain and trypsin (Lowey, 1971) are summarized in Fig. 6.1. Two main regions that are susceptible to proteolysis exist in the molecule. Cleavage in one region about 200 Å long of the myosin tail divides the molecule into two subfragments called light meromyosin (LMM), which represents the bulk of the tail, and heavy meromyosin (HMM), which comprises the remainder of the tail together with the two globular heads. Heavy meromyosin can be cleaved further at the second susceptible region to yield the two separate heads each termed HMM subfragment-1 (or S-1) and a short segment of the myosin tail termed HMM subfragment-2 (S-2). These and other proteolytic fragments of myosin will be discussed further in Section 6.2.3 of this chapter. Suffice it to say here that, by studying the

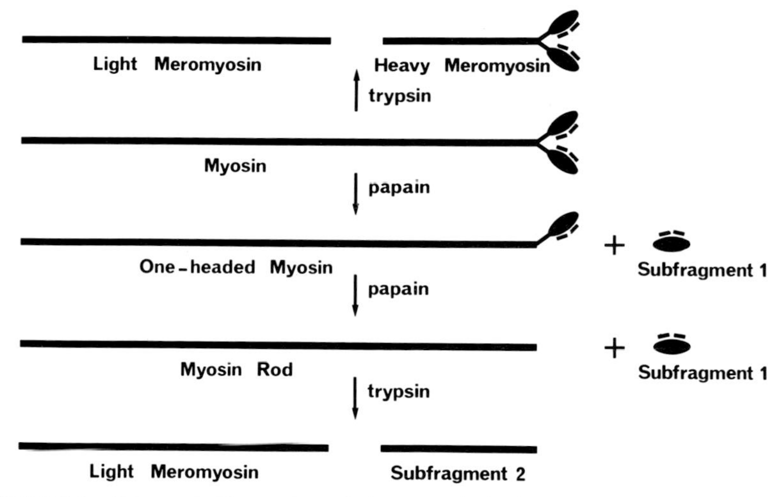

FIGURE 6.1. Schematic illustration of the proteolytic cleavage of myosin (two globular heads with light chains on the end of a long rod) by trypsin and papain. Trypsin cleaves the molecule into light meromyosin (LMM) and heavy meromyosin (HMM). Papain can cleave the molecule close to the globular heads to give either single-headed myosin and one separated head (called HMM subfragment-1 or S-1) or two S-1 moieties and the myosin rod. The myosin rod can be further cleaved into LMM and the head end of the myosin rod which is termed HMM subfragment-2 (S-2). (After Offer, 1974.)

properties and interactions of such fragments and also of the true subunits of myosin (the heavy and light chains), it has been possible to characterize more fully the properties of the whole myosin molecule.

Unfortunately, neither the amino acid sequence nor the structure of the myosin moelcule is known in detail, primarily because of the sheer size and complexity of the protein. As a consequence, any discussion of the myosin molecule can only be carried out, as yet, at a relatively low resolution. Notwithstanding this limitation, the purpose of this chapter is to describe the known features of the myosin molecule and its subunits insofar as they seem relevant to a structural discussion of muscular contraction. An excellent review on myosin has recently been written by Lowey (1979).

6.2. Characterization of the Myosin Molecule

6.2.1. Studies of the Molecular Size and Shape

Early solution studies of myosin indicated that the length of the myosin molecule was about 1600 to 1700 Å and that it had a diameter in the range 20 to 30 Å. As early as 1960, it was suggested that the molecule might be thickened at one end to account for various results of the time (Kielley and Harrington, 1960). Although the evidence used was later shown to be inaccurate, the conclusion itself was confirmed by later studies, especially on LMM and HMM (Holtzer *et al.*, 1961; Lowey and Cohen, 1962). It was found that considerable disagreement occurred between the observed and predicted values for the radius of gyration of the whole myosin molecule if both HMM and LMM were taken to be uniform rods. Lowey and Cohen (1962) therefore proposed a formal model for myosin consisting of a 20 Å diameter rod 800 Å long (LMM) linked to an ellipsoid of major axis 800 Å and minor axis 28 Å. More direct evidence obtained by Rice (1961a,b) confirmed the basic features of this kind of myosin model. He studied shadowed myosin molecules (using the mica replication technique mentioned in Chapter 3; Hall, 1956) and obtained electron micrographs in which match-shaped particles were seen. Further similar experiments confirmed this basic form for the myosin molecule (see Perry, 1967) and established that the molecule consists of a rod of length about 1500 Å and diameter 20 Å terminating in a globular region of diameter about 30 to 60 Å and length about 200 Å (H. E. Huxley, 1963; Zobel and Carlson, 1963; Rice 1964; Rice *et al.*, 1966).

Later work of a similar kind but using specimens that had been rotated during the metal-shadowing procedure (Slayter and Lowey,

1967; Lowey *et al.*, 1969), revealed that the globular region at the end of the myosin rod was actually composed of two similar globular heads that seemed to be rather flexibly joined to the rod and that were estimated to have a diameter of about 70 Å (see Fig. 6.2).

With this structure for the myosin molecule established, two further questions remained to be answered. First, could improvements in this type of technique reveal details of the shape and size of the two myosin heads? And second, do the myosin molecules in different muscles have a similar structure? We shall return to the first of these questions in Section 6.2.4. The second has been the subject of a recent detailed study by A. Elliott *et al.* (1976) who have used a modification of the mica replication technique to obtain much improved micrographs of myosin prepared from a variety of sources. The special feature of their technique (also described in Section 3.3.2) was that the specimens were first rapidly cooled in a solution containing the antifreeze glycerol and then frozen, dried, and shadowed while cold. This reduced the damage caused by the action of surface tension forces during the normal procedure of drying protein solutions at room temperature. Using this technique, A. Elliott *et al.* (1976) found that good visualization of the two myosin heads could be observed even if unidirectional shadowing were used, and this was taken to be because of the improved preservation of the myosin molecules resulting from the reduced surface tension forces. These forces could otherwise tend to clump the two heads together.

In their comparative study, A. Elliott *et al.* (1976) used myosins prepared from adult and embryonic vertebrate skeletal muscle, vertebrate cardiac and smooth muscle, invertebrate striated and smooth muscles, from platelets, and from the embryonic brain. Although these various myosins are known to have similar molecular weights (all about 500,000) and similar chain compositions, they do differ significantly in their chemical properties, in their amino acid sequences, and in their aggregation properties. Despite this, it was found that in the electron microscope they all appeared to have the same characteristic structure of a long thin tail terminating in two globular heads (Fig. 6.2). Estimates of the tail length, which were complicated slightly by the difficulty in defining the junction point between the rod and the heads, were very similar in each case and were taken to be identical to within ± 100 Å. The estimated common length was 1500 Å (±200 Å). In another study, Hatano and Takahashi (1971) showed that slime mold myosin has a similar size and shape.

That this characteristic shape is conserved despite the differences that exist between other features of different myosin molecules seems to be clear evidence that the shape itself is important to the proper function of the myosin molecule. The fact that no single-headed myosins seem to

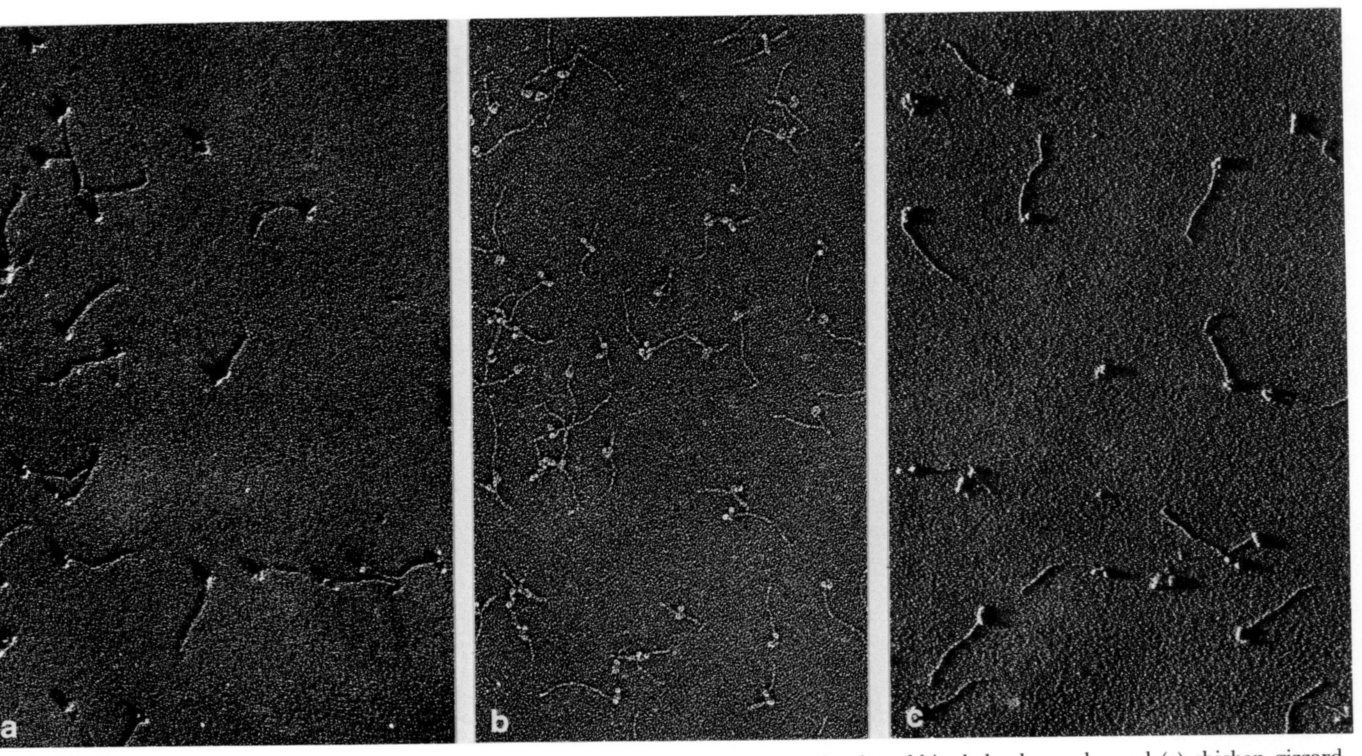

FIGURE 6.2. Shadowed isolated myosin molecules from (a) chicken cardiac muscle, (b) rabbit skeletal muscle, and (c) chicken gizzard muscle, all prepared by the mica replication technique (see text). In each case the molecules appear to have the same shape and dimensions. Magnification $\times$ 100,000 (reduced 43% for reproduction). From A. Elliott *et al.* (1976) and A. Elliott and Offer (1978). Note that (a) and (c) were shadowed unidirectionally and (b) was rotary shadowed.

occur in nature may mean, as A. Elliott *et al.* (1976) suggested, that two heads are needed for the contractile event. But it could simply mean that it is easier to synthesize two identical heavy chains than it is to synthesize two different ones.

6.2.2. Proteolytic Fragments of Myosin

The cleavage of myosin into its characteristic proteolytic fragments LMM and HMM, and of HMM into subfragments S-1 and S-2 has already been described (Fig. 6.1). It has also been mentioned that myosin itself possesses three characteristic properties, the ability to aggregate, the ATPase activity, and the ability to bind actin. Studies of the individual fragments have shown that LMM possesses the ability to bind strongly to itself under physiological conditions but does not possess ATPase or actin-binding properties. On the other hand, HMM will not aggregate readily, but it does bind to actin and also has an ATPase per mole equal to that of the intact myosin molecule. In fact, the ATPase and actin-binding properties reside in the S-1 subfragments of HMM, the S-2 moieties having a more structural role.

Analysis of the amino acid composition and α-helix content of HMM, LMM, and HMM S-1 and S-2 has shown that both LMM and HMM S-2 are highly α-helical (both greater than 95%), whereas HMM S-1 is only 37% α-helical. Table 6.1 lists the molecular weights, dispersion constants, and estimated helix content of these various myosin preparations. The values are also given for the intact myosin rod which can be prepared from precipitated myosin by papain digestion. Like LMM, the rod has a very high α-helix content (greater than 95%). It is interesting to note that single-headed myosin (Fig. 6.1) appears to behave in a very similar way to intact myosin from vertebrates (Margossian and Lowey, 1973; Lowey and Margossian, 1974; Cooke and Franks,

TABLE 6.1. Molecular Weights and Helix Contents of Myosin and Its Proteolytic Derivatives

Parameter	Myosin	HMM	S-1	S-2	LMM	Rod
Mol. wt. $\times 10^{-3}$	470 ± 10[b]	340 ± 10	115 ± 5[b]	62 ± 2[b]	140 ± 5	220 ± 10
Dispersion constant (b_0)	400 ± 10	320 ± 10	230 ± 10	610 ± 10	630 ± 10	660 ± 10
Helix content (%)[a]	63	51	37	>95	>95	>95

[a] Calculated assuming $b_0 = -630$ is equivalent to 100% alpha helix. From Lowey *et al.* (1969) and Weeds and Pope (1977).

[b] Estimates of these molecular weights vary according to the particular preparation involved. In the text, the myosin molecular weight is quoted as 500,000 which represents a rounded up figure that is probably close to the true value. Weeds and Pope (1971) quote a higher value for the S-2 molecular weight.

1978). In particular, single-headed myosin will produce contraction in an actomyosin thread—a result that leads one to wonder about the usefulness of two heads.

The amino acid compositions of myosin, S-1, and the myosin rod (given in Table 6.2) are clearly consistent with the high α-helix content in the rod (high content of Glu and low content of Gly and Pro) and the globular nature of the S-1 heads (much lower Glu content, etc.). Comparison of the molecular weights and lengths (~900 Å and 1300 Å, respectively) of LMM and myosin rod indicated that they must contain two α-helical chains, since single α-helices of the same molecular weight would have calculated lengths almost twice the value observed. The sequences of parts of the myosin chain have recently been determined by Elzinga and Collins (1977) and Trus and Elzinga (1980).

6.2.3. The Subunit Structure of Myosin

Vertebrate Skeletal Muscle. Although there were considerable difficulties in defining the subunit structure of myosin in early studies, it can now be established readily by using SDS gel electrophoresis. Rabbit skeletal muscle myosin consists of two heavy chains of 200,000 daltons

TABLE 6.2. Amino Acid Compositions of Myosin, HMM, and Myosin Rod[a]

Amino acid	Myosin	HMM Subfragment I	Papain rod
Alanine	9.0	7.8	9.7
Arginine	5.0	3.8	6.4
Aspartic acid[b]	9.9	9.4	10.1
Glutamic acid[b]	18.2	13.0	25.0
Glycine	4.6	6.8	2.3
Half-cystine[c]	1.0	1.2	0.5
Histidine	1.9	2.0	1.7
Isoleucine	4.9	5.9	4.1
Leucine	9.4	8.3	11.3
Lysine	10.7	9.2	12.2
Methionine	2.7	3.1	2.6
Phenylalanine	3.4	5.8	0.8
Proline	2.5	4.1	0
Serine	4.5	4.5	4.5
Threonine	5.1	5.4	4.3
Tyrosine	2.3	3.8	0.7
Valine	5.0	6.1	3.8

[a] Lowey *et al.* (1969). Given in residues/100 residues, rounded to nearest 0.1 residue. Table after Fraser and MacRae (1973).
[b] Includes amide.
[c] Estimated as cysteic acid.

and four light chains, two of 19,000 daltons, one of 21,000 daltons, and one of 17,000 daltons. The 19,000-dalton subunits can be dissociated from myosin by the action of Ellman's reagent, 5,5′-dithiobis-(2-nitrobenzoate), which is usually known as DTNB (Weeds, 1969; Gazith *et al.*, 1970; Weeds and Lowey, 1971). These light chains are therefore termed the DTNB light chains. Although their removal from myosin seems to remove the ability of myosin to bind calcium, this appears to have a negligible effect on the ATPase of the molecule *in vitro*. So their role is, as yet, undefined even though there is a strong indication that they may be involved in some way in thick filament regulation (see below).

Further treatment of the myosin with alkaline solutions (pH 11) causes dissociation of the remaining two light chains (Kominz *et al.*, 1959). These are therefore known as the alkali light chains, A-1 (21,000 daltons) and A-2 (17,000 daltons) (Frank and Weeds, 1974). Their removal from myosin results in a loss in the ATPase and actin-binding properties of the molecule. The A-1 and A-2 light chains are, in fact, closely related since much of their sequence is common (Fig. 6.3), and the part that is not is termed the *difference peptide.* It has been shown that these two light chains occur in differing amounts, there being about 50 to 100% more of the A-1 light chain in preparations of whole rabbit muscle (Weeds *et al.*, 1975; Sarkar, 1972; Lowey and Risby, 1971). This suggested (but did not prove) that the alkali light chains are probably not both present in a single myosin molecule. More likely, there are two populations of myosin molecules, one having two A-1 light chains per molecule and the other having two A-2 light chains. In addition, Gauthier *et al.* (1978) have shown that in developing fast-twitch mammalian muscle (rat diaphragm), two distinct classes of myosin heavy chain are present ("slow" and "fast" myosin) in all fibers. However, as differentiation into characteristic fiber types proceeds, each fiber becomes associated with a predominant myosin type.

Early on in the study of different myosins, it seemed to be likely that particular heavy chain types were associated with a particular type of alkali light chain and that these occurred in different fiber types within the same muscle so that individual fibres contain only a single myosin species. However, it was later shown by Weeds *et al.* (1975) that both of the alkali light chains are present in single fibers of histochemically homogeneous muscles. The different abundance of the two light chains strongly suggested that there are myosin isoenzymes in each of which there is only one kind of light chain. In theory, if there were two different heavy chain types (A and B), there would be three possible heavy-chain structures for a myosin molecule, AA, AB, and BB, each of which could bind with two DTNB light chains and any combination of

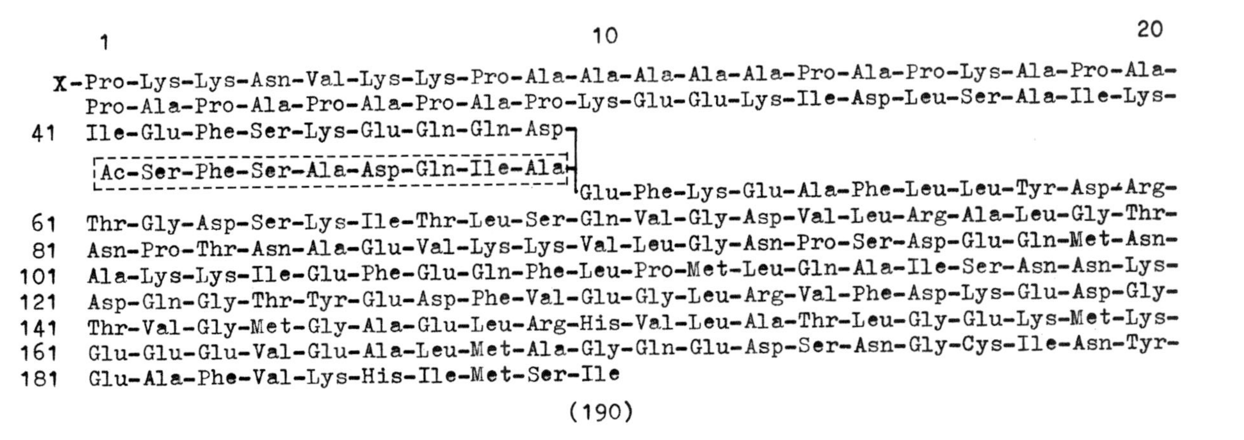

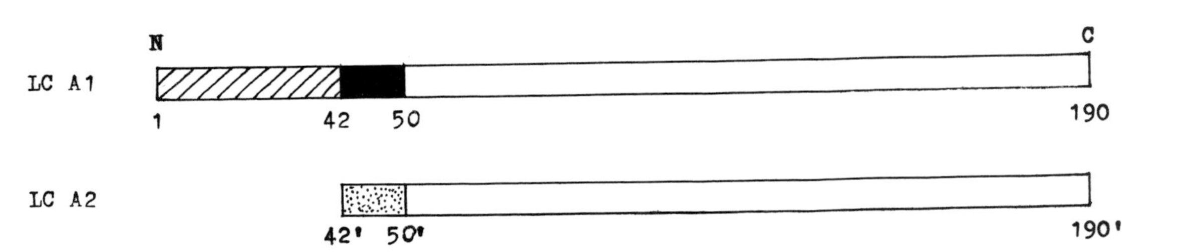

FIGURE 6.3. The amino acid sequences of the two alkali light chains in rabbit skeletal muscle myosin. The lower part of the diagram indicates that the light chains have a common sequence from residue 50 to the C-terminal end at residue 190. Residues 42 to 49 in both chains have different but related sequences. The A1 light chain has a further 41 residues at its N-terminal end, but the A2 sequence starts at residue 42. The top of the diagram shows the details of both sequences starting with residues 1 to 49 of A1, then residues 42 to 49 (boxed in) of A2, and finally listing the common sequence from residue 50 to residue 190. (From Frank and Weeds, 1974.)

the two alkali light chains A-1–A-1, A-1–A-2, or A-2–A-2, thus giving a total of nine different alternatives. Since it seems that particular adult fibers probably contain only one predominant heavy-chain type, structures of the kind A + A-1,A + A-1 (homodimer), A + A-1,A + A-2 (heterodimer), or A + A-2,A + A-2 (homodimer) would be predominant in a single fiber.

Striking evidence about the properties of the alkali light chains has been obtained by Weeds and Taylor (1975) who managed to prepare homogeneous S-1 preparations from solutions of synthetic myosin filaments using α-chymotrypsin. They found that two types of S-1 could then be separated by chromatography on DEAE–cellulose, and these were found to contain either the A-1 or the A-2 light chains (the DTNB chains being removed during the treatment with α-chymotrypsin). Subsequent tests (Weeds and Taylor, 1975; Wagner and Weeds, 1977) showed that, in the presence of actin, the two S-1 preparations gave different ATPase rates, that of the A-2 S-1 being higher. However, this difference was only observable at very nonphysiological ionic strengths, so its significance is unclear.

The problem remained to determine if both light chains could be present in a single myosin molecule. In fact, the results reported recently by Holt and Lowey (1977) and Wagner and Weeds (1977) have shown that most of the myosin in chicken and rabbit does occur as homodimers, but Lowey *et al.* (1979) have shown that heterodimers probably do exist and may represent up to 30% of the myosin population. In other experiments, Gauthier and Lowey (1977) and Gauthier *et al.* (1978) made use of the difference peptides betwen the A-1 and A-2 light chains (Fig. 6.3); they prepared antibodies to these, coupled the antibodies to a fluorescent marker, and used the coupled antibodies to label rat diaphragm. It was then found that in transverse sections, the fibers of developing muscles showed little evidence of the presence of the A-2 light chain (see also Chi *et al.*, 1975), but adult fibers contain both light chain types. In general, it was concluded that although most adult fibers do contain both light chains, the proportions of these vary characteristically with fiber type. Note finally that it has not yet been totally ruled out that the light chain populations are associated with different myosin heavy chains. Hoh and Yeoh (1979) have demonstrated several myosin isoenzymes in muscles containing only a single light chain type, and these contained both homodimers and heterodimers of the heavy chains.

Recent work by Perry and his collaborators (Perrie *et al.*, 1973; Pires *et al.*, 1974; Frearson and Perry, 1975; Frearson *et al.*, 1976a,b; Perrie and Perry, 1970; Perry *et al.*, 1975) has revealed an interesting property of the DTNB light chains from rabbit skeletal muscle. Although on SDS gels the light chains always ran as three bands corresponding to the molecular weights of 22,800 (A-1), 18,500 (DTNB), and 15,500 (A-2), respectively, if corresponding preparations were run on polyacrylamide gels in the presence of 8 M urea, four bands were seen (Fig. 6.4). Both of the two central bands were shown to be from the DTNB light chains, the difference being that some of these light chains contained organic phosphate (they were phosphorylated), whereas the others did not. It was later shown that the nonphosphorylated DTNB light chains could be converted to the phosphorylated form by the action of a crude protein kinase preparation from skeletal muscle in the presence of ATP. Similarly, the phosphorylated chains could be dephosphorylated by the action of the alkaline phosphatase of *E. coli.* It was also found that the proportion of phosphorylated light chains in tissue preparations gradually decreased as the preparations were stored—a result taken to indicate that in intact resting muscle the DTNB light chains may be predominantly in the phosphorylated form.

Later work by Pires *et al.* (1974) showed that in vertebrate skeletal muscle there is a 77,000-dalton kinase specific to the myosin DTNB light chains, and this was termed myosin light chain kinase. This important result became even more interesting when the action of the kinase (with ATP as substrate) was found to be as Ca^{2+}-dependent as that of troponin in the thin filaments. These results imply that the kinase will be fully active during muscle contraction when the ATPase activity is high.

In a similar way, a myosin light chain phosphatase (monomer mol. wt. 70,000) has been discovered by Morgan *et al.* (1976). This may possibly be complexed *in vivo* either with myosin light chain kinase or with other components of the sarcoplasm. The combined system comprising myosin light chain kinase, myosin light chain phosphatase, and the

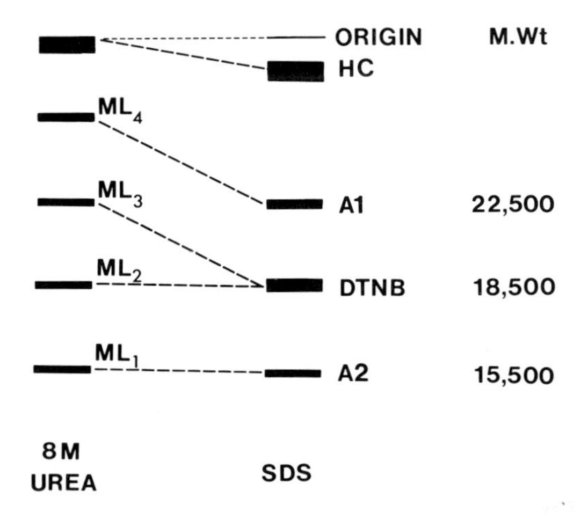

FIGURE 6.4. Schematic representation of the separation of the myosin light chains on SDS gels and in the presence of 8 M urea. Three light chain bands were seen on SDS gels, but urea gels separated the light chains into four bands. In the latter, the middle two bands (ML_2 and ML_3) were both found to contain the DTNB light chain, but only some of these chains (in the ML_3 band) were phosphorylated. (After Perrie *et al.*, 1973); the molecular weights quoted are their values.)

DTNB light chains was found to behave as a Ca^{2+}-sensitive ATPase, but this was probably too slow to confuse investigations of the actin-activated myosin ATPase. However, the actomyosin ATPase was found to be little affected by phosphorylation of the DTNB light chains, although there is evidence that some kind of involvement in this interaction must occur.

From structural studies (to be described in Chapter 10), it seems probable that vertebrate muscle thick filaments, like the thin filaments, are "switched on" in some way when the muscle is activated. It is clear that the DTNB light chains may be involved in the process, and some indications that this may be so have already been obtained (Werber *et al.*, 1972; Lehmann, 1977a,b). According to Morgan *et al.* (1976), such a role for the DTNB light chain is suggested by the fact that actin activates the Mg^{2+}-stimulated ATPase of chymotrypsin-prepared HMM much more effectively than that of typsin-prepared HMM in which the DTNB light chain is modified. Similarly, Wagner and Weeds (1977) have suggested that this light chain can affect the proteolytic susceptibility of the S-1-to-S-2 link and that it may thus have a structural role in regulating the thick filament rather than a direct effect on the myosin ATPase. This idea will be discussed further in Chapters 10 and 11. Note that Morgan *et al.* (1976) actually referred to the DTNB light chain as the P light chain, since similar light chains in myosins from sources other than rabbit skeletal muscle were found to be similar in their interactions with myosin light chain kinase and phosphatase even though they often had slightly different molecular weights and could not be dissociated from myosin by the action of DTNB. The term P light chain is therefore more generally applicable to the 18,000- to 20,000-dalton light chain found in different vertebrate muscles. In this context, it is worth noting that, unlike the negative results on the effect of phosphorylation of the P light chain in muscle, it was shown by Adelstein and Conti (1975) that phosphorylation of this light chain in preparations of human platelet myosin leads to a dramatic increase in the actin-activated hydrolysis of ATP in the presence of Mg^{2+}. Phosphorylation also seems to be involved directly in the regulation of vertebrate smooth muscle (see below).

Invertebrate Muscle. The subunit composition of the myosins in invertebrate muscles is rather different from that in vertebrate muscles, and the light chains have distinct properties. We have already seen that in vertebrate muscle Ca^{2+} regulation of the thin filaments occurs through the action of the troponin–tropomyosin system, but the thick-filament-regulating system has yet to be clearly identified. However, in molluscan muscle there is little troponin on the thin filaments, and tropomyosin, although present in the thin filaments, is not needed for regulation (Lehman *et al.*, 1972). In this case, the Ca^{2+} dependence of the actomyosin interaction is definitely associated with the myosin (Kendrick-Jones *et al.*, 1970, 1976) and is conferred on the myosin by one type of light chain.

The light chains of scallop myosin are of two distinct types: the so-called EDTA light chains obtained by reducing the divalent cation concentration with EDTA (ethylenediaminetetraacetate) and the thiol light chains that contain one S–H group and that are only released when the myosin is denatured. The thiol light chains are thought to be analogous to the alkali light chains in vertebrate muscle since they are essential for the ATPase activity of the myosin. On the other hand, the EDTA light chains are associated with Ca^{2+} binding, and the removal of one of the two chains from myosin causes the complete loss of the myosin Ca^{2+} sensitivity; they are clearly strongly implicated in the Ca^{2+} regulation of the myosin ATPase.

The thiol and EDTA light chains have similar molecular weights (17,000) and run as a single diffuse band on SDS gels. But they have different sequences and can be separated on the basis of charge by gel electrophoresis in the presence of urea. Each myosin contains two moles of each type of light chain (Kendrick-Jones *et al.*, 1976), each myosin head being associated with one thiol and one EDTA light chain. Kendrick-Jones *et al.* (1976) have also shown that different molluscan muscles contain myosin molecules each with two regulatory light chains (although these are not released by EDTA) and two "alkali" light chains (i.e., light chains analogous to the EDTA and thiol light chains of the scallop).

These results are complicated by the fact that EDTA treatment of scallop myosins only yields one mole of EDTA light chain per mole of myosin (leaving one mole still bound), but Ca^{2+} sensitivity of the myosin is totally lost. This result could be explained if removal of one EDTA light chain causes a change in the myosin structure that "locks" the second EDTA chain into position (Kendrick-Jones, 1975). The complete loss of Ca^{2+} sensitivity could be taken to mean that calcium regulation requires cooperativity between the two myosin heads in a single molecule, an idea which is supported by observations that HMM S-1 with a bound EDTA light chain is not Ca^{2+} sensitive, but HMM with two moles of EDTA light chain bound is fully sensitive.

Another twist to the light chain story was reported by Kendrick-Jones (1974). He showed that scallop myosin desensitized by the removal of one mole of EDTA light chain could be resensitized either by the readdition of purified EDTA light chain, or, strangely, by the addition of the DTNB light chain (P light chain) from vertebrate muscle. This is despite the fact that the P light chain has not yet been shown to have a similar role in its parent muscle (even though it is associated with some calcium binding; Bremel and Weber, 1975).

Insect Flight Muscle. It was shown by Lehman *et al.* (1973), Bullard *et al.* (1973), and Lehman *et al.* (1974, 1977a,b) that in insect flight muscle Ca^{2+} regulation is associated both with the thin filaments (as in vertebrate muscle) and the thick filaments (as in molluscan muscle). Annelids and several examples (other than insect flight muscle) of arthropod muscles also contain this dual regulation system (Lehman *et al.*, 1973; A. G. Szent-Gyorgyi, 1975; Lehman and Szent-Gyorgyi, 1975). In addition, it was shown (Lehman *et al.*, 1974) that the presence of tropomyosin enhances the ATPase activity in the presence of calcium whether the myosin is Ca^{2+} dependent or has been rendered independent of Ca^{2+}.

It has been shown (Winkelman and Bullard, 1977) that there are two myosin light chains in insect flight muscle: one of 30,000 daltons and one of 20,000 daltons. The former can be phosphorylated, and it appears that this phosphorylation, as well as Ca^{2+} binding, is necessary for myosin regulation in this muscle.

Vertebrate Smooth Muscle. Like insect muscles, vertebrate smooth muscles may be Ca^{2+} regulated through the myosin light chains (Sobieszek and Small, 1976; for a completely contrary view, see Mikawa *et al.*, 1977). But it is now becoming apparent that this regulation takes place by activating a myosin kinase so that one of the myosin light chains (20,000 daltons) is phosphorylated. Once phosphorylated, the system no longer appears to be calcium sensitive (Sobieszek and Small, 1976; Adelstein *et al.*, 1976; Chacko *et al.*, 1977). Dephosphorylation of the light chain by a smooth muscle phosphatase is associated with a loss in actin-activated ATPase activity and therefore with relaxation of the system.

Comparative Study of Myosin Regulatory Light Chains. In a more general survey of the properties of the regulatory light chains from a variety of sources, Kendrick-Jones *et al.* (1976) combined these light chains with desensitized scallop myosin (i.e., myosin lacking 1 mole of EDTA light chain) and tested for calcium binding and Ca^{2+} sensitivity. They found that two classes of light chain could be distinguished. Regulatory light chains from several molluscs and from vertebrate smooth muscles restored full calcium binding to pure desensitized scallop myosin. On the other hand, similar light chains from vertebrate skeletal, cardiac, and fast decapod muscles did not restore calcium binding or sensitivity to scallop myosin unless the myosin were complexed with actin. They noted that it was this last group of muscles in which no myosin-linked regulatory system had yet been discovered by *in vitro* tests. In this case, it may well be that the myosin-linked regulation is rather more subtle, as discussed in Chapter 10.

One final point worth making is that it seems very likely from the results of Kendrick-Jones *et al.* (1976) that the regulatory light chains may be attached to the myosin heads reasonably close to the actin binding site. This is indicated by the effect that actin has on the properties of the regulatory light chains of vertebrate skeletal muscle when added to desensitized scallop myosin. But they cannot be close enough to block actin attachment, and the recent results of Weeds and Pope (1977) suggest that they may also have an effect on the proteolytic cleavage site of S-1 from HMM.

6.2.4. Shape and Size of the Myosin Head

One feature of the myosin molecule that is clearly of vital importance in our understanding of muscle contraction is the size and shape of the two myosin heads. But in addition, as will be discussed in Chapter 10, Section 10.2, the knowledge of the size and shape of the myosin head is also crucial for a realistic interpretation of the X-ray diffraction evidence on relaxed and contracting muscles, and to date this is the most useful way of studying the movements of the myosin head involved in the contractile event. Unfortunately, there is still considerable uncertainty about this aspect of myosin morphology. Indeed, it is not even known if the overall shape of the head remains more or less constant during the contractile cycle. On the contrary, myosin S-1 is already known to be rather labile (Lowey *et al.*, 1969), and the observed S-1 shape may actually be a function of the technique used to study it.

The aim of this section is to summarize what evidence there is about this important aspect of the myosin molecule and to outline what appears to be the most likely shape of the myosin head. At the same time, the relative configuration of the two heads in the myosin cross bridge (analogous to HMM) will also be considered. The types of methods that have been used to tackle this problem fall into two main categories. In one, the physical properties of HMM S-1 in dilute solutions have been studied (e.g., by measurements of viscosity, sedimentation, light scattering, fluorescence depolarization, and X-ray and neutron scattering), and axial ratios for the equivalent ellipsoids or in some cases values for the radius of gyration of the molecules have been determined. In the other, the problem has been approached more directly by studying electron micrographs either of intact isolated myosin molecules or of actin filaments decorated with S-1 (see Chapter 5, Section 5.2.4). As a rule, these two approaches have tended to lead to somewhat different conclusions about the form of the myosin head. This may mean either that the shape is genuinely variable or that there is a single shape which is not immediately apparent but which will account for all of the evidence.

It has already been mentioned that early electron micrographs of rotary-shadowed isolated myosin-molecules showed two apparently globular regions on the end of the myosin rod, and such pictures gave the first indications of the size of the myosin heads. They were said to resemble spheres of apparent diameter 70 Å (Lowey *et al.*, 1969), although possibly with a slight asymmetry. According to these authors, spherical molecules of molecular weight 115,000 would have a calculated diameter of about 64 Å, a value reasonably close to the result obtained from measurements of electron micrograph images. But their value for the viscosity of S-1 solutions (0.06 dl/g) was much higher than typical viscosities of proteins known to be fairly spherical (0.035 dl/g). They concluded that a prolate ellipsoid of axes 90 Å and 54 Å (giving the right volume for a molecule of molecular weight 115,000) might be a better approximation to the true shape of the myosin heads.

Since that early work, various groups have tackled the problem, and their conclusions are summarized below.

Three-Dimensional Reconstruction. The results of Moore *et al.* (1970) on the three-dimensional reconstruction of S-1-decorated thin filaments have already been described (Chapter 5), Section 5.24). These authors concluded (Fig. 6.5d,e) that S-1 is banana-shaped with approximate dimensions 150 Å × 45Å × 30 Å, (the latter being in the plane of curvature). It will be remembered that this result depended heavily on the assignment of the zero contour level in the reconstructed density map and on the assumption of helical symmetry.

Fluorescence Anisotropy Decay. Mendelson *et al.* (1973) used the method of fluorescence anisotropy decay to measure the rotational diffusion of S-1 in solution and by this means concluded that S-1 must be "longer than 165 Å," (assuming it was a prolate ellipsoid).

X-Ray and Neutron Scattering. These methods rely on the predicted scattering distribution at low angles from a collection of identical, randomly oriented, noninteracting, homogeneous particles of radius of gyration R_G. The scattered intensity I at a small angle θ is given by:

$$I \propto I_0 \exp\left[-R_G{}^2\left(\frac{4\pi^2S^2}{3}\right)\right]$$

where S is $(2/\lambda)\sin(\theta/2)$ and λ is the wavelength of the radiation used. From this a plot of $\ln I$ against $4\pi^2S^2$ should be linear with a slope $-R_G^2/3$ (see Guinier, 1963).

In this way, X-ray scattering measurements of S-1 in solution by Kretzschmar *et al.* (1976, 1978) gave a value for the radius of gyration

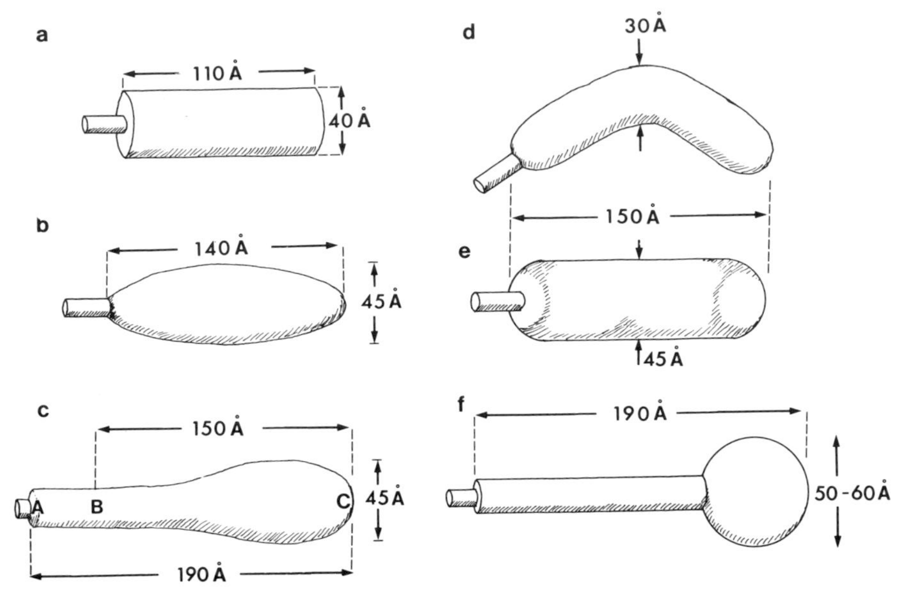

FIGURE 6.5. Illustrations of various models for the shape of a single myosin head. (a) A cylinder with a radius of gyration (R_G) of 34.76 Å. (b) A prolate ellipsoid of R_G value 34.4 Å. (c) Approximation of the new pear-shaped model of A. Elliott and Offer (1978). Only region BC is thought to correspond to the S-1 portion of the myosin head, and this has an R_G value of about 35 Å. (d and e) Two orthogonal views of the banana-shaped model of Moore *et al.* (1970). One of the most significant features here is the marked curvature of the myosin head. (f) A simplified version of (c) in which the structure is modeled using only sphere and a cylinder (Offer and Elliott, 1978). On the left of each model is a hypothetical connecting link to myosin S-2.

(R_G) of S-1 equal to 32 Å, and analogous experiments using thermal neutrons instead of X-rays (D. L. Worcester, G. Offer, R. Starr, and J. M. Squire, unpublished) gave similar results with a value of R_G equal to 35 Å. These R_G values are hard figures, but to interpret them in terms of S-1 shape requires assumptions to be made. Unlike other solution methods where the effects of hydration can be significant, hydration effects are not significant in this case and can be ignored.

In interpreting the R_G value, the first assumption that one needs to make is about the volume of the particle. It has already been mentioned that S-1 has a molecular weight of about 115,000 and, knowing its amino acid composition and hence its approximate density (about 1.36 gm/cm^3), the volume (V_p) of the molecule can be calculated. It is V_p = 115,000/(1.36 × 6.02 × 10^{-1}) = 140,500 Å^3. The second assumption concerns the shape of the molecule. If the heads are taken to be spherical, then an R_G value of 35 Å would require the sphere radius to be about 45 Å, and hence the sphere volume would be about 390,000 Å^3.

This is clearly much too large and indicates directly that the myosin heads must be elongated. Satisfactory values of both R_G and V_p could be obtained from a cylindrical rod 110 Å long and 40 Å in diameter (Fig. 6.5a) or from an ellipsoid of axes 140 Å, 45 Å, and 45 Å (Fig. 6.5b). But in fact the myosin head may really be much more complex in shape as indicated below. In order to get more direct evidence about shape from X-ray and neutron scattering studies of the myosin head, it is necessary to record intensities at higher angles of diffraction where the effects of the interference of the beams scattered by different parts of the same molecule can be observed. A first step in this has been reported by Mendelson and Kretzschmar (1979), who suggest a structure like that in Fig. 6.5f but 130 Å long. It is hoped that such studies will help to evaluate the results described in the next section.

Shadowed Myosin Molecules. The clearest visual demonstration of the myosin head shape to be obtained so far has been provided by the recent work of A. Elliott and Offer (1978) who used the mica replication technique described earlier (Chapter 3, Section 3.3.2). In their beautiful electron micrographs, the two myosin heads on an individual myosin molecule could often be seen clearly, and this allowed Elliott and Offer to measure carefully the relative dimensions of different regions of a single head right back to its point of attachment to the myosin rod. By averaging similar measurements on a number of exceptionally well preserved heads and allowing for the thickness of the metal shadow, A. Elliott and Offer concluded that each head was about 180 to 190 Å long and was pear-shaped, being larger towards the tip than near the junction with the rod. Figure 6.5f shows a first approximation to this kind of shape (a 50- to 60-Å diameter sphere on the end of a 20-Å diameter rod; A. Elliott and Offer, 1978). But Fig. 6.5c represents more realistically the details of the measurements made in Elliott and Offer's latest work. The head is about 45 Å in diameter at its widest point and tapers down to about 20 Å at its narrow end. In cross section it is not circular but is slightly flattened.

With this shape in mind, it is clearly appropriate to ask how it fits in with the results from the other methods described above. A. Elliott and Offer claim that this shape will account well for the earlier observations of near-spherical heads in previous studies using less well-developed shadowing methods, since it is bulbous at one end. Because of its length, it can also account to some extent for the results of Moore *et al.* (1970). Although the shape shown in Fig. 6.5d,e approximates the head shape used in the model of Moore *et al.* shown in Fig. 5.11c, their original reconstructions (Fig. 5.11a) do actually show that the head could be larger at its actin-bound end than at its free end. In this sense, the new results of Elliott and Offer are consistent with the three-dimensional reconstructions. However, A. Elliott and Offer have only rarely found any evidence for a marked curvature along the length of the head, and this is one of the features thought by Moore *et al.* to be crucial for producing the characteristic arrowhead appearance on S-1-decorated thin filaments. Some uncertainty therefore remains about this feature of the head.

A further problem becomes apparent when the radius of gyration of the head shown in Fig. 6.5c is compared with the results from X-ray and neutron scattering studies of S-1 in solution. The predicted R_G value of about 50 Å is very much higher than the observed value of 32 to 35 Å, and this has led A. Elliott and Offer to suggest that myosin S-1 may represent only part of the whole intact myosin head that they see in their micrographs. For example, if the S-1 fragment consisted only of the outer 140 to 150 Å of the myosin head shown in Fig. 6.5c (i.e., the region BC), then the R_G value would reduce to only 35 to 36 Å, which is in good agreement with the experimental R_G values. This S-1 size and shape would also account reasonably well for the hydrodynamic properties of S-1 (Lowey *et al.*, 1969; Yang and Wu, 1977) and for the length (if not shape) observed by Moore *et al.* (1970).

A further interesting feature of the results of Elliott and Offer is that the inner end of the myosin head (i.e., region AB in Fig. 6.5c) is reported to be larger in diameter than would be a single extended protein chain or even a single α-helix. Like the remainder of the myosin head, it must also represent a region of folded polypeptide chain about 20 Å or more in diameter.

6.2.5. Flexibility of the Myosin Molecule

It has already been shown that the myosin molecule is susceptible to proteolysis by papain and trypsin in two main regions (Fig. 6.1). Region 1 on the myosin rod divides the molecule into LMM and HMM, and Region 2 divides the HMM fragment into HMM S-1 and HMM S-2. Because of the proteolytic susceptibility of Region 1, it has long been supposed that the myosin chain is not regularly α-helical in this region (Mihalyi and Harrington, 1959; Segal *et al.*, 1967) and that it could therefore be flexible (Pepe, 1967b; H. E. Huxley, 1969). Indeed, the existence of a hinge region along the myosin rod is crucial to most models of cross bridge action during contraction (H. E. Huxley and Brown, 1967; Pepe, 1967b) in which the heads move out towards the actin filaments by different amounts at different sarcomere lengths (see Chapter 7, Section 7.1.2). Flexibility of the myosin heads on the end of the myosin rod has previously been inferred from electron micrographs that showed that the heads could lie at a variety of positions relative to the rod (Slayter and Lowey, 1967). This too has been incorporated into a model of cross bridge action (H. E. Huxley, 1969) in which it was suggested that the heads could then interact with actin in an identical manner whatever the separation between the actin and myosin filaments. The fluorescence depolarization results of Mendelson *et al.* (1973) supported the idea that the myosin heads are flexibly linked to the myosin tail.

The new results of A. Elliott and Offer (1978), as well as giving a clearer idea about the shape of the myosin head, have also provided direct evidence for the existence of hinge regions in the myosin molecule. From measurements of the angles α and β (Fig. 6.6a–c) between the two heads and the myosin tail, it was found that there was little correlation between the position of one head and that of the other. On the whole, there was fairly uniform distribution of angular positions except that few heads were seen with angles α and β less than about 60°. These results indicated clearly that the heads in isolated myosin molecules must be flexibly joined to the myosin rod.

Apart from this, A. Elliott and Offer frequently observed molecules that were bent sharply back onto themselves (Fig. 6.6d) at a well-defined position along the myosin tail. Their measurements showed that the

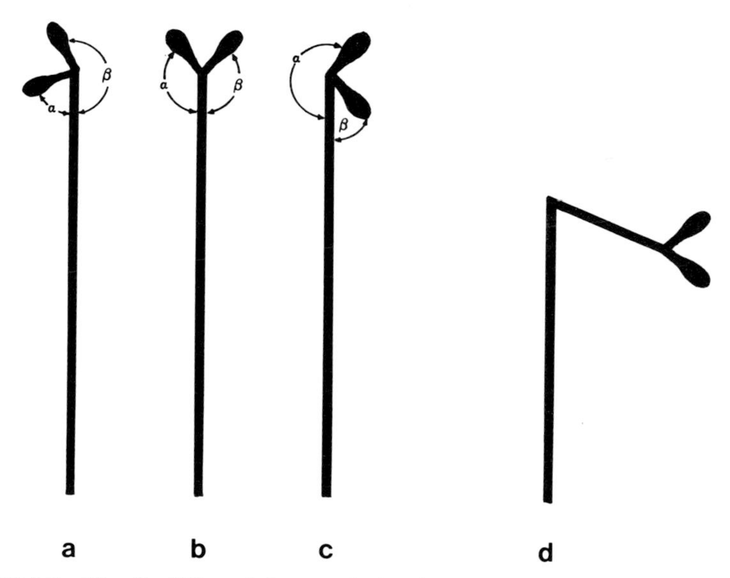

FIGURE 6.6. The flexibility of the myosin heads on the end of the rod (a–c) and the definition of their orientations in terms of the angles α and β. Observed values of α and β lie fairly uniformly in the range 60° to 300°, but a few occur outside this range. (d) The observed "kink" in the myosin rod of some molecules at a position that is probably close to the cleavage site between LMM and HMM. (After A. Elliott and Offer, 1978.)

bend occurred at about 1100 to 1150 Å from the free end of the tail (50% of the measurements in the range 1110 ± 20 Å) and at about 400 to 450 Å from the head end of the tail (50% of the measurements in the range 430 ± 25 Å). These values were also consistent with the estimated total tail length of straight molecules which was about 1550 to 1600 Å (50% of the measurements in the range 1560 ± 50 Å). In itself this is, to date, the most accurate direct determination of the length of the myosin tail.

These new results have therefore confirmed the existence of a hinge along the myosin tail and between the tail and the myosin heads. Since in their micrographs the two heads on one molecule are often quite well separated and since the heads themselves are quite long, Offer and A. Elliott (1978) have suggested that it might be possible for one single myosin molecule to interact simultaneously with two different actin filaments. Direct evidence for this has been obtained by Freundlich *et al.* (1980) as discussed in Chapter 10, Section 10.2.4.

6.3. Aggregation of Myosin and Its Subfragments

6.3.1. Introduction

A preliminary description of the thick filaments in vertebrate skeletal muscle was given in Chapter 1. It was shown that these filaments are about 1.6 μm long and comprise an approximately cylindrical backbone about 150 Å in diameter which is almost entirely covered in projections. The only region free of projections (the bare zone) occurs halfway along the filament and is about 1500 to 2000 Å long. The filaments also taper at each end and are clearly bipolar.

It was H. E. Huxley who originally suggested that the myosin molecules that aggregate to form these filaments might be packed together in an antiparallel fashion at the bare zone and in a parallel fashion elsewhere along the filament, thus giving rise to such a bipolar structure. This idea is shown schematically in Fig. 1.9. It also seemed to be very likely that it was the insoluble LMM portion of the myosin molecule that aggregated to form the cylindrical filament backbone, leaving the soluble myosin heads on the surface of the filament to form the projections. This basic structure for the thick filaments in vertebrate skeletal muscle has not been faulted and has been used as the basis for a variety of detailed packing models of bipolar thick filaments (Pepe, 1967a; Harrison *et al.,* 1971; Squire, 1973).

It will be shown in the next chapter, section 7.2, that the myosin filaments in vertebrate skeletal muscle exhibit two pronounced axial periodicities, one of 143 Å and the other of 429 Å (3 × 143 Å). The presence of these periodicites has been deduced mainly from X-ray diffraction patterns of intact muscle (H. E. Huxley and Brown, 1967). The reasonable deduction can be made that these two related axial repeats are associated in some way with the packing arrangement of the myosin molecules within the myosin filaments. It will be shown later that some synthetic aggregates of myosin have the same periodicities.

Since the myosin tails are about 1500 Å long, it is clear that they must be lined up almost parallel to the long axis of a filament that is only 150 Å in diameter. That this is so is clearly indicated by the high-angle X-ray diffraction patterns of a variety of muscles all of which show clear evidence of α-helical structures lined up parallel to the fiber axis. These patterns usually show a meridional arc at an axial spacing of about 1/5.1 $Å^{-1}$ together with a pair of near equatorial relfection (see Fig. 2.22). As mentioned in Chapter 4, this type of appearance is typical of a coiled-coil α-helical structure (MacArthur, 1943; Astbury, 1947; Fraser and MacRae, 1961, Cohen and Holmes, 1963; Fraser *et al.*, 1965).

In this section, descriptions are given of the various types of synthetic aggregates that can be produced from myosin or from the subfragments of myosin (such as LMM, HMM, S-2, and myosin rod). It will be shown that these aggregates have given valuable indications of the way myosin molecules may pack together to form thick filaments. It should be noted, however, that myosin molecules from different sources, although similar in overall shape and size, do have different properties and can often form characteristically different kinds of synthetic aggregates. One must, therefore, be cautious when applying results of this kind in a general way.

6.3.2. Formation of Synthetic Myosin Filaments

When the ionic strength of a solution of myosin molecules is lowered to 0.2 or 0.1 M, then the myosin molecules aggregate to form synthetic myosin filaments (H. E. Huxley, 1963) (Figs. 6.7 and 6.8f). These filaments were found to be rather similar to myosin filaments separated directly from muscle (Fig. 6.8a–e) except that the native filaments always had a length of about 1.5 to 1.6 μm, whereas the lengths of the synthetic filaments were very variable. Indeed, synthetic filaments only 0.25 to 0.30 μm long were often seen (Figure 6.7). In addition, the synthetic filaments were rather less regular in appearance than the native filaments. But like the native filaments, they always showed a very clear bare zone, about 0.15 to 0.2 μm long and placed halfway along the filament.

In a number of detailed investigations of synthetic filament forma-

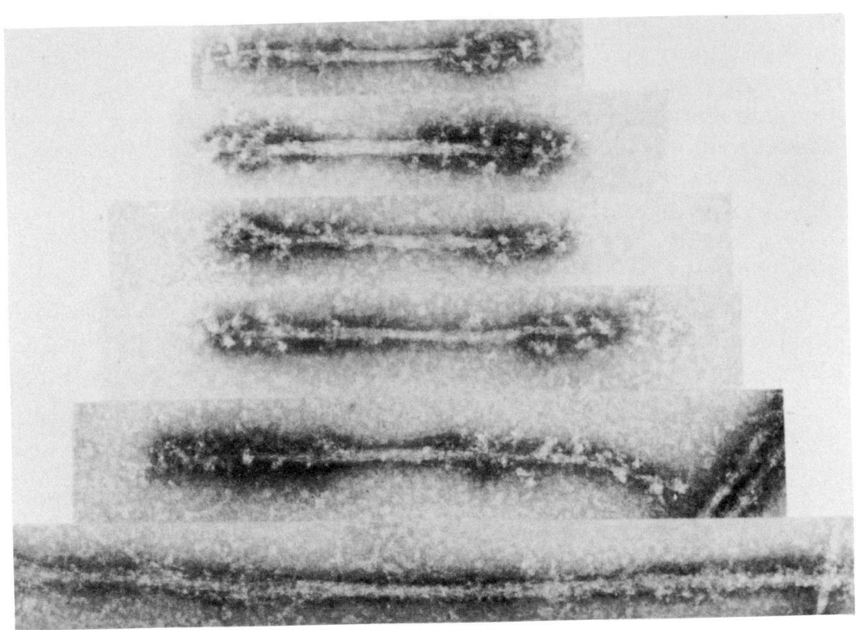

FIGURE 6.7. Synthetic myosin filaments visualized by negative staining. Despite the marked variation in length (from 0.5 to more than 1.4 μm), all of the filaments show the characteristic bipolar structure with a central bare zone about 1500 to 2000 Å long. Magnification ×145,000 (reduced 43% for reproduction). (From H. E. Huxley, 1963.)

tion, the effects of altering the pH, ionic strength, cation composition, dielectric constant, and general preparative procedure have been studied (Kaminer and Bell, 1966; Sanger, 1971; Kaminer *et al.*, 1976). It was found that systematic variations in mean filament length could be obtained by varying many of these parameters. But in each case there remained a considerable spread in the distribution of observed filament lengths. A reasonable conclusion from these results is that myosin molecules have the intrinsic ability to pack to form bipolar filaments but that there is nothing intrinsic that can determine exactly the length of the filament that is formed. Rather, under particular environmental conditions, the filament length will only be defined approximately by the changes in packing energy that occur as the filaments grow. The results of Kaminer *et al.* (1976) that showed that under similar conditions the myosin from vertebrate striated muscle and vertebrate smooth muscle formed filaments of different mean length seem to support this view. Different filament lengths might be expected, since the two myosins most probably have different packing energies. As would be expected, filaments of intermediate length were obtained by Kaminer *et al.* (1976) from solutions containing mixed myosin from smooth and striated mus-

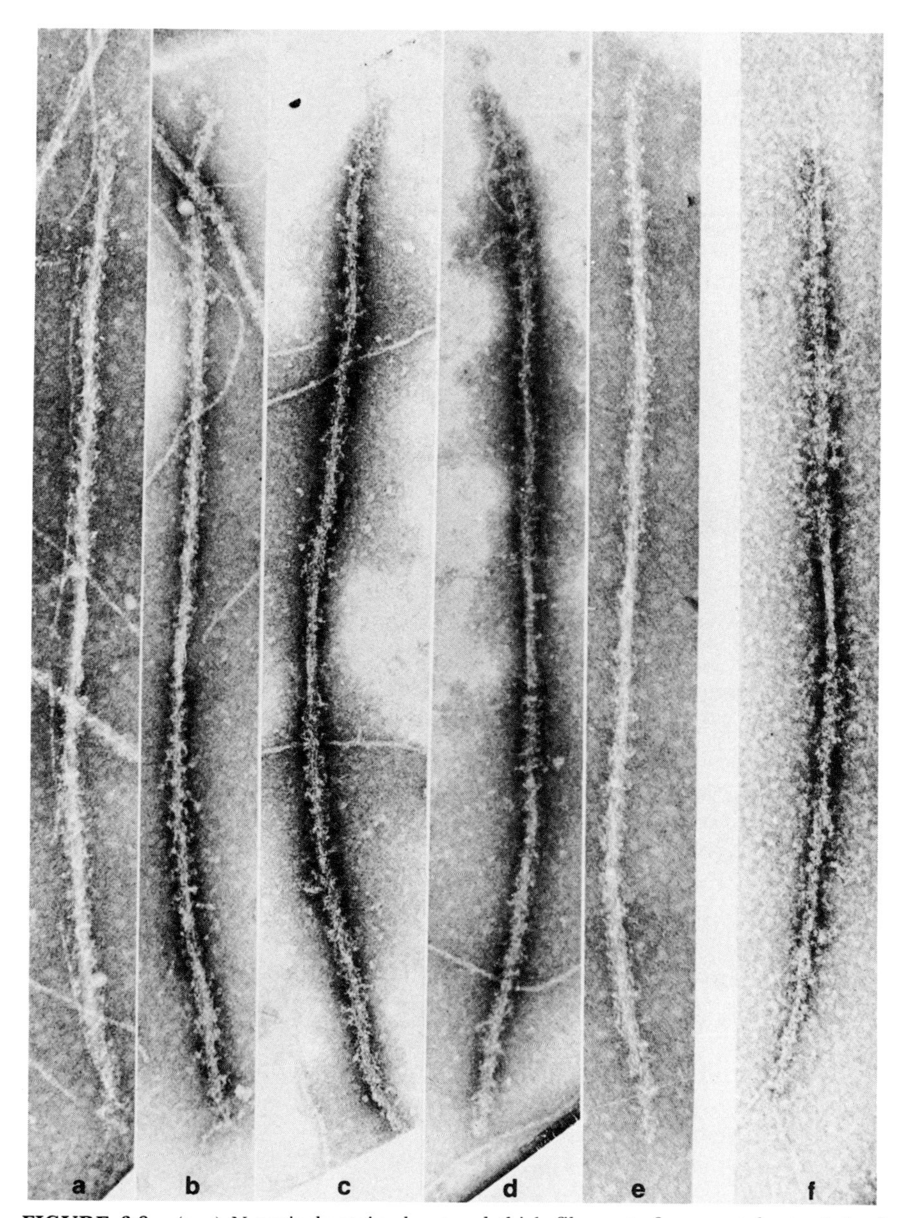

FIGURE 6.8. (a–e) Negatively stained natural thick filaments from vertebrate skeletal muscle. (f) A negatively stained synthetic myosin filament of comparable length. As in Fig. 6.7, all of the filaments have a bipolar structure with a central bare zone. Magnification ×105,000 (reduced 20% for reproduction). (From H. E. Huxley, 1963.)

cle. Since the native thick filaments are always about 1.6 μm long in vertebrate skeletal muscle, it is clear from these studies of synthetic filaments that some additional mechanism (not intrinsic to myosin itself) must be involved in determining the filament length *in vivo*. It is also clear from the observation that the bare zone (when it can be seen clearly) is always near the center of the filament whatever its length, that the antiparallel interaction at the bare zone may well be the initiating interaction involved in filament formation. The filaments would then grow by the addition of molecules on each side of this central zone.

Very few of the synthetic filaments of myosin have shown evidence of any axial periodicities. Even isolated native filaments have until recently shown little evidence of a well-ordered arrangement of myosin heads. But there are exceptions (Eaton and Pepe, 1974; Sobieszek, 1972, 1977; Reedy and Garrett, 1977). Eaton and Pepe (1974) prepared synthetic aggregates of skeletal muscle myosin by dialyzing myosin preparations against a 0.3 M KCl solution. The filaments they obtained had no bare zone and exhibited a periodicity of about 430 Å along their entire length. This periodicity seemed to be caused by repeating rows of myosin heads. As will be described in Chapter 8, Section 8.4, the myosin from smooth muscle can form bipolar filaments. But it can also form aggregates that have no central bare zone. Instead, they have a smooth bare region at one end (Sobieszek, 1972, 1977; Sobieszek and Small, 1973). The remainder of the filament shows a periodicity of about 140 Å (one of the characteristic myosin repeats). This is apparently also produced by projecting myosin heads.

Since there is good evidence that in relaxed vertebrate skeletal muscle, the myosin cross bridges are regularly arranged on the surfaces of the thick filaments, the usual lack of any evidence of a periodicity on isolated thick filaments (and on the bipolar synthetic filaments) must be attributed to a lack of preservation of the native structure of the filaments. The myosin heads are thought to be flexibly linked to the myosin tail (since the HMM S-1-to-S-2 junction is susceptible to proteolysis), and it is clear that they could easily become disordered by the staining or dehydration of the isolated filaments prior to their being viewed in the electron microscope. For this reason, efforts have been made to prepare thick filaments in such a way as to avoid these procedures. One promising approach was developed by Trinick (1973) who obtained filaments by the critical point drying method and then shadowed them. The filaments obtained did show evidence of regularly arranged projections on the thick filaments. Unfortunately, the exact arrangement of these projections could not be determined unambiguously. But Trinick and Elliott (1979) have applied the mica replication technique to isolated filaments and have obtained some beautiful images of the cross bridges and bare zones. The filament symmetry, however, remains elusive.

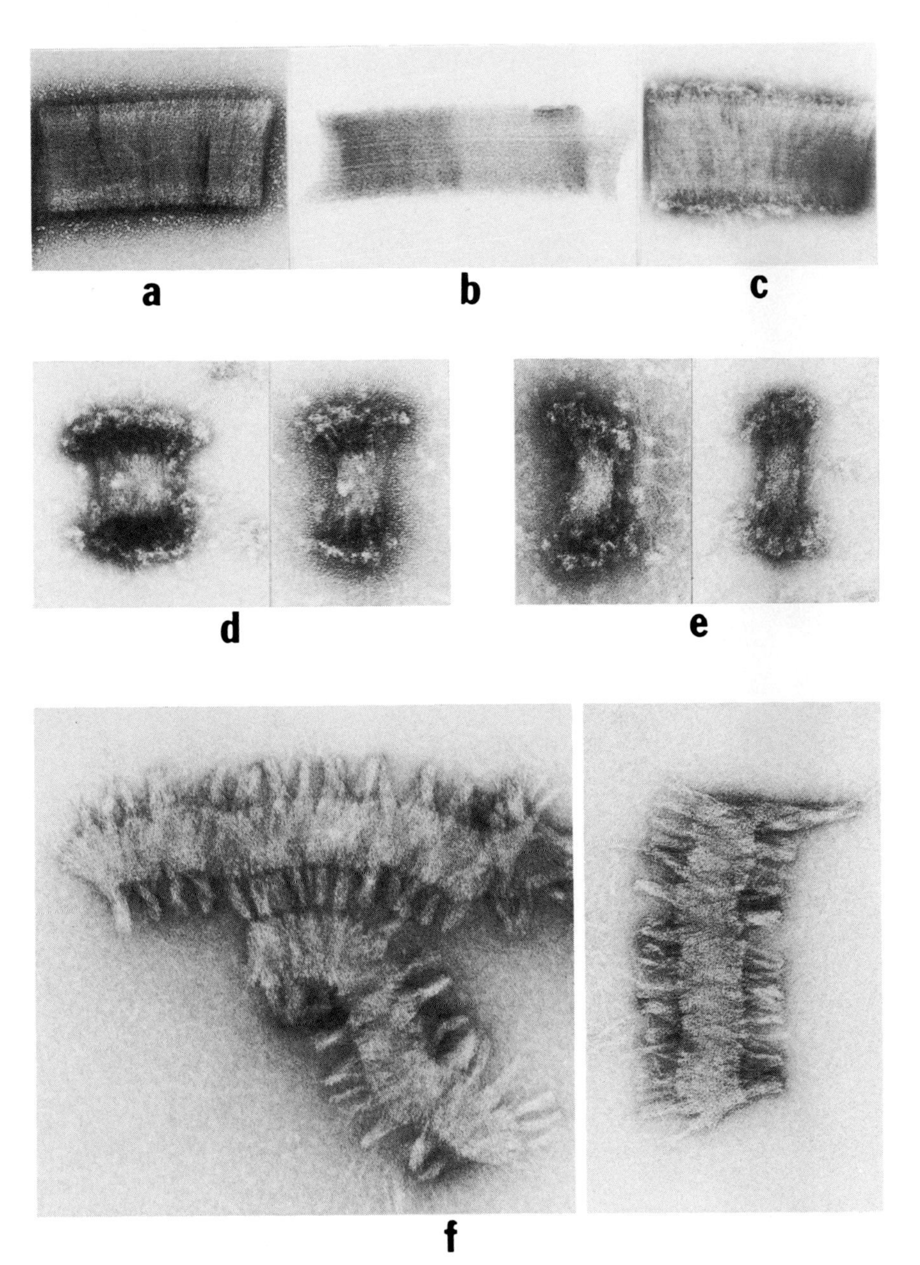

FIGURE 6.9. Segments from myosin and from myosin rod. The protein was dispersed in 1.0 M KCl, 0.05 M Tris-HCl (pH 8.2) and precipitated by dialysis against various concentrations of KSCN, 0.05 M Tris-HCl, 0.05M $CaCl_2$ (pH 8.2). (a) Negatively stained Type I rod segment; 0.05 M KSCN. (b) Positively stained Type I rod segment as in (a); 0.05 M KSCN. (c) Type I rod segment showing incorporation of myosin molecules; 0.05 M KSCN. (d) Type II segments from chicken myosin; 0.10 M KSCN. (e) Type II segments from rabbit myosin; 0.085 M KSCN. (f) Type II segments from chicken rod; 0.075 M KSCN. All preparations except (b) were negatively stained with 1% uranyl acetate. Magnifications: (a) to (c) ×100,000; (d) to (f) ×105,000 (From Harrison *et al.*, 1971.)

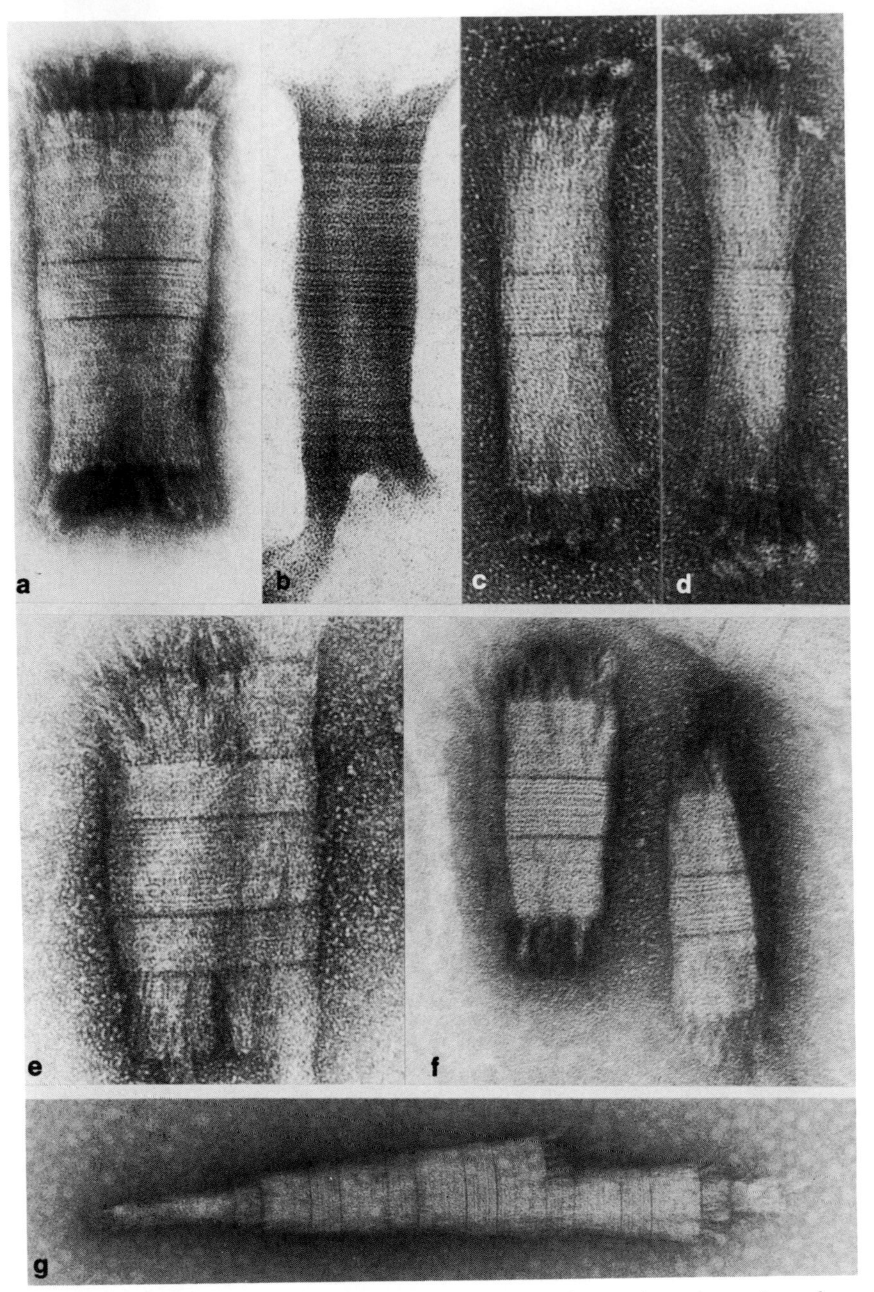

FIGURE 6.10. Segments from vertebrate smooth muscle myosin and myosin rod prepared under a variety of conditions. All preparations from chicken gizzard myosin. (a) Segment from chicken gizzard myosin rod prepared by papain digestion of myosin (Lowey

6.3.3. Formation of Myosin Segments

Myosin, myosin rods, and LMM, when precipitated in the presence of divalent cations, form characteristic "segment" structures. These are bipolar structures of well-defined width formed by molecules with antiparallel overlaps of about 1300 Å (Type I) or about 900 Å (Type II). These two forms are illustrated in Fig. 6.9.

Type I segments have only been prepared from the myosin rods of chicken breast muscle or from a mixture of chicken myosin and rods in the ratio of about 1 : 5 by weight (Cohen *et al.*, 1970). The segments have an overall width of about 1800 Å including the lightly stained outer fringes. The segments were interpreted by Cohen *et al.* (1970) to comprise antiparallel molecules with a tail-to-tail overlap of about 1300 Å (Fig. 6.9a–c). From this they estimated the minimum rod length. If the segment width (excluding the terminal weak fringes on each side of the segment) is about 1600 Å (1958 ± 44 Å), then from Fig. 6.9, the rod length is ½(1600 + 1300) = 1450 Å.

Type II segments (Fig. 6.9d–f) are produced either by myosin alone, by myosin rods, or by LMM (Harrison *et al.*, 1971), and similar results are obtained with myosins from chicken breast muscle and from rabbit. In this case, an antiparallel overlap of about 900 Å is involved, giving rise to a lightly stained region 900 Å wide in the middle of the segments. Segments from myosin (d) show heads about 625 Å from the edge of the light region, giving a minimum value for the length of the whole molecule of about 1525 Å. The segments from LMM (not shown) show fringes about 100 Å wide on each side of the lightly stained overlap zone, giving a length of about 980 to 1000 Å for LMM (Harrison *et al.*, 1971).

---

et al., 1969). Segment formed by precipitation with 50 mM Ca^{2+} from 50 mM Tris-HCl (pH 8.0), 100 mM KSCN. The central zone is about 430 Å across and the dark terminal fringes are about 400 Å long. (×200,000). (b) Positively stained segment as in (a) (×200,000). (c) Segment from myosin coprecipitated with myosin rods. Equal weights of each were mixed in 0.6 M KCl at neutral pH and then dialyzed against a large volume of 50 mM Tris-HCl (pH 7.7), 70 mM KSCN (×200,000). (d) Another segment as in (c). (e) Segments from a myosin fragment obtained by brief tryptic digestion (15 min) precipitated with 50 mM Ca^{2+} from 50 mM Tris (pH 8.0), 75 mM KSCN (×200,000). (f) Segments from myosin fragment after prolonged tryptic digestion (40 min) precipitated with 50 mM Ca^{2+} from 50 mM Tris (pH 8.0), 75 mM KSCN (×200,000). (g) Multiple segments from myosin fragments prepared by brief digestion with chymotrypsin (15 min) precipitated with 50 mM Ca^{2+} from 50 mM Tris (pH 8.0), 70 mM KSCN (×100,000). All figures reduced 16% for reproduction. All preparations except (b) were negatively stained with 1% uranyl acetate. Note that the 430-Å central zone in (a) has the same striation pattern as the 430-Å zones in (e). Alternate 430-Å zones in (f) also show the same pattern, but the intervening zones which correspond to the fringe overlap region have a different pattern. (From Kendrick-Jones *et al.*, 1971.)

Bipolar segments have also been obtained from vertebrate smooth muscle myosin rod (Kendrick-Jones *et al.*, 1971). But in this case, compound segments were obtained which were about 3500 Å wide. These segments (shown in Figs. 6.10 and 6.11) were taken as corresponding to a molecular arrangement (Fig. 6.11b) involving both parallel and antiparallel molecular overlaps of 430 Å. Similar segments were obtained

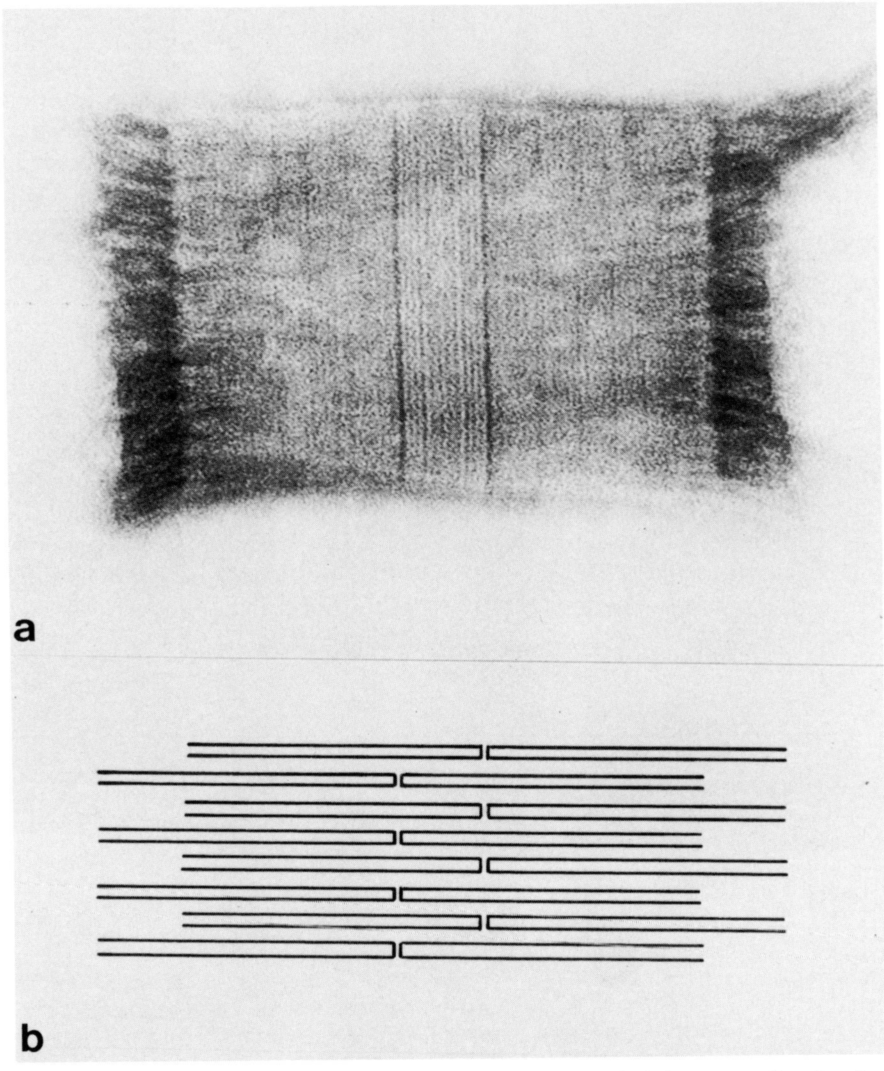

FIGURE 6.11. Interpretation of the segments as in Fig. 6.10 in terms of molecular packing. Magnification ×280,000. The "open" end of the rods indicates the head end of the molecule. Similar assemblies using rods 1130 Å or 1000 Å in length will account for the segments in Fig. 6.10e–g. (From Kendrick-Jones *et al.*, 1971.)

from a mixed preparation of myosin and myosin rod (Fig. 6.10c) and from rods of different lengths produced by digesting the myosin for different times. In each case, a central zone 430 Å wide was always observed. In addition, some segments were occasionally observed with a central antiparallel overlap of about 700 Å (Fig. 6.10e,g) as well as the 430 Å overlaps.

From the dimensions of the various smooth muscle myosin segments, the rod length was estimated to be about 1560 Å. This is slightly greater than the estimated rod length of striated muscle myosin (~1450 Å; Cohen *et al.*, 1970), but whether this difference is significant or not is difficult to say, especially in view of the recent observations of A. Elliott *et al.* (1976) in which no difference in length between the two intact myosins could be detected.

6.3.4. Paracrystals of the Myosin Rod and Its Subfragments

Light meromyosin will precipitate at low ionic strength to form a variety of tactoids or paracrystals. Micrographs of some of these are shown in Figs. 6.12 and 6.13. The majority of the tactoids described in the literature have dihedral symmetry and show a repeating pattern with a periodicity of either about 140 Å (Podlubnaya *et al.*, 1969, Cohen *et al.*, 1970; Katsura and Noda, 1973) or about 430 Å (Philpott and Szent-Gyorgyi, 1954; A. G. Szent-Gyorgyi *et al.*, 1960; H. E. Huxley, 1963; Lowey *et al.*, 1967; Podlubnaya *et al.*, 1969; Cohen *et al.*, 1970; Nakamura *et al.*, 1971; Katsura and Noda, 1973). Cohen *et al.* (1970) also described a fibrous long-spacing aggregate of repeat about 760 Å that seemed to contain a molecular overlap of about 100 Å. Podlubnaya *et al.* (1969) described an LMM paracrystal with a repeat close to 48 Å (143/3 Å). From these studies, it appears that the paracrystals with a 430-Å repeat can appear in at least three different forms. In one, there is a marked subrepeat of about 143 Å (Podlubnaya *et al.*, 1969). In the second, the main feature of the 430-Å repeating unit is a band edged by two darkly stained lines separated by about ⅜ × 430 Å (Fig. 6.13a) (Katsura and Noda, 1973; Bennett, 1976). The third form has dark transverse lines 430 Å apart but little apparent substructure (Katsura and Noda, 1973). The tactoids with a 143-Å repeat (Podlubnaya *et al.*, 1969) sometimes show a subrepeat of about 72 Å. Chowrashi and Pepe (1977) have prepared tactoids from various types of LMM preparations that show characteristically different appearances. Some of these are shown and described in Fig. 6.13.

Because of their dihedral symmetry, it is rather difficult to interpret the appearance of the various LMM paracrystals in terms of molecular packing, although some approach towards this will be made in

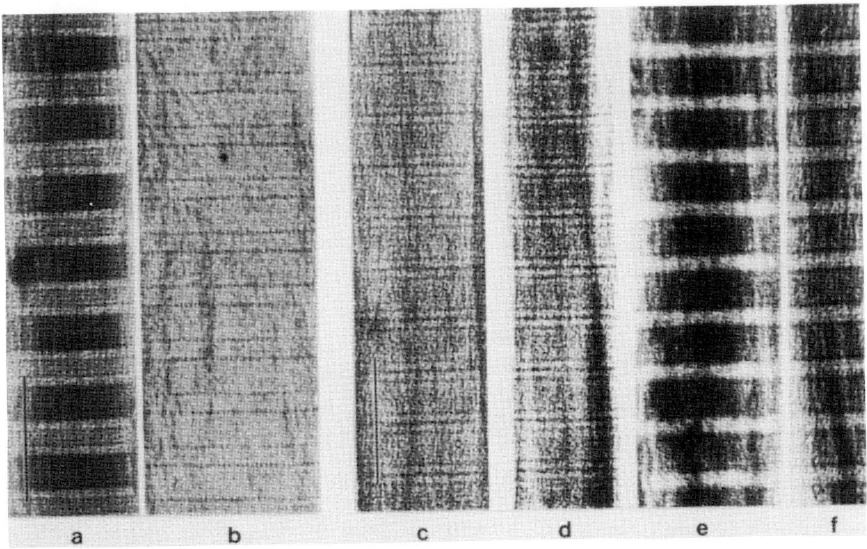

FIGURE 6.12. Light meromyosin paracrystals formed using various types of LMM preparation. (a and b) From column-purified LMM; both ×300,000. That in (b) is from a 40-min papain digestion of myosin and contains two components of lengths 880 and 620 Å as estimated from the observed chain weights assuming 100% α-helix content. The repeat of about 440 Å consists of pairs of dark lines about 150 Å apart. The paracrystal in (a) is from LMM prepared by 25-min trypsin digestion of a column-purified preparation obtained by 15-min papain digestion of myosin. The repeat is about 600 Å and consists of alternating dark and light bands about 300 Å wide. The bar represents 0.1 μm. (c to f) The effect of the presence of a specific RNA fraction on paracrystal formation. (f) From an ethanol-resistant fraction (A. G. Szent-Gyorgyi *et al.*, 1960) of a 5-min trypsin digestion of myosin. The pattern of repeat 440 Å consists of alternating light (130 Å wide) and dark (310 Å wide) bands. LMM length about 960 Å. (e) From an LMM preparation as in (f) after column purification and the subsequent readdition of the separated RNA-containing fraction. (d) From an ethanol-resistant fraction of an 18-min papain digestion of myosin. The repeat of about 440 Å contains three prominent dark lines, the outer two in each group being about 160 Å apart. (c) From a preparation as in (d) but after column purification and redigestion for 15 min with trypsin and readdition of the nucleic acid fraction. The pattern obtained is as in (d). Note that in the absence of nucleic acid here and in (e), the column-purified preparations gave a simple 150-Å axial repeat. Bar in (c) 0.1 μm. Magnifications ×300,000 (reduced 45% for reproduction). These results from Chowrashi and Pepe (1977) were taken to indicate that different portions of the myosin molecule code for different axial staggers between adjacent interacting molecules, since different repeats were obtained either from LMM fragments of different lengths prepared using the same enzyme or from LMM fragments of the same length prepared using different enzymes. Clearly the presence of the nucleic acid fraction also has a marked effect on paracrystal formation. All preparations negatively stained with 1% uranyl acetate.

Chapter 9. However, Bennett (1976) has recently analyzed one of these paracrystalline forms in some detail. The paracrystal involved was the 430-Å form with the dark lines separated by ⅜ × 430 Å (Fig. 6.13). It

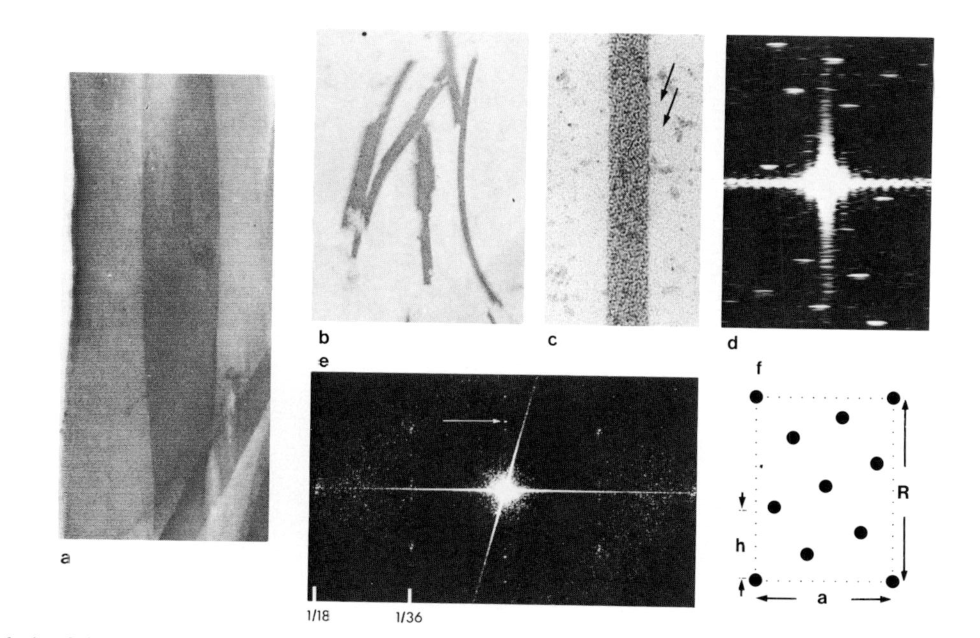

FIGURE 6.13. Analysis of the structure of LMM paracrystals by Bennett (1976). (a) A 430-Å repeat LMM paracrystal negatively stained with 2% uranyl acetate. The characteristic 430-Å repeat with the prominent subbands separated by ⅜ × 430 Å can be seen (×110,000). (b) Part of a section from a pellet of LMM paracrystals in near transverse view (×100,000). (c) Part of a longitudinal section perpendicular to the plane of the flat paracrystal. Note the prominent oblique stripes that can be seen in this and other similar sections (×250,000). (d) An optical diffraction pattern from the section in (c) which clearly demonstrates the two-dimensional order visible in the section. (e) An optical diffraction pattern from a negatively stained paracrystal such as that in (a). The meridional reflections index as orders of a 430-Å repeat. Of the higher orders, fifth, eighth (arrowed), and 11th are consistently strong. The lateral order in the paracrystal gives rise to a row line at a spacing of 1/36 $Å^{-1}$ and a true equatorial reflection at a spacing of 1/18 $Å^{-1}$. The latter probably indicates the lateral separation of adjacent molecules, and the former suggests that the crystallographic repeat across the paracrystal spans two molecules. (f) The form of the lattice arrangement visible in transverse views of the paracrystals such as (c); h is ⅜ × 430 Å, R is 430 Å and a is 330 Å. (Courtesy of P. Bennett; from Bennett, 1976.)

had already been shown by Katsura and Noda (1973) that 430-Å paracrystals can adhere to each other with a ⅜ × 430 Å stagger between them, indicating that an interaction between molecules with a relative axial translation of ⅜ × 430 Å or ⅝ × 430 Å might be possible. Bennett studied such paracrystals in both face and edge view, the latter being obtained by embedding pellets of the paracrystals, cutting sections through them, and then viewing these sections at an appropriate tilt. It appeared from such edge views that a structure involving a regular molecular stagger of ⅜ × 430 Å was present right through the paracrystal (Fig. 6.13). Face views of these paracrystals were analyzed by optical diffraction, and it was found that axial reflections that were orders of the 430-Å repeat were clearly seen, with the orders 5, 8, and 11 relatively strong. These orders are also relatively strong in patterns from native myosin structures (see Chapter 7, Section 7.3). Also seen were an equatorial reflection of spacing 1/18 $Å^{-1}$ and a row line of spacing 1/36 $Å^{-1}$ (with no intensity on the equator). This suggested that the repeating unit in the assembly comprised a pair of molecules each of diameter 18 Å (see Section 4.3.4).

In addition to the tactoids described above, two netlike aggregates of LMM have been observed. One of these had the form of a square net of side about 400 Å (Philpott and Szent-Gyorgyi, 1954; Lowey *et al,* 1967), and the other was a hexagonal net of side about 600 Å (H. E. Huxley, 1963; Lowey *et al.,* 1967). The thickness of the strands in these nets was estimated to be about 70 Å and was claimed to consist of substrands of width about 35 Å (Katsura and Noda, 1973). These might possibly be related to pairs of molecules of diameter 18 Å.

Tactoids have also been prepared from HMM S-2, the α-helical subfragment of HMM (Fig. 6.14). Only one type has been seen, and it has a repeat of about 145 Å (Lowey *et al.,* 1967). But these tactoids are not formed at normal pH, suggesting that *in vivo* the S-2-to-S-2 interaction may be relatively weak. Myosin rod prepared by papain digestion (Lowey, 1971) gives rise to tactoids with a 145-Å repeat under normal pH conditions.

Another highly α-helical derivative of myosin, prepared by treat-

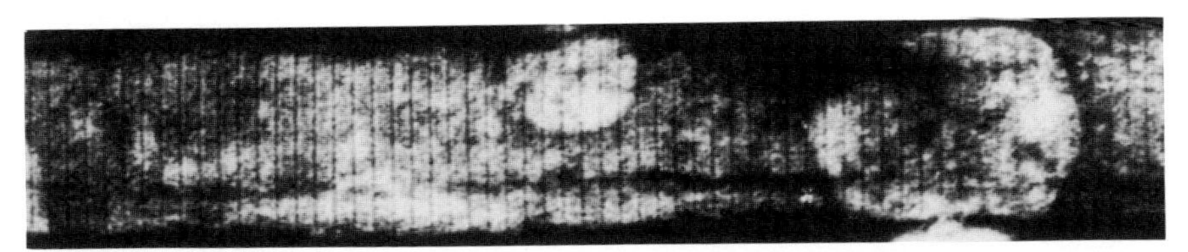

FIGURE 6.14. A negatively stained paracrystal of the S-2 fragment of myosin. The repeat is about 145 Å. (From Lowey *et al.,* 1967.)

ment of myosin with cyanogen bromide and termed LMM-C (Young *et al.*, 1968), forms tactoids with repeats of 143 Å or 429 Å and square nets of side 685 Å and 390 Å (King and Young, 1972). The action of cyanogen bromide is said to be more specific and to produce a more homogeneous rod preparation than do proteolytic methods.

Bipolar segments have also been produced from LMM-C, and these are similar to the Type II segments of Cohen *et al.* (1970) with a 900-Å overlap between antiparallel molecules (Fig. 6.9). From these segments, the molecular length was estimated to be 1000 Å (King and Young, 1972), a value consistent with the estimated molecular weight of 173,000.

King and Young (1972) and Young *et al.* (1972) also described the formation of an unusual aggregate from LMM-C. This was in the form of tubes of molecules (Fig. 6.15) in which the molecules are tilted relative to the axes of the tubes and follow helical patterns within the tube walls. These are the only synthetic assemblies of myosin in which flexibility of the myosin rod has been directly demonstrated.

6.3.5. Studies of Myosin Aggregates in Solution

Aggregation Properties. A novel approach toward the determination of myosin aggregation properties has been used by Harrington and his collaborators (Godfey and Harrington, 1970; Burke and Harrington, 1971, 1972; Harrington and Burke, 1972; Harrington *et al.*, 1972; Reisler *et al.*, 1973) who carried out solution studies of myosin under conditions approaching those suitable for filament formation and found that myosin existed in a mixed monomer–dimer state in which there was a rapid and reversible monomer–dimer equilibrium. From analysis of the viscosities of similar preparations of myosin, myosin rod, and LMM, they were able to show that the most abundant dimer form is one in which the two myosin molecules are parallel and have a relative axial stagger of about 400 to 500 Å. This was taken to indicate that a parallel dimer of overlap about 430 Å was likely to be involved in myosin filament formation. Since the bare zone appears to be the first part of a bipolar filament to be formed, it was suggested that parallel myosin dimers might interact in an antiparallel fashion to give a simple bipolar aggregate and that further parallel dimers might then build up on each end of this central structure to form the whole filament.

In a further paper (Reisler *et al.*, 1973), an account was given of aggregates of myosin existing in solution near conditions for filament formation. These had been stabilized by cross-linking with glutaraldehyde or dimethyl suberimidate. Analysis of the cross-linked species confirmed that the parallel dimer form of myosin increased in abundance as the filament-forming conditions were approached. But a

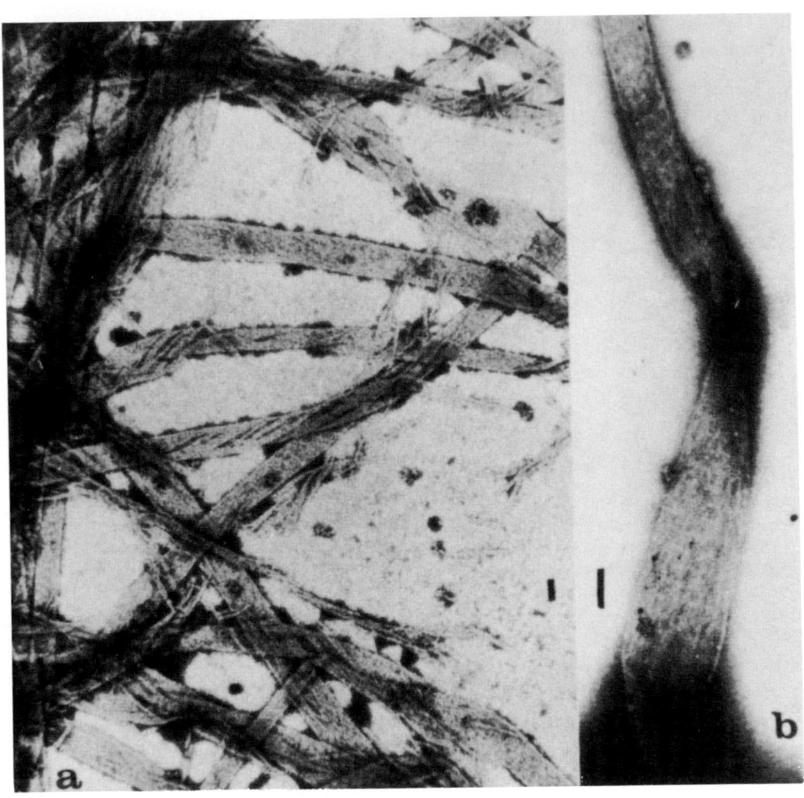

FIGURE 6.15. Negatively stained "tubes" of myosin formed from a preparation of LMM-C (see text). The tubes contain molecules that run slightly obliquely to the tube axis, and the myosin rods must therefore curve slightly along their length. The scale bars in (a) and (b) are both 1000 Å. (From Young *et al.*, 1972.)

number of other aggregates were also observed including antiparallel dimers and several multimers of myosin. Analysis of these multimers may lead to a further insight into myosin filament formation.

Thick Filament Regulation. Morimoto and Harrington (1974) carried out a different kind of solution study on myosin. They worked on intact thick filaments from vertebrate skeletal muscle and attempted to observe changes in either the sedimentation coefficient of these filaments or in the viscosity of the solutions brought about by the presence of Ca^{2+}. Their results clearly indicated that as the Ca^{2+} concentration changed from about 10^{-7} to 10^{-4} M, a gradual change in sedimentation coefficient and viscosity occurred. The sedimentation coefficient increased by about 3%, whereas the viscosity of the solution decreased in a similar fashion. The observed changes both followed a sigmoid curve (Fig. 6.16) comparable in shape to a plot of the ATPase activity of ac-

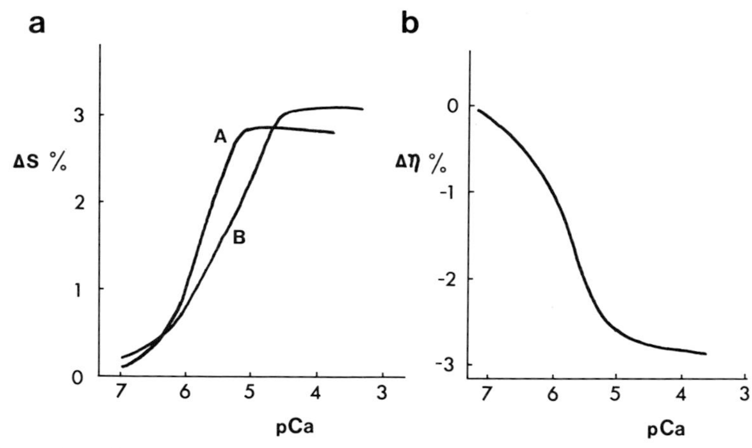

FIGURE 6.16. Variations of sedimentation coefficient (a) and viscosity (b) as a function of Ca^{2+} concentration in solutions of myosin filaments. The shapes of these sigmoid curves are similar to the change in ATPase activity of myosin as a function of Ca^{2+} concentration (Fig. 1.23). In (a), curve A is from natural thick filaments, and curve B is from synthetic filaments. (After Morimoto and Harrington, 1974.)

tomyosin as a function of calcium concentration (Fig. 1.23). In each case the midpoint of the transition was close to a calcium level of $10^{-5.5}$ to $10^{-5.7}$ M.

These results indicated that a change in calcium concentration typical of that which occurs when a muscle is activated can cause the myosin filaments to change their structure in some way. However, it is not yet clear that this change occurred on a sufficiently fast time scale to be involved in thick filament regulation. Morimoto and Harrington (1974) and Bremel and Weber (1975) also showed that the DTNB light chain (P light chain) is associated with calcium binding. However, this binding is now known to involve a nonspecific divalent ion binding site (Bagshaw, 1977; Bagshaw and Reed, 1977; Bagshaw and Kendrick-Jones, 1979) from which bound Mg^{2+} ions at low calcium levels are only slowly displaced when the calcium level is increased. These binding sites are clearly not likely to be associated with thick filament regulation. However, as mentioned earlier in Section 6.2.3, it is still possible that the DTNB light chain is involved in thick filament regulation even though *in vitro* studies have not yet detected any change in the actin-activated ATPase of myosin with Ca^{2+} concentration. It is quite possible, as suggested by Wagner and Weeds (1977), that thick filament regulation is a structural event that does not directly involve the myosin ATPase. It may, for example, simply lock the myosin heads onto the surfaces of the thick filaments at low calcium concentrations so that they are physically unable to interact with actin. This lock would then be released by the binding of calcium to some as yet undiscovered specific calcium binding site on the DTNB light chain (possibly in association with the myosin head), and interaction between myosin and actin could then occur. *In vitro* studies in which this constraint is lacking would clearly be unable to detect any change in the myosin ATPase because the actin could diffuse up to the locked-in myosin heads. Further work is obviously required to check this possibility, but it should be noted that studies using fluorescence depolarization (Mendelson and Cheung, 1976), cross-linking (Sutoh and Harrington, 1977), and electron microscopy (A. Elliott and Offer, 1978) have so far failed to detect any specific structural effect of calcium ions.

6.4. Conclusion

It is appropriate to conclude with a brief summary of the general properties of the myosin molecule as described in this chapter.

1. Most myosin molecules are about 1600 to 1700 Å long including the myosin heads, and they consist of a long two-chain α-helical tail terminating in two large globular heads, each of which is associated with two light chains.

2. Two main kinds of light chain are present. One type, comparable to the alkali light chains in vertebrate myosins, is essential for the myosin ATPase. The other type, the DTNB, P, or regulatory light chain, is associated in some way with regulation of the thick filament. In molluscan, insect, and possibly vertebrate smooth muscle, the regulatory light chain directly affects the ATPase of the myosin depending on the [Ca^{2+}]. In vertebrate skeletal muscle, the regulatory light chain can bind Ca^{2+} and can be phosphorylated, but it probably does not act directly on the myosin ATPase. It very likely controls the arrangement of the myosin heads. Fig. 6.17 summarizes the known distribution of actin- and myosin-linked regulation in different species.

3. Myosin molecules have been shown to form aggregates with antiparallel overlaps of about 1300 Å, 900 Å, and 430 Å, the latter only being observed with vertebrate smooth muscle myosins that also show end-to-end butting (zero overlap).

4. Parallel overlaps between myosin molecules of 430 Å have been demonstrated directly in myosin segments and in solution studies. Other parallel overlaps may be inferred from the existence of paracrystals with 48-Å, 143-Å, 430-Å, 600-Å, and 760-Å repeats.

5. Bipolar myosin filaments are probably built up from the bare zone outwards, possibly using a 430-Å dimer as the building block. Myosin molecules themselves contain the required information to build a bipolar filament but not to form filaments of well-defined length.

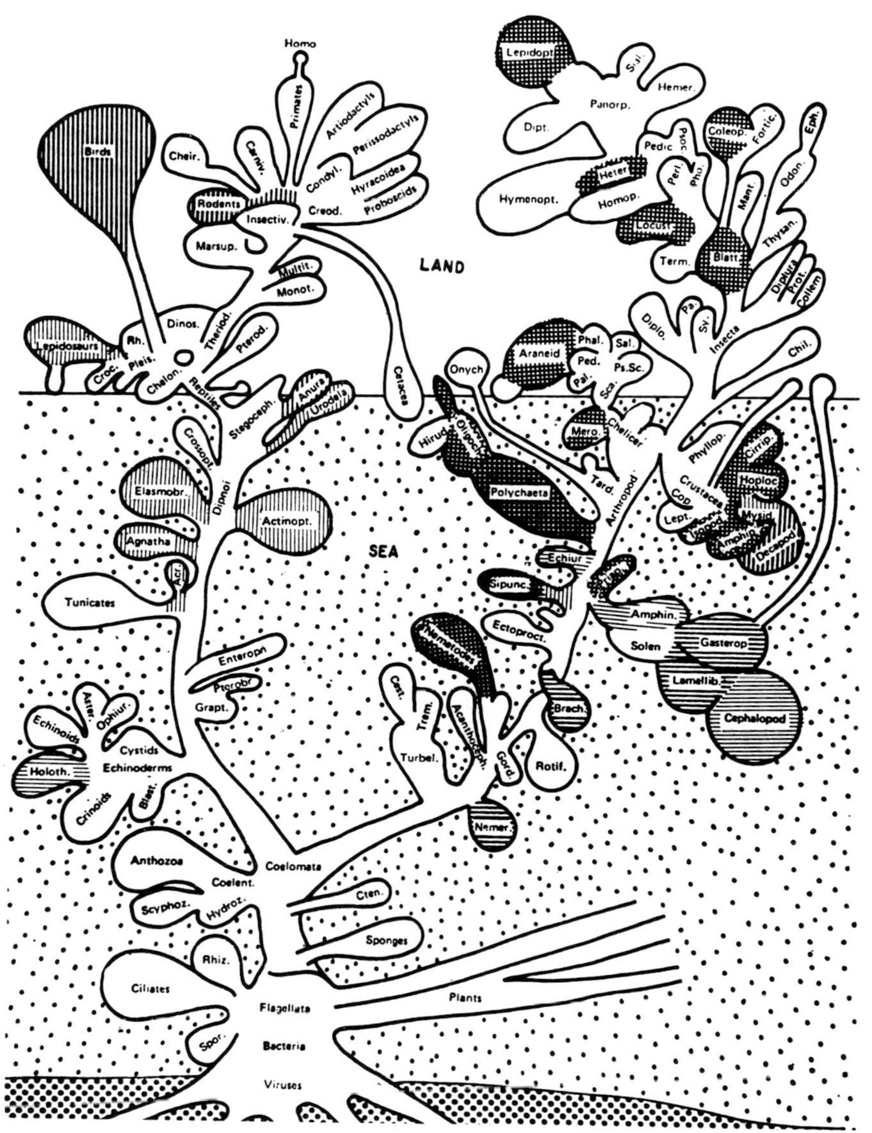

FIGURE 6.17. The distribution of actin-linked and myosin-linked Ca^{2+} regulation in the animal kingdom. Vertical lines indicate the presence of actin-linked regulation, and the horizontal lines, myosin-linked regulation. In several cases, both types of regulation have been demonstrated. Some kind of myosin-linked regulation in vertebrates can be inferred from the results shown in Fig. 6.16, but this has not yet been demonstrated directly. See discussion in Chapter 10. (From A. G. Szent-Gyorgyi, 1975.)

6. The myosin heads approximate in shape to an elongated pear about 180 to 190 Å long and ranging in width from about 45 Å at its widest down to about 20 Å at its narrow end. S-1 may only represent the terminal 140 to 150 Å of the intact myosin head.

7. The ability of proteolytic enzymes to preferentially cleave the myosin molecule in two specific regions suggests that at the LMM–HMM S-2 junction and at the HMM S-2–HMM S-1 junction the tail may not be regularly α-helical but may be less regular and therefore slightly flexible. This possibility is supported by fluorescence depolarization studies of myosin in solution and by the apparent flexibility of the head–rod junction that is suggested by the general disorder in the head arrangement in electron micrographs of myosin molecules and myosin aggregates. Distinct kinks in the myosin rods have been observed in these regions by A. Elliott and Offer (1978).

More details of the likely packing arrangements of myosin molecules in a variety of thick filaments, will be given in the next few chapters in which the structures of individual muscle types are described.

Finally, Fig. 6.17 provides a summary of the distribution of actin-linked and myosin-linked regulation throughout the animal kingdom as determined in an elegant series of experiments by A. G. Szent-Gyorgyi (1975) and his colleagues. Actin-linked and myosin-linked regulation are indicated in Fig. 6.17 by vertical and horizontal lines, respectively. On the whole, the invertebrates show either myosin-linked or dual regulation, whereas vertebrates have, so far, only shown clearly the existence of actin-linked regulation (except in one case).

7

Vertebrate Skeletal Muscle

7.1. Introduction: Structure of the Sarcomere

7.1.1. Introduction

In the main part of this chapter an account is given of the information available on the structure of the sarcomere in vertebrate striated muscle. The particular example most commonly quoted here is the sartorius muscle of the frog. It must be recognized, however, that this muscle, although probably the most extensively studied vertebrate skeletal muscle, is actually rather a specialized example. Different skeletal muscles in the same animal, different fibers in the same muscle, and analogous muscles in different animals can have markedly different properties, and many of these appear to be associated directly with the ultrastructure of the sarcomere. The sarcomere of the frog sartorius muscle will therefore be used as a kind of standard against which the structures of other vertebrate skeletal muscles (where these are known) can be compared.

Frog sartorius muscle is a strap-shaped muscle running from the knee to the pelvis and is often about 2–3 cm long, 5 mm wide, and up to 1 mm thick. It is composed of long parallel fibers that extend from the tendon at the knee to the bone connection at the pelvis. Because of the regular organization of the component fibers, this particular muscle has been extensively used for X-ray diffraction and electron microscope studies. The flat form of the muscle is actually ideal for X-ray diffraction work since it is thin enough to allow satisfactory X-ray transmission but is large enough in the other dimensions to allow full use to be made of the extended beams that are produced by mirror monochromator cameras (see Chapter 2, Section 2.5.2). In addition, the fact that the muscle is thin allows fast penetration of fixatives prior to electron microscopy.

The fibers in frog sartorius are tens of millimeters long and about 50 to 100 μm in diameter, and they comprise parallel myofibrils about 1 to 2 μm across (see Fig. 1.4). The myofibril is composed of repeating structures, the sarcomeres, which are about 2 to 3 μm long depending on the amount of overlap of the thick and thin filaments. A "rest length" has been defined in a variety of ways for this muscle and corresponds to a sarcomere length of about 2.2 to 2.4 μm. "Rest length" as defined by H. E. Huxley (1968) corresponds to the length of the muscle *in situ* with the upper leg at right angles to the body and to the lower leg.

The general form of the sarcomere in vertebrate striated muscle was described in Chapter 1. The central region of the sarcomere (Fig. 7.1a) consists of myosin-containing filaments arrayed side by side to form the A-band. These filaments are mechanically linked at their centers through a structure called the M-band. At each end the thick filaments interdigitate with an ordered array of thin actin-containing filaments which in turn are linked by the Z-band to a similar actin array in the next sarcomere.

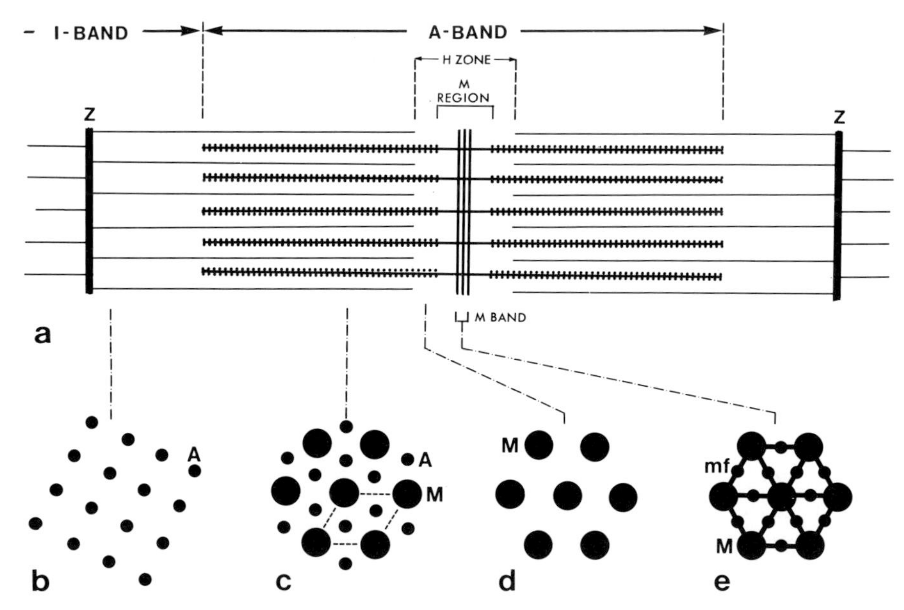

FIGURE 7.1. Schematic diagram to show the general layout and structure of the vertebrate sarcomere. Longitudinal view (a) and transverse views of different parts of the sarcomere showing the I-band in the region next to the Z-band (b), the overlap region (c), the H-zone (d), and the M-band (e). Abbreviations: Z, Z-band; A, Actin filaments; M, Myosin filaments; mf, M-filaments (see text).

7.1.2. Lateral Order in the Sarcomere

Early electron micrographs of vertebrate skeletal muscle in transverse section (H. E. Huxley, 1953b, 1957) showed that the myosin filaments are arranged in a hexagonal lattice (Fig. 7.1d). In the region of overlap between the thin and thick filaments, it was found that the thin filaments were also regularly arranged: they were located at the trigonal points of the thick filament array (Fig. 7.1c) so that there were two thin filaments and one thick filament in each unit cell in cross section. This particular arrangement of thick and thin filaments is actually a characteristic of all vertebrate striated muscles. Later it was shown that near to the Z-band the thin filaments are arranged in a nearly square lattice (Fig. 7.1b). It was concluded that the thin filament array must change through the I-band from being square at the Z-band to being hexagonal in the overlap region of the A-band.

Confirmation of such an arrangement was obtained by X-ray diffraction studies of this type of muscle at various sarcomere lengths. The early X-ray diffraction work of H. E. Huxley (1951, 1952, 1953a) had shown that vertebrate skeletal muscles gives rise to a characteristic set of strong equatorial reflections (Fig. 7.2). These were interpreted (before the electron micrographs showing the filament array were available) as arising from a hexagonal array of spacing in the range 350 to 480 Å, depending on sarcomere length. The reflections were later shown to be caused by the hexagonal array of thick and thin filaments (Fig. 7.1c) observed in electron micrographs.

Figure 7.2 shows how the form of the diffraction pattern from the filament array can be determined. A hexagonal lattice can be defined in terms of two equal vectors **a** and **b** with an angle of 120° between them (Fig. 7.2a). Thick filaments located on this lattice can be thought of as lying in planes parallel to the filament axis which can be indexed according to the scheme outlined in Chapter 2, Section 2.2.3. In Fig. 7.2a, the positions of the 10 and 11 planes are illustrated. X-ray patterns of the form shown in Fig. 7.3c are recorded with the X-ray beam perpendicular to the long axis of the muscle. In terms of Fig. 7.2b, the incident beam can be thought of as lying in the plane of the page. Whether or not diffraction will occur from the various planes of filaments will depend on the orientation of these planes relative to the X-ray beam. If they are at an angle (θ) from the X-ray beam that satisfies Bragg's Law for that particular interplanar spacing

$$d = a/[4/3(h^2 + hk + k^2)]^{\frac{1}{2}}$$

then diffraction will occur. In general, for a particular orientation of the lattice, diffraction will occur from very few, if any, of the different lattice

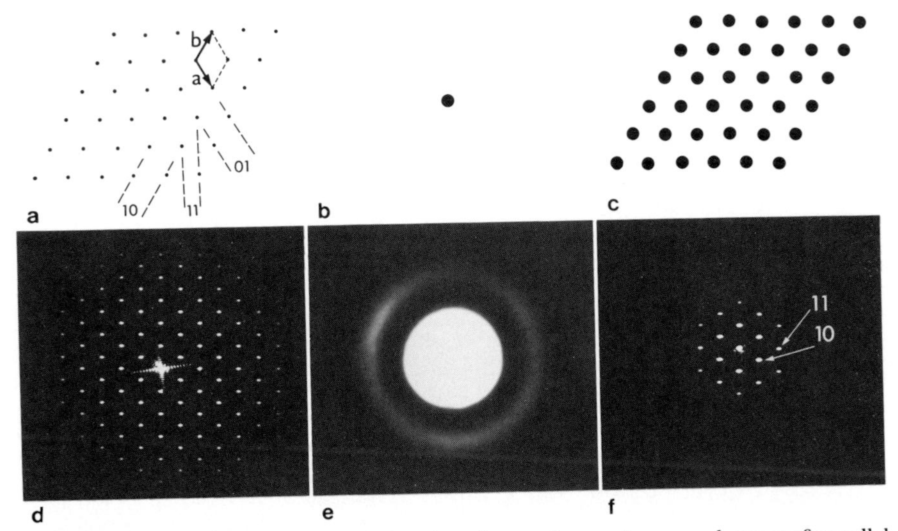

FIGURE 7.2. Illustration of the diffraction pattern from a hexagonal array of parallel cylindrical objects as a first approximation to the myosin filaments in the A-band. (a) A simple hexagonal array of points; (d) its optical diffraction pattern. (b) A cylindrical structure viewed down its axis; (e) its optical diffraction pattern (the Airy disk). (c) A hexagonal array of objects as in (b); this can be thought of as a convolution of (b) with the hexagonal array (a). (f) The diffraction pattern of (c); from the convolution theorem (Chapter 2) this is the product of (d) and (e). Only the first three reflections ($10\bar{1}$, $11\bar{2}$, and $20\bar{2}$) are strong in the final pattern.

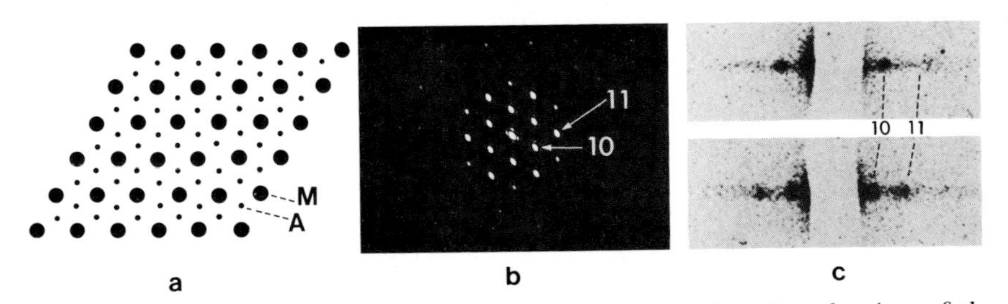

FIGURE 7.3. The effect of adding actin filaments (A) to the trigonal points of the hexagonal array of myosin filaments (M) as in (a) is to increase the intensity of the $11\bar{2}$ reflection relative to the $10\bar{1}$ as in (b) which is the optical diffraction pattern from (a) (this should be compared with Fig. 7.2f to see the effect). (c) Equatorial X-ray diffraction patterns from toad sartorius muscles at two different sarcomere lengths (top, S is 3.4 μm; bottom, S is 2.9 μm) (From G. F. Elliott *et al.* 1967.) It is clear that at the longer sarcomere length the intensity of the $11\bar{2}$ reflection is considerably weakened relative to the $10\bar{1}$ reflection compared with the shorter sarcomere length, an effect consistent with the withdrawal of the actin filaments from the regular A-band lattice (see text).

planes. However, in a whole muscle, the many different component fibrils will have many different orientations about the long axis of the muscle. There will, therefore, always be some fibrils in the correct orientation for diffraction from a particular set of planes to occur. Equatorial reflections can therefore be recorded from each of these diffraction planes. The reflections closest to the center of the pattern will clearly come from the planes with the largest d spacings starting with the $10\bar{1}$, the $11\bar{2}$,* etc. (Fig. 7.2)

In the muscle, each lattice point is occupied by a thick filament of diameter about 100 to 150 Å. The array can be thought of as a convolution of a single filament with the hexagonal lattice. From the convolution theorem, the diffraction pattern (Fourier transform) of the filament must be multiplied by the diffraction pattern of the lattice to give the diffraction pattern of the array of filaments. The filament profile, in projection down its long axis, is almost circular, and the corresponding diffraction pattern will approximate in form to an Airy disk pattern. The diffraction pattern of the lattice is just the reciprocal lattice. When multiplied by the Airy disk pattern, the intensity of the peaks in the reciprocal lattice will be modulated, and in particular the intensities of all but the $10\bar{1}$ and $11\bar{2}$ reflections will be markedly reduced. This effect is demonstrated in Fig. 7.2 by means of optical diffraction. The diffracting masks represent the lattice (a), a single filament (b), and an array of filaments (c) in projection down the fiber axis. Here the illumination can be thought of as being oriented along the fiber axis direction, so diffraction can be observed in two dimensions. It is clear from the optical diffraction patterns (d), (e), and (f) that the $10\bar{1}$ and $11\bar{2}$ reflections dominate the pattern from the array of filaments. The observed equatorial X-ray diffraction patterns from muscle (Fig. 7.3c) show comparable features, with the $10\bar{1}$ and $11\bar{2}$ reflections being relatively very strong.

It can be seen in Fig. 7.3a that the effect of adding actin filaments to the trigonal points of the myosin filament array is to increase markedly the density in the $11\bar{2}$ planes of the lattice, whereas the added actins lie between the $10\bar{1}$ planes. The result of this is that the intensity of the $11\bar{2}$ reflections increases relative to that of the $10\bar{1}$ reflections (Fig. 7.3b). This kind of effect was observed by G. F. Elliott *et al.* (1963) when they recorded the equatorial diffraction patterns of striated muscles at different sarcomere lengths. They found that at very long sarcomere lengths the intensity ratio of the $10\bar{1}$ reflection to the $11\bar{2}$ was high, but that it gradually decreased as the sarcomere length was reduced (Fig.

*In this indexing system, the third index i is such that $h + k + i = 0$. It helps to recognize equivalent planes in a hexagonal lattice.

7.3c). They took this to be evidence that the actin filaments are relatively disordered in the I-band but become ordered in the overlap region because of the regular thick filament array. On this basis, as the sarcomere length decreases, more filament overlap will occur. Therefore, the effective density of the thin filament contribution at the trigonal points of the myosin array will increase, thus strengthening the $11\bar{2}$ reflection relative to the $10\bar{1}$. The lack of order of the thin filaments in the I-band is clearly consistent with a changeover from a hexagonal arrangement in the A-band to a tetragonal array at the Z-band (Fig. 7.1). G. F. Elliott *et al.* (1963) also confirmed an earlier result of Huxley (1953b) that the spacings as well as the intensities of the $10\bar{1}$ and $11\bar{2}$ reflections changed with sarcomere length (S). Their results indicated that the thick filament array (of spacing a) became larger at short sarcomere lengths and smaller as the muscle was stretched. This happened in such a way that the total volume of the system ($\sim a^2S$) remained virtually constant—an observation known as the "constant volume effect." Further mention of this phenomenon will be made in Chapters 10 and 11. Suffice it to say here that any mechanism of force generation in muscle must clearly allow for the changing separation of the thick and thin filaments as the sarcomere length changes.

7.1.3. Axial Periodicities in the Sarcomere

Early Electron Microscopy. In electron micrographs of longitudinal sections, the vertebrate sarcomere can be seen to comprise several different bands. The A-band, I-band, H-zone, Z-band, and M-band have already been described (Fig. 7.1). The H-zone is commonly less densely stained than the remainder of the A-band and, as would be expected from the sliding filament model, it changes length with changing sarcomere length such that the distance from the H-zone edge to the Z-band remains constant. The M-band is located at the center of the H-zone. On each side of the M-band are regions even less densely stained than the remainder of the H-zone. These lightly stained bands together with the M-band were originally termed the "pseudo H-zone," but since this name is of more historical than practical value, an alternative name, the M-region, has been suggested (Sjöström and Squire, 1977a) and will be used here.

The pseudo H-zone or M-region, in fact, corresponds to the extent of the bare zones in the middles of the myosin filaments that line up to form the A-band. Although isolated filaments do appear to be truly bare in this region, the existence of extra material at the M-band in the intact A-band renders the name bare zone inappropriate for the description of this part of the sarcomere.

The regions of the sarcomere between the edges of the H-zone and the edges of the A-band are conveniently termed the overlap region. The regions between the edges of the M-region and the edges of the A-band have been termed the bridge regions since this is where the myosin cross bridges occur.

Early electron microscopy of muscle (Page and Huxley, 1963; H. E. Huxley, 1957) showed that both the A-bands and the I-bands of the vertebrate sarcomere are crossed by fine transverse striations approximately 400 Å apart. Later work by O'Brien *et al.* (1971) showed that in the I-band the axial periodicity was about 380 to 385 Å, and this was taken to be caused by the repeat of the tropomyosin–troponin structure along the thin filaments that form the I-band. The work of H. E. Huxley (1957) showed that the fine transverse striations in the A-band appeared to be caused by cross connections between myosin and actin filaments. It was concluded that such cross bridges occurred between any pair of myosin and actin filaments at axial intervals of about 400 Å. Since each myosin filament was known to be covered with projections and to be surrounded by six nearest-neighbor actin filaments, Huxley deduced that there were probably about six cross bridges projecting from the surface of the thick filaments in a filament length of about 400 Å. That these projections were associated with the thick rather than the thin filaments was clearly indicated by their existence both at the edge of the myofibril and in the H-zone. In neither place were actin filaments present.

X-Ray Diffraction Evidence. Relaxed vertebrate skeletal muscles give a very characteristic low-angle X-ray diffraction pattern (G. F. Elliott, 1964; G. F. Elliott *et al.,* 1967; H. E. Huxley and Brown, 1967). Apart from the equatorial reflections which have already been described (Fig. 7.2), the diffraction patterns also contain a very beautiful arrangement of layer lines and meridional reflections, all of which are associated with the axial periodicities in the sarcomere (Fig. 7.4a). The schematic diagram in Fig. 7.4b shows in a simplified way the various contributions to the total diffraction pattern. Layer lines with *d* spacings that are orders of a 360- to 370-Å repeat are thought to be caused by the thin filament structure. They can be compared with the patterns shown in Chapter 5 (Figs. 5.5 to 5.7). According to H. E. Huxley and Brown (1967) and Haselgrove (1975a), these layer lines have the spacings given in Table 5.4. A reflection on the meridian at a spacing of 385 Å is also thought to be associated with the thin filaments (see Section 7.4).

Another system of layer lines associated with *d* spacings that are orders of a 429-Å axial repeat (H. E. Huxley and Brown, 1967) is thought to be associated with the myosin filament. A particular feature of this layer line system is that layer lines 3, 6, 9, etc. have strong intensity

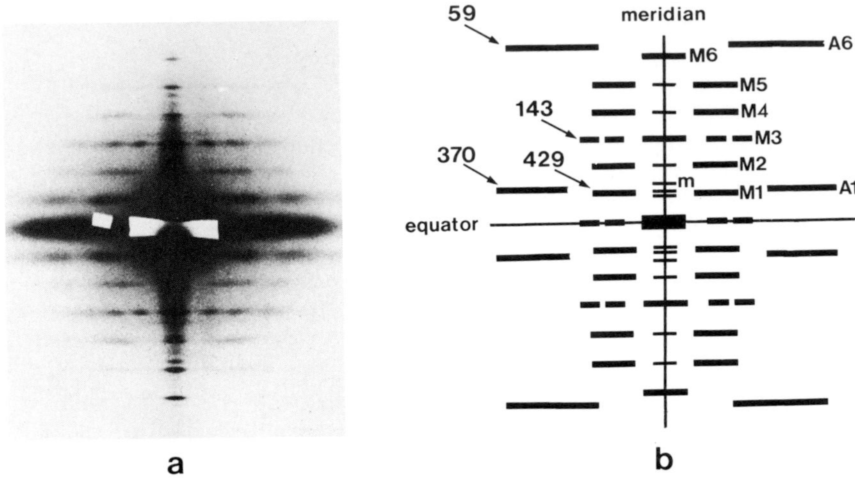

FIGURE 7.4. (a) Low-angle X-ray diffraction pattern from frog sartorius muscle in the relaxed state showing the characteristic layer line patterns caused by the thick and thin filaments (courtesy Dr. J. C. Haselgrove; muscle axis is from top to bottom of page). The spacings and probable origin of these layer lines are summarized in (b) where the layer lines from the myosin filaments are labelled M1, M2, etc., and those from actin are labelled A1 and A6 (for details of the actin pattern see Chapter 5). The myosin layer lines are orders of a 429-Å repeat, and the actin layer lines are orders of a repeat of about 2 × 370 Å. Some of the so-called "forbidden" meridional reflections (see Table 7.1) are labeled m.

on the meridian, whereas all of the other layer lines have greater intensity off the meridian (Fig. 7.4b). This was interpreted by H. E. Huxley and Brown (1967) to mean that the structure giving rise to the layer line pattern was a helical array of myosin projections with a repeat (R) of 429 Å and a subunit axial translation (p) of 143 Å (see Figs. 2.12 and 2.14). The fact that the intensity on layer lines 1,2,4,5 etc. was generally closer to the meridian than the reflections in the actin pattern suggested that the helix of diffracting units had a much larger radius than the actin helix, and this supported the view that it was the myosin heads on the thick filaments that were arranged with the 429-Å helical repeat.

Further support for this conclusion seemed to come from a quite different feature of the X-ray diffraction patterns. Use of a virtually point focus double-crystal monochromator camera (H. E. Huxley and Brown, 1967) to record X-ray diagrams from frog sartorius muscle appeared to show that the system of layer lines based on the 429-Å repeat was sharply "sampled" by row lines (i.e., reflections on lines parallel to the meridian). Sampling of this kind gives information about the lateral packing of the diffracting units. The most interesting feature of the apparent sampling was that the intensity maxima on the first and second layer lines were on different row lines than the reflections on the equator and on the third layer line (Fig. 7.5).

An effect analogous to this was described in Chapter 4 where the packing of two-strand coiled-coil molecules was considered. It was shown there that an axial shift between adjacent molecules would generate a "superlattice," and the effect on the diffraction pattern would be that reflections on the equator would be in different positions from those on the first layer line as observed by A. Elliott *et al.* (1968a). In fact, the observed reflections followed a selection rule that in that particular case was $h + k + l = 2N$ (where N is integral).

A similar kind of analysis was carried out by Huxley and Brown on

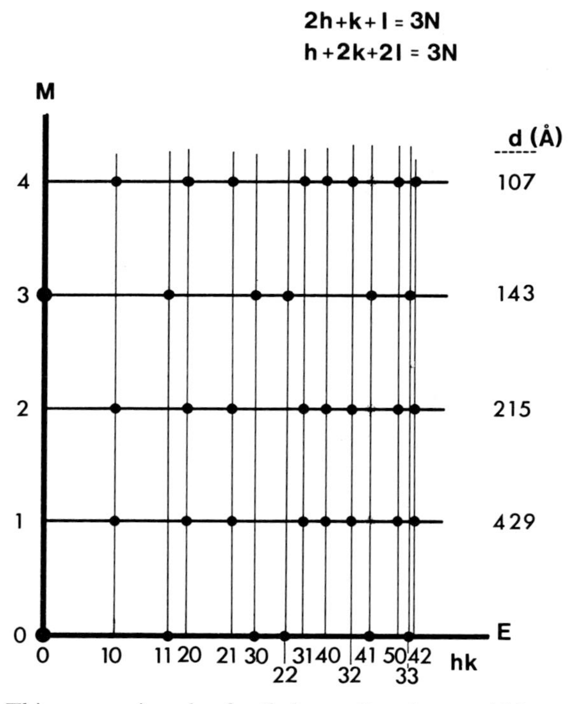

FIGURE 7.5. This summarizes the detailed sampling that would be predicted for the myosin layer lines in low-angle X-ray diffraction patterns from frog sartorius muscles using the selection rules $2h + k + l = 3N$, $h + 2k + 2l = 3N$ (see text). Some of these reflections were observed using X-ray diffraction cameras with good two-dimensional resolution by H. E. Huxley and Brown (1967). Note that the reflections seen on layer lines 1, 2, and 4 are on different row lines (parallel to the meridian, M) from the reflections on the equator (E) and the third layer line. That the observed sampling pattern is consistent with the selection rules at the top of the diagram may indicate that the myosin filaments in the A-band are organized into a particular kind of superlattice (see Fig. 7.6).

the sampled layer lines associated with the 429-Å repeat. It was found that a superlattice arrangement might occur such that the unit cell of the superlattice has a side that is $\sqrt{3}$ times the side of the lattice deduced from the equatorial reflections alone. In this case, the selection rule $2h + k + l = 3N$
or $h + 2k + 2l = 3N$ seemed to apply, and it was deduced from this that the filaments giving rise to the layer line pattern must be located on the lattice points of the hexagonal array and must have a threefold screw axis. The reason for this was that the existence of the superlattice was thought to be caused by an effective translation of adjacent filaments by ⅓ of the axial repeat of 429 Å (i.e. by 143 Å). The resulting scheme is shown in Fig. 7.6. That there was no evidence of this larger lattice on the equator meant that all of the filaments in the superlattice must appear identical in projection down their axes. The conclusions that H. E. Huxley and Brown reached from these results were that the 429 Å repeat

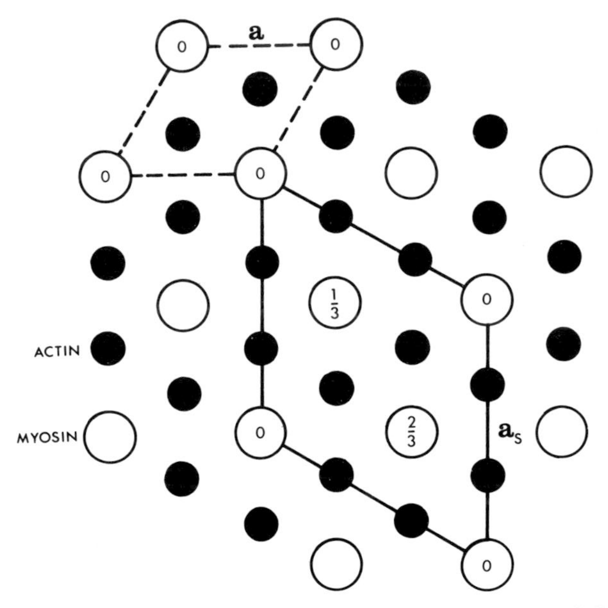

FIGURE 7.6. Diagram of the A-band lattice in the overlap region of the A-band as viewed down the filament axis. The dotted lines indicate the shape of the simple unit cell of the hexagonal A-band lattice, and this contains one myosin filament and two actin filaments. The side a varies with sarcomere length between about 350 and 480 Å. The bold outline indicates the size and shape of the superlattice proposed by H. E. Huxley and Brown (1967) to account for the layer line sampling in X-ray diffraction patterns from resting frog sartorius muscles (Fig. 7.5c). The numbers for each myosin filament indicate the effective axial positions of the different filaments in the superlattice in multiples of 143 Å that would be demanded by the proposed selection rules (for details see text). The superlattice side a_s is $a\sqrt{3}$.

belonged to the thick filaments and was caused by the myosin projections or cross bridges and that the thick filaments are arranged to give the superlattice shown in Fig. 7.6. We shall see later that although the first conclusions are probably correct, the real superlattice may be different.

Some other features of the observed X-ray diffraction patterns are also worth noting. First, the meridional reflections that are orders of 143 Å are very narrow in an axial direction, indicating that the structure that gives rise to them (the helical array of myosin cross bridges) is very well ordered for several thousand angstroms and possibly for the whole length of the thick filaments.

The second feature is that the width of these same reflections across the meridian, in patterns recorded using the double-crystal monochromator (which has very good resolution in that direction), is also consistent with an array that is ordered laterally over several thousand angstrom units. Clearly, individual filaments do not have such dimensions. But a whole myofibril is of this size, and the observation suggests that the thick filaments in the A-band are in precise axial register over most, if not all, of the width of the myofibril.

Finally, it should be noted that in addition to the 385-Å tropomyosin–troponin reflections and to the meridional reflections that are orders of 143 Å, there are many other meridional reflections that cannot be accounted for either by the known thin filament structure or by a simple helix of myosin cross bridges. The spacings of these "forbidden" reflections according to H. E. Huxley and Brown (1967) and Haselgrove (1975a) are given in Table 7.1. Their origin will be discussed fully in Section 7.3 and in Chapter 10.

TABLE 7.1. Axial Spacings of Low-Angle Meridional X-Ray Reflections from Relaxed Frog Sartorius Muscles at Rest Length[a]

600 ± 20[d]		394.9[b]	doublet	229.7		176.9	
529.4		383.0[b]		222.5		153.1	
494.0		365.3		214.3	doublet	150.8	
441.8	doublet	352.2		209.8		148.0	
418.4		271.3		187.4		143.4[c]	doublet
		237.7		181.9		141.4	

[a]From Haselgrove (1975a). The reflections were recorded on a mirror/monochromator camera (see Chapter 2) with a specimen-to-film distance of 2 m. All wavelengths in ångstroms. Unless stated otherwise the probable error on all reflections is about ±0.5%.

[b]These are thought to be reflections caused by the tropomyosin–troponin complex on the thin filaments.

[c]The spacing of this reflection was calculated absolutely from measurements using the 2-m camera and using Cu Kα = 1.5405 Å. This reflection was then used to calibrate the diffraction pattern.

[d]This is a very broad, diffuse reflection.

7.2. Thick Filament Symmetry and the Transverse Structure of the A-Band

7.2.1. Introduction

So far we have discussed three lines of evidence about the arrangement of the myosin cross bridges on the surface of the thick filaments. (1) H. E. Huxley's electron micrographs (H. E. Huxley, 1957) were interpreted to mean that there were six cross bridges in the repeat of about 400 Å (which could clearly be taken to be equivalent to the 429 Å repeat *in vivo*). (2) The X-ray diffraction evidence of H. E. Huxley and Brown (1967) showed that the cross bridges were very likely arranged on a helix with a 429-Å repeat and 143-Å subunit axial translation. (3) The existence of the myosin filament superlattice was taken to mean that the helix involved had a threefold screw axis (H. E. Huxley and Brown, 1967; H. E. Huxley, 1972a).

H. E. Huxley and Brown (1967) put these results together and derived a model for the cross bridge array consisting of two helical strands of cross bridges, each strand being a helix with six cross bridges per turn and having a pitch of 2 × 429 Å and subunit axial translation 143 Å. This model is illustrated in Fig. 7.7 where its radial projection, cross section, and view in perspective are given. It is commonly termed a two-strand, six residue per turn (6/1)* helix. It will be seen that even though each helix has a pitch of 2 × 429 Å, since there are two helical strands, the repeat of the structure is only 429 Å. Figure 7.7c shows the suggested arrangement of the superlattice in terms of this model for the filament symmetry. It will be seen from this diagram that an effective axial shift of 143 Å between adjacent filaments, as required to give the observed superlattice, can in fact be produced from filaments with no axial shift simply by rotating adjacent filaments by 60° as has been done in Fig. 7.7c. This is because the filaments have helical symmetry with successive subunits being related by a 60° rotation plus a 143-Å axial shift.

To H. E. Huxley and Brown, this model seemed to be very satisfactory except for one kind of observation. This was that from quantitative estimations of myosin content there seemed to be many more myosin molecules in the 429-Å repeat than six. Since the A-band is about 1.6 μm long and the bare zone about 1600 Å long (see later section), then the two bridge regions on a single thick filament should together contain very roughly 14,400/143 levels of cross bridges (i.e., about 100). H. E. Huxley and Brown's model has two cross bridges on each level, so there

*In crystallographic terms, it is a 6_2 helix.

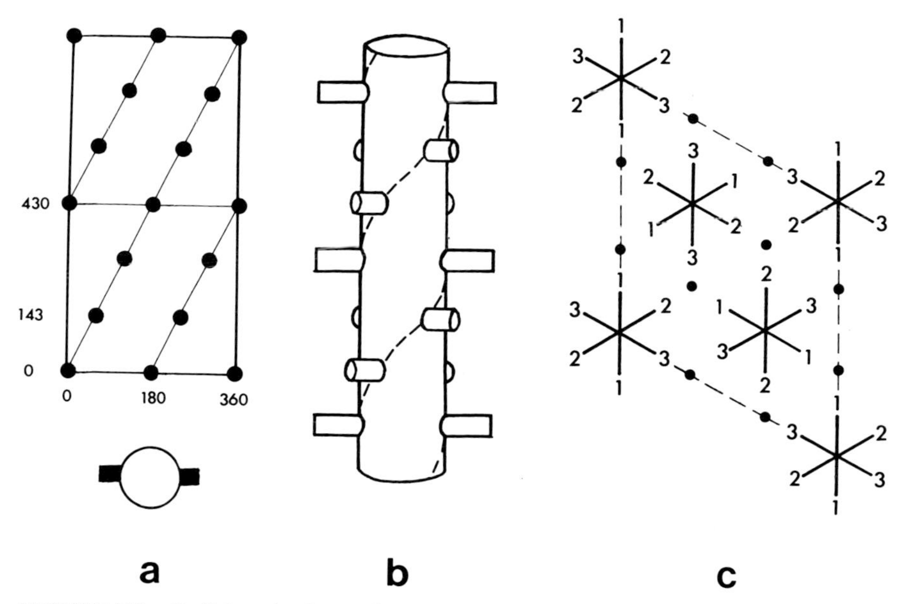

FIGURE 7.7. Radial projection and cross-section (a), schematic three-dimensional view (b), and the required superlattice arrangement (c) for thick filaments with their cross bridges arranged on a two-stranded helix as proposed by H. E. Huxley and Brown (1967). In (c) the superlattice is generated by rotations between filaments of 60° rather than by axial shifts of 143 Å. Since the filaments have helical symmetry, the two operations are equivalent.

should be about 200 ($=N_M$) molecules in the whole filament. However, early estimates of the myosin content in the thick filament suggested that there were really 300 to 400 molecules per filament. For example, H. E. Huxley (1960, 1972a) used data from Hanson and Huxley (1957) on rabbit psoas muscle to estimate the myosin content in a given volume of muscle, and knowing the geometry of the sarcomere, he obtained a value for N_M of about 384 molecules. An alternative approach more commonly used nowadays is to estimate the relative amounts of myosin and actin in the muscle and, knowing the structure of the thin filament and the geometry of the sarcomere, to deduce N_M.

Figure 7.8 illustrates the measurements needed to carry out this second approach (Tregear and Squire, 1973). Since (1) the molecular weight of actin is 41,800, (2) the thin filament length is about 1 μm, and (3) the thin filament subunit axial repeat is 27.5 Å, there must be a mass (M_A) of actin in each thin filament given by $M_A = (10{,}000)/(27.5) \times 41{,}800$. Similarly, since the molecular weight of the whole myosin molecule is 470,000 (Table 6.1), the mass of myosin in the thick filaments must be $M_M = N_M \times 470{,}000$. Each thick filament interacts with two

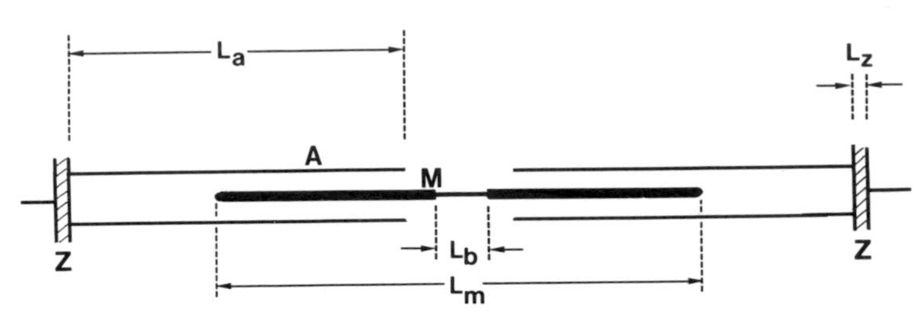

FIGURE 7.8. Schematic diagram to show the geometry of the vertebrate sarcomere and the dimensions required to predict the weight ratio of myosin to actin. After Tregear and Squire (1973); for details see text.

arrays of actin filaments, and in each array there are two thin filaments to every thick one. The ratio of myosin mass to actin mass for the whole sarcomere must therefore be $M_M/4M_A = (N_M \times 470{,}000)/(4 \times 15{,}200{,}000) = 0.00773\ N_M$. H. E. Huxley and Hanson (1957) estimated the mass ratio of A substance to I substance to be 1.53 using the interference microscope. If this is corrected for the tropomyosin/troponin content (~30%) in the thin filaments (Spudich *et al.*, 1972) and for the presence of about 5% of proteins other than myosin in the thick filaments (e.g., C protein, Offer, 1972), then the estimated myosin-to-actin mass ratio is about 2.1, and hence $N_M = 2.1/0.00775 = 270$ molecules. Both this value of 270 molecules, and H. E. Huxley's (1972b) estimate of 384 molecules are significantly greater than the value of 200 expected from the two-strand 6/1 helical model for the cross bridge arrangement. H. E. Huxley (1972a) concluded that there could be two possible explanations for this. One was that each cross bridge position in the two-strand helix might correspond to the heads of two myosin molecules (Fig. 7.9a), thus giving about 400 molecules per filament. The second explanation was simply that for some reason a proportion of the myosin cross bridges present on the filament were not visualized in electron micrographs of the overlap region.

With this kind of uncertainty in mind, Squire (1971, 1972) proposed two alternative models for the cross bridge arrangement. The evidence on myosin content in the thick filament seemed to suggest that the heads of either three or four myosin molecules were located at 143-Å intervals along the thick filament. Squire therefore suggested that alternative cross bridge arrangements might be three-stranded or four-stranded helical arrays of cross bridges with the helix pitches equal to 3 × 429 Å and 4 × 429 Å, respectively. These models are illustrated in Fig. 7.9b,c. In theory, any multistrand helix of pitch $n \times 429$ Å with n helical strands will have a true repeat of only 429 Å whatever the value of n, provided that there is an integral number of subunits on each strand in the 429-Å repeat (see Section 2.3.7). The alternatives with $n = 3$ and $n =$

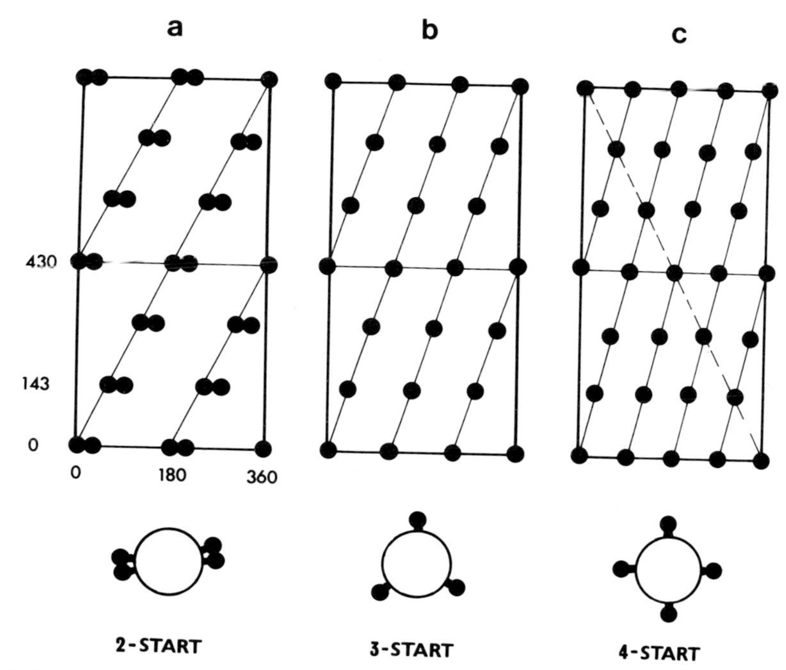

FIGURE 7.9. Radial projections (above) and cross-sectional views (below) of the three alternative regular helical distributions of cross bridges that account for the layer line pattern from resting frog sartorius muscle (Fig. 7.5) and with the requirement that, at each 143-Å-spaced axial level, there are the heads of either three or four myosin molecules. Each projection in the lower figures corresponds to one myosin molecule or to two myosin heads (see Chapter 6).

4 were chosen since only these values seemed to be compatible with the estimated myosin content.

That the values $n = 3$ and $n = 4$ should be accepted as realistic alternatives to the two-stranded helix ($n = 2$) clearly required that they account satisfactorily for the observed X-ray diffraction and electron microscopic evidence on the cross bridge arrangement. It will be shown below (Section 7.2.6) that all three arrangements give diffraction patterns with layer lines of repeat 429 Å and meridional reflections of repeat 143 Å. The main difference between the patterns is that the position of the off-meridional intensity on layers 1, 2, 4, 5, etc. does depend on the model. In order that each model give a peak at the same position on a particular layer line, the radial position of the cross bridge from the filament axis in the two-stranded model would have to be relatively smaller than it is in the three-stranded model, and that in the four-stranded model would need to be relatively larger. A detailed discussion of this effect will be given later.

Figs. 7.10–7.12 show how the three different models account for the observed myosin filament superlattice (assumed to be perfect). The effect on the superlattice of altering the number of strands in the helix is to alter the relative orientations of adjacent filaments in the superlattice. The two-strand model has sixfold rotational symmetry in projection down its axis. A rotation of 60° (= 360°/6) between adjacent filaments therefore produces the required superlattice. The symmetries in projection of the three- and four-stranded helices are ninefold and 12-fold, respectively. The superlattice is therefore generated by rotations between adjacent filaments of either 40° (= 360°/9) or 30° (= 360°/12) (Squire, 1974).

The additional evidence that needs to be explained is Huxley's observation (H. E. Huxley, 1957) of one cross bridge from myosin to each actin filament about every 400 Å. Squire (1974) has accounted for this observation in the case of the three-stranded filament in terms of the possible interactions that myosin cross bridges can make with the adja-

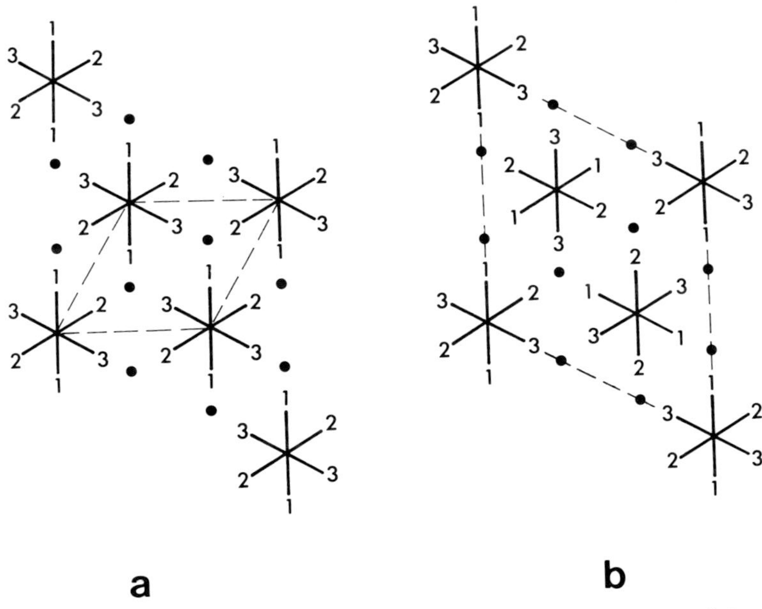

FIGURE 7.10. Schematic diagrams of the simple lattice arrangement (a) and the superlattice arrangement (b) for a two-start (two-stranded) myosin filament (produced by systematic 60° rotations). The diagram shows 429 Å thick transverse slices through the overlap region, and the numbers by the projections indicate the axial levels of the cross bridges in multiples of 143 Å. Note that in (a) the actin filaments (black dots) are all surrounded by projections at different levels (1, 2, and 3), whereas in (b) some actin filaments have 1, 2, and 3 around them and others have all the surrounding bridges at the same level (e.g., 2,2,2). In this case, different actin filaments have nonequivalent surroundings.

cent actin filaments. The argument was based on the conclusion that only a fraction (⅔) of the total number of cross bridges could attach to actin in a rigor muscle. If only those cross bridges that were attached to actin were sufficiently stabilized to be preserved and visualized in the electron microscope, then H. E. Huxley's observations could be explained. The details of this argument require some knowledge of the structure of the I-band, so we shall return to it at the end of the chapter (Section 7.4.6).

To summarize, it was shown in a series of papers (Squire, 1971, 1972) that preliminary analysis of the available evidence on the cross bridge array could not choose among three alternative models. The cross bridge arrangement could be described as a three-stranded or four-stranded helical array with the heads of one myosin molecule as the helix subunit in each case; alternatively, the structure could correspond to the H. E. Huxley and Brown (1967) two-stranded helical array but with the heads of two myosin molecules as the subunit in the helix (Fig. 7.9a). The problem remained to decide among these alternatives.

In principle, there are several ways that this could be done and some of these are summarized below. The value of n could be determined by:

1. Determining accurately the number N_M of myosin molecules in the thick filament using physical and biochemical methods.
2. Analyzing the intensity distribution on the 429-Å, 215-Å, 143-Å, etc. layer lines in the observed X-ray diffraction patterns, since these should in theory carry all of the information needed to give the rotational symmetry of the helix.
3. Studying the symmetry of the myosin filament superlattice in order to obtain clues about the symmetry of the component filaments.
4. Studying the structure, shape, and symmetry of the myosin filament backbone since this must be closely related to the surface cross bridge arrangement.
5. Studying isolated myosin filaments in order to visualize directly the myosin cross bridge array.
6. Extending Huxley's original electron microscope observations using improved techniques in order to count directly the number of cross bridges in a thin cross section of the thick filaments.

All of these different approaches are, in fact, being used to determine the value of n, and fortunately the bulk of the evidence available so far seems to be in favor of the value $n = 3$. The results and arguments that seem to support this conclusion are summarized in the next few pages. It should be mentioned, however, that the results described are all very much part of "work in progress." At present, many of the useful approaches to the problem have not yet been fully worked through. This section of the chapter should therefore be taken as a description of work in hand and of the conclusions that appear to be emerging from it.

7.2.2. Biochemical Evidence on Myosin Content

Once it had been realized that the symmetry of the thick filaments was still uncertain, several attempts were made to determine accurately the number of myosin molecules in a single filament using biochemical or physical chemical methods. The technique of SDS gel electrophoresis had already been introduced, and it therefore became the most commonly used method.

Tregear and Squire (1973) applied the technique to glycerinated preparations of rabbit psoas muscle. The muscle was washed to remove glycerol and was separated into myofibrils by homogenization. The myofibril preparation was then centrifuged (to remove mitochondria, etc.), washed, and dissolved in a 1% SDS solution. After the usual preparation procedure for electrophoresis, the solution was run on 5 cm of 5% gel layered on top of 5 cm of 10% gel. In this way, the proteins moved rapidly through the upper gel, thus allowing the slow myosin heavy chain to penetrate adequately into the gel, while the relatively fast actin molecules were slowed down by the lower 10% gel so that they did not run off the gel too soon. The resulting gels were stained with Coomassie Blue, and the densities of the different peaks were determined by scanning along the gel with a densitometer. In this way, the ratio of the area of the myosin heavy chain peak (M_{HC}) to the actin peak (M_A) could be determined. After correction for differential staining, Tregear and Squire (1973) estimated that the ratio of myosin to action (M_{HC}/M_A) in their preparations was close to 1.51. They then used a calculation based on the geometry of the sarcomere and similar to that described earlier (see Fig. 7.8) to estimate the number of myosin molecules per filament. The value they obtained for N_M was about 243 molecules per myosin filament. This was thought to be consistent with a filament symmetry that was three-stranded (Fig. 7.9b), since on their assumptions, the number of myosin molecules per 143-Å repeat was estimated to be 2.66. In reality, this figure should clearly be integral, and it was thought likely that the experimental value probably slightly underestimated the myosin and that the number (n) should really be 3.

Two groups have published results of comparable SDS gel electrophoresis experiments on vertebrate skeletal muscle. Morimoto and Harrington (1974) obtained a myosin heavy chain to actin weight ratio of 2.65 ± 0.04, and Potter (1974) obtained a value for this ratio of 1.44. The values N_M calculated from these results are about 386 ± 6 and 254, respectively.

In addition, Morimoto and Harrington (1974b) used another approach in which they estimated the molecular weight of intact thick filaments by determining both the weight concentration of a filament suspension and the number of filaments per unit volume. The latter was estimated by a particle-counting method in which filaments were centrifuged onto an electron microscope grid and the number in a given area of the grid estimated from electron micrographs of the negatively stained filaments. In this way Morimoto and Harrington obtained a value for the molecular weight of the thick filament of $204 \pm 11 \times 10^6$ After making a small allowance for the presence of nonmyosin proteins in the thick filaments, they estimated the number of myosin molecules per filament to be about 430.

It has already been mentioned that the results of H. E. Huxley and Hanson (1957) gave a value for N_M of 270 molecules. Similarly, using a value of 170 μM for the absolute myosin concentration and using the known sarcomere geometry to determine the number of myosin filaments per unit volume of muscle, H. E. Huxley (1972a) estimated that N_M must be about 384 molecules. However, more recent nucleotide-binding studies by Marston and Tregear (1972) have cast some doubt on the validity of H. E. Huxley's calculation based on the myosin concentration of 170 μM (H. E. Huxley, 1960). Their value for this concentration was 120 μM which gives $N_M = 305$ molecules using the same method of calculation. More recently, Lamvik (1978) has attempted to "weigh" thick filaments directly in the scanning transmission electron microscope. This gave $n = 3$. It is unfortunate, but nonetheless clear from these very varied results (Table 7.2) that present physical and biochemical methods have so far been unable to determine conclusively the number of myosin cross bridges in a thick filament. But taken together, the results do at least confirm that n must be either 3 or 4 and is not 2 or 5.

The main problem with the method involving SDS gel electrophoresis is that, although one probably obtains quite a good estimate of the proportion of myosin and actin in the preparation that is put onto the gel (unless some unidentified protein happens to run together with the myosin or actin peak; see Pepe and Drucker, 1979), the sample preparation may not itself be representative of the situation in the intact muscle. This could easily be the case if either the actin or the myosin is incompletely extracted by the SDS dissolving medium or if proteolysis occurs at any stage during the preparative procedure.

7.2.3. Symmetry Evidence from the Myosin Superlattice

An analysis of the myosin filament superlattice in frog sartorius muscle (assumed to be perfect) and the implications of its symmetry have been given by Squire (1974). It was shown that the alternative two-strand

TABLE 7.2. Estimates of the Number of Myosin Molecules (N_M) in Vertebrate Skeletal Muscle Myosin Filaments and the Number (n) of Molecules on Each 143-Å Separated Level along the Myosin Filaments[a]

Reference	Method	N_M	n[b]
Huxley and Hanson (1957)	Interference microscopy	318	3.18
Huxley (1972a)	Calculated from myosin concentration (170 μM)	432	4.32
Marston and Tregear (1972)	Calculated from myosin concentration (120 μM)	305	3.05
Tregear and Squire (1973)	Quantitative SDS gel electrophoresis	266	2.66
Morimoto and Harrington (1974)	SDS gels	386 (±6)	3.86
Morimoto and Harrington (1974)	Density of suspension plus particle counting	430	4.3
Potter (1974)	SDS gels	254	2.54
Lamvik (1978)	Filament weighing in scanning transmission electron microscopy	—	2.7

[a] All data from rabbit psoas muscle.
[b] Calculated on the assumption that there are about 100 levels (crowns) of myosin cross bridges in each myosin filament.

6/1, three-strand 9/1, and four-strand 6/1, helical symmetries of the thick filaments would give characteristically different cross bridge environments for the actin filaments at the trigonal points in the lattice. Only in the case of the myosin filaments with three-strand helical symmetry would there be any apparent advantage in arranging the myosin filaments in a superlattice.

This was deduced in the following way. Figure 7.10 shows how myosin filaments with two-strand helical symmetry would fit into either a simple lattice in which all of the myosin filaments have the same orientation (a) or into the superlattice (b). As mentioned earlier, the required superlattice is produced in this case if there is a 60° rotation between adjacent myosin filaments. It can be seen that the cross bridge environments that the actin filaments see in the simple lattice (Fig. 7.10a) are all equivalent. Each actin filament is approached by cross bridges equally spaced at 143-Å intervals (as indicated by the 143-Å-spaced cross bridge levels 1, 2, and 3). On the other hand, when the myosin filaments are arranged to give the superlattice (Fig. 7.10b), the cross bridge environments of half of the actin filaments in the lattice are quite different from those of the other half. In one case, the cross bridges are evenly spaced at 143-Å intervals; in the other, all three myosin filaments surrounding the actin filament contribute cross bridges at the same level, and there is no interaction for 429 Å. If it is remembered that the two-strand helix would have the heads of two myosin molecules as the repeating unit on each lattice point of the helix (to give four molecules per 143-Å interval), then in Fig. 7.10b the actin filaments would have 12 myosin heads directed towards them at the same level, and presumably these would compete for interaction with the same short length of the actin filament. Taking this apparently unlikely situation together with the nonequivalence of the actin filaments, it would seem that a superlattice arrangement of myosin filaments would be a disadvantage for a two-stranded filament. The simple lattice arrangement (Fig. 7.10a) would appear to be much more satisfactory.

The symmetries of the two-stranded and four-stranded helical models for the thick filament are very closely related. For this reason, similar results to those described above are obtained when the four-stranded filaments are arranged in either a simple lattice (Fig. 7.11a) or a superlattice (Fig. 7.11b). The main difference is that in this case the superlattice is produced by a 30° rotation between adjacent filaments. Once again, the simple lattice gives equivalent actin environments, but the superlattice does not.

Turning now to the three-stranded model for the thick filament, quite a different result is obtained. This time the thin filaments can have virtually equivalent environments whether the superlattice exists or not. But in the simple lattice, the actin filaments would be approached by

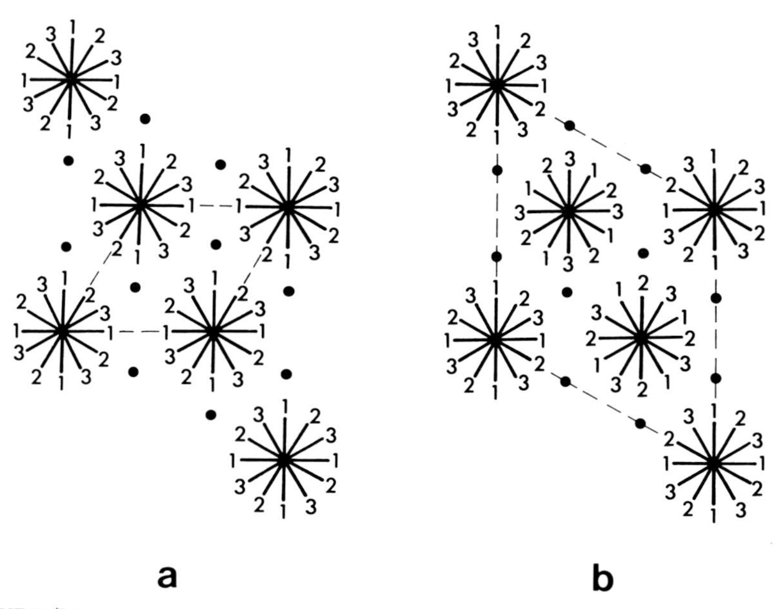

FIGURE 7.11. A comparable figure to Fig. 7.10 but for a four-start (four-stranded) myosin filament. This time the superlattice (b) is generated by systematic 30° rotations. Note that just as for the two-stranded filament, the actin environments are equivalent in (a) but nonequivalent in (b).

cross bridges at the same level (Fig. 7.12a), whereas they would be evenly spaced at 143-Å axial intervals in the superlattice (Fig. 7.12b). It should be noted, however, that the superlattice is such that the cross bridges directed towards one half of the actin filaments would follow a left-handed screw axis (i.e., the sequence 1–2–3 in Fig. 7.12b would go round the actin in an counterclockwise direction), whereas those around the other half would follow a right-handed screw. The two situations are therefore only quasiequivalent. Note also that the thick filaments would need to have a specific orientation in the hexagonal lattice (with some cross bridges lying along the 1,1 planes of that lattice) to give the actin filaments this quasiequivalence.

The conclusion from this discussion, as it stands, is that only in the case of the three-stranded structure does it seem advantageous for there to be a superlattice. If the ideal superlattice does occur in frog sartorius muscle, then this conclusion would argue strongly that the thick filaments in this muscle must be three-stranded. Unfortunately, this simple conclusion is complicated to some extent by recent results from both frog sartorius muscle and from fish skeletal muscle. As shown in the next

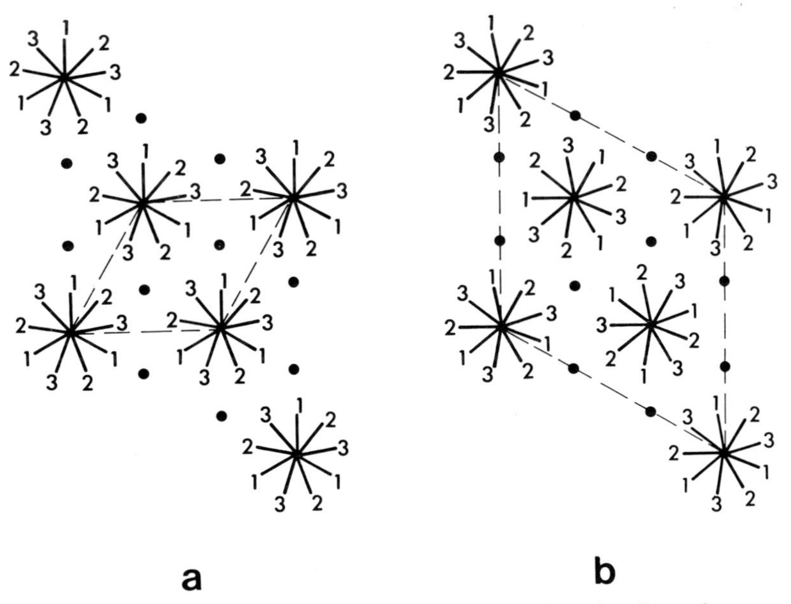

FIGURE 7.12. A diagram comparable to Figs. 7.10 and 7.11 but for a three-stranded myosin filament. The superlattice is produced by systematic 40° rotations between adjacent filaments. Unlike the other two cases, the thin filament environments are equivalent in (a) and still at least quasiequivalent in (b). Here only the sense of the crossbridge helix around the different thin filaments varies. Note that this result is only true for the particular absolute orientation of the thick filaments relative to the lattice that is illustrated.

section, a superlattice arrangement of thick filaments does not seem to occur in fish muscle. All of the thick filaments seem to have identical orientations. If the thick filaments in frog sartorius muscle and fish skeletal muscle have equivalent symmetries (which has not been proved conclusively but is commonly assumed to be the case and is supported by recent electron micrographic evidence), then it could be argued that the filament symmetry must be such that, in either the simple lattice or the superlattice, the thin filament environments should be equivalent. The only cross bridge symmetry for which this is true is the three-stranded 9/1 helical symmetry. It could then be argued that it is only a minor disadvantage to have all of the cross bridges approaching the thin filament at the same level as would occur in the simple lattice in fish muscle. (Note that in none of the cases considered is the filament symmetry such that staggered cross bridge arrangements occur in both the simple and superlattices.) It is worth noting that the actual number of myosin heads approaching the few available actin sites at the same level would be less for the three-stranded structure (6) than it would for the two- or four-stranded structures (12).

It would be a tempting speculation to suggest that, if it exists, the superlattice in frog sartorius muscle (and probably in many other vertebrate skeletal muscles) would be a slightly advantageous development of the analogous structure in fish muscle in that while still preserving the equivalence of the thin filaments, it allows the myosin cross bridges to interact much more satisfactorily with actin by spreading them out at 143-Å axial intervals around the thin filaments. However, new results to be described in Section 7.2.5 suggest that contrary to expectation, the superlattice in frog sartorius muscle is not as well developed as had previously been thought.

7.2.4. Evidence from Electron Microscopy

It has long been realized that direct visualization of the cross bridges on the thick filaments using the electron microscope is likely to be the best method of determining the filament symmetry and structure. Two different approaches spring to mind. One is to isolate thick filaments from muscle, to stain them or shadow them, and then to analyze the structure that is seen. The other is to study embedded muscle, to count the number of heads emanating from the thick filaments in thin cross sections of the muscle, and to relate successive levels of cross bridges by using either longitudinal sections of muscle or serial transverse sections.

Isolated Filaments. Unfortunately, as mentioned in the last chapter, isolated thick filaments have rarely shown regularly arranged projections. At best, they have tended to show an axial periodicity of about

140 Å but little evidence of helical order. To date, the most promising results have been obtained by Trinick (1973) who used the critical point drying method to dry isolated thick filaments prior to electron microscopy. Use of this procedure, which reduces the surface tension forces that usually flatten biological structures onto the grid, resulted in the preparation of filaments that showed at least some evidence of helical order. Unfortunately, it was reported that it was not possible to decide from the available pictures whether the helix was three-stranded or four-stranded, but it seemed not to be two-stranded. Some recent results of Davey and Graafhuis (1976) seem to indicate that the thick filament backbone has threefold rotational symmetry, but these results will need to be followed up before their true significance can be evaluated. Finally, Trinick and Elliott (1979) have applied the mica replication technique (Chapter 3) to isolated rabbit myosin filaments, and they obtained some very beautiful appearances with remarkably clear cross bridges radiating from the filament shaft. Unfortunately, no helical symmetry was evident.

This and other comparable approaches are being continued and should eventually provide decisive evidence on the structure of the thick filaments. In the meantime, recent evidence from embedded muscle has started to provide solid evidence about thick filament symmetry and morphology. This new evidence has come mainly from transverse sections of frog sartorius muscle and of fish muscle.

Sections. The appearance of a transverse section clearly depends on its location in the sarcomere, and, as it turns out, transverse sections through many different parts of the A-band have given information on the filament symmetry (Luther, 1978; Luther and Squire, 1978, 1980). It is convenient to consider five specific types of transverse section, the locations of which are indicated in Fig. 7.13. Type A is through the M-band alone, type B is through the region (the bare region) between the outer edge of the M-band and the inner edge of the bridge region, type C is through the part of the bridge region that is not overlapped by actin, type D is from the middle of the overlap region, and type E is from the very tip of the A-band (i.e., close to the A–I junction).

The two crucial requirements of a good transverse section of muscle are (1) that the native structure of the muscle be well preserved and (2) that the section be cut in a plane that is accurately perpendicular to the long axis of the muscle (Luther, 1978; Luther and Squire, 1978). It has been found that greatly improved transverse sections can be prepared if the muscle (e.g., frog sartorius) is fixed *in situ* in the animal in question. This avoids the longitudinal disruption that often occurs if a muscle is dissected before being fixed. In addition, if the piece of muscle that is embedded is also fairly thin (e.g., a single fiber), then, by viewing the

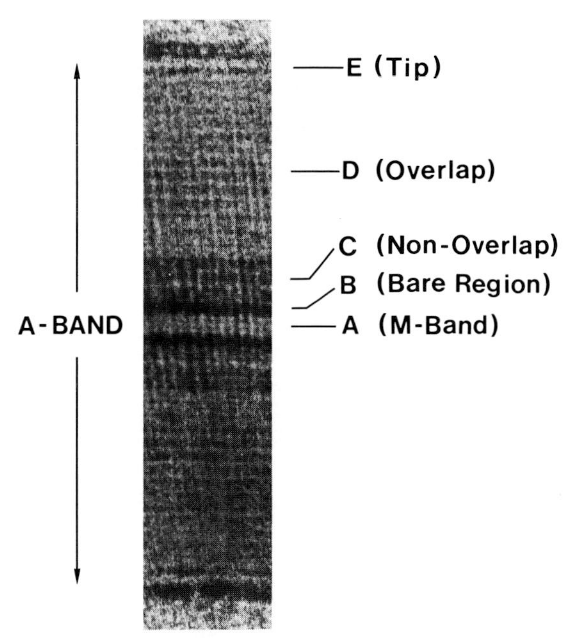

FIGURE 7.13. Longitudinal negatively stained cryosection of the vertebrate A-band (frog sartorius muscle) indicating the various regions in the A-band that are discussed in detail in the text. Unpublished micrograph of Dr. A. Freundlich. (Magnification ×74,000; reduced 40% for reproduction).

block in the light microscope, it is possible to see the cross striations in the muscle and hence to orient the block in the specimen holder of the ultramicrotome so that the striations are aligned parallel to the edge of the microtome knife. Very good transverse sections have been obtained in this way. The appearances of such sections through the different parts of the A-band indicated in Fig. 7.13 are described below. It will be shown that these sections, as well as giving information on the filament symmetry, have also revealed many new details of the ultrastructure of the A-band.

Transverse Sections through the M-Band and Bare Regions. The first published details of the M-band structure were given by Pepe (1967a) and Knappies and Carlsen (1968). It was shown that at the M-band each thick filament appears to be approximately circular in profile and to be cross-linked to its six nearest neighbor thick filaments by bridging structures commonly termed M-bridges. This arrangement is shown in Fig. 7.14a. Knappies and Carlsen (1968) also noted that the M-bridges were often seen to be thickened halfway along their length. Since then, the recent work of Luther and Squire (1978) has shown that,

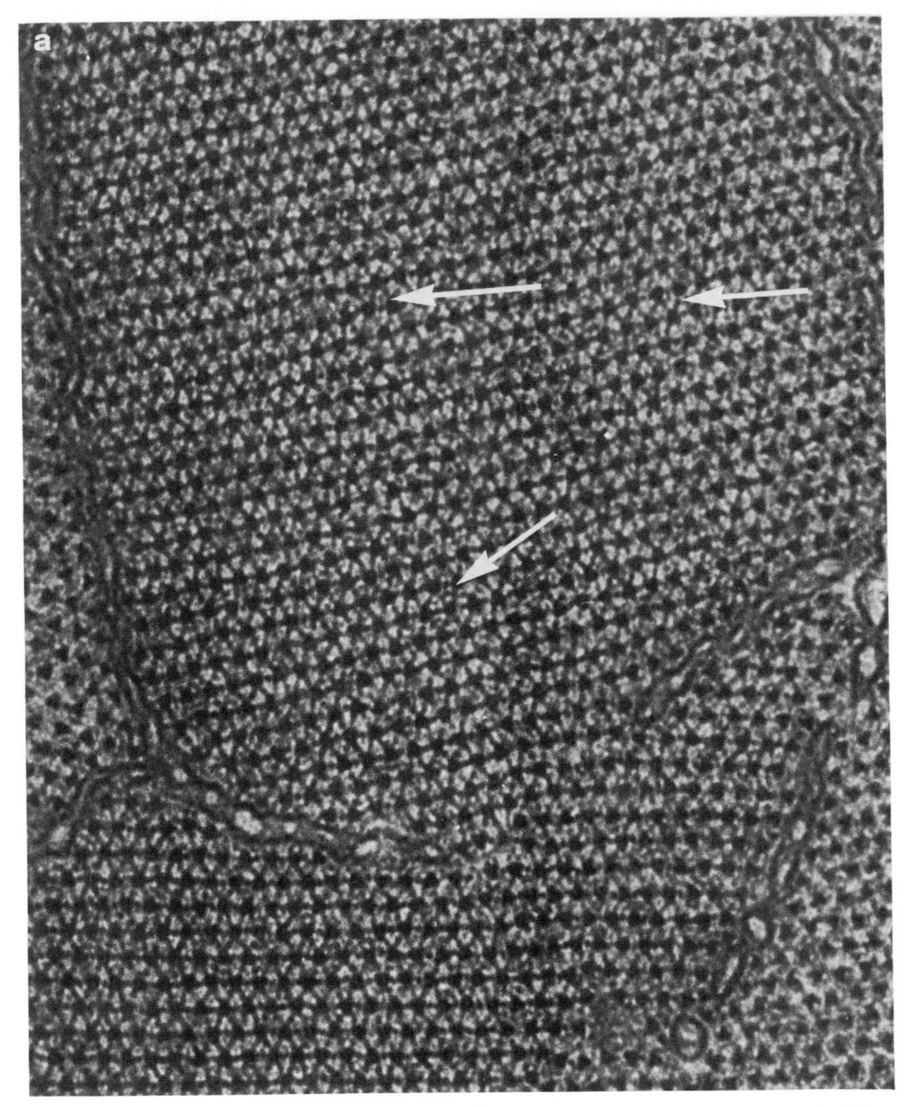

FIGURE 7.14. (a) A moderately thin (500 Å) transverse section of the M-band of frog sartorius muscle. As well as the hexagonal network of myosin filaments and M-bridges (see Fig. 7.1e), finer secondary M-bridges are apparent. Three regions with clear secondary M-bridges are arrowed. Magnification ×85,000. (b) An electron micrograph (shown in negative contrast) of a very thin (~200 Å) transverse section of the M-band as in (a). The order in the section is very good as evidenced by the sharpness of the spots in its optical diffraction pattern (c). An enhanced image of (b) obtained by filtering (c) is shown in (e). This image clearly shows the main M-bridges, the secondary M-bridges [as in (a)], and the myosin filament profiles which consistently have a hollow, roughly circular appearance. The diagram (d) summarizes what can be seen in (e). Magnifications of (b) and (d) ×180,000. (From Luther, 1978; Luther and Squire, 1978.)

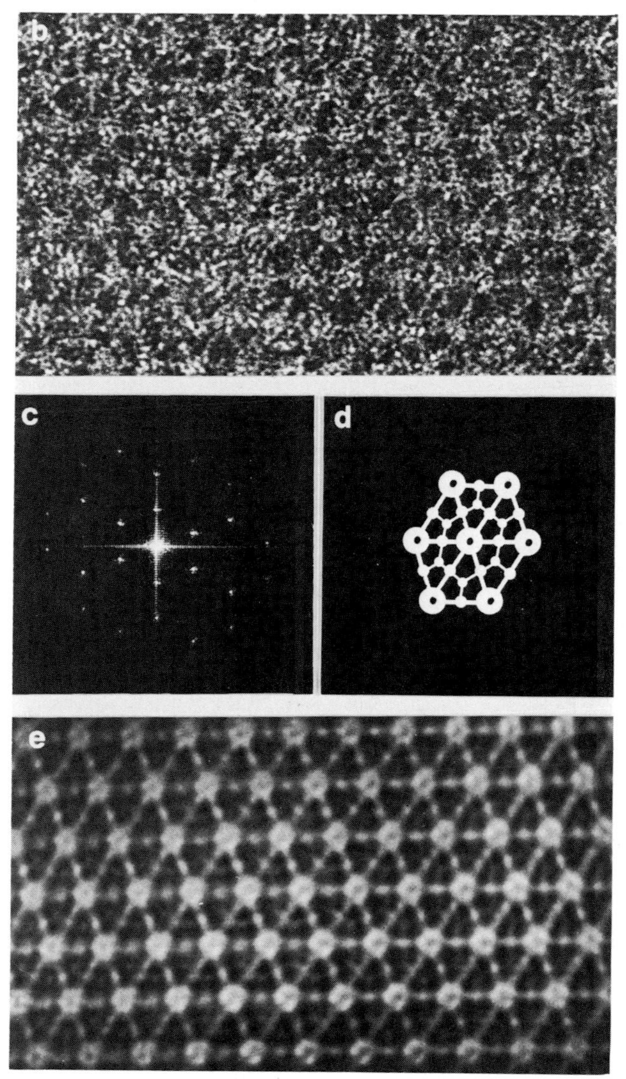

FIGURE 7.14 *Continued*

in addition to the thick M-bridges described above, there are smaller bridging structures in the M-band that run from the middle of each M-bridge (the thickened part reported by Knappies and Carlsen) to the trigonal positions of the myosin lattice (i.e., to the centers of the triangles formed by the M-bridges). This is clearly seen in Fig. 7.14b where an enlarged micrograph of the M-band is shown together with its optical diffraction pattern and filtered image. The structure that is visible in these micrographs is illustrated diagrammatically in Fig. 7.14c. Even in such very thin transverse sections (Fig. 7.14b), each thick filament can be seen to be joined to all six of the neighboring filaments by the main M-bridges. This result is clearly incompatible with the model for the M-band in which the three prominent levels of M-bridges seen in longitudinal sections (Fig. 7.1) are each caused by M-bridges that point in one direction only and thus link a thick filament to just two of its six neighbors (Pepe, 1967a, 1971, 1972, 1975).

Transverse sections taken in the bare regions adjacent to the M-band (type B in Fig. 7.13) are quite different in appearance from M-band sections. Here, no bridging structures can be seen, but the thick filaments are very clearly triangular in profile (Pepe, 1971) as shown in Fig. 7.15 for frog sartorius muscle and fish muscle. The fact that the filaments have such a profile, which clearly is one with threefold rotational symmetry, is itself evidence that the whole filament, including the cross bridge arrangement, probably has threefold symmetry. Indeed, as will be shown later, close to the M-band itself the whole thick filament M-bridge arrangement sometimes shows threefold rather than sixfold symmetry.

It was Pepe (1971) who showed that in the bare regions of different vertebrate skeletal muscles the filaments are arranged in different ways. This is illustrated in Fig. 7.15 for the frog sartorius and fish muscles. It will be seen that in the latter all of the triangular profiles of the thick filaments are pointing in the same direction. By following a slightly oblique section of this type of muscle from one side of the M-region to the other, Pepe was able to show that through the M-band the thick filament structure changes so that the triangular profiles in the bare region on one side of the M-band point in different directions to those on the other side (Fig. 7.15c). An effective rotation of the triangular profile by about 40° through the M-band would produce this appearance (Luther *et al.*, 1981).

→

FIGURE 7.15. (a) Slightly oblique thin transverse section through the M-region of relaxed frog sartorius muscle showing the transition from the M-band (M) to the bare region (B) where the myosin filaments show a characteristic triangular appearance. The triangles do not all point in the same direction. (b) A micrograph comparable to (a) but for fish muscle (roach:rutilus). This time the whole of the M-region is traversed, and the overlap regions on each side of the H-zone can be seen (left and right edges). The M-band (M) can be seen between the two adjoining bare regions (B). As in frog, the bare region filament profiles are clearly triangular, but here all the triangles on one side of the M-band point in the same direction and all those on the other side in a different direction (see Pepe, 1971). This is seen even more clearly in (c) which is an optically filtered image of (b). Protein is white in (c). Note that the change in direction of the triangles across the M-band is about 40° (see arrows). Magnifications: (a) ×90,000; (b) ×90,000; (c) ×175,000 (reduced 36% for reproduction). (From Luther, 1978; Luther *et al.*, 1981.)

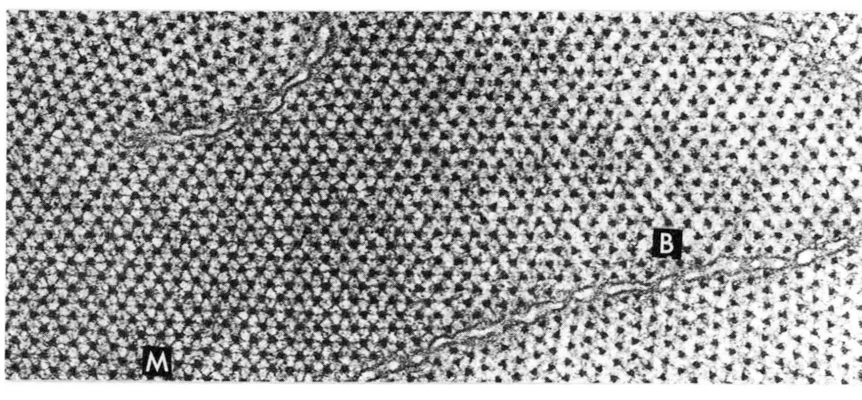

a

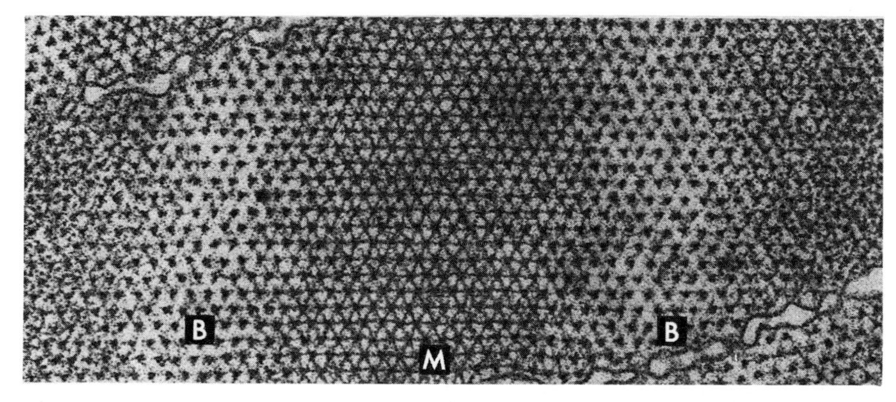

b

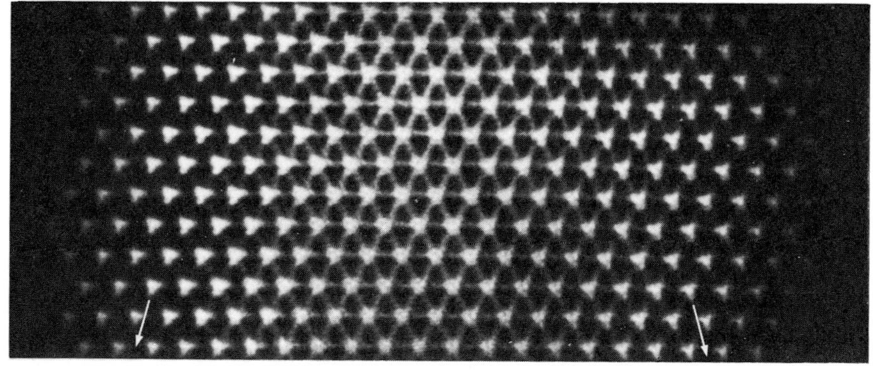

c

The appearance of the bare region in frog muscle is rather different. Here, adjacent triangular filament profiles can be seen to have different orientations even on the same side of the M-band (Fig. 7.15a). As described earlier, it is now thought that in this muscle the myosin cross bridges are arranged to give a superlattice. The differences in the orientations of adjacent triangular profiles in the bare region may therefore be associated with the relative rotation between filaments that is needed to produce the cross bridge superlattice. If each thick filament backbone behaves as a rigid structure, then this must be true. It is shown below that there is evidence from the observed cross bridge array in transverse sections of fish muscle that no superlattice arrangement occurs in this muscle. The fact that the triangular filament profiles in the bare region in fish muscle have identical orientations is consistent with the absence of such a superlattice. But, as discussed in Section 7.2.5, unless the cross bridge array has threefold rotational symmetry, it does not itself prove that there is no cross bridge superlattice.

The Edge of the M-Region. Before leaving the description of the M-region, one further characteristic structure must be mentioned. In very thin (about 200 to 400 Å), slightly oblique transverse sections of well-ordered muscles, it is possible to follow the structural changes that occur through the M-region as shown in Fig. 7.16. Here the transition from the M-band (a) to the bare region (b) can be seen quite clearly. Further away from the M-band, an appearance characteristic of the bridge region can be seen (d,e). But between the bare region and the bridge region, a different structure can be recognized (Fig. 7.16c; Luther and Squire, 1978). Here the myosin filament backbones have very clear threefold rotational symmetry. However, they are not triangular but have more the shape of a thick Y. In addition, from the ends of each arm of the Y, thin curving strands can be observed which can often be seen to link across to other filaments. Like the M-band, this structure appears to be a protein other than myosin forming some kind of interfilament bridge. The axial position of this structure can be estimated if it is known at what angle the section (of known thickness) has been cut and if the lateral separation in the electron micrograph between the center of the M-band and the center of the region in question can be measured. The sectioning angle can be determined approximately from a knowledge of the axial extent of the M-band (about 440 Å) and the section thickness (estimated from section folds). On this basis, the axial position of these extra protein bridges was estimated by Luther and Squire (1978) to be between 600 and 900 Å from the center of the M-band. They were also estimated to lie in a narrow band less than 200 Å thick axially. It will be shown later in this chapter that longitudinal sections show a strong transverse band, previously thought to correspond to an extra protein,

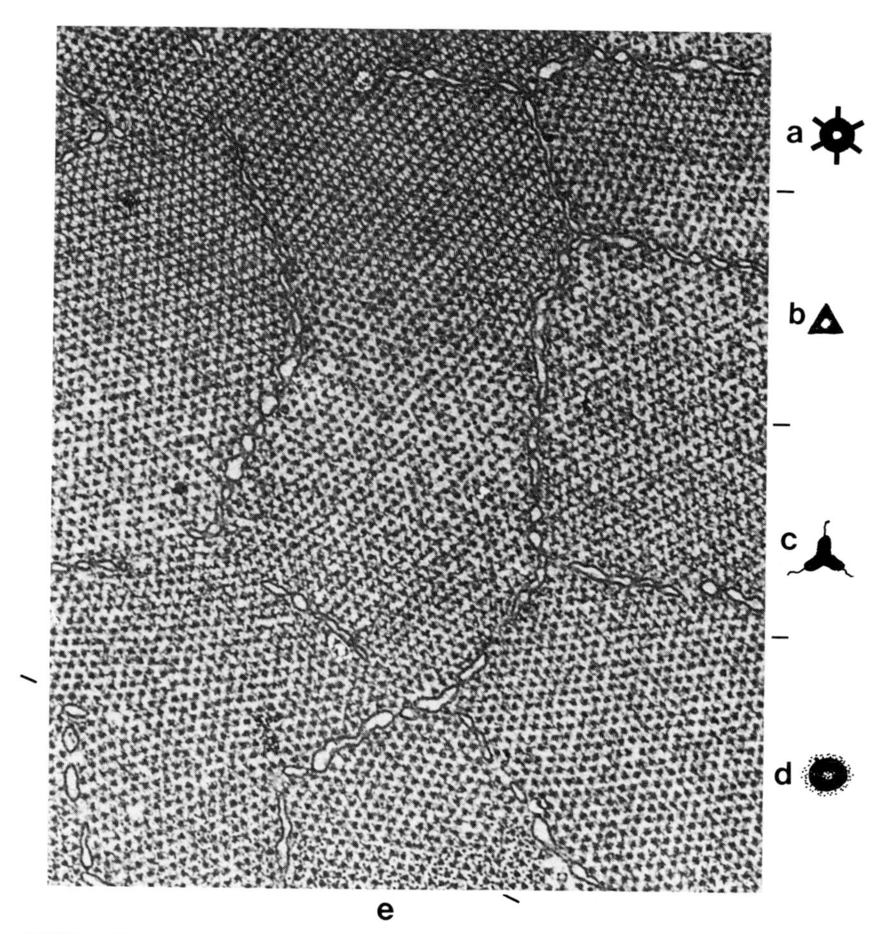

FIGURE 7.16. Very slightly oblique section about 400 Å thick showing the profiles of the myosin filament in different parts of the M-region of relaxed frog sartorius muscle. (a) The M-band, (b) the bare region, (c) the M9 region (showing a Y-shaped profile and thin curving bridges from the tips of the Y), (d) the start of the bridge region, and (e) the start of the overlap region. Magnification ×65,000 (reduced 25% for reproduction). (From Luther, 1978; Luther and Squire, 1978.)

at a distance of 692 ± 13 Å from the center of the M-band (Sjöström and Squire, 1977a). This line was labeled M9 (see Section 7.3.2), and it appears very likely that it corresponds to the axial position of the curving bridges now visible in thin transverse sections. These bridges were therefore termed the M9-bridges by Luther and Squire (1978).

The fact that the Y-shaped filament profiles and the locations of the M9-bridges on each filament are so clearly related by threefold rotational symmetry argues once again that the thick filaments are three-stranded. Note also that sections of the kind shown in Fig. 7.16 can also be used to estimate the axial position of the start of the bridge region. Luther and Squire (1978) estimated this position to be about 800 to 1000 Å from the center of the M-band, a value that is close to measurements made directly on longitudinal sections as described below.

Transverse Sections through the Bridge Region. Transverse sections taken through the bridge region might be expected to reveal directly the cross bridge arrangement on the thick filaments. But if the section is taken through the nonoverlap portion of the bridge region (type C section in Fig. 7.13), very little evidence of regularly arranged cross bridges can normally be seen. The thick filaments usually appear to have a more or less circular backbone profile (but sometimes with a hint of triangularity), and according to H. E. Huxley (1969), this is surrounded by a rather fuzzy ring of density which presumably represents the location of disorganized or overlapping myosin cross bridges. In other micrographs (e.g., Fig. 7.16d, e), the thick filaments do have structures projecting from them, but these are too variable in appearance to allow direct analysis.

Transverse sections taken through the overlap region (type D sections in Fig. 7.13) show a rather different appearance. But this appearance depends on the physiological state of the muscle prior to fixation. Muscles that are originally in rigor (e.g., glycerinated muscles) show a few clear cross bridges in transverse sections. That few cross bridges can be seen in such sections may be because the myosin heads are tilted relative to the long axis of the actin filaments as in the decorated thin filaments studied by Moore *et al.* (1970) (Fig. 5.11). When viewed in projection down the filament axis, these bridges may appear to be associated closely with the thin filaments, and they may not therefore appear as dense bridges between the thick and thin filaments assuming that the S-2 linkage is normally too thin to be seen clearly.

On the other hand, if the muscle is nominally relaxed prior to fixation, then cross-links from myosin to actin are sometimes more apparent in transverse sections. This paradoxical result seems to come about because the fixative itself modifies the thick filaments, thus allowing the myosin cross bridges to interact in some way with actin. Indeed, fixation occasionally results in the generation of tension in a relaxed muscle. Whatever its nature, the actin–myosin interaction induced by the fixative seems to be sufficient to stabilize the cross bridges during the processing procedures for electron microscopy, and, as a result, thin transverse sections of vertebrate skeletal muscle can show the presence of numerous cross bridges. Note also that X-ray diffraction patterns from fixed relaxed muscles sometimes show evidence of myosin labeling of the thin filaments (Sjöström and Squire, 1977b).

The results from relaxed fish muscle are probably among the most striking so far obtained (cf. micrographs of insect flight muscle shown in Chapter 8). Here, in extremely thin transverse sections of the start of the bridge region, it can be seen that each myosin filament seems to make cross-links to three of its six neighboring actin filaments (Fig. 7.17). In addition, every thick filament in a myofibril makes cross-links in the same three directions (120° apart). This indicates clearly that there is no rotation between adjacent filaments in this case and, therefore, presumably that a cross bridge superlattice does not occur in fish muscle. At the same time, it explains the unique orientation of the triangular filament

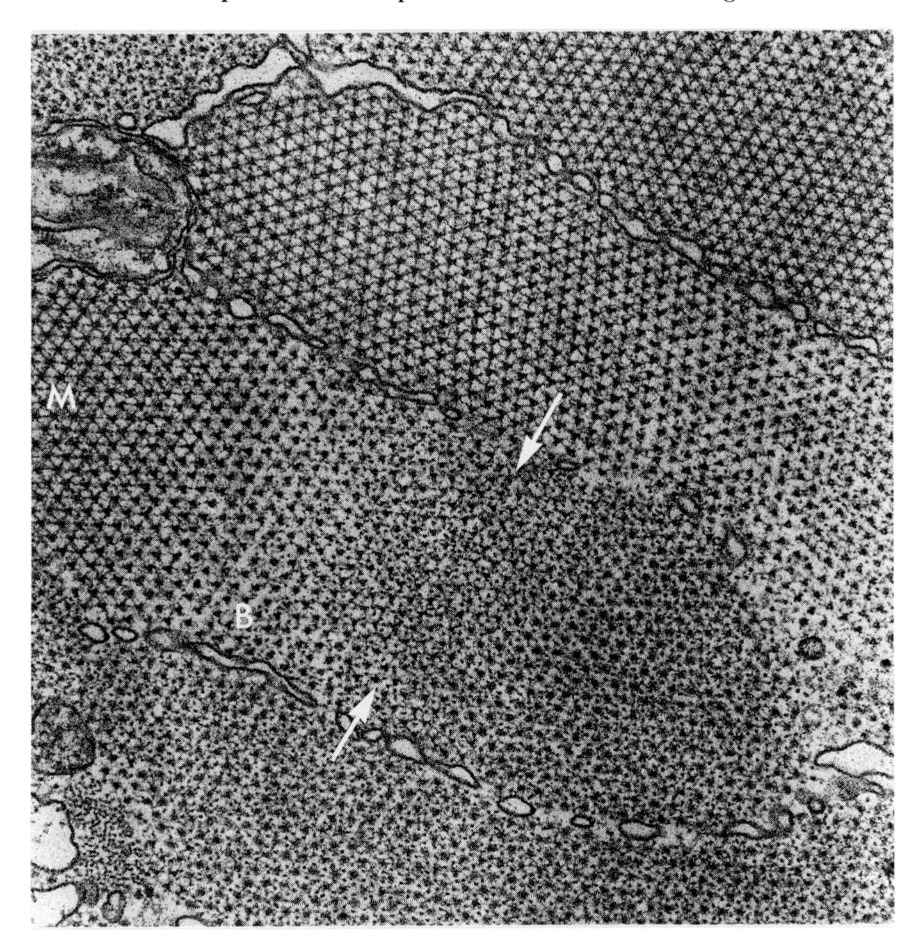

FIGURE 7.17. Micrograph of a thin transverse section of the M-region in fish muscle (roach:rutilus). M9 (Y-shaped) filament profiles (arrowed) have a single orientation as expected from Fig. 7.15b. M, M-band; B, bare region. Magnification ×80,000 (reduced 18% for reproduction). (Courtesy of P. M. G. Munro.)

profiles in the bare region (Fig. 7.15). The single orientation of the thick filaments and of the cross-links to actin allows very easy interpretation of the structure present in the sections. But it is not clear whether the cross-links are myosin heads or structures analogous to the M9 bridges in frog muscle. But once again, these pictures indicate very strongly that the thick filaments have threefold rotational symmetry.

Equivalent transverse sections of frog sartorius muscle are much more complicated. This time one would expect that the cross bridges would ideally be in a three-filament superlattice arrangement, but analysis of this region in exceptionally well-preserved muscles has shown that the expected superlattice is only partially developed. The observed transverse sections (e.g., Fig. 7.18a) sometimes appear to show clear evidence of cross bridges, but at best these appear to be regularly arranged only over a relatively small area of a myofibril. Such micrographs have yet to be analyzed in detail, but very little evidence of any form of superlattice can be seen in optical diffraction patterns. The existence of this superlattice will also depend on how regularly the actin filaments are arranged. Evidence will be presented later in this chapter and also in Chapter 10 that might suggest that in frog muscle the thin filaments are arranged systematically at the Z-band and possibly also in the overlap region as in fish muscle. But this evidence is not conclusive, and some thin filament irregularity in the overlap region may in fact occur. If it does not, then it would be hard to account for the observed irregularity in the overlap region unless the myosin filament superlattice is only poorly developed. As shown later, this does, in fact, seem to be the case.

The A-Band Tip. The final type of transverse section to be discussed is the type E section of Fig. 7.13 which is located at the tip of the A-band close to the A–I junction. Particularly interesting transverse sections of fixed relaxed muscle have been obtained by Luther (1978) and Freundlich *et al.* (1980) from this region. It will be shown later in this chapter that there is now clear evidence of a gap in the cross bridge array right at the tip of the thick filaments (Section 7.3.4). At the filament tip there is probably only one (or at most two) rows of cross bridges beyond the gap. Thin transverse sections that include this terminal cross bridge array have a very characteristic appearance. First of all, it can occasionally be seen that there are cross bridges between the thin filaments and the thick filaments (which appear very small in diameter here because of the filament taper). Sometimes the cross bridges appear as three spokes, 120° apart, to three of the six surrounding thin filaments (Fig. 7.18e). In other places in Fig. 7.18e, an even more remarkable structure is visible (boxed). This time it appears that there are three (120°-spaced) myosin projections radiating from the thick filament backbone toward the gaps between the neighboring thin filaments. But then at a distance of 150 to

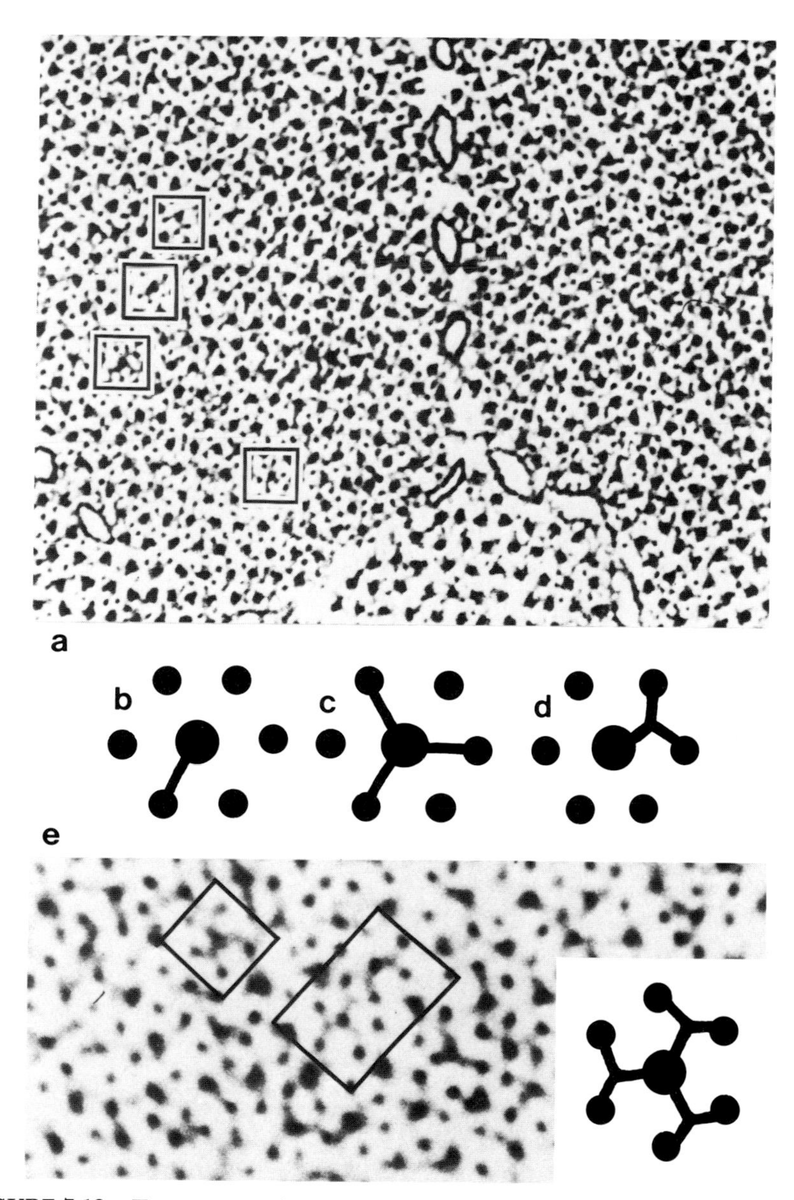

FIGURE 7.18. Transverse sections through the start of the bridge region (a) and the tip region (e) of frog sartorius muscle (fixed relaxed) showing a variety of cross bridge configurations. (a) Some single links (boxed) straight from myosin to actin as in (b), groups of single links as in (c), and some dividing links (d) where a single thick filament links through one dividing link to two actin filaments. In (e) groups of dividing links occur on a single thick filament. In some cases (boxed) up to six actin filaments are linked by dividing bridges to a single thick filament. Magnification: (a) ×125,000; (e) ×300,000 (reduced 33% for reproduction). (From Freundlich *et al.*, 1980.)

250 Å from the thick filament axis, these projections divide into two arms that link to the two thin filaments on each side of the gap. In this way, all six thin filaments are cross-linked to a single thick filament. The appearance is as if the two myosin heads in a particular myosin molecule are attaching to two different thin filaments—in itself a fascinating possibility which will be considered again later (cf. Reedy, 1967; Trinick and Offer, 1979).

One final noteworthy feature of these transverse sections through the edge of the A-band where the filament taper occurs is that the filament backbones can clearly be seen to be triangular in profile just as in the bare region. These triangles often appear to comprise three subunits as shown in Fig. 7.19.

Summary. The full implications of these observations are only now being analyzed. On the whole it is clear that there is overwhelming evidence for the threefold nature of the rotational symmetry of the thick filaments. The accumulated data now available would be very hard to explain away by any model for the thick filament other than one with a threefold rotation axis. It seems almost beyond doubt that the myosin cross bridges are arranged, at least to a first approximation, on a three-stranded 9/1 helix.

7.2.5. The Nature of the A-Band Superlattice

In earlier parts of this chapter, two apparently conflicting results have been described. One is that H. E. Huxley and Brown (1967) interpreted their X-ray diffraction patterns from frog sartorius muscle in terms of a regular superlattice of myosin filaments (Fig. 7.6), and the other is that in electron micrographs of the bridge region in very well-ordered muscles, there is little or no direct evidence for such a superlattice (Fig. 7.18). These contradictory results are discussed more fully in this section; it is shown that although a superlattice of myosin filaments does occur, it does not appear to be the arrangement described by H. E. Huxley and Brown.

The evidence that led to this conclusion was obtained from a systematic study of the bare region of frog sartorius muscle (Luther, 1978; Luther and Squire, 1980). Since in this region the myosin filament profiles are clearly triangular in shape (Fig. 7.15), the relative orientations of adjacent filaments in a transverse section should provide an indication of the nature of any superlattice of myosin filaments that might exist. To analyze the possible appearances of this region, Luther and Squire (1980) prepared models of the predicted bare region appearances of A-bands containing either two-stranded, three-stranded or four-stranded filaments (Figs. 7.20 and 7.21). It was assumed

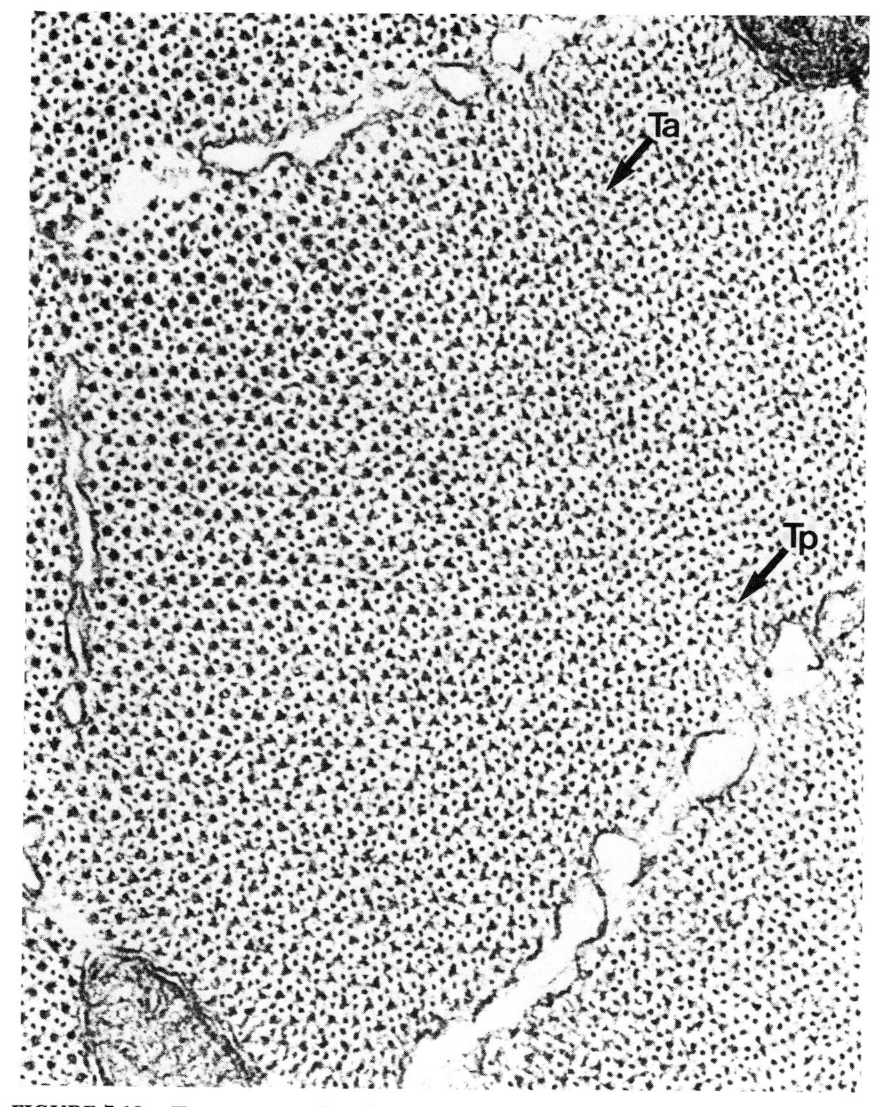

FIGURE 7.19. Transverse section through the taper (Ta) and tip (Tp) regions (E in Fig. 7.13) of Roach red muscle (fixed relaxed) showing the triangular profiles of the thick filaments in this region and an indication that the myosin filament backbones comprise three circular subunits (arrows). Magnification ×97,000 (reduced 28% for reproduction). (Courtesy of P. M. G. Munro.)

throughout that the thick filaments were arranged in such a way that a perfect Huxley–Brown superlattice was generated by the cross bridges on the filaments. The ways in which this superlattice can be formed in

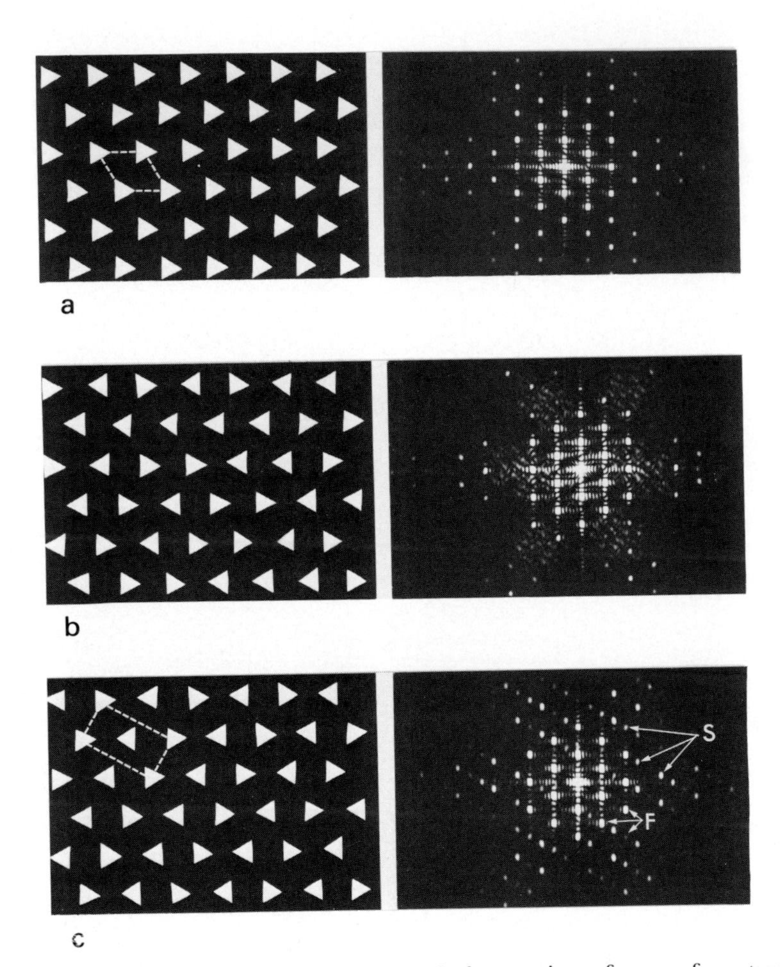

FIGURE 7.20. Some of the possible models for the bare regions of two- or four-stranded myosin filaments assuming that the cross bridges on these filaments are arranged to give a perfect Huxley and Brown superlattice of the kind shown in Fig. 7.10b and Fig. 7.11b. All of these possible arrangements comprise triangular filament profiles in one of two orientations 60° apart (four-stranded, 2 × 30° apart). In (a), all the triangles point in the same direction. This means that there is a simple hexagonal unit cell of triangles and a simple optical diffraction pattern. In (b), the two filament orientations are randomly distributed. The optical diffraction pattern has reflections in the same positions as in (a), corresponding to a simple statistical unit cell with 'average' contents, but the background in the diffraction pattern is much increased because of the disorder in the structure. In (c), filaments are regularly arranged to give a two-filament (rectangular) unit cell. This two-filament superlattice gives rise to superlattice reflections (S) in the optical diffraction pattern, and these are between the fundamental reflections (F) which relate to the simple unit cell as in (a) and (b).

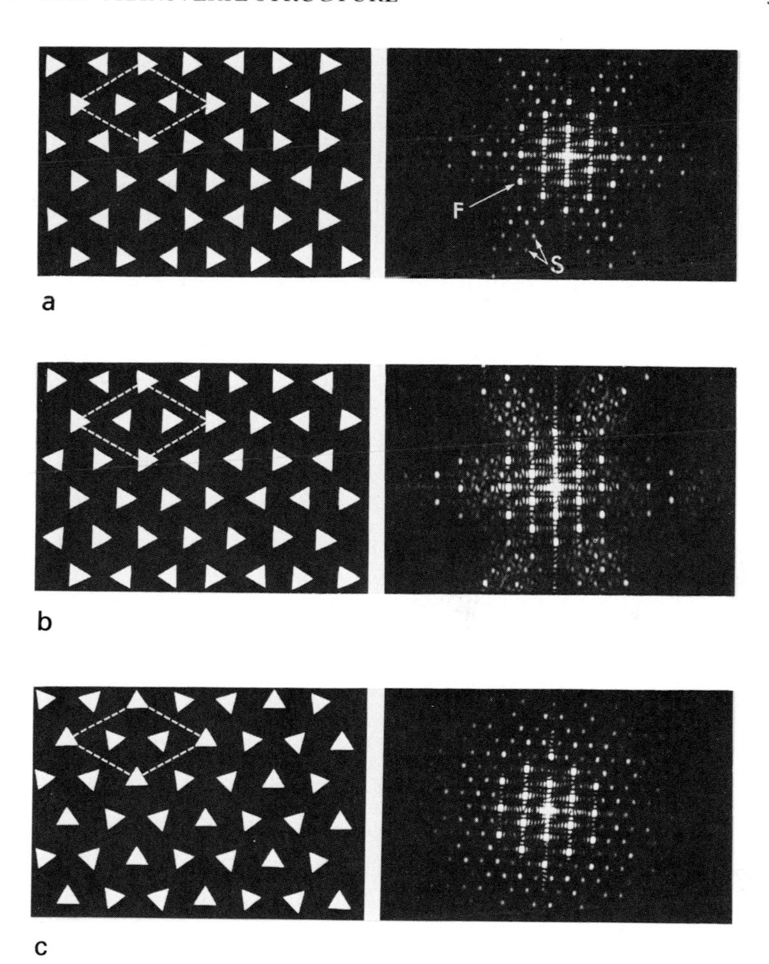

FIGURE 7.21. Further models of the frog bare region based on the assumption that the Huxley–Brown (1967) cross bridge superlattice is correct; (a) and (b) are for two- or four-stranded filaments, and both contain a three-filament superlattice of the size of the Huxley–Brown lattice. (a) A regularly arranged superlattice; (b) the corner filaments in the superlattice are regularly arranged, but the two other filaments in the unit cell are in arbitrarily chosen positions either pointing in the same direction or opposite to the triangles at the corners. The diffraction patterns in (a) and (b) both show superlattice reflections (S) in the middles of the triangles formed by any three adjacent fundamental reflections (F). Because of the randomness in (b), the diffraction pattern in that case also shows considerable background intensity. In (c) is the only possible type of bare region triangle arrangement for three-stranded filaments with cross bridges in the Huxley–Brown superlattice. Three-filament orientators 40° apart occur systematically across the lattice to generate a regular superlattice. Superlattice reflections at the same positions as those in (a) and (b) can be seen clearly in (c). But note that in the patterns (a) and (b) and also in the patterns of Fig. 7.20 there is an overall distribution of intensity which shows six-fold symmetry. This is because there are only two orientations of triangles 60° apart. However, in (c) where there are three filament orientations, the sixfold nature of the intensity distribution has disappeared.

each case have already been described (Figs. 7.10 to 7.12). Relative rotations between adjacent filaments of 60°, 40°, and 30°, respectively are required for the three filament symmetries. If the filaments behave as rigid structures, then the triangular filament profiles in the bare region would be expected to have comparable relative orientations. Possible appearances of the bare region can therefore be deduced in each case.

It is simplest to start with the three-stranded model for the thick filaments. Here a 40° rotation between filaments is required to give the superlattice. In this case, the triangular profiles in the bare region must also have 40° rotations between them, and the result is shown in Fig. 7.21c. Fortunately, a 40° rotation between equilateral triangles makes them look quite different, and if the filaments are three-stranded and in a perfect superlattice, then it should be possible to see, in transverse sections through the bare region, an arrangement of triangular profiles similar to that in Fig. 7.21c. It is also clear that there is no way in which a cross bridge superlattice could be produced if, as in fish muscle, the triangular filament profiles have identical orientations (Fig. 7.15) and if there is no axial stagger between filaments.

The superlattice arrangement in Fig. 7.21c can be compared with the simple lattice arrangement of the kind shown in Fig. 7.20a. It will be seen that the presence of the superlattice introduces extra diffraction spots into the pattern (superlattice reflections) that lie at the centers of the triangles in the hexagonal arrangement of diffraction spots from the simple lattice. The presence of superlattice reflections in equivalent optical diffraction patterns from electron micrographs of the bare region would be direct evidence for the existence of the superlattice.

Turning now to what happens with a two- or four-stranded filament, the possibilities are much more numerous and complicated. This time, rotations of $n \times 60°$ or $m \times 30°$, respectively, can occur between adjacent filaments. But if n is even (or if m is a multiple of four) then the triangles will be rotated by 120° or a multiple, and they will therefore appear unchanged. In addition, on any one lattice point, it is not certain what orientation a triangle should be given. If the cross bridge array has two-fold (or four-fold) symmetry, then there are two (or four) different orientations of the triangles 180° (or 90°) apart which will give equivalent cross bridge orientations in the superlattice.

In the two-stranded model only two different appearances of the triangles (60° or 180° apart) would be possible, and in theory either of these two orientations on any lattice point could give the required cross bridge superlattice. Some of the alternative possibilities are shown in Fig. 7.20b,c and 7.21a,b. In the case of the four-stranded structure, any lattice point could be occupied by triangles with any one of four orientations 30° apart. Some of the resulting arrangements are equivalent to those shown for the two-stranded filament. Other possibilities are unique to the four-stranded structure, and these involve all four of the filament orientations. Throughout Figs. 7.20 and 7.21 the optical diffraction patterns from the models are also shown. In some cases, only the simple lattice is present (Fig. 7.20a), some show a regular superlattice (Figs. 7.20c and 7.21a), and some display a measure of randomness in their structure (Figs. 7.20b and 7.21b). This shows up in the diffraction pattern as increased background intensity and a weakening of the superlattice reflections.

In sections of the bare region of frog sartorius muscle, the triangular profiles of adjacent thick filaments generally do have different orientations (Figs. 7.15). In light of the background given above, it is clear that if the premises are correct and there is a perfect Huxley and Brown superlattice of cross bridges, then the relative rotation between adjacent triangles should be close to 30°, 40°, or 60°. It should then be possible in theory to decide directly whether the filaments had a threefold rotation axis (40°) or a two- or fourfold rotation axis (60° or 30°). Unfortunately, it is no easy task to distinguish orientation differences of 10° or 20° between triangles that are somewhat ill-defined because of the inevitably imperfect preservation in the section and the limited resolution of the electron microscope image. But conclusive results have now been obtained by Luther (1978) and Luther and Squire (1980), and these are described below.

First, optical diffraction patterns of micrographs of the bare region (sometimes together with part of the M-band) do show evidence of the existence of a superlattice with the same shape and size as the Huxley–Brown lattice (Fig. 7.22). However, the superlattice reflections from the micrographs are not as strong and as numerous as might have been expected from the model structure for a three-stranded filament (Fig. 7.21c). In addition, it is clear from micrographs such as that in Fig. 7.22 that despite the remarkable regularity in the hexagonal array of filaments, there is a certain amount of irregularity in the arrangement of the triangles within the array. This could still mean that the native structure has been imperfectly preserved, but more likely it indicates that the superlattice in the bare region is not regularly organized. There could be two reasons for this: either the filaments are really two- or four-stranded, in which case any of the random arrangements of filaments shown in Figs. 7.20 and 7.21 would be possible, or there is some ambiguity about the way three-stranded filaments are linked at the M-band.

These possibilities have been investigated by more detailed studies of the bare region (Luther, 1978; Luther and Squire, 1980). The bare regions structure has been subjected to image averaging on the optical diffractometer. The filtering masks used were such that the average

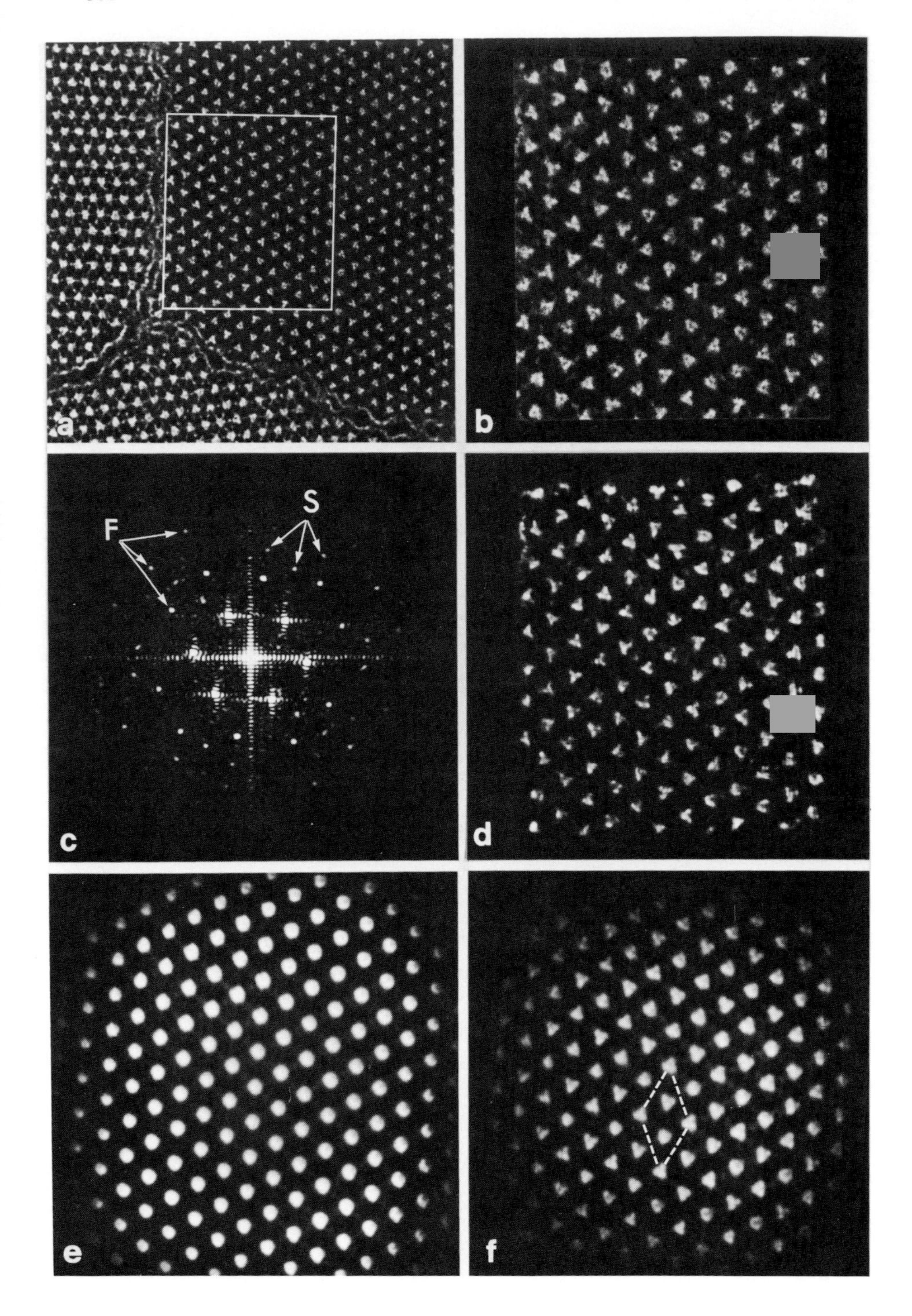
a
b
S
F
c
d
e
f

produced was a convolution of the micrograph with a small array of points on the superlattice. The average images produced by this process (e.g., Fig. 7.22f) are such that a number of important features of the structure in the micrographs can be recognized. First, where a superlattice of filaments can be seen, it only seems to extend for a few unit cells in any direction. In the adjacent few unit cells, the superlattice is normally evident as well, but the triangle orientations are different from those in the first region. The structure therefore appears as one in which a two-dimensional crystal contains numerous packing defects that define limited domains a few unit cells in width. Where the packing defects occur, one very occasionally sees a two-filament superlattice (cf. Fig. 7.20c). Second, a number of well-ordered micrographs that showed an exceptionally well-developed three-filament superlattice clearly showed in their averaged images only two orientations of triangles 60° (or 180°) apart. But in some cases the observed average filament profiles were triangular, and in some cases they were more or less circular. Since the average image was produced by convoluting the micrograph with a small number of repeats of the superlattice, each filament in the average image must represent the average of a few filaments placed in equivalent positions in adjacent unit cells of the superlattice in the original micrograph. Observation of a triangular profile in the final image therefore indicates that all of the filaments contributing to that particular image had similar (probably identical) orientations. Observation of a circular image, on the other hand, means that the contributors to the average had different orientations.

FIGURE 7.22. Analysis of the structure of the bare region in frog sartorius muscle. (a) A micrograph of a thin transverse section (~500 Å) of the frog bare region (×75,000). (b) An enlargement of the area outlined in (a). The triangular filament shapes and their three-subunit structure are clearly visible (×140,000). (c) The optical diffraction pattern from (b). This clearly shows the presence of superlattice reflections (S) in addition to the fundamentals (F). (d) The reconstructed unfiltered image of (b) obtained using the optical diffractometer. This provides a test of the optical system. (e) A filtered reconstructed image of (b) obtained by allowing only the fundamentals to contribute to the image. (f) The filtered reconstructed image obtained with both the fundamental and superlattice reflections included. Because of the size of the circular holes in the filtering mask, the final image in (f) is effectively an image of (b) averaged over a perfect superlattice of points that extends a few unit cells in any direction (see Chapter 3). As indicated by the dashed lines in (f), a superlattice of filament orientations is readily apparent, and only two triangle orientations are present in the image. Note that some filaments appear circular in (f) but that this means that in that position the image is an average of several filaments with mixed orientations in (b). There is, therefore, some randomness in the superlattice contents in (b). The optical diffraction pattern of (b) when seen at much higher exposure shows a sixfold symmetric intensity distribution characteristic of the presence of two triangle orientations rather than three (see Fig. 7.21). Figures reduced 24% for reproduction. (From Luther, 1978; Luther and Squire, 1980.)

The conclusions from these observations can be summarized in the following way.

1. A superlattice of myosin filaments does occur, and this has the same size and shape as that suggested by H. E. Huxley and Brown (1967).
2. The filament superlattice is not perfectly developed. Even in well-ordered muscles it only extends for a few unit cells in any direction.
3. Even though the weight of evidence (described earlier in the chapter) is clearly in favor of the myosin filaments being three-stranded, only two filament orientations are evident in the bare regions of otherwise very well-ordered muscles.

This last result could be taken to mean that the conclusions about the threefold symmetry of the myosin filament are wrong and that the filaments are two- or four-stranded, in which case only two filament orientations might be expected in the bare region even with a perfect Huxley–Brown superlattice of cross bridges. On the other hand, and almost certainly more realistically, it could be taken to mean that the myosin filaments are three-stranded but that the superlattice does not have the perfect symmetry in which adjacent filaments are rotated by 40°. In fact, the structure appears to be one in which three-stranded filaments can have one of only two different orientations 60° (or 180°) apart. It will be shown later that such an arrangement makes very good structural sense.

Feature 2 above is crucial to an understanding of A-band structure, and it leads one to ask whether the aggregation of myosin filaments is rather haphazard or is controlled regularly at the M-band in such a way that a superlattice of limited extent is automatically produced. Not surprisingly, it seems that the latter possibility is the correct one, and this can be shown in the following way. We have seen that in electron micrographs of the bare region there are two orientations of the triangular filament profiles. Referring to one of these orientations as "up" and to the other orientation as "down," then, in any group of three filaments on the corners of a triangle in the hexagonal filament array, it is found almost invariably that if two of the filaments are "up" filaments then the third filament points "down." It is relatively rare to find all three filaments with the same orientation. Since three adjacent filaments like this are all linked directly to each other by the M-bridges at the M-band, it is perhaps not surprising that there is a restriction on the way the filaments can interact. What is at first sight surprising is that if one takes a hexagonal array of points and, building from a particular starting point, one places "up" and "down" triangles on these points in such a way that no three adjacent filaments have the same orientations, then the structure developed need not be completely regular, but the large three-filament superlattice almost always occurs over a significant area of the array (Fig. 7.23). Note that in generating such a structure it often happens that a group of three filaments is being completed in which the first two filaments have opposite orientations. The orientation of the third filament in the group is therefore ambiguous, and this is the source of the randomness in the lattice. In Fig. 7.23, the appropriate orientation in such a situation was chosen by tossing a coin. The superlattice appears in this type of array in the sense that many triangles of the same orientation lie at the corners of the superlattice, even though the orientations of the two other filaments within the unit cells may vary. The simple "no three alike" rule (rule 1) therefore accounts directly for the main features of the appearance of the bare region as seen in electron micrographs of frog sartorius muscle. However, it is possible that a second but less rigid aggregation rule (rule 2) occurs that makes it unlikely for more than three filaments in a straight line to have the same orientation. The relative importance of this second rule is not yet clear, and exceptions to it are not uncommon.

It is of interest to consider how the "no three alike" rule can be interpreted in terms of possible molecular interactions that might occur

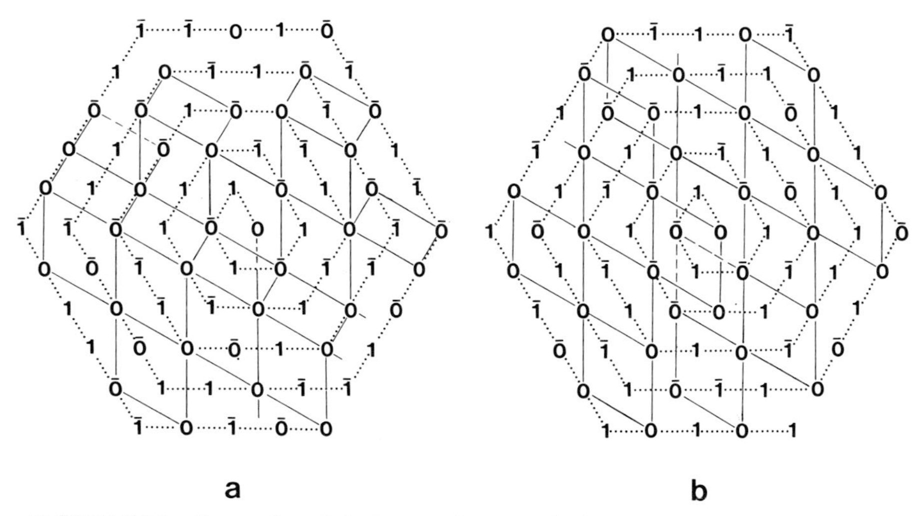

FIGURE 7.23. Generation of the bare region superlattice (as in Fig. 7.22) using (a) the "no three alike" rule (1) and (b) the "no three alike" rules (1) and (2) together. For definition of rules see text. Note that 0 and 1 represent the two triangle orientations 60° apart and that the bar over any number (e.g., $\bar{0}$) indicates that the choice of 0 or 1 at that position was made arbitrarily by tossing a coin.

in the M-band. It has been shown that there are good reasons for believing that the myosin filaments have threefold symmetry. They also seem to have a bipolar structure, which probably means, on any model for the thick filament, that at the middle of the M-band, the filament possesses three twofold rotation axes perpendicular to the filament axis and related one to the other by a threefold rotation axis coincident with the filament axis. In this case, it is likely that at the middle of the M-band where a filament makes six M-bridge connections to its neighbors, the filament does not have true sixfold symmetry. Rather, there are three equivalent M-bridges 120° apart (A in Fig. 7.24) and related by a threefold axis, together with three more M-bridges (B in Fig. 7.24) identical in structure to the first three but obtained from them by applying

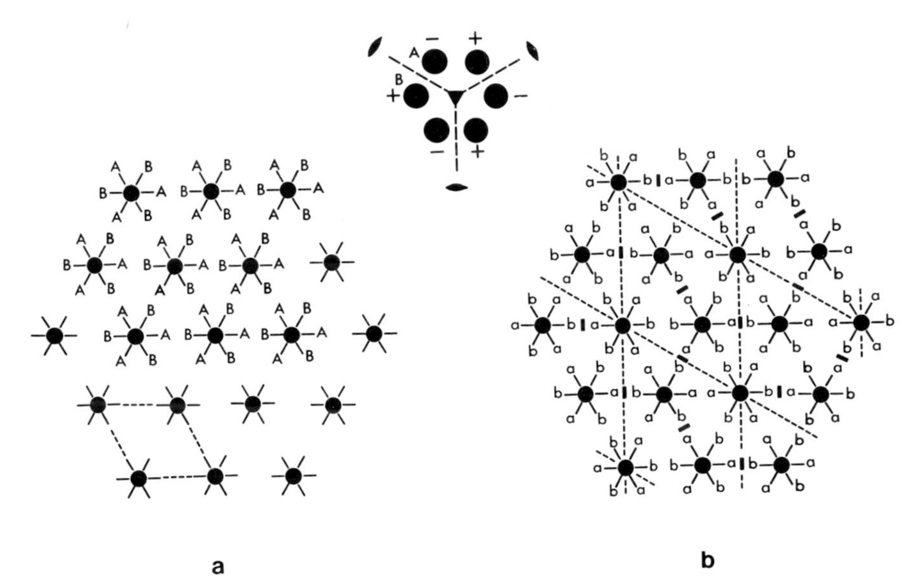

FIGURE 7.24. (Inset) Illustration of the probable symmetry of the thick filament backbone at the center of the M-band (M1). The axis of the filament is a threefold rotation axis, and perpendicular to this are three twofold rotation axes 120° apart. The filled circular objects marked + (and denoting an arbitrary asymmetric unit) are equivalent to those marked − except for their polarity with respect to the filament axis (i.e., + points up out of the page, and − points down). This means that whatever the interaction sites for the M-bridges are at this level there will be three A-type positions and three B-type positions of opposite polarity. Illustrated in (a) and (b) are two ways that A and B sites can interact in a systematic way. In (a), all of the A-type sites interact through M-bridges to a B-type site, and this inevitably means that all of the filaments have identical orientations as in fish muscle (Fig. 7.15b). On the other hand, mixed interactions [A to B in (b)] might be less easy to make than like-with-like interactions (A to A or B to B). In this case, like interactions are maximized if there can be some 60° rotations between adjacent filaments, and the "no three alike" rule (1) is automatically generated as seen in the A-band of frog sartorius muscle.

the twofold rotation operation once only (Fig. 7.24). The whole arrangement has the symmetry of the dihedral point group 32. When adjacent filaments in the M-band interact, they can therefore do so either via an A-to-A (like) interaction, by a B-to-B (like) interaction, or by an A-to-B(mixed) interaction. Were all of these possible interactions to be equally likely, then a random A-band structure would result, since all filament orientations would be equally likely. But if, for example, the mixed interactions were dominant, there would be preferred ways of arranging adjacent filaments at the M-band: they would all have identical orientations as in fish muscle (Fig. 7.24a). Alternatively, if the M-band lattice is built up using the "no three alike" rule (rule 1), the number of possible like-with-like interactions is automatically maximized. The analysis of the nature of the bare region superlattice has therefore led directly to an understanding of the possible mode of assembly of the A-band: it is probably controlled by the optimization of certain types of M-bridge interaction. In frog sartorius muscle where there is a superlattice, the like interactions are dominant, whereas in fish muscle where all the myosin filaments have identical orientations, the mixed interactions must dominate.

Although at the middle of the M-band, the M-bridge structure as seen in transverse sections will appear to have pseudosixfold symmetry, there is no reason to believe that on each side of the center the filament symmetry would appear other than threefold. Indeed, in the bare region, this fact is evident in the triangular shape of the filament profiles. It has recently been found (Luther, 1978; Luther *et al.*, 1981) that this symmetry is also evident in the two M-bridge rows on each side of the central M-bridge line in the M-band.

Figure 7.25 shows the original and the averaged image of a very slightly oblique ultrathin transverse section (about 200 Å) containing the central M-bridge line in one part and one of the two outer M-bridge lines in the other part. The optically averaged image of this micrograph (Fig. 7.25d) obtained using the superlattice as the averaging function (Chapter 3) shows an interesting difference between the middle and outer M-bridge lines. The middle line shows rather little evidence of the superlattice, thus indicating that the structure has at least pseudosixfold symmetry in the M-bridge array and approximately circular myosin filament backbones. On the other hand, the outer M-bridge line where the filament profiles are starting to become triangular shows very clear evidence of the superlattice both in the orientations of the triangular filament profiles and, more interestingly, in the appearance of the M-bridges. On each of the myosin filaments, three of the M-bridges 120° apart are clearly different (they appear denser) from the other three. But, in addition to this threefold M-bridge structure, the M-bridges on

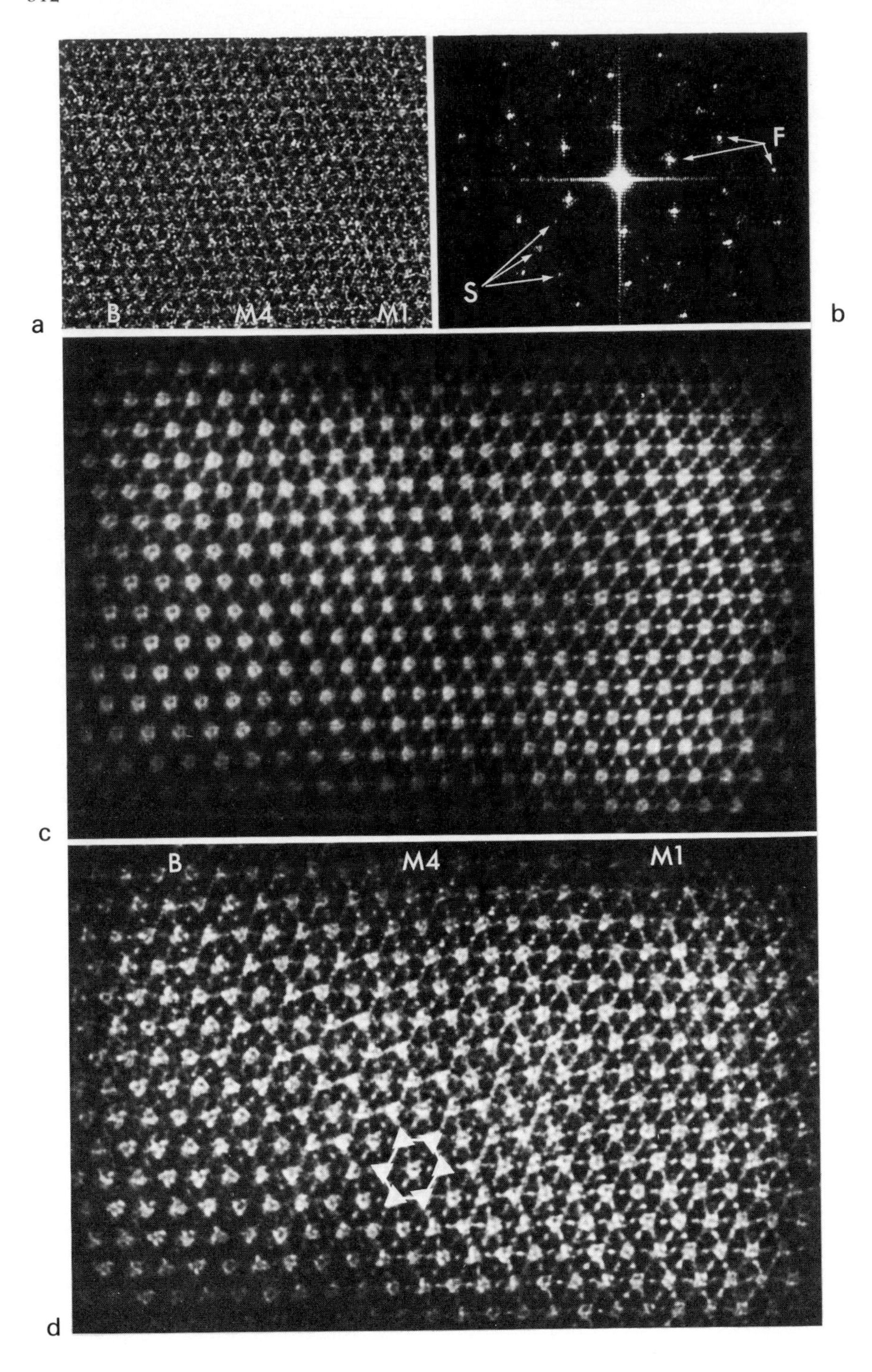
a
B
M4
M1
b
F
S
c
d
B
M4
M1

one filament are related to those on the adjacent filaments by the "no three alike" rule just as the filament profiles. This observation in itself could reasonably be taken as a proof of the fact that the myosin filaments have threefold symmetry and yet are arranged with just two orientations to give the three-filament superlattice structure.

Having reached this very clear conclusion, it is crucial to consider the implications of this structure in the bridge region. Inevitably, the cross bridges themselves must be arranged in exactly the same superlattice if, as must be assumed, the filament backbones behave as rigid structures. Since this superlattice clearly does not have the symmetry of the lattice predicted by H. E. Huxley and Brown (1967) (i.e., as in Fig. 7.6), the selection rule for the sampling on the myosin layer lines will not apply. The intensity distribution on the first layer line of the pattern therefore needs to be reinterpreted.

The data used by H. E. Huxley and Brown to deduce the existence of a superlattice is summarized in Table 7.3. It was found that the observed equatorial reflections could be indexed well as orders $10\bar{1}$, $11\bar{2}$, $20\bar{2}$, $21\bar{3}$, $30\bar{3}$, etc. of a simple hexagonal lattice of side 412 Å (in the particular case quoted). But on the first myosin layer line (Fig. 7.5), intensity peaks could be seen at positions measured from the meridian that were different from the positions of the equatorial reflections. As mentioned earlier, these first layer line peaks could be accounted for reasonably well in terms of the hexagonal superlattice of side $\sqrt{3} \times 412$ Å (i.e., 713 Å) as indicated in the appropriate column in Table 7.3. H. E. Huxley and Brown accounted for these peaks in terms of either of the selection rules $2h + k + l = 3N$ and $h + 2k + 2l = 3N$ as required by a

←

FIGURE 7.25. (a) a very thin slightly oblique transverse section of the M-band in frog sartorius muscle showing the transition from M1 to M4 to the bare region. The micrograph is printed in negative contrast. (b) The optical diffraction pattern of (a) clearly shows superlattice reflections and very good order. In (c) and (d) are reconstructed images (as in Fig. 7.22f) obtained using either (c) only the fundamental reflections [F in (b)] or (d) both the fundamentals and the superlattice reflections (S). The details of these images are described in the text, but note in particular that the middle part of (d) which corresponds to M4 shows an obvious superlattice arrangement of M-bridges in which each myosin filament has three dense M-bridges and three weaker M-bridges to the six adjacent filaments. The threefold symmetric arrangements of dense M-bridges themselves follow the "no three alike" rule (1), and they clearly generate open hexagons of dense M-bridges. An exaggerated example of one such open hexagon has been drawn on the image in (d). The open hexagons are formed by dense M-bridge interactions from the tip of the triangular profile of one filament to the tip of an adjacent filament. It has been shown by Luther (1978) and Luther *et al.* (1981) that observance of rule (1) together with the maximization of these tip-to-tip interactions at M4 generate a filament distribution indistinguishable from the observed distribution in the frog bare region. (a) ×120,000; (c) and (d) × 235,000 (reduced 35% for reproduction). (From Luther, 1978; Luther *et al.*, 1981.)

TABLE 7.3. Lattice Spacings on the Equator and on Layer Lines of a Rest-Length Frog Sartorius Muscle[a]

Index on 713-Å cell (superlattice)	Index on 412-Å cell (simple lattice)	Calc. spacing (Å)	Observed spacings on equator	If allowed by rules	Observed spacings on first layer line	If allowed by rules	Observed spacings on third layer line[b]
$10\bar{1}$	—	618	Absent	No	—	Yes	—
$11\bar{2}$	101	357	361 Å	Yes	Absent	No	362 Å
$20\bar{2}$	—	309	Absent	No	V. weak[c]	Yes	Absent
—	—	—	260 Å[d]	—	—	—	—
$21\bar{3}$	—	232	Absent	No	232.5 Å	Yes	Absent
$30\bar{3}$	$11\bar{2}$	206	206 Å	Yes	Absent	No	205 Å
$22\bar{4}$	$20\bar{2}$	179	177.4 Å	Yes	Absent	No	—
$31\bar{4}$	—	172	Absent	No	172.0 Å	Yes	—
$40\bar{4}$	—	154.5	Absent	No	—	Yes	—
$32\bar{5}$	—	141.5	Absent	No	—	Yes	—
$41\bar{5}$	$21\bar{3}$	135	137 Å	Yes	Absent	No	—
$50\bar{5}$	—	123.5	Absent	No	—	Yes	—
$33\bar{6}$	$30\bar{3}$	119	119 Å	Yes	Absent	No	—

[a] The 206-Å equatorial reflection ($11\bar{2}0$) was used to determine cell size. Superlattice spacing is $3 \times 412 = 713.6$ Å. The selection rules used here are $2h + k + l = 3N$ and $h + 2k + 2l = 3N$. These results are from one particular muscle, but a large number of other experiments gave similar results: the two characteristic lattice reflections on the first layer line (near to 172 and 232 Å, depending on the characteristic lateral spacing of the muscle which varies with sarcomere length) did not lie on a line with any equatorial reflections. The actual value of the spacings sometimes differed from the theoretical value by 2 or 3%, probably because of the finite breadth of the reflections and only partial layer line sampling. From H. E. Huxley and Brown (1967).
[b] The selection rule on the third layer line allows the same reflections as on the equator.
[c] This reflection could sometimes be picked up in densitometer tracings of the first layer line.
[d] This reflection is probably caused by the Z-band and the part of the I-band nearby (see Section 7.4).

superlattice with the symmetry of the sort shown in Fig. 7.6. Since the intensities and positions of the observed peaks were very variable [the position often being up to 3% away from the predicted spacing according to H. E. Huxley and Brown (1967) and sometimes rather more as indicated in Fig. 7.26] and since the peaks on the first layer line were often more diffuse than the equatorial reflections in the same diffraction pattern, it is apparent that an extensive superlattice of cross bridges as in Fig. 7.6 might not be adequate to account completely for the observations.

Table 7.3 lists the reflections that would be expected for the three-filament superlattice in the absence of any selection rules. If a superlattice exists but there are only two filament orientations 180° apart and the filaments are 3-stranded, then all of these superlattice reflections should be seen in the region where the first layer line peak is strong. The predicted peaks on this layer line would therefore be the $21\bar{3}$, $30\bar{3}$, $22\bar{4}$, $31\bar{4}$, $40\bar{4}$, $32\bar{5}$, and $41\bar{5}$ reflections. Since the extent of each area of superlattice in the A-band structure is limited (being only a few unit cells

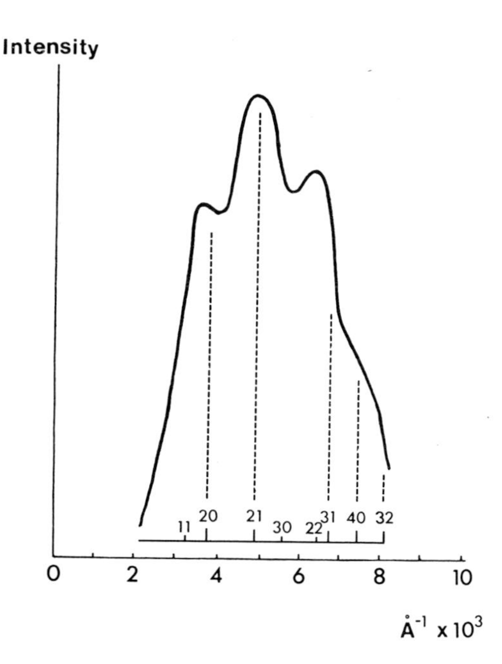

FIGURE 7.26. The intensity profile on the first myosin layer line (1/429 Å^{-1}) from a resting frog sartorius muscle (see Fig. 7.5; Huxley and Brown, 1967). The peaks in the profile lie close to the $20\bar{2}$, $21\bar{3}$ and $31\bar{4}$ reflections of the A-band superlattice, and these peaks are predicted by the selection rules $2h + k + l = 3N$ and $h + 2k + 2l = 3N$ as noted by H. E. Huxley and Brown. However, as shown in Fig. 7.27, there may be another interpretation of this profile.

in any direction), then the peaks that sample the layer line intensity will be rather broad and will overlap to a considerable extent (Fig. 7.27). But, more significantly, different reflections will have a different number of crystal planes contributing to them (i.e., they have different multiplicities). The effect of this is to give the $21\bar{3}$, $31\bar{4}$, $32\bar{5}$, and $41\bar{5}$ reflections twice the weight (multiplicity 12) of the other reflections (multiplicity 6). Taking this into account, the general form of the sampling profile for such a lattice can be predicted and is shown in Fig. 7.27. It will be seen that the prominent peaks in this sampling profile correspond closely to the peaks observed by H. E. Huxley and Brown (1967) on the first layer line from frog sartorius muscle; it seems, therefore, that their observations can be accounted for without the need to introduce any special symmetry into the superlattice, and no selection rules need apply.

From all the evidence given above, it is fair to conclude that the myosin cross bridges are arranged in the same "no three alike" superlattice (Fig. 7.23) as appears in micrographs of the bare region and M-band and that none of the X-ray diffraction data conflicts with this conclusion. The randomness within this structure would indeed provide a direct explanation for the variability in the sampling on the first layer line since, apart from cross bridge disorder, the presence of different statistical populations of the two filament orientations within the superlattice arrangement in different muscles would give rise to slightly different sampling effects. Some variation would therefore be expected even in

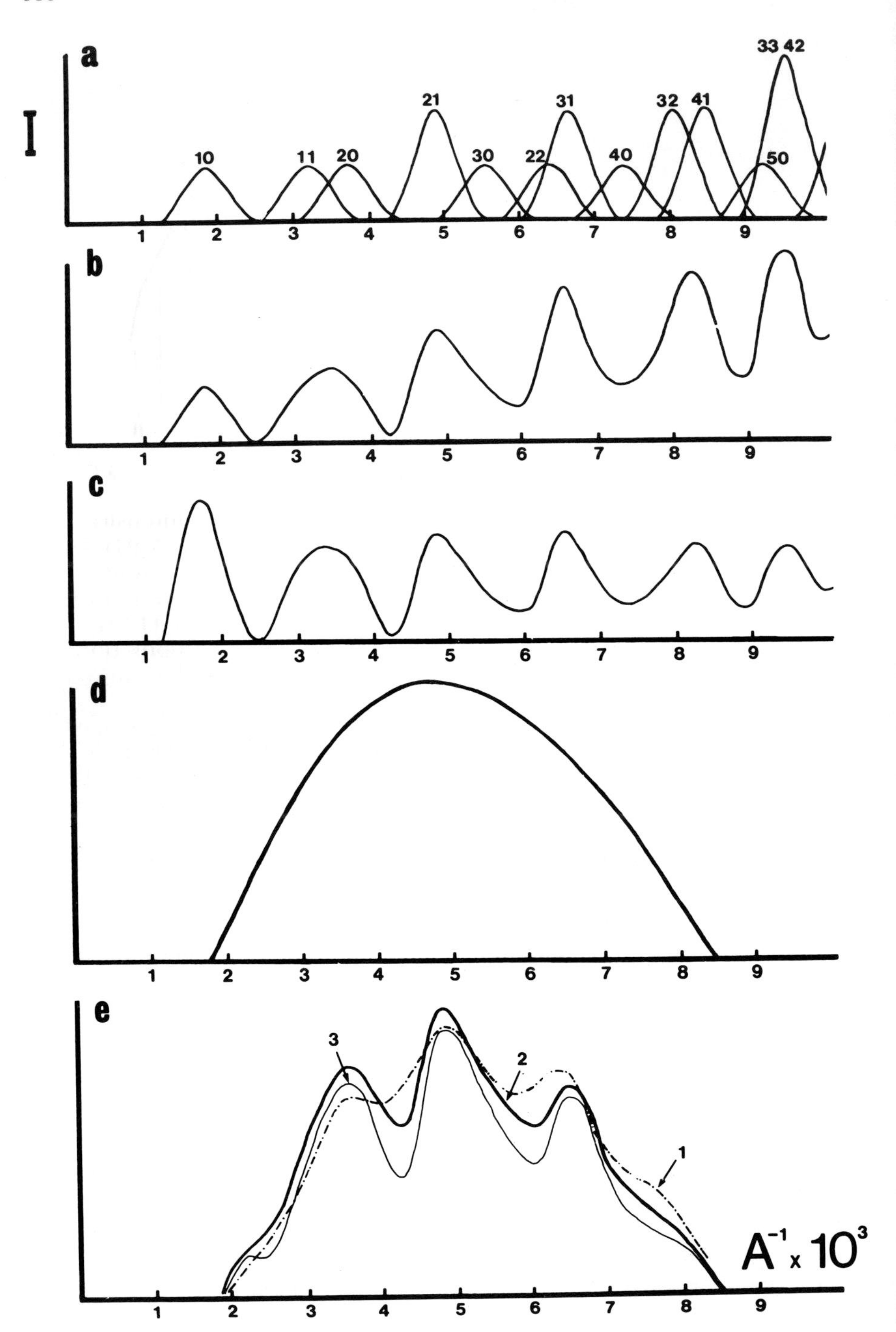
a
b
c
d
e
I
10
11
20
21
30
22
31
40
32
41
50
33 42
1
2
3
4
5
6
7
8
9
$A^{-1} \times 10^3$

patterns with good cross bridge order on each filament. It has already been shown that electron micrographs of the bridge region support the conclusion that in frog sartorius muscle there is a large amount of randomness in the cross bridge array.

To conclude this section on A-band structure, two further points must be made. The first is that, although the two orientations of myosin filaments might in principle be evident at high resolution when the A-band is viewed down the fiber axis, no evidence for the superlattice has been seen on the equator of the X-ray diffraction pattern from frog sartorius muscle. This must mean that the first few equatorial reflections are at too low a resolution to be able to detect the difference between "up" and "down" orientations of the cross bridge array. It must therefore be concluded that in this view of relaxed myosin filaments, the cross bridges form a circularly symmetric shelf of electron density around the filament backbone. The significance of this result together with more detailed analysis of the observed equatorial diffraction patterns will be given in Chapter 10.

The second point to make concerns the conclusions reached earlier about the possible form of the cross bridge environment around an actin filament in the overlap region of the A-band (cf. Fig. 7.12). As it happens, the new type of superlattice with only two orientations of three-stranded filaments cannot produce at any particular level even pseudoequivalent actin environments of the kind in Fig. 7.12. But what happens is that some actin filaments will interact with a single cross bridge in any 143-Å axial repeat, whereas others will interact with two cross bridges (Fig. 7.28). In no situation will more than two cross bridges interact with a single actin filament at the same level, since this would require the three adjacent myosin filaments to have identical orientations, a situation forbidden by the "no three alike" rule dictated by the M-bridge interactions between filaments. Note also that a thin filament interacting with a single cross bridge at one level will tend to interact with

←

FIGURE 7.27. (a) All of the reflections to be expected on the first layer line from a cross bridge arrangement following the "no three alike" rules. No reflections are systematically absent, but some reflections are stronger than others because of differences in multiplicity (see Chapter 2). It has been assumed that the coherent unit in the "no three alike" superlattice is a few unit cells across and hence that the reflections will be broad. (b) A simple summation of the intensities in (a). (c) The reflections in (b) after the application of a Lorentz correction. (d) A reasonable first layer line profile in the diffraction pattern from a single myosin filament. (e) Comparison of the profile observed by Huxley and Brown (1967) (broken line) with the product of (c) and (d) with either 50% (weak line) or 75% (bold line) of the unsampled filament peak (d) added in to allow for the effect of randomness in the filament lattice. The calculated profiles have peak positions in exactly those positions seen by H. E. Huxley and Brown.

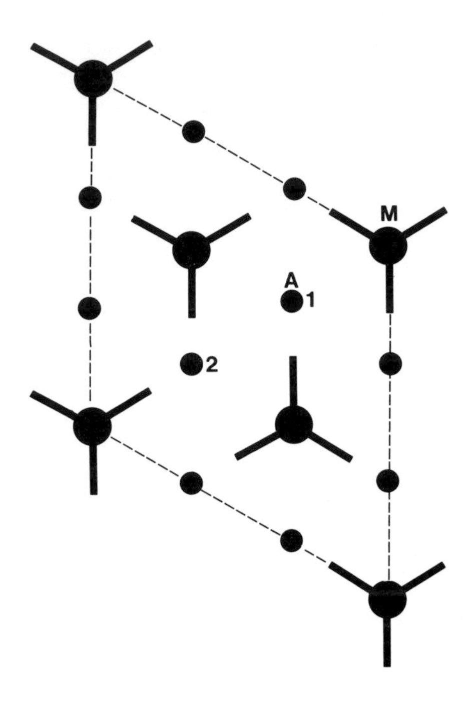

FIGURE 7.28. Diagram of a 143 Å thick cross section of the overlap region of an A-band with a "no three alike" cross bridge superlattice. Two types of actin environment are produced (labeled 1 and 2). Some actin filaments are approached by two bridges at the same level (actin 2) and others by only one (actin 1). But elsewhere along these actins, the situation changes so that in the 430-Å A-band repeat actin 2 is approached by four bridges and actin 1 by five. The actin filaments are therefore nearly equivalent even in this arrangement. (Cf. Fig. 7.12; M is myosin filament; A is actin filament.)

two cross bridges at other levels along its length. On average, each thin filament will interact with the same total number of myosin cross bridges at a given amount of filament overlap, so the quasiequivalence of the actin environments does exist for the filaments as a whole. And despite the fact that the actin filaments are not entirely equivalent at a single axial level in this type of superlattice arrangement, the situation is still favorable relative to that in fish muscle where the thin filaments have three cross bridges directed toward them at the same axial level. Since the thin filaments have two similar strands of actin monomers with which cross bridges can interact (see Chapter 5), it does not seem unreasonable that at any level they should, with advantage, interact with two rather than three cross bridges.

7.2.6. X-Ray Diffraction Evidence on the Myosin Cross Bridge Arrangement in Relaxed Muscle

It was shown earlier (Fig. 7.9) that each of the three alternative cross bridge models can account qualitatively for the observed X-ray diffraction patterns from relaxed vertebrate skeletal muscle (H. E. Huxley and Brown, 1967). However, the intensity distribution in the diffraction patterns of the three models would be expected to be different. This can be seen in the following way. Each of the models is a multistranded helix of cross bridges. For example, the model proposed by H. E. Huxley and Brown (1967) is a two-stranded model in which two identical helical arrangements of cross bridges, with six cross bridges per turn in each helix, twist around each other to give the whole structure. Let us consider the diffraction pattern of just one of these strands. From the account of helical diffraction theory given in Chapter 2, we should expect the diffraction pattern of a helix with six residues per turn (of pitch 858 Å) to consist of a helix cross at the center of the diffraction pattern together with other similar helix crosses spaced evenly along the meridian. The layers in each helix cross would have spacings measured along the meridian of $n/858$ Å^{-1} (where n is the number of the layer in the cross), and different crosses would be at spacing $m/143$ Å^{-1} along the meridian. The result is shown schematically in Fig. 7.29a,e. As mentioned in Chapter 2, the value of n clearly gives a measure of the distance from the meridian of the diffraction pattern at which the peak on the nth layer occurs. For a particular value of n, the peak position on a particular layer can be shifted by altering the radius of the helical diffracting object (See Fig. 2.14).

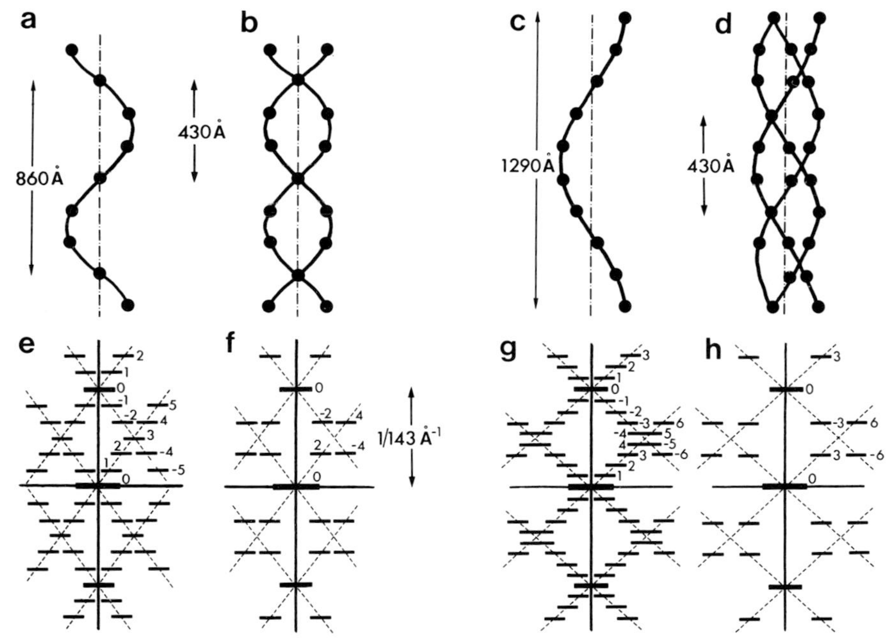

FIGURE 7.29. (a) A 6/1 helix and its diffraction pattern (e). (b) A two-stranded 6/1 helix and its diffraction pattern (f). (c) A 9/1 helix and its diffraction pattern (g). (d) A three-stranded 9/1 helix and its diffraction pattern (h). The numbers on any layer in (e) to (h) indicates the n value (Bessel function order) associated with that particular layer (see Chapter 2). For detailed discussion see text.

The effect of adding a second helix to complete the two-strand helix is to halve the axial repeat from 858 Å to 429 Å, but to leave the subunit axial translation (143 Å) unaltered. In the diffraction pattern, this results in the removal of all layers in a particular helix cross except those for which n is even, as illustrated in Fig. 7.29f. These layers will have spacings that are orders of the 429-Å repeat. For this reason, the layer $n = 1$ disappears, and the first layer line (1) observed in the diffraction pattern now has a spacing 1/429 Å^{-1} which corresponds to an n value equal to 2.

In the case of the three-stranded model, there are three equivalent strands of 9/1 helices of cross bridges, each of pitch 3 × 429 Å (Fig. 7.29c, d). The diffraction pattern from one such helix therefore consists of helix crosses spaced at intervals of $m/143$ Å^{-1} along the meridian in which each helix cross has layers of spacing $n/1287$ Å^{-1}. This is shown in Fig. 7.29g. The effect of adding two additional identical helices related by a threefold rotation axis is to cut the repeat of the structure to one-third of 1287 Å (i.e., 429 Å), but the subunit axial repeat of 143 Å remains unchanged. The effect on the diffraction pattern is to cut out all layers in a particular helix cross except those for which n is a multiple of 3. In this case, layers $n = 1$ and $n = 2$ disappear, and the first layer line (1) occurs at a spacing 3/1287 Å^{-1} (i.e., 1/429 Å^{-1} as before), and this time it corresponds to $n = 3$ (Fig. 7.29h).

In a completely analogous way, the four-stranded 12/1 structure of pitch 4 × 429 Å would give a diffraction pattern in which the first layer line in the central helix cross will have a spacing of 4/4 × 429 Å^{-1} (i.e., 1/429 Å^{-1}), and this would correspond to an n value of 4.

To summarize, the diffraction patterns from the three models will have layer lines at the same axial positions (orders of 1/429 Å^{-1}), but the intensity distribution on the first layer line (and indeed, any layer line not an order of 1/143 Å^{-1}) will be related to n values of either 2, 3, or 4, depending on the symmetry involved. This means that if all of the models had helix radii that were the same, the intensity peak on the first layer lines of their respective diffraction patterns would have different positions according to the value of n. Alternatively, since the position of this peak in the observed X-ray diffraction pattern is known, it should be possible to choose for each model a helix radius such that the layer line peak of the appropriate n value is brought to the correct position. In fact, when rigorous diffraction theory is applied, it is found that the radius of the helix (r) and the position of the intensity peak from the meridian (R) are related by the simple expression $2\pi Rr = X$, where X varies with n as shown in Fig. 7.30 (Fraser and MacRae, 1973). In the case being considered here, the intensity distribution on the first layer line in the observed diffraction pattern from relaxed frog muscle (H. E.

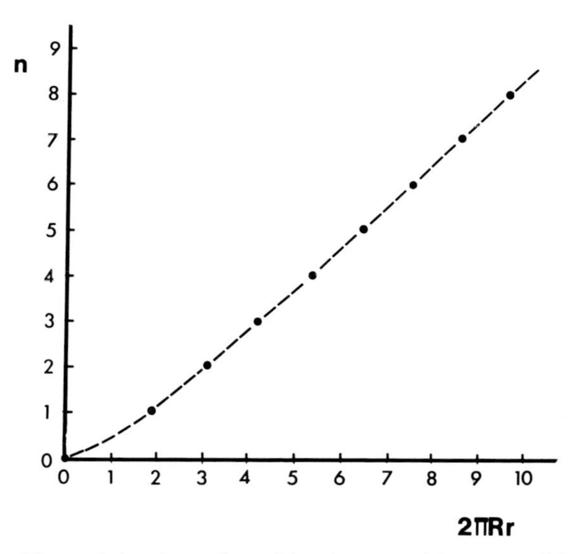

FIGURE 7.30. The variation in peak position (measured in terms of $2\pi Rr$) associated with different n values in helical diffraction patterns (see Fig. 7.29). R is the helix radius in real space (measured in Å), and r is the radial position of the peak along a layer line (measured in $Å^{-1}$). The plot shows the peak positions of nth order Bessel functions in terms of the argument $2\pi Rr$.

Huxley and Brown, 1967) is as shown in Fig. 7.26. It can be seen that it has a peak at about $R = 0.005$ $Å^{-1}$. If $n = 2$, as for the two-stranded helix, then $2\pi r \times 0.005 = 3.1$, and hence $r = 98$ Å. Similarly if $n = 3$, $r = 134$ Å, and if $n = 4$, $r = 169$ Å.

But the argument given above assumes that the helices involved are helices of point scatterers. In fact, in the case of the thick filaments, the repeating units are the myosin cross bridges which are clearly far from being point scatterers. The simple argument given above has merely given an indication of the approximate positions of the centers of mass of the myosin cross bridges for a particular model.

Because the myosin cross bridges are large and have an elongated shape (Fig. 6.5c), the observed intensity distribution on the different layer lines will depend not only on the number of strands in the helix and on the helix radius but also on the cross bridge shape and the orientation of the cross bridge in space. An added complication is that each myosin cross bridge comprises two myosin heads, and the cross bridge shape will depend on how these two heads are spatially related to each other.

Since there are so many parameters involved and there is still so much uncertainty about some of these (e.g., the arrangement of the two myosin heads in one cross bridge), it is clear from the outset that unambiguous analysis of the observed X-ray diffraction pattern in terms of a unique thick filament structure is not yet going to be possible. But this does not mean that preliminary analysis is not worthwhile provided that the results are treated with caution.

The problem can be approached in two different ways. One approach is to ask if it is possible, by making plausible assumptions about the myosin cross bridge shape and geometry, to use the observed X-ray results to give some indication of the correct value for the rotational symmetry of the cross bridge helix. The other approach is to accept the evidence already outlined in earlier parts of this chapter that the cross bridge arrangement is three-stranded and then to use the observed X-ray diffraction results to find the most probable configuration of the myosin cross bridges on the myosin filament surface in relaxed muscle. Preliminary attempts to apply both of these approaches have been carried out by Squire and Roberts (quoted in Squire, 1975) and more recently by Haselgrove (personal communication) and by Poulsen and Lowy (personal communication), and these will be described very briefly here.

The first step in the procedure is to establish what experimental evidence can be used in the analysis. Squire and Roberts made use of the published results of H. E. Huxley and Brown (1967) on the intensity distribution along the first myosin layer line (at 1/429 $Å^{-1}$) shown in Fig. 7.26. In addition, the evidence about the relative intensities of the first three layer lines was also taken into account. It was estimated that the intensity of the second layer line was less than one-half that of the first layer line on average. The relative intensities of the first layer line and the third layer line (the meridional reflection) were estimated both by applying approximate correction factors to the intensity ratio ($I_3/I_1 = 2.56$) quoted by H. E. Huxley and Brown (1967) and by direct measurements of the integrated intensities on these layer lines carried out by Dr. J. C. Haselgrove (personal communication). Both of these methods led to an estimate of the relative peak heights of the first and third layer lines (after being corrected approximately for sampling) of between 1 : 1 and 10 : 1. The main uncertainty about these estimates is caused by the limited superlattice order in the sarcomere which would affect the correction factors to be applied. As mentioned earlier, efforts have recently been made to obtain unsampled diffraction patterns by using frog muscles that have been stretched to destroy the longitudinal order in the A-band (J. C. Haselgrove, F. R. Poulsen, and J. Lowy, personal communications). These have tended to confirm that $I_3 < I_1$ and that $I_2 \approx I_4 \approx I_5 \approx \frac{1}{4}I_1$. It has also been shown that the peaks on all of these layer lines have similar positions (0.005 $Å^{-1}$: J. C. Haselgrove, personal communication) and that, in addition, layer lines 1 and 2 have distinct minima and then subsidiary peaks at about 0.010 Å^{-1}. It is therefore reasonable to start the analysis with this basic evidence.

The second step in the procedure is to define a model for the myosin cross bridge. In the analysis of Squire and Roberts, each myosin head was taken to approximate to a cylinder about 120 Å long and about 40 Å in diameter (for which R_G = 37.5 Å; Fig. 6.5a). The cylindrical shape (which is quite a good first approximation to the shape observed by Elliott and Offer, 1978; Fig. 6.5c) was in turn modeled as six overlapping spherical scattering units of radius 20 Å and equally spaced with their centers 20 Å apart (Fig. 7.31). This particular model was shown to be a realistic approximation (at the resolution involved) to a cylinder of length 120 Å, by test computations carried out by Squire and Luther (unpublished results). A computer program was then set up to calculate the intensity distributions on the first, second, and third layer lines of the diffraction pattern as a function of the number of strands in the helix and the cross bridge configuration as defined in terms of the radial position R of the innermost sphere in the myosin head and the angles ϕ and θ between the axis of the myosin head and, respectively, the filament axis and a tangent to the filament surface perpendicular both to the radius R and to the myosin filament axis.

An additional feature of the computation was that the myosin heads in one cross bridge were assumed to lie side by side as in Fig. 7.31b and either to lie with their inner ends in the same plane parallel to the helix

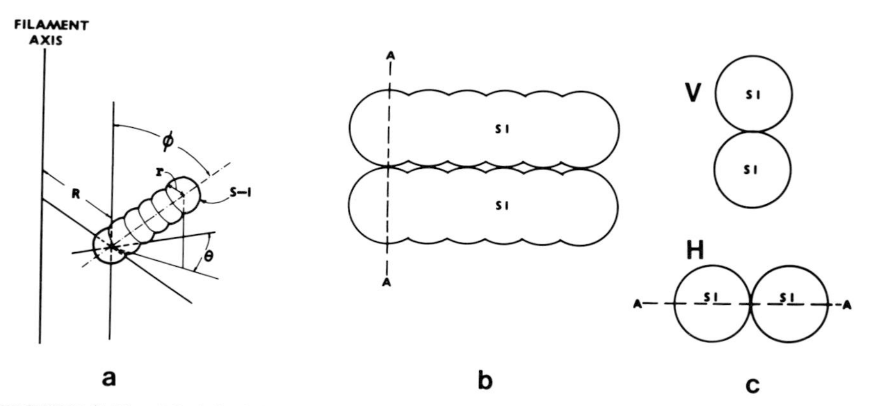

FIGURE 7.31. Model of the myosin cross bridge used for computation of the first layer line profiles in Fig. 7.33 and the relative layer line peak heights in Fig. 7.32. (a) The shape of one myosin head (here termed S-1) and the definitions of the parameters R, ϕ, and θ. (b) A model of one cross bridge comprising two myosin heads side by side. (c) Two modeled configurations of the cross bridge in space relative to the filament axis as in (a). The inner ends of the two heads were either taken to be in a plane parallel to the helix axis (V) or perpendicular to it (H).

axis (the V position) or perpendicular to the helix axis (the H position). Placing the two heads together like this is probably the most arguable assumption in the computations, and it is as well to remember that the two heads may in fact be angled to each other. However, it is an order of magnitude more involved to consider models in which the two heads are positioned independently in space, and with the data available at present, it is probably premature to try to model the structure in such detail.

First Layer Line Fitting. The first significant feature of the observed intensity distribution on the first layer line is that it peaks at a spacing of about 0.0045 to 0.0055 $Å^{-1}$ (H. E. Huxley and Brown, 1967). Even assuming that this layer line is sampled by the Fourier transform of a "no three alike" superlattice, the unsampled peak will still remain in about this position (see Fig. 7.27). The question can then be asked what parameters are needed in our model to give the right peak position and approximately the right intensity distribution? Table 7.4 summarizes the values of R, θ, and ϕ that give sensible peak shapes and positions accord-

TABLE 7.4. Computer Modeling of the Intensity Distribution in the X-Ray Diffraction Pattern from Resting Frog Sartorius Muscle[a]

Number of strands	Cross bridge configuration (H or V)	Radius (R) (Å)	Axial tilt (ϕ)	Slew angle (θ)	I_2/I_1[b]	I_3/I_1[b]
2	H	70	30°	75–90°	0.36	0.47*
2	H	80	(15°)	120°	0.36	0.20*
2	H	90	(15°)	0°	0.50	0.20*
2	V	80	(15°)	60°	0.39	0.20
2	H	90	(0°)	0–30°	0.40	0.13*
3	V	100	60°	75°	0.61	1.76*
3	V	100	45°	75°	0.52	1.10*
3	V	90	45°	90°	0.54	1.03*
3	H	100	60°	45°	(1.19)[c]	(4.93)[c]
4	H	130	75°	45°	(1.27)	(10.9)
4	H	140	60°	45°	(1.38)	(6.64)
4	V	(140)[c]	60°	75°	0.67	(2.40)
4	V	(140)	45°	75°	0.57	1.48
4	V	(160)	15°	60°	0.46	0.42
4	V	(160)	15°	120°	0.37	0.40

[a] The cross bridge model used was as in Fig. 7.31, and the parameters R, ϕ, and θ are as defined there. The best fits listed here all give a very close fit to the first layer line profile shown in Fig. 7.26 except for the four-stranded filaments where even the best fit is rather poor (J. M. Squire and E. C. Roberts, unpublished calculations; see Squire, 1975).

[b] The quoted relative intensities are actually relative peak heights. Asterisks indicate that the fit to the first layer line profile was very good.

[c] Figures in parentheses seem to be outside the probable range of the parameter involved. For discussion, see text.

ing to this modeling. It will be seen that for the two-strand helix (here taken to have one myosin molecule per cross bridge), the best fit is obtained either with R very small or with the cross bridge tilted to be almost parallel with the filament axis (especially for the H configuration of the cross bridge). On the other hand, for the four-stranded filament, the best fit is rather poor and could only be obtained with large values for R (120 to 140 Å). In the case of the three-stranded model, the best fit is obtained with R about 80 to 110 Å.

Relative Layer Line Intensities. If the relative intensities of the layer lines are now taken into account, it becomes clear that satisfactory agreement is only obtained if the cross bridge axis is tilted towards the filament axis. To demonstrate this, Fig. 7.32 shows the variation in the computed value of I_3/I_1, as a function of ϕ for the three models and the two cross bridge configurations (V and H). It is clear from these plots that even if I_3/I_1 is only required to be less than 1, then ϕ must be in the range 0 to about 45°. It is also clear that it is easier to produce the required low value of I_3/I_1 if the cross bridges are in the V configuration (Fig. 7.31b).

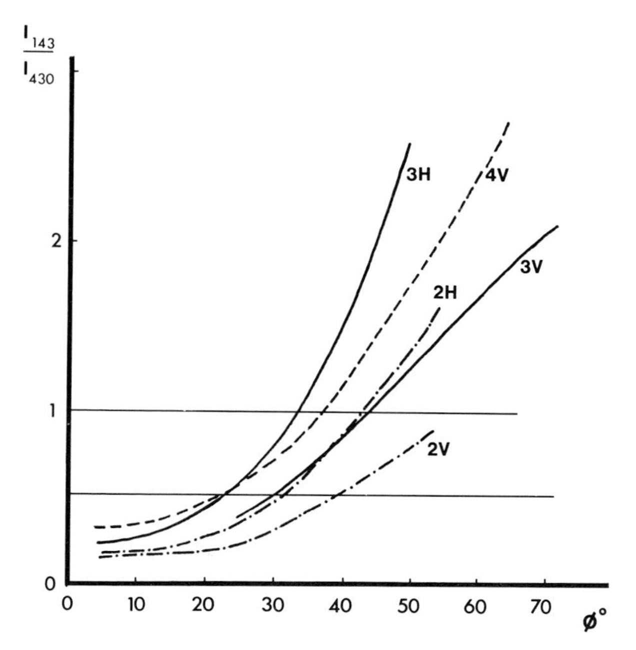

FIGURE 7.32. Variation of the ratio of the peak heights on the first and third layer lines (I_{430} and I_{143}, respectively) as a function of axial tilt (ϕ) in diffraction patterns from two-, three-, or four-stranded helices of cross bridges modeled as in Fig. 7.31. The values were taken from models that gave a reasonable fit to the first layer line profile. The observed peak height ratio is probably less than 1.

Combining the conclusions from the first layer line fitting and the estimates of I_3/I_1 for particular cross bridge positions, it is reasonable to conclude that the two-stranded model gives a best fit to the observations provided R is about 80–90 Å and ϕ is 0 to 45° (V and H equally good); and that the three-stranded model gives a best fit if R is about 100 Å (V better than H) and ϕ is between 30° and 60° (Haselgrove puts it as 50° to 70°). The four-strand model gives at best only a poor fit on the first layer line if R is very large (160 Å) or alternatively if ϕ is so large that the intensities on the second and third layer lines are relatively too great. The kind of fit obtained in these cases is illustrated in Fig. 7.33.

Some further qualitative arguments can be used to define the cross bridge configuration more closely (after Wray *et al.*, 1975; J. C. Haselgrove, personal communication). Figure 7.34 shows a radial projection of a three-stranded helix, and on it have been drawn different sets of helices. The density distribution in projection along one of these helices determines to a large extent the intensity distribution that will be seen on a particular layer line in the diffraction pattern. Thus, the helices marked A give rise to the first layer line, those marked B to the second layer line, and so on. It can be seen that in each case the density profile projected along each helix set has been sketched in (Fig. 7.34) for the particular cross bridge position illustrated. The relative intensities of the different layer lines will be determined to a large extent by the differences between these density profiles. For example, if the cross bridges are lined up more or less along the A helices (as in Fig. 7.34), then the

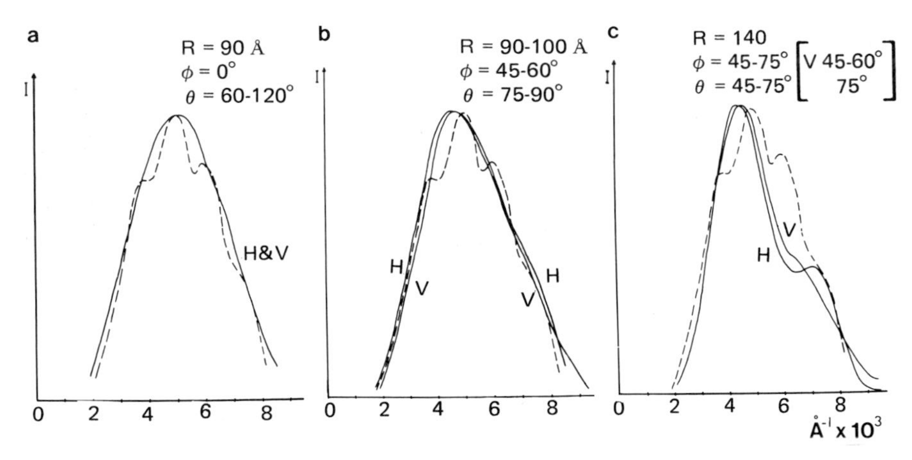

FIGURE 7.33. The best fits to the observed first layer line profile (Fig. 7.26) as computed for (a) two-stranded, (b) three-stranded, and (c) four-stranded helices of cross bridges modeled as in Fig. 7.31. Even the best fit for a four-stranded filament is rather poor. Solid lines, computed profiles; dashed lines, H. E. Huxley and Brown's (1967) results.

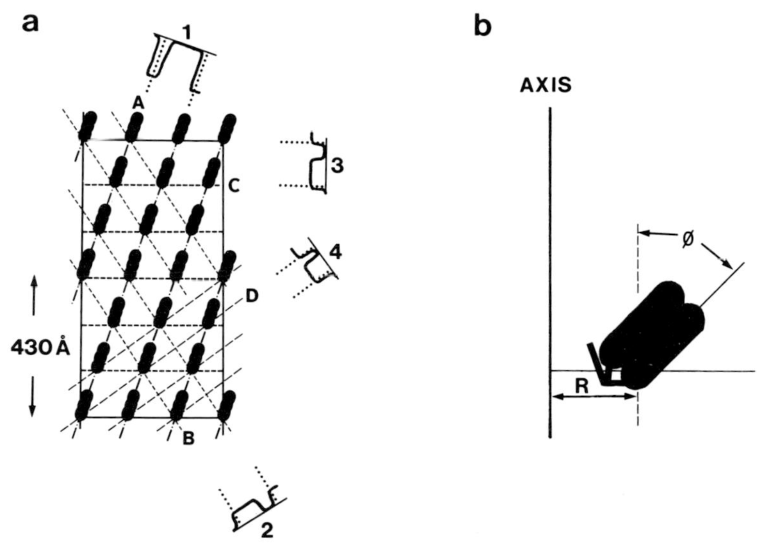

FIGURE 7.34. (a) Radial projection of a three-stranded helix of cross bridges showing the probable slewing of the cross bridges along the helices of pitch 3 × 430 Å. For details see text. (b) Probable axial tilt of the bridges (ϕ is probably 45° to 60°).

first layer line will tend to be strong, since there is a very marked density variation in the projected profile. On the other hand, the 143-Å repeat (c) will be relatively weak, since the bridges lie across this repeat (this is another way of looking at the results summarized in Fig. 7.32). Diffraction from helices B, D, and E will also be relatively weak, since the bridges would lie across these repeats also. A model in which the crossbridges are slewed azimuthally to lie more or less along the A helices would therefore explain qualitatively why layer line 1 is stronger than layer lines 2, 4, and 5. Note that this qualitative argument will also apply equally well if the filaments are two-stranded or four-stranded since the radial projections for all three symmetry models are very similar (Fig. 7.9). More detailed arguments can be used to suggest a more precise location for the "relaxed" cross bridge (J. C. Haselgrove, personal communication), but since there is still some doubt about the true shape of the whole cross bridge, it is probably too early for such arguments to be taken as more than useful indications of the factors involved.

With these general trends in mind, it is appropriate to ask if there is any additional information that might be used to help choose among the models.

One additional piece of evidence is that when a muscle is activated it is possible that the myosin cross bridges may attach to the thin filaments at a tilt ϕ of about 90° (see Chapter 10). If ϕ is only 0 to 15° in relaxed muscle, the cross bridges would be required to swing through about 75 to 90° before attaching to actin. Such a swing might be expected to increase the intensity of the 143-Å meridional reflection by a factor of 3 to 5 (Fig. 7.32). But experimentally, no such increase is observed. Although it is likely that some countereffect is involved that will reduce the intensity of the 143-Å reflection (such as increased axial disorder), it seems unlikely that this could counteract sufficiently an increase by a factor of 3 to 5. On the other hand, a tilt of about 60° might be possible, and this is the situation modeled for the three-strand filament. The two-stranded model with ϕ about 0 to 15°, therefore, may be unsatisfactory in this respect.

Conclusions. The analysis described above clearly depends heavily on the assumptions that are made about the size and shape of the myosin heads and their arrangement in a cross bridge. However, it could be argued that since the analysis leads to the same conclusions about filament symmetry as the other quite independent lines of evidence described earlier in the chapter, the assumptions made might have been reasonable. The results can then be used to indicate the cross bridge position in relaxed muscle, assuming that they are arranged on a three-stranded helix. The computations do clearly show that the correct relative layer line intensities can only be produced if the projected length of the cross bridges onto the filament axis is ½ to ⅔ of the 143 Å repeat (Fig. 7.34b). This can be produced with a long thin cross bridge of the sort modeled here if the cross bridge is tilted significantly towards the filament axis ($\phi \approx 60°$). Since the only other sensible explanation of this kind of result would be that the cross bridges themselves are much more globular in shape than had been assumed in the computations described earlier or than has now been observed in the electron microscope by A. Elliott and Offer (1978; Fig. 6.5c), then it is fair to conclude that in relaxed muscle the cross bridges do not project in a direction perpendicular to the filament surface but may be tilted up to 40° to 50° from this position. It remains to be seen exactly how great the tilt is. Some early evidence from neutron diffraction studies of vertebrate striated muscle (Worcester *et al.*, 1975) and also fluorescence depolarization studies (Tregear and Mendelson, 1975) tend to support the view that in relaxed muscle the cross bridges are indeed tilted. It will be seen in Chapters 10 and 11 of this book that this result is important for a complete understanding of the nature of the cross bridge movements that are involved in muscle contraction. Unfortunately, the diffraction evidence can give no indication of the direction in which the tilt occurs: it could be towards or away from the M-band. This too is crucial to an understanding of the contractile event. Haselgrove (personal communication) has suggested that the two heads in one cross bridge might actually point in opposite directions, one toward and one away from the M-band, but whether this is so remains to be seen. If not, it seems likely for reasons given in Chapter 10 that the relaxed tilt of both heads might well be toward the M-band.

7.2.7. Discussion

The conclusions that have been reached in this section can easily be summarized as follows. First, the myosin cross bridges on the thick myosin filaments are probably arranged on a three-stranded helix of pitch 3 × 429 Å and subunit axial translation 143 Å. Second, the thick filaments in the frog A-band are arranged in a "no three alike" superlattice with just two filament orientations 60° apart, and this lattice is defined by specific M-bridge interactions in the M-band. And finally, the myosin cross bridges are probably arranged in space so that there is a certain amount of tilt toward the filament axis and slew around the filament axis.

However, it would be wrong to leave this section without drawing attention to the fact that there is some disagreement about the number of strands of cross bridges on the surface of the myosin filaments. In particular, Pepe and his collaborators (Pepe, 1967a, 1971; Pepe and Drucker, 1972, 1979; Pepe and Dowben, 1977) have a particular model for the structure of the vertebrate thick filament in which the filament has a more or less triangular profile in any cross section and yet has its cross bridges arranged on a two-stranded helix. This model was originally derived from a suggested structure for the M-band and, as shown earlier in this section, does not account for the structure of the M-band as seen in thin transverse sections. There is, therefore, no compelling evidence that supports the two-stranded model, since there is little disagreement about the fact that the filament profile often has a triangular shape, and the simplest interpretation of that is clearly that the crossbridges are arranged on a three-stranded helix. Any other interpretation, such as Pepe's, must mean that myosin molecules at the same axial position along the thick filament are in completely nonequivalent environments. Indeed, Pepe's revised model (Pepe and Dowben, 1977) requires different molecular environments for almost all of the myosin molecules in one half of the thick filament, and this must remain one of the most serious objections to the model. A further discussion of this point will be given in Chapter 9. Pepe and Dowben's clear observations of triangular profiles in thick transverse sections of the bridge region will also be considered in Chapter 9.

7.3. Components and Axial Structure of the A-Band

7.3.1. A-Band Components

It is clear from the description of the A-band given so far that in addition to myosin some other proteins must be responsible for forming the M-band structure that cross-links the thick filaments. A considerable effort has been spent in characterizing these M-band proteins, and various groups have worked on the problem (see reviews by Pepe, 1975; Lowey, 1979; Eaton and Pepe, 1972; Morimoto and Harrington, 1972; Turner *et al.*, 1973; Masaki and Takaiti, 1972, 1974; Landon and Oriol, 1975; Trinick and Lowey, 1977; Walliman *et al.*, 1977, 1978). So far, the general conclusions are that M-band extracts contain four main components. One of molecular weight 40,000 appears to be creatine kinase, an enzyme known to catalyze the reaction ATP + creatine → ADP + phosphocreatine (Eaton and Pepe, 1972; Morimoto and Harrington, 1972; Turner *et al.*, 1973). Walliman *et al.* (1978) have evidence that creatine kinase is a principal structural component of the M-band. The second protein of molecular weight 90,000 has phosphorylase activity [it is glycogen phosphorylase b (Eaton and Pepe, 1972; Masaki and Takaiti, 1972, 1974; Pepe, 1975; Lowey, 1979)]. The last two components are proteins of molecular weight about 165,000, and only one of these seems to act primarily as a structural protein (Masaki and Takaiti, 1972, 1974; Landon and Oriol, 1975; Trinick and Lowey, 1977). The other is glycogen debranching enzyme, and this and the 90,000-molecular-weight protein are not structural M-band components (Trinick and Lowey, 1977; Heizmann and Eppenberger, 1978).

It will be shown later that the M-bands in different vertebrate fiber types and muscle types have characteristically different structures. It is natural to conclude that they probably contain some common proteins and some proteins that are characteristic M-proteins for that particular fiber or muscle. It is clear then that much work has yet to be done to determine the composition of particular M-bands.

The M-proteins are not the only nonmyosin A-band proteins. This was shown clearly by Starr and Offer (1971), who analyzed a nominally "pure" preparation of rabbit myosin (Perry, 1955) using the new technique of gel electrophoresis in the presence of SDS. They ran a heavily loaded 6% gel, stained it with Coomassie blue dye, and the result, shown in Fig. 7.35, was a series of unexpected protein bands in addition to those of the myosin heavy and light chains and a trace of actin. Starr and Offer labeled these various bands with letters, the three most abundant—those with molecular weights 180,000, 150,000 and 110,000—being termed the B, C, and F proteins, respectively. B protein may be the same as Trinick and Lowey's structural H-protein (Offer, personal communication). Since these could not be identified as thin filament or

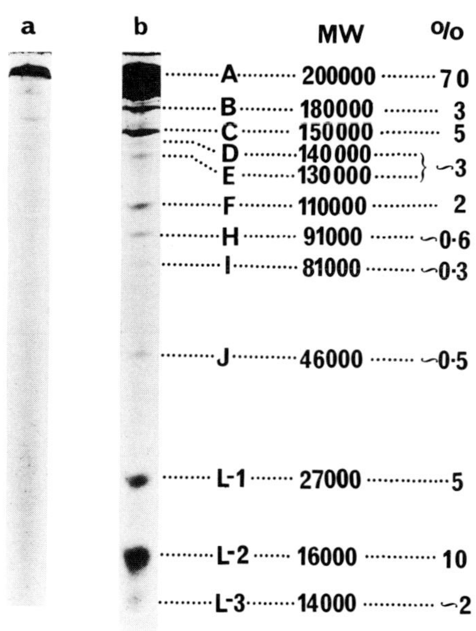

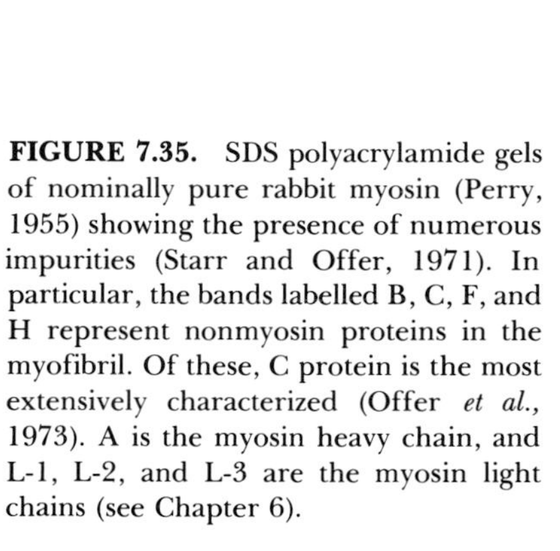

FIGURE 7.35. SDS polyacrylamide gels of nominally pure rabbit myosin (Perry, 1955) showing the presence of numerous impurities (Starr and Offer, 1971). In particular, the bands labelled B, C, F, and H represent nonmyosin proteins in the myofibril. Of these, C protein is the most extensively characterized (Offer *et al.*, 1973). A is the myosin heavy chain, and L-1, L-2, and L-3 are the myosin light chains (see Chapter 6).

Z-band proteins and were abundant in myosin preparation, it seemed likely that they were, in fact, additional components of the A-band. That they were true myofibrillar proteins was indicated by electrophoresis of myofibril preparations. The gels showed similar bands (Sender, 1971), two of which were later found to be enhanced if C and F proteins were added (Starr and Offer, 1971).

The most abundant of these extra proteins, C protein, has been the most extensively studied. It has been described and characterized largely by Offer and his collaborators (Offer, 1972; Offer *et al.*, 1973). It binds strongly to myosin and to LMM; it is almost completely devoid of α-helix (see also Morimoto and Harrington, 1973) and probably contains 40 to 60% of β structure, with the remainder random coil. Its amino acid composition is given in Table 7.5.

Electron microscopy of isolated C protein molecules as well as solution studies and low-angle neutron scattering studies have indicated that it is an elongated molecule about 350 to 400 Å long (assuming it is rod-shaped) and probably 20 to 30 Å across (Offer *et al.*, 1973). Its radius of gyration estimated from low-angle neutron scattering studies of solutions is about 125 Å (D. Worcester, G. Offer, R. Starr, and J. M. Squire, unpublished results).

The interactions of C protein and myosin and the location in the A-band of the M-proteins, C protein, and some other nonmyosin proteins will be described in the remainder of this section when the axial structure of the A-band is considered in detail.

TABLE 7.5. Comparison of the Amino Acid Composition of C Protein with That of Myosin and Its Fragments[a]

Amino acid	Residues per 140,000 g C protein	Residues per 10^5 g					
		C protein	Myosin[b]	HMM[b]	LMM[b]	S-1[c]	Rod[c]
Lys	103	74	92	86	94	83	107
His	17	12	16	14	21	18	15
Arg	53	38	43	34	60	34	56
Cya[d]	17	12·0	8·8	7·4	4·0	11	4·6
Cys[e]	17	12·5	—	—	—	—	—
Asp	116	83	85	82	83	85	88
Thr	71	51	44	44	33	49	38
Ser	69	49	39	39	34	41	39
Glu	141	101	157	137	210	117	219
Pro	85	61	22	32	0	37	0
Gly	85	61	40	50	18	61	20
Ala	80	57	78	73	81	70	85
Val	126	90	43	48	38	55	33
Met	21	15	23	26	19	28	23
He	61	44	42	44	39	53	36
Leu	84	60	81	73	96	75	99
Tyr	35	25	20	21	9	34	5·9
Phe	45	32	29	36	4	52	7·0
Try	27	19					
Amide	(52)	(37)					

[a] From Offer *et al.* (1973).
[b] Lowey and Cohen (1962).
[c] Lowey *et al.* (1969).
[d] From performic acid-oxidized C protein.
[e] From dithiobis-(2-nitrobenzoate) assays.

7.3.2. Structure of the M-Region

Earlier in this chapter, the transverse structures of the M-bands in frog and fish muscle were described. These were both shown to consist of M-bridges joining adjacent thick filaments, the M-bridges often showing thickening halfway along their length. In addition, in frog muscle the M-bridges often appeared to be linked by secondary M-bridges from their thickened central regions to the trigonal points of the thick filament lattice (Fig. 7.14). Here a description is given of the structure of the M-band and the remainder of the M-region as seen in longitudinal sections of muscle. This structure is then related to the known transverse M-band structure to produce a tentative three-dimensional structure for the M-band.

Early longitudinal sections of embedded muscle showed that the M-band usually appeared as either three or five strong lines of density, the lines being separated axially by about 220 Å (Pepe, 1967a,b; Knappeis and Carlsen, 1968). In suitable sections it could be seen that these lines were caused by bridging structures between adjacent thick filaments (Fig. 7.36), and they were therefore taken to represent the axial positions of the M-bridges which could clearly be seen in transverse sections. Pepe (1971) compared micrographs from several muscles, and these showed that in longitudinal sections of a particular muscle, the M-band can have quite different appearances. These are all shown in Fig. 7.36 and can be described as three centrally placed lines, five lines with the central three more prominent, four lines, two lines of which one is centrally placed in the M-region, and two lines placed about 220 Å on

FIGURE 7.36. A variety of M-band appearances in longitudinal sections of vertebrate muscle (From Pepe, 1971). The fibril axis is horizontal. From left to right the appearances are: three centrally placed line; five lines, generally with the central three more distinct; four lines; two lines, one of which is centrally placed; and two lines placed one on each side of the missing central one. (a) Chicken breast muscle. (b) Lateral muscle of the black molly (Mollienesia sphenops). (c) Lateral muscle of the freshwater killifish (Fundulus diaphanus). Magnifications ×70,000 (reduced 16% for reproduction).

each side of a missing central line. Pepe (personal communication) has stated that different patterns were observed within a single fiber, and all of these appearances were attributed by Pepe (1971) to the differences in appearance that would occur when a particular M-band model is viewed in different directions. The structure, described briefly earlier, was one in which a thick filament was only linked to two of its six neighbors at one of the M-bridge levels seen in Fig. 7.36, and it has already been shown that thin transverse sections do not support such a model (Fig. 7.14). In addition, it was shown by Sjöström and Squire (1977a) that there might be quite another reason for seeing different appearances in longitudinal sections. Negatively stained sections of frozen muscle (obtained as described in Chapter 3) showed some of the different M-band appearances described by Pepe (1971). However, it was noticed that all of the M-bands in a particular fiber appeared to have similar structures, whereas the M-band structures in adjacent fibers could have quite distinct appearances. Since many different myofibril orientations will occur in a given fiber, it seemed to be clear that some of the differences in M-band appearance were related to the particular fiber types involved and not to differences in viewing direction. Support for this conclusion was obtained by comparing M-band appearances with other features of the fiber (such as Z-band width, mitochondrial content, glycogen content, and so on) that were already known to be characteristic of different fiber types. In this way it was evident that fibers of a particular type always appeared to show the same M-band structure. For example, in human M. tibialis anterior, an appearance with five strong M-bridge lines was always seen in fibers identified histochemically as being Type I (slow twitch) fibers, whereas Type II (fast twitch) fibers commonly showed an M-band of three strong lines (Fig. 7.37).

Since then, Sjöström *et al.* (1981) have extended the study of M-band structure to different histochemically defined fiber types in a variety of muscles. Apart from human M. tibialis anterior, three muscles from the rat were studied: M. extensor digitorum longus, M. soleus, and cardiac muscle. It was found that three quite different M-band appearances could be seen, a three-line, five-line, or four-line pattern (the central line being missing here), and that these could be related to fiber type (Fig. 7.38). Table 7.6 summarizes these results in terms of M-band appearance and the ultrastructural, histochemical, and physiological differences among the corresponding fiber types. It is fair to conclude that sometimes different M-band appearances similar to those reported by Pepe (1971) may correspond to fiber type differences. But it is also possible that some of these appearances can be accounted for in terms of the threefold M-bridge arrangement illustrated in Fig. 7.25.

As well as relating M-band appearance to fiber type, the results of Sjöström and Squire (1977a) also revealed many previously unobserved

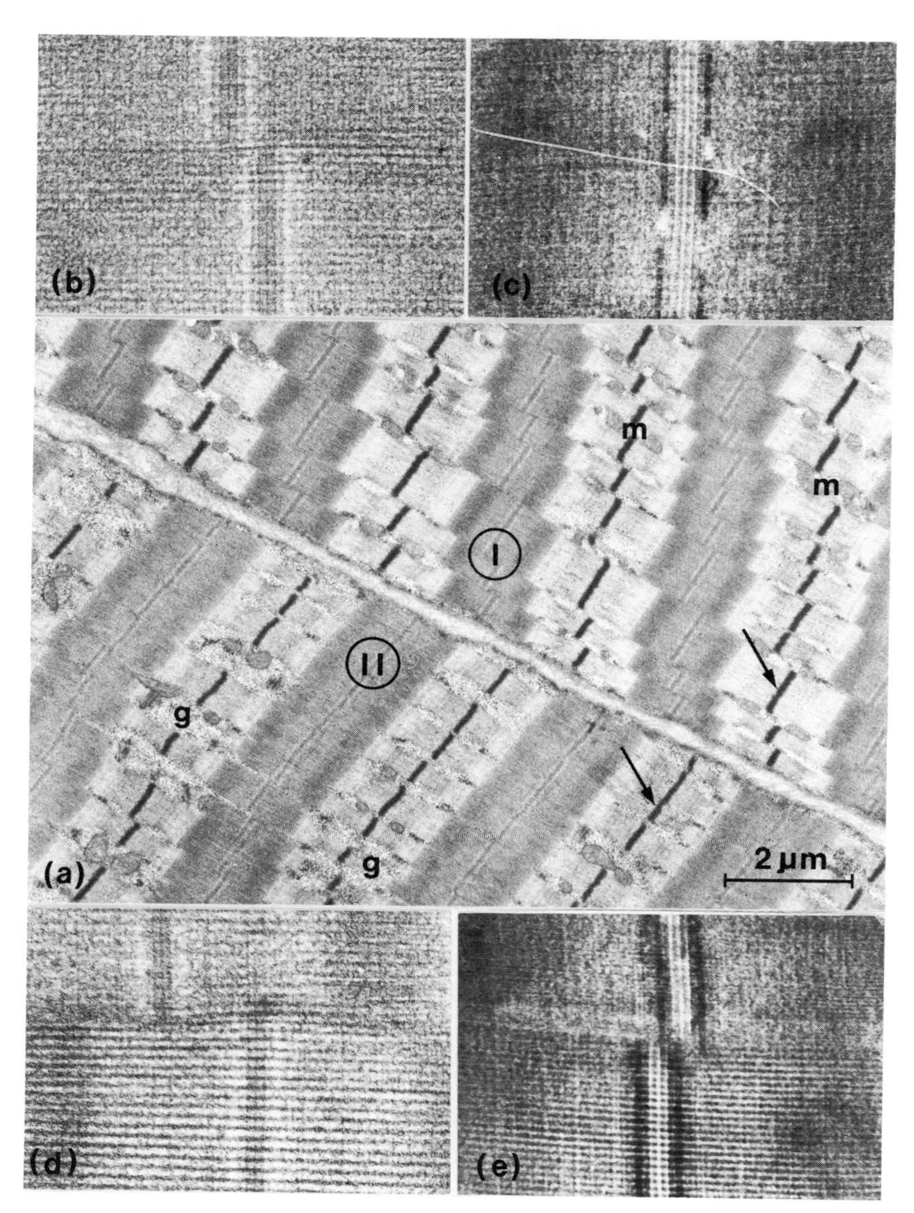

FIGURE 7.37. Different M-band appearances in different fiber types. (a) A longitudinal plastic section of human M. tibialis anterior showing two neighboring fibers of different types as revealed by their Z-band width (arrows) and, for example, their glycogen (g) and mitochondrial contents (m). They have been identified as Type I (slow oxidative) in the top right and Type II (fast) in the lower left. In addition to these differences, the fibers have consistently different M-band structures. The Type I fiber has wide M-bands comprising five strong M-bridge lines as shown in (b) (plastic section) and (c) (negatively stained cryosection). The Type II fiber has narrower M-bands, showing only three strong M-bridge lines as in (d) (plastic section) and (e) (cryosection). Note that (b) and (d) are enlargements of parts of (a). Magnifications: (a) ×12,000; (b) to (e) ×60,000 (reduced 25% for reproduction). (Courtesy of M. Sjöström.)

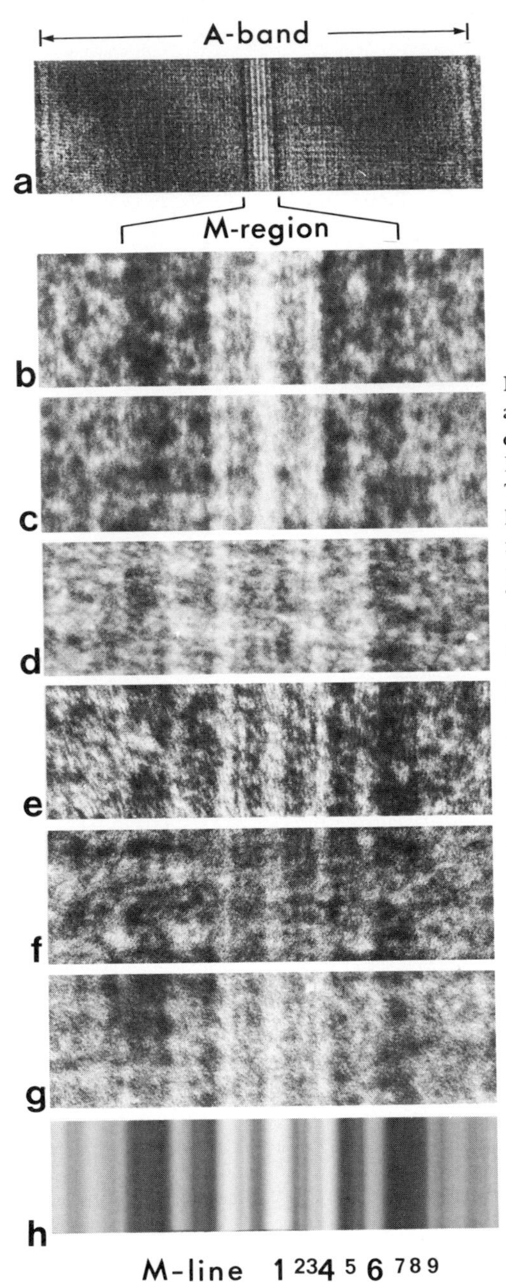

FIGURE 7.38. (a) Typical appearance of an A-band in longitudinal cryosection showing the location of the M-region (magnification ×52,000). This and all the other micrographs here are negatively stained with ammonium molybdate, so protein appears white and stain black. (b to g) The appearance of the M-regions in different muscles and fiber types (all ×380,000; reduced 30% for reproduction). (b) Human M. tibialis anterior Type II fiber (three-line pattern). (c) Rat M. extensor digitorum longus Type II B fiber (three-line pattern). (d) Rat M. soleus Type I fiber (four-line pattern). (e) Rat M. extensor digitorum longus Type IIA fiber (five-line pattern). (f) Rat cardiac muscle (five-line pattern). (g) Human M. tibialis anterior Type I fiber (five-line pattern). A five-line pattern is also seen in rat M. soleus Type II fibers. (h) A laterally averaged image of the M-region in a cryosection of human M. tibialis anterior showing improved visibility of the characteristic striations; these are numbered according to Sjöström and Squire (1977a). The averaging was achieved during printing by moving the photographic paper laterally during the exposure. (From Sjöström *et al.*, 1981.)

TABLE 7.6. Correlation between M-Band Appearance and Other Ultrastructural, Histochemical, and Physiological Properties of Rat Striated Muscle Fibers[a]

Types of muscles and of muscle fibers		Ultrastructural properties				Histochemical properties		Contractile properties	
		M-band pattern	Z-band width	SR content	Mitochondrial content	Myosin ATPase	Succinate dehydrogenase	Speed of contraction	Resistance to fatigue
Ext. dig. long.	Type IIA	5-line	Thick	Rich	Rich	Alkali-resistant	High	High	Moderate
	Type IIB	3-line	Thin	Rich	Poor	Alkali-resistant	Low	High	Low
Soleus	Type I	4-line	Thick	Poor	Moderate	Alkali-sensitive	Moderate	Low	High
	Type II	5-line	Thick	Rich	Rich	Alkali-resistant	High	High	Moderate
Cardiac		5-line	Thick	Poor	Rich				

[a] From Sjöström *et al.* (1981). Also based on data from Edström and Kugelberg (1968), Schiaffino *et al.* (1970), Edgerton and Simpson (1971), and Kugelberg (1973).

ultrastructural features of the M-region. Negatively stained cryosections of human M. tibialis anterior showed that the whole of the M-region was crossed by weak striations in addition to the strong M-bridge lines. This is shown in Fig. 7.39a. These various lines were numbered and their positions measured. The central strong M-bridge line was termed M1, and successive striations were then numbered M2, M3, etc., out to M9 at the outer edge of the M-region. The spacings and relative densities of these lines (visually estimated) are given in Table 7.7.

Occasionally the strong M-bridge lines (M1, M4, and M6 in Sjöström and Squire's nomenclature) were seen to be split into two sublines about 50 to 60 Å apart. Table 7.7 shows that if this splitting of the strong M-bridge lines is included, then all of the weak lines in the M-region (except M3) seem to lie on one of two periodic arrays of spacing about 145 Å. It seems to be very likely that these weak striations are caused by some feature of the thick filament backbone. If this is so, then the occurrence of two 145-Å repeats in the M-region is consistent with H. E. Huxley's suggestions (H. E. Huxley, 1963) that in this part of a thick

TABLE 7.7. Analysis of the Line Pattern in the M-Region of Human M. Tibialis Anterior[a]

Line	Strength[d]	Width	Distance from M1	Number of obs.[b]	Calculated spacing on 145-Å repeats	Possible origin of lines
			−30[c]		−33	Rod
M1	v.v.v.s.	Broad (100 Å)	0	40		M-bridge
			30[c]		33	Rod
M2	w.	Narrow (30 Å)	96(±12)	33	112	Rod
M3	m.s.	Narrow	151 (±6)	35		?
			190[c]		178	Rod
M4	v.v.v.s.	Broad (80 Å)	220 (±5)	40		M-bridge
			250[c]		257	Rod
M5	w.	Medium	340(±11)	32	323	Rod
			407[c]		402	Rod
M6	v.v.v.s.	Broad (80 Å)	437 (±6)	40		M-bridge
			467[c]		468	Rod
M7	v.w.	Narrow	544(±13)	33	547	Rod
M8	v.w.	Narrow	608(±10)	33	613	Rod
M9	m.s.	Medium	692(±13)	39	692	Rod
P1	s.	Medium	810(±18)	38		Myosin head

[a] The measurements were performed on microdensitometer tracings along the A-band. On the tracings the main periodicity over the C-zone (assumed to be 438 Å as in Table 7.9) was taken as the reference distance. From Sjöström and Squire (1977a).

[b] Two measurements were performed on four sarcomeres from each of five people.

[c] The peaks corresponding to the broad lines were sometimes split into 2 minor peaks separated by 48·2 (±4·3) Å ($n = 18$). However, measurements on the best specimens often tended to separate the peaks by a slightly larger distance (55 to 60 Å).

[d] Abbreviations used: v., very; m., moderate; s., strong; w., weak.

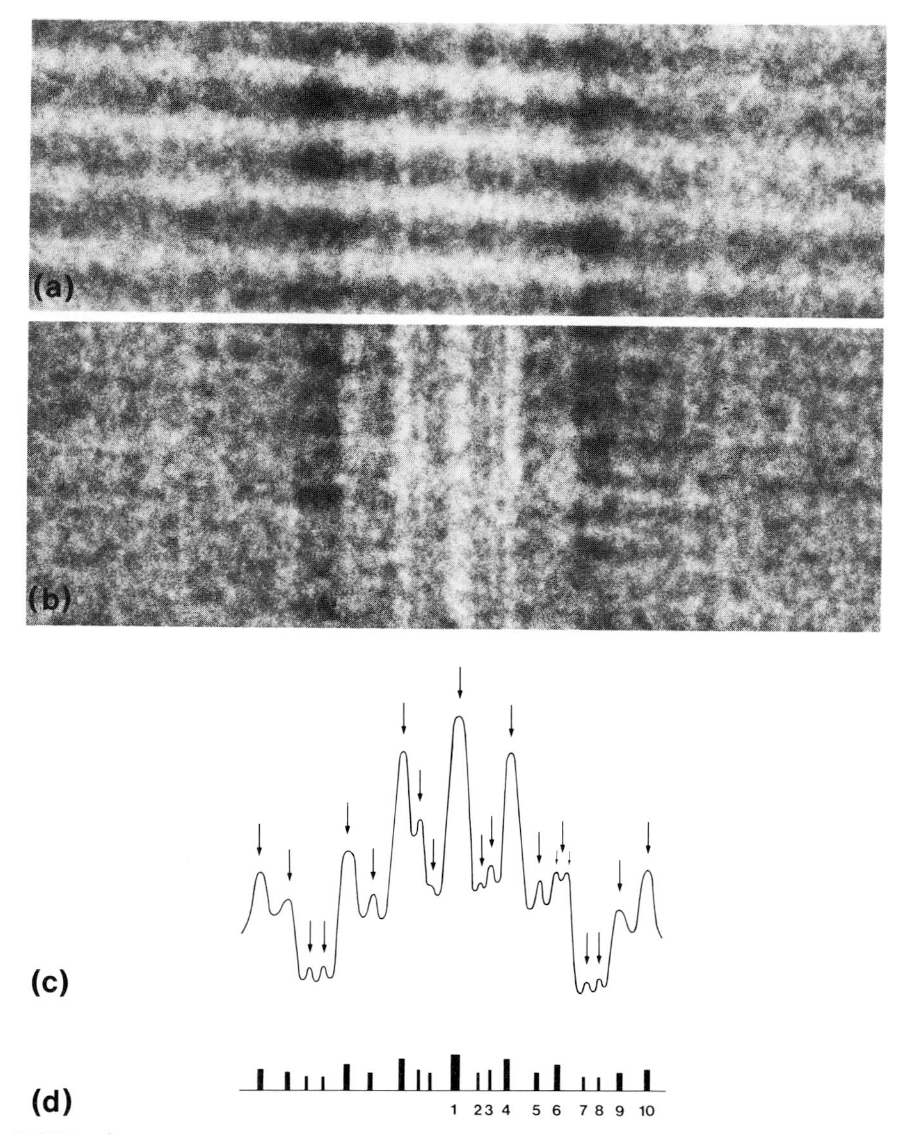

FIGURE 7.39. (a and b) Negatively stained cryosections of the M-region of human M. tibialis anterior (protein white). The section in (a) appears to show density linking between thick filaments at the prominent M-bridge positions. (c) A typical microdensitometer trace across such M-regions. The five strong M-bridge lines M6, M4, M1, M4′, and M6′ are clearly seen. These lines sometimes appear to be split into two lines about 50 to 60 Å apart (e.g., line M6; see Table 7.7 for details). Numerous fine lines can also be identified between and outside the M-bridge lines. (d) The numbering scheme used to define the various lines seen in the M-region. Magnifications of (a) and (b): ×352,000 (reduced 25% for reproduction). (From Sjöström and Squire, 1977a.)

filament antiparallel packing of myosin molecules occurs. Since the crossbridge array is known to have a 143-Å repeat, the presence of a similar repeat in the M-region seems to suggest that the myosin rods in the M-region are arranged with a similar helical symmetry to the cross bridges as originally suggested by Harrison *et al.* (1971). The existence of two overlapping sets of striations is clearly consistent with the presence of two overlapping antiparallel sets of myosin rods. That the weak M-region striations are indeed caused by staining features on the myosin rods seems to be indicated by the results of Sjöström *et al.* (1981) on M-region structure in a variety of muscles and fiber types. It was found that, despite the differences in M-band structure that were seen to occur (Fig. 7.38), the weak M-region striations always seemed to be present in the same positions. Their existence in the bare regions on each side of the M-band (where extra proteins did not seem to be present) clearly supported this view. Sjöström *et al.* (1981) therefore summarized their results in terms of the diagram shown in Fig. 7.40. It was concluded that

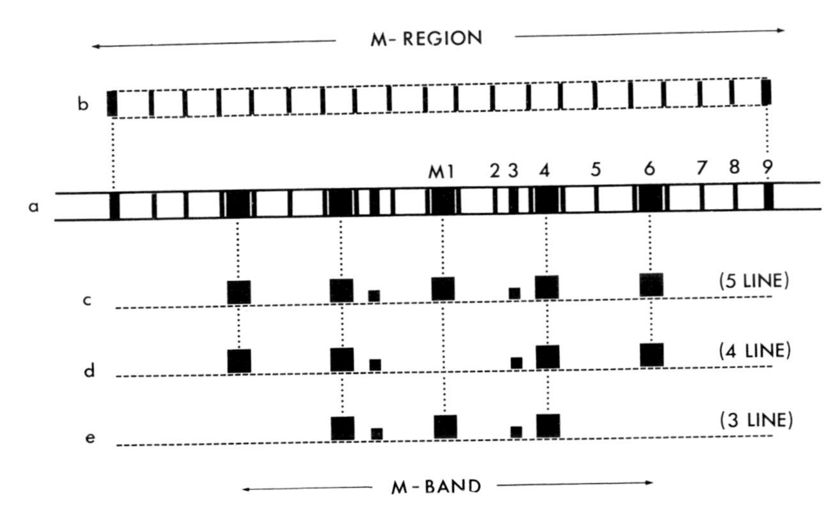

FIGURE 7.40. A schematic diagram to illustrate the different M-band appearances and their interpretations. Row (a) shows all of the observed lines in the M-region including five strong M-bridge lines. These are labeled as in Fig. 7.39. The splitting of the strong lines, which is often seen, has also been indicated here. Row (b) shows the positions of the lines in the two overlapping 145-Å arrays that are thought to be caused by stain-excluding features on the myosin rods which pack antiparallel to form the myosin filament bare zone. Rows (c) to (e) represent the different M-bridge appearances that can be seen in Fig. 7.38 as a function of muscle and fiber type. These are appearances with five, four, and three strong lines of M-bridges, respectively. The line M3 appears to be a constant feature of these M-bands but does not fit onto the 145-Å arrays. It is therefore taken as a part of the M-band structure but is not included in the M-bridge number. The possible origin of M3 is indicated in Fig. 7.41. (From Sjöström *et al.*, 1981.)

M-bridges could occur at some or all of positions M1, M4, and M6 and that the line M3, which did not lie on one of the two 145 Å repeats given in Table 7.7, might also be the location of M-protein. That this line has an origin different from the other weak striations is indicated by results from plastic sections where M3, in many muscles, is the only weak striation that can be seen. All of the other lines were attributed to myosin packing, including the weak doublets superimposed on the strong M-bridge lines. It is worth noting that this doublet is very clear in the middle of the four-line M-band where the central M-bridge line is absent (Fig. 7.38d).

These new results on the longitudinal structure of the M-band can be combined with the evidence on M-band structure in transverse section described earlier in the chapter to give a tentative model for the M-band. Since the line M3 has not as yet been accounted for, it is proposed that this line represents the axial location of the secondary M-bridges described by Luther and Squire (1978). On this assumption, the M-band structure would be as shown diagrammatically in Fig. 7.41 for a three-line M-band (as in frog sartorius and fish muscles). The existence of some longitudinal element in the M-band in the form of the M-filaments described by Knappeis and Carlsen (1968) and Sjöström

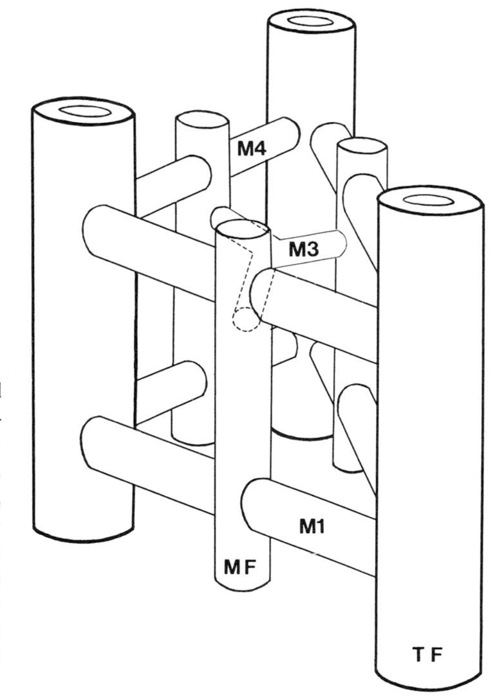

FIGURE 7.41. A three-dimensional drawing of part of the M-band illustrating the possible arrangement of the secondary M-bridges (see Fig. 7.14) at M3. The relationship among these bridges, the M-filaments (MF), the myosin filaments (TF), and the M-bridge lines M4 and M1 is indicated. In a three-line M-band, the structure as visualized here would be mirrored in the M1 plane to yield the whole M-band assembly. (From Luther and Squire, 1978.)

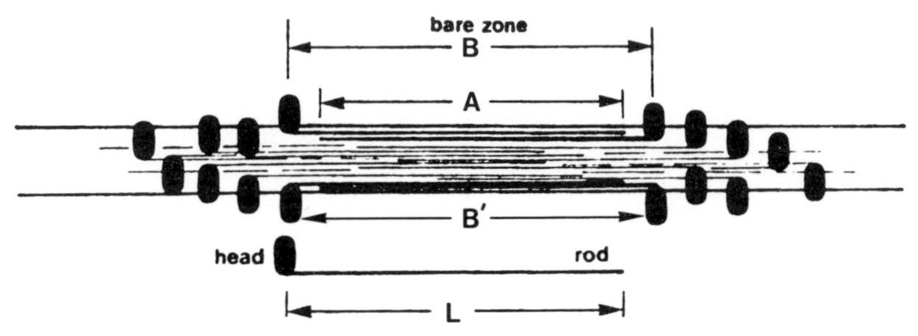

FIGURE 7.43. Schematic diagram of the H. E. Huxley (1963) model for the myosin filament bare zone showing the meaning of the parameters B, B', A, and L used in the text. (From Sjöström and Squire, 1977a.)

7.3.3. Location of C Protein in the A-Band

In an elegant series of experiments, Offer and his collaborators (Offer, 1972; Craig and Offer, 1976b) and Pepe and Drucker (1975) have determined the location of C protein in the A-band. This was done by preparing antibody to rabbit C protein and then allowing a solution of the resulting anti-C to diffuse into glycerinated myofibrils of rabbit psoas muscle. Early attempts at this technique using whole antiserum to C protein revealed nine stripes in the middle of each half of the A-band together with some M-band labeling. But with partially purified anti-C preparations (Offer, 1976; Craig and Offer, 1976b), it was sometimes found that in the inner parts of a particular fiber only seven stripes could be seen in each half of the A-band (Fig. 7.44) and that the M-band labeling was reduced. At the outer edges of these fibers, labeling of nine stripes and of the M-band could still be seen. This kind of result was consistent with the presence in the anti-C preparation of antibodies to minor contaminants of the original C-protein preparation. It was concluded that the seven stripes represented the location of C protein.

These C-protein stripes were about 430 Å apart, and the stripes nearest to the M-region were at a measured separation from the center of the M-band of about 2500 Å (Craig and Offer, 1976b). Similar stripes caused by antibody labeling were first reported by Pepe (1967a,b) who used antibody to a myosin preparation. Pepe and Drucker (1975) have clearly shown that the stripes he observed were, in fact, caused by the presence of C protein in his myosin preparation and hence to anti-C in his antibody preparation.

Estimates of the amount of C protein present in a single thick filament have been obtained by SDS gel electrophoresis of myofibrils (Offer *et al.*, 1973; Morimoto and Harrington, 1974). Using the ratio of C protein to actin in myofibrils and the known sarcomere geometry, Offer

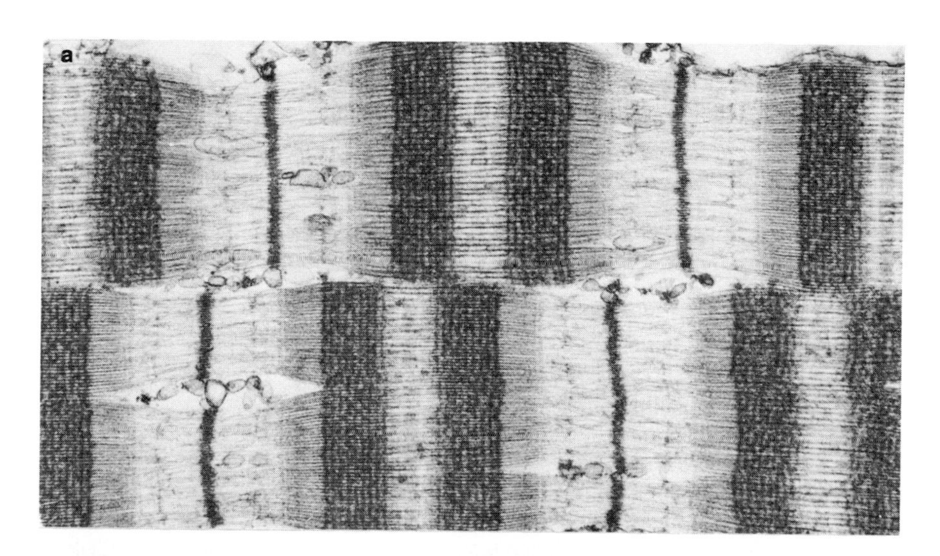

FIGURE 7.44. Labeling of rabbit psoas muscle with relatively pure antibody to C protein. (a) A sarcomere near to (but not at) the surface of a fiber labeled with anti-C showing nine stripes in each half of the A-band. (b) A sarcomere deeper into a fiber showing seven stripes in each half of the A-band. Magnification ×30,000. (From Craig and Offer, 1976b.)

et al. obtained a value of 37 (±7) for the number of C-protein molecules in each thick filament. This gives a value of 2.6 (±0.5) C-protein molecules per stripe if there are 14 stripes on one filament (Craig and Offer, 1976b). Morimoto and Harrington (1974) estimated the molar ratio of myosin to C protein to be 8.3 ± 0.3. If the thick filaments are three-stranded, then there would be very roughly 300 myosin molecules per filament. There would therefore be 300/8.3 = 36 C-protein molecules per filament or 2.5 molecules per stripe (which agrees quite well with the conclusions of Craig and Offer). Alternatively, if the thick filament is four-stranded, there would be about 400 myosin molecules and hence about 48 C-protein molecules per filament. In this case there would be about 3.4 C-protein molecules per stripe. Of course, it is very likely that this number is really integral and is probably exactly 3 if the filaments are three-stranded.

Further antibody-labeling work (e.g., Fig. 7.44b) has shown that the eighth and ninth stripes originally seen in sections labeled with impure anti-C on the M-region side of the seven stripes now known to be C protein are in fact caused by two other A-band proteins (Craig and Offer, 1976b; Pepe and Drucker, 1975). In addition, two further stripes (one of them being M9) often seen in isolated A-band preparations (A-segments; Hanson *et al.*, 1971; Craig, 1977) and situated even closer in to the M-region than the eighth and ninth stripes are also thought to be caused by additional A-band proteins (Craig and Offer, 1976b; Sjöström and Squire, 1977a). A series of about 11 stripes more or less evenly spaced therefore extends in each half of the A-band from the edge of the M-region (the first is at M9) towards but not up to the outer end of the A-band. More details of this arrangement are given in the next section where the structure of the bridge region is considered in detail.

7.3.4. General Structure of the Bridge Region

Early studies of longitudinal sections of skeletal muscle showed that the A-band outside the M-region is crossed by striations about 400 Å apart (H. E. Huxley, 1957; Draper and Hodge, 1949), but few details of the structure of the bridge region were obtained. More recently much improved micrographs of plastic sections, isolated A-bands, isolated myofibrils, and cryosections have provided a wealth of new detail on A-band structure. The new details of the M-region have already been described. In this section the ultrastructure of the remainder of the A-band will be considered.

The first significant details of A-band structure were obtained by O'Brien *et al.* (1971) and Hanson *et al.* (1971) who studied electron micrographs of negatively stained A-segments. Figure 7.42a is from a similar preparation to theirs. They described the A-segment structure on each side of the M-region as comprising ten bands, each about 420 Å wide (probably corresponding to the 429-Å repeat *in vivo*), extending from the edge of the M-region to a position about 0.3 μm from the outer edge of the A-band. Few details of the A-band structure could be seen outside these ten bands on each side of the M-region. However, each of these bands showed a clear pattern of weaker striations, and it could be seen that the weak striation patterns in the three bands nearest to the M-region (bands 1 to 3) were different from each other and from the seven remaining bands. The outer seven bands (4 to 10) showed very similar striation patterns. Beyond band 10, little structural detail could be seen.

Craig and Offer (1976a) studied the bridge region by labeling sections of glycerinated rabbit psoas muscle with antibodies to rabbit subfragment-1. The kind of result they obtained is illustrated in Fig. 7.45. It was found that the whole of the A-band was densely labeled except in the M-region (about 1500 Å long) and at the outer edges of the A-band. Here a single unlabeled line could be seen in each half of the A-band about 200 to 300 Å away from the filament tip. Evidence has already been given that the myosin cross bridges are located at 143 Å axial intervals along the bridge regions of the A-band. This 143-Å cross bridge repeat could not be seen in the anti-S-1-labeled preparations of Craig and Offer (1976a), but from the position of the unlabeled stripes at the A-band tips, it was concluded that the filament probably terminates with two rows of myosin cross bridges beyond the gap. The gap itself was thought to correspond to a single missing row of cross bridges. Between these gaps and the edges of the M-region, it was concluded, the array of cross bridges must be complete (i.e., with no further gaps). But since the 143-Å cross bridge periodicity could not be seen, this conclusion must be treated with caution even though it is the most obvious one.

Many further details of A-band structure have recently been obtained. Sjöström and Squire (1977a,b) have analyzed the structure of the A-band as seen in negatively stained cryosections of muscle, and Craig (1977) has followed up the work of Hanson *et al.* (1971) on negatively stained A-segments. The results so obtained, although from different muscles, are complementary and very revealing about thick filament structure. For convenience, the analysis by Sjöström and Squire will be described first.

At high magnification, the A-band in negatively stained cryosections of human M. tibialis anterior can be seen to be crossed by many closely spaced striations (Fig. 7.46). The measured spacing between these striations is about 140 to 150 Å and is clearly associated with the myosin cross bridge repeat known from X-ray diffraction work to be 143 Å. But these striations do not lie on an exactly regular 143-Å periodicity, and they have characteristically different densities. In particular, in the middle portion of each bridge region, every third striation is prominent, thus

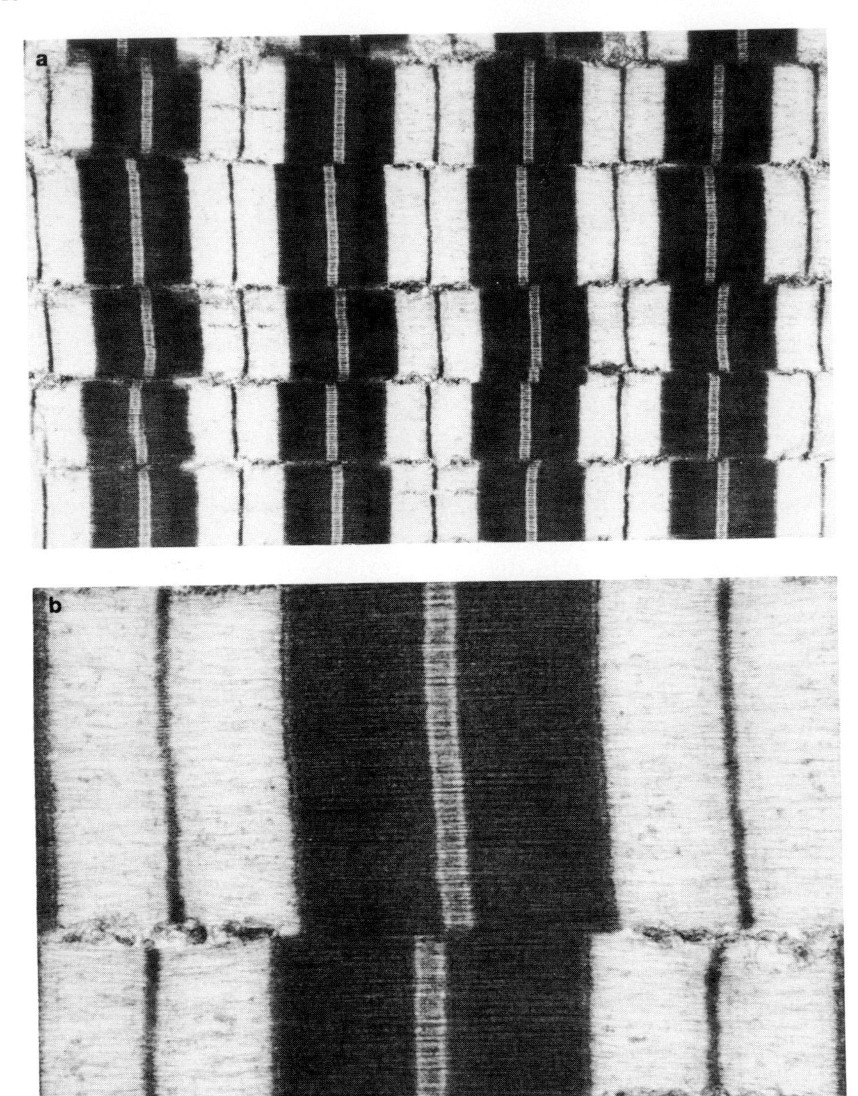

FIGURE 7.45. Glycerinated rabbit psoas muscle labeled with antibody to rabbit HMM S-1 (see Chapter 6). Dense staining is seen everywhere in the A-band except in the M-region and at single lines at each end of an A-band. Magnifications: (a) ×11,200; (b) ×33,000 (reduced 29% for reproduction). (From Craig and Offer, 1976a.)

giving rise to a repeat of about 3 × 140 to 3 × 150 Å. The region where this occurs is just that region in the A-bands of rabbit psoas muscle where C protein is known to be located.

For ease of description, the bridge region can be thought of as

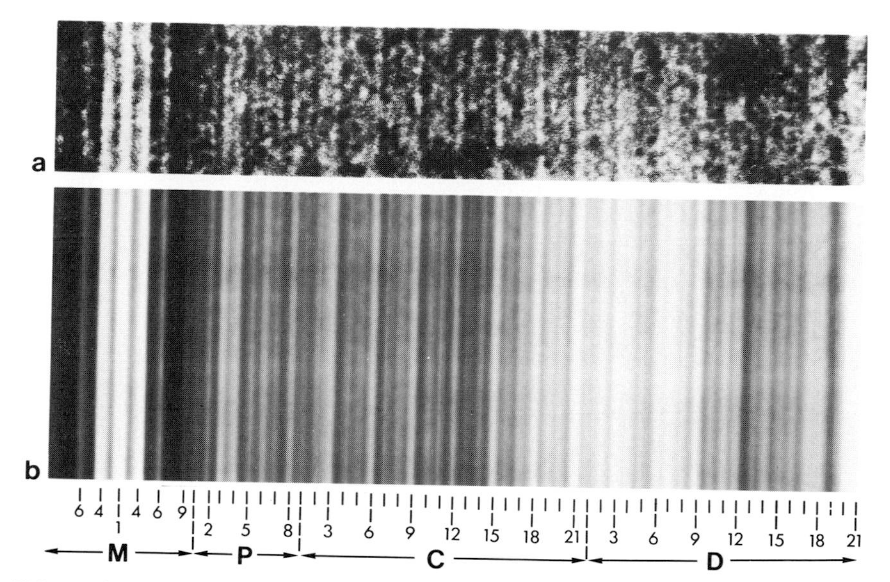

FIGURE 7.46. Part of a negatively stained cryosection of human M. tibialis anterior (a) showing details of the striation pattern in one half of the A-band together with a laterally averaged image of the same micrograph (b) and an indication of the numbering of the bridge region striations used in the text. Magnification ×230,000 (reduced 45% for reproduction).

comprising 17 sections about 430 Å wide. Each section comprises three unstained lines about 140 to 150 Å apart. Sections 1 to 3, which lie just outside the M-region (Figs. 7.42d and 7.46), are termed the proximal zone (or P-zone). Sections 4 to 10, which correspond to the region where C protein is located, are termed the C-zone. Sections 11 to 17 are termed the distal zone (or D-zone). Sections 1 to 10 correspond approximately to the ten bands in the A-segments described by Hanson *et al.* (1971). As in these A-segments, sections 1 to 3 in the cryosections show striation patterns that are different from each other and from the striation patterns in sections 4 to 10 of the C-zone. Within the C-zone the striation patterns in the seven sections are rather similar, but slight variations in pattern do occur, and these will be considered later. Sections 11 to 17 in the D-zone are also very different from each other and from the patterns in the C-zone. However, sections 15 to 17 share some common features with the P-zone sections.

Table 7.8 from Sjöström and Squire (1977a) summarizes the appearances of the many bridge region striations in terms of their estimated densities and their visual grouping. Each individual striation has been numbered. In the P-zone, the first section comprises the terminal M-region line M9 and two 140- to 150-Å-spaced lines P1 and P2. Sec-

TABLE 7.8. Intensity and Grouping of the Striations in the Bridge Region of Negatively Stained Cryosections of Human M. Tibialis Anterior According to Sjostrom and Squire (1977a)

				Possible origin	
Zone and section	Line number		Relative intensity and grouping[a]	Myosin contribution[b]	Extra protein positions
P-zone	1	(M9)	m.s.[c]	Rod	Yes
		P1	s.	Head + Rod	
		P2	s.	Rod	Yes[d]
	2	P3	(v.)v.s.	Rod	Probably
		P4	(v.)v.s.	Head + Rod	
		P5	s.	Rod	Yes
	3	P6	m.s.	Head + Rod	
		P7	m.s.	Head + Rod	
		P8	v.s.	Rod	Yes[d]
C-zone	4	C1	m.s.	Head + Rod	
		C2	m.s.	Head + Rod	
		C3	v.s.	Head + Rod	C protein
	5	C4	m.s.	Head + Rod	
		C5	m.s.	Head + Rod	
		C6	v.s.	Head + Rod	C protein
	6	C7	m.s.	Head + Rod	
		C8	m.s.	Head + Rod	
		C9	v.s.	Head + Rod	C protein
	7	C10	m.s.	Head + Rod	
		C11	m.s.	Head + Rod	
		C12	v.s.	Head + Rod	C protein
	8	C13	m.s.	Head + Rod	
		C14	m.s.	Head + Rod	
		C15	v.s.	Head + Rod	C protein
	9	C16	m.s.	Head + Rod	
		C17	m.s.	Head + Rod	
		C18	v.s.	Head + Rod	C protein

(*continued*)

tions 2 and 3 comprise lines P3 to P5 and P6 to P8, respectively. The seven C-zone sections comprise lines C1 to C21, and the seven D-zone sections comprise lines D1 to D21.

It seems to be very likely that the prominence of the seven distal striations in the C-zone (i.e., C3, C6, etc.) is caused by the presence of C-protein molecules in addition to the underlying 143-Å myosin cross bridge repeat. In a similar way, lines P8 and P5, which are also relatively strong, may represent the locations of the two other "extra" protein stripes identified by Craig and Offer (1976a) and by Pepe and Drucker (1975). In particular, line P5 may be caused by the H protein component

TABLE 7.8 *Continued*

Zone and section	Line number		Relative intensity and grouping[a]	Possible origin	
				Myosin contribution[b]	Extra protein positions
	10	C19	m.s.	Head + Rod	
		C20	m.s.	Head + Rod	
		C21	(v.)v.s.	Head + Rod	C protein
D-zone	11	D1	m.s.	Head + Rod	
		D2	m.s.	Head + Rod	
		D3	v.s.	Head + Rod	Possibly
	12	D4	m.s.	Head + Rod	
		D5	m.s.	Head + Rod	
		D6	v.s.	Head + Rod	Possibly
	13	D7	m.s.	Head + Rod	
		D8	m.s.	Head + Rod	
		D9	s.	Head + Rod	Possibly
	14	D10	m.s.	Head + Rod	
		D11	m.s.	Head + Rod	
		D12	s.	Head + Rod	Possibly
	15	D13	m.s.	Head + Rod	
		D14	m.s.	Rod	
		D15	m.s.	Head + Rod	
	16	D16	m.s.	Head + Rod	
		D17	m.s.	Rod	
		D18	m.s.	Head + Rod	
	17	D19	—	—	
		D20	v.s.	Rod	
		D21	v.s.	Head (+ Rod)	

[a] The grouping mentioned here is the visual grouping of lines which is caused by a combination of stain distribution and line separation.
[b] This refers to the interpretation by Squire (1973). In the model suggested by Pepe (1967a,1971), all of lines P1 to D21 have myosin heads on them including D19.
[c] For definition of abbreviations see legend to Table 7.7.
[d] See text.

in Fig. 7.33 (Offer, personal communication). This accounts for nine of the 11 stripes of nonmyosin protein known to be present in each half of the A-band. The other two stripes on the inner end of the bridge region would be located close to lines P2 or P3 and M9.

Apart from these known locations of extra proteins, lines P3 and P4 are exceptionally strong; indeed, they are often the most prominent unstained lines in the whole A-band. Also, the distal striations in sections 11 to 14 (i.e., D3, D6, etc.) are slightly more prominent than their neighbors. This could either mean that further "extra" proteins (as yet unidentified) label the A-band in these positions (especially on P3 and P4) or, alternatively, that the myosin packing in the thick filaments has an intrinsic 430-Å axial periodicity. However, even in this latter case, the prominence of the lines would be hard to explain.

Apart from these characteristic density differences between the striations of the bridge region, there are also differences in the way that successive striations appear to be grouped. This is especially clear in the laterally averaged image of a cryosectioned A-band shown in Fig. 7.46b. It is clear that in the P-zone the striations appear to be grouped in pairs. Similarly, in sections 4 to 6 the striations are grouped in threes, but elsewhere in the C-zone the grouping is more complex. In sections 11 to 14 of the D-zone, the grouping is once more in threes, and in sections 15 to 17, the striations are grouped in pairs as in the P-zone. However, in section 17, the line D19 is always absent, a result consistent with the results from the anti-S-1-labeled muscle studied by Craig and Offer (1976a) and described earlier (Fig. 7.45). By means of optical diffraction (Fig. 7.47), Sjöström and Squire (1977a) showed that the axial distribution of density in the cryosectioned A-band must be very similar to the distribution of density in intact living muscle. Comparison of the optical diffraction patterns with the meridional X-ray diffraction data obtained by H. E. Huxley and Brown (1967) from frog sartorius muscle (Table 7.9) showed that most of the observed meridional X-ray reflections had their counterparts in the optical diffraction patterns. This indicated both that the preservation of structure in the cryosections must be quite good and that many of the "forbidden" meridional reflections in the X-ray diffraction pattern can be attributed solely to the axial distribution of density in the A-band. It also suggested that the A-band structures in frog sartorius muscle and human M. tibialis anterior must be rather similar (but not necessarily identical).

The results of Craig (1977) on A-segment structure are largely consistent with those of Sjöström and Squire (1977a), but he makes some additional points that are worth considering. First, A-segments negatively stained with 1% uranyl acetate (Fig. 7.48) showed 11 bands about 430 Å wide starting at the edges of the M-regions and extending towards the outer edges of the A-segment. The proximal end of each band was defined by a strong unstained line thought to be associated with the extra proteins described earlier, including C protein. Craig noted two points about these A-segments. He said that no change in band pattern could be seen through bands 4 to 11 where C protein is located and that within each band, the subpattern of striations thought to be caused by the packing of the myosin rod into the filament backbone was not such that a subrepeat of ⅓ × 430 Å (i.e., 143 Å) could be seen. Respectively he took these to mean that (1) the C protein and myosin periodicities are

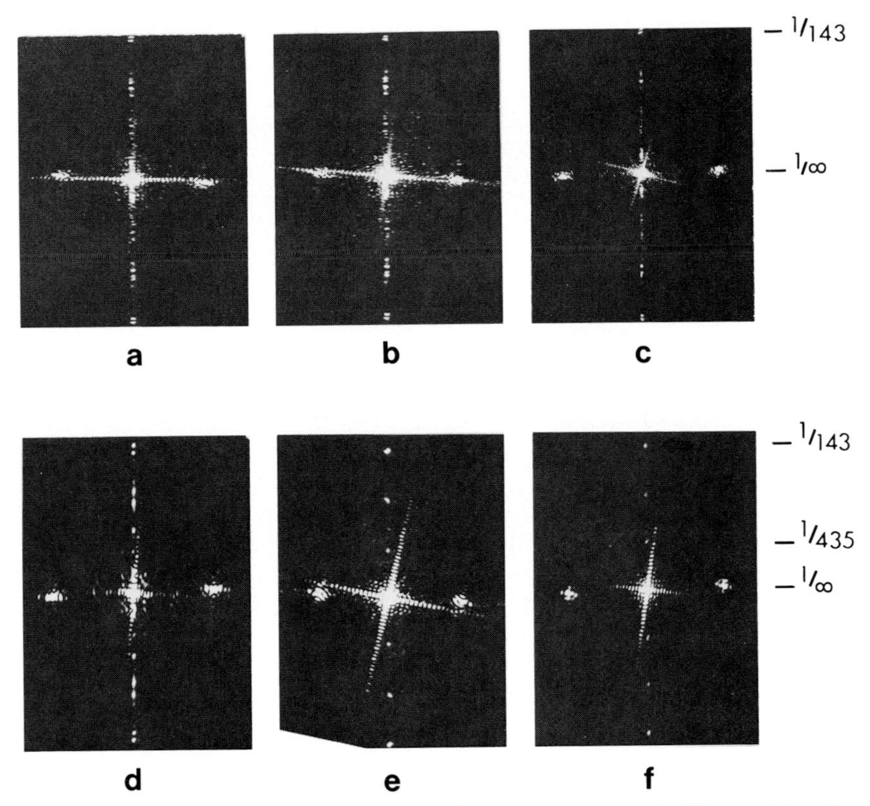

FIGURE 7.47. Optical diffraction patterns from various parts of different A-bands as visualized in negatively stained cryosections of human M. tibialis anterior. (From Sjöström and Squire, 1977a). Patterns in (a) and (b) are from whole A-bands (fiber axis top to bottom of page), (c) is from one-half of an A-band, (d) is from the M-region, and (e) and (f) are from different C-zones. Approximate spacings calibrated from the 143-Å reflection around in (f) are given in $Å^{-1}$. For accurate spacings, see Table 7.9.

the same and (2) that the true myosin periodicity is not 143 Å but is 430 Å. Conclusion (1) conflicts with the conclusions of Sjöström and Squire (1977a); the problem will be considered in more detail in the next section. Point (2) could be right, but it could surely mean that the C-protein stripes do not coincide axially with the 143-Å myosin rod repeat. A "thirding" of the C-protein repeat would not then be expected. However, it is clear, despite these comments, that it has not yet been decided whether the myosin repeat is 143 Å or 430 Å.

Craig's A-segments, when negatively stained (without prior fixation) with ammonium molybdate (Fig. 7.48b) showed an appearance very similar to that of the cryosections shown in Fig. 7.46 (Sjöström and Squire, 1977a) in which 143-Å-spaced striations could clearly be seen.

TABLE 7.9. Optical Diffraction Spacings in Patterns from Human M. Tibialis Anterior A-Bands in Negatively Stained Cryosections[a]

Meridional X-ray reflections[b]	Optical diffraction patterns[c]		
	Whole A-band	M-region plus P-zones	C-zone plus D-zone
		(110)	
139	139·3 (±0·4)	138 (±2)	
143·2	143	143	143
183·5	(186)	183 (±3)	
191·5	198·9 (±1·6)	198 (±3·5)	
—	205·9 (±1·2)		
214	218·2 (±0·5)		216·25 (±1·42)
222		223 (±2)	
231·5	228·3 (±0·8)		
237·5	239·9 (±2)		
—	(310)		
—	327·9 (±2·3)	329 (±5)	
383	373·2 (±4·6)		(370)
395·5			
416	(414)		
441	434 (±3)		438 (±7)
495	475 (±10)		
528	(510)		
870	(840)		
1028	(1030)		
1228	(1190)		

[a] From Sjöström and Squire (1977a).
[b] From H. E. Huxley and Brown (1967).
[c] Calculated standard deviations are given in parentheses after each figure, except where the figures themselves are in parentheses. The latter indicates that the reflections were either very weak or were not consistently seen. The reflection at 143 Å was used to calibrate the optical diffraction patterns.

Pairing of these lines in the P-zone, grouping in threes in the C-zone, and further pairing at the end of the D-zone could be seen in segments from both frog and rabbit muscle. This agrees with the results of Sjöström and Squire summarized in Table 7.8 and supports their view that the presence of thin filaments in their sections was not contributing greatly to the observed striation pattern.

7.3.5. Analysis of the C-Protein Periodicity

Introduction. It was shown in the last section that most of the striations visible in the bridge region either in cryosections of the A-band or in A-segments can be accounted for in terms of the locations of myosin cross bridges and a number of additional proteins such as C protein. But the true axial repeat of the myosin molecules could not be

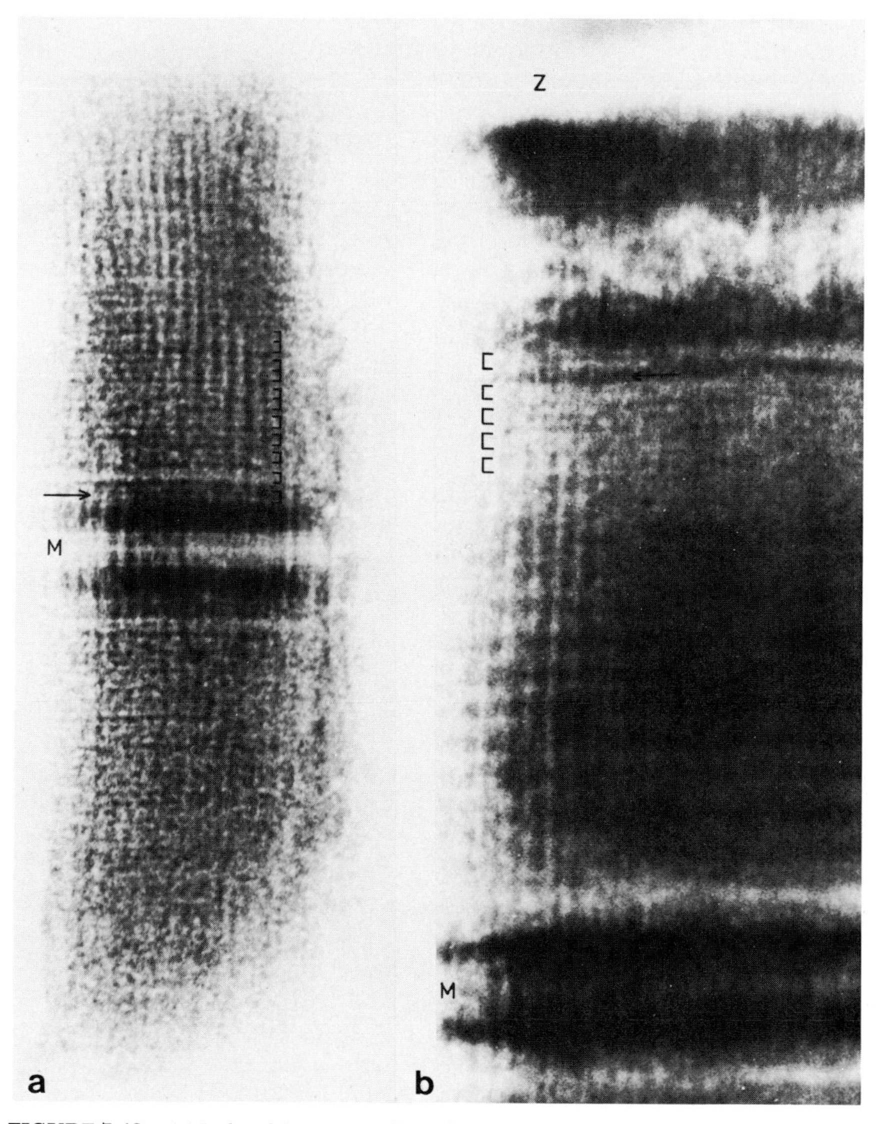

FIGURE 7.48. (a) Isolated A-segment from frog thigh muscle and (b) rabbit psoas muscle myofibril, both negatively stained with ammonium molybdate without prior fixation. The appearances are rather similar to negatively stained cryosectioned A-bands (e.g., Fig. 7.13 and Fig. 7.42d) and show a clear 143-Å periodicity. Magnifications: (a) ×84,000; (b) ×112,000 (reduced 22% for reproduction). (From Craig, 1977.)

defined. In most models for the structure of the thick filaments, it is commonly assumed that in the bulk of the bridge region the myosin cross bridges are arranged on a helix with a regular repeat of 143 Å or possibly of 429 Å.

It has also been claimed by many researchers (such as Craig, 1977) that the axial separation between the C-protein stripes in the bridge region is exactly 3 × 143 Å. This has been indicated by optical diffraction studies of antibody-labeled muscle, by studying paracrystals of LMM labelled with C protein, and by studying A-segments (Offer, 1972; Moos, 1972; Craig, 1975; Moos *et al.*, 1975; Pepe *et al.*, 1975; Craig and Offer, 1976b; Craig, 1977). If, in the C-zone, the myosin cross bridges have a regular 143-Å or 429-Å repeat and the C-protein stripes have a repeat of exactly 429 Å, it would be expected that every section in the C-zone would show an identical striation pattern.

But in cryosections this usually is clearly not the case; the striation patterns do change through the C-zone. This can be seen especially clearly in the laterally averaged image shown in Fig. 7.46b. Since in cryosections there are always thin filaments overlapping the A-band, it could be argued that in some way these thin filaments are perturbing the arrangement of the myosin cross bridges in such a way that the equivalence of the C-zone sections is lost. However, similar C-zone appearances can be seen in cryosections of muscles fixed at different sarcomere lengths (Sjöström and Squire, 1977a and unpublished results). It therefore seems very unlikely that it is actin which is causing the systematic variation in pattern through the C-zone.

Two alternative possibilities remain: one is that the packing of the myosin molecules might be changing right through the C-zone, and the other is that the C-protein and myosin periodicities may, in fact, be slightly different. With these possibilities in mind, Sjöström and Squire (1977a), and Squire *et al.* (1976, 1981) analyzed in detail the periodicities present in the C-zones of cryosections such as that shown in Fig. 7.46. They used a variety of methods as detailed below.

Optical Diffraction Analysis. Optical diffraction patterns from the C-zone were of the form shown in Fig. 7.47. Three prominent axial reflections were commonly observed, and these corresponded to *d* spacings of 438 ± 7 $Å^{-1}$, 216.3 ± 1.5 $Å^{-1}$, and 143 $Å^{-1}$. The latter reflection was used to calibrate the diffraction pattern, and so all of the error is included in the standard deviations quoted for the other two reflections. Occasionally a weak reflection at a spacing of about 370 Å was also observed. This was very likely caused by the thin filaments.

If the myosin and C-protein repeats are identical, then it would be expected that the first meridional reflection would have a spacing (429 Å) exactly equal to three times the spacing of the 143-Å reflection. In fact, the first meridional reflection has a spacing of 438 ± 7 Å which is significantly greater than 3 × 143 Å. Since this reflection must be primarily caused by the spacing of the prominant C-zone striations (C3, C6, etc.) which are thought to be the locations of C-protein molecules, it is not unreasonable to conclude that the C-protein stripes may have a separation that is slightly greater than 429 Å, say about 435 Å.

Direct Analysis of Electron Micrographs. If the C-zone really comprises C-protein stripes spaced 435 Å apart and superimposed on an underlying 143-Å myosin cross bridge repeat, then the appearance of the C-zone striations would be of the form shown in Fig. 7.49. It will be

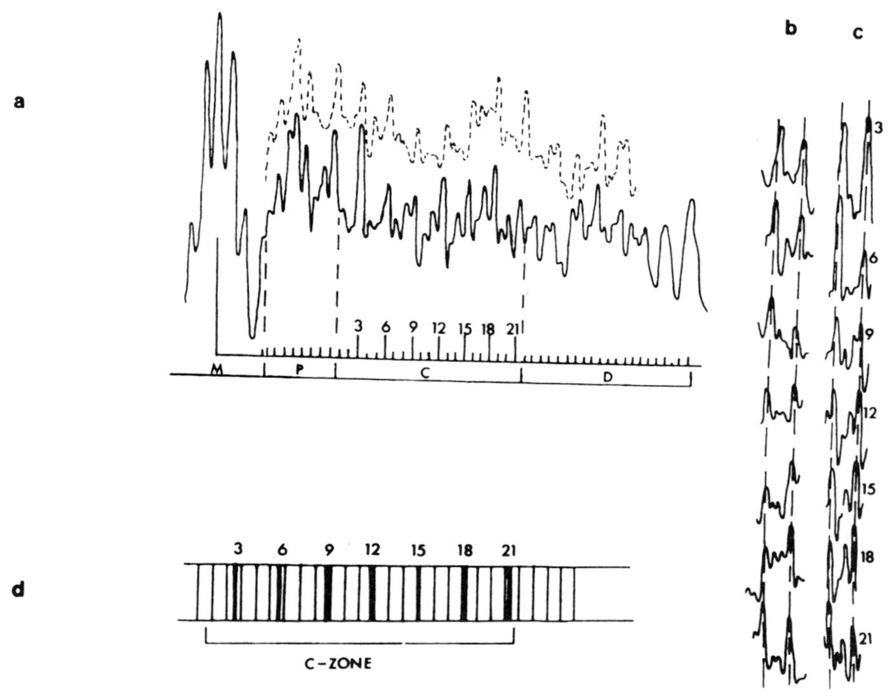

FIGURE 7.49. Microdensitometer trace through the micrograph shown as Fig. 7.42d. The lower trace (—) is from the left of Fig. 7.42d, and the upper trace (.....) is from the right. The striation pattern in the micrograph is much clearer on the left than the right as confirmed in the traces. The traces in (b) and (c) from different C-zone sections are respectively from the upper and lower traces in (a). In this latter series of traces (i.e., from the clearer half of Fig. 7.42d, a systematic change of pattern in successive C-zone sections can be seen. This is present but not so clear in (b). (d) A possible scheme that might explain in part the change in the C-zone striation pattern is the superposition of a 440-Å C-protein repeat (**—**) on top of a cross bridge repeat of 143 Å (—) as illustrated here. It would give rise to an apparent systematic shift of the few weak lines in each section through the C-zone. No attempt has been made to reproduce exactly the type of profile seen in (a). (From Sjöström and Squire, 1977a.)

seen that as required every third striation is relatively strong, and the difference in periodicity is manifested as a slight systematic shift of the two weak cross bridge lines in successive sections through the C-zone. Close inspection of the original electron micrographs and especially of the laterally averaged image shown in Fig. 7.46b reveals that just such an appearance can be followed through the C-zone. For example, in section 4 the space between lines D8 and C1 is relatively greater than that between C2 and C3. Similarly, C3 and C4 are more widely separated than C5 and C6. This gives rise to the grouping of the lines in sections 4, 5, and 6 into threes. However, at the other end of the C-zone in section 10, lines C20 and C21 are slightly more widely separated than lines C18 and C19, and in sections 7, 8, and 9 the two cross bridge lines are almost centrally placed so that the space between C9 and C10, C11 and C12, C12 and C13, C14 and C15, etc. are approximately equal.

Microdensitometry. Movement of the two weak lines within the C-zone relative to their neighboring prominent lines was investigated by making direct measurements both on average images of the kind shown in Fig. 7.46 and on microdensitometer traces of the optical density across contrasting negatives of the C-zones in micrographs (or their average images) (Fig. 7.49). To obtain the maximum accuracy, the measurements were made on the strong striations by measuring, for example, from C3 to C12, C6 to C15, and C9 to C18 (or C21), so that each line was only used once. These measurements were then divided and averaged by 3 (or 4) to give the required C-protein spacing (l_c). Measurements on lines C1, C4, C7, C10, C13, C16, and C19 were made in a similar way, as were measurements between lines C2, C5, C8, C11, C14, C17, and C20. The results from these two sets of weak lines were then averaged to give a single value for the myosin repeat (l_m). Results from seven C-zones gave a mean value for the ratio l_c/l_m of 1.014 (±0.006). On this basis if the myosin cross bridge repeat is 429 Å, then the C-protein repeat would be 435 (±3) Å which is in good agreement with the optical diffraction evidence that gives a spacing of 438 ± 7 Å.

X-Ray Diffraction Evidence. One of the most prominent of the "forbidden" meridional reflections in the X-ray diffraction patterns from intact muscle is one at a *d* spacing of 442 Å (H. E. Huxley and Brown, 1967). H. E. Huxley and Brown considered the possibility that this might be caused by a nonmyosin A-band protein that interacts with myosin in such a way as to control the length of the thick filaments. The two different repeats (429 Å and 442 Å) would be involved in a kind of vernier length-determining mechanism. In an elegant experiment performed by Rome *et al.* (1973a), glycerinated rabbit muscle was labeled with antibody to C protein and its X-ray diffraction pattern recorded. It was then found that the intensity of the 442-Å meridional reflection was considerably enhanced relative to that from unlabeled muscle, a clear indication that this reflection is at least partly caused by C protein. In addition, a very slight shift in spacing was noted from 442 ± 2 Å to 444 ± 2 Å. A second reflection at a measured spacing of 418 ± 2 Å was also recorded. This could only be seen under optimum conditions in diffraction patterns from unlabeled muscle.

Rome *et al.* (1973a) believed that the C-protein and myosin periodicities were identical and explained their results in terms of a concept originally suggested by E. J. O'Brien. They pointed out that in each A-band the two C-zones would be at a specific separation and, therefore, that the diffraction patterns from the two C-zones would interfere with each other, the form of the interference function being related to the separation of the centers of the two C-zones.

The system can be thought of as a set of seven C-protein lines convoluted with a function consisting of two points separated by the same separation as the centers of the C-zones (Fig. 7.50). From the convolution theorem (Chapter 2), the total diffraction pattern will then consist of the diffraction pattern of one C-zone multiplied by the diffraction pattern of the array of two points. The latter is just a simple cosine function. The result has the form shown in Fig. 7.51. Here it has been assumed that the C-protein and myosin repeats are both 429 Å. The measured separation of the two C-zones is approximately 7100 Å, and the 16th and 17th orders of this repeat would have spacings close to 1/442 $Å^{-1}$ and 1/418 $Å^{-1}$ as required. It can be seen in Fig. 7.51c that these two peaks are almost symmetrically placed around the center of the C-protein peak at 1/430 $Å^{-1}$. The results is therefore to produce two peaks of almost equal intensity at 1/442 $Å^{-1}$ and 1/418 $Å^{-1}$. But in the observed diffraction patterns, the 442-Å reflection is very much stronger than the reflection at 418 Å (ratio about 3 : 1: Haselgrove quoted in Rome *et al.,* 1973a). This discrepancy can be overcome to some extent if the C-protein peak that is being sampled has a spacing nearer to 1/442 $Å^{-1}$. Rome *et al.* considered the situation that would result with a C-protein repeat of 440

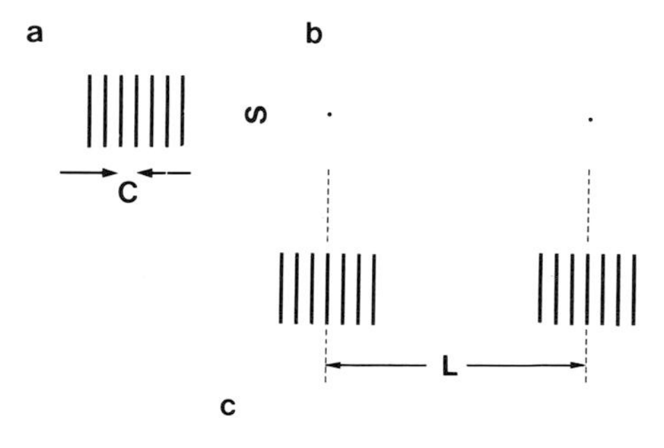

FIGURE 7.50. Illustration of the idea that the two C-zones in an A-band (c) can be thought of as the convolution of a seven-line array of C-protein molecules (spaced C apart) with two points (b) separated by the separation (L) of the centers of the C-zones.

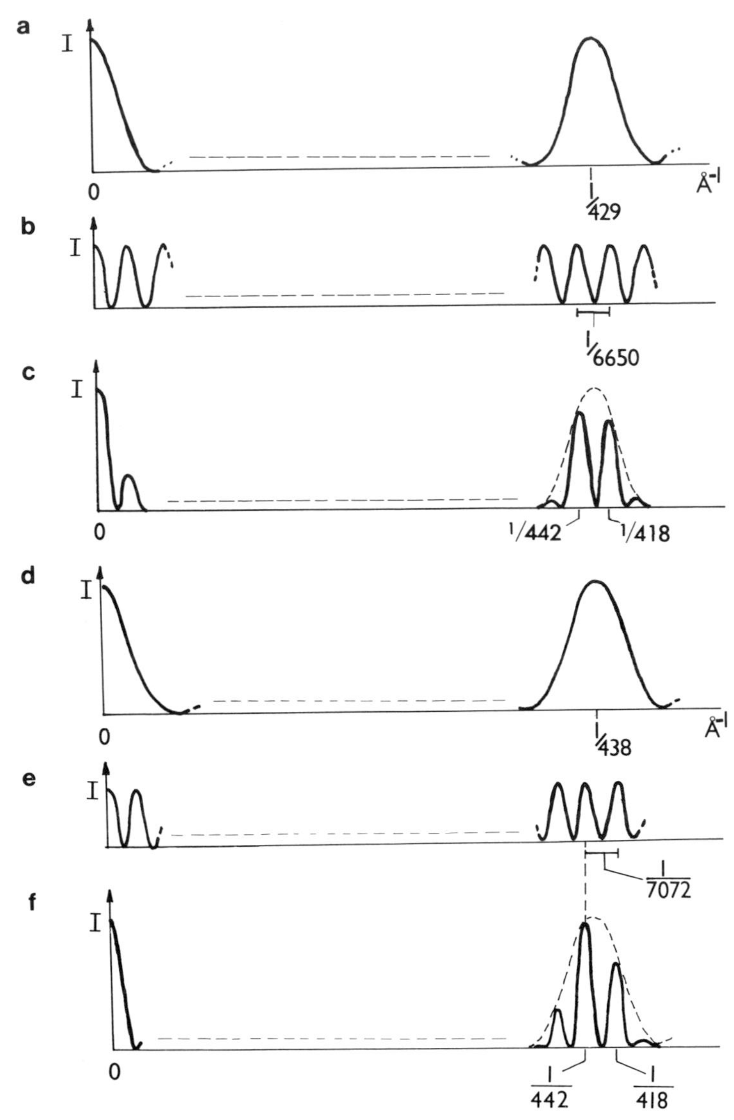

FIGURE 7.51. Explanation of the "forbidden" meridional reflection at 1/442 $Å^{-1}$ and 1/418 $Å^{-1}$ in X-ray diffraction patterns from frog sartorius muscle (see Table 7.1) in terms of diffraction from the C-zones in the A-band. (a) the expected diffraction pattern from a seven-line array of spacing 429 Å (*C* in Fig. 7.50). (b) The expected interference function for two points 6650 Å apart (*L* in Fig. 7.50). The total effect of two seven-line arrays centred *L* apart is here to produce lines of almost equal intensity at 1/442 $Å^{-1}$ and 1/418 $Å^{-1}$ as required. But the 418-Å peak is rather too strong. (d–f) A to similar set (a),(b), and (c) except that here *C* is 438 Å and *L* is 7072 Å. The result (f) is to give a reflection at 1/442 $Å^{-1}$ rather stronger than that at 1/418 $Å^{-1}$. A trace of reflection of about 1/472 $Å^{-1}$ is also seen, but this is not observed in X-ray diffraction patterns. (After Rome *et al.*, 1973a.)

Å. The effect of this is that the interference peaks at 1/442 Å^{-1} and 1/418 Å^{-1} are now asymmetrically placed in the peak because of the seven-line C-protein array. This results in an intensity ratio of about 1 to 0.43 between these two reflections. But, in addition, the 15th order reflection at 1/473 Å^{-1} has become stronger, and yet this has never been observed in the recorded X-ray diffraction patterns.

Setting the C-protein repeat at 1/435 Å^{-1} provides a compromise between these two situations. The reflection at 1/442 Å^{-1} would be stronger than that at 1/418 Å^{-1} (ratio 1 : 0.87), but the reflection at 1/473 Å^{-1} would still be very weak (0.23 on the same scale) and might well not be observed. The presence of a C-protein repeat different from that of myosin might therefore be consistent with, but not proved by, the observed X-ray diffraction evidence. This conclusion has been checked by optical diffraction analysis of a model of the double C-zone structure. The diffraction pattern from the model of a single C-zone shows a peak at about 1/435 Å^{-1}, and when both C-zones are included, this peak is split into two components of unequal intensity as required. However, there are several other features of the A-band such as the striations at P5, P8, D3, D6, D9, etc. which are also on a repeat of about 430 to 435 Å. The M-band too might contribute to this region of the diffraction pattern, and for this reason, the interpretation summarized above may well only be part, but not all, of the story.

Further Analysis of the C-Zone. As a final comment on the C-zone structure, it is instructive to compare the results obtained (1) from plastic sections and cryosections of M. tibialis anterior, (2) from plastic sections, cryosections, and A-segments of frog sartorius muscle, (3) from plastic sections of rabbit psoas muscle, and (4) from anti-C-labeled rabbit psoas muscle. Measurements show that with the possible exception of frog muscle, the prominent C-zone stripes in all of the preparations of unlabeled muscle lie in approximately the same positions. In fact, the first C-protein line, C3, is consistently at a measured separation from the center of the M-band of about 2300 to 2400 Å. But in the anti-C-labeled rabbit muscle, the first anti-C stripe seems to be located slightly further from the M-band. Its position is estimated to be about 2500 Å from the center of the M-band (see also Pepe and Drucker, 1975; Craig and Offer 1976b). If these stripes all represent features of the same C-protein molecules, then it is clear that, contrary to previous assertions, the axial extent of C-protein is limited (Craig and Offer, 1976b). The C-protein molecules each seem to extend for a distance of at least 200 Å along the axis of the thick filaments. The significance of this will be considered in Chapter 9 where the structure of myosin filaments is considered in detail.

Suffice it to say here that the existence of a C-protein repeat different from that of myosin would clearly make C-protein a strong contender for the role of the length-determining protein in the thick filaments.

7.4. Structure of the I-Band

7.4.1. General Description of the I-Band

Despite the detailed information now available on the structure of the component thin filaments, the structure of the I-band of vertebrate skeletal muscle is relatively poorly understood. The I-band, in fact, contains a number of proteins in addition to actin, tropomyosin, and troponin, but it is only recently that some of these proteins have been isolated and characterized.

For the purposes of this discussion, the I-band is taken to be the whole of the thin filament assembly from the edge of the H-zone in one sarcomere through the Z-band and out to the H-zone in the next sarcomere. This is illustrated in Fig. 7.52a. It has already been mentioned that in sections of muscle the I-band can be seen to be crossed by transverse striations originally estimated to be about 400 Å apart. These striations can also be seen in isolated negatively stained I-segments (O'Brien *et al.*, 1971). As described in Chapter 5, the striations are now attributed to the 385-Å repeat of the tropomyosin–troponin complex along the thin filament grooves, an idea that has been confirmed by antibody-labeling studies. Rome *et al.* (1973b) prepared antibody to troponin-C and then diffused this antibody into glycerinated rabbit psoas muscles (as was done for anti-C-protein (Rome *et al.*, 1973a). X-Ray diffraction patterns were then recorded from both labeled and unlabeled muscles. The un-

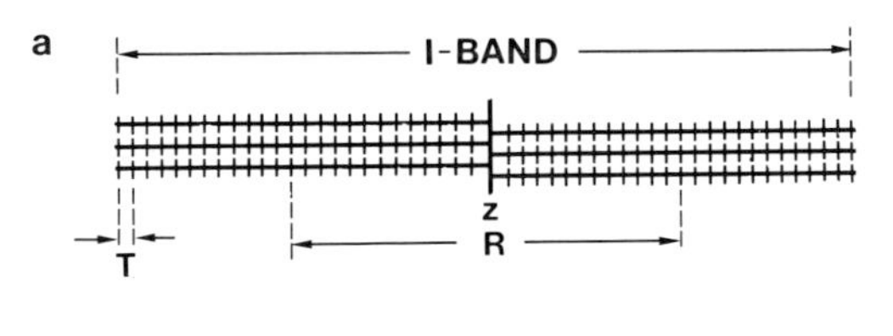

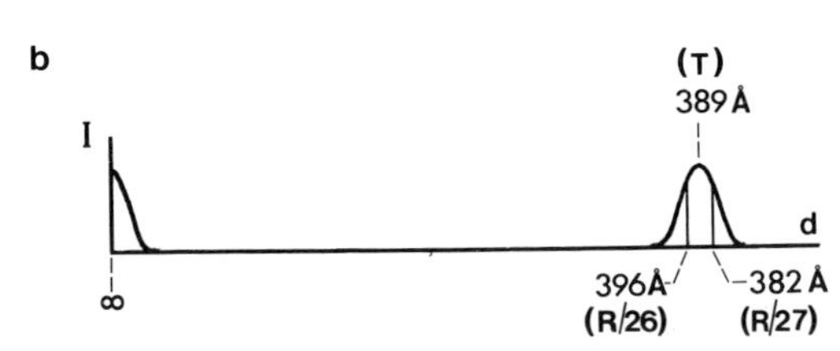

FIGURE 7.52. (a) Representation of the vertebrate I-band, like the C-zones, as two sets of lines of periodicity T centered a distance R apart. The lines would be caused by the tropomyosin–troponin assembly on the thin filaments (see Chapter 5). (b) The net result of taking T as 389 Å and R as 10,300 Å. Two axial reflections close to 1/396 Å^{-1} and 1/382 Å^{-1} are generated (see text and Table 7.1).

labeled muscle gave rise to a meridional X-ray reflection at a measured spacing of 380 ± 2 Å (cf. reflection at 385 ± 2 Å reported by H. E. Huxley and Brown, 1967, and attributed by them to a component of the thin filaments). Diffraction patterns from rabbit muscle labeled with anti-troponin C showed considerable enhancement of the intensity of this reflection (by a factor of three to five times), and the reflection had a measured spacing of 383 ± 3 Å. This clearly indicated that troponin was contributing to the reflection. Rome *et al.* noted that the measured spacing is almost exactly seven times the actin repeat of 54.6 Å (H. E. Huxley and Brown, 1967: 7 × 54.6 = 383.3) as would be expected if there is a functional unit in the thin filaments of the sort described in Chapter 5 (i.e., seven actins to one tropomyosin to one troponin).

However, with very-long-focus X-ray diffraction cameras, the 380- to 385-Å reflection can be resolved into two closely spaced reflections (Table 7.1; H. E. Huxley and Brown, 1967; Haselgrove, 1970, 1975a) at measured spacings of 394.9 Å and 383.0 Å. This doublet has been attributed, in a similar way to the 442-Å and 418-Å C-protein reflections, to an interference effect between the diffraction patterns from the thin filament arrays in each half of a single I-band. Since they have approximately equal intensity, the repeat (R) of the interference function must be equal to about $(n + \frac{1}{2})T$, where T is the troponin–tropomysin repeat (O'Brien *et al.*, 1971). Reflections at 383 Å and 395 Å would very likely be successive orders of the interference function of repeat R [i.e., $395 = R/N$, $383 = R/(N + 1)$]. It can easily be shown that the repeat distance R must then equal one of a series of values ranging from about 12640 Å ($N = 32$) down to about 10,300 Å ($N = 26$). If R is 12,640 Å, then orders 32 and 33 have spacings of almost exactly 395 Å and 383 Å. However, the values of R closer to 10,300 may be more appropriate for reasons given below. The 26th and 27th orders of this repeat are 396.2 Å and 381.5 Å, respectively. Although these values are not quite so close to the observed spacings as are the orders of 12,640 Å, their shifts to slightly shorter and slightly longer spacings respectively are attributable to the fact that these peaks would be sampling the sloping sides of the underlying peak corresponding to the troponin–tropomyosin repeat (T) (Fig. 7.52b). If this is the case, then this repeat T must be such that $T = 10{,}300/(n + \frac{1}{2})$. Since T must lie between 383 Å and 396 Å, it would have to be close to 388.7 Å ($n = 26$). This is slightly different from seven times the actin repeat. The latter is determined most accurately from the first meridional reflection from the actin helix at a measured spacing of 27.3 (±0.1) Å (H. E. Huxley and Brown, 1967). This is half of the spacing of the actin molecules along one of the long-pitch strands (Fig. 5.2). Groups of 14 of such actin molecules would therefore be separated axially by 14 × 27.3 (±0.1) Å = 382.2 (±1.4) Å. If the 383- and 395-Å reflections are really caused by I-band interference, then it must be concluded either that the actin repeat is closer to 2 × 27.76 Å or that the repeats of actin and of tropomyosin–troponin in the thin filaments are not, after all, exactly integrally related. The latter possibility might conceivably be associated with a thin filament length-determining mechanism, but consideration of such a possibility must await further evidence. Support for the view that the observed 383-Å and 395-Å reflections may be caused by I-band interference is obtained when it is realized that the midpoints of the two thin filament arrays in one I-band are at a separation R given by $R = L_I/2 + L_z/2$, where L_I is the total length of the I-band (about 1.9 to 2.0 μm) and L_z is the Z-band width (very approximately 1000 Å in frog and rabbit muscle). These figures yield a value of R of 10,000 to 10,500 Å, a figure that agrees well with the value 10,300 Å estimated from the spacings of the 395-Å and 383-Å reflections. Each half of the I-band would consist of about 25 periods of 388 Å (i.e., 9700 Å), and an array of this extent would give a diffraction peak extending from 1/404 $Å^{-1}$ to 1/373 $Å^{-1}$. This easily spans the interference lines at 1/395 $Å^{-1}$ and 1/383 $Å^{-1}$ as shown in Fig. 7.52b.

7.4.2. Structure of the Z-Band

In transverse sections through the overlap region of vertebrate skeletal muscle, the thin filaments can be seen to occupy the trigonal positions of the hexagonal lattice of myosin filaments. But outside the A-band, the thin filament lattice gradually changes from being hexagonal to being square as the Z-band is approached. Just to one side of the Z-band the thin filament array has a spacing of about 200 to 250 Å. But transverse sections through the Z-band itself have one of two characteristic appearances (Fig. 7.53; Knappeis and Carlsen, 1962; H. E. Huxley, 1963; Franzini-Armstrong and Porter, 1964; Reedy, 1964; Kelly, 1967, 1969; Kelly and Cahill, 1972). One appearance (Fig. 7.53b) is of a square lattice of exactly half the spacing of the actin array, and this appears to be formed by a crossing set of bridging structures (Fig. 7.54d). The other appearance (Fig. 7.53a) is a kind of "basket-weave" appearance based on the same size array as the actin array outside the Z-band but this time containing twice as many filaments (Fig. 7.54b). Slightly oblique transverse sections indicate that the basket-weave appearance is caused by two overlapping square arrays of filaments linked together by cross connections, each array being continuous with the thin filament arrays on one side only of the Z-band. The profiles of the filaments and their cross connections viewed from the two sides of the Z-band both show approximately fourfold rotational symmetry, but they have opposite senses as indicated in Fig. 7.55c. This is consistent with the thin filament polarity

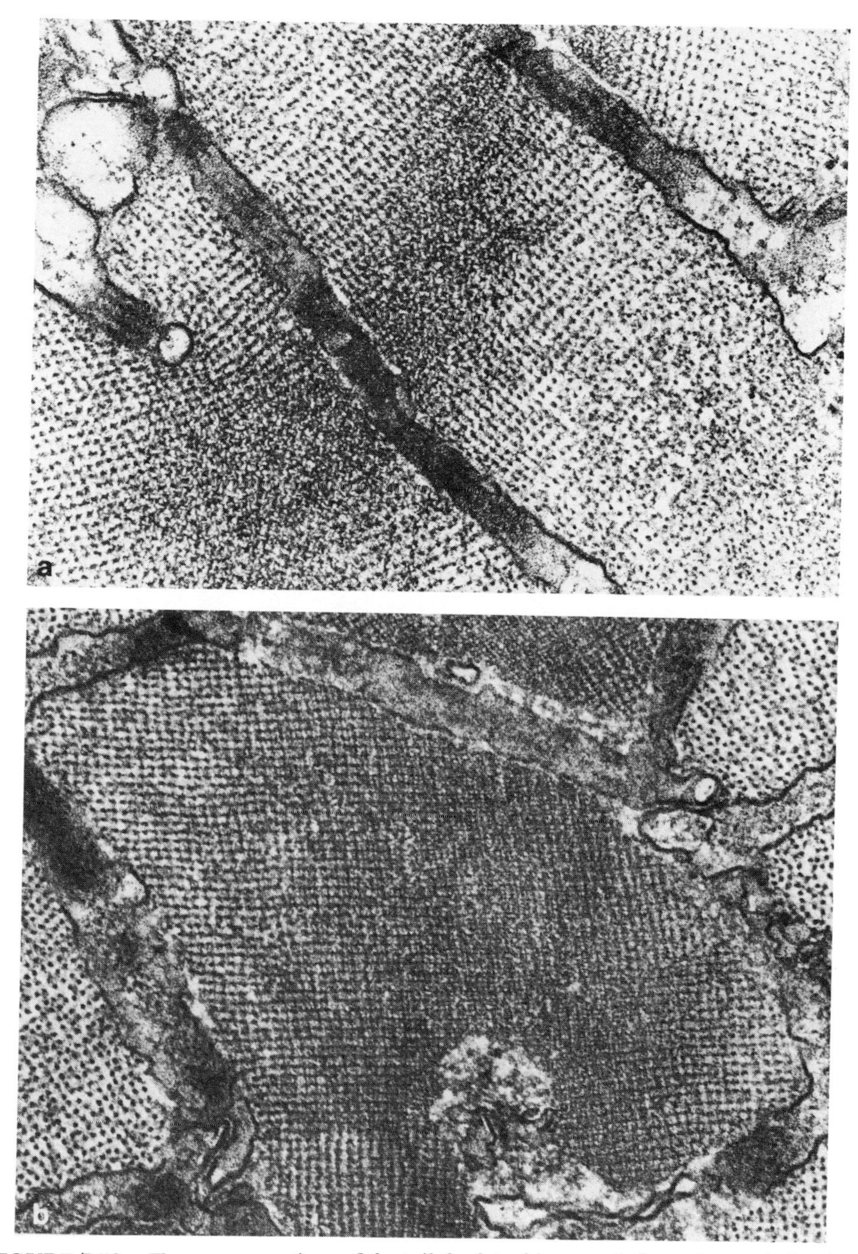

FIGURE 7.53. Transverse sections of the tail (body) white muscle from the roach (leuciscus) (a) and of the frog sartorius muscle (b), both fixed in glutaraldehyde in phosphate buffer and stained with uranyl acetate and lead citrate. The sections are both through the Z-band, and they show the characteristic Z-band appearances of (a) the basketweave and (b) the small and large square lattices. For details see text and Fig. 7.54. Magnifications: (a) ×120,000; (b) ×133,000 (reduced 38% for reproduction). (Courtesy of Dr. P. K. Luther.)

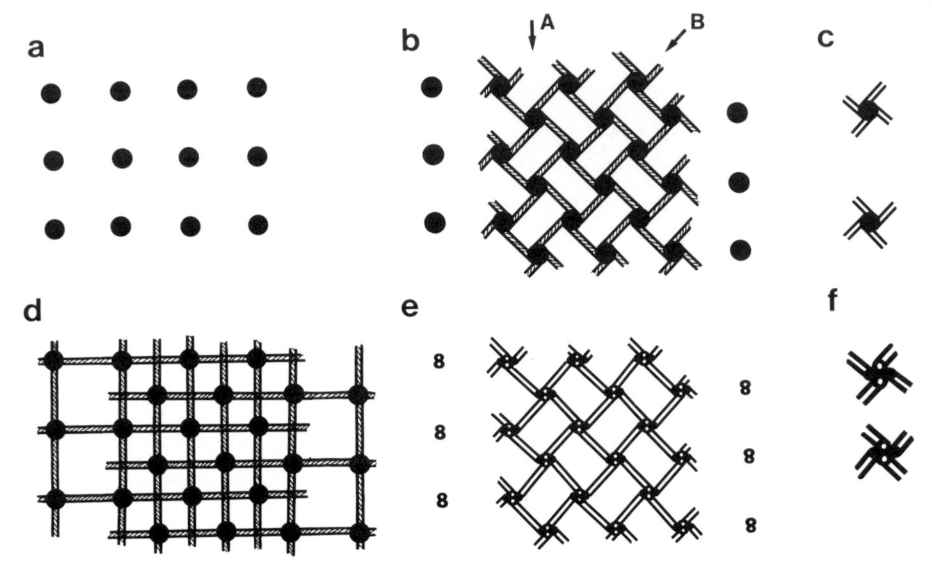

FIGURE 7.54. Representations of (a) the I-band lattice of thin filaments close to the Z-band and (b) the basketweave appearance generated by sets of Z-filaments (shaded) tangentially linked to thin filaments with alternating sense (polarity). These are the thin filaments from each side of the Z-band. The two thin filament appearances are illustrated in (c). (d) The large square lattices (left and right edges) and the small square lattice (center) commonly seen in certain Z-band sections (see text and Fig. 7.53b). (e) when treating actin filaments as structures with approximate twofold symmetry (represented by two touching open circles), then giving all of the thin filaments the same orientations only demands the use of one type of Z-filament interaction (i.e., open circle to filled circle) to generate the basketweave appearance. Other more complicated filament arrangements (e.g., one generated from the structure in Fig. 7.62b) would require several different types of interfilament interaction. (f) The two types of filament appearance used in (e).

being different on each side of the Z-band as shown by H. E. Huxley (1963).

The appearance of longitudinal sections of the Z-band depends to some extent on the particular muscle and fiber type that is being studied. In the past it has been shown that in rat plantaris muscle the Z-bands in white (Type II) fibers are thinner than those from intermediate fibers and much thinner than those in red (Type I) fibers (Schiaffino *et al.*, 1970; Rowe 1973). Similarly, the recent cryosection results on human M. tibialis anterior have shown that thick Z-bands are present in Type I fibers and thin Z-bands in Type II fibers (or a subgroup of this category). Thick and thin Z-bands have also been correlated with M-band structure in different fiber types from the rat (Sjöström *et al.*, 1980). Of course, the terms "thick" and "thin" are very vague, and Z-band structure clearly has to be quantified in some way. To see how this might be

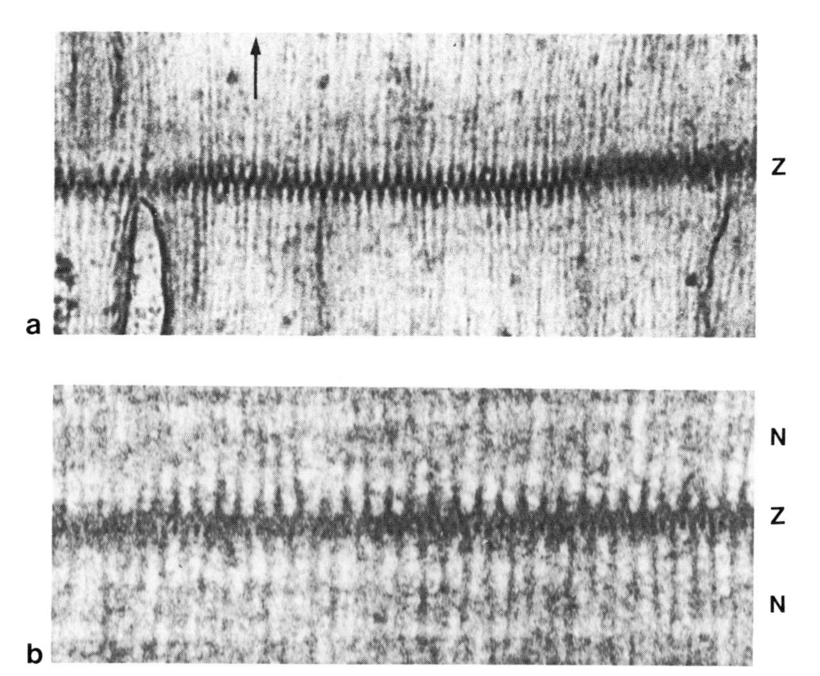

FIGURE 7.55. Two different Z-band appearances in longitudinal sections of roach (*Rutilus rutilus*) white fibres. The two appearances are probably two different views of the same Z-band structure (cf. Figs. 7.5b and 7.59). Micrographs courtesy of P. M. G. Munro. (a) ×133,000; (b) ×200,000 (reduced 30% for reproduction); Z, Z-band; N, N-lines (see text); muscle long axis vertical (arrow).

done, it is most convenient to start with the appearance of what appears to be the thinnest and simplest Z-band so far studied, that in fish muscle (Franzini-Armstrong, 1973). These Z-bands show only the simple basket-weave appearance in cross section. In longitudinal section, a number of appearances can be seen depending on the orientation of the myofibril in the section, but a characteristic appearance is one in which the Z-band consists of cross connections linking the two thin filament arrays and giving rise to a striking zigzag pattern (Fig. 7.55). This kind of Z-band appearance originally led Reedy (1964) to propose a model for the Z-band in which each thin filament was connected by four tangential links to the nearest four thin filaments in the opposite half of the I-band (Fig. 7.54b). Since these thin filaments have opposite polarity, such a system will automatically generate the basket-weave appearance in transverse section.

Rowe (1973) has shown details of the Z-bands in a number of other fibers. He used red, white, and intermediate fibers of rat plantaris muscle, and the Z-band micrographs he obtained showed clear zigzag profiles in longitudinal sections. These were somewhat similar to the fish Z-band but clearly contained more layers of zigzag structure. White fibers showed a double zigzag structure, intermediate fibers a triple zigzag, and the red fibers a quadruple zigzag as illustrated in Fig. 7.56(b–d). Adjacent zigzag layers were said, in all cases, to be separated axially by about the same distance as the tropomyosin–troponin repeat. Rowe at-

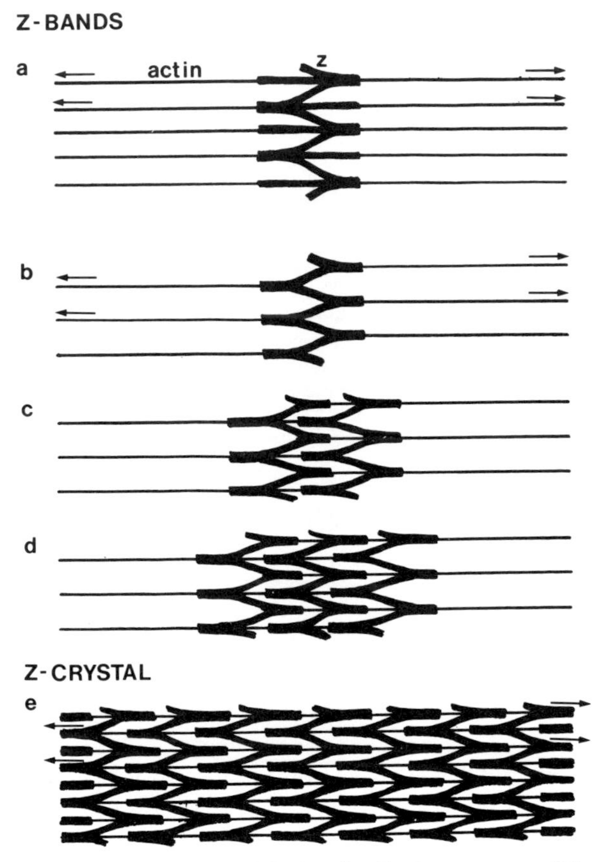

FIGURE 7.56. (a and b) Illustrations of the longitudinal appearance of the Z-band structure shown in Fig. 7.58 when viewed in different directions. (cf. Fig. 7.55) (a) The appearance of view B in Fig. 7.54b. (b) The appearance of view A (see also Katchburian *et al.*, (1973). (b–d) Corresponding views of Z-bands containing different numbers of layers of Z-filaments as seen in Fig. 7.55. Thus, (b) would correspond to the Z-band structure in fish muscle, (c) would correspond to rat plantaris white fibers, (d) would correspond to rat plantaris intermediate fibers, and red fibers would have four levels of Z-filaments (not shown). (e) The same structure extended into many layers; the result is remarkable similar to the longitudinal structure of the anomalous Z-bands in Figs. 7.57 and 7.58.

tempted to interpret his observations in terms of a very complex looping filament structure (see the original paper for details), but this model does not seem to account for the basket-weave Z-band appearance or for the structure of the simple fish Z-band. On the other hand, most of the present appearances can be accounted for by a much simpler model to be described below.

An interesting extension of Rowe's results can be seen in the micrographs of Stromer *et al.* (1976) who studied the structure of the abnormal Z-bands (nemaline rods) present in diseased muscle (Afifi *et al.*, 1965; Gonatas, 1966; Shafiq *et al.*, 1967; Shy *et al.*, 1963). Comparable structures called Z-crystals have also been observed in heart muscle (Goldstein *et al.*, 1974; Fawcett, 1968; Thornell, 1973). Longitudinal sections of these nemaline rods (Fig. 7.57) and Z-crystals (Fig. 7.58) show structures that are remarkably similar to those of the various Z-band appearances reported by Rowe (1973). Transverse sections also have similar appearances. Only preliminary analyses of these various structures have so far been published, but their obvious similarities make it tempting to conclude that they are structurally very closely related.

One possible general scheme that would account for most of these appearances is described briefly below. It is a logical extension of Reedy's Z-band model illustrated in Fig. 7.54b. Here each thin filament is linked to the four closest thin filaments in the next sarcomere by four bridging structures. Let us suppose that the four bridging structures have diameters comparable to that of one of the strands of actin monomers in the thin filament. These four strands would be packed together at the Z-band end of the thin filament and would then divide and link across to the appropriate four thin filaments in the next sarcomere. This type of structure (Fig. 7.59) would in itself be sufficient to explain the appearance of the fish Z-band (Fig. 7.56b). If it is then suggested that the bridging structures interact with a specific site in the 385-Å thin filament repeat, then similar bridging structures could link to actin 385 Å along the thin filament from the first set of bridges. Allowing the two abutting thin filament arrays in adjacent sarcomeres to overlap (Fig. 7.56c) would therefore permit a double Z-band bridge array to be formed as in the case of the white fibers of rat plantaris muscle studied by Rowe (1973). Increasing the thin filament overlap by further multiples of 385 Å would then produce the three-layer and four-layer structures seen by Rowe (1973) in intermediate and red fibers of rat plantaris muscle (Fig. 7.56c,d). Finally, a very large amount of thin filament overlap may explain many features of the nemaline bodies (Fig. 7.57) and the Z-crystals of heart muscle (Fig. 7.58) in a similar way (Fig. 7.56e). It is clear then that a relatively simple type of structure might explain most of the vari-

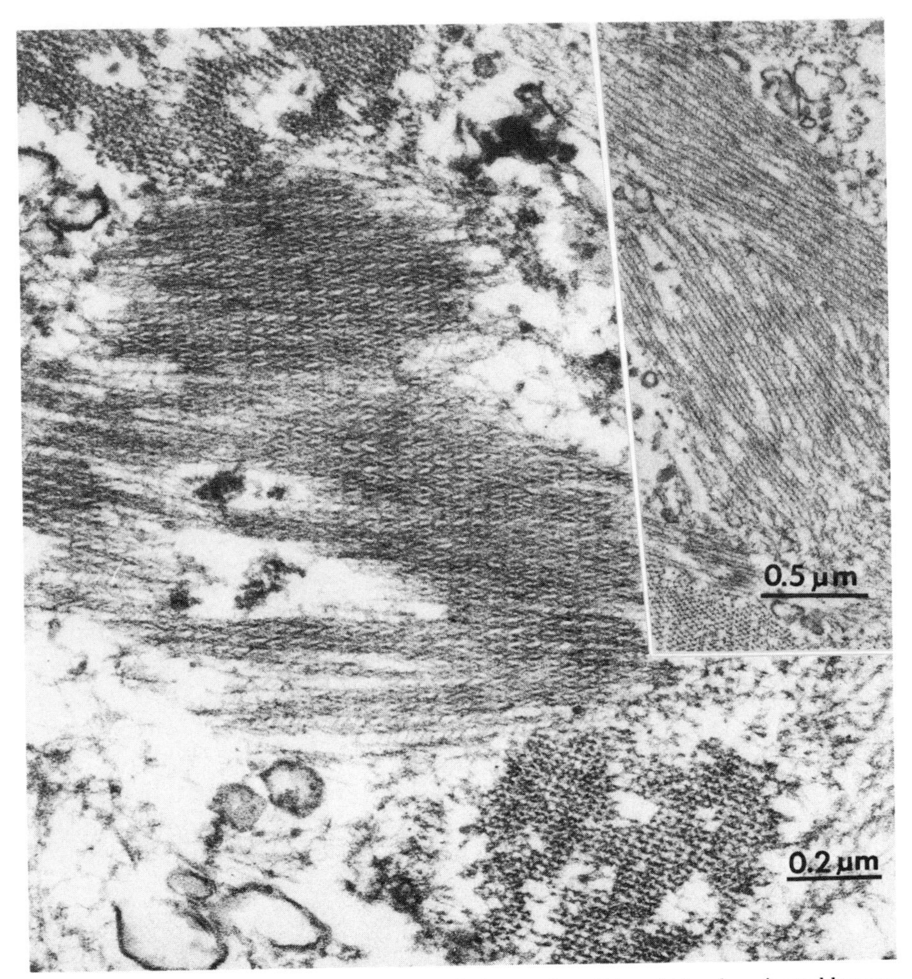

FIGURE 7.57. Appearance of nemaline rods (anomalous Z-bands) in glycerinated human skeletal muscle after low-ionic-strength extraction for 9 days. The center part shows a nemaline rod in longitudinal section, and the top and bottom parts of the micrograph show transverse views. The similarity between the rod structure and the appearances of conventional Z-bands is striking. The inset (top right) shows the appearance of a similar rod extracted for 28 days. Much of the structure in the 9-day rod has disappeared. For details of the methods used, see Stromer *et al.* (1976) from which the figure is reproduced.

ous Z-band appearances that have so far been seen in longitudinal sections of muscle. The Z-band thickness could then be quantified in terms of multiples of the 385-Å repeat.

The next step is to ask what the bridging structures consist of. In fact, it is not yet clear which proteins form these bridges, but the obvious contenders are actin, tropomyosin, and α-actinin. The latter is a protein

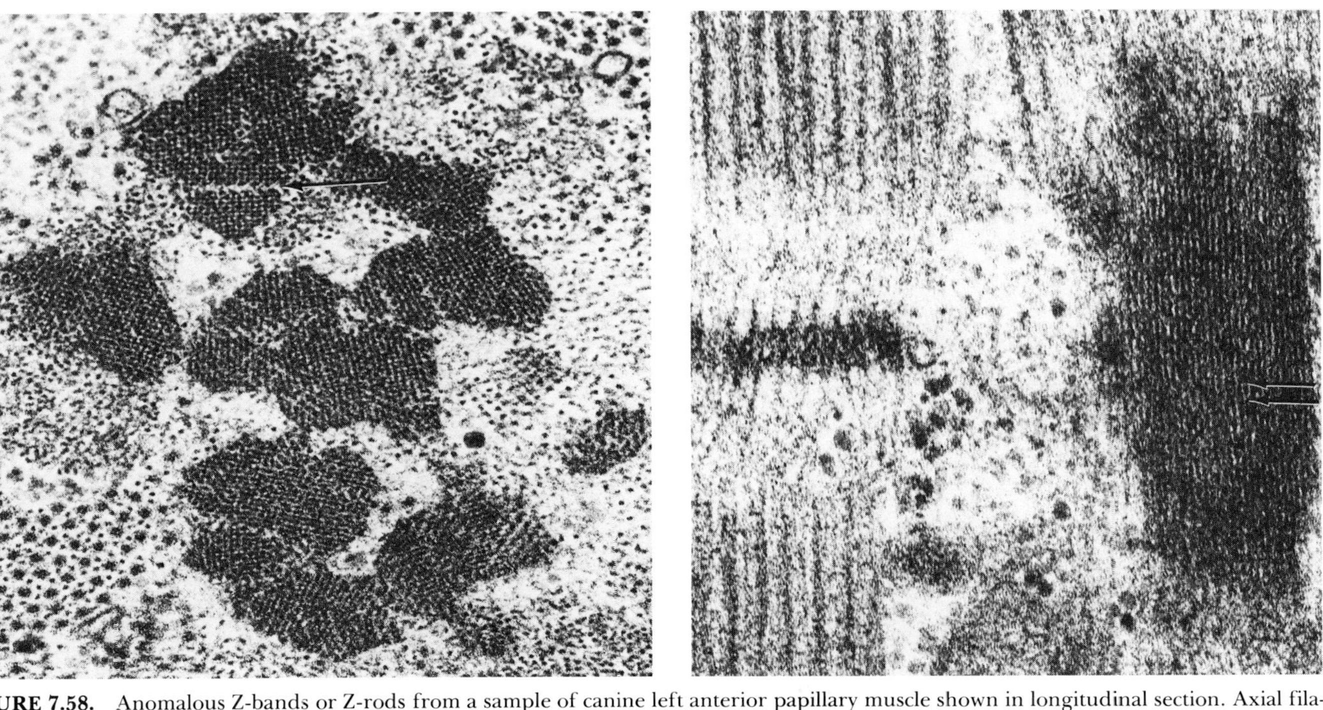

FIGURE 7.58. Anomalous Z-bands or Z-rods from a sample of canine left anterior papillary muscle shown in longitudinal section. Axial filaments parallel to the fiber axis are spaced 115 Å apart and have a transverse repeat every 387 Å (arrows). In some regions the axial filaments vary in apparent thickness from 30 to 100 Å. A normal Z-band is shown to the left. (b) Transverse sections of Z-rods as in (a). This micrograph exhibits distortion of the square Z lattice to a rectangular lattice because of compression shown also in the A-band lattice. Large (240 Å) and small lattice squares oriented in the same direction as the I squares are seen. An occasional region (arrow) is suggestive of the basketweave pattern, another distinct lattice form also observed in the normal cardiac Z-band. Magnifications ×96,000 (reduced 21% for reproduction). (From Goldstein *et al.*, 1977.)

FIGURE 7.59. Perspective drawing of the possible structure of the Z-band. The Z-filaments shown schematically as shaded or white rods would account for both the zigzag structure seen in longitudinal sections and the basketweave appearance in transverse sections. The black rods merely indicate the twofold symmetric nature of the thin filaments but clearly not their helical twist.

that has been characterized largely by Goll and his collaborators (see Suzuki *et al.*, 1976, for the many references). It has a molecular weight of about 200,000, consists of two similar chains each of weight 100,000, has a moderately high α-helix content (74%), and is rod-shaped with approximate dimensions 400 to 500 Å long by 40 Å in diameter. It can bind strongly to actin, can compete with tropomyosin under low temperature conditions for actin binding sites, and may even induce aggregation of actin into filaments under conditions where actin would not normally polymerize. That it is located at or near the Z-band has been demonstrated by specific extraction and analysis of the Z-band and by partial Z-band reconstitution with purified α-actinin. Since this protein has also been found in smooth muscles and in nonmuscle motile systems, it may conceivably have a common key role in organizing actin in these various contractile systems. This may be true even though, at most, only rudimentary Z-bands (the dense bodies—see next chapter) seem to occur in smooth muscles, and no such structures have been reported in nonmuscle motile systems.

The dimensions of both α-actinin and tropomyosin might make them suitable candidates for the bridging proteins that form the Z-filaments, but antibody-labeling studies (Pepe, 1966) have suggested that tropomyosin may not be present in the Z-band. Extraction, reconstitution, and antibody labeling of α-actinin have clearly shown that this protein is associated with the Z-band even if it is not the bridging protein (Suzuki *et al.*, 1976). It may possibly be associated with a matrix material as described below.

There is one more feature of the vertebrate Z-band that has not been accounted for. This is that frequently in transverse sections of some muscles (e.g., frog sartorius muscle) an appearance is seen (Fig. 7.53b) in which a square bridging lattice of half the spacing of the thin filament array is predominant. The basket-weave appearance can hardly be seen, if at all. It is also clear in such micrographs that there is material linking adjacent thin filaments on the same side of the Z-band to give a large square cross-linked lattice. The problem is that these appearances would not be accounted for directly in terms of the Z-filament model that accounts well for the basket-weave appearance.

Two classes of explanation have been put forward to overcome this problem. In one type of Z-band model (Macdonald and Engel, 1971; Ullrick *et al.*, 1977), the basket-weave and square lattice appearances are both explained in terms of Z-filament structures that have the filaments "kinked" by different amounts (Fig. 7.60a,b). In the other model, the two appearances are attributed to different components of the Z-band (Kelly and Cahill, 1972). Here the square lattices are attributed to a matrix material (α-actinin?) that stretches between thin filaments on the same side of the Z-band (Fig. 7.60c). As before, the basket weave is attributed to Z-filaments. These two types of Z-band appearance were said by both Macdonald and Engel (1971) and Kelly and Cahill (1972) to be caused by the use of either osmium tetroxide or glutaraldehyde as primary fixative. But Franzini-Armstrong (1973) could not detect any comparable differences in the fish Z-band with these two primary fixatives or alternatively with changes in sarcomere length. The simple Z-filament scheme accounts well for the fish Z-band but clearly not for the frog Z-band. However, this may not in itself be a problem. It must be borne in mind that these Z-bands are probably different in structure anyway, just as the M-bands in these muscles are different. For example, it could be that in frog muscle the matrix material that gives rise to the square lattices (Fig. 7.60c) is very densely stained and stable to the fixatives used and that in fish this material is either completely absent or is not preserved by fixation. The structure in frog could still have Z-filaments as in fish, but these could be largely obscured in moderately thick transverse sections by the dense matrix material. Even in very thin

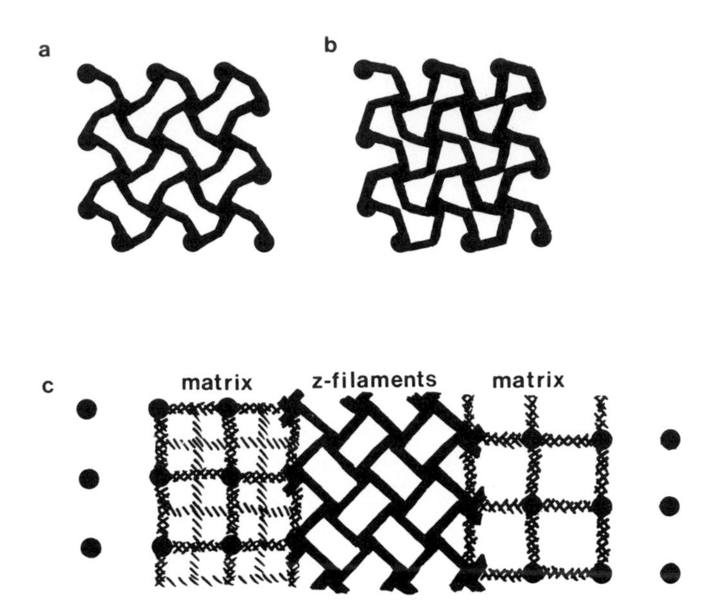

FIGURE 7.60. Alternative explanations of the basketweave and square lattice Z-band appearances (see Fig. 7.54). (a and b) the scheme of Macdonald and Engel (1971) in which the Z-filaments kink by different amounts to give the two appearances. Here, (a) is the basketweave, and (b) is the small square lattice. It is not clear how the large square lattice can be generated by this. (c) The alternative scheme of Kelly and Cahill (1972) in which there are Z-filaments (center) which form the basketweave together with a matrix material that may form a small square lattice (left) or a large lattice (right). In this case, the basketweave would appear after extraction of the matrix, otherwise it would be obscured by the denser matrix material.

transverse sections (up to, say, 300 Å), the Z-filaments could still be somewhat obscured by the matrix since it is not known how extensive axially the matrix is. It is clearly not possible at present to come to any definite conclusions about the structures of different Z-bands. What is needed is a study of variety of Z-bands in thin transverse sections of well-preserved muscles, a study comparable to the M-band study described earlier in this chapter (Section 7.3.2) and which has obviously been so fruitful. No doubt such a study will be carried out in the near future.

To conclude this section it is appropriate to mention a further feature of the Z-band that has not yet been touched on. The discussion above on Z-band structure has obviously been based on the kind of model described by Kelly and Cahill (1972). But none of that discussion rules out the possibility that the Z-filaments themselves may change their shape under different conditions, just as in the model of Macdonald and

Engel (1971). It has already been shown by Davey (1976) and Yu *et al.* (1977) that the dimensions of the Z-band and of the I-band regions just next to the Z-band probably vary as a function of sarcomere length in the same way as the dimensions of the hexagonal lattice in the A-band vary. The main evidence for this is that there is an equatorial reflection in the X-ray diffraction patterns from, for example, frog or toad sartorius muscle (G. F. Elliott *et al.,* 1967; H. E. Huxley and Brown, 1967; see Table 7.3) that does not appear to come from the A-band but which varies in spacing with the A-band reflections. Kinked Z-filaments (Fig. 7.60a,b) could be associated with this change in spacing. But whatever its cause, the apparent flexibility of the Z-band could contribute significantly to the mechanical properties of the sarcomere.

7.4.3. The N-Lines

Before we leave this discussion of the I-band, two transverse stripes in each half of the I-band must be briefly mentioned. They are referred to as the N-lines (N_1 and N_2) and can be seen in Fig. 7.61. The N_1 lines occur about 0.17 μm from the center of the Z-band (Franzini-Armstrong, 1970) and are about 400 to 500 Å wide. The N_2 lines are further away from the Z-band, their exact position being dependent on the sarcomere length (Page, 1968; Franzini-Armstrong, 1970). In a given sarcomere, the N_2 lines have a constant separation of about 1.6 μm at sarcomere lengths below about 2.25 μm, but their separation increases linearly to 2.15 μm as the sarcomere length increases to about 3.6 μm. The width of the N_2 lines is said to be about twice that of the N_1 lines (i.e., 800 Å to 1000 Å).

The protein components of these two kinds of N-line have not been determined, and their role in the sarcomere is unknown. But there is some evidence (1) that they serve to divide the sarcomere into structural compartments and (2) that they may be in some way involved in storing Ca^{2+} in the I-band (Maunder *et al.,* 1976).

In a recent series of papers on highly stretched beef muscle, Locker and Leet (1975, 1976) and Locker *et al.* (1976) have provided evidence that there may be longitudinal filaments linking the N-lines, and possibly the A-bands too, from one sarcomere to the next. Mention has been made of the apparent mechanical continuity of myofibrils after extraction of myosin and actin (Hanson and Huxley, 1953). A number of other references to continuous filaments have been made (see Hoyle, 1968), but no conclusive evidence for their existence has really been published. The recent results of Locker and Leet are very suggestive and may well lead to a clarification of this important aspect of muscle structure. It

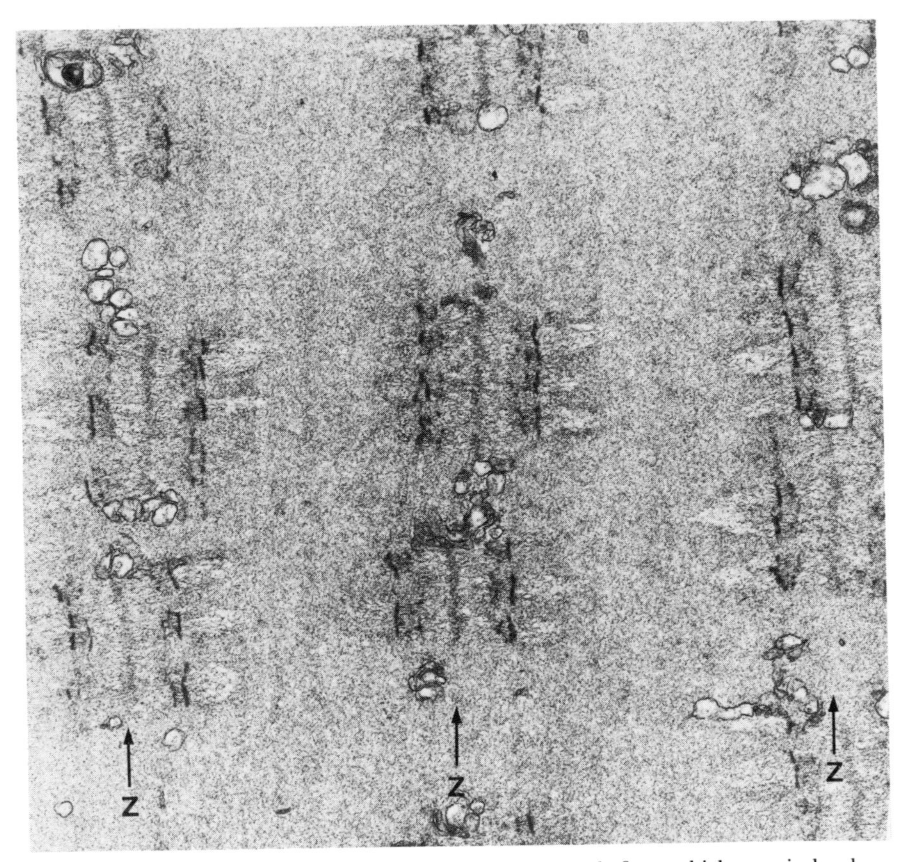

FIGURE 7.61. Longitudinal section of rabbit psoas muscle from which myosin has been exhaustively extracted. The remnants appear to be I-bands containing Z-bands and N-lines whose relative positions are remarkably well preserved. This maintained alignment may indicate the presence of some mechanical continuity between the Z-bands at the ends of a sarcomere. Magnification: ×33,000 (reduced 34% for reproduction). (Courtesy of P. M. G. Munro.)

seems possible that extensions of the original work of Hanson and Huxley (1953) might also be fruitful. For example, the micrograph in Fig. 7.61 shows a section of a rabbit muscle from which the myosin has been thoroughly extracted (P. M. G. Munro and J. M. Squire, unpublished results). Although longitudinal filaments cannot be seen in the A-band, it is evident that the remnants of the I-band, including the Z-band and possibly the N-lines, remain and that they are in good longitudinal register across the micrograph. The existence of some physical continuity from Z-band to Z-band is strongly suggested by results of this kind.

7.5. The Three-Dimensional Structure of the Sarcomere

The aim of this final section of the chapter is to integrate the results and conclusions (already described) about A-band and I-band structure and thus see how the structure of the whole sarcomere must appear in three dimensions. The approach follows closely that given in Squire (1974).

We have already seen that the actin filaments are arranged in a more or less square array near to the Z-band and in a hexagonal array (defined by the thick filaments) in the overlap region of the A-band. It is instructive to see how the thin filaments might be arranged in the square lattice and how these thin filament arrangements would fit into the myosin filament superlattice in the A-band. For the purposes of this discussion, it will be assumed that the thick filament superlattice is a three-filament superlattice of "up" and "down" filaments as shown in Fig. 7.23.

Since the thin filaments have a pseudotwofold rotation axis (i.e., there are two coaxial strings of actin monomers; Fig. 5.2), there are, in principle, two regular ways of placing these filaments on a square array. In one, all of the filaments have identical orientations; in the other, adjacent filaments are systematically related by a 90° rotation so that a two-filament superlattice is generated. These two possibilities are illustrated in Fig. 7.62. It was shown by Pringle (1968) that it is possible to transform an almost square lattice to a hexagonal one by equal displacements of all the lattice points as shown in Fig. 7.62 so the two thin

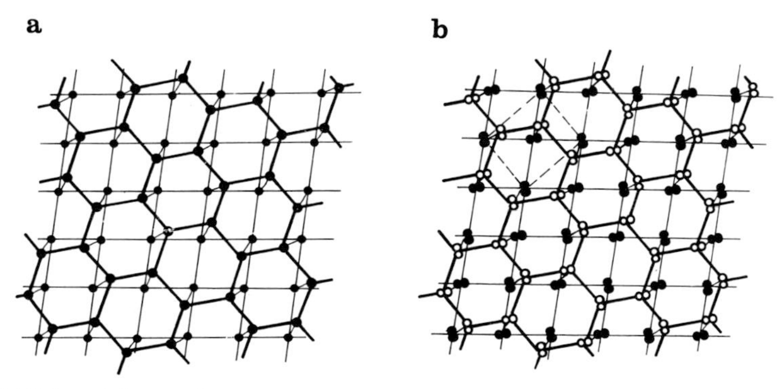

FIGURE 7.62. The possible transformation from the almost square (rhomboidal) Z-band lattice to the hexagonal actin arrangement in the A-band (as suggested by Pringle, 1968) for situations in which (a) all the actin filaments have identical orientations in the Z-band and (b) adjacent actin filaments are rotated by 90°, thus giving a Z-band superlattice (dashed lines). (From Squire, 1974.)

filament arrangements in (a) and (b) could be transformed to the hexagonal arrangements in the same figures, assuming that the relative orientations of the thin filaments are preserved during this transformation. Strictly, for this transformation to include exactly equal shifts of all the thin filaments, the Z-band lattice would need to be rhomboidal with a lattice angle of 83° (Pringle, 1968). With this in mind and also the fact that the thin filaments do not have an exact twofold rotation axis, it does appear that the arrangement (Fig. 7.62a) in which the thin filaments have exactly identical orientations is more likely to occur than the superlattice arrangement. But, unfortunately, this is by no means absolutely certain, and, in the absence of direct evidence, it is worth considering both of the suggested thin filament arrangements while bearing in mind that the simple arrangement may be the more likely.

Figure 7.63 shows how these two arrangements would fit into the three-filament myosin superlattice. As it happens, these two arrangements are the only two thin filament arrangements that could fit regularly into the myosin superlattice. In Fig. 7.63, this superlattice is shown for a very thin transverse slice through the A-band in which only one level of myosin cross bridges has been drawn. This is approximately to scale assuming that the myosin cross bridges are roughly elliptical in shape and about 120 Å long.

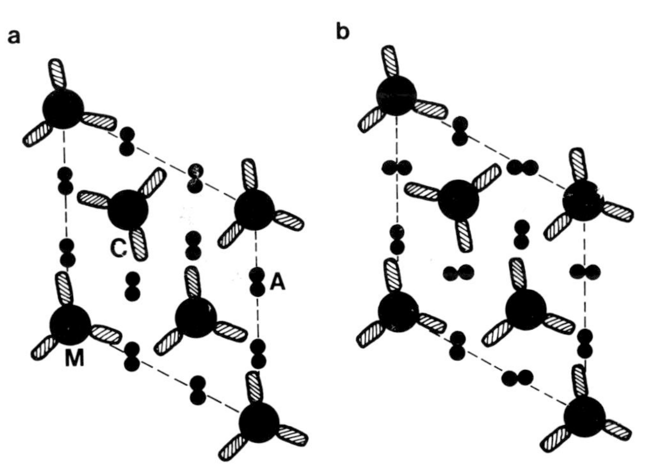

FIGURE 7.63. This shows that the two hexagonal arrangements of actin filaments in Fig. 7.62 are also the only two actin arrangements that would fit systematically into the thick filament (no three alike) superlattice. In (a), all the thin filaments have identical orientations; in (b), nearest neighbors are rotated by 90° just as in Fig. 7.62. The diagram corresponds to a transverse section of the A-band about 143 Å thick (i.e., it shows only one cross bridge level) with the myosin backbones shown as solid circles (large), the cross bridges shaded, and the actin filaments as pairs of small solid circles. (After Squire, 1974, Fig. 5.)

With this arrangement in mind, it is instructive to consider what kind of interactions these cross bridges could make with the neighboring actin filaments. It can be assumed that the myosin heads make roughly the same kind of interaction as that found by Moore *et al.* (1970) on S-1-decorated thin filaments (Fig. 5.11), since this interaction is thought to mimic the rigor interaction (Chapter 10). Note, however, that the arguments do not depend heavily on this particular kind of interaction. The result is illustrated in Fig. 7.64 for the two types of thin filament arrangement. Here it can be seen that some interactions are readily made, whereas others require considerable contortions of the S-2 link between the attached S-1 heads and the backbone of the thick filaments. In Fig. 7.64 three kinds of interaction have been distinguished: in the first (good) the myosin head does not need to move out of the 60° sector defined by the two nearest thin filaments, and the attachment site on actin is not outside a circle through the centers of the six thin filaments surrounding a single thick filament. This is taken to be a favorable interaction. The second type (possible), which is taken to be possible but less favorable than the first, is one in which the cross bridge needs to move outside the 60° sector or the actin circle for attachment to occur but which requires little distortion of the S-2 link. The third category (poor) is such that all possible interactions seem to be sterically unreasonable, either needing considerable contortions of the S-2 link or having to move beyond a cross bridge from another myosin filament in order to interact. It turns out that whichever of the two thin filament arrangements is taken, about one-third of the cross bridges can make each type of interaction. This could mean that in a rigor muscle only about two-thirds of the available cross bridges would be attached to the thin filaments. This could well explain why H. E. Huxley (1957) only saw

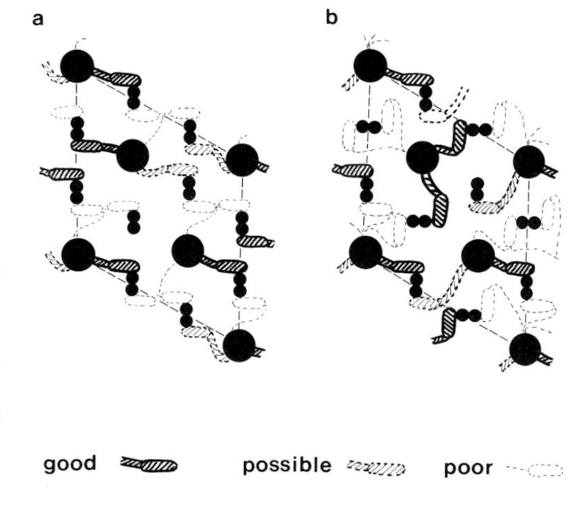

FIGURE 7.64. (a and b) Possible cross bridge interactions for the two arrangements in Fig. 7.63. Three categories of interaction are distinguished (lower illustrations) as good, possible, and poor. For detailed definitions of these, see text. In both (a) and (b), the three types of interaction occur in equal proportions with three out of the nine cross bridges in the superlattice forming interactions of each type. (After Squire, 1974, Fig. 6.)

six cross-links to actin from one thick filament per 430-Å repeat even though it seems likely that there are nine cross bridges in this repeat.

The implications of this result are far-reaching in that they suggest that a given myosin cross bridge can only interact with particular regions of a thin filament as the thin filament moves past it during shortening of a muscle. The idea that there are particular thin filament target areas for attachment of a given cross bridge was originally suggested by Reedy (1967) and was spelled out in detail by Squire (1972). It will be considered again in the next chapter.

8

Comparative Ultrastructures of Diverse Muscle Types

8.1. Introduction

In the last chapter a very detailed description was given of the ultrastructure of the vertebrate skeletal muscle myofibril. At the same time many of the general ideas and methods concerning the analysis of muscle structure were introduced. Fortunately, this makes it much easier to consider the structures of other types of muscle. For this reason, the present chapter provides short and, it is hoped, concise descriptions of what is known about the ultrastructures of a number of other important muscle types. However, this division of space should not be taken as an indication of the relative usefulness of these muscles as tools in muscle research. Results from all of these muscles, but in particular from insect flight muscle, have been instrumental in forming the currently held views on how muscles work. This should be apparent from the discussion in Chapter 10.

8.2. Arthropod Muscles

8.2.1. General Description

An excellent review of the biology of arthropod muscles has been written by Pringle (1972), and the reader is referred to that for a summary of the comparative anatomy and physiology of these muscles. For more detailed reading two recent books on insect muscles are recommended: *Insect Muscle* edited by P. N. R. Usherwood (1975) and *Insect Flight Muscle* edited by R. T. Tregear (1977). Here the discussion is primarily concerned with a description of the comparative ultrastructures of different arthropod myofibrils—in particular, those in insect flight muscles.

A brief account of the classification of arthropod muscles was given in Chapter 1. As a general rule all of these muscles are cross striated and have sarcomeres that are comparable in general appearance to those in vertebrate skeletal muscles. A-bands, I-bands, Z-bands, and sometimes M-bands can all be recognized in these muscles, as can an H-zone of variable width. It was realized soon after the original postulation of the sliding filament model by H. E. Huxley and Hanson (1954) and A. F. Huxley and Niedergerke (1954) that arthropod muscles must shorten by a similar mechanism. The fact that arthropods and vertebrates have evolved such similar myofilament arrangements despite their very early divergence in the evolutionary tree (Fig. 6.17) has been commented on by many writers (see Elder, 1975). It may well be a demonstration of the fact that the cross-striated myofibril represents a very efficient way of organizing the contractile machinery.

Despite their common cross-striated form, the detailed structures of different arthropod muscles show considerable variations. Differences occur in myofibril diameter, in sarcomere length, in filament geometry and number, and in structural regularity. These differences clearly permit a wide range of physiological behavior as required by the fact that in arthropods the skeletal and visceral muscles are both striated. This contrasts markedly with the situation in vertebrates where two distinct kinds of muscles, the striated and the smooth muscles, are used for these two functions.

The fibers in arthropod muscles can vary in diameter from about 1 to 20 μm in visceral muscles, 10 to 100 μm in nonfibrillar insect muscles, up to 1 mm in some fibrillar insect muscles, and as much as 4 mm in giant barnacles. The enormous size of these barnacle fibers has made them extremely useful for work on single fibers since they are much more manageable than vertebrate fibres only 20–100 μm in diameter. As a result, many interesting results have been obtained from them (Ashley and Ridgway, 1970).

Within these different types of arthropod fibers, several distinct patterns of myofibrillar arrangement occur (Elder 1975). In particular, in the insect fibrillar muscles with which we are primarily concerned, the myofibrils are large (1–5 μm in diameter) and are well separated from the neighboring myofibrils by mitochondria (~40% of the cell volume) and other cell organelles (Fig. 8.1a). Fibers in other muscles often contain myofibrils that are only partially separated from each other

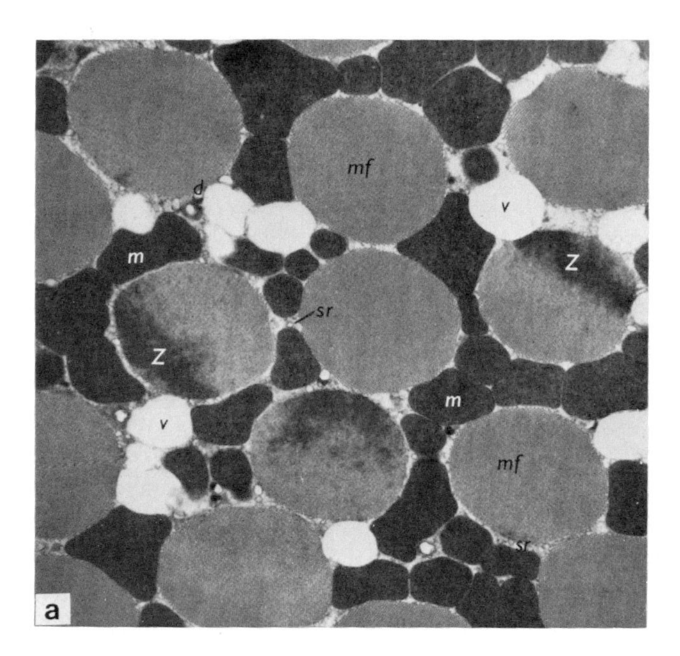

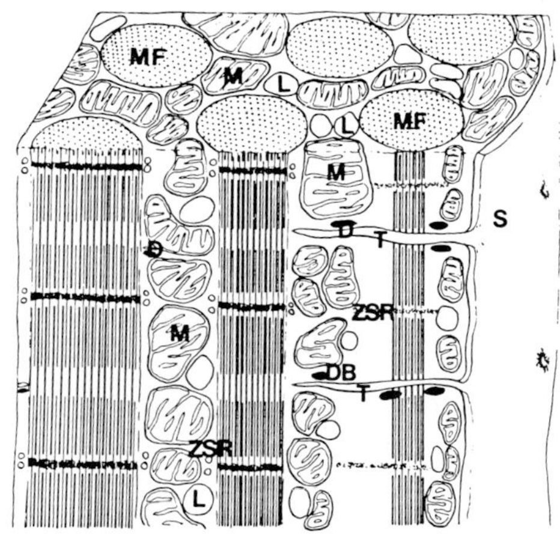

FIGURE 8.1. (a) Electron micrograph of a transverse section of flight muscle from *Lethocerus maximus.* The myofibrils (mf) are discrete and approximately circular in profile. Most are sectioned at the level of the A-band, but in some (Z) the Z-band is in the section. Many mitochondria (m) and some large vesicles (v) are present between the myofibrils. A few vesicles of the sarcoplasmic reticulum (sr) and a few dyads (d) are visible. (Magnification ×9,000; from Ashhurst, 1967a.) (b) Diagrammatic representation of the 3-D structure of the muscle shown in (a). Key: myofibrils, MF; dyad, D; dense body of dyad, DB; T-tubule, T; sarcoplasmic reticulum vesicles at Z-band, ZSR; sarcolemma, S; mitochondrion, M; lipid droplet, L. (From Ashhurst and Cullen, 1977.)

(Fahrenbach, 1964). The sarcomere lengths in insect fibrillar muscles are usually about 3.0 μm (Fig. 8.1b), but in other arthropod muscles this length can range from 3.0 μm to about 13 μm, and the thick and thin filament lengths increase accordingly. As a general rule, the sarcomere length is greater in slow tonic muscles. Note also that the thick filaments tend to be larger in diameter than those in vertebrate muscles, and they often show a clearly hollow core (see Auber and Couteaux, 1963; Reedy, 1967).

We have already seen in the last chapter that in vertebrate skeletal muscle each thick filament is surrounded by six nearest neighbor thin filaments located at the trigonal points of the hexagonal thick filament array. This filament arrangement (Fig. 8.2a) can be compared with those that occur in different arthropod muscles. Fibrillar insect flight muscles (Fig. 8.18) also have six thin filaments regularly arranged around each thick filament (Fig. 8.2b), but here the thin filaments are located at positions midway between two adjacent myosin filaments. Each thin filament is therefore "shared" by two thick filaments, and since there are

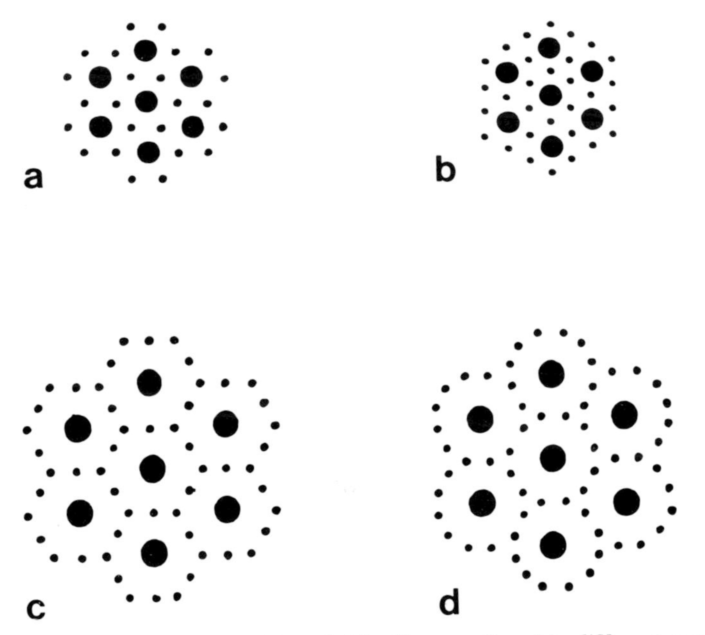

FIGURE 8.2. Arrangements of thick and thin filaments found in different muscles. (a) Vertebrate striated muscles. (b) Insect fibrillar flight muscle. (c and d) Two possible arrangements of 12 actin filaments around each myosin as in arthropod leg and trunk muscles and some flight muscles. The overall ratio of thin to thick filaments is (a) 2 : 1, (b) 3 : 1, (c) 5 : 1, and (d) 6 : 1. (From Tosselli and Pepe, 1968.)

six of them around one thick filament, the thin-to-thick filament number ratio is 3 : 1. This compares with the ratio 2 : 1 that occurs in vertebrate muscles.

The filament arrangements in the leg and trunk muscles of arthropods are quite distinct. Here there can be as many as 10 to 12 thin filaments in a ring around each thick filament. The thick filament diameter is sometimes greater as well, but a more or less hexagonal thick filament lattice is invariably present. It was shown by Toselli and Pepe (1968) that there are two distinct ways of arranging rings of 12 actin filaments to fit symmetrically into the hexagonal thick filament array. These two arrangements are illustrated in Fig. 8.2c, and d; they give thin to thick filament ratios of 5 : 1 and 6 : 1 respectively. Despite the obvious regularity of these arrangements, filament counts on a number of muscles have given filament ratios of anywhere between 3 : 1 and 6 : 1. For example the flight muscles of *Vanessa* (Auber, 1967), which have a low wing beat frequency, gave an estimated filament ratio of 4 : 1 with seven to nine actin filaments around each myosin. Similarly crab leg and eyestalk muscles gave ratios 3.6 : 1 and 5.5 : 1 (Franzini-Armstrong, 1970), and insect leg muscles gave 5.3 : 1 (Tregear and Squire, 1973). Insect nonfibrillar flight muscles with wing beat frequencies greater than about 10/sec tend to have the same regular 3 : 1 filament ratio as in fibrillar flight muscles, a clear indication that this arrangement is adapted for speed (Fig. 8.2b).

The Z-band in arthropods serves a similar purpose to that in vertebrate skeletal muscle. Z-band structures in muscles with a 3 : 1 filament ratio tend to be well defined; the structure of that in insect fibrillar flight muscles will be described later. In other muscles, the Z-band often appears to be thicker and less well defined (Reger, 1967). In some extreme cases [e.g., giant fibers of the barnacle (Hoyle *et al.,* 1965), trunk muscles of the blowfly larva (Osborne, 1967), slow fibers from a crab antennal muscle (Hoyle and McNeill, 1968)], the Z-band is unusual in being perforated to allow penetration of the thick filaments and hence to allow supercontraction to occur. A similar structure is also observed in visceral muscles where the Z-band appears to consist of only partially aligned "dense bodies." It will be seen later in this chapter that such an arrangement does bear some resemblance to the mode of thin filament organization in certain smooth muscles.

Clear M-band structures can be seen in fibrillar insect flight muscles, but they are often much less evident in other insect and crustacean muscles. Even the M-band structures in different insect flight muscles are very varied (Pringle, 1972). A detailed description of that in the giant water bug (*Lethocerus*) will be included in the next section.

8.2.2. The Thick Filaments in Insect Flight Muscles

Early Evidence. The particular insect flight muscles that have been most extensively studied are the fibrillar flight muscles of the giant water bugs (*Lethocerus;* Fig. 8.18). That these flight muscles might be suitable subjects for biophysical studies was originally suggested by Professor J. W. S. Pringle, and most of the published work on these muscles is from Pringle's laboratory in the Zoology Department at Oxford. It will be seen in Chapter 10 that it is work on vertebrate striated muscle and on these insect flight muscles that has provided most of the current evidence on the nature of the cross bridge cycle. In fact, it was the early electron microscopy of insect flight muscle by Reedy *et al.* (1965) that gave the first indication that the generation of force in muscle might involve a change in the configuration of the myosin cross bridges while attached to the thin filaments. Figure 8.3, taken from their original paper, shows longitudinal sections of rigor and relaxed muscle, and here it can be seen that the cross bridges between the thick and thin filaments have different tilts in the two muscle states. What is also clear from these micrographs is that the structure of the insect sarcomere is beautifully regular. Reedy (1967, 1968) took full advantage of this regularity to determine the structure of the rigor muscle, and he discovered many interesting features about the geometry of the thick and thin filaments and of their interaction. Some of the details of his results are discussed in the following pages.

The sarcomere in insect fibrillar flight muscles is about 2.5 μm long, and the thick filaments extend for almost the whole of this length (2.2 μm). The I-band is therefore very short indeed. This arrangement is apparently related to the specific function of this muscle. It oscillates when activated (see Chapter 1, Fig. 1.22), and only very small changes in sarcomere length (1–2%) occur during this oscillation. Contraction therefore occurs with the bridge regions of the A-band being totally overlapped by the thin filaments. As described later there is some evidence that the thick filaments are actually mechanically linked to the insect Z-band, another feature that may be closely linked to the oscillatory function of the muscle. The ability of the muscles to oscillate is a direct consequence of the phenomenon known as stretch activation in which the muscle pulls harder after a stretch and a slight delay than it does after being released. The mechanism of stretch activation remains one of the main unsolved problems in muscle research, although a number of possible mechanisms come to mind and will be detailed in Chapter 11.

Insect muscle thick filaments have a similar form to those in vertebrate skeletal muscle. They are roughly cylindrical in shape, about 180

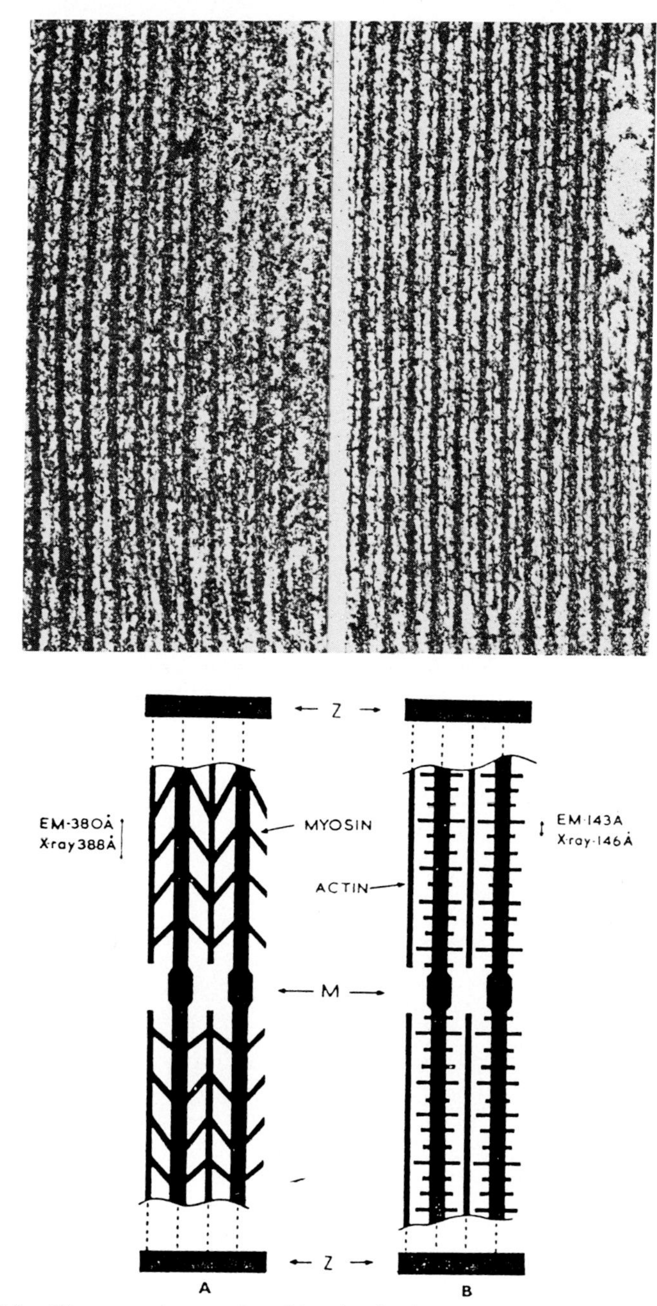

FIGURE 8.3. Electron micrographs of longitudinal sections of insect flight muscle (*L. Maximus*) either in rigor (top left) or fixed while relaxed (top right) together with the interpretation by Reedy *et al.* (1965) of the cross-bridge configurations that are present in the two muscle states (A, rigor; B, relaxed). The micrographs are from Reedy (1968); magnifications ×70,000.

to 200 Å in diameter, about 2.2 μm long, have projections over most of their surface except in the bare zone (M-region) halfway along their length, and are tapered at each end. This basic form is as illustrated in Fig. 1.9. The main difference from vertebrate thick filaments, apart from that of size, is that in the M-region of the insect thick filaments the backbone becomes slightly wider. In this region, the backbone is also oval in profile as shown in Fig. 8.4b. Elsewhere in the A-band, the thick filaments are circular in profile and appear to be hollow (Fig. 8.4a) in the sense that the centers of the thick filaments often appear less densely stained than their peripheries.

Optical diffraction analysis of longitudinal sections of rigor insect flight muscle by Reedy (1967, 1968) showed up strong axial periodicites of about 145 Å, 385 Å, and 1155 Å (Fig. 8.5). All of these are orders of the 1155-Å repeat. It was evident from these repeats, from the early X-ray diffraction evidence of Reedy *et al.* (1965), and from the general regularity of the A-band that the axial repeats of the thick and thin filaments were very closely related. Unlike vertebrate striated muscle, where the axial repeats of the thin and thick filaments are quite different, in insect muscle the thin filament repeat (385 Å) is almost exactly one-third of the 1155-Å thick filament repeat. This helps to explain why the rigor interaction between the thin and thick filaments repeats so regularly through the A-band (Fig. 8.5).

A remarkably regular structure was also seen by Reedy (1967, 1968) in ultrathin transverse sections of rigor insect flight muscle. These sections were thin enough to show predominantly only a single level of cross

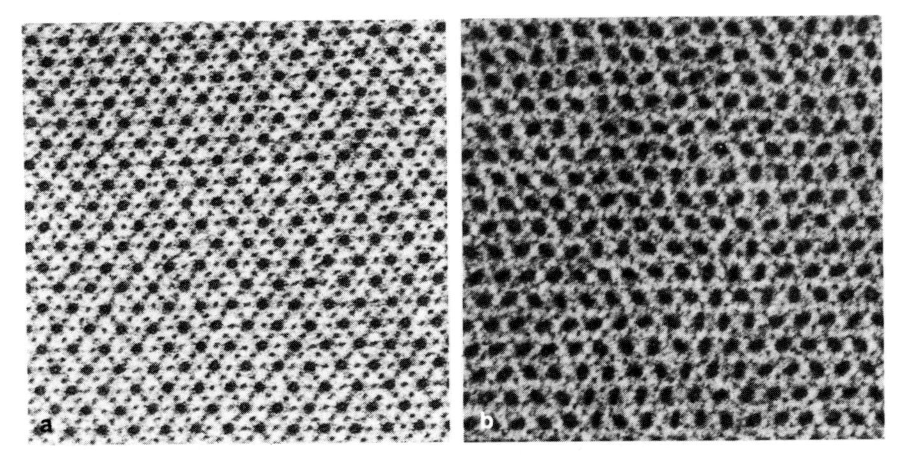

FIGURE 8.4. Transverse sections of *L. maximus* flight muscle. (a) The overlap region of the A-band showing the hexagonal arrays of actin and myosin filaments. (b) At the M-band where the profiles of the thick filaments are oval in shape. In both (a) and (b) the thick filaments have a lightly stained core. (From Ashhurst, 1967a.) (a) ×121,000 (b) ×108,000.

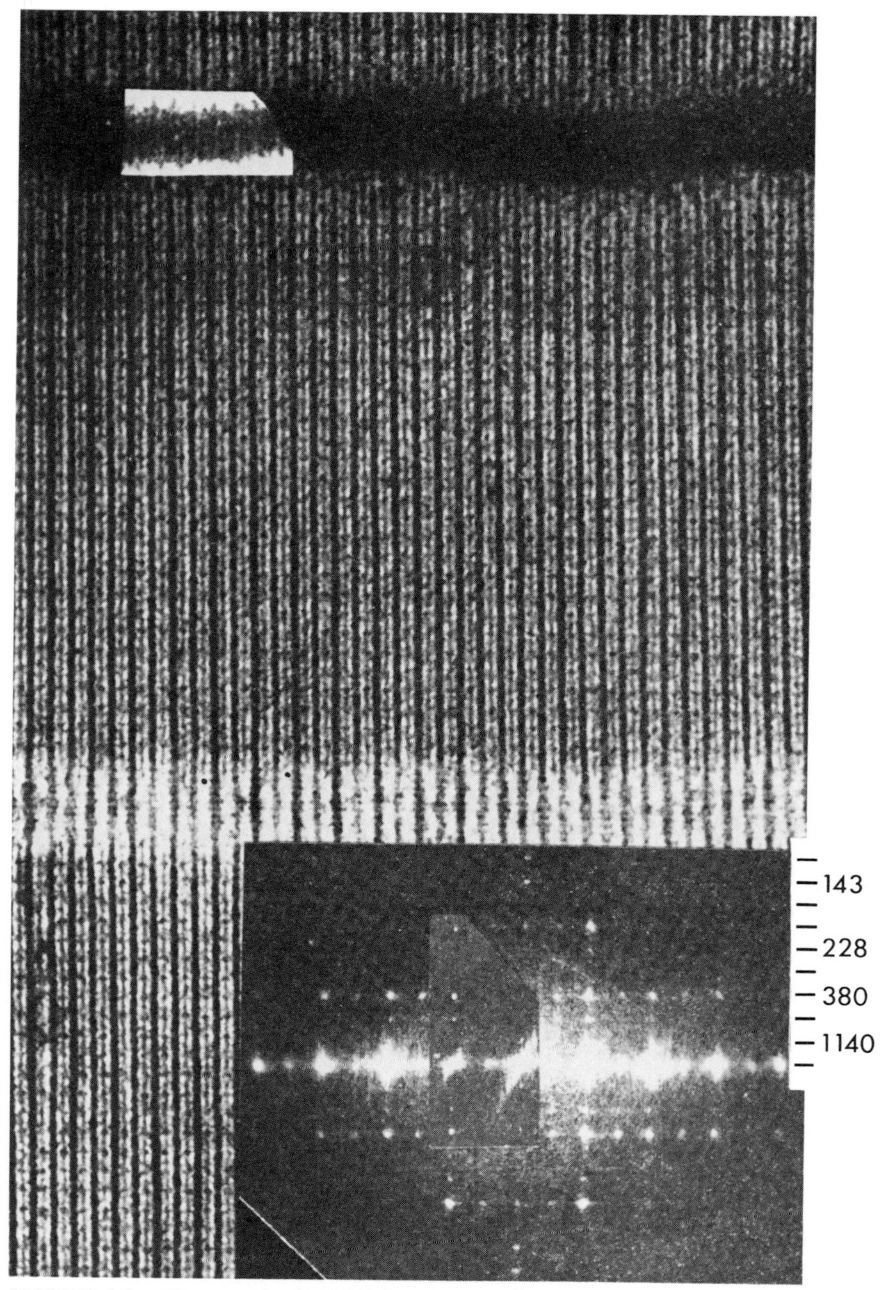

FIGURE 8.5. Micrograph of a thick longitudinal section of rigor flight muscle from *L. maximus* together with its optical diffraction pattern. Many orders of the 1140-Å repeat can be seen, but the fourth and seventh orders are weak. For discussion see text. (From Reedy, 1968.) ×92,000.

bridge origins (spaced 145 Å axially), and a very characteristic arrangement of cross connections from the thick to the thin filaments could be seen. In fact, four cross bridges could be seen originating from a single thick filament and connecting to four of the six neighboring thin filaments. Figure 8.6 (a–h) shows some of Reedy's electron micrographs with this characteristic "flared X" appearance. Figure 8.6i summarizes this flared X appearance, and (j) illustrates one or two variations of the flared X that Reedy saw in his micrographs.

Slightly oblique transverse sections showed that a systematic rotation of the flared X appearance occurred between successive cross bridge levels 145 Å apart along the thick filaments. This was evident from the sudden concerted change of orientation of the flared Xs from one region of the fibril to the next—these changes gave rise to a clear "herringbone" effect across the section (Fig 8.6k).

By correlating the cross bridge appearance in both transverse and longitudinal sections, Reedy (1967, 1968) was able to show that the flared X arrangements at successive 145-Å levels along the thick filaments were related by a helix of pitch 770 Å. He took this to mean that the cross bridge origins on the thick filaments followed a similar symmetry. The kind of cross bridge arrangement he suggested is illustrated in Fig. 8.7a; it is formally described as a two-strand, 16/3 helix of pitch 770 Å and true repeat 1155 Å.

With this cross bridge arrangement in mind, the next question to answer was whether each of the four arms of the flared X represented the whole globular part of one myosin molecule or just one of the two component myosin heads. Evidence available at the time (Chaplain and Tregear, 1966) suggested that the myosin content in this muscle was at least sufficient for each arm to represent a separate myosin molecule. This would mean that the heads of four myosin molecules would be located at each 145-Å level (or crown) along the bridge regions of the thick filaments. Actually, the results of Chaplain and Tregear (1966) suggested that there was enough myosin in each thick filament for there to be six molecules per crown, but in the face of Reedy's striking flared X images, little attention was paid to this discrepancy at the time.

A New Interpretation. The problem was reconsidered by Squire (1971, 1972) in the light of the evidence then coming to hand on myosin packing in other thick filaments (Small and Squire, 1972; discussed later in this chapter, Section 8.4). It was suggested that the insect thick filaments might not really have a two-stranded arrangement of double cross bridges (Fig. 8.7a,b) but rather might have a six-stranded arrangement of single cross bridges (Fig. 8.7c,d). This would then allow the myosin molecules in the thick filaments of vertebrate skeletal muscle, insect flight muscle, and vertebrate smooth muscle (see Section 8.4) to have a

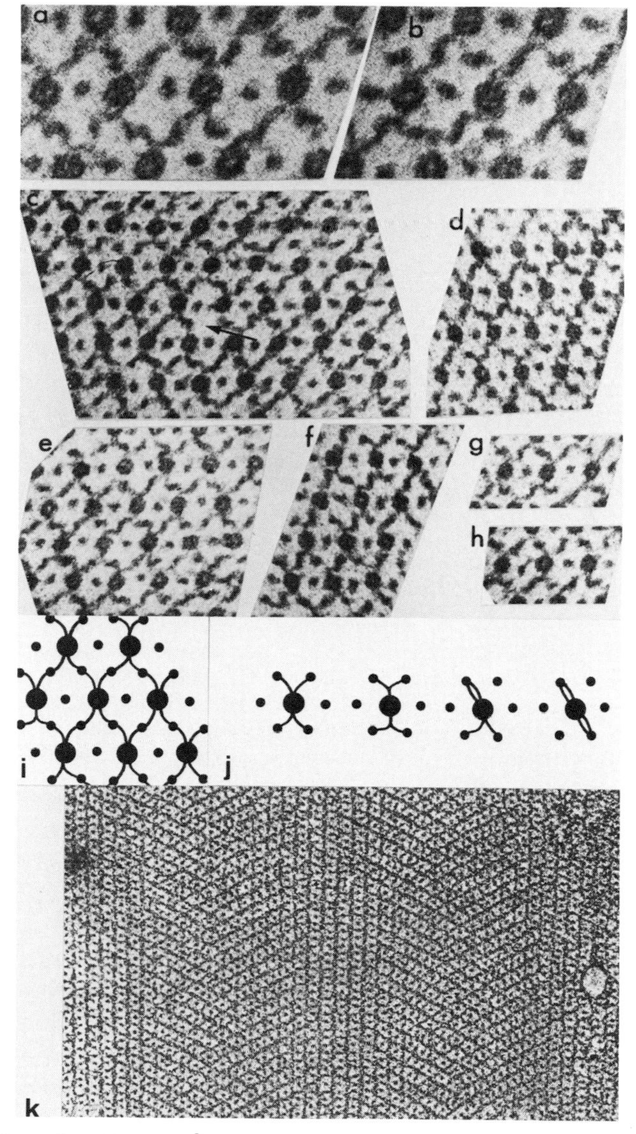

FIGURE 8.6. (a/h) Thin (200-Å) transverse sections of *L. maximus* flight muscle in rigor showing the flared X appearances produced by the cross bridge interactions with actin. (From Reedy and Garrett, 1977). Scale bars 1000 Å. Note that the flared Xs on adjacent thick filaments all point in the same direction as illustrated in (i). (j) Some alternative forms of the flared X seen by Reedy (1967; 1968) shown diagrammatically. (k) Slightly oblique thin transverse section of the same preparation showing the herringbone appearance of the section caused by the transitions from one axial level of flared Xs to another level where they are differently oriented. (From Reedy, 1967, 1968.) ×60,000. Note the missing actin filament (arrowed) in (c).

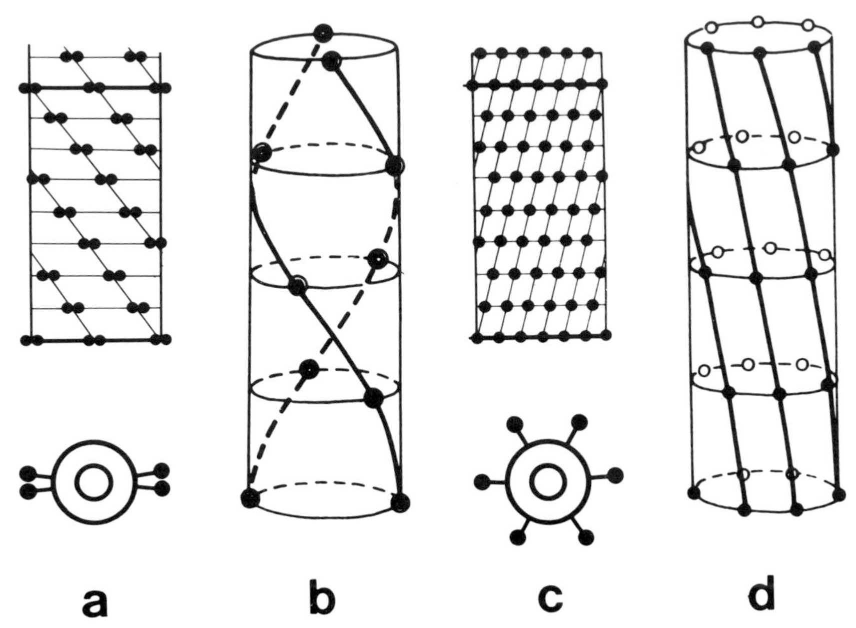

FIGURE 8.7. Illustrations of two alternative models for the arrangement of the myosin cross bridges on insect thick filaments. (a and b) A radial projection, section, and perspective view of a two-stranded arrangement (two-start, 16/3 helix) where each repeating unit comprises the heads of two or three molecules. (c and d) Comparable views of a six-stranded 16/1 helix of single myosin cross bridges. (a and c) From Squire (1977); (b and d) after Tregear (1975). Note that the helix senses in (c) and (d) are different; the true sense is not known (see text).

common mode of aggregation. This idea formed the basis of the general model of myosin filament structure which will be considered in detail in Chapter 9.

With this new six-stranded model in mind, it was clear that the available structural data on insect flight muscles would need to be reinterpreted. In particular, for the six-stranded model to be acceptable, it was clearly necessary to show that it could account both for the X-ray diffraction evidence from this muscle (Miller and Tregear, 1972) and for Reedy's flared X images.

The X-ray diffraction evidence from *Lethocerus* that required interpretation was of the kind shown in Fig. 8.8 (Miller and Tregear, 1972). It was thought that the parts of this diffraction pattern (from relaxed muscle) that corresponded to the structure of the thick filaments were the layer lines of spacing 385 Å, 235 Å, and 145 Å indicated in Figure 8.8e. The 145-Å layer line contained a strong meridional component, whereas the other layer lines had intensity predominantly off the meridian. From the arguments given in Chapter 2, it is clear that this

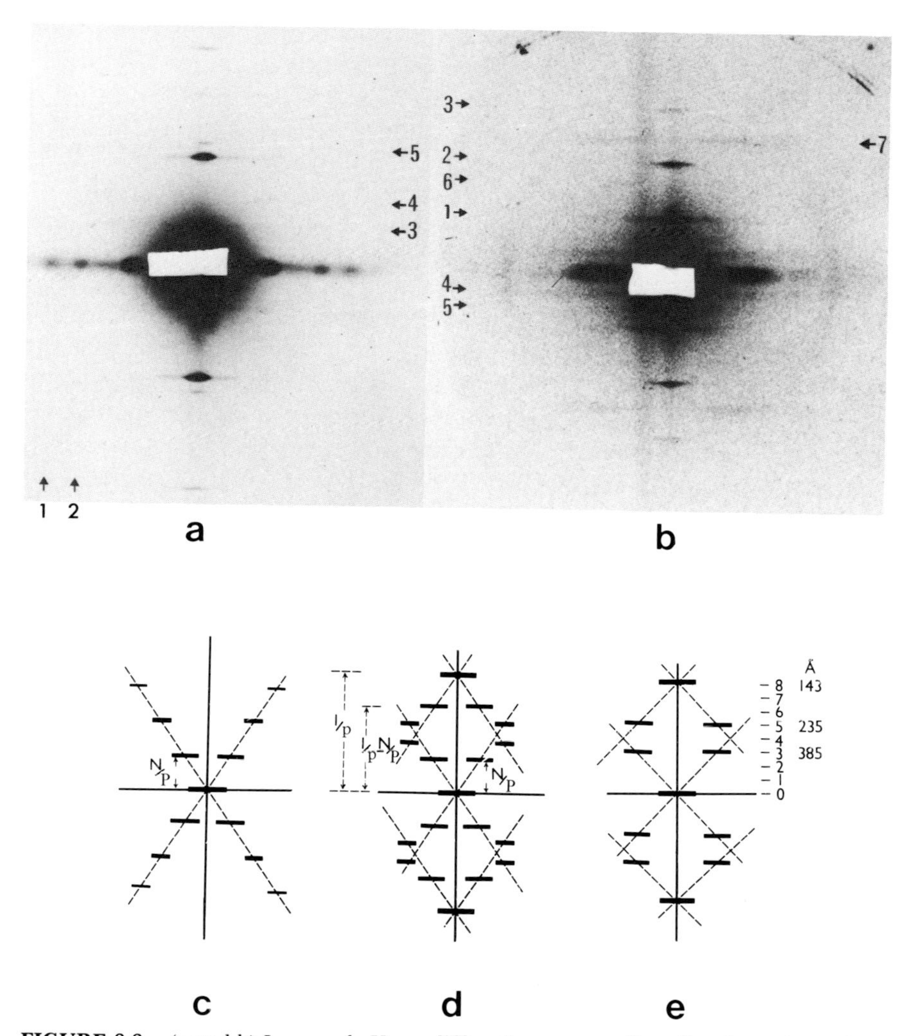

FIGURE 8.8. (a and b) Low-angle X-ray diffraction patterns from live *L. maximus* fibers (fiber axis vertical) showing the layer line pattern typical of this muscle state. In (a) reflections 1 and 2 are the 101 and 20$\bar{2}$ equatorial reflections caused by the hexagonal filament array (see Fig. 7.2), and reflections 3, 4, and 5 are the characteristic layer lines of spacing 385 Å, 235 Å, and 145 Å, respectively, which are orders of the 1150-Å repeat probably associated with the cross bridge array on the thick filaments. In (b) reflections 1, 2, and 3 are the appropriate orders of the 145-Å meridional repeat, and reflections 4, 5, and 6 are the "myosin" layer lines with spacings 385 Å, 235 Å, and 88 Å. Reflection 7 is the 59-Å layer line associated with the structure of the actin filaments (see Chapter 5). (c–e) The probable origin of the myosin layer line patterns in (a) and (b) and the expected appearances of diffraction patterns from (c) a continuous N-stranded helix of the pitch P, (d) a discontinuous N-stranded helix of pitch P and subunit axial translation p, and (e) a two-stranded 16/3 helix or six-stranded 16/1 helix. (a and b) From Miller and Tregear (1972); (c–e) from Squire (1977).

diffraction pattern is very likely caused by a helical structure of unit axial translation 145 Å. The layer line spacings (385, 235, and 145 Å) can be indexed as orders 3, 5, and 8, respectively, of a repeat of 1155 Å (just as in Reedy's micrographs). Since the axial distance (in $Å^{-1}$) from the origin to the 385-Å reflection is the same as the distance from the 145-Å layer line down to the 235-Å layer line, it is clear that these off-meridinal layer lines could correspond to layers defined by equal n values (Figs. 2.13 and 8.8c,d) in two helix crosses, one centered on the origin and one centered on the 145 Å reflection.

On this basis, the pitch (P) of the helix must be given by the two relationships explained and notated in Fig. 8.8d (Squire, 1972): $1/145 - n/P = 1/235$ and $n/P = 1/385$. In this case n (N in Fig.) corresponds both to the number of the layer in the helix cross and to the number of strands in the helix; the reasons for this are explained in Chapters 2, Section 2.3.7 and 7, Section 7.2.6 for the equivalent analysis of the X-ray diffraction evidence from vertebrate skeletal muscle. The equations above summarize the whole of the information on the axial periodicities in the insect thick filament that can be gleaned from the axial positions of the layer lines. Unfortunately, for any value of n that is chosen, a suitable value of P can be determined to satisfy these equations. In particular, if $n = 2$ (Fig. 8.7a) then $P = 770$ Å as deduced by Reedy (1967, 1968) and by Miller and Tregear (1972); if $n = 6$, then P must be 2310 Å (Squire, 1972). A six-stranded cross bridge arrangement must therefore have the symmetry of a helix of pitch 2310 Å as illustrated in Fig. (8.7c). On the basis of the X-ray diffraction evidence alone, other structures with different values of n, such as 3, 4, 5, etc., could not be excluded. But Reedy's micrographs (i.e., the flared X images) do seem to show clearly that the thick filaments have a twofold rotation axis which appears to rule out $n = 3$ and $n = 5$. Taking at its face value Chaplain and Tregear's result that there may be six myosin molecules per crown (i.e., every 145 Å), then the value $n = 6$ is clearly preferred.

We come now to the flared X appearances: how can they be explained if n is 6? One possible answer was put forward by Squire (1972) in terms similar to the arguments given at the end of the last chapter concerning the likely cross bridge interactions with actin that might occur in vertebrate skeletal muscle. In short, it was suggested that in rigor muscle only four of the six cross bridges in any crown (i.e., a single axial level of cross bridges) will interact with the thin filaments; the other two would require too much distortion of the S-2 linkage for actin attachment to occur. On the assumption that cross bridges attached to actin would be better preserved and stabilized during fixation, dehydration, and embedding than would unattached bridges, then electron micrographs would tend to show only the four attached cross bridges. Evidence that this might be the case in insect flight muscle was obtained by Reedy (1971). He found that occasionally, in some of his sections, a thin filament would be absent from its normal location in the hexagonal filament array (Fig. 8.6c). In such a case it was invariably found that if the thin filament would normally have been linked to one arm of the flared X, then the relevant arm of the flared X was not visible. The implication that cross bridges not attached to actin are not normally preserved and visualized in electron micrographs is hard to escape. Such a view is reinforced by Reedy's recent pictures of rigor muscle (Fig. 8.9) showing longitudinal sections of regions with occasional missing thin filaments (Reedy and Garrett, 1977). Once again, the cross bridges are not visible in the region where the actin filaments are missing.

One of Reedy's results which has not yet been mentioned is that the filaments surrounding any thick filament are themselves arranged on a helix. He found in his electron micrographs of ultrathin longitudinal sections that contained only thin filaments that the actin ends of the rigor flared Xs followed the pattern illustrated in Fig. 8.10a. Here adjacent thin filaments are related by an axial stagger of 128 Å. When this was considered in terms of the six thin filaments around one thick one, it was realized that these thin filaments must be on a helix of unit axial translation 128 Å and pitch 2 × 385 Å as illustrated in Fig. 8.10b. Since the thin filament itself repeats approximately after 385 Å, the arrangement can be thought of as a two-stranded 6/1 helix of cross bridge interactions. This is actually produced by systematic 60° rotations among the six thin filaments (1 to 6 in Fig. 8.10b), thus giving the arrangement shown in Fig. 8.11a. It was shown by Squire (1972) that the six-stranded cross bridge arrangement would interact with this helical array of thin filaments to give flared X patterns that followed closely those observed by Reedy (1967, 1968). Reedy's results are summarized in Fig. 8.12a,b, and the appropriate cross bridge interactions according to Squire (1972) are shown in Fig. 8.12c on the assumption that these interactions follow the observations of Moore *et al.* (1970). It is instructive to compare this with Fig. 7.64 of the last chapter for the corresponding interactions in vertebrate muscle where a similar conclusion is reached: that in rigor muscle only about two-thirds of the myosin heads will be linked to actin. Once again, it is evident that there are well-defined areas on the thin filaments that define where cross bridge attachment can occur.

Are There Six Molecules per Crown? After the six-stranded model had been shown to be a possible alternative to Reedy's two-stranded model, various attempts were made to find out which of the two was correct. The technique that was generally chosen was to follow up the approach of Chaplain and Tregear (1966) and to determine by other methods the number of myosin molecules per thick filament.

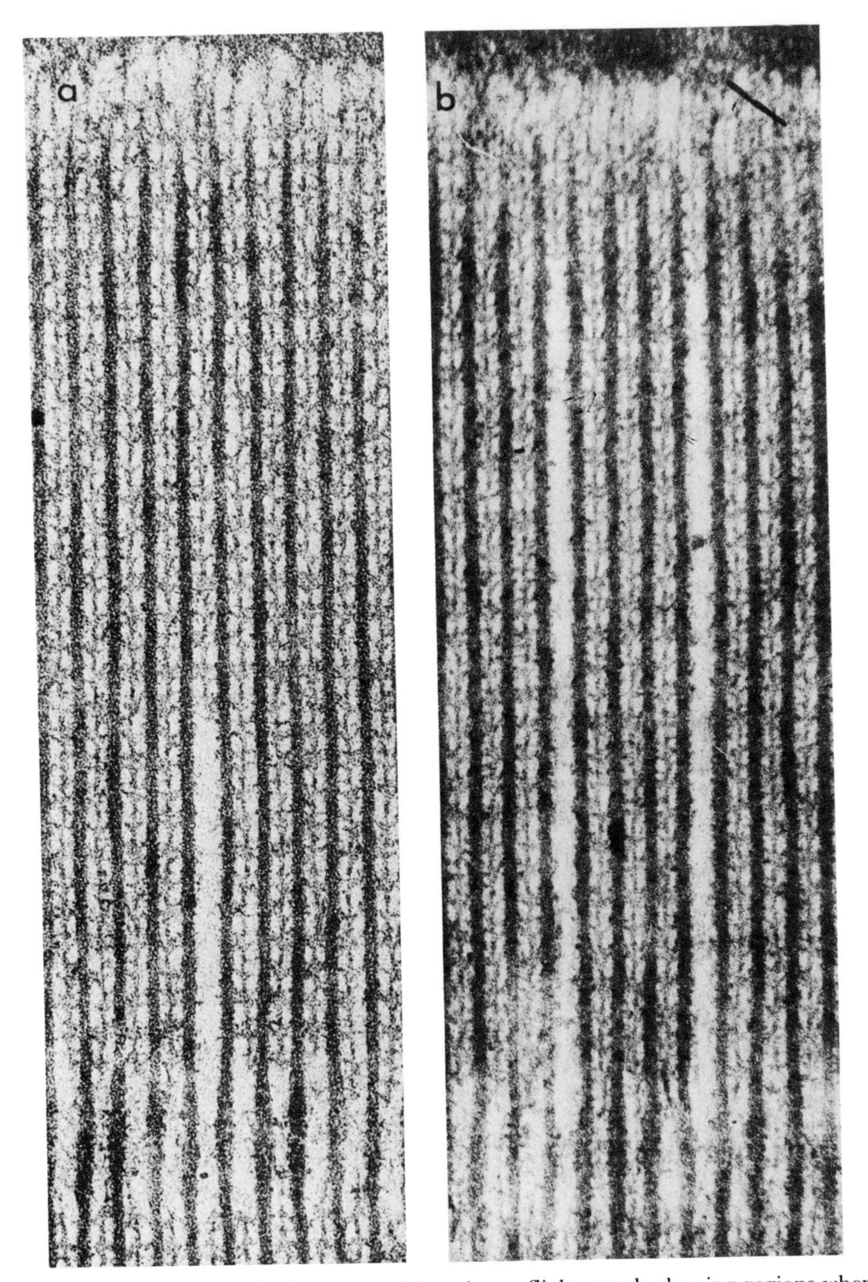

FIGURE 8.9. Longitudinal sections of rigor insect flight muscle showing regions where actin filaments are missing from the normal filament array. In these positions no cross bridges are seen on the thick filaments (from Reedy and Garrett, 1977). As Reedy puts it, "Rigor cross bridges don't show up without an actin to hang on to." ×150,000.)

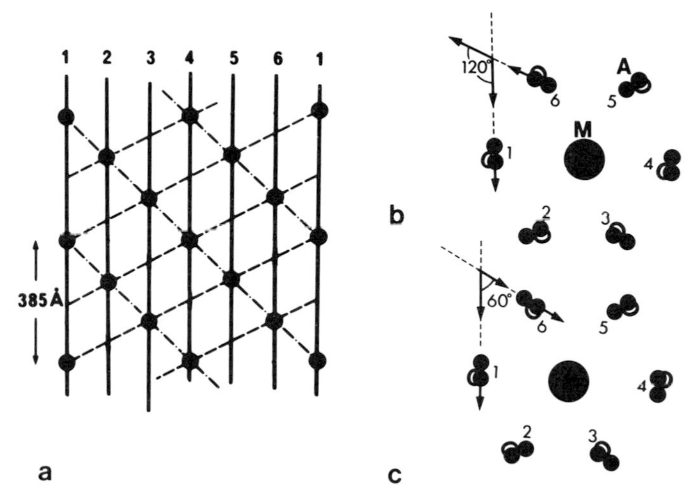

FIGURE 8.10. (a) The helix of six actin filaments that surrounds each thick filament in insect flight muscle. The helix can be thought of as approximating to a two-stranded 6/1 helix of actin target areas of pitch 2 × 385 Å as indicated. This helix can be generated at least approximately in two different ways: successive thin filaments (A) around one thick (M) are related by 120° rotations in (b) and by 60° rotations in (c). (c) An exact helix around the thick filament shown, but with diagonally related actins (e.g., 1 and 4) not equivalent. (b) Only an approximate helix, but opposite actins (1 and 4) are identically oriented and would thus allow a simpler arrangement of thin filaments in the Z-band. It is known that the long-period strands in the actin helix (see Fig. 5.22) follow a right-handed helix and that the helix of actins as in (a) is left-handed (Reedy, 1967, 1968).

Reedy *et al.* (1972) used quantitative microscopy, and Bullard and Reedy (1972) and Tregear and Squire (1973) both used SDS gel electrophoresis to give values for the actin-to-myosin mass ratio. Fortunately, all of these techniques led to the same conclusion, namely, that there are probably six myosin molecules per crown. This result tends to exclude Reedy's two-stranded model with two molecules per lattice point (giving four molecules per crown) but leaves open the possibility that the cross bridge arrangement is two-stranded but with three molecules per lattice point (thus giving six molecules per crown). We are therefore left with two models: a six-stranded model with one molecule per lattice point and a two-stranded model with three molecules per lattice point. The problem remains to decide between these alternatives. Unfortunately, the more spread out azimuthally the three molecules are in the two-stranded model, the more it will look like the six-stranded model. In fact, the six-stranded structure is a special case of the two-stranded structure in that in the latter, the three molecules per lattice point are separated azimuthally by 60°. In the next few pages arguments are presented that

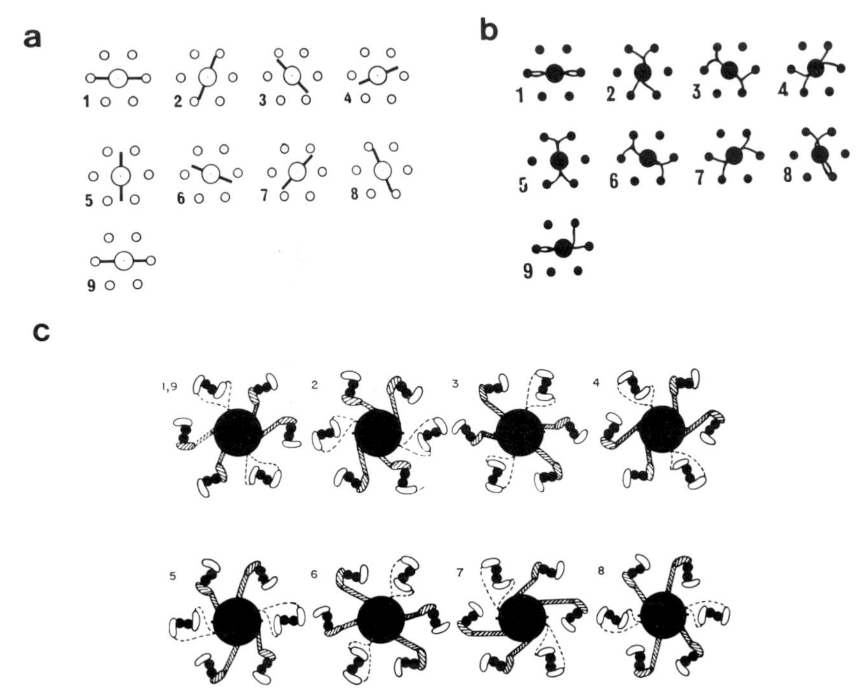

FIGURE 8.11. (a) The positions relative to the six surrounding actin filaments of successive cross bridge origins on a thick filament with the symmetry of a two-stranded 16/3 helix of pitch 770 Å. The levels shown are at the appropriate multiples of 145 Å from level 1, and levels 1 and 9 are identical and show the helix repeat of 1150 Å. (b) The same scheme as in (a) but in terms of the change in azimuth of the flared X cross bridge arrays that would be expected from the helix in (a). (From Reedy, 1967.) (c) A diagram corresponding to (b) but for a six-stranded helix of cross bridges and with actin filament profiles in approximately the correct relative orientations. Note that here for the sake of simplicity the thin filaments are assumed to approximate to structures with two-fold rotational symmetry, in which case the two actin arrangements in Fig. 8.10 would be indistinguishable. In this diagram it can be seen that, as in Fig. 7.64 for vertebrate muscle, it is easier for some cross bridges to interact with actin than it is for others. The "good" interactions in (c) produce the flared X interactions just as in (b). (After Squire, 1972.)

favor a six-stranded structure rather than a two-stranded one with closely bunched molecules, but a pseudo-six-stranded structure could not be excluded. The arguments are based on the observed structure of the M-region in *Lethocerus*.

It has already been mentioned that in transverse sections through the M-region, the *Lethocerus* thick filaments appear oval in profile. It can also be seen (Fig. 8.4b) that the oval profiles of adjacent filaments appear to point in different directions as illustrated in Fig. 8.12c which also shows the arrangement of the M-bridge connections between adjacent filaments (Pringle, 1972). In fact, there seem to be three orientations of

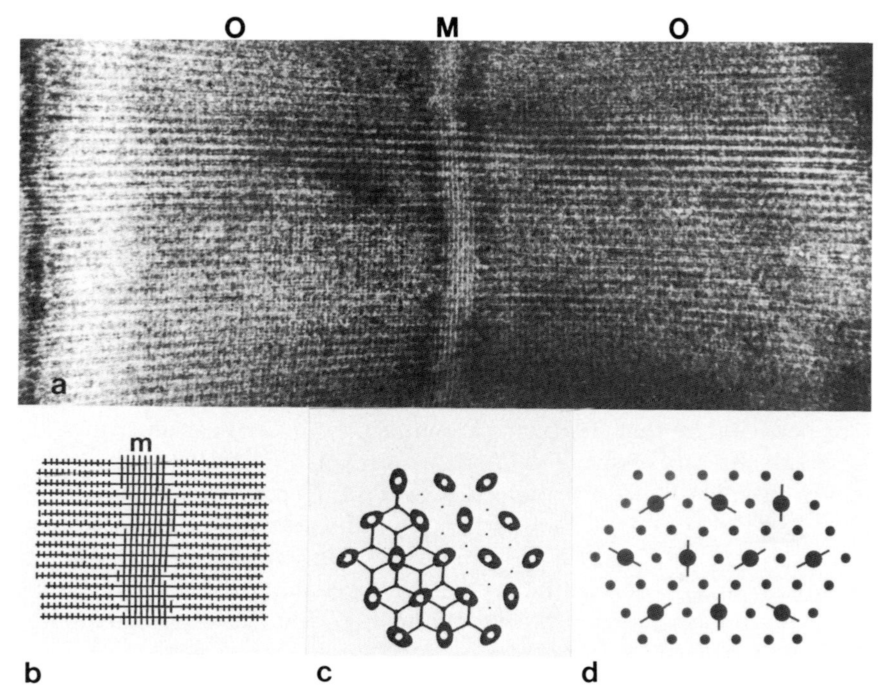

FIGURE 8.12. (a) Longitudinal cryosection (negatively stained) of relaxed *L. maximus* flight muscle showing clear 145-Å repeats both in the overlap regions (O) and in the M-region (M). It can be seen that in the M-region axial shifts of 145 Å or a multiple sometimes occur between adjacent thick filaments as illustrated in (b). (a) Courtesy of Dr. A. Freundlich; ×87,000. (c) The appearance of the insect M-band in transverse section (as in Fig. 8.4b) showing three orientations of the oval thick filament profiles 60° apart. (d) The kinds of orientations in the overlap region that the cross bridge origins on adjacent two-stranded filaments would have if positioned randomly in one of three orientations 60° apart as in (c). To satisfy Reedy's evidence (Fig. 8.6), these filaments would be required to interact with the surrounding actin filaments to produce flared Xs that all point in the same direction. (b–d) After Squire (1977).

these oval profiles separated by about 60°, but these different orientations appear to occur randomly through the fibril. If the thick filaments are assumed to behave as rigid rods (and if there is no axial stagger between filaments), then it is clear that in a transverse section through the overlap region of the A-band similar random orientations of the thick filaments must occur. If the thick filaments are two-stranded, then the cross bridge origins in such a transverse section might well be arranged as illustrated in Fig. 8.12d (from Squire, 1977). But one of Reedy's results (strikingly obvious in Fig. 8.6) is that in a given transverse section, the flared Xs on neighboring filaments always tend to point in the same direction. One might reasonably ask how cross bridges originating in as many different orientations as in Fig. 8.12d could interact with actin to produce identically oriented flared X structures. Clearly, whatever the thick filament orientations, the positions of the actin target areas (assuming the thin filaments are regularly arranged) will still be in the same place in each unit cell. But one could reasonably question whether such things as target areas could occur if the myosin cross bridges possessed sufficient flexibility for them to interact with thin filaments up to 90° away. This argument can be extended still further when it is also realized that axial shifts between thick filaments sometimes occur. That this is so can be seen in Fig 8.12a where, in the M-region, it can be seen that the M-band is not perfectly straight. However, it can also be seen that the 145-Å-spaced M-bridge lines on each thick filament remain in register right across the M-region. This can only mean that the axial shifts that occur between the thick filaments are multiples of 145 Å as illustrated in Fig. 8.12b. But successive 145-Å-spaced crowns in a two-stranded filament are related by 67.5° rotations between the cross bridge origins (as defined by the 16/3 helix shown in Fig. 8.7a), and this would mean that adjacent filaments could easily have cross bridge origins either 67.5° or possibly 2 × 67.5° apart at the same axial level along the A-band. Once again, such an arrangement would not be expected to give rise to identically oriented flared X structures.

The conclusion can therefore be reached that, provided the random rotation and axial shifts between adjacent filaments are not systematically related so as to preserve the orientation of the cross bridge origins in the overlap region (and they do not seem to be), then either the two-stranded filament model or the idea that there are actin target areas must go. Since the target area idea is supported by results on vertebrate muscle and by other results on insect muscle to be described in Chapter 10, it seems that it is the two-stranded model that must be dropped.

On the other hand, it is easily seen that the six-stranded model is quite compatible with the results on the M-region. Random rotations of 60° between adjacent filaments will leave the cross bridge origins unaltered, and at no time will an actin target area be more than about 30° away azimuthally from the nearest cross bridge whatever the axial shift between adjacent thick filaments. It is fair, therefore, to conclude that the cross bridge array in relaxed insect flight muscle may well be described by the six-stranded 16/1 helix illustrated in Fig. 8.7c.

The Remaining Uncertainties. Two fundamental uncertainties remain about the thick filament structure and the interpretation of the flared X images. Although the six-stranded 16/1 helix is the most straightforward interpretation of the layer line pattern in X-ray diffraction patterns from relaxed insect muscle (Squire, 1972), it has been realized by a number of those involved (e.g., Reedy and Garrett, 1977) that the structure giving rise to the relaxed X-ray pattern may actually be a result not of the thick filaments alone, but of the perturbation of the cross bridge array by some kind of interaction with the thin filaments. Thus, such patterns from relaxed muscle may not reveal the intrinsic thick filament geometry. For example, if there are some cross bridges attached to actin even in relaxed muscle, and such relaxed attachment follows the arrangement of actin target areas, then the perturbed myosin cross bridge array on the thick filaments will display repeats of 385, 240, and 1150 Å as observed. Electron micrographs of isolated thick filaments from insect muscle (Fig. 8.13; Reedy and Garrett, 1977) do indeed

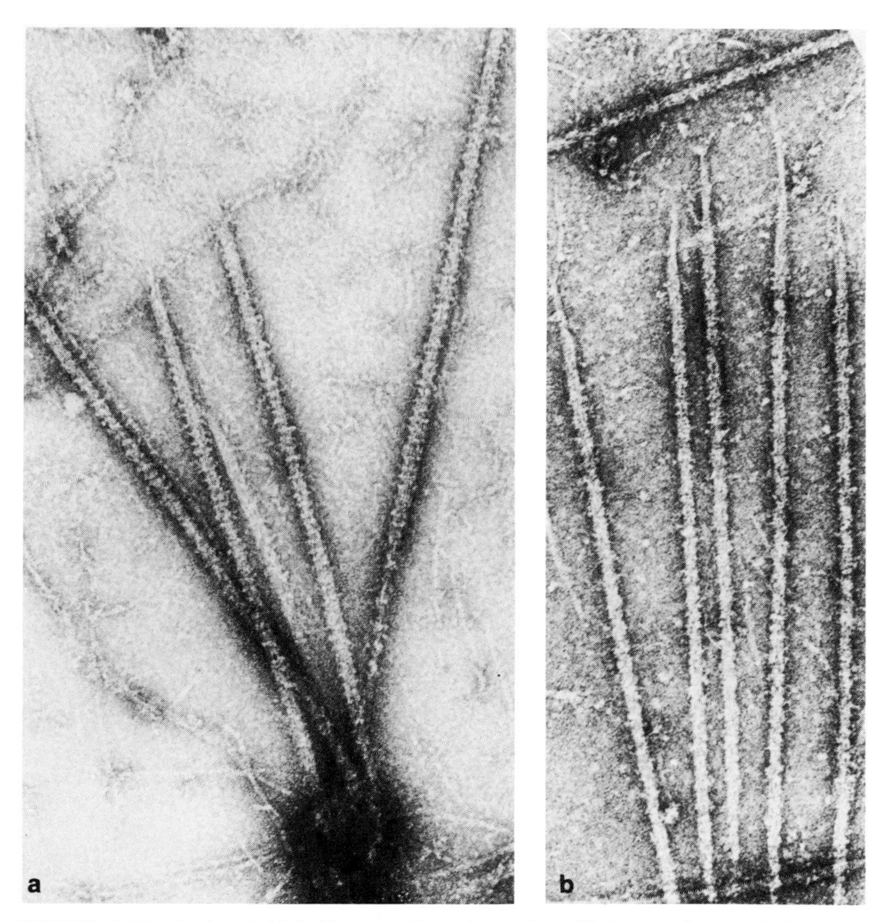

FIGURE 8.13. Isolated thick filaments from *L. maximus* flight muscle negatively stained with uranyl acetate. In (a) the thick filaments show a clear 145-Å periodicity but little evidence of helical order; the presumed cross bridges produce a uniform shelf of density every 145 Å. In (b), where the thick filaments have been treated with papain (briefly) to remove the myosin heads, little evidence of the 145-Å striping is apparent. Magnifications: (a) ×110,000; (b) ×90,000. (From Reedy and Garrett, 1977.)

seem to show a shelf of density every 143 Å and little evidence of helical order or a repeat of 1150 Å. It therefore seems likely but not certain that the cross bridge array is as in Fig. 8.7c, but in any case it seems to be the arrangement of actin target areas, not the cross bridge array on the thick filaments, that determines the probability of cross bridge attachment to actin in rigor and probably also in active muscle (Squire, 1972). Provided the actin filaments see shelves of cross bridges every 143 Å axially on the thick filaments, then the actin target areas will pick out those cross bridges with which they need to interact (Squire, 1972; Reedy and Garrett, 1977).

The second uncertainty, raised by Offer and Elliott (1978), concerns the number of myosin heads that are seen in each flared X structure. They suggested that each of the four arms of the flared X may represent one myosin head and that each half of the flared X is a Y-shaped structure (cf. Fig. 7.19) produced by the heads on a single myosin molecule separating and linking to two adjacent thin filaments (Fig. 8.14). This would mean that in rigor only four of the 12 available heads on one crown would be attached to actin. This important possibility will be considered in detail in Chapter 10.

Paramyosin. Apart from myosin, insect thick filaments also contain a significant amount of the protein paramyosin. This protein is a two-chain α-helical coiled-coil molecule similar in shape and size to the rod part of myosin. It is found most abundantly in molluscan muscles

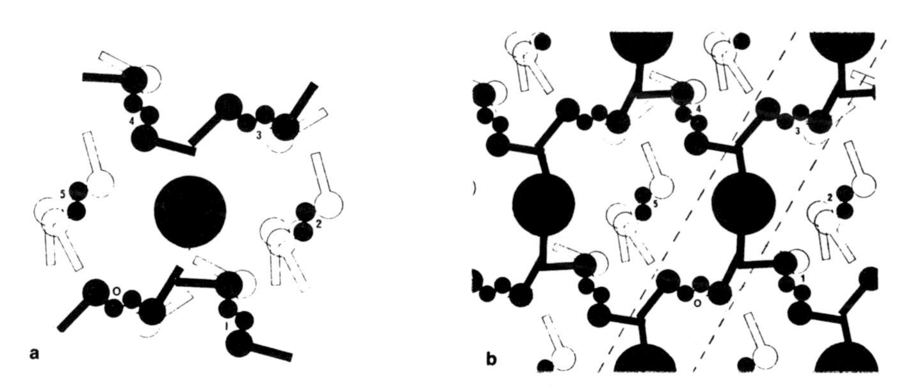

FIGURE 8.14. Possible pattern of myosin–actin interaction in rigor insect flight muscle according to Offer and Elliott (1978). The diagram shows six actin filaments as in Fig 8.10b surrounding one thick filament. With the particular absolute actin orientations shown, a number of myosin heads on adjacent thin filaments have their distal ends close together. They could belong to the same myosin molecule. (b) Continuing this possibility over a large area and linking each pair of close heads to the neighboring thick filament produces a cross bridge pattern reminiscent of Reedy's flared X apperances. Here each arm of the flared X is a single myosin head, not two as in Squire's model (Fig. 8.11c).

and, since these are the subject of a later part of this chapter, the properties of paramyosin will be considered at length there. Suffice it to say here that paramyosin commonly has the role of forming a core in the backbones of myosin filaments of large diameter and is frequently found in arthropod muscles.

In *Lethocerus,* paramyosin accounts for about 11% of the total thick filament mass, and this is sufficient to fill up the hollow core of the thick filaments, assuming that the core is about 100 Å in diameter (Squire, 1973). But in other insects, there is much less paramyosin (about 2% of the myosin mass in blowfly: Bullard *et al.,* 1973), and it has been found that the thick filaments sometimes appear more hollow near the periphery of a myofibril than at the center (Auber, 1969; Reedy *et al.,* 1972). It has also been found by Levine *et al.* (1976), who studied a variety of invertebrate muscles, that paramyosin content seems to correlate well with thick filament length but less well with thick filament diameter. It seems clear from this that paramyosin does not merely fill up the hollow parts of the thick filaments. On the other hand, Bullard *et al.* (1977) have shown by antibody-labeling studies of *Lethocerus* muscle that paramyosin is located in the M-region and the inner parts of the bridge region. It may therefore serve as a nucleating structure for filament growth and may also be involved in the thick filament length-determining process in invertebrate muscles. Insect thick filaments are considered in more detail in Chapter 9.

Finally, some comments on the tapered regions of the thick filaments are worth noting. One is that occasionally a slight thickening of the myosin filaments seems to occur partway down the taper (White, 1967). The second is that in transverse sections through this part of the A-band the thick filament profiles sometimes appear to be rather triangular, suggesting that the six-stranded structure might possibly taper down to a structure with only threefold symmetry at the filament tip, possibly with only three myosin molecules per crown (Squire, 1977). The third is that the hollow centers of the thick filaments appear to reduce in size in proportion to the outer diameter as the filament tapers (Squire, 1977). These points will be considered further in Chapter 9 where myosin filament structure is considered in detail.

8.2.3. I-band Structure in Insect Flight Muscle

The Z-Band. Unlike the Z-band in vertebrate skeletal muscle which is square and requires a redistribution of the thin filaments from their hexagonal arrangement in the overlap zone, the Z-band in insects is hexagonal and seems to require no displacements of the thin filaments. It was Ashhurst (1967b) who defined in detail the Z-band structure in

Lethocerus following the early work of Auber and Couteaux (1963) who used the flight muscle of two *Diptera*. Figure 8.15 shows a slightly oblique section through the *Lethocerus* Z-band and includes parts of the two adjacent sarcomeres. Here it can be seen that on each side of the Z-band

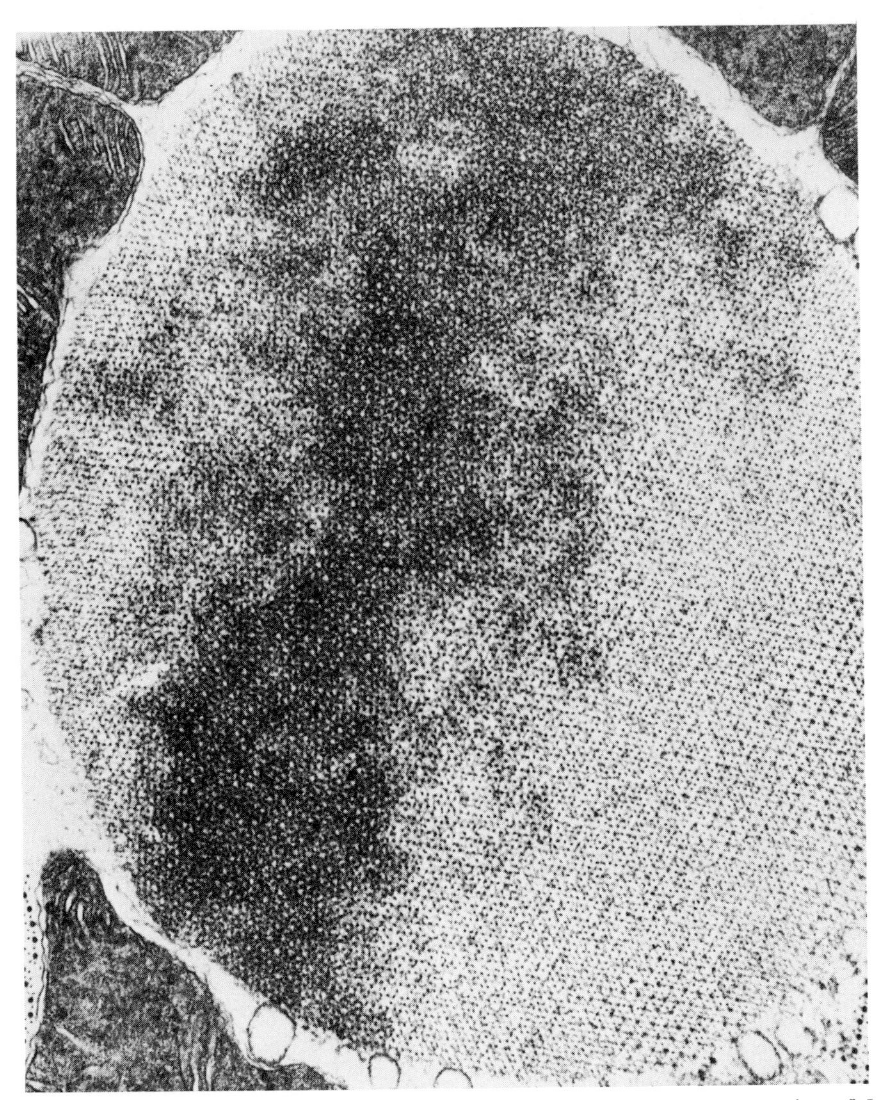

FIGURE 8.15. Electron micrograph of a very slightly oblique transverse section of *L. maximus* flight muscle showing the appearance at and near the Z-band. Top left and lower right are the filament arrays of successive sarcomeres. In the dense region between these areas in the Z-band itself. The thin filament lattices on each side can be followed into the Z-band where they can be seen to interdigitate to form part of the Z-band matrix as illustrated in Fig. 8.16a. (From Ashhurst, 1967b.) × 60,000.

the thick filaments have apparently terminated and an open hexagonal thin filament array is left. If the two open thin filament arrays on each side of the Z-band are followed into the Z-band itself, it can be seen that they are slightly displaced from each other but fit together regularly to form the hexagonal arrangement illustrated in Fig. 8.16a. Figure 8.16b shows a schematic three-dimensional view of the structure of one of the resulting small hexagonal Z-band units according to Ashhurst (1967b). Note that, although it is likely that adjacent thin filaments are cross-linked by some kind of Z-band bridge, the hexagonal structures drawn at each edge of the Z-band in Fig. 8.16b are not supposed to represent these cross-links. Figure 8.16c shows a remarkable micrograph of the insect Z-band in a longitudinal cryosection, and the observed image correlates

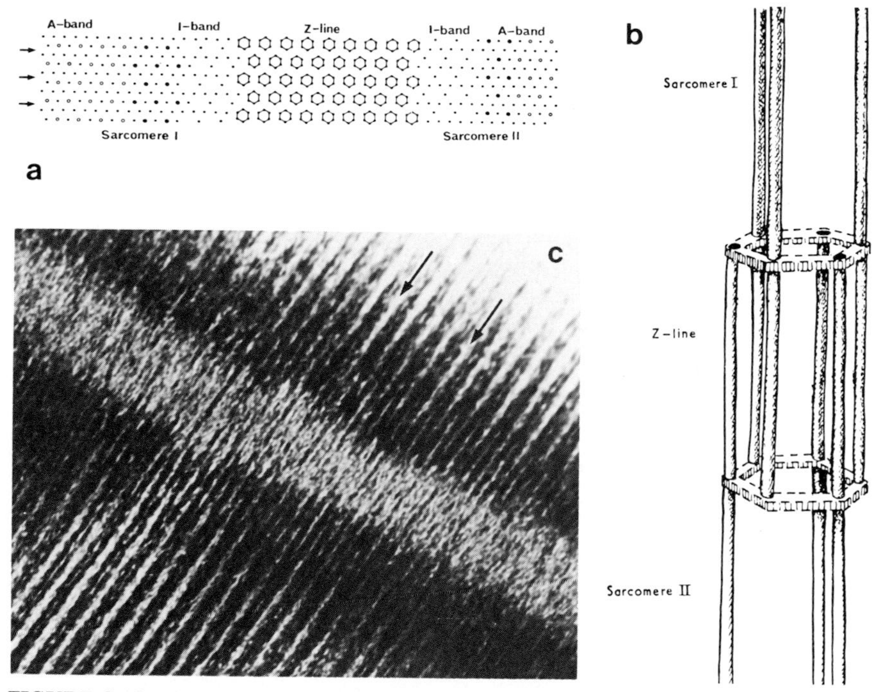

FIGURE 8.16. (a) Diagram to illustrate the precise interdigitation of the I-filament lattices of adjacent sarcomeres in the Z-bands of insect fibrillar flight muscle. In the Z-band one of the hexagonal units is thought to be as illustrated in (b), although the hexagonal cross-links are not supposed to represent any structural Z-band component. The general validity of this model is confirmed by the longitudinal negatively stained cryosection shown in (c) (courtesy of Dr. A. Freundlich). This is entirely consistent with the requirement that "no filament of one sarcomere is exactly opposite another in the adjacent sarcomere" (to quote Ashhurst, 1977). Viewing the micrograph along the direction of the filament axis (arrow shows this to be true (note that protein is white). (a) From Ashhurst (1977); (b) from Ashhurst (1967b); (c) magnification ×90,000.

well with the view through the Ashhurst model (Fig. 8.16a) in the direction of the arrow. The Z-band thickness is about 1100 to 1400 Å.

A slight variant of the Ashhurst model has been proposed by Saide and Ullrich (1973) to explain their electron micrographs of muscles from *Apis mellifica,* and another model for the honey bee Z-line has been proposed by Trombitas and Tigyi-Sebes (1975). For a discussion, see Ashhurst (1977).

It has recently been shown by Bullard and Sainsbury (1977) that the main constituents of isolated Z-bands (termed Z-disks) from *Lethocerus* flight muscle are α-actinin (Hammond and Goll, 1975), actin, and tropomyosin, but the roles of these proteins in the Z-band remain unclear. The marked density of the insect Z-band is electron micrographs seems to suggest that some kind of matrix material is present there; this may possibly be α-actinin or another protein that has recently been identified by Bullard *et al.* (1977). Some details of the latter are described below.

C Filaments. Brief mention was made earlier in the chapter of the fact that there is evidence for the existence of mechanical connections (C filaments) between the insect thick filaments and the Z-band. This idea was originally suggested by Auber and Couteaux (1963), and more recent evidence for it has been published by D.C.S. White (1967) and Reedy (1971). The most striking evidence is from longitudinal and transverse sections of highly stretched rigor muscle. In such sections (Fig. 8.17), it can be seen that the thin filaments in some sarcomeres have been pulled away from the Z-band but that thin strands can still be seen joining the thick filaments to the Z-band (D.C.S. White, 1967; Reedy, 1971). Despite these appearances, sections of unstretched muscle such as those published by Ashhurst (1967b) and shown in Fig. 8.15 show little evidence of continuity of the thick filaments between the A-band and the Z-band.

However, Bullard (1977) has recently obtained an antibody to a hitherto unknown insect protein, and this antibody specifically labels the insect I-band on each side of the Z-band. Trombitas and Tigyi-Sebes (1979) have also reported thick filament continuity in bee muscle. It has been suggested (Pringle, 1978) that these various results might be explained if the unknown I-band protein forms a matrix with the properties of a hydrated polymer gel between the A-band and the Z band. Sections of unstretched muscle might then not be expected to reveal the presence of "filaments" as such, but "filaments" could well be drawn out of the matrix if the sarcomere is stretched as in Fig. 8.17. This attractive idea might help to account for the mechanical continuity through the sarcomere that seems to be demanded by the stretch activation phenomenon. Further discussion of this property of insect muscle will be given in

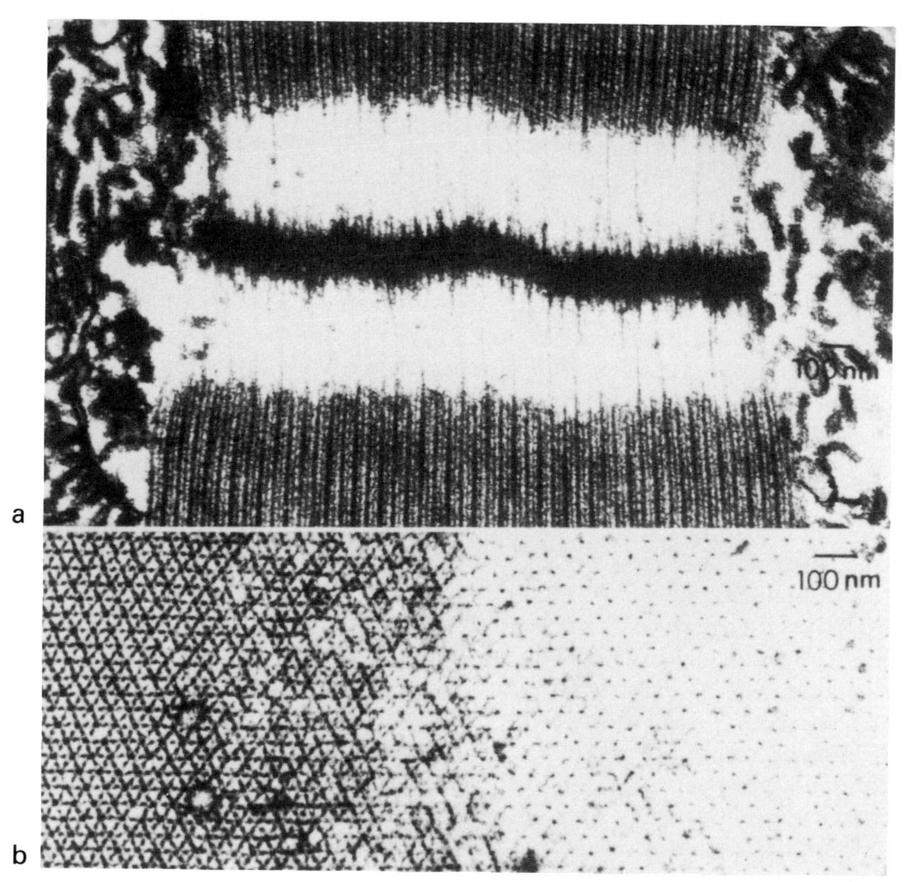

FIGURE 8.17. Electron micrographs of insect flight muscle (*L. maximus*) that was stretched after being put into rigor. The thin filaments appear to have detached from the Z-band, and filamentous material (the C filaments?) is left linking the thick filaments to the Z-band. (a) Longitudinal section; (b) slightly oblique transverse section; both courtesy of Dr. D. White.

Chapter 10, where the nature of the cross bridge cycle is considered in detail. The structure of insect rigor muscle will also be discussed more fully there.

8.2.4. Discussion: Details of Other Arthropod Muscles

Insect Muscles. So far the structure of *Lethocerus* flight muscle has been considered in detail, but little has been said about other insect muscles. The reason for this is simple: very little is known about the detailed ultrastructures of other insect muscles. Tregear and Squire

(1973) have studied the myosin content of insect leg muscle and have concluded that there are probably at least five myosin molecules per crown. Little else is known about the filament geometry in that type of muscle. Rodger (1973) has studied the structures of a number of insect muscles by X-ray diffraction, and all show the same axial periodicities of 385 Å and 145 Å. But generally, the trunk and leg muscle of insects are sufficiently irregular to make it difficult, as yet, to determine their structures in detail. However, it does seem likely that the cross bridge arrangements on the thick filaments of these muscles will be similar to those in the insect flight muscles provided that the filament diameters are similar (see Chapter 9, Section 9.1.1).

Crustacean Muscles. Until recently relatively little work had been carried out on the detailed ultrastructure of the myofibrils in crustacean muscles. An interesting review of the biology of crustacean muscles has been written by Atwood (1972), and the general morphology of the muscle fibers and fibrils was described earlier in the chapter. One of the most interesting features of crustacean muscles is the variability of ultrastructure that occurs within a fiber or even within a single myofibril.

A detailed analysis of sarcomere structure in the walking legs of the crab *Portunus depurator* was carried out by Franzini-Armstrong (1970). She showed that three groups of fibers could be distinguished. They contained sarcomeres of average length about 4 μm, 5 μm, and 7μm. She also found that within a single myofibril in any of the fibers in these groups a variation of sarcomere lengths of as much as 25% could sometimes be observed. In such a case, where the lengths of successive sarcomeres in one myofibril were clearly different, the thin filaments on each side of the common Z-band were clearly of different lengths as well. Franzini-Armstrong (1970) also noted that the M-region of the A-band in this muscle (which she termed the L-line) did not contain any structure equivalent to an M-band and was only about 600 Å long. This is much less than the M-region in vertebrate skeletal muscle (Sjöström and Squire, 1977a; Craig, 1977). Like the insect thick filament, the thick filaments in the crab (normally about 210 Å in diameter) become slightly larger at the M-region and often appear less hollow there than elsewhere along their length.

April *et al.* (1971) reported evidence from combined electron microscopy, light diffraction, and X-ray diffraction methods that the A-band lattice in single fibers from the walking legs of crayfish (*Orconectes*) exhibits the same constant volume behavior as the vertebrate A-band (Chapter 7). This was found to be true even when the fibers had been skinned and demonstrated that the constant volume behavior is an intrinsic property of the myofilament lattice in the A-band of this muscle. Subsequently, April (1975) and April and Wong (1976) showed that this was true only under certain conditions.

Wray *et al.* (1975) carried out an X-ray diffraction study of lobster abdominal extensor muscle and found that the diffracted intensity pattern from relaxed muscle could be accounted for by a helix of cross bridges that was six-stranded. Insufficient evidence was available to define the geometry of the cross bridge lattice. In another X-ray diffraction study, Yagi and Matsubara (1977) obtained good evidence that the thick filaments in crustacean muscles are probably hollow.

More recently, Wray (1979a) has carried out a systematic X-ray diffraction study of various crustacean muscles and has found that they give rise to very beautiful and simple X-ray diffraction patterns. Wray found that the layer line patterns could not be interpreted solely in terms of arrays of myosin cross bridges; it was also necessary to take into account the structures of the thick filament backbones. These patterns are now being analyzed in detail, and they should prove to be very informative. The main conclusions from them about thick filament backbone structure are described in Chapter 9, Section 9.3.4. But two points are already clear. One is that the thick filaments in different crustacean muscles, although not identical, are very closely related indeed. In fact, it seems that the cross bridge arrays can be described in terms of helices with four, five, or six strands of cross bridges but that the axial repeats of these helices vary slightly from muscle to muscle. The other important point, which arises directly out of the potential usefulness of Wray's results, is that it is well worth studying in detail the structures of muscles other than vertebrate skeletal muscle, insect flight muscle, and the smooth muscles. There is little doubt that in the foreseeable future crustacean muscle will have made its mark as a very useful tool for ultrastructural research.

Limulus Muscle. Arthropod muscles that have been the subjects of a considerable amount of study are the muscles in *Limulus,* the horseshoe or king crab (Fig. 8.18). Strangely, this is not a crustacean but an arachnid closely related to the spiders. For many years it has been reported that contraction of *Limulus* muscle involves a shortening of the A-band in addition to the normal sliding filament mechanism (de Villafranca, 1961; Dewey *et al.,* 1973). The *Limulus* telson muscles have thick filaments of normal length about 5 μm (containing paramyosin), thin filaments about 2.4 μm long, orbits of about 9 to 13 thin filaments around each thick filament (Fig. 8.19), and no evidence of an M-band.

Dewey *et al.* (1973) made two main observations about A-band length. One was that at long sarcomere lengths the apparent A-band length increases. This was thought to be caused by relative axial sliding

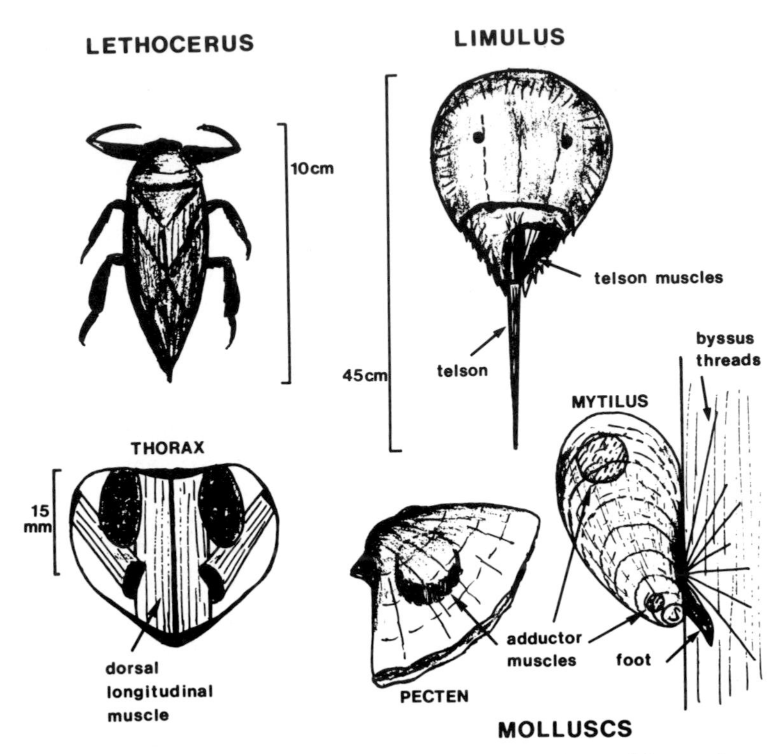

FIGURE 8.18. Sketches of some important invertebrate species used in muscle research. Top left shows a dorsal view of the giant water bug *Lethocerus*. *Lethocerus maximus* can grow to 10 cm in length (see Cullen, 1977). Bottom left is a schematic diagram of the *Lethocerus* thorax showing the dorsal longitudinal (flight) muscles, which are good subjects for insect flight muscle research (after Pringle, 1977b). Top right shows a dorsal view of the horseshoe crab *Limulus*. Part of the carapace is removed to show the dorsal telson muscles. For detailed dissection of *Limulus* telson muscles see Dewey *et al*. (1973). Bottom right shows two examples of molluscan bivalves, and the locations of their adductor muscles have been "ghosted" in. *Mytilus* (mussel), with its characteristic byssus threads and foot, has the usual complement of two adductor muscles but the anterior adductor is much reduced. *Pecten* (scallop) has a single (posterior) adductor muscle that is nearly centrally placed.

of adjacent thick filaments without there being a change in filament length. The second observation was that at sarcomere lengths below about 7.4 μm the A-band length appeared to reduce linearly with sarcomere length (down to about 4.5 μm). This was thought to be evidence of shortening of the thick filaments. Dewey's results are summarized in Fig. 8.20.

Partly in order to test this idea, Wray *et al.* (1974) recorded X-ray diffraction patterns from *Limulus* muscle and measured the spacings of the observed layer lines as a function of sarcomere length. They found that even at the shortened sarcomere lengths the myosin reflections

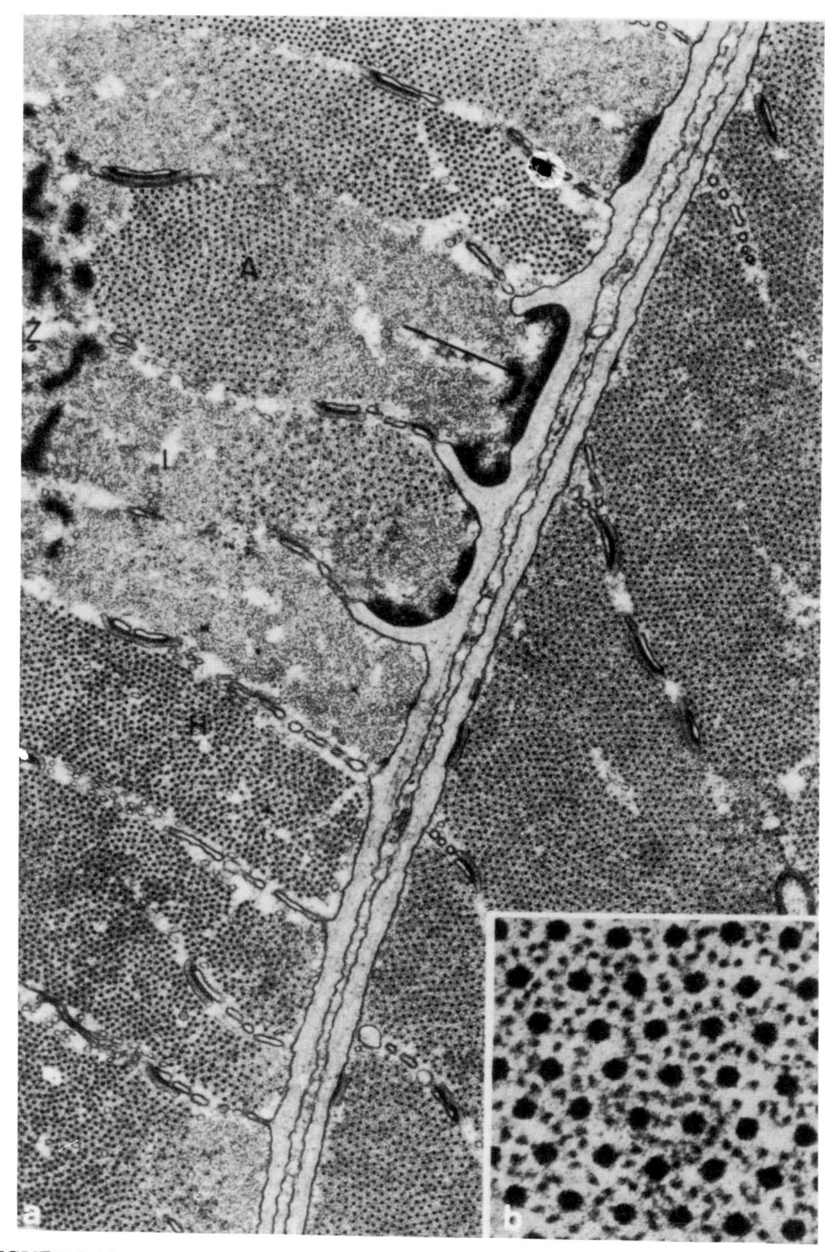

FIGURE 8.19. (a) Transverse section of lengthened fibers of *Limulus* telson muscle. A-, H-, I-, and Z-bands are visible as are myofibrillar attachment regions (arrow). Tubules arise from both the muscle fiber surfaces and clefts, and they form dyadic and triadic junctions with elements of the sarcoplasmic reticulum (×35,000). (b) Region of overlap of thick and thin filaments showing the characteristic filament arrangement (×170,000). (From Dewey *et al.*, 1973.)

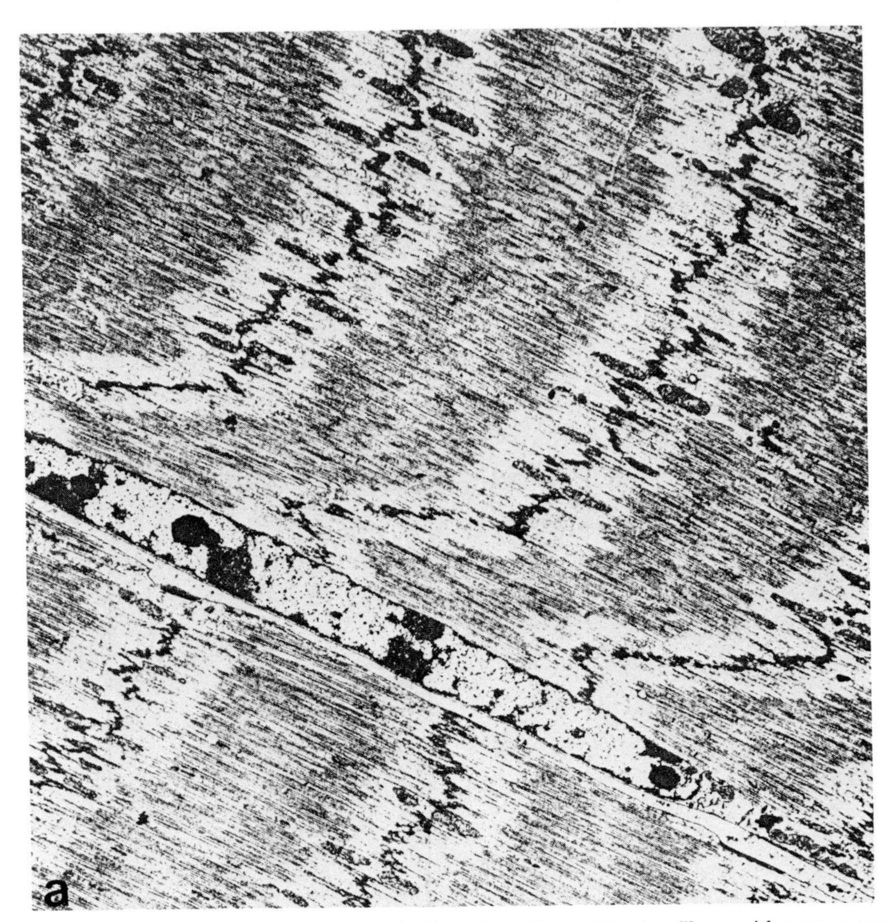

FIGURE 8.20. (a) Part of a longitudinal section of two *Limulus* fibers with sarcomere length of 7.6 μm. Thick filaments are misaligned, giving an irregular A–I boundary. An indistinct H-zone is apparent, but there is no M-band. Z-bands are irregular and discontinuous. Mitochondria occur near the Z-band. (from Dewey *et al.,* 1973.) (b) Low-angle X-ray diffraction pattern from glycerinated *Limulus* telson muscle in relaxing medium of ionic strength 0.08. Intensity on the 146-Å layer line falls to a minimum at a radius of 360 Å and rises to a weak subsidiary maximum at 270 Å. The 438-Å and 110-Å layer lines have maxima at a radius of 215 Å, and there is a minimum on the 438-Å layer line at 165 Å. The pattern is less well developed when recorded from muscles in a relaxing solution of higher ionic strength. (From Wray *et al.,* 1974.) (c) Graph showing the relationship of both A-band (solid line) and thick filament (dashed line) lengths as a function of sarcomere length in *Limulus* telson muscle according to Dewey *et al.* (1973). Ranges *a, b,* and *c* correspond to the telson maximally elevated (a), straight out (b), and maximally depressed (c).

remained at the same positions (146 Å and 73 Å) as in patterns from muscles at longer sarcomere lengths. On the other hand, diffraction patterns from muscles put into rigor at different lengths showed very

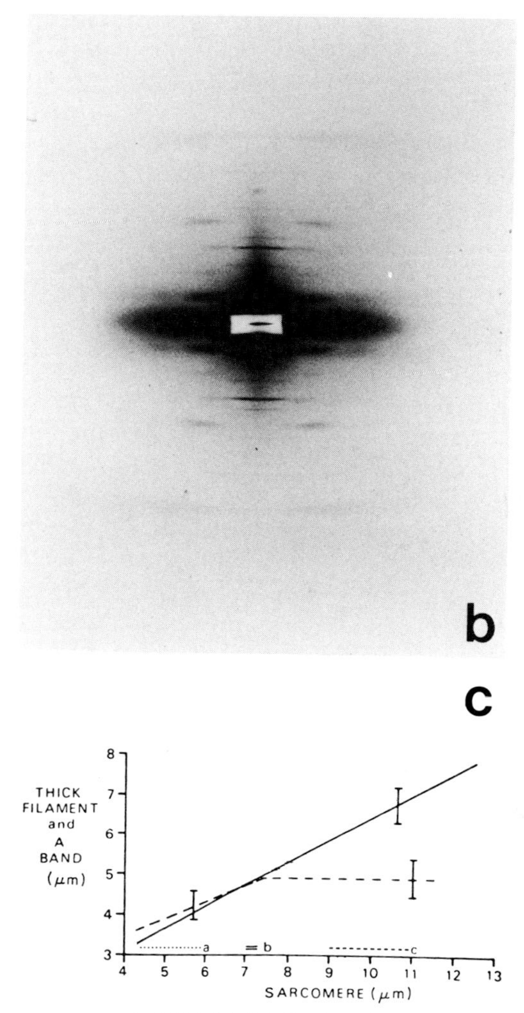

FIGURE 8.20 *Continued*

little change in the intensity of these reflections which are known to be caused by the cross-bridge-labeled thin filaments. It therefore appeared either that the relative axial sliding of the thick filaments was tending to preserve the overlap length of the thick and thin filaments or that the thin filaments must have been very long.

Dewey and his collaborators (Dewey *et al.*, 1973, 1977; Brann *et al.*, 1979) have suggested that their observations of shortening A-bands at short sarcomere lengths may be compatible with the X-ray diffraction evidence either if the tips of the thick filaments dissolve progressively as the sarcomere shortens (an idea not strictly compatible with the constancy of intensity of the 146-Å and 72-Å myosin reflections: Wray *et al.*, 1974) or if the myosin filaments shorten by a stepwise movement of molecules from one 146-Å-spaced configuration to another. It is clear that further direct evidence is required to clarify these points; at this moment their true significance is obscure.

Another aspect of the X-ray diffraction studies of *Limulus* muscle by Wray *et al.* (1974) was that the layer line pattern caused by the myosin cross bridges in relaxed muscle was very clear and unsampled (Fig. 8.21). It was therefore possible to carry out a direct analysis of the intensity distribution in the diffraction pattern in terms of the likely cross bridge arrays on the thick filament surface. They found that the layer lines (which had almost the same axial repeat as those from frog muscle, i.e., 438 Å, 219 Å, and 146 Å) could be modeled reasonably well by either three- or four-stranded helical arrangements of the kind shown in Fig. 7.9 for vertebrate muscle. Since the thick filaments in *Limulus* have a slightly larger backbone diameter (Levine *et al.*, 1976) than those in vertebrates, it was suggested by Wray *et al.* (1974) that if the cross bridge array in vertebrates is three-stranded, then that in *Limulus* is very likely four-stranded. Another important feature of the results of Wray *et al.* (1974) was that the experimenters found that changes in ionic strength had a marked effect on the cross bridge order in relaxed glycerinated muscle. Muscles in a relaxing medium with ionic strength 0.08 gave much better diffraction patterns (Fig. 8.20b) than those at ionic strength 0.20. It was suggested that activation of the thick filaments might be associated with a change in charge distribution on the thick filaments. This attractive hypothesis will be considered again in Chapter 10. Regulation in *Limulus* is myosin-linked as well as actin-linked, and the alteration of charge distribution could, in this case, be associated with Ca^{2+} binding through the myosin light chains.

The observations of Wray *et al.* (1974) on *Limulus* muscle in the rigor state will be considered in the general discussion on rigor muscle in Chapter 10.

8.3. Molluscan Muscles

8.3.1. Introduction

The muscles of molluscs provide a fascinating study in themselves since they are in some cases cross-striated, sometimes obliquely striated, and sometimes smooth. As a general rule, molluscan muscles display the unusual property of *catch,* in which tensions can be maintained for very long times with relatively little expenditure of energy. But not all

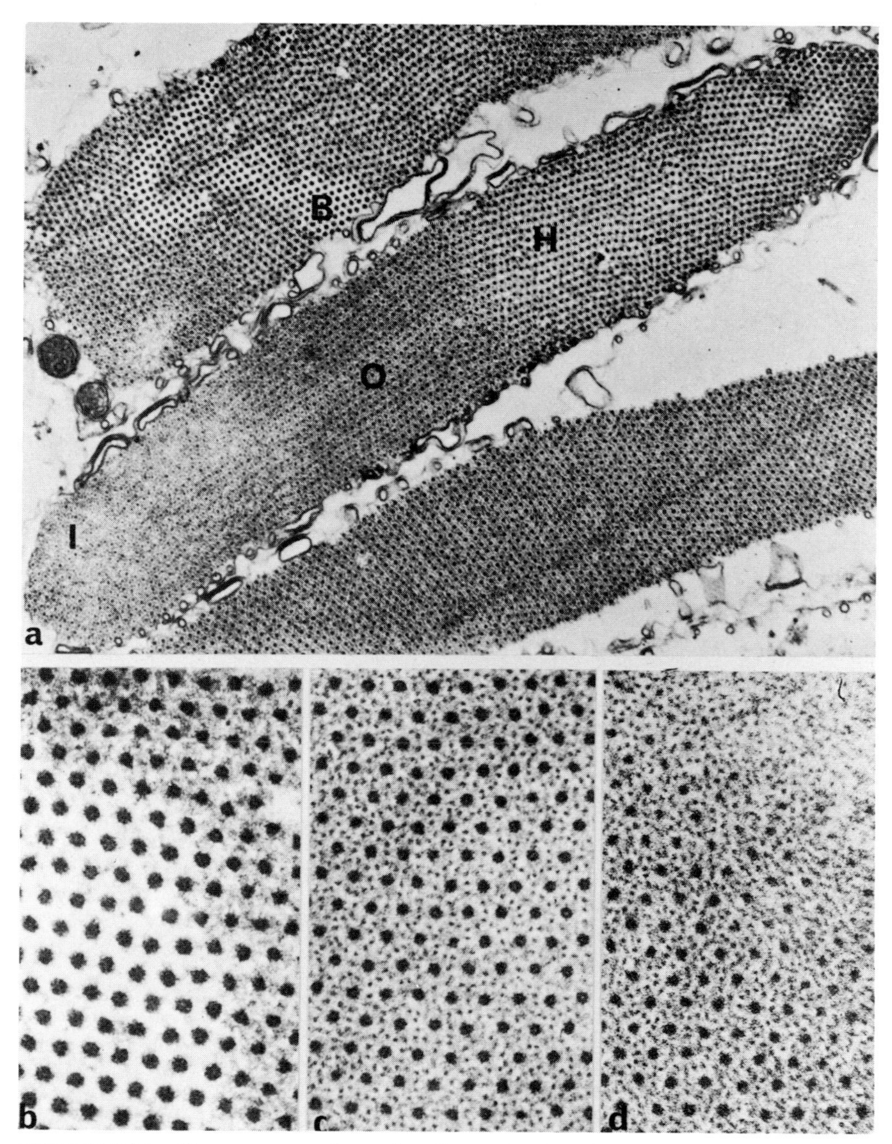

FIGURE 8.21. Transverse sections of scallop striated adductor muscle. (From Millman and Bennett, 1976.) (a) Low-magnification view showing the elongated profile of the muscle fibers and the appearance of the filaments in different regions of the sarcomere; B, bare region; H, the part of the H-zone where the thick filaments have projections; O, the overlap region; I, the I-band. Magnification ×25,000. (b–d) High-magnification views (×110,000) of different regions of the sarcomere; (b) the bare region, (c) the overlap region showing the filament lattice, and (d) the overlap region near the A–I junction where the thick filaments taper.

molluscan muscles are catch muscles. Just as in the case of insect flight muscles where the stretch activation phenomenon has been the driving force behind much of the insect muscle research, in the study of molluscan muscles the main driving force has been the exploration of the catch phenomenon. Unfortunately, just as with stretch activation in insects, molluscan catch is not yet completely understood.

The molluscs that are usually studied are the bivalves, which include clams, mussels, oysters, and scallops (Fig. 8.18). As outlined briefly in Chapter 1, the particular muscles that are suitable for biophysical studies are the adductor muscles (which open and close the two shells of the bivalve) and also in the case of the mussel (*Mytilus*) the muscle that controls the foot and byssus threads of the animal (the anterior byssus retractor muscles or ABRM). In most bivalves, the adductor muscles are smooth catch muscles that are well-suited to their usual role of holding the two shells tightly shut for long periods. But in the case of the scallop (*Pecten*), the adductor (Fig. 8.18) is largely striated and is not a catch muscle. In many ways this muscle is similar to vertebrate skeletal muscle (as described in Section 8.3.2) and is adapted to open and close the shells of the scallop fairly rapidly so that it can "swim" along by expelling water. A small part of the scallop adductor is a normal catch smooth muscle.

Another type of molluscan muscle is known as obliquely striated muscle and occurs in, for example, the translucent adductor of the oyster. Since obliquely striated muscles are predominant in the annelids and nematodes, a brief discussion of this type of muscle will be given in a later part of this chapter (Sections 8.5.1, and 8.5.2).

8.3.2. Structure of Scallop Striated Adductor Muscle

Recently Millman and Bennett (1976) have carried out a detailed study (by electron microscopy and X-ray diffraction) of the ultrastructure of the cross-striated adductor of the scallop (Fig. 8.18), and they and Wray *et al.* (1975) have analyzed the X-ray diffraction pattern from this muscle in the relaxed state to determine the cross bridge symmetry on the thick filaments. Some details of these results are summarized here.

In some ways this muscle (Fig. 8.21) is similar to the slow arthropod muscles, since it comprises thick filaments containing paramyosin and arranged in a hexagonal lattice, each surrounded by rings of about 12 actin filaments (Lowy and Hanson, 1962; Sanger and Szent-Gyorgyi, 1964; Lowey *et al.,* 1966). On the other hand, the thick and thin filament lengths (about 1.76 and 1.0 μm) and sarcomere lengths (Fig. 8.22) seem to be constant right through the muscle (Millman and Bennett, 1976) and are similar to those in vertebrate skeletal muscle. The fibers in this

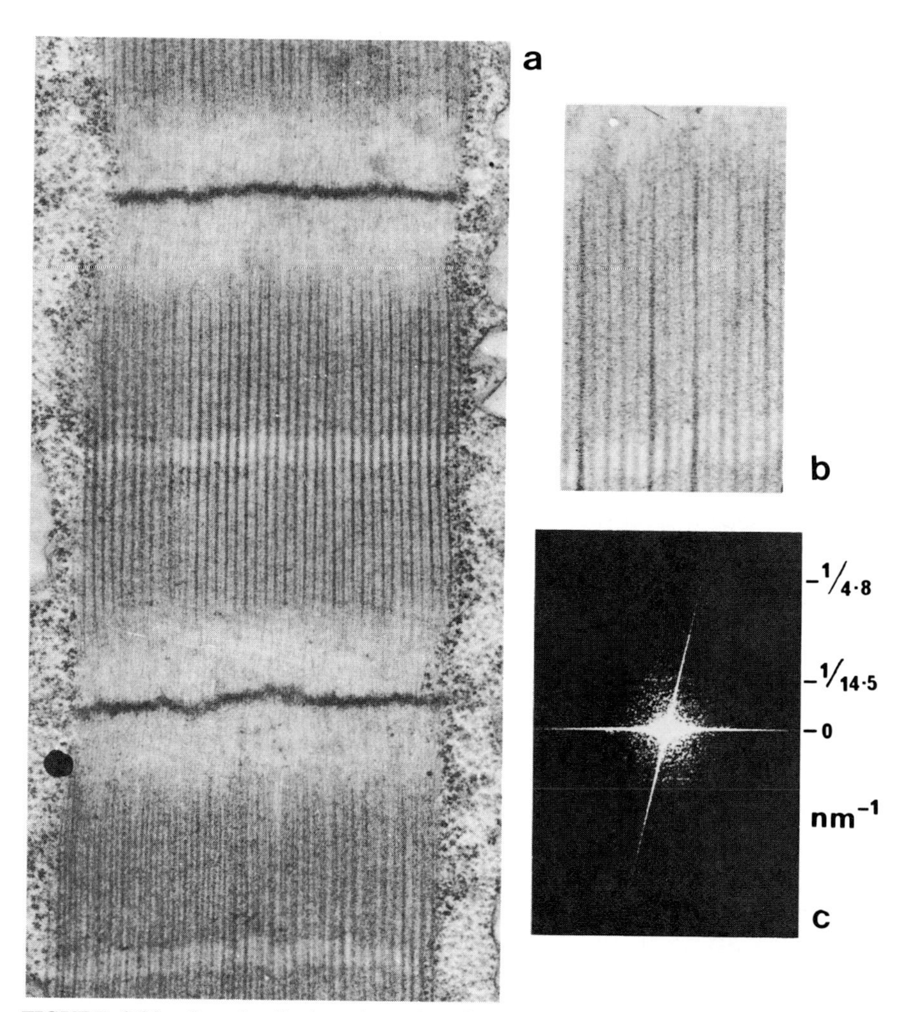

FIGURE 8.22. Longitudinal section of scallop striated adductor muscle. (a) General view; (b) selected area of the overlap region; (c) optical diffraction pattern from (b). Note the transverse striping that can be seen on some filaments in (b). (From Millman and Bennett, 1976.)

muscle are elongated and have dimensions about 1 μm × 10 μm (Millman and Bennett, 1976). No structure equivalent to the vertebrate M-band has been seen, but there are well-defined Z-bands about 400 Å wide. The Z-band is extremely dense and shows little of the regular structure seen in sections of vertebrate muscle. In the M-region the thick filaments are thicker (220 Å in diameter) than elsewhere (170 Å diameter) and do not appear hollow, although hollowness has occasionally been seen in other parts of the A-band.

X-ray diffraction patterns and optical diffraction patterns from electron micrographs indicate that the same thick filament repeat of 145 Å occurs in scallop as in other muscles and that the thin filaments have the usual structure with a pitch of about 2 × 385 Å. Meridional reflections at spacings of about 380 Å, 190 Å, and sometimes 128 Å were attributed to the tropomyosin repeat along the thin filaments despite the virtual absence of troponin in this muscle.

Analysis of the intensity distribution in the X-ray diffraction patterns from relaxed muscle (Fig. 8.23) by Millman and Bennett (1976) and by Wray *et al.* (1975) seemed to indicate that the cross bridge arrangement on the scallop thick filaments could be described by a six-stranded helix of pitch 2900 Å. This would account for strong layer lines of spacings that are orders 3, 7, 10, 13, 17, 20, etc. of a 1450-Å repeat. For example, the third order at 464 Å would be the sixth layer (n = 6) in the helix cross from a helix of pitch 2900 Å (2900/6 = 483 Å). Figure 8.23b shows a radial net for this cross bridge arrangement. Evidence that the thick filaments are approximately six-stranded was also obtained by Millman and Bennett from electron micrographs of scallop thick filaments.

Additional information about the myosin cross bridge arrangement

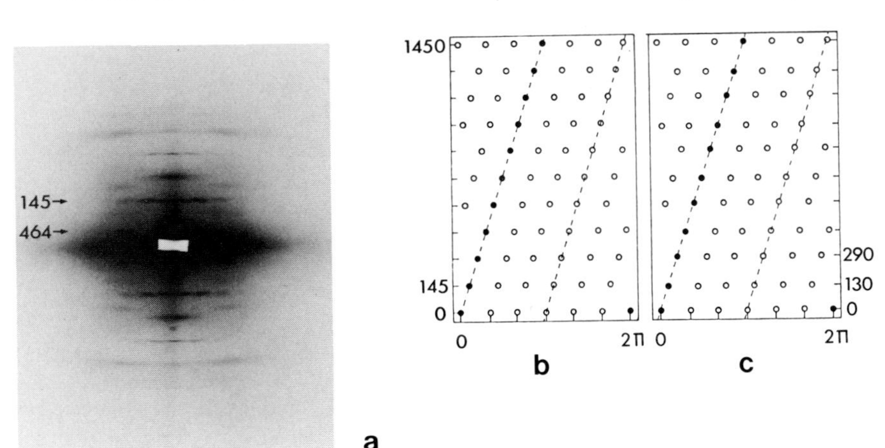

FIGURE 8.23. (a) Low-angle X-ray diffraction pattern from scallop striated adductor muscle showing the characteristic myosin layer line pattern with prominent reflections at 464 Å and 145 Å. (From Wray *et al*, 1975.) (b and c) Radial projections of possible models for the cross bridge arrangement in the scallop thick filaments. Both have axial repeats of 1450 Å and are based on six-stranded helices of pitch 2 × 1450 Å. In (b) successive cross bridge levels are 145 Å apart, whereas in (c) the levels are spaced alternately by 130 Å and 160 Å, giving a 290-Å subunit repeat. (b and c) After Millman and Bennett (1976). In both lattices a filament diameter of 210 Å would have the cross bridge origins spaced 110 Å apart.

in scallop was the presence in X-ray diffraction patterns from relaxed muscle (Millman and Bennett, 1976) of meridional reflections that were orders 3 and 5 of a 290-Å repeat. It was thought that these reflections must be caused by the cross bridge arrangement rather than any other feature of the muscle since they only appeared when the other myosin layer lines were present too. Formally, "forbidden" meridional reflections from a helical structure can be produced by any periodic disturbance of the repeating units in the helix. Millman and Bennett (1976) attributed the 290-Å repeat to a disturbance of the 145-Å myosin cross bridge repeat so that successive crowns were spaced alternately by about 130 Å and 160 Å (Fig. 8.23c). In this way, a 290-Å axial repeat is generated, and the displacement is such that the unobserved first-order reflection at 290 Å would be weak, but the third order at 97 Å would be strong (Fig. 8.24). Clearly, this interpretation may well be correct, but it is probably as well to remember that other periodic perturbations of the cross bridge array (with a 290-Å repeat) might possibly explain the observations.

Discussion of the structure of *Pecten* striated adductor muscle in rigor will be given in Chapter 10, Section 10.2.2.

8.3.3. Structure of Molluscan Smooth Muscles

Molluscan smooth muscles that exhibit the catch response (Baguet and Gillis, 1968; Lowy and Millman, 1963) have a characteristic structure which is markedly different from other muscles. As the name implies, the myofilaments in this type of muscle are not regularly organized in three dimensions in such a way that a striated appearance is produced. On the other hand, contraction does seem to depend on the relative sliding of thin actin-containing filaments and thick myosin-containing filaments. But even within one muscle cell, the thick filaments can have very different diameters. Filament diameters in the range 280 to 1800 Å, have been observed in a variety of molluscan muscles (G. F. Elliott, 1960, 1964). As in other invertebrate muscles, these thick filaments contain the protein paramyosin in addition to myosin. But unlike other invertebrate thick filaments which are composed predominantly of myosin, molluscan thick filaments can contain as much as 90 to 95% paramyosin. It is thought that paramyosin forms the bulk of the central core of the thick filaments and that myosin forms a thin surface layer on top of this core. The evidence for this will be given largely in Section 8.3.4.

In addition to the thick and thin filaments, molluscan muscles also contain amorphous-looking structures called dense bodies. It now seems reasonably clear that these dense bodies carry out a role similar to that of

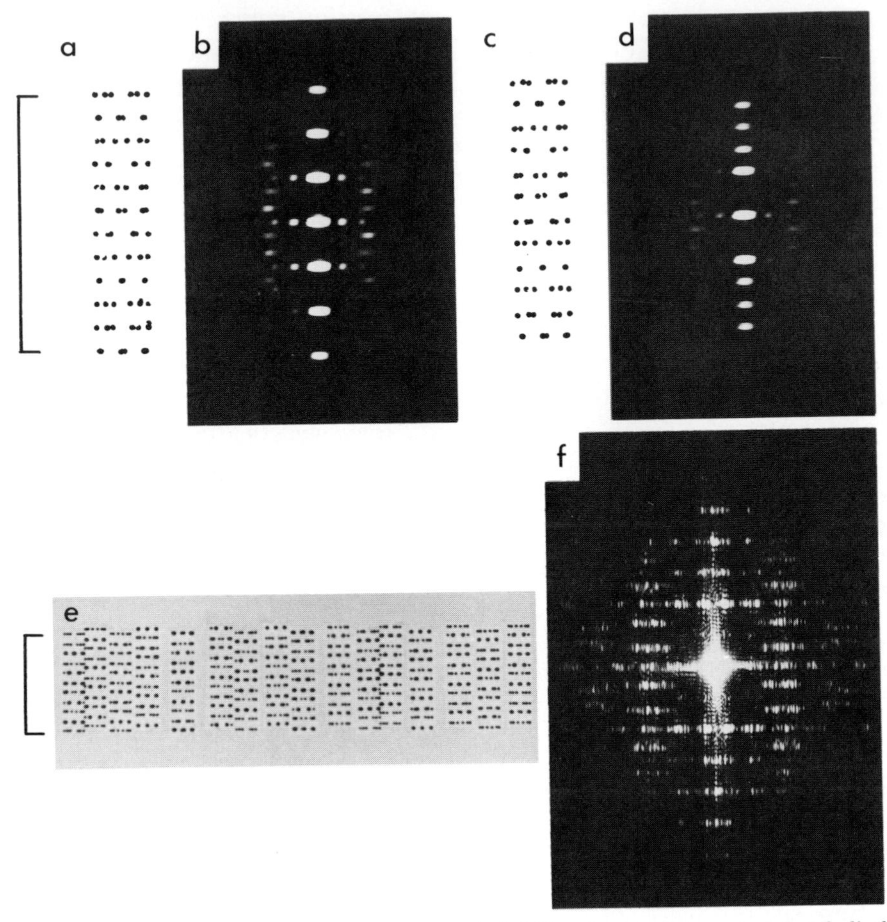

FIGURE 8.24. (a and c) Photographs of optical diffraction masks representing helical projections of the two alternative models for the cross bridge arrays on scallop thick filaments shown in Fig. 8.23, b, c. (e) Optical diffraction mask of a series of projections like (c) alternately staggered by about 72 Å and with centers at randomly selected lateral separations between 300 and 600 Å. (b, d, and f) Optical diffraction patterns of (a), (c), and (e), respectively. In (b) the meridional reflections are integral orders of 145 Å. (d) The pattern is similar to (b) but has additional reflections that are the third and fifth orders of the 290-Å repeat. (f) This is like (d), but it shows the lattice-sampling effects that include a splitting across the meridian of the 145-Å meridional reflection as observed in the X-ray diffraction patterns (Fig. 8.23a).

the Z-bands in striated muscles in that they act as mechanical linkages between two thin filament arrays of opposite polarity (Szent-Gyorgyi *et al.*, 1971).

Two different hypotheses have been presented to explain the origin of the catch phenomenon. One of these explains the mechanism solely in terms of cross bridge behavior so that the slow turnover rate of energy is a direct reflection of a slow rate of breaking of actin–myosin cross bridges (Lowy and Millman, 1959, 1963; Lowy *et al.*, 1964; Szent-Gyorgyi *et al.*, 1971; Parry and Squire, 1973). The other idea, which now seems to be less favored, is that catch occurs by direct fusion of neighboring thick filaments (Johnson, 1962; Johnson *et al.*, 1959; Rüegg, 1958, 1961). The first of these ideas will be considered further in Chapter 10. The second is mentioned briefly later in this section.

Sobieszek (1973) has made a careful study of the arrangements of thick and thin filaments and dense bodies through the cells of the ABRM of *Mytilus*. A number of his results, described briefly below, are probably typical of molluscan smooth muscles in general.

First of all, Sobieszek estimated the cell size to be about 1.6 mm long and 4 to 4.5 μm at its widest part. The myofilaments in the cell were estimated to be extremely long, the thin filaments' length being about 11 μm and the thick filament length about 25 μm. Dense body sizes were about 1.8 μm axially and about 0.12 μm in diameter.

Figure 8.25b,c shows typical cross-sectional views of fibers from ABRM. Here the large, dense, almost circular objects are the thick filaments (usually called paramyosin filaments), which have an average center-to-center separation of about 700 (±100) Å. The thin filaments can be seen as clusters of small dots around the thick filaments. In addition, the large "fuzzy" objects are the dense bodies.

In regions of the cell where thin filaments are grouped together, a regular hexagonal thin filament lattice sometimes occurs (Fig. 8.25). This is commonly known as the actin lattice and is usually such that the inter-thin-filament spacing is close to 120 Å. Evidence for a regular thin filament separation in living molluscan muscle (and also in vertebrate smooth muscle—see Section 8.4) has come from X-ray diffraction patterns (Lowy and Vibert, 1967) where a diffuse equatorial reflection of spacing 120 Å is commonly seen. This supports the results from electron micrographs of the kind shown in Fig. 8.25b. From his studies of ABRM, Sobieszek concluded that the contractile material, although not being perfectly ordered, did form a number of parallel contractile units consisting of a "sarcomere" of thick filaments overlapping two sets of thin filaments that were linked to the next "sarcomere" by dense bodies at each end. Each sarcomere was thought to comprise about three or four thick filaments and about 60 to 80 thin filaments per dense body. Thick filaments could often be seen to be ringed by about 20 thin filaments. The whole contractile unit would then be linked to the cell membrane by structures analogous to dense bodies (Fig. 8.25c). Such an arrangement

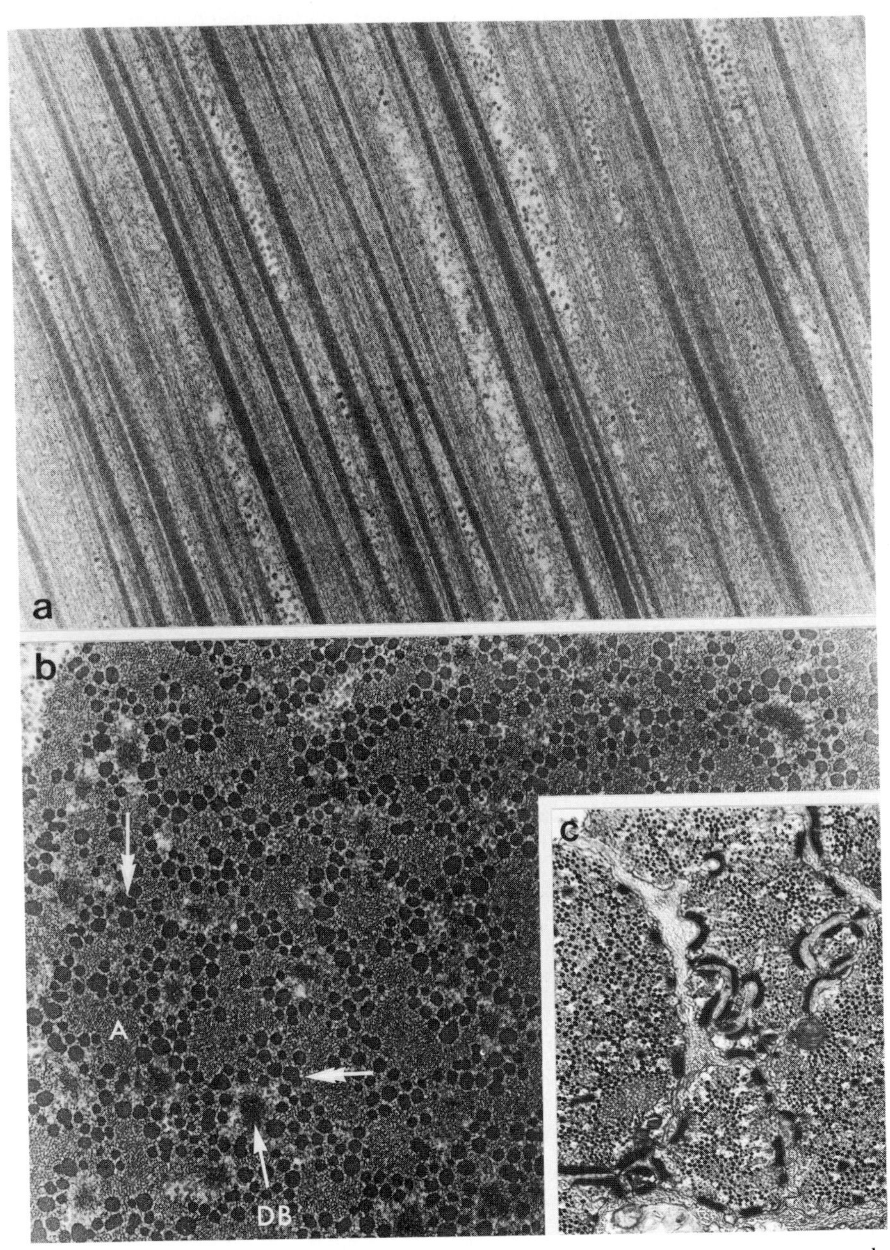

FIGURE 8.25. Electron micrographs of sections of the anterior byssus retractor muscle (ABRM) of *Mytilus*. (From Sobieszek, 1973.) (a) Longitudinal section of muscle fixed while fresh, showing the thick paramyosin filaments and intervening well-ordered arrays of actin filaments. This is considered by Sobieszek to correspond to the *in vivo* filament arrangement in this muscle (×50,000). (b) Transverse section of glycerinated muscle which

is said to account satisfactorily for the known length–tension curve for this muscle (Abbott and Lowy, 1958; Jewell, 1959; Lowy and Millman, 1963; Sobieszek, 1973).

Sobieszek (1973) also studied the structure of glycerinated ABRM and occasionally found an appearance in cross section that clearly showed aggregated thick filaments (Fig. 8.25b) as previously seen by Johnson and his collaborators (1959, 1962) and Ruegg (1958, 1961). Sobieszek found that the preservation in glycerinated muscle was extremely varied, some cells showing only rosettes of thin filaments around the thick filaments and no evidence of dense bodies (Hanson and Lowy, 1959). But of the better preserved cells, most showed appearances similar to that of relaxed muscle (Fig. 8.23), and relatively few showed the fused thick filament appearance. Those cells in which thick filament aggregation was observed also showed what were thought to be abnormally extensive actin lattices. It was concluded that the large lattices and fused thick filaments were preparative artifacts, but the point still remains to be proved conclusively.

8.3.4. Structure of Paramyosin Filaments

The Paramyosin Molecule. The structure and interactions of paramyosin molecules have recently been reviewed in depth by Fraser and MacRae (1973). Here space limitations allow only a limited account of those properties of paramyosin that are clearly relevant to a discussion of muscle contraction.

As mentioned earlier, paramyosin is a two-chain α-helical coiled-coil molecule rather like the rod part of myosin. It has a molecular weight of about 200,000 to 220,000 (Lowey *et al.*, 1963; Woods, 1969; Szent-Gyorgyi *et al.*, 1971), very high α-helical content (Cohen and Szent-Gyorgyi, 1957, 1960) and an amino acid composition rather similar to those of myosin rod and tropomyosin in being rich in α-helix-favoring residues (Chou and Fasman, 1974). It is rodlike in shape and is about 1300 Å long by about 20 Å in diameter (Hodge, 1952, 1959; Lowey *et al.*,

---

shows the apparent variable diameter of the paramyosin filaments (some of it because of the filament taper) together with large dense bodies (DB) and an almost crystalline organization of actin filaments. These are even more extensive than the actin lattices seen in transverse sections of fresh muscle and are considered by Sobieszek to be artifacts (×63,000). (c) Transverse sections showing dense areas on the cell membrane. These are particularly numerous in cell profiles of small diameter presumed to represent regions close to the cell ends. Both these and the dense bodies are thought to be actin attachment sites analogous to the Z-bands in striated muscles (×17,000). Note that in (b) some aggregation of adjacent thick filaments can be seen (double arrows), whereas in (c) the thick filaments are well separated. For the relevance of these appearances, see text.

1963; A. Elliott *et al.*, 1968a). This length, taken together with the molecular weight and α-helix content, suggested that paramyosin molecules must contain two chains, an idea supported by the hydrodynamic and light-scattering studies of Olander *et al.* (1967) and Olander (1971), by estimates of the chain weight (100,000 to 110,000; Szent-Gyorgyi *et al.*, 1971; Weisel and Szent-Gyorgyi, 1975), and by X-ray diffraction studies (Cohen and Holmes, 1963; Fraser *et al.*, 1965; A. Elliott *et al.*, 1968a; see Chapter 4). Weisel and Szent-Gyorgyi (1975) have presented evidence that suggests that the two chains in the paramyosin from *Mercenaria* (and probably some other muscles too) are very similar if not identical. On the other hand, Walker and Stewart (1975) have analyzed the sequences of certain peptide fragments from scallop paramyosin and have found evidence for chain heterogeneity. To be really certain about these results, complete sequencing of the molecules is desirable, but unfortunately the enormous size of the chains makes this a formidable task.

Synthetic Aggregates of Paramyosin. A large number of different sythetic aggregates can be formed from paramyosin under a variety of conditions of pH and ionic strength. These aggregates often show a periodic banded pattern in which the repeat can be 1800 Å (Hodge, 1959; Locker and Schmitt, 1957), 1400 Å (Hodge, 1952), 360 Å (Locker and Schmitt, 1957), 145 Å and 725 Å (Hanson *et al.*, 1957; Hodge, 1959), or about 70 Å (Hodge, 1959). All of these repeats are multiples of about 72 Å, and subrepeats of 72 Å within the major repeats were also evident. Since the aggregates appeared to possess dihedral symmetry, the molecular packing arrangements involved could not easily be defined.

More recently, Cohen *et al.* (1971a) obtained a number of different aggregates by precipitation with divalent cations, and two of these showed polar band patterns of period 725 Å. They are shown in Fig. 8.26 together with the three dihedral structures that Cohen *et al.* also obtained. The dark (stain-filled) bands in all of these negatively stained aggregates were taken to be the location of gaps in the molecular array, and the simplest polar pattern (Fig. 8.26a), which was termed the PI structure, was thought to be caused by systematic axial shifts of 725 Å between parallel molecules 1275 Å long. This scheme is shown alongside the micrograph in Fig. 826a. The second polar pattern (Fig. 8.26b) was thought to be caused by two overlapping PI arrays with an axial stagger of 725 Å × 2/5. For this reason it was termed the PI(2/5) pattern. The proposed molecular arrangements for this aggregate and the three dihedral structures are shown under Fig. 8.26b.

Figure 8.27 shows a micrograph from Cohen *et al.* (1971a) in which the paramyosin aggregate has a changing band pattern along its length.

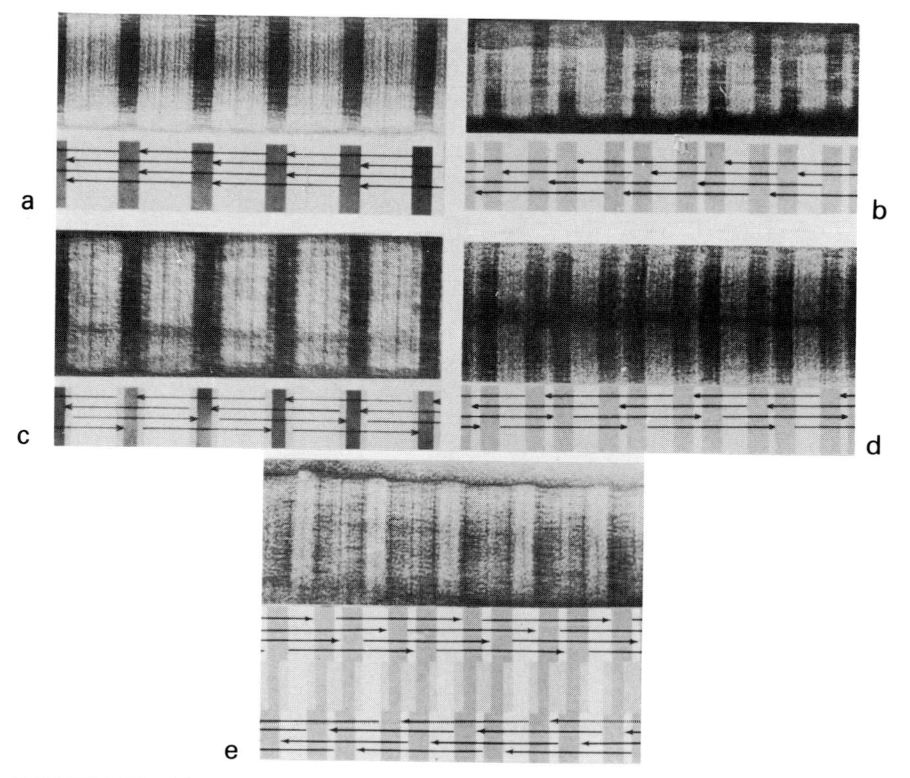

FIGURE 8.26. Electron micrographs illustrating various forms of ordered aggregates of paramyosin obtained by precipitation with divalent cations. All are negatively stained. (a) The PI form; (b) the PI(2/5) form; (c–e) various forms with dihedral symmetry. (From Cohen *et al.,* 1971a.) These micrographs are interpreted in terms of possible molecular packing arrangements as indicated beneath each picture. The aggregate in (e) is considered to be a sum of two PI(2/5) forms as in (b). The axial repeat in each case is about 720 Å.

In the center of this aggregate, one period shows the dihedral pattern of Fig. 8.26c, and on each side of this repeat the banding pattern gradually changes to oppositely directed PI structures. An additional kind of synthetic aggregate that shows a surface pattern similar to that on the paramyosin filaments prepared from molluscan muscles (Cohen *et al.,* 1971a) will be considered in a later part of this section.

Recently Weisel (1975) obtained synthetic segments of paramyosin, this time not from molluscan muscle but from the earthworm. These aggregates included antiparallel overlaps of about 470 Å and 700 Å and indicated a molecular length of 1235 Å for this type of paramyosin.

Structure of Native Paramyosin Filaments. When isolated from molluscan muscles, the thick paramyosin filaments often show a regular

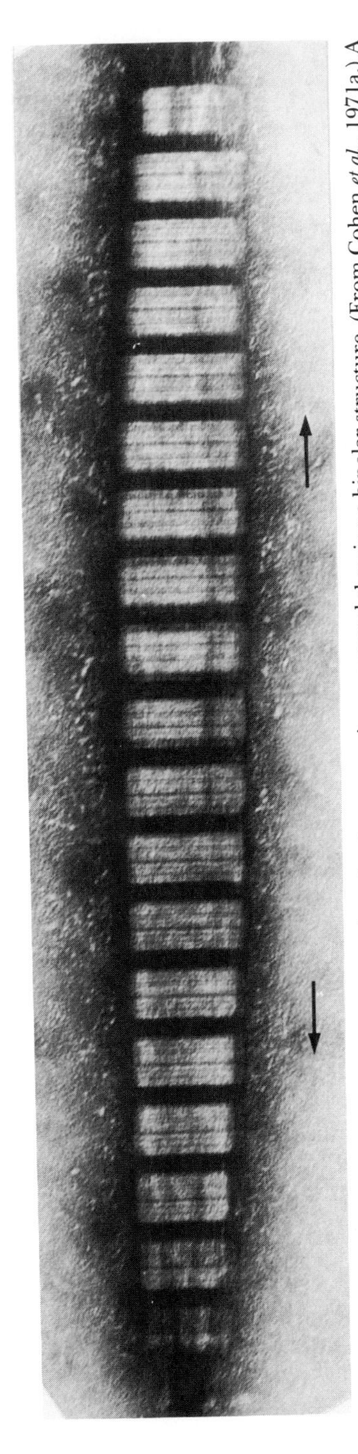

FIGURE 8.27. Electron micrograph of a negatively stained paramyosin paracrystal showing a bipolar structure. (From Cohen *et al.*, 1971a.) A single period of the dihedral form in Fig. 8.26c occurs just to the right of center, and this changes gradually in each direction to give two oppositely directed PI forms.

two-dimensional surface net pattern when negatively or positively stained (Fig. 8.28; Hall *et al.*, 1945; Hanson and Lowy, 1964). It was also found by Bear and Selby (1956) that low-angle X-ray diffraction patterns from wet or dry molluscan muscles could be interpreted in terms of the same net pattern. This pattern, which is commonly known as the Bear–Selby net, is illustrated in Fig. 8.28b. It is a two-dimensional arrangement of lattice points in which the **c** axis is constant at about 725 Å and the *a* axis is variable among different thick filaments and different muscle types. This rectangular net is divided into smaller (primitive) unit cells by lattice points spaced axially at intervals of 725/5 = 145 Å (the usual thick filament repeat) and laterally at intervals of *a*/5. In fractional coordinates, the lattice points are at (0,0), (3/5,1/5) (1/5,2/5), (4/5,3/5), and (2/5,4/5).

The diffraction pattern from the Bear–Selby net has a characteristic appearance and can be described in terms of layer lines that have spacings that are successive orders of 1/720 $Å^{-1}$. The strongest layer lines are at 2/720 $Å^{-1}$ = 1/360 $Å^{-1}$, 3/720 $Å^{-1}$ = 1/240 $Å^{-1}$, and 5/720 = 1/144 $Å^{-1}$, the latter having strong intensity on the meridian of the diffraction pattern. Figure 8.29a shows a model of the Bear–Selby net together with its optical diffraction pattern which clearly shows the features described above. Also shown is an X-ray diffraction pattern from an oyster white adductor muscle that had been treated with acetone. It will be seen here

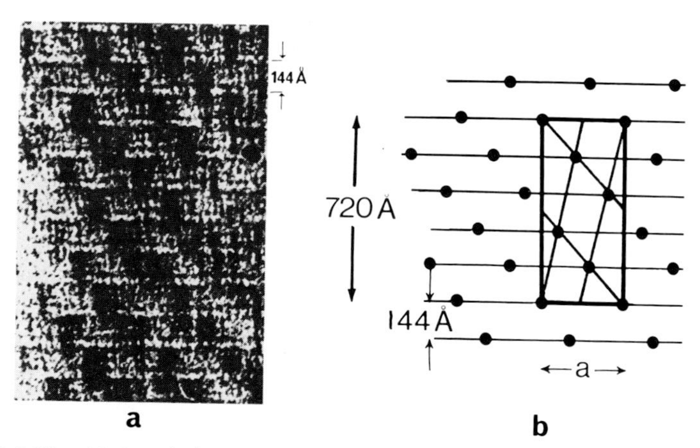

FIGURE 8.28. (a) A typical appearance of the negatively stained surface of an isolated paramyosin filament from the adductor muscle of the oyster *Crassostrea angulata* (courtesy Dr. J. Lowy). Clear transverse stripes can be seen at axial separations of 145 Å (filament axis vertical), and there is regular two-dimensional pattern of almost triangular staining features as indicated in (b) which shows the form of the lattice. It has a 720-Å axial repeat and a lateral spacing (the *a* spacing) that varies among filaments and among muscles. This characteristic lattice is known as the Bear–Selby net (see text).

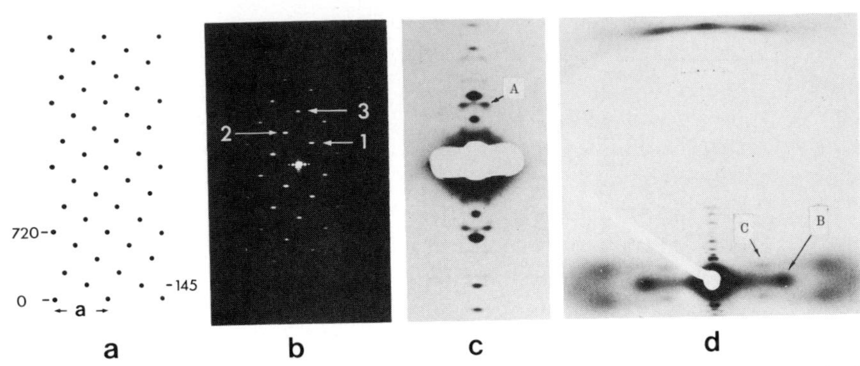

FIGURE 8.29. (a) Representation of the Bear–Selby net pattern (Fig. 8.28) used to obtain the optical diffraction pattern shown in (b). Here the prominent layer lines at 360 Å (720/2), 240 Å (720/3), and 145 Å (720/5) are indicated. These layer lines are typical of diffraction patterns from molluscan muscles except, of course, that these are symmetrical about the meridian. (c) Low-angle X-ray diffraction pattern from the opaque portion of the adductor muscle of *Crassostrea angulata* showing the characteristic layer lines which are orders of the 720-Å repeat. The radial position of the off-meridional reflections such as A can be related to the *a* spacing of the Bear–Selby net. (d) High-angle X-ray diffraction patterns from the same muscle as (c) but after soaking in 50% aqueous acetone. The specimen has been tilted to the X-ray beam to reveal detail around the 5.1-Å meridional reflection (top edge). The equatorial reflection (B) has a spacing of about 19.3 Å, and the near-equatorial layer line of spacing about 70 Å is sampled at C. (c and d from A. Elliott and Lowy, 1970.)

that comparable layer lines occur in both the X-ray diffraction pattern and the optical diffraction pattern of the models (even though the relative intensities are different).

It is generally assumed that the striking Bear–Selby net pattern which can be seen on isolated paramyosin filaments (Fig. 8.28a) must be an expression of the underlying arrangement of the component paramyosin molecules. On that basis, one of the first questions to answer was whether the paramyosin filaments were helical in nature or, alternatively, were an assembly of flat layers parallel to the filament axis, each of which possessed a Bear–Selby net structure. Bear and Selby (1956) and G. F. Elliott (1964) favored the planar layer structure, but in a very elegant experiment, A Elliott (1971) shadowed both sides of a single isolated paramyosin filament (Fig. 8.30a) and was able to show by optical diffraction that the surface Bear–Selby net on the back of the filament had the opposite hand to that on the front, apparently a clear indication that the filament must be helical. This conclusion was based on the fact that the optical diffraction pattern from a single Bear–Selby net (Fig. 8.29) is asymmetrical about the meridian, whereas the pattern from the filament shadowed on both the front and back surfaces (Fig. 8.30) was

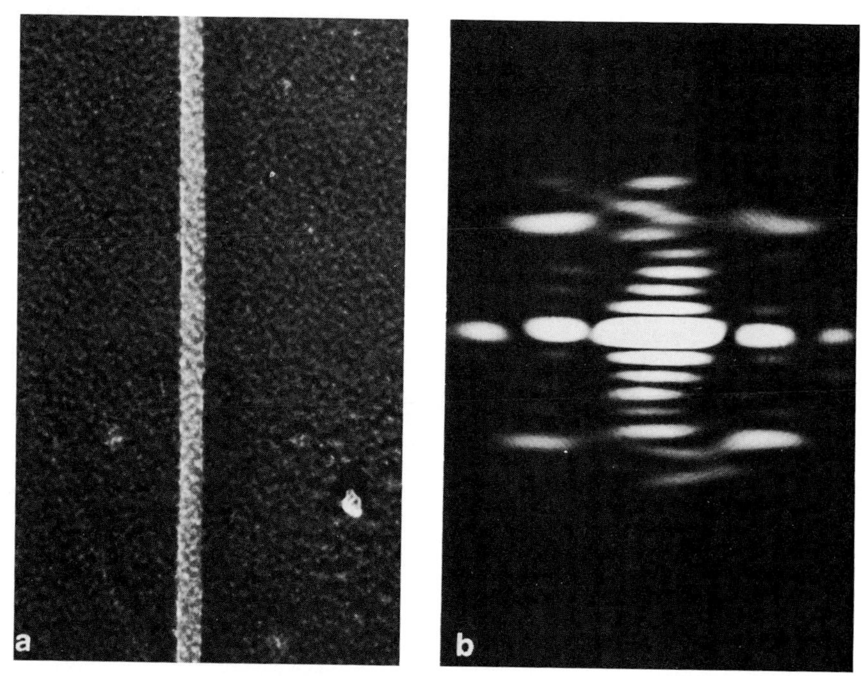

FIGURE 8.30. (a) Electron micrograph of a paramyosin filament from the adductor muscle of *Crassostrea angulata* that has been shadowed with heavy metal on both sides. (b) Optical diffraction pattern of part of the filament in (a) showing an essentially symmetrical appearance of the 360-Å layer line about the meridian (vertical). (From A. Elliott, 1971, with permission of the Royal Society, London.)

symmetrical and therefore must have been caused by a superposition of two Bear–Selby net patterns of opposite handedness.

It was mentioned in Chapter 4, Section 4.3.4 that A. Elliott *et al.* (1968) obtained evidence from the high-angle X-ray diffraction pattern from oyster white adductor muscle that the paramyosin molecules in the thick filaments were probably arranged in a centered tetragonal lattice (Fig. 4.14). One might therefore expect the *a* spacing of the Bear–Selby net to be related directly to the molecular packing distances involved in this tetragonal lattice. A. Elliott and Lowy (1970) argued that in such a case the equatorial reflection (B in Fig. 8.29d) at a spacing of about 1/20 $Å^{-1}$ should be on the 20th row line created by the *a* spacing repeat. It should therefore have a reciprocal coordinate that is exactly 20 times larger than that of the first row line (A in Fig. 8.29c) that would appear on layer lines 2, 3, 7, 8, etc., of the 720-Å *c*-axis repeat (Fig. 8.29b). By changing the bathing solution of the oyster white adductor muscle using various concentrations of acetone in water, A. Elliott and Lowy (1970) found that the spacing $(1/R)$ Å^{-1} of the reflections A and B gradually changed (they increased as the acetone concentration increased). But a plot of $(1/R)_B$ against $(1/R)_A$ was found not to be linear, thus indicating that the paramyosin filaments could not be considered as structures formed by placing paramyosin molecules into a regular three-dimensional crystalline space lattice.

A further interesting feature of these paramyosin filaments was deduced from an analysis of the *a* spacings of individual negatively stained filaments of different diameters. A. Elliott and Lowy (1970) found that as a general rule the larger the filament diameter, the smaller was the *a* spacing of the surface lattice as deduced from measurements of the optical diffraction patterns (Fig. 8.31) from the filaments (see also Fig. 8.29b).

These various observations on the structure of paramyosin filaments led A. Elliott and Lowy (1970) to suggest a structural model for these filaments in which the paramyosin molecules were arranged in an extensive two-dimensional layer exhibiting the Bear–Selby net pattern which rolled up on itself (like a rolled carpet) to give a cylindrical filament (Fig. 8.32). Clearly, in such a situation, the inner layers would be bent more than the outer layers, and it was thought that this bending would open out the surface Bear–Selby net so that the observed *a* spacing would change. Small-diameter filaments in which the surface layer would be highly curved would therefore display larger *a* spacings than larger filaments which would have less distorted surface layers.

This rolled layer model would actually have the repeating units in the Bear–Selby net arranged not on a true helix with constant pitch and radius (and therefore constant pitch angle) but on a helicoid in which the pitch angle remains nearly constant as the radius gradually changes from a large value at one end of the helicoid to a small value at the other. The opposite handedness of the front and back surfaces of the filament (Fig. 8.30) would be equally well explained by such a helicoidal structure as by a helix. This helicoidal arrangement would help to explain another feature of the X-ray diffraction patterns from molluscan muscles. As pointed out by A. Elliott and Lowy (1970), an unusual feature of these diffraction patterns is that the intensity profiles of the peaks on layer lines 2, 3, 7, 8, etc. of the 720-Å repeat terminate abruptly at their outer ends (i.e., away from the meridian) but taper off gradually towards the meridian (Fig. 8.29c) (see also Bear and Selby, 1956). Such an observation could be explained by a helicoidal structure provided that the pitch angle of the helicoid became distorted in such a way that it reduced slightly as the layers became more tightly curved towards the middles of the filaments. As described in Chapter 2, helices of smaller pitch angle

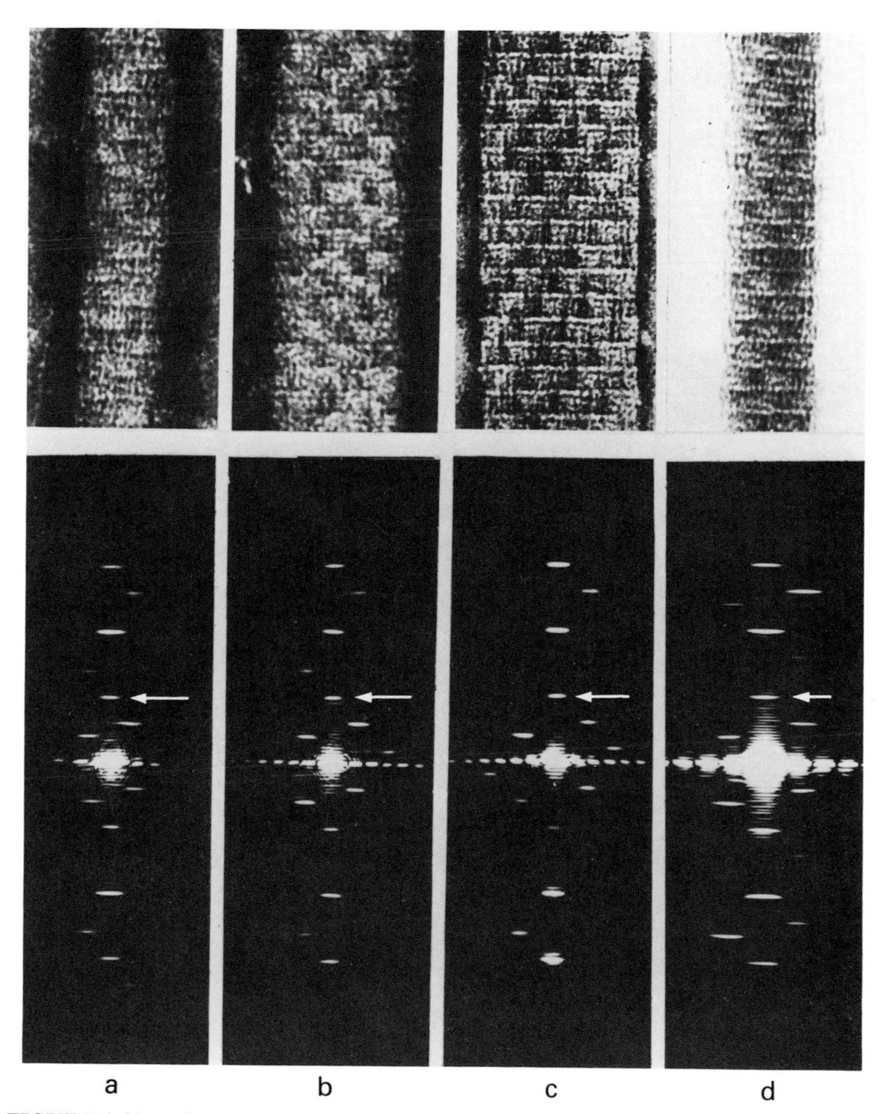

FIGURE 8.31. Electron micrographs of negatively stained paramyosin filaments of different diameters from the oyster white adductor muscle together with their optical diffraction patterns. Note that the diffraction spots from the widest filaments have the greatest horizontal distance from the meridian (and hence the smallest *a* spacing of the net). The structure on the right may be a flat ribbon of paramyosin. The diffraction patterns can be compared with Fig. 8.29b. The 145-Å reflection is arrowed. (From A. Elliott and Lowy, 1970.)

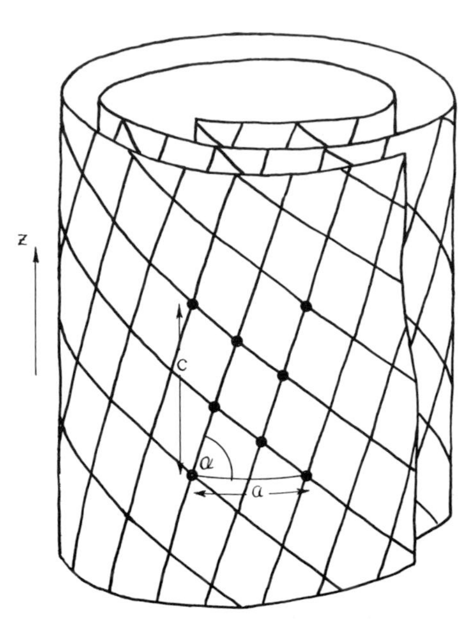

FIGURE 8.32. Representation of the helicoidal model of the paramyosin filament proposed by A. Elliott and Lowy (1970). A two-dimensional Bear–Selby net pattern is rolled up to form a spiral cylindrical lattice.

(large radius for a given pitch) give rise to intensity closer to the meridian than do helices of larger pitch angle. Since the bulk of the paramyosin would be located in the relatively undistorted surface layers of the paramyosin filaments, diffraction from these layers would dominate the observed patterns, and the layers with a smaller radius of curvature would form the observed intensity tail towards the meridian. Also, since successive layers of the helicoid would not pack together in a systematic way, the structure would be virtually a two-dimensional one with significant a and c axes but a relatively small b axis. Only two indices, h and l, would then be required to index the diffraction patterns, as has been found to be the case (Bear and Selby, 1956).

On the assumption that paramyosin filaments contain some sort of rolled layer structure, the next question to be asked concerns the nature of the molecular packing arrangement within the layer that gives rise to the Bear–Selby net pattern. As mentioned earlier, one of the synthetic paramyosin aggregates obtained by Cohen *et al.* (1971a) had the appearance of a synthetic paramyosin filament since it showed a lattice similar to the Bear–Selby net pattern. This is shown in Fig. 8.33. Cohen *et al.* (1971a) interpreted this structure as being generated by the side-by-side packing of subfilaments each of which comprised the PI type of molecular arrangement shown in Fig. 8.26a. Here molecules in each

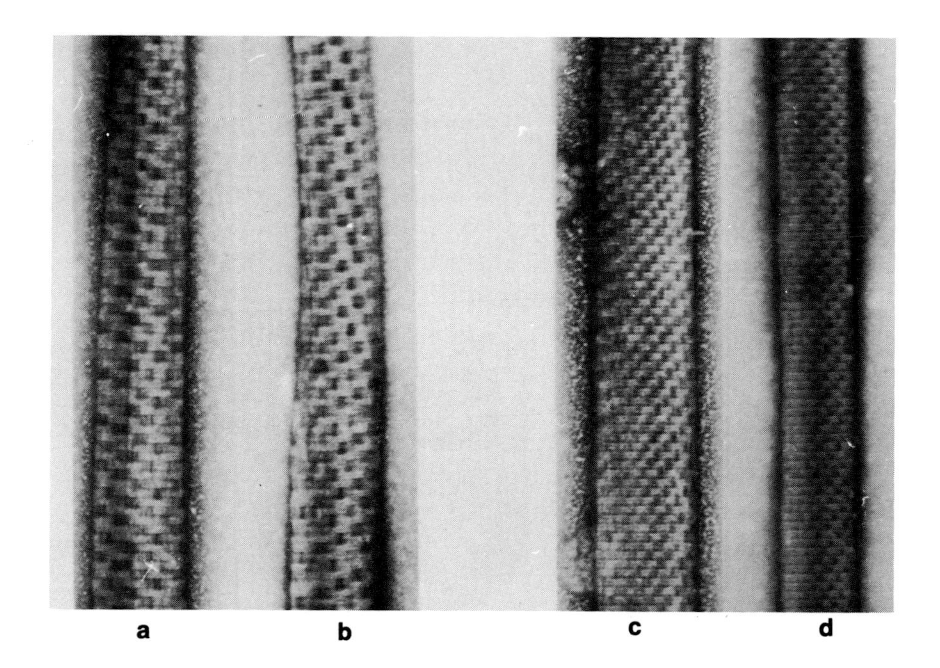

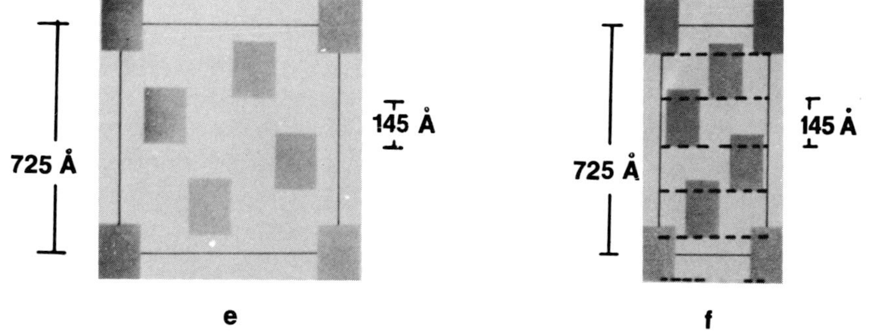

FIGURE 8.33. Electron micrograph of negatively stained synthetic and natural paramyosin filaments. (From Cohen *et al.*, 1971a.) Synthetic filaments are from *Atrina rigida* translucent muscle paramyosin (a) and *Placopecten megellanicus* smooth muscle paramyosin (b), and native filaments are from *Mercenaria mercenaria* white muscle (c) and *Crassostrea virginica* translucent muscle (d). (e and f) The lattice arrangements in the synthetic and native filaments, respectively.

subfilament would be related by an axial shift of c, and adjacent subfilaments would be related by axial staggers of $2c/5$ where c is about 725 Å. It seems very likely that this is indeed the structure of the synthetic paramyosin filaments. But various arguments, mainly those of A. Elliott and Lowy (1970), suggest that native paramyosin filaments may have a different structure. These arguments are summarized briefly below.

The most striking evidence about paramyosin packing is from the appearance of negatively stained native paramyosin filaments (Fig. 8.28a). First, it can be seen in Fig. 8.28a that the densely stained feature on every lattice point of the Bear–Selby net has a well-defined shape. It is not rectangular as in the synthetic paramyosin filaments (Fig. 8.33a,b) but is spade-shaped (almost triangular), being straight at one end and sharply curved or pointed at the other. This is clearly visible in the image-averaged picture shown in Fig. 3.14. Also shown up well by this average image is the fact that the white (protein) features form a very well-defined zigzag edge that runs along the short diagonal of the unit cell of the Bear–Selby net (Fig. 8.28). To explain this feature, A. Elliott and Lowy (1970) suggested that the layer in their helicoidal model might be composed of small planar aggregates (plates) of paramyosin molecules (as in Fig. 8.34a) that would pack together with the symmetry of the Bear–Selby net. They showed that in principle four kinds of

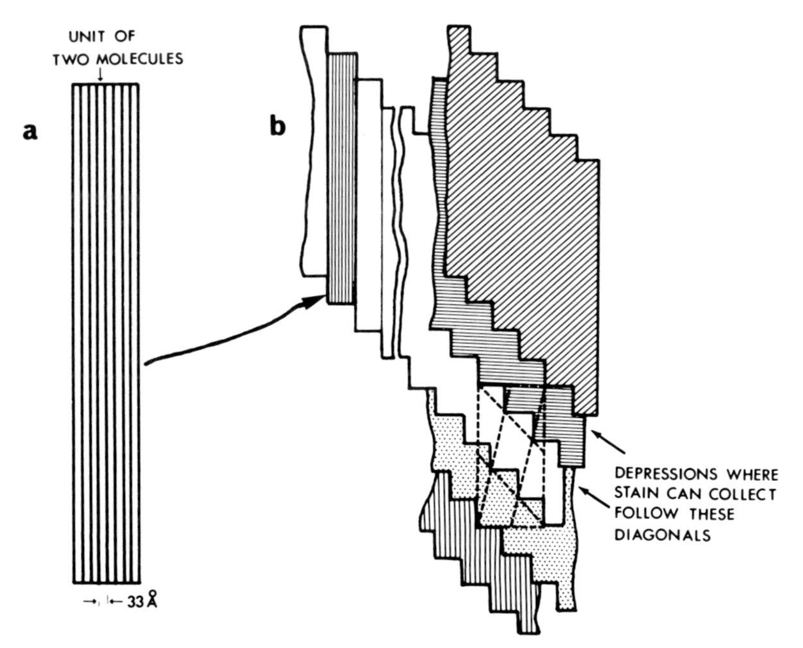

FIGURE 8.34. Diagrams illustrating (a) the structure of one plate of paramyosin molecules (see Fig. 8.35) and (b) the organization of these plates to form stepped planar sheets of molecules that overlap as in Fig. 8.35c to give the Bear–Selby net. As indicated, stain would collect in the diagonal depressions to give the appearance most clearly seen in the average image of Fig. 8.28a which is shown as Fig. 3.14.

arrangements of these molecular plates could generate the Bear–Selby net pattern. These four possibilities are shown in Fig. 8.35.

In model 1, the plates of width $a/5$ are joined edge to edge to other plates shifted axially by $^{3c}/5$, and several of the extended sheets so produced are layered on top of each other to give the right net pattern. In a similar way, the adjacent plates have relative axial shifts of $^{2c}/5$, $^{c}/5$, and $^{c}/5$, respectively in models 2, 3, and 4, and the plates have widths $^{a}/5$, $^{2a}/5$, and $^{3a}/5$.

By an ingenious technique analogous to the rubbing of monumental brasses, A. Elliott and Lowy (1970) investigated the likely staining patterns of these alternative structures. They made cardboard models of each, covered them with tracing paper, and then rubbed over them with a greasy black crayon. The results, shown in Fig. 8.35, are very striking, and only one model (No. 3) resembles the appearance of the averaged image shown in Fig. 3.14. This model is illustrated in Fig. 8.34b. It is fair to conclude that if the negative staining pattern in Fig. 8.28 really represents the nature of the surface topography of the paramyosin filaments, then model 3 could explain the observations extremely well. Unfortunately, it is conceivable that the appearance of the negatively stained

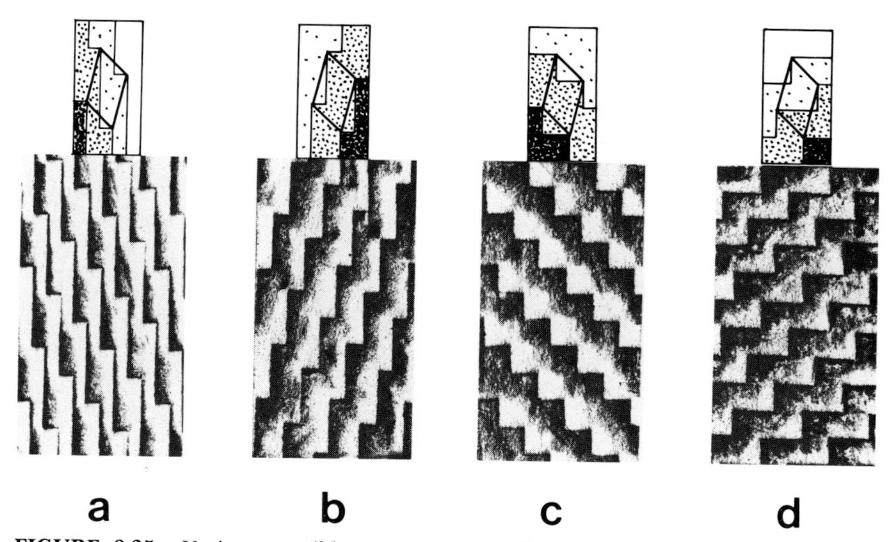

FIGURE 8.35. Various possible arrangements of rectangular plates of paramyosin molecules that would give a surface lattice with the symmetry of the Bear–Selby net. Adjacent plates are related by axial shifts of (a) 3 × 145 Å, (b) 2 × 145 Å, (c and d) 145 Å. The plate widths in (c) and (d) are respectively twice and three times the width of those in (a) and (b). Only (c) gives a surface appearance (lower figures) obtained by a kind of brass-rubbing technique (see text) that resembles that on negatively stained paramyosin filaments. Details of (c) are given in Fig. 8.34. (From A. Elliott and Lowy, 1970.)

filaments may be caused by overlapping staining effects through successive molecular layers in the filament surface, in which case another paramyosin packing scheme might be involved. This might help to explain the rather variable appearance of stained filaments.

Clearly, one way to test this is to study paramyosin filaments that have been shadowed with heavy metals, in which case only the surface features will appear. G. F. Elliott (1964) and A. Elliott and Lowy (1970) have shown pictures of such preparations, and, although these show good evidence for the presence of the Bear–Selby net pattern on the filament surface, the detailed structure of each repeat in this lattice is not really clear. However, one thing that does seem apparent from these micrographs is that since very marked ridges parallel to the filament axis are not seen, a subfilament-packing model of the kind suggested by Cohen *et al.* (1971a) may not be permissible.

It is clear that a considerable amount of work still needs to be done to define the true structure of paramyosin filaments. Indeed, the whole of this work has recently been questioned by A. Elliott (1979) who has tilted isolated paramyosin filaments about their long axes and recorded electron micrographs of the tilted views (Fig. 8.36a). Here, both the Bear–Selby net and a simple 145-Å periodicity can be seen. Elliott concludes that, after all, these filaments may have a planar layer structure similar to that suggested by G. F. Elliott (1964).

Fortunately, one feature of these filaments does seem unambiguous. Lowy (personal communication) and Szent-Gyorgyi *et al.* (1971) have obtained micrographs of isolated paramyosin filaments that clearly show an abrupt change of polarity along their length (Fig. 8.36b). This is evident from the opposite directions in which the triangular stained features point on opposite ends of the filament. It therefore seems clear that paramyosin filaments are bipolar because of the bipolar paramyosin packing arrangement.

Finally, some features of the local packing arrangements of paramyosin molecules are worth noting. It has already been mentioned that under certain conditions paramyosin molecules seem to form a centered tetragonal packing arrangement with an axial shift of about 35 Å (or an odd multiple of this) between adjacent molecules (A. Elliott *et al.*, 1968a). As explained in Chapter 4, this could be related to the way in which neighboring two-chain coiled-coil molecules could fit into each other to allow close packing (A. Elliott *et al.*, 1968a; Longley, 1975).

Electron micrographs of native paramyosin filaments have also given information on the lateral packing arrangements of individual molecules. A. Elliott *et al.* (1968b) used the technique of self-convolution (Chapter 2) to show that a lateral repeat of about 33–34 Å (a/10) was present in a micrograph of a filament of a spacing 330 Å from oyster

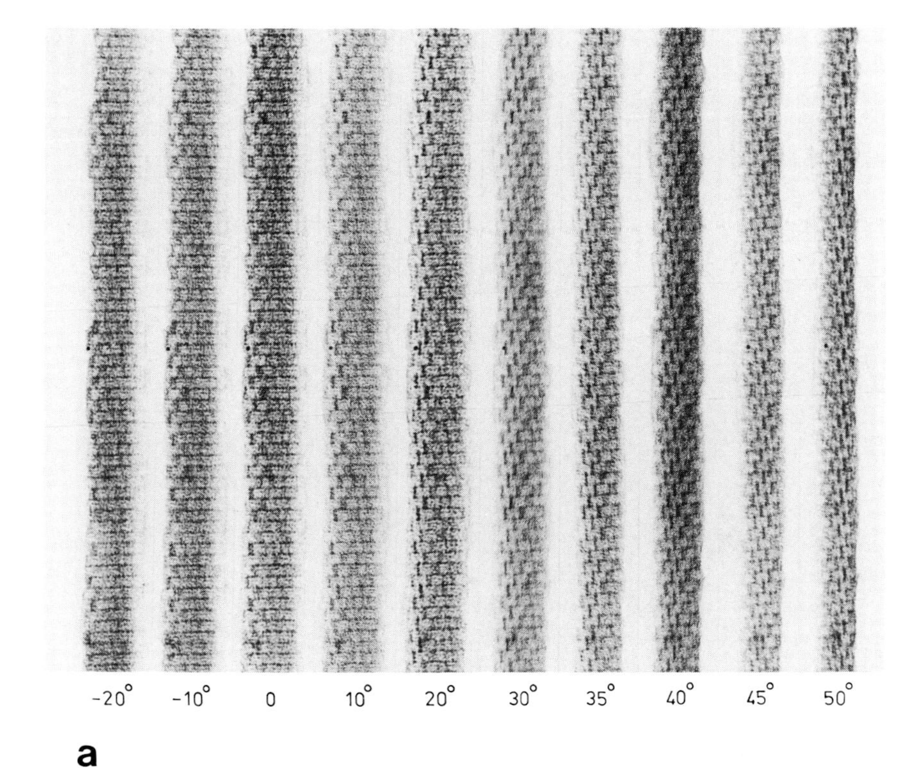

144 Å

b

FIGURE 8.36. (a) Paramyosin filament from *Ostrea edulis* rotated around its long axis by the amount shown. Negatively stained with ammonium molybdate (×115,000) (From A. Elliott (1979). (b) An isolated negatively stained paramyosin filament from *Crassostrea angulata* (courtesy Dr. J. Lowy) showing a change in polarity halfway along its length. As indicated, the triangular staining features on each side of the center point towards the middle.

white adductor muscle (an axial subrepeat of 48 Å was also evident; Fig. 3.14). It was concluded that the 33- to 34-Å lateral repeat represented a grouping of pairs of molecules (each of diameter 16 to 17 Å; A. Elliott *et al.*, 1968a). This pairing could well correspond to the slightly different axial positions (35 Å apart) of adjacent molecules.* Since the width of the molecular plate in model 3 of Fig. 8.35 is $^{2a}/5$, and a comprises about 20 molecular widths, each plate in model 3 would be about 8 molecules wide. A. Elliott and Lowy (1970) noted that the relative radial positions of the off-meridional peaks on layer lines 2, 3, 7, 8, etc. in the low-angle X-ray diffraction pattern from *Venus, Crassostrea,* and *Mytilus* adductor muscles were approximately in the ratio 3 : 4 : 5. In other words, the thick filament a spacings in different smooth muscles can be very different as they can on different thick filaments in the same muscle. A. Elliott and Lowy suggested that a structure like model 3 (Fig. 8.35) but with 6, 8, and 10 molecules, respectively, in each plate might explain the appearances and the differences between the filaments in these three molluscan muscles. This possibility is considered further in Chapter 9, Section 9.3.3.

Location of Myosin on Paramyosin Filaments. It has long been supposed that myosin molecules form a surface layer on top of the paramyosin core in the thick filaments of molluscan muscles (Hanson and Lowy, 1961). Since then a number of observations of projections on the surfaces of these thick filaments have been reported (Hanson and Lowy, 1964; A. Elliott, 1974; Sobieszek, 1973). In Fig. 8.25, it can be seen that the thick filaments are separated from the neighboring thin filaments by a semidense halo region thought to be caused by the myosin cross bridges. This idea has also gained support from an analysis of the relative amounts of myosin and paramyosin in molluscan muscles that contain different average filament diameters (Szent-Gyorgyi *et al.*, 1971; Squire, 1971). However two factors complicate this evidence: one is that it is difficult to assess how accurate the biochemical evidence really is (see Levine *et al.*, 1976), and the other is that it is difficult to relate the biochemical evidence to the known filament dimensions since the latter are so variable even within a single muscle (G. F. Elliott, 1964). In theory, with a given filament diameter and a given paramyosin-to-myosin ratio, it is possible to calculate the thickness of the surface myosin layer assuming that it approximates to a uniform cylindrical shell. Unfortunately, with molluscan muscles, it is only possible to take average filament diameters for use in such a calculation.

Squire (1971) attempted to show on the basis of biochemical evi-

*68 Å spacing on one filament.

dence available at the time that all of the observations were consistent with there being a uniform layer of myosin about 40 Å thick (not including the myosin heads) on every type of myosin-containing filament considered. This idea will be discussed fully in Chapter 9. Similarly, Szent-Gyorgyi *et al.* (1971) concluded from their work that as a general rule there was sufficient myosin in various molluscan muscles to give one myosin molecule per 15,000 Å^2 of filament surface. Several recent papers have provided more direct evidence of the location of myosin on the filament surface. A number of reports have suggested that virtually intact thick filaments can be separated from molluscan muscle and that these show either a very rough appearance consistent with the presence of surface projections (Hanson and Lowy, 1964) or a simple band pattern with a 144 Å periodicity (Hardwicke and Hanson, 1971; Szent-Gyorgyi *et al.,* 1971; A. Elliott, 1974).

It was also shown by Szent-Gyorgyi *et al.* and Hardwicke and Hanson that myosin can be extracted from the thick filaments leaving virtually intact paramyosin filament cores. This is a clear indication that myosin resides on the filament surface. Such extracted filaments were then thought to show predominantly the Bear–Selby net pattern, suggesting that the core has this structure and that the surface myosin layer tends to obscure everything except the typical 145-Å myosin repeat (Szent-Gyorgyi *et al.,* 1971; Sobieszek, 1973). However, the recent results of A. Elliott (1979) showing that sometimes a 145-Å repeat may be just another view of the Bear–Selby net (Fig. 8.36a) obviously complicate this simple conclusion.

Furthermore, A. Elliott (1974) and Sobieszek (1973) had already reached different conclusions about the arrangement of myosin on the filament surface. A. Elliott first studied both intact and extracted oyster thick filaments that had been negatively stained and found that in both cases a very variable appearance was obtained; some filaments showed the 145-Å repeat and others the Bear–Selby net. This was attributed largely to the variation in stain penetration in different filaments. For this reason he then prepared filaments using a variety of methods and contrasted them by tungsten shadowing so that only surface features would appear. It was found that all of the methods used produced some paramyosin filaments (Fig. 8.37) that showed "a very regular two-dimensional pattern of raised rounded objects on the surface." The pattern did, in fact, have the same symmetry as the familiar Bear–Selby net pattern as confirmed by optical diffraction. Other filaments showed rather less surface detail.

Filaments similar in appearance to the shadowed ones were obtained by A. Elliott by negatively staining these after they had been

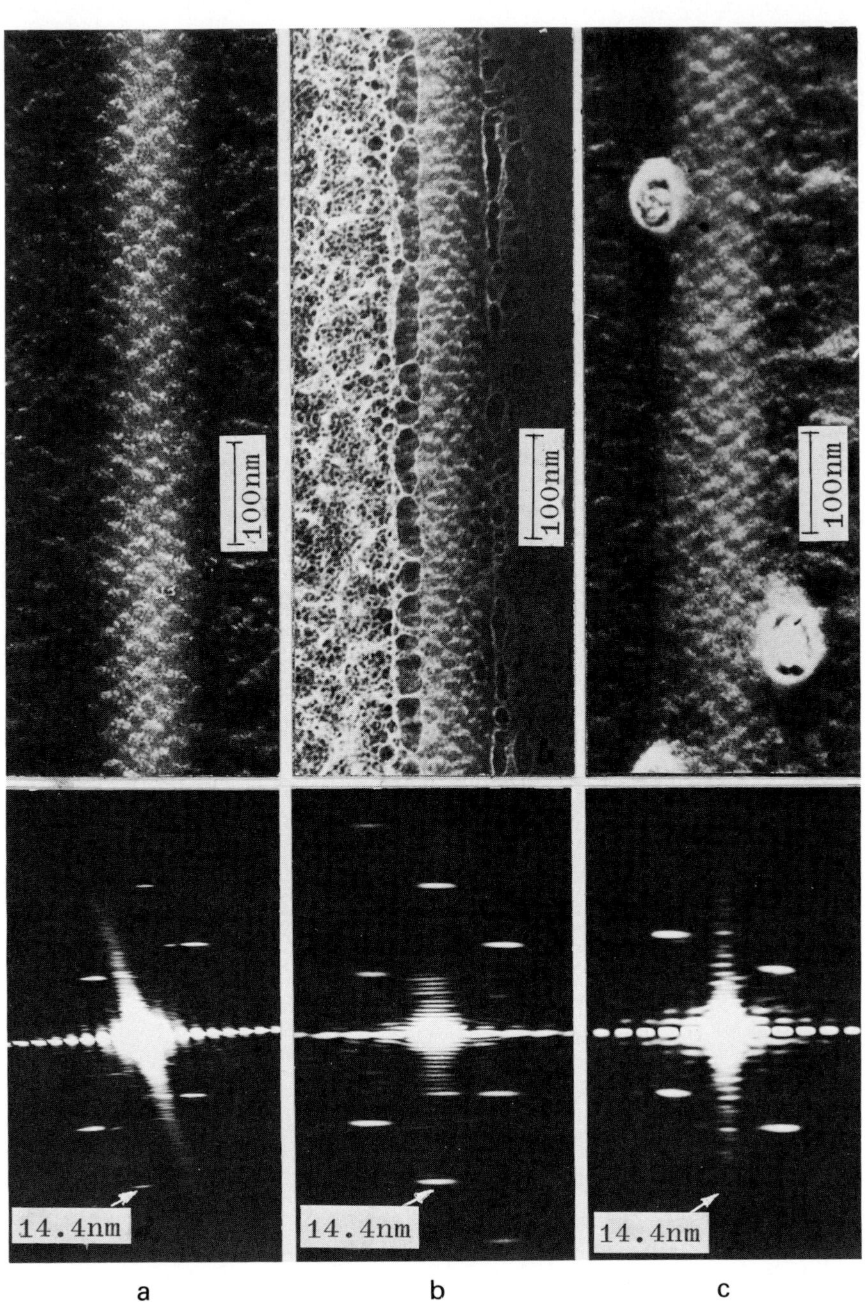
100nm
100nm
100nm
14.4nm
14.4nm
14.4nm
a
b
c

"waterproofed" by brief examination in the electron microscope. It then appeared that a true surface appearance was being seen. A. Elliott also estimated the paramyosin-to-myosin ratio in the oyster using SDS gel electrophoresis and obtained a value of about 8 : 1. This was interpreted in terms of the population of filament diameters measured by G. F. Elliott (1964) together with the known *a* spacings to give an estimate of the number (n) of myosin molecules per lattice point on the Bear–Selby net. It was concluded that n must be either 2 or 3. Each of the "raised rounded objects" in the shadowed preparations was therefore interpreted as being a clump of the heads of two or three myosin molecules. Sobieszek (1973) studied sections of ABRM and found a variety of thick filament appearances including the Bear–Selby net pattern, 145-Å-banded patterns, and occasionally patterns of dots that appeared to be the myosin projections. Some of his sections had similar appearances to those of A. Elliott's shadowed filaments, and they can presumably be interpreted in a similar way. Other sections indicated that projections from the thick filament surface were spaced about 100 Å apart circumferentially.

Squire (1971) analyzed an unusual paramyosin filament from the ABRM and showed that surface "blobs" which could be clearly seen in this filament divided the *a* spacing regularly into intervals of $a/5$ and $c/5$. This is shown in Fig. 8.38. Once again, the separation of the blobs (thought to be myosin projections) was about 120 to 130 Å. Finally, both A. Elliott (1974) and Sobieszek (1973) showed micrographs of paramyosin filaments of rather small diameter that seemed to have the structures of two-stranded helices of pitch 1450 Å (10 × 145 Å). The filaments would therefore have two *a* spacings around their circumference, and the normal Bear–Selby net arrangement would be twisted into a cylindrical layer to give the required structure.

In conclusion, it is clear from this discussion that there is still a great deal to be learned about the molecular packing arrangements in paramyosin filaments. Some further consideration of these filaments is included in Chapter 9 where a general discussion of myosin filament structure is given.

The Control Mechanism in Molluscan Muscle. Before leaving this discussion of molluscan muscle it is worth noting that there is evi-

←

FIGURE 8.37. Shadowed native paramyosin filaments from oyster white adductor muscle which are thought to have an intact layer of myosin molecules on their surface. In each case very clear evidence of the Bear–Selby net pattern can be seen as indicated by the optical diffraction patterns beneath each micrograph. Note that the filament in (c) has the opposite hand to those in (a) and (b). (From A. Elliott, 1974, by permission of the Royal Society, London.)

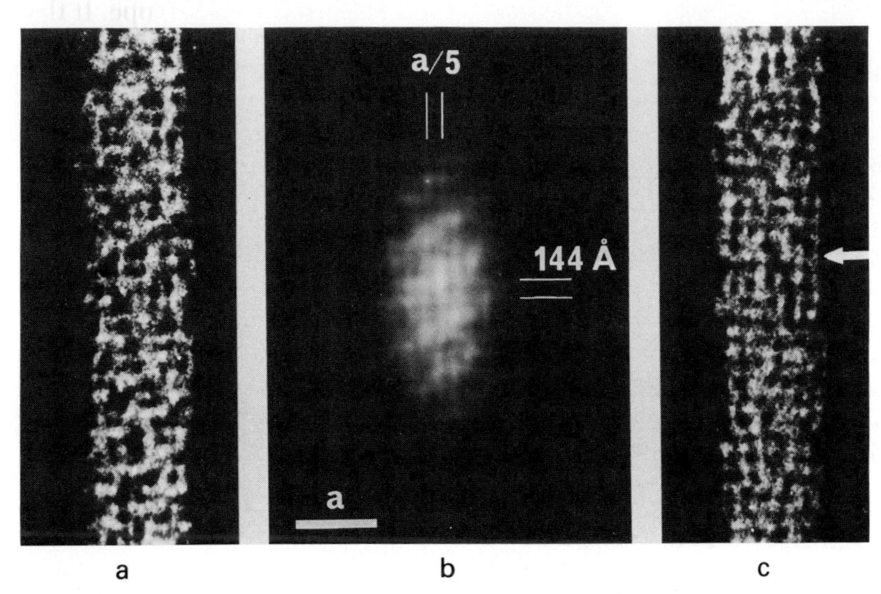

FIGURE 8.38. (a) Electron micrograph of part of a negatively stained paramyosin filament from Mytilus ABRM (courtesy of Dr. J. Lowy). (b) Self-convolution function of part of the filament in (a) showing that the *a* spacing of the Bear–Selby net is clearly divided into five subperiods. (c) Another region of the same filament as in (a) showing a regular near-rectangular arrangement of subunits (arrowed). (From Squire, 1971.)

dence that paramyosin molecules may have more than just a structural role in the thick filaments. It has already been mentioned in Chapters 5 and 6 that molluscan muscles contain little or no troponin and that regulation of contraction is mediated by the myosin light chains (Kendrick-Jones *et al.,* 1970; Szent-Gyorgyi *et al.,* 1973). In addition, it was found by Szent-Gyorgyi *et al.* (1971) that when paramyosin was added to solutions of actomyosin and coprecipitated, "the actin-activated ATPase of myosin was inhibited progressively with an increasing paramyosin to myosin ratio, while the Ca^{2+}-activated ATPase remained unaffected." In such an experiment, relatively low paramyosin concentrations were needed to give "total" inhibition of the actin-activated ATPase, but if actomyosin were added to the cores of native paramyosin filaments or to synthetic paramyosin filaments, much higher paramyosin concentrations were required. Szent-Gyorgyi *et al.* suggested that paramyosin might in fact have a regulatory role in tension maintenance and that in the catch mechanism a phase change in the paramyosin may be coupled to the movement of the cross bridges between the thick and thin filaments.

8.4. Vertebrate Smooth Muscles

8.4.1. Introduction

As described in Chapter 1, the involuntary visceral muscles of vertebrates are smooth muscles. Their component cells are spindle-shaped (Fig. 1.19), have a single nucleus, and are about 50 to 400 μm long and 1.5 to 8 μm in diameter (Huddart, 1975). These cells are arranged differently in different organs. In blood vessels, they are largely circumferential so that the vessel size can be regulated. In the intestines, where peristaltic contractions are required, there are two layers of cells, one circumferential and one longitudinal. Finally, in the uterus and bladder, which need to contract uniformily in all directions, the cells are arranged tangentially but with different orientations within the tangential surface.

A number of reviews of the ultrastructure and biochemistry of smooth muscle have recently been written including Burnstock (1970), Needham and Shoenberg (1968), and Small (1977b). The main ultrastructural features of this muscle are outlined briefly here.

From the discussion of this chapter and the last, it is clear that almost all muscles must operate by the sliding filament mechanism. But for a long time it was found that ultrastructural studies failed to reveal unambiguous evidence for the existence of thick myosin filaments in vertebrate smooth muscle (Hanson and Lowy, 1964; Shoenberg *et al.*, 1966; G. F. Elliott, 1964, 1967; Panner and Honig, 1970). Only thin filaments and so-called 100-Å filaments were consistently seen (Choi, 1962; Needham and Shoenberg, 1964; Lane, 1965; Ellis, 1965; Fawcett, 1966). Since the latter were similar to neurofilaments and to other cytoskeletal filaments, they were thought not to be myosin. This conclusion is supported by the recent results of Cooke and Chase (1971), Cooke (1975), Small and Sobieszek (1977), and Small (1977a). That myosin was present in abundance in vertebrate smooth muscle cells was not, however, in doubt. But it was found to be rather different in solubility and in enzymic activity from striated muscle myosin (Barany *et al.*, 1966; Hamoir, 1969; Huriaux *et al.*, 1965; Hamoir and Laszt, 1962; Hamoir, 1973). In fact, smooth muscle myosin is much more soluble at low ionic strength than are other myosins, and the ATPase activity is generally lower (Needham and Williams, 1963a,b,c). But, like these other myosins, smooth muscle myosin will also aggregate under suitable conditions (see Chapter 6) to form synthetic filaments (H. E. Huxley, 1963; Hanson and Lowy, 1964; Kaminer, 1969; Sobieszek, 1972; Pepe *et al.*, 1975). It was therefore a puzzle that myosin filaments could not be satisfactorily identified in sections. To explain this, various possibilities were considered. For example, it was suggested that antiparallel myosin dimers might be involved in contraction or that myosin filaments might form as a result of activation. Another possibility was that the myosin was so labile that the fixation methods that had been used might have disrupted the thick filaments' structure.

It was the X-ray diffraction evidence of Lowy and his collaborators (Lowy *et al.,* 1970, 1973; Vibert *et al.,* 1971) that settled at least part of this question. First, it was shown that X-ray diffraction patterns from living relaxed guinea pig taenia coli muscles (Fig. 8.39) contain a meridional reflection at a spacing of 143 Å which is clearly diagnostic of the presence of aggregated myosin. In addition, an equatorial reflection of spacing 120 Å (first seen by G. F. Elliott and Lowy, 1968) indicated that here, as in molluscan muscles, the thin filaments are organized into some kind of actin lattice. The conditions for recording these diffraction patterns were that the muscles were held at about 12 to 15°C in normal Ringer's solution (Bülbring and Golenhofen, 1967) aerated with 97% O_2 and 3% CO_2. The relatively low temperature was used here to prevent the muscles from contracting, since at their physiological temperature (37°C), they exhibit spontaneous contractile activity as required for their function *in vivo.*

On thing that seemed at first to be surprising was the size of the 143-Å reflection. As mentioned in Chapter 2, the size of a reflection depends on the extent of the lattice that is giving rise to it. The larger the lattice, the smaller is the reflection. In X-ray diffraction patterns from taenia coli, the width along the meridian of the 143-Å reflection was thought to be consistent with a 143-Å array extending for at least 0.5 μm axially (Lowy *et al.,* 1973), and its width across the meridian (which was rather difficult to estimate accurately) was thought to suggest that the diffracting object was several hundred angstroms wide (Lowy *et al.,* 1970), wider than might be expected for a single myosin filament.

When it was realized that there must be an actin lattice in this muscle, various attempts were made to preserve this structure so that it could be seen in electron micrographs. When this was achieved (Lowy and Small, 1970; Rice *et al.,* 1970; Heumann, 1970), it was found that thick myosin filaments were also present, a result that tended to confirm the X-ray diffraction evidence. However, it was found that the use of different preparative methods produced thick filaments that were very different in shape and size. In one case (Fig. 8.40), they were roughly cylindrical and about 150 to 250 Å in diameter (Rice *et al.,* 1970, 1971; Heumann, 1970), and in the other (Fig. 8.41), they were flat, ribbon-shaped structures about 100 to 150 Å thick, several hundred angstroms wide, and several micrometers long (Lowy and Small, 1970; Small *et al.,*

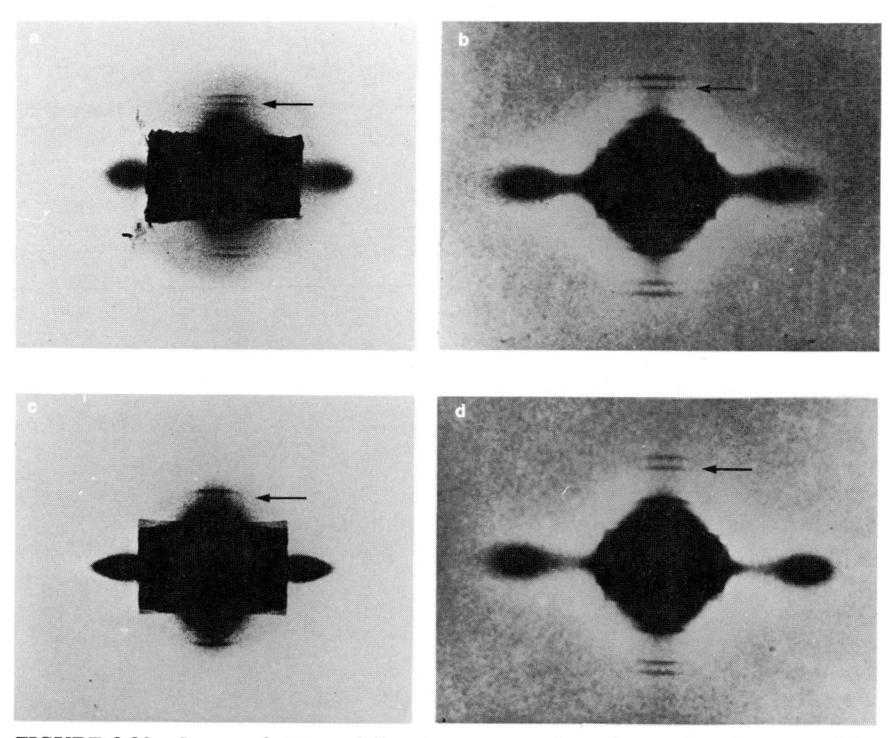

FIGURE 8.39. Low-angle X-ray diffraction patterns from the taenia coli muscle of the guniea pig recorded under various conditions. (a) Resting muscle at 9°C. (b) Contracting muscle at 32°C. Here the 143-Å reflection (arrowed) is much weaker than in (a), but it has more or less the same shape. (c and d) Patterns recorded on a longer camera (36 cm muscle-to-film distance) than (a) and (b). (a) Resting muscle at 9°C in isotonic Ringer solution of composition given by Bulbring and Golenhofen (1967). (d) Muscle in resting state at 35°C in Ca-free Ringer's solution containing 2 mM EGTA and also sucrose to give a hypertonic Ringer's of tonicity 480 mOs. Note that in both (c) and (d) the intensity of the 143-Å reflection (arrowed) is comparable to that of the adjacent (130-Å) reflection caused by collagen. The strong meridional reflection on the edge of the central scatter (260 Å) is also caused by collagen, but the strong equatorial peaks are thought to be due to regions in the muscle cells where there are regular actin lattices. The spacing of the equatorial reflection in (c) is about 115 Å, and that in (d) is about 95 Å. All patterns are from Lowy *et al.* (1973).

1971; Small and Squire, 1972). It is clear from the large width of these ribbons that they could explain directly the apparent width of the 143 Å reflection, but the cylindrical filaments appeared to be too small in diameter to explain the X-ray data unless two or more filaments were aligned axially to produce a coherent diffracting unit (Vibert *et al.*, 1972; Small and Squire, 1972). However, the ribbons were found to be most consistently seen and at their widest (Small and Squire, 1972) when the

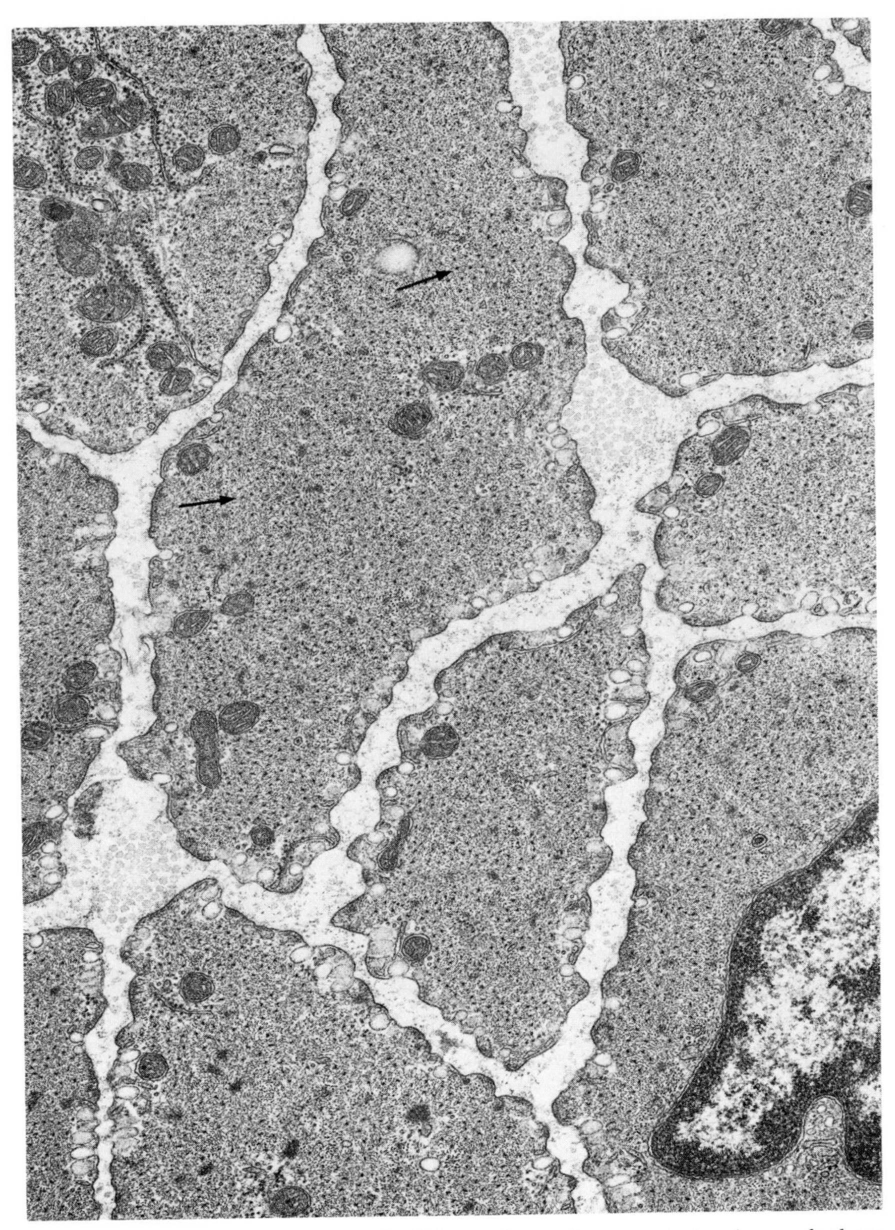

FIGURE 8.40. Transverse section of rabbit portal–anterior mesenteric vein muscle showing the presence of regular actin lattices and a fairly even distribution of round thick filaments (arrowed). Collagen fibers can be seen in cross section in the intercellular space. Dense structures rather similar in appearance to the dense bodies in molluscan muscle (Fig. 8.25) can also be seen (×35,000). (From Somlyo *et al.*, 1973.) Note also the 100-Å filaments associated with the dense bodies.

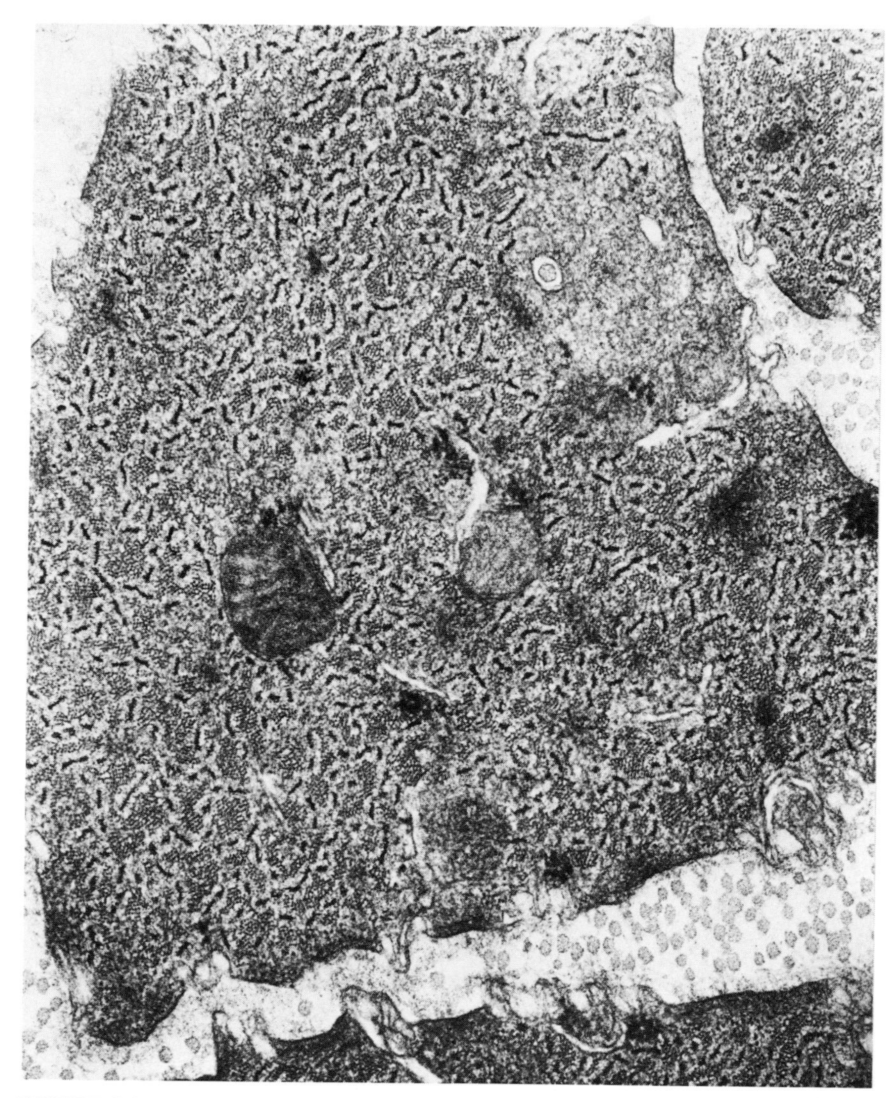

FIGURE 8.41. Transverse section of an apparently well-preserved cell from guinea pig taenia coli muscle according to Small and Squire (1972). The appearance is much as in Fig. 8.40, with actin lattices and dense structures clearly evident, but here the thick filament profiles are elongated. These are the so-called myosin ribbons. They can be seen to be surrounded by a less dense halo region about 140 to 180 Å wide. Magnification ×80,000.

muscle was fixed cold (0–4°C) or in a normal Ringer's solution made hyperosmolar by the addition of sucrose (hypertonic Ringer's solution). In the same way it was found that the 143-Å meridional X-ray reflection was strongest under the same conditions (Shoenberg and Haselgrove,

1974). This result under nonphysiological conditions was taken by many to be an indication that wide ribbons might be a preparative artifact. However, it should be noted that none of the chemical preparative treatments for electron microscopy can be said to be remotely physiological whatever the initial treatment of the muscle; the X-ray diffraction evidence should, on the other hand, be more reliable although less easy to interpret.

The effect of contraction on the X-ray diffraction pattern was rather different(Lowy *et al.*, 1970, 1973).Here, the intensity of the 143-Å reflection was always less than that in patterns from relaxed muscle (Fig. 8.39), and this was attributed (as in other muscles) to movement of the myosin cross bridges from their regular relaxed positions. Since the 143-Å reflection did not disappear and had about the same width as in diffraction patterns from relaxed muscle, it seemed reasonable to Lowy *et al.* (1973) to conclude that some form of aggregated myosin must be present in the active muscle. This has recently been confirmed by Small (1974) who obtained preparations of isolated cells from which the 100-Å filaments had been extracted using collagenase and who found that these contained well-organized contractile units and would still contract. It would be expected that any myosin molecules that were not in an aggregated form would have been extracted by the preparative treatment.

Since, as mentioned above, very wide ribbons seem to be most evident in muscles fixed under nonphysiological conditions (Shoenberg and Haselgrove, 1974), it has been argued in several papers that ribbons must be preparative artifacts and that the true *in vivo* myosin elements are the round filaments (Rice *et al.*, 1970; Heumann, 1970). The ribbons would then be explained as products of the side-by-side aggregation of the round filaments. In addition, it is becoming evident that measurement of the width of the 143-Å reflection is a very uncertain method of determining the coherent size of the thick filaments (Squire, unpublished calculations). It therefore appears that the disagreement about the *in vivo* form of the smooth muscle thick filaments must reduce to the question "Is it round filaments or ribbons of small width (say 150–200 Å) that exist *in vivo* and aggregate under various conditions, some nonphysiological, to form very wide ribbons?"

Whether or not wide ribbons are artifacts, what is clear is that they represent an extremely well-ordered form of myosin aggregate (Small and Squire, 1972). As described in the next section, the structure of these ribbons has been analyzed in detail and the results of this analysis help to provide an answer to the question just posed. They also help to show how the contractile machinery in vertebrate smooth muscle might be organized.

8.4.2. Structure of the Myosin Ribbons

The remarkably regular structure of the wide myosin ribbons in vertebrate smooth muscle was analyzed in detail by Small and Squire (1972) using electron microscopy and optical diffraction. Since the ribbons are planar structures, three distinct orthogonal views of them can be recognized. In transverse sections (Fig. 8.42) the ribbons can be seen in projection down their long axes, and here they appear as lines 80 to 120 Å thick and of variable length between 200 and 1000 Å. As in the case of the paramyosin filaments in molluscan muscle (Fig. 8.25), these ribbons can be seen to be surrounded by a semidense halo region which separates them from the surrounding actin arrays.

In longitudinal section, two distinct orthogonal views of the ribbons can be recognized as well as some intermediate views. The characteristic views are edge views parallel to the filament surface and perpendicular to its length and face views normal to the plane of the ribbon. In edge views, the ribbons appear as very long dense lines about 80 to 120 Å thick and edged by the semidense halo material seen in transverse sections (Fig. 8.43). In certain preparations it could be seen that the halo region in edge view actually comprised projections from the filament surface that had an axial spacing of about 140 Å [Fig. 8.43(a–d)]. This same feature could be recognized in face views as lateral stripes of density with the same axial spacing of 140 Å [Fig. 8.43(i–l)]. It was evident from the presence of this 140-Å spacing (which can be taken to correspond *in vivo* to the 143-Å spacing measured from X-ray diffraction patterns) that the projections seen in edge views must correspond to the heads of myosin molecules aligned across the ribbon width, a clear indication that the ribbons are indeed myosin-containing filaments.

Optical diffraction analysis of a number of ribbon edge views revealed a consistent but unusual feature of these myosin aggregates. It was found that, as expected, the optical diffraction patterns commonly showed a reflection at a spacing of about 140 to 150 Å (taken to be 143 Å), but in addition many patterns showed an off-meridional reflection on one side of the meridian only [Fig. 8.43(a–d)]. By analyzing a number of model structures by optical diffraction [Fig. 8.43(e–h)], it was concluded that the cross bridges on the ribbons must be arranged in such a way that all those on one face of the ribbon are tilted axially in a direction opposite to those on the other face as indicated in Fig. 8.44a. Myosin ribbons are therefore not bipolar as are the thick filaments in other muscles but have "face polarity." It is evident that this face polarity of the cross bridge arrangement must reflect the underlying molecular packing arrangement in the ribbon backbones so that the rods in one face of the ribbon must point in a direction opposite to those in the other face.

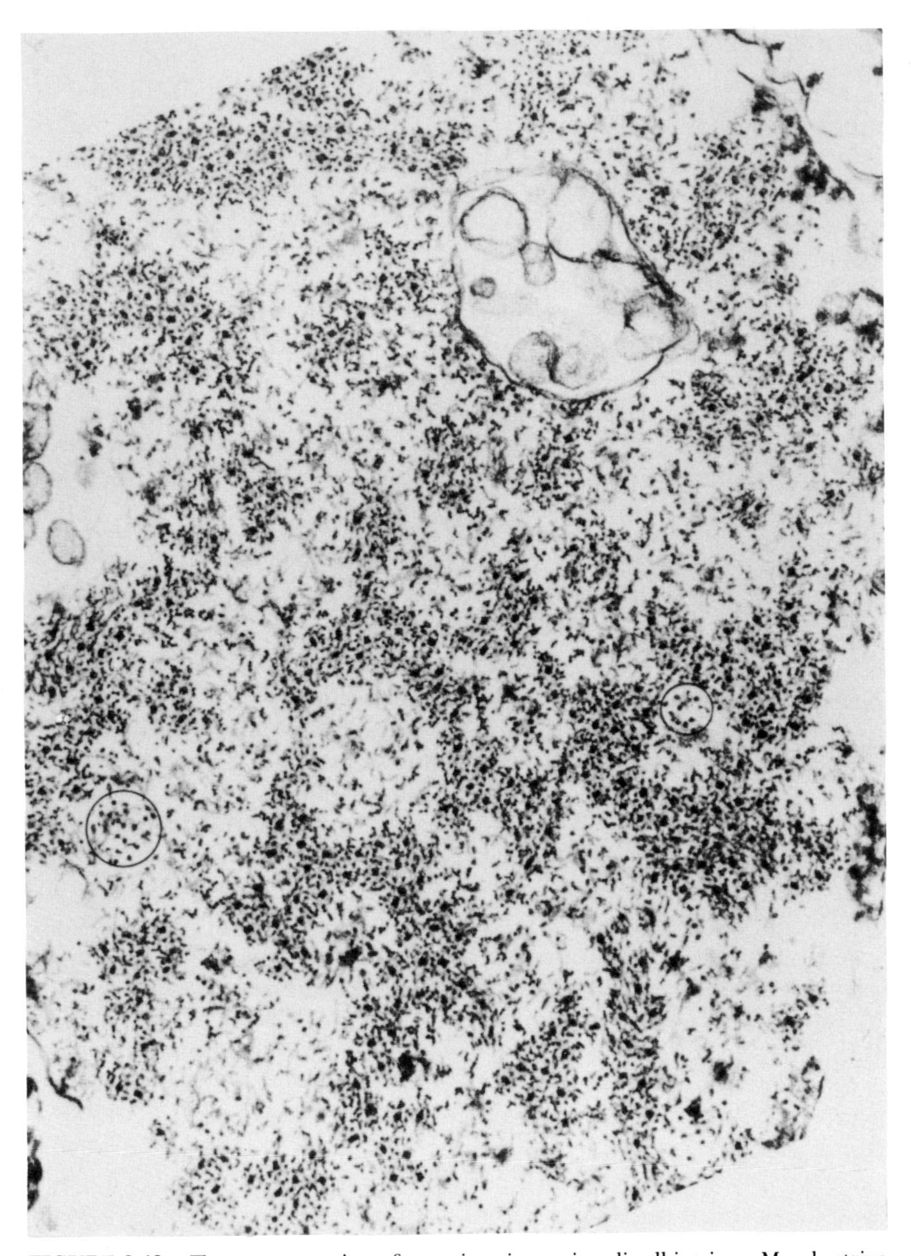

FIGURE 8.42. Transverse section of a guniea pig taenia coli cell in rigor. Muscle strips, prepared by treatment of the muscle with collagenase, were bathed in a solution containing 0.03 to 0.05% Triton X-100 plus dithiothreitol but no ATP for 12 to 24 hr at 4°C. 100-Å filaments (circled) lie between the groups of actin and myosin filaments. The myosin filaments exhibit an approximately square profile in cross section. (From Small, 1977a.) ×80,000.

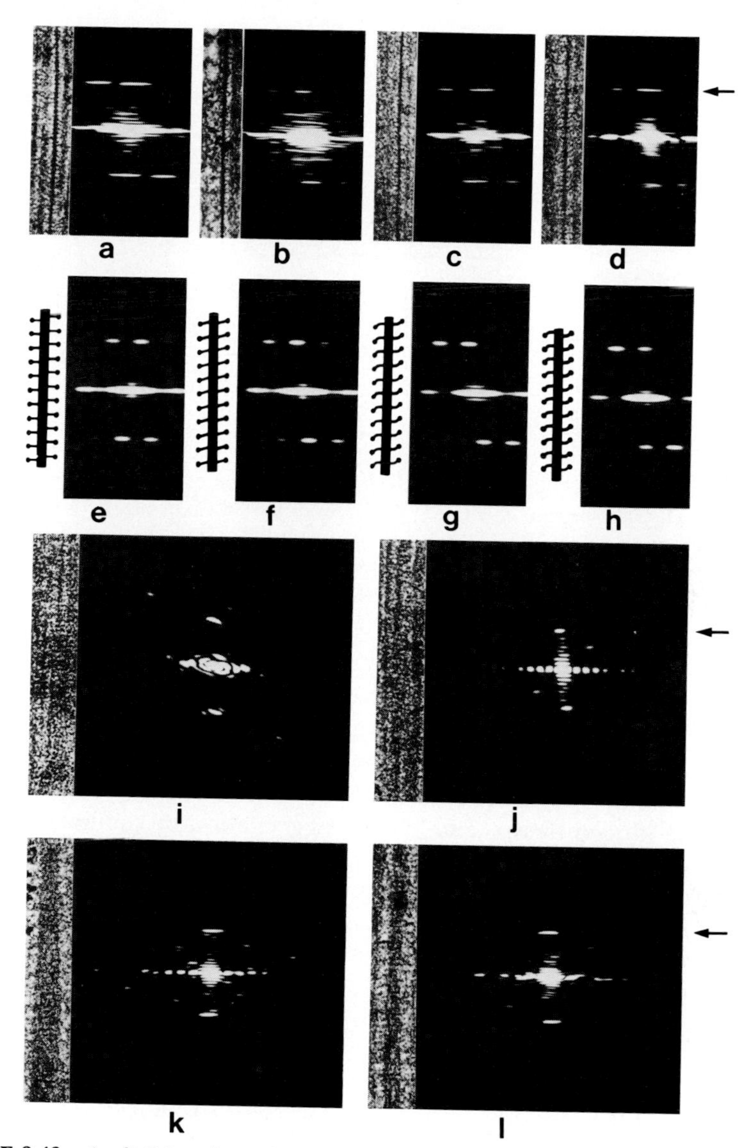

FIGURE 8.43. (a–d) Edge-view micrographs of myosin ribbons in guinea pig taenia coli muscle together with their optical diffraction patterns. The dominant feature of the patterns is a layer line (arrowed) at a spacing of about 140 Å which comprises a meridional reflection and an off-meridional reflection on one side of the meridian only. (e–h) Four possible models (from several) of the edge-view ribbon structure with their optical diffraction patterns. All models contain face polarity (see text), but only (h) seems to mimic accurately the patterns in (a) to (d). Details of model (h) are given in Fig. 8.44. (i–l) Myosin ribbons seen in face view and their optical diffraction patterns. In addition to the prominant meridional reflections of spacing about 144 Å (arrowed) other reflections off the meridian seem to relate to a cross bridge lattice of axial repeat about 720 Å. (From Small and Squire, 1972.)

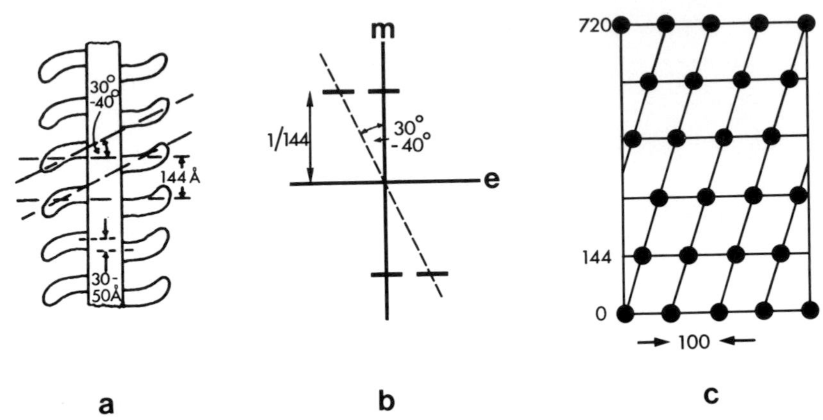

FIGURE 8.44. (a) Diagram illustrating the form of model (h) in Fig. 8.43 for the edge views of the myosin ribbons. The face-polarized cross bridge array illustrated here explains directly the origin of the asymmetric diffraction patterns as in (b) (m, meridian; e, equator). (c) Representation of the probable array of myosin cross bridges on each face of the myosin ribbons that will account for patterns such as those in Fig. 8.43 (i–l). Dimensions are given in Å. The lattice has an axial repeat of 720 Å and is like a Bear–Selby net (as on paramyosin filaments) but with an *a* spacing of only about 100 Å. Each lattice point probably represents the location of the heads of a single myosin molecule. (a and c) After Small and Squire (1972).

From this result alone it can be concluded that if wide ribbons are artifacts produced by the side-by-side aggregation of smaller filaments, then these smaller filaments must themselves have face polarity, and they must aggregate in a very specific manner so that their cross bridges are in axial alignment. They cannot therefore be cylindrical filaments of the conventional bipolar type as has been argued by Rice *et al.* (1971) and Somlyo *et al.* (1970, 1971), but they must be small ribbons. That ribbons can break up into smaller units was shown by Small and Squire (1972) using serial transverse sections. It could then be seen that what started off in one section as a rather wide ribbon became a number of smaller filaments looking rather like the round filaments of Rice *et al.* (1970) and Somlyo *et al.* (1970, 1971, 1973) in subsequent sections further along the muscle cell. From this it is evident that ribbons are not formed from a reaggregation of the dissolved components of the round filaments but, rather, that one is the direct derivative of the other. It is hard to escape the conclusion that if myosin does exist in an aggregated form *in vivo* then it must be formed into at least small ribbons of the form described above and that these narrow ribbons may be similar to the round filaments observed by Rice *et al.* (1970) and Heumann (1970).

Many more details of the structure of the ribbons were obtained by Small and Squire (1972) from an analysis of face views. It was found that the face views gave optical diffraction patterns [Fig. 8.43(i–l)] that could be satisfactorily interpreted in terms of an array of cross bridges of the kind shown in Fig. 8.44c. Here, the cross bridges are arranged about 100 to 120 Å apart to form rows across the ribbon surface, and cross bridges in successive rows, 143 Å apart axially, are arranged to produce a surface lattice rather like a Bear–Selby net with a repeat of about 720 Å. Since the diffraction patterns were invariably asymmetric about the meridian, it was evident that the cross bridge array on the back surface of the ribbon must have the same hand as that on the front, as might be expected from such a planar structure. In addition to the 720-Å repeat, some occasional evidence for a thick filament repeat of between 400 and 500 Å was also obtained.

Apart from thick filaments and ribbons, transverse sections of guinea pig taenia coli showed evidence of 100-Å filaments (mentioned earlier) and of dense areas (Fig. 8.41). The latter may not be equivalent to the dense bodies in molluscan muscles, since in longitudinal sections, it can be seen that actin filaments often go around these rather than through them (Small and Squire, 1972). On the other hand, they do seem to be associated with 100-Å filaments which can sometimes to seen within the dense areas. On the basis of circumstantial evidence, Small and Squire (1972) originally suggested that the 100-Å filaments and dense areas were breakdown products of the ribbons resulting from inadequate preservation of the muscle, and since the ribbon backbones sometimes seemed to show a subunit structure similar to that of the 100-Å filaments (which comprised a square array of four subfilaments, each of diameter 20 to 40 Å), it was suggested that these filaments might form a core structure to the ribbon backbones and that myosin might form layers on the two faces of this core. The dense areas would then be pools of myosin produced by disruption of this surface array. This idea has since been disproved by Small and Sobieszek (1977) and Small (1977a,b) who demonstrated that myosin and actin can be totally extracted from smooth muscle cells, leaving a matrix of 100-Å filaments containing a 55,000-dalton protein subunit which they termed skeletin (see also Cooke, 1975). These filaments were thought to serve together with the dense areas as a cytoskeleton within the muscle cells. A similar conclusion was reached by Ashton *et al.* (1975). In addition, as mentioned earlier, using Triton X-100 and collagenase to preferentially extract the 55,000-dalton protein from isolated cells, Small (1974) and Small and Sobieszek (1977) found that 100-Å filaments were largely absent but that the cells that resulted could still contract in response to ATP.

It is natural to conclude from this that the filaments containing the 55,000-dalton protein are definitely not responsible for forming the

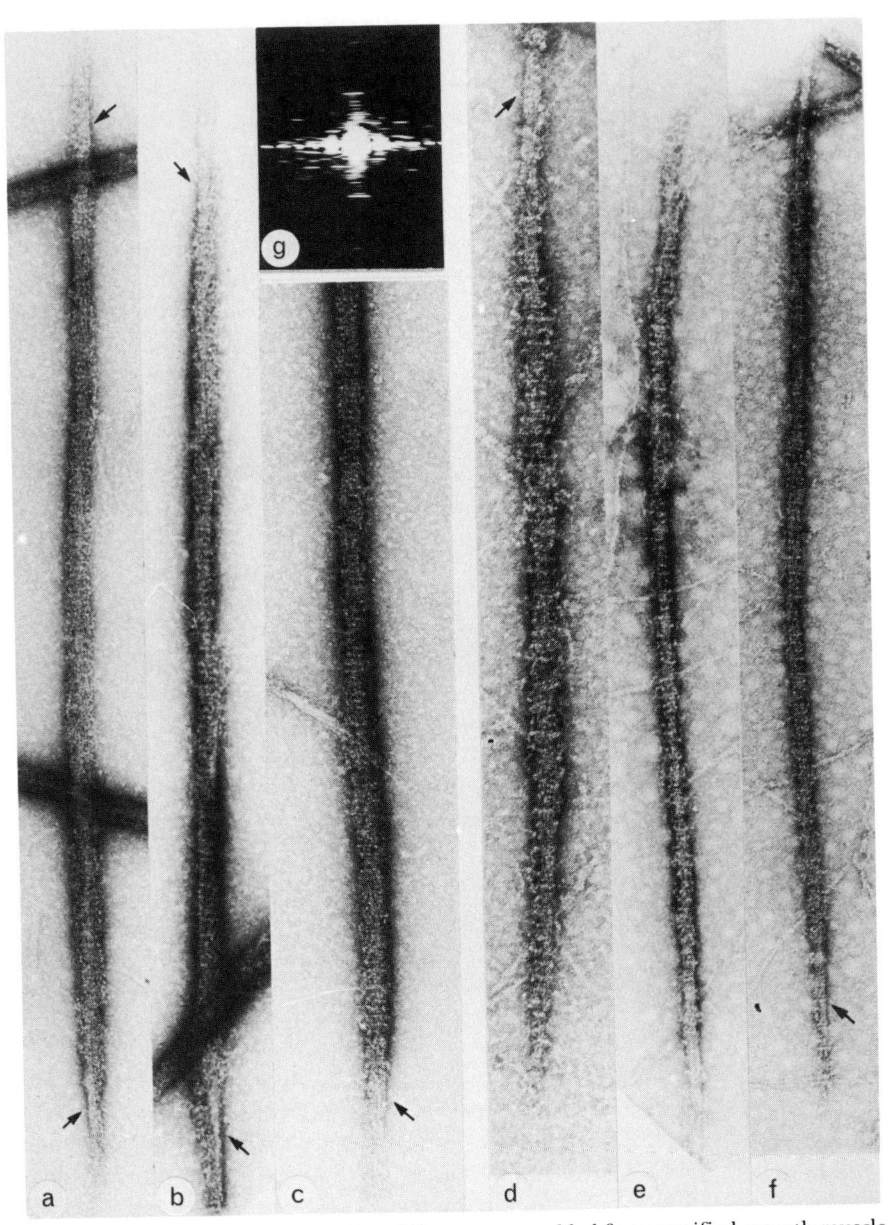

FIGURE 8.45. Structural similarity of filaments assembled from purified smooth muscle myosin (a–c) and the native filaments released from isolated cells (d–f). (From Small, 1977b; micrographs a–c supplied by A. Sobieszek.) A periodicity of about 140 Å, probably from the cross bridge array, can be recognized along the entire length of the filaments, and bare edges occur at the filament ends (arrows). (g) Optical diffraction pattern from the native filament shown in (e). Note that (g) and similar patterns from other filaments

cores of the ribbons. But, with that result clear, a number of questions then come to mind about the nature of the ribbons. (1) Do the myosin filaments or ribbons contain any protein other than myosin? (2) Why do transverse sections of ribbons sometimes show a clear subunit structure? (3) Why do some ribbon face views show evidence of both a 720-Å repeat and a repeat about 400 to 500 Å?

With regard to these questions, there is one point that does seem to be clear. This is that preparations of purified smooth muscle myosin can form filaments that appear rather similar to the ribbons (Sobieszek, 1972; Sobieszek and Small, 1972; Wachsberger and Pepe, 1974; Craig and Megerman, 1977; Fig. 8.45). Although short bipolar filaments can occur with the characteristic bone-shaped appearance and a central bare zone (Fig. 8.46a), longer filaments (Figs. 8.45 and 8.46b,c) do not have the central bare regions sometimes observed in other kinds of sythetic filaments (H. E. Huxley, 1963). They can, however, show a regular 145-Å periodicity along most of their length. If bare regions are evident, these tend to be at the ends of the long filaments rather than the middle, and they thus give the filaments in some views a flat parallelogram shape. This has been interpreted in terms of a specific face-polar myosin arrangement. Evidence for this is that such filaments can show the same asymmetric diffraction pattern indicative of face polarity (Fig. 8.46e) as the ribbon edge views in sections (Craig and Megerman, 1977). Sobieszek (1972) concluded that the filaments might be formed by the staggered aggregation of a number of short bone-shaped filaments, but clearly there are other possibilities as outlined in the next chapter.

Since pure myosin can form ribbons on its own, one can question whether a core structure is really necessary. For example, if one calculates the volume that would be occupied by the myosin rods in the ribbons assuming that on the two ribbon faces the myosin heads are arranged on the lattice shown in Fig. 8.44c, then it is found that rods alone could almost account for the measured thickness of 80 to 120 Å for the ribbons (Small and Squire, 1972). The volume occupied by one myosin rod would be about $20 \times 20 \times 1500 \text{Å}^3$, and since the surface area per myosin cross bridge on the surface lattice in Fig. 8.44c is about 100 to 120 Å × 143 Å, then assuming that the rods in the filament backbone are close-packed, they would form a layer of thickness about

---

(Sobieszek, 1977) can be accounted for in terms of the same basic cross bridge lattice as that shown in Fig. 8.44c for the intact myosin ribbons seen in face view. Sobieszek interprets his patterns in terms of a helical array of cross bridges (for which Fig. 8.44c would be a radial projection), but it is by no means clear that these filaments are helical. They could equally well be planar like the ribbons.

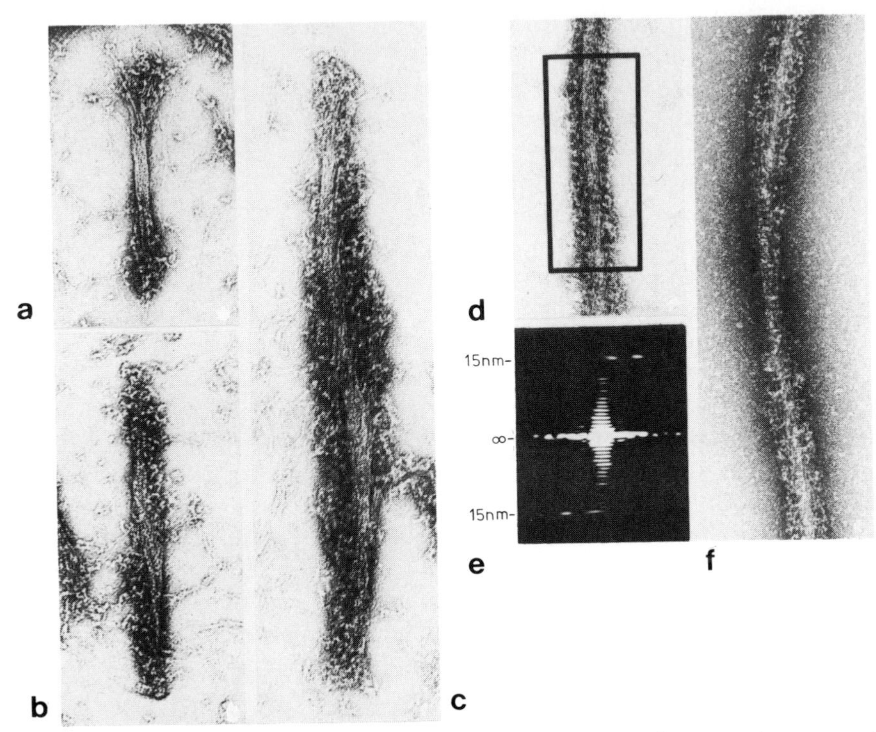

FIGURE 8.46. All except (e) are synthetic aggregates of myosin from vertebrate smooth muscle (calf aorta negatively stained). (a) A bone-shaped filament typical of such preparations using striated muscle myosin. (b and c) Face polar filaments with two bare edges and two edges covered in projections. (e) The optical diffraction pattern of the filament in (d) showing the asymmetric 144-Å layer line profile characteristic of ribbon edge views (Fig. 8.43). (From Craig and Megermann, 1977.)

600,000/(14,300 to 17,160) Å or 35 to 42 Å. Two of these layers would give a filament backbone of thickness 70 to 84 Å. This agrees fairly well with the original measured value of the ribbon thickness of 80 to 120 Å but is slightly less than more recent estimates (145 to 160 Å), and this may suggest that a small core exists. Certainly it does not appear to be the 100-Å filaments that form the ribbon core as originally thought (Small and Squire, 1972).

An early alternative idea was that some of the tropomyosin in vertebrate smooth muscles might form a core structure (Small and Sobieszek, 1972, Sobieszek and Small, 1973). A low-ionic-strength extract from vertebrate smooth muscle in the presence of ATP was found to form long ribbon-shaped aggregates. The aggregates were formed either if $CaCl_2$ was added to the extract or simply if the extract was left to stand in the cold for one or more days. These ribbons were found to have the same repeats of

391 ± 4 Å and 56 Å as in other tropomyosin aggregates (Chapter 5). That they might be involved in forming the myosin ribbons observed in sections seemed to be suggested by two results. One was that when left in solution in the presence of smooth muscle myosin, ribbons could be obtained that seemed to show both the characteristic tropomyosin repeat and the 143-Å repeat typical of myosin. The other was that estimates of tropomyosin content in vertebrate smooth muscles (Sobieszek and Bremel, 1975; Tregear and Squire, unpublished results) seemed to indicate that these muscles contained up to twice the amount of tropomyosin that could be accommodated by the thin filaments in the muscle. It should be noted, however, that a difference in the staining properties of actin and tropomyosin in SDS gels might account for some of this discrepancy (Sobieszek and Bremel, 1975).

Apart from this, Sobieszek and Bremel (1975) have identified two previously unknown protein components in vertebrate smooth muscle in addition to the 55,000-dalton protein, skeletin (Cooke, 1975; Small and Sobieszek, 1977). These two unknown proteins of chain weights 110,000 and 130,000 have not been characterized. An alternative possibility is, therefore, that if a ribbon core exists it is composed of one of these proteins. However, it must be concluded that much more evidence is needed to decide these questions. It may well be that the ribbons have no core at all. The recent evidence of Small (1977a,b) that shows myosin filaments in isolated cells from which the 100-Å filament cytoskeleton had been extracted now seems to provide confirmation that the contractile filaments in vertebrate smooth muscle exist in the form of small filaments. These filaments are said to be square and about 160 Å across. When isolated, they show 145 Å striations along their entire length except at their ends where bare edges occur. They also seem to be able to aggregate side by side and in axial register to form wider structures. Although Small (1977a,b) and Sobieszek (1977) seem to believe that these filaments have a four-stranded helical arrangement of cross bridges, there seems to be little direct evidence for this, and the evidence already described seems to be very suggestive that these structures are small face-polar ribbons. In such a case, their width of 160 Å might suggest that a core component is necessary. Estimates of myosin content in this muscle suggest that there are about four molecules per 143-Å repeat along the square filaments (Small, 1977a,b).

8.4.3. Discussion

It must be clear from the discussion so far that even now there remains some uncertainty about the structure of myosin filaments in active vertebrate smooth muscles. However, some concluding remarks made by H. E. Huxley at the close of the Royal Society Meeting in 1973 in "A discussion on recent developments in vertebrate smooth muscle physiology" do appear to remain valid. Among other things, he said, "A myosin component is present. . . . This myosin can form either filaments or ribbons, the latter at least showing a remarkably well organised fine structure." Later on he said, "I do not really think that the evidence really proves that tension is actually generated by these large aggregates rather than by some other form of myosin, but it would be very surprising to me if this ability to form large aggregates were an accidental property of the myosin."

On that basis, it is instructive to see how face-polar myosin ribbons whatever their width might fit into a contractile assembly. The analysis was originally carried out for the taenia coli of the guinea pig by Small and Squire (1972) in terms of wide ribbons. But, in principle, it applies equally well to any myosin filament (such as a narrow ribbon) that has mixed cross bridge polarity along its length.

On the assumption that none of the filaments changes length during contraction, then the contractile apparatus must be arranged so that relative sliding of the thick and thin filaments can be produced. Assuming that the myosin filaments (ribbon) are face-polarized, it is natural to suggest that thin filaments of one polarity will interact with one face of a ribbon and those of the opposite polarity will interact with the other face. But to provide a continuous contractile unit through the cell, it clearly must be necessary for the thin filaments to change their polarity somewhere along their length as occurs at the Z-band of striated muscles. But in this muscle it is not definite that the dense bodies are structures analogous to Z-bands (despite the observations of Ashton *et al.*, 1975, on vascular smooth muscle), and no site of thin filament attachment has yet been unambiguously identified other than the clear attachment sites at the cell surfaces. It must therefore be postulated either that thin filament polarity does not matter and the direction of movement depends only on the polarity of the cross bridges (a seemingly very unlikely possibility) or that some structure, as yet unidentified (but possibly analogous to dense bodies), links thin filaments of opposite polarity.

If the latter is taken to exist, then contractile units of the kind illustrated in Fig. 8.47 can be envisaged. As in molluscan muscle, these units will form a kind of myofibril in the muscle cell. In this arrangement it is envisaged that each ribbon face of width $n \times 100$ Å to 120 Å would interact with a row of about n thin filaments spaced 100 to 120 Å apart (seen edge on in Fig. 8.47). The dots on the thin filaments indicate the site of the polarity change. In the model shown in Fig. 8.47, it is assumed that the terminal thin filaments have only half the length of the other

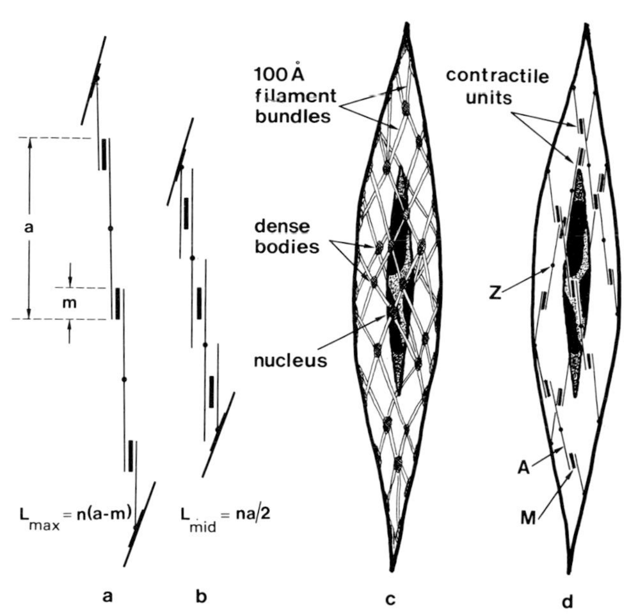

FIGURE 8.47. (a and b) Schematic diagrams illustrating the nature of contractile units (rudimentary myofibrils) that can be formed using thick filaments (such as the face-polar ribbons) that have mixed cross bridge polarity along their entire length. (a) The greatest length at which 100% tension is generated. (b) The middle length for 100% tension generation. It is assumed that the actin filaments are linked to structures (as yet to be clearly defined) analogous to the Z-bands or dense bodies in other muscles. These are shown as dots halfway along the filaments where the filament polarity is assumed to change. Such a structure is assumed to link the thin filaments to the cell surface. For explanation see text and Small and Squire (1972). (c) Schematic illustration of the probable arrangement of the 100-Å filaments and dense bodies in vertebrate smooth muscle cells to form a kind of internal structural network. Within this network the contractile units as in (a) would be located in a kind of spiral array around the nucleus. A, actin; M, myosin; Z, the location of the polarity change on the thin filaments. (c) After Small (1977a); (d) after Small and Squire (1972).

bipolar thin filaments assemblies. This difference has very little effect on the conclusions given below.

The behavior of these contractile units at different muscle lengths clearly depends on the relative lengths of the thick and thin filaments. Small and Squire (1972) estimated that the ratio of the thick to thin filament lengths was probably in the range between 1 to 4 and 1 to 6. With the absolute lengths taken to be about 6 to 8 μm (see also Small, 1977) and 32 to 36 μm, respectively, and the whole contractile unit to be 200 μm long, it was found that a number of physiological results from these muscles could be explained rather well. In particular, it has been shown by a number of workers (Aberg and Axelson, 1965; Mashima and Yoshida, 1965; Lowy and Mulvany, 1973) that the variation of tension with length is much broader than that in vertebrate striated muscle (see Fig. 1.8). It is such that if L_0 is the muscle length at which maximum tension is developed, then tensions above 90% of this value occur in vertebrate smooth muscle in a range that is $0.50L_0$ to $0.65L_0$ wide. This compares with a range of $0.28L_0$ for frog semitendinosus muscle (Gordon *et al.*, 1966).

Similarly, vertebrate smooth muscle will work over a vast range of lengths, so that the ratio of the maximum working length (L_{max}) to the minimum (L_{min}) is about 4 or 5 : 1 compared with 1.5 or 2 : 1 in striated muscles. Using the estimated values of 6 to 8 μm and 32 to 36 μm for the thick and thin filament lengths, respectively, in the contractile units shown in Fig. 8.47, Small and Squire (1972) estimated that maximum tension (taken to correspond to the recorded 90% tension range) could be developed over a range 0.50 to $0.67L_0$ and that L_{max}/L_{min} would be about 5 or 6 : 1. Each unit would comprise four or five ribbons. These results are clearly comparable to the observed values, and they encourage one to believe that face-polar ribbons may have some relevance to the contraction of vertebrate smooth muscle.

It is instructive to consider how such contractile units in vertebrate smooth muscle might be arranged to produce the observed spindle-shaped cells. Figure 8.47d shows how Small and Squire (1972) thought their ribbon units might be arranged. (Note that the recent scheme suggested by Small, 1977a,b is very similar.) Here the width of the cell has been greatly exaggerated to illustrate the point that the contractile units are joined to different parts of the cell membrane. Since they are oriented slightly obliquely to the cell axis, the structure automatically leaves a central area which in practice is found to contain the nucleus and other cell organelles. In fact, the contractile units would be only about 1° or 2° from the long axis of the muscle in extended cells. The lack of perfect alignment among different contractile units would, therefore, not be evident in sections of reasonably long cells. In fact, direct evidence for the presence of slightly oblique contractile units has been obtained by Small (1974) in his studies of isolated cells.

The results of Ashton *et al.* (1975) on the contractile apparatus of rabbit portal anterior mesenteric vein smooth muscle seem to differ considerably from the scheme described above. In particular, the thick filaments are reported to be only 2 to 3 μm long, and the dense bodies are claimed to be actin attachment sites and to contain the protein α-actinin usually associated with Z-bands (Schollmeyer *et al.*, 1973). It is possible that these results may reflect a genuine difference in structure between taenia coli and vascular smooth muscles, and there is little reason as yet to believe that the ultrastructures of different vertebrate smooth muscles must be identical. Here the thick filaments were said to have diameters of about 145 Å and to taper at each end, but no direct evidence was obtained that could distinguish between round, bipolar filaments and narrow ribbons.

As a final comment, it is worth noting that future studies of the contractile apparatus of vertebrate smooth muscle will very likely be facilitated by the recent results (some described above) of Small (1974, 1977a,b and Sobieszek and Bremel (1975) who have been able to produce functional preparations of intact smooth muscles or of isolated cells that may be similar in their application and advantages to the isolated myofibrils of striated muscles. Already these preparations have provided several valuable results, and their future application in both biochemical and structural studies is extremely promising. But it seems to be sensible to bear in mind, as Huxley suggested, that the ability of the small myosin filaments to aggregate to form wider ribbons may have a functional significance. These new smooth muscle preparations may help to make it possible to explore this fascinating possibility.

8.5. Obliquely Striated Muscles

8.5.1. Introduction

The characteristic appearance of a number of muscles in annelids, nematodes, and certain molluscs is not cross-striated but is obliquely or helically striated. A number of reviews of this muscle type have been published (Hoyle, 1964; Rosenbluth, 1972), and the intention is to present only a brief discussion here.

It was Englemann in 1881 who first recognized that the oblique striations in this muscle type were not primarily because of obliquely pointing myofilaments (as for example seems to occur in vertebrate smooth muscle, Fig. 8.47) but were caused by a particular arrangement of axially oriented filaments. Since then, this conclusion has been largely confirmed by electron microscopy (for detailed references see Rosenbluth, 1972). The arrangement of the contractile material in the body wall of the round worm *Ascaris lumbricoides* is shown in Fig. 8.48a, and that in fibers of annelid muscles (segmented worms) is shown in Fig. 8.48b. Following the discussion of Rosenbluth (1972), it is convenient to consider three orthogonal sections through the ribbon-shaped muscle structure in Fig. 8.48a. Sections cut transversely to the filament and muscle long axis are here termed *XY* sections, and the two longitudinal sections are the *XZ* section cut parallel to the ribbon surface and the *YZ*

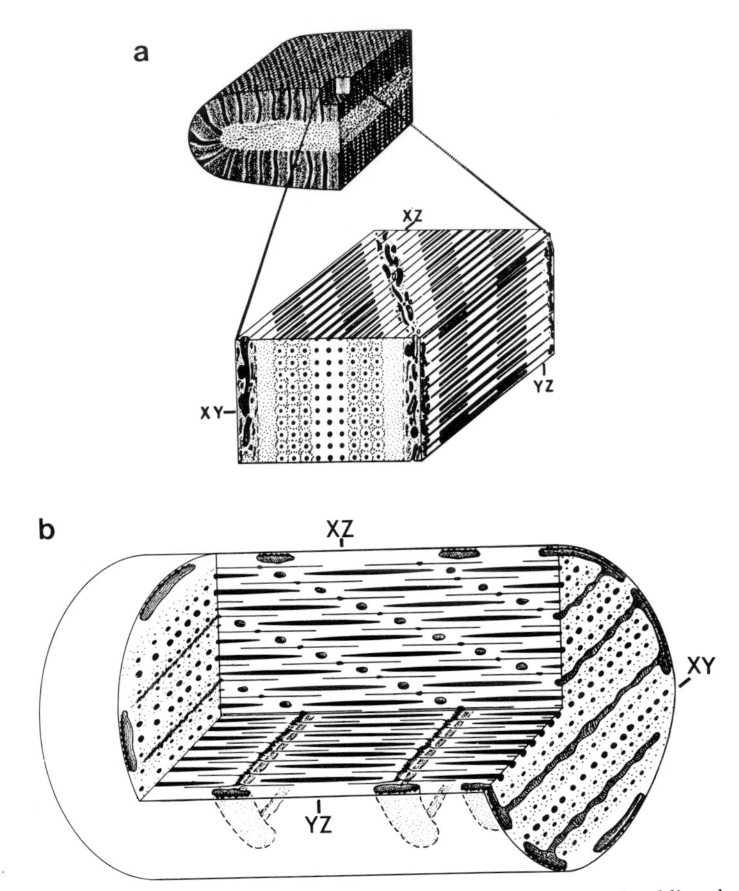

FIGURE 8.48. Schematic diagrams of the myofibrillar arrangements in obliquely striated muscles. (a) Diagram of part of a muscle fiber of *Ascaris*, showing the U-shaped tip of the fiber which lies close to the body wall of the animal. The enlarged block has three characteristic faces. The front facing *XY* plane is the standard transverse view of the contractile material. *YZ* is a longitudinal view in which the structure appears cross striated, and in the *XZ* plane the systematic staggering of the myofilaments produces an obliquely striated appearance. In reality, the oblique lines are much more nearly parallel to the muscle axis than has been indicated here. (b) Equivalent diagram to (a) but for the cylindrical fibers in annelid muscle. (From Rosenbluth, 1965a, 1968.)

section cut perpendicular to the ribbon surface. Three similar sections can be drawn through the annelid fibers as in Fig. 8.48b.

From the appearance in Fig. 8.48, it can be seen that in the *XZ* sections adjacent thick filaments are staggered axially in a systematic way so that the A-band moves obliquely across the section. On the other hand, no such stagger occurs in the *YZ* longitudinal section. The effect of this arrangement is to produce in the *XY* transverse sections a striped appearance caused by sampling of the sarcomere successively through the Z-band, I-band, overlap region, H-zone, next overlap region, I-band, Z-band, and so on.

8.5.2. Myofilament Structure and Arrangement

The *XY* transverse sections show that the thick filaments are relatively large in diameter (250 to 700 Å), and they can sometimes (not in nematodes) vary in shape and size within one muscle. This variation is distinct from the changes in diameter that occur through the A-band because of tapering of the thick filaments. In the A-band, the thick filaments generally tend to be arranged in a more or less hexagonal lattice, and they are commonly surrounded by orbits of nine to 12 thin filaments. The precise location of these thin filaments in the A-band lattice has not yet been determined.

Estimates of the lengths of the thick and thin filaments are difficult because of the obliquity of the A-band, but from the appearance of transverse sections, it can be deduced that the relative filament lengths in the earthworm are similar to those in vertebrate skeletal muscle, whereas in *Ascaris* the thin filaments are probably relatively longer. As a rough guide, the absolute A-band length in *Ascaris* is about 6 μm, and that in the earthworm is about 3 μm.

The thin filaments in adjacent sarcomeres are linked together by structures analogous both to the Z-bands of cross-striated muscles and to the dense bodies in molluscan muscles. As in molluscan muscles, these structures are also termed dense bodies, but here they appear to be rod-shaped so that in *YZ* longitudinal sections and in *XY* transverse sections, they appear continuous across the fiber, whereas in *XZ* sections they are discontinuous. In the latter case, they can be seen to alternate with regularly spaced membrane-limited sarcotubules in annelid muscles (Fig. 8.49). The whole oblique layer of dense bodies and sarcotubules, if they occur, is analogous to the Z-band of other muscles. Figure 8.49 also illustrates the point that the striation pattern in these muscles is very much more oblique than was indicated in Fig. 8.48. In this diagram the striations in the *XZ* section have been drawn for clarity at an angle of about 30 to 40° to the muscle long axis. In fact, this angle is commonly only a few degrees, being very roughly 1 to 2° in annelids and about 6° in nematodes and molluscan muscles. Exact values for these angles cannot really be given since the striation angle varies with sarcomere length and muscle length. As the muscles shorten, the relative stagger of neighboring thick filaments is reduced, and the striation angle therefore increases. This relative shearing of the thick filaments seems to be similar to that which occurs in *Limulus* muscle and probably in other

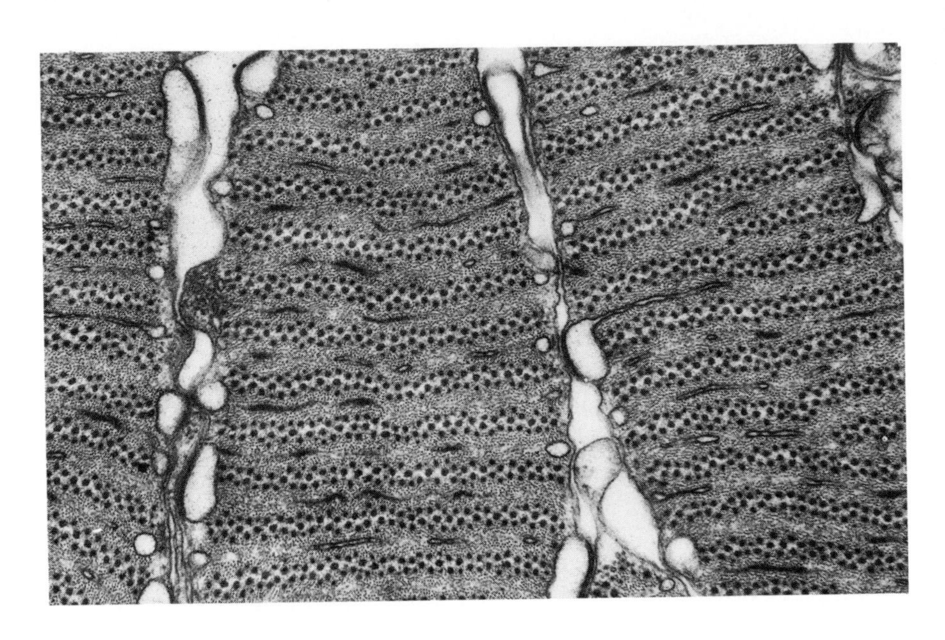

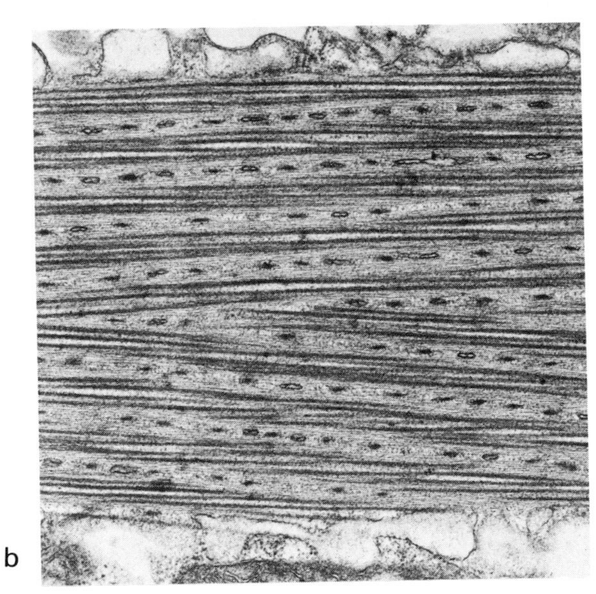

FIGURE 8.49. Electron micrographs of *Ascaris* muscle viewed (a) in the *XY* (transverse) plane showing the transition from A to I to Z to I to A, and so on (×58,000). (b) A longitudinal (*XZ*) section of an *Ascaris* fiber at rest length showing the staggered filament arrangement and the alternating dense Z rods and hollow sarcotubules in the Z-band. A change in the direction of stagger can be seen halfway down the fiber (×34,000). (Both micrographs from Rosenbluth, 1972.)

muscles as well and seems to be a means of allowing the muscle to work over a very large range of lengths.

Very little is known about the detailed structures of the myofilaments in this muscle type. The thin filaments seem to be very similar in form to all other thin filaments, and they give rise to the usual 380-Å period in the I-band. The thick filaments are known to have the usual myosin periodicity of 143 Å (Hanson and Lowy, 1961) and to contain paramyosin (Ruegg, 1961; Lowy *et al.*, 1964), and so they may have structures similar to other paramyosin-containing filaments. There is evidence that a "material of moderate density" (as yet unidentified) may surround and cross-link the thick filaments in the H-zones of some muscles (Rosenbluth, 1967), and this might tend to oppose the relative shearing of the thick filaments. It was suggested by Rosenbluth that a strong interaction among thick filaments might in some way help to account for the catch state in these muscles but not in oyster obliquely striated muscle where there is no true catch state (Millman, 1963). However, as in molluscan smooth muscles, a slow rate of breaking of actomyosin crosslinks would also account for this. Rosenbluth also suggested that the presence of a fibrillar cytoskeletal network, which seems to occur right through the muscle cells of *Ascaris* (Apathy, 1894; Reger, 1964; Rosenbluth, 1965a) and which may be associated in some way with the Z-bands, might be an efficient means of transmitting tension to the surrounding connective tissue during tonic contractions. These ideas must clearly await further evidence before they can be tested.

8.6. Discussion

It is appropriate at this point to summarize the differences and similarities among the various muscle types that have been described and to make some generalizations (albeit tentative) about muscle design.

One of the most crucial conclusions from the evidence given in this chapter and the last (and dealt with again in Chapter 10) is that all known muscles appear to operate by the relative sliding of thick and thin filaments. This confirms the original suggestions of H. E. Huxley and Hanson (1954) and A. F. Huxley and Niedergerke (1954) and emphasizes the far-reaching nature of their results. The precise organizations of these filaments in different muscles can, however, be very different, and since these muscles have presumably developed from a single common ancestor, it seems clear that organizational differences that occur must have evolved because of their suitability for different functional purposes. It is therefore of interest to distinguish those features of the myofilament arrangement and architecture that seem to be common, and which can therefore be presumed to be basic to the contractile mechanism, from those features that clearly are specializations for particular functions.

The common features of all muscle types seem to be the following.

1. They contain actin filaments (of actin and tropomyosin molecules) that are virtually identical in structure but which may or may not contain troponin. The actin monomers seem to have amino acid sequences and presumably three-dimensional conformations that are highly conserved throughout the animal kingdom. (In this respect it is interesting that molluscan thin filaments, which normally do not contain troponin, can bind functionally to troponin from other muscles.)

2. All muscles (and many nonmuscle contractile systems) contain myosin molecules that are similar in their shape and size and in having an actin-activated ATPase. The common shape indicates that this shape must itself be a fundamental requirement of filament formation and hence of contraction. Different myosins differ in their light chain components and hence in their functional properties, but the fact that light chains from one myosin type (e.g., vertebrate skeletal muscle myosin) can interact functionally with myosin from a completely different source (e.g., scallop myosin) may suggest that at least part of the myosin head is also highly conserved. One can speculate that this conserved region may be the part of the head that interacts with actin, since the actin attachment site is probably itself highly conserved.

3. Despite the considerable variability in the organization of the thin filament in different muscles, it does seem clear that adjacent thin filaments in nearly all muscle types (except those where the actin positions are very precisely located) are approximately 100 to 140 Å apart. It has been shown by Spencer (1969) and Lednev (1974) that actin filaments in actin gels can also pack with a similar interfilament distance (the exact distance depending on the conditions used).

4. As shown in detail in the next chapter, it is becoming clear that the lateral separation of myosin heads on a single thick filament crown is also about 100 to 140 Å. That this cross-bridge separation and the separation of the adjacent thin filaments are so similar seems to be more than coincidental.

The main differences among the contractile mechanisms of different muscles seem to be of two types. One is simply that different muscle types have thick filaments of different symmetry, diameter, and length and also thin filaments of different length, and that these filaments are organized in different ways. The other is that the regulatory proteins—the troponin complex, the myosin light chains, and possibly the phosphatase/kinase system—are different in different muscles.

With respect to the structural geometries of the various contractile units, certain generalizations can be made (on the assumption that one is always dealing with cross-bridge cycles that have similar rates and with thick and thin filaments that have similar properties).

1. Muscles that tend to contract fast generally have rather short sarcomere and myofilament lengths and small thick filament diameters, and they tend to have regular arrangements of thin and thick filaments, usually with a 2 : 1 or 3 : 1 thin to thick filament number ratio. A myofibril with many short sarcomeres in series will shorten faster than a myofibril of similar length but containing a small number of long sarcomeres.

2. The shortness of the sarcomeres in these fast muscles means that reasonable tensions can be generated over only a relatively short range of sarcomere lengths. Maximum total shortening of a muscle can be achieved if the individual myofibrils are parallel to the muscle axis and run the whole length of the muscle. But the tension generated by a single myofibril is the same as that generated by a single sarcomere in that myofibril (unless there is some connective tissue right through the muscle), and so a large total range of shortening can only be achieved using very long myofibrils at the expense of the level of tension that can be produced. A larger effective number of myofibrils in parallel can be produced by a feathered arrangement, but with this arrangement only a relatively small amount of total shortening can occur.

3. Longer sarcomeres permit higher tensions to be generated (assuming that tension is related to the amount of overlap of the thick and thin filaments), but the muscles must inevitably shorten more slowly since there are fewer sarcomeres in series. Since the presence of longer thick filaments allows more tension to be generated per filament, these filaments are often larger in diameter and they often interact with more actin filaments so that the tensions per thin filament or per unit cross-sectional area of thick filament do not become prohibitively large (i.e., large enough to disrupt the filaments).

4. The different sarcomere arrangements in different muscles are tailored to particular functional demands. Highly organized sarcomeres with rigid A-band and I-band structures (e.g., vertebrate skeletal muscle and insect flight muscle) are required to optimize the probability of attachment of myosin cross bridges to actin.

The lack of an M-band or a well defined A-band makes the filament geometry less optimal for cross-bridge attachment, but it permits the relative shearing of adjacent thick filaments, which in turn allows significant tension to be generated over a much wider range of sarcomere lengths (e.g., in some muscles of arthropods such as *Limulus*). Obliquely striated muscles represent an extreme case of this in which the sarcomeres are permitted to shear in a precisely defined manner during contraction (hence the change in obliquity in the striations) and thus to act over a very wide range of lengths. The same problem has been overcome in vertebrate smooth muscles by the use of face-polar myosin filaments which also allow tensions to be developed over a wide length range.

Of course, the original assumption that all of these systems would involve identical cross-bridge interactions is not valid. Although it seems reasonable to assume that the actual force-generating process in the cross-bridge cycles in different muscles must be similar, the rate at which these cycles occur is clearly different for different myosins. Even within a single group (e.g., vertebrate striated muscles) where the sarcomere lengths and the lengths and arrangements of the thick and thin filaments are similar, there can be a wide variety of behaviors. These will correlate with specific differences in the ATPase rates of the myosin (possibly because of differences in the light chains), with differences in the other regulatory proteins (e.g., troponin and the phosphatase/kinase system), and with other more subtle features of the sarcomere (e.g., Z-band and M-band structure, etc.). These differences are manifested by the presence of different fiber types in the same muscle and by the different responses of different muscles of the same basic type (e.g., cross-striated vertebrate muscle).

The regulation of different muscles clearly depends on a variety of factors. The primary cause of these differences lies in the way the muscles are innervated and in the way nerve impulses are transmitted to the myofibrils (as described in Chapter 1, Section 1.4.2). Cross-innervation experiments have indicated that it is the mode of innervation that actually determines to some extent the detailed structure of the myofibrils. Fibers that are initially of one type (e.g., slow) can be converted to respond differently by cross-innervation, and the change in response is attributable to the synthesis in the muscle of different proteins. It has been reported (see Chapter 1) that it is actually the nature of the nervous stimulation that triggers this change. However, whether this directly suppresses the synthesis of some protein and activates the synthesis of others or whether there is some transmitter substance is not known.

Finally, we can consider the variations in filament regulation that occur in different muscles. This has been dealt with in various parts of Chapters 5, 6, 7, and 8, and the results will only be summarized here. We have seen that regulation of the thin filaments can occur in all muscles (except molluscan muscles and a few others) via the troponin–tropomyosin system. The process involved seems to be primarily a steric blocking mechanism involving tropomyosin movement but also involving some cooperative behavior of the whole thin filament assembly.

Thick filament regulation appears to be of two kinds. In most muscles it has been found that one of the light chains (the P light chain) has a regulatory role and acts directly on the myosin ATPase presumably by blocking or structurally modifying the site on the myosin head that normally interacts with actin. This may be controlled either by calcium concentration or by phosphorylation or by both mechanisms. The exception seems to be the vertebrate skeletal muscle in which the DTNB light chain does not act directly on the myosin ATPase in *in vitro* experiments. However, the accumulating evidence (described in detail in Chapter 10) that in this muscle too there is some form of myosin-linked regulation suggests that the DTNB light chain has an indirect controlling effect. It could well be that *in vivo* it binds the myosin heads to the thick filament backbone in the absence of Ca^{2+} and frees the heads to interact with actin when the calcium level is increased. The light chains would not directly modify the myosin ATPase, and their effect might not therefore be evident in *in vitro* studies.

A. G. Szent-Gyorgyi and his collaborators have summarized their studies of the actin- and myosin-linked control systems in terms of the evolutionary tree shown in Fig. 6.17. Here the horizontal lines indicate the demonstrated presence of direct myosin control, the vertical lines the presence of actin control, and the two sets of lines together indicate the presence of both systems. It is evident from this diagram that actin control is present in almost all of the species studied whereas direct myosin control is limited markedly to the right-hand (invertebrate) branch of the tree. As described above, myosin control in vertebrates does seem to occur, but it is different in nature from the invertebrate systems in that it is indirect in the skeletal muscles and associated directly with phosphorylation in smooth muscle. Since these regulatory mechanisms must clearly have been developed extremely early in the evolutionary process, and since there are obvious functional advantages, it seems very likely that similar regulatory proteins will be found in all of the nonmuscle contractile systems that involve actin and myosin. The further study of these nonmuscle systems should be extremely informative about the basic requirements of any contractile mechanism.

9

Molecular Packing in Myosin-Containing Filaments

9.1. Introduction

9.1.1. The General Model of Myosin Filament Structure

In Chapters 7 and 8 the detailed structures of a variety of muscles were described. Here the structures of the various thick filaments in these muscles are considered collectively to give some insight into the general design of myosin filaments. A number of physically sensible alternative models for packing of myosin molecules are put forward, and an attempt is made to choose between them.

Towards the end of the 1960s the structures of two kinds of myosin filament, those in vertebrate skeletal muscle and those in insect flight muscle, seemed to be fairly well understood. As described in Chapters 7 and 8, both were thought to be such that the myosin cross bridges were arranged on two-stranded helices. However, by 1971 the cross-bridge arrangement on the myosin ribbons in vertebrate smooth muscle had been worked out, and comparison of this arrangement with the surface cross-bridge arrays on the two two-stranded models for the vertebrate and insect thick filaments (Fig. 9.1) made it very obvious that these different models must of necessity require the component myosin molecules to pack together in completely different ways. On the other hand, it was thought at the time, and has been confirmed since, that all of these different myosin molecules are very similar (but not identical) in size, shape, and properties. The marked differences in the three myosin filament models shown in Fig. 9.1 were, therefore, a real puzzle.

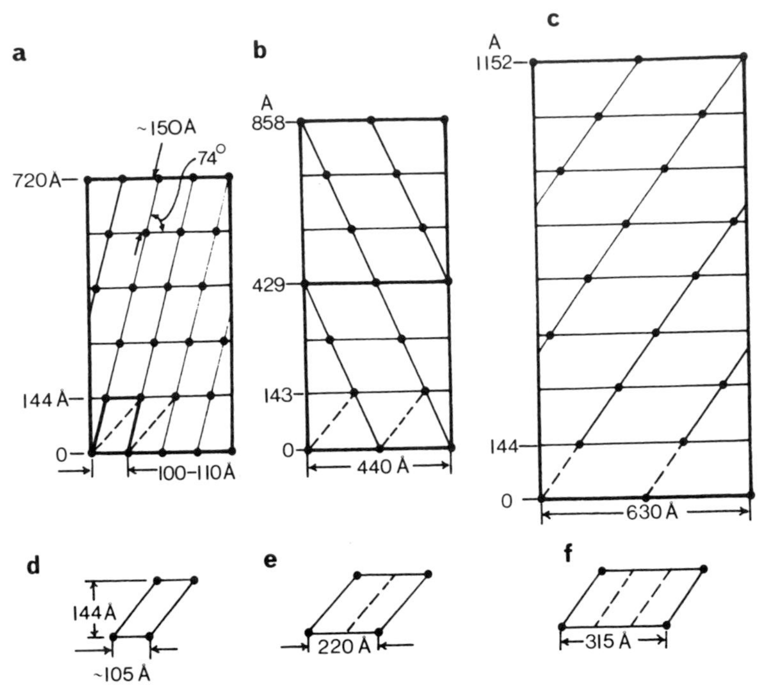

FIGURE 9.1. Radial projections of the early published cross-bridge lattices of models for the myosin filaments in (a) vertebrate smooth muscle (Small and Squire, 1972), (b) vertebrate skeletal muscle (H. E. Huxley and Brown, 1967), and (c) insect flight muscle (Reedy, 1967). (d–f) The repeating units in (a) to (c), respectively, demonstrating how these units may be related. (From Squire, 1971.)

But when the data were viewed in the light of the biochemical evidence on the myosin content in these three muscle types, a possible solution to the puzzle becamse apparent (Squire, 1971). In the vertebrate smooth muscle ribbons, it was thought that there was one myosin molecule per lattice point in the array in Fig. 9.1a. The models for the vertebrate and insect thick filaments (Fig. 9.1b,c), to be consistent with the myosin content, would need to have, respectively, between one and a half and two molecules and about three molecules per lattice point. Since the circumference of the insect thick filaments was larger than that of the vertebrate skeletal muscle filaments in the ratio of about 3 to 1.5 or 2, a natural suggestion to make was that the myosin cross bridges on these filaments might really be arranged on lattices similar to that on the myosin ribbons. As described in Chapter 7, it seemed likely on this basis that the thick filaments in vertebrate skeletal muscle might be three- or four-stranded and that those in insect flight muscle might be six-stranded (Squire, 1972).

A direct extension of these ideas led to the formulation of a general model of myosin filament structure (Squire, 1971) which, it was thought, might be applicable to all types of myosin-containing filament. The basic idea behind this general scheme was that myosin molecules, being rather similar in all muscles, would tend to pack in all muscles so that they have virtually equivalent environments. The cross-bridge arrays on the thick filament surfaces would then have one myosin molecule per lattice point (i.e., no clumping of the molecules in twos and threes), and the myosin rods would be packed in similar ways so that the surface area per cross bridge on different myosin filament types would be approximately constant. From the available evidence, it seemed very likely that adjacent cross-bridge origins on one crown (i.e., one 144-Å repeat) would be spaced at lateral intervals of 100 to 120 Å, giving a surface area per cross bridge of about 14,400 to 17,300 A^2. A natural consequence of this idea was that all myosin filaments of the same diameter would have the same number of cross bridges per crown and that larger filaments would have more cross bridges per crown than smaller ones.

Since the original postulation of this idea in 1971, much more evidence has been obtained on the structures of the thick filaments in a number of different muscles types (Chapters 7 and 8). These have served to test the model, and so far its predictions have been largely confirmed. It has been shown that the thick filaments in vertebrate striated muscle are probably three-stranded and that those in insect flight muscle are probably six-stranded. In addition, a number of other muscles with thick filament diameters similar to those in insect flight muscle (e.g., *Pecten* striated and lobster muscles) may also be six-stranded (see discussion in Section 9.2.2). And the thick filaments in the horseshoe crab (*Limulus*), which have a diameter slightly larger than vertebrate thick filaments and slightly smaller than insect thick filaments, may well be four-stranded.

A very striking feature of all of these recent results is evident from a direct comparison of the cross-bridge arrays on these filaments (Fig. 9.2). It is immediately obvious that the arrays on different filament types, although having slightly different long repeats, are reasonably similar. In fact, one surface lattice can easily be transposed into any of the others by a very slight twisting or untwisting of the helix, an effect that could easily be produced by very slight differences in the amino acid sequence along the myosin rods. It is the mode of interaction of these rods that presumably dictates the basic geometry of the cross-bridge array.

The general model can be extended further on the assumption of equivalent myosin molecules to show how the myosin rods might be packed together to form the backbones of myosin filaments. In fact, it

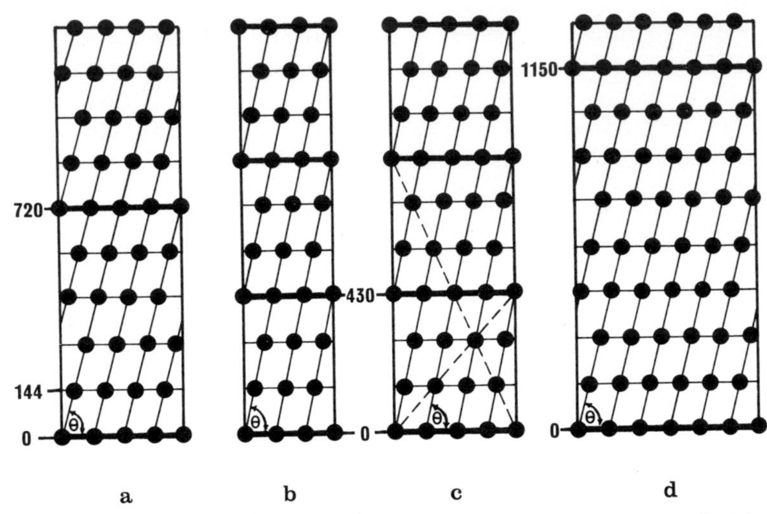

FIGURE 9.2. Radial projections of current models for the thick filaments in (a) vertebrate smooth muscle, (b) vertebrate striated muscle (three-stranded filament), (c) *Limulus* muscle (four-stranded), and (d) insect flight muscle (six-stranded). In each case the repeating unit is very much the same except for slight changes in the angle θ. (From Squire, 1973.)

has been shown that two classes of packing arrangement are conceivable. One of these packs the myosin rods to form a continuous surface layer in the thick filaments (Squire, 1973), and the other packs groups of molecules into subfilaments and then packs these subfilaments together to form the full filament backbone (Squire, 1975). These two alternatives will be described and compared in this chapter to reveal the likely packing arrangements in a variety of myosin filaments. The chapter then presents a brief discussion of the effects of allowing the equivalence of the myosin molecules to break down. The early and historically important myosin model of Pepe (1967a) will be considered at this point, where it fits logically into the structure of this chapter. Finally, a brief analysis is given of the structure and interactions of the coiled-coil α-helical myosin rod. But first, the known structural properties of myosin rods are briefly summarized, since it is these which must form the basis of any packing model for myosin filaments.

9.1.2. Summary of the Structural Properties of Myosin

A detailed account of the structural properties of myosin molecules was given in Chapter 6 and the appropriate references are given there. These properties are summarized briefly here since the discussion in this chapter depends heavily on them.

1. Myosin molecules consist of a rod (LMM plus HMM S-2) about

1550 Å long on one and of which are attached the two myosin heads (HMM S-1).

2. The myosin rod is a two-chain α-helical coiled-coil molecule, and such molecules are known to pack together with a center-to-center intermolecular packing distance of about 20 Å *in vivo,* although this may be reduced to about 17 Å in dehydrated specimens.

3. The LMM portions of the myosin rod are insoluble under physiological conditions and are thought to pack together within the backbones of myosin filaments.

4. Heavy meromyosin S-2 is much more soluble than LMM and is therefore thought to reside at the filament surface in order to allow the myosin heads attached to it to move out and interact with the neighboring actin filaments.

5. Various synthetic aggregates of LMM, myosin rod, HMM S-2, and whole myosin molecules have provided information on the interactions of myosin. Antiparallel overlaps of about 430, 900, and 1300 Å and end-to-end butting have been observed in synthetic bipolar segments, the 430-Å overlap being observed only in segments of vertebrate smooth muscle. Paracrystals of LMM have predominantly shown repeats of 145 Å and 430 Å, but occasionally repeats of 48 Å and 72 Å have also been seen.

6. Studies of myosin molecules in solution have suggested that under conditions close to those that promote filament formation, myosin molecules tend to aggregate as parallel dimers with a 430 Å axial shift between molecules.

7. Synthetic aggregates from modified rod portions of myosin (LMM-C) have been found to have the form of cylindrical tubes in which the molecules are tilted slightly to the tube axis and bend slightly around the tube surface.

8. The appearance of high-angle diffraction patterns from a variety of muscle types in which the 5.1-Å meridional reflections from the myosin coiled-coil structure (Fig. 4.13) are fairly well localized across the meridian suggest that the α-helical molecules giving rise to them are aligned more or less parallel to the meridian (i.e., to within about 5°).

It will be seen in the next section that this list of properties is sufficient, when taken together with the known geometry of the surface arrays of cross bridges, to define the most likely kinds of model for the packing of the myosin rods in various myosin filament backbones.

9.2. Myosin Packing in Uniform Layers

9.2.1. Packing in a Planar Sheet

In considering possible myosin packing schemes, it will be assumed to start with that the myosin rods approximate uniform cylinders about

1500 Å long that are slightly flexible. At the end of the chapter, the effect of considering the α-helical nature of these rods will be briefly assessed (Section 9.6.3).

It is most convenient when establishing a general packing model to consider first the structure of a specific myosin filament type and then to extend the conclusions to other known filament types. For this purpose, any of the filaments illustrated in Fig. 9.2 could be used as the starting point, and the result would be the same. Originally, Squire (1973) used the structure of the myosin ribbons in vertebrate smooth muscle as a starting point (it being immaterial whether these are the *in vivo* myosin filaments or not). The main advantage of using the ribbons for this is that they are planar structures. This is still an advantage, and so the initial analysis described here will be concerned with the structure of planar myosin arrays.

To start the analysis, two very reasonable working assumptions need to be made. One is that the myosin rods will tend to close pack so as to minimize the potential energy of the system. The other is that the myosin rods of different myosin molecules may tend to be in exactly equivalent environments, a feature that itself would appear to be necessary if the myosin cross bridges are all equivalent in the surface arrays shown in Fig. 9.2. On the basis of these assumptions, a number of deductions can be made right away about the rod packing arrangement. First, if the rods behave as cylinders, then they can pack together tightly in a regular lattice with a packing angle (α in Fig. 9.3) of between 60° and 90°. The volume occupied by one myosin rod will therefore be about $1500 \times 20 \times 20 \times (0.87$ to $1)$ or 500,000 to 600,000 Å^3. In all of the cross-bridge arrays shown in Fig. 9.2, the surface area per cross bridge is about 144×100 to 120 Å (i.e., 14,400 to 17,000 Å^2). The myosin rods

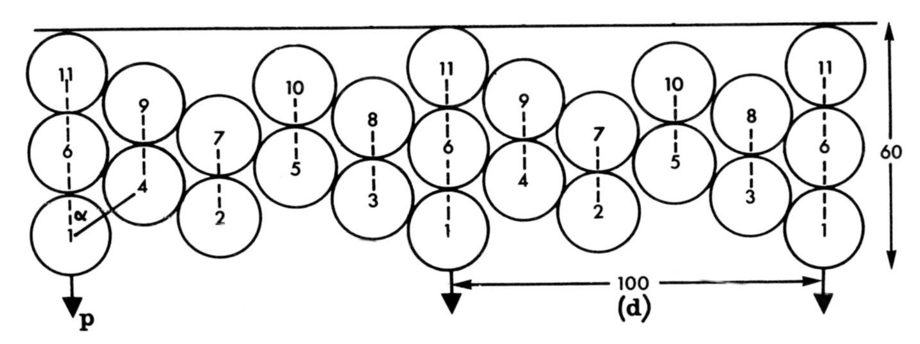

FIGURE 9.3. A possible myosin packing scheme for a planar myosin layer. Each circle represents the rod part of one myosin molecule. The dashed lines indicate the direction in which the myosin molecules must tilt to produce a repeat of 720 Å. The numbers in the circles represent the different axial levels in multiples of 143 Å at which the cross bridge on the end of that rod will appear (see Fig. 9.4). (After Squire, 1973.)

must therefore pack to form a layer of thickness equal to at least 500,000 to 600,000/17,000 Å or 30 to 40 Å. Since the myosin rods are only 20 Å in diameter, there must be at least two thicknesses of myosin rods in the myosin layer. The number of myosin rods that form any cross section through the layer can also be deduced. Since the rod length is about 1500 Å and the axial repeat in the layers is 144 Å, then there must be about 10 or 11 (i.e., 1500/144) rods in a width of 100 to 120 Å in any cross section of the layer.

On the basis of these deductions, a packing scheme of the sort shown in Fig. 9.3 can be deduced for a planar myosin layer. Here it has been assumed that the lateral separation (d) of the myosin heads (arrows) is exactly 100 Å. The consequence of this is that the packing angle (α) between adjacent myosin rods (circles) is about 72°. This packing scheme is such that interactions occur between myosin rods staggered axially by 430 Å and 720 Å (i.e., circles 1 and 4, and 1 and 6 are in contact). The 430-Å interaction seems to be one that appears again and again in synthetic aggregates of myosin, and so its presence in the model seems to be necessary. Its occurence may mean that the myosin rod itself possesses a 430-Å repeat. The other interaction of 720 Å is compatible with the appearence of LMM paracrystals of repeat 145 Å provided that the LMM molecules possess the 430-Å repeat or stagger mentioned above. The largest common factor of these two repeats is clearly 145 Å, and this would therefore be the observed repeat on the paracrystals.

The cross section of the packing scheme shown in Fig. 9.3 is taken at level X–X in the face view of the model shown in Fig. 9.4. Here the lateral scale has been greatly exaggerated to make it possible to show the relative positions of the axes of the myosin rods (the near-vertical lines in the diagram). Figures 9.3 and 9.4, taken together, show how the molecules are arranged in the model. In Fig. 9.3, the number in any circle represents the level along the filament (spaced axially at intervals of 145 Å; Fig. 9.4) at which the cross bridge on the end of that particular rod will appear. Thus, rod 2 in Fig. 9.3 will have its crossbridge 145 Å along the axis from rod 1. It can be seen from this that the position of rod 1 represents the position of the part of the rod that is near to the cross bridge, whereas position 11 is the position of the far end of the LMM part of another myosin rod. This means that if Fig. 9.3 is taken not as a true cross section but as a diagram showing the progress of one molecule down through the layer, then the cross-bridge end of the rod will be at position 1, and the rod will gradually move away from the surface and towards the back of the layer as it goes through levels 1 down to 11. Since the axial repeat in the myosin ribbons is 720 Å, the path followed by one molecule will be along one of the dotted lines in Fig. 9.3. This joins circles 1, 6, and 11 separated by 5 × 145 or 720 Å.

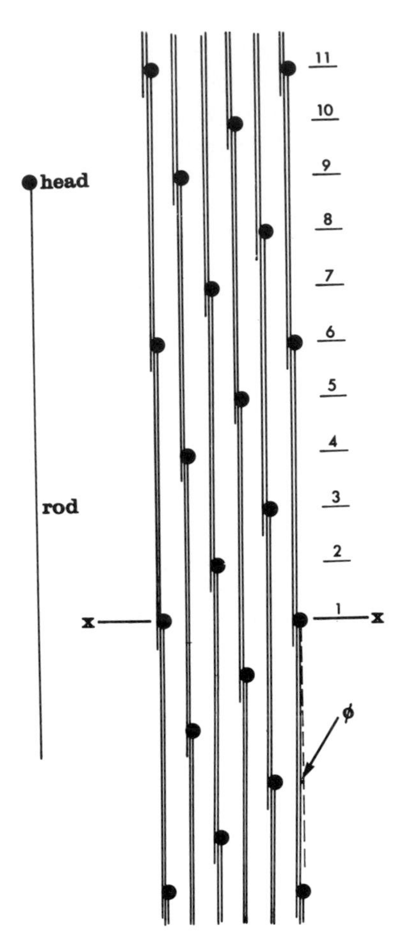

FIGURE 9.4. A view of the packing scheme in Fig. 9.3 in a direction normal to the surface of the planar myosin layer. Fig. 9.3 corresponds to a section through X–X in this diagram. The lateral scale is exaggerated, and only the positions of the rod axes are shown. (From Squire, 1973.)

This model for myosin packing can be extended indefinitely either laterally or axially to produce a very extensive layer in which all of the myosin molecules are in exactly equivalent environments. It is also a close-packed structure, it makes use of the known axial staggers between myosin molecules, and it generates the required cross-bridge lattice. It can therefore be said to be a satisfactory model structure. But there are other possibilites, and it is instructive to see how these compare with the first model. The way this can be done is to assume, to start with, that the cross-bridge separation (d) has a specific value (say 100 Å) and then to consider alternative packing models that will produce this cross-bridge separation.

A study of the arrangement in Fig. 9.4 shows that if the myosin head lattice has a defined shape then the only independent parameter that can be altered is the angle ϕ between the rod axis and the filament long axis in this face view. In Fig. 9.4 the value of ϕ would be close to 0.5°. The effect of altering ϕ is illustrated in Fig. 9.5. Here it has been altered to $-1°$ in (a) and to $+1°$ in (b), and the cross-section views of these are shown in (c) and (d), respectively. It can be seen that the effect is to open out the layer considerably so that it is no longer tightly packed, and the layer thickness is increased to 70 to 90 Å. A 288-Å axial stagger between adjacent molecules occurs in (c), and one of 430 Å in (d). It seems to be inevitable that such structures would be unstable and would collapse to a more closely packed structure as in Fig. 9.3. As ϕ increases beyond $\pm 1°$, it can be shown that the myosin layer increases in thickness to about 120 Å until ϕ reaches $\pm 4°$, and it then reduces to a second minimum thickness at $\pm 8°$. However, such a large angular tilt seems to be unlikely because of the appearance of the high angle 5.1-Å meridional reflection in X-ray diffraction patterns from muscle. It is reasonable to conclude, therefore,

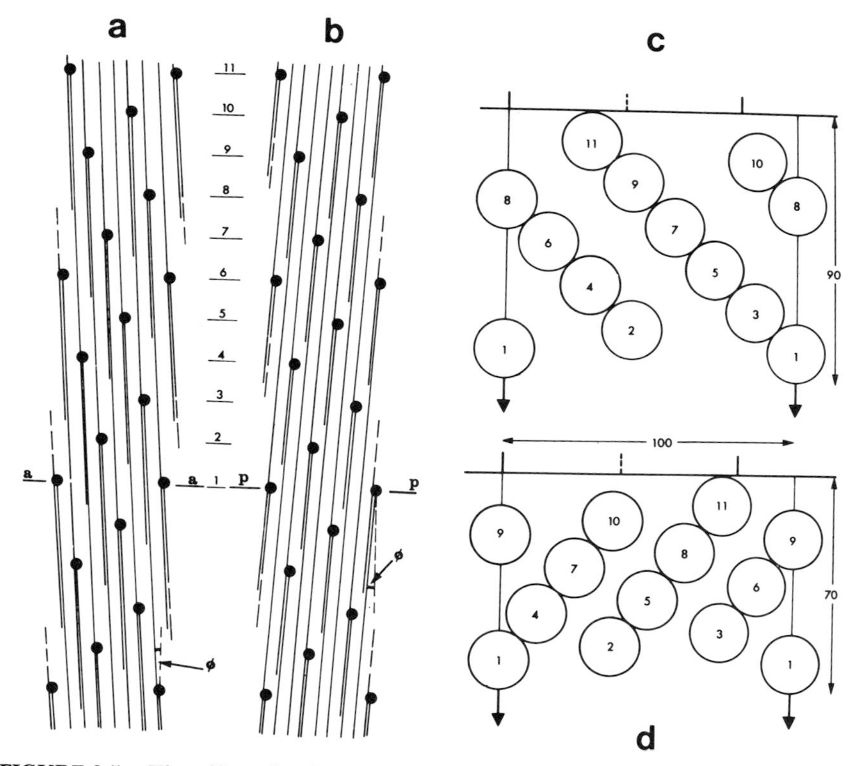

FIGURE 9.5. The effect of variations in the tilt (ϕ) of the myosin rods in the planar layer structure. (a and c) The equivalents of Figs. 9.4 and 9.3, respectively, for a molecular tilt (ϕ) equal to $-1°$. (b and d) Comparable diagrams when ϕ is $+1°$. The section in (c) corresponds to a–a in (a), and that in (d) to p–p in (b). For details see text. (From Squire, 1973.)

that if d is 100 Å, then the packing arrangement shown in Figs. 9.3 and 9.4 is probably the most reasonable planar layer structure.

The next step in the analysis is to alter the value of d in Fig. 9.3 and to see what happens. In fact, all that happens is that the packing angle (α) changes, but the overall form of the model remains much the same. Figure 9.6 shows the type of result that is obtained. The effect of increasing d is to increase α. In fact, it is easy to show that $d = b\sqrt{34 - 30 \cos \alpha}$ where b is the rod diameter. In Fig. 9.6a, d has become 115 A, and α has become about 90°. In (b), α has been reduced to 60° by reducing d to

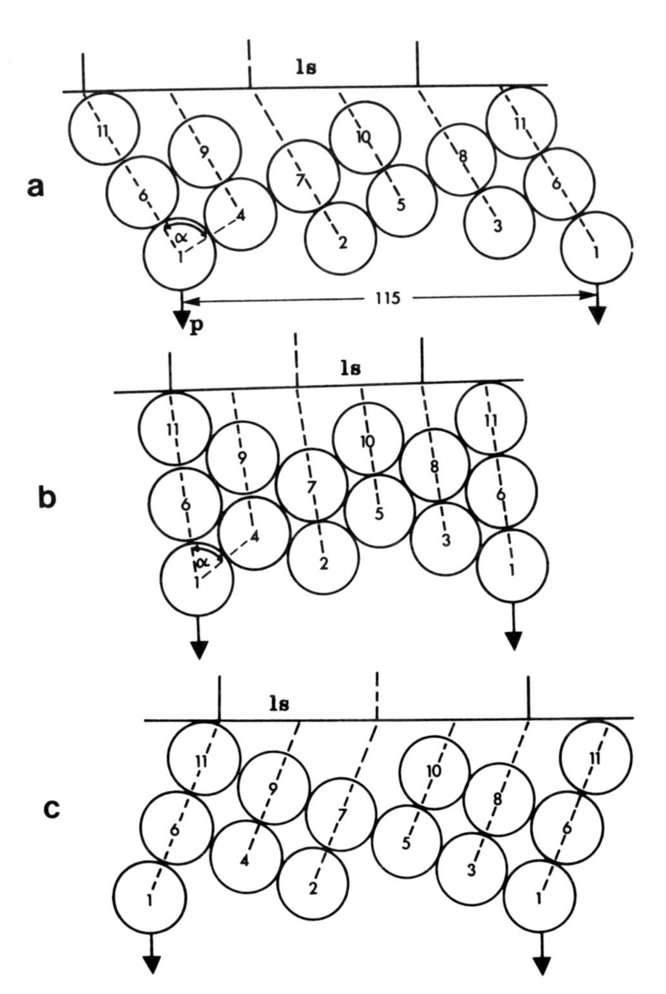

FIGURE 9.6. The effect on the packing arrangement in Fig. 9.3 of changing the value of the separation (d) between the myosin projections. (a) d = 115 Å, α = 90°; (b) d = 90 Å, α = 60°; (c) d = 115 Å, α = 90° (but here, no 430-Å overlap occurs). (From Squire, 1973.)

about 90 Å. Shifting the arrangement still further in the same direction opens out the packing again to $\alpha \simeq 90°$ when $d = 115$ Å, but here interactions of 720 Å and 288 Å occur, whereas in (a) it is 720 Å and 430 Å, and in (b) all three interactions occur. Since the value of d is not known exactly, it is not possible to choose among these alternatives with certainty. However, various observations suggest that d may be slightly greater than 100 Å. Measurements on the myosin ribbons give $d \simeq 100$ Å, but since lateral shrinkage of up to 15% is known to occur in dehydrated specimens (A. Elliott *et al.*, 1968a), it seems likely that this value underestimates the true value of d. Similarly, in other muscles the value of d generally seems to be larger than 100 Å (Section 9.2.2). But if d is about 110 to 120 Å, it is evident that the type of model in Fig. 9.6a is better than that in Fig. 9.6c since it includes the prominent 430-Å molecular overlap.

9.2.2. Packing in Cylindrical Myosin Filaments

The planar myosin packing model discussed in the last section can easily be applied to cylindrical myosin filaments simply by taking n lateral repeats of the planar structure and bending them into a cylindrical shell. An n-stranded helical arrangement of myosin cross bridges will automatically be produced. Thus a three-stranded filament (as may occur in vertebrate skeletal muscle; Chapter 7) would appear as in Fig.

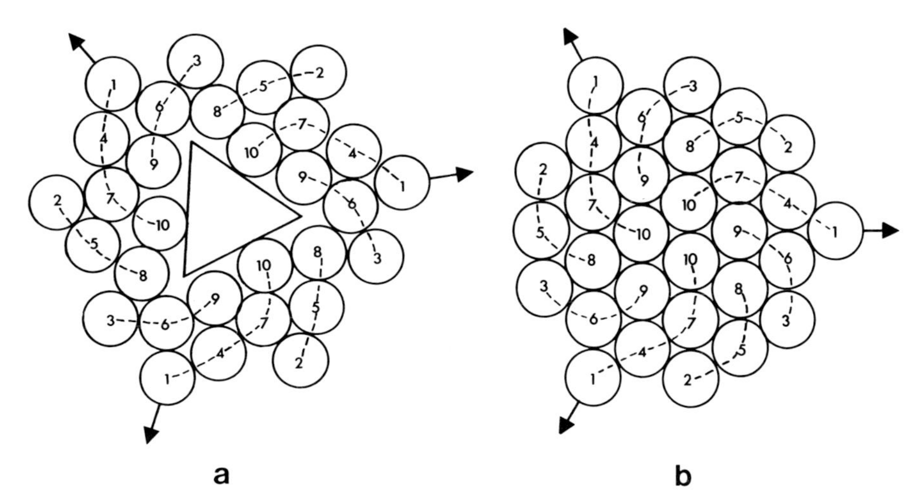

FIGURE 9.7. Cross sections of models for three-stranded myosin filaments that are derived from the planar layer packing arrangement in Fig. 9.3. The configuration in (a) has a small hollow core, and that in (b) is more compact and contains no core. The numbers have the same meaning as in Fig. 9.3, and the dotted lines indicate the molecular tilt necessary to generate a repeat of 430 Å. (From Squire, 1973.)

9.7a, a four-stranded filament (a possibility for *Limulus* muscle; Chapter 8) as in Fig. 9.8, and a six-stranded filament (as may occur in insect muscle and *Pecten* striated muscle; Chapter 8) as in Fig. 9.9. In each of these, the dominant intermolecular interactions are all the same and are 430 Å and 720 Å. Because the molecular layers are bent, it is inevitable that the rod packing must become slightly looser at the filament surfaces and slightly more close-packed in the middles of the filaments. Note also that to produce the correct axial repeat along these filaments, the dotted lines that indicate the progress of one rod down through the filament would not be as in Fig. 9.3 but would join circles 1, 4, 7, 10, etc. in Figs. 9.7 and 9.8, where a 3 × 145-Å repeat is required, and would join circles 1 and 9 in Fig. 9.9 for insect muscle where a repeat of 8 × 145Å (1150 Å) is required. In *Pecten* striated muscle, the repeat is even longer (10 × 145 Å), and the packing arrangement in this case would be as in Fig. 9.9 but with the dotted lines going almost radially inwards from circle 1.

Figure 9.7b shows a similar scheme to that in Fig. 9.7a for a three-stranded filament except that the packing has been made truly close-packed, and the filament no longer has a hollow center. Since these filaments often show a less dense central region in transverse sections (Fig. 7.14), it seems likely that the structure in Fig. 9.7a would be more appropriate. In this case, taking the rod diameter to be 20 Å, the three-stranded filament would have an outer diameter of about 155 Å, with the density falling off rapidly outside a diameter of about 130 Å. This is

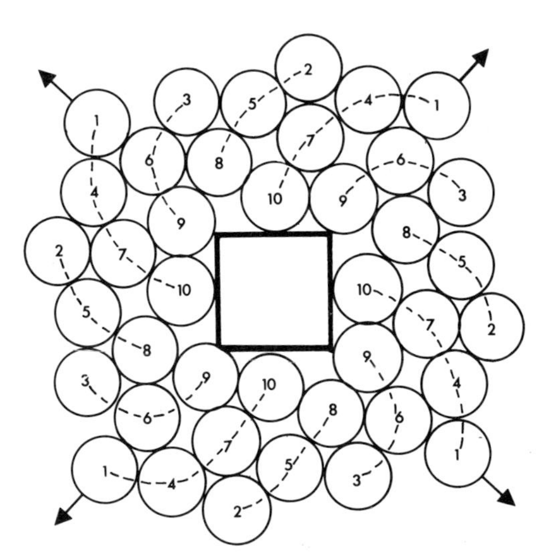

FIGURE 9.8. As for Fig. 9.7 but for a four-stranded filament of repeat 430 Å (as may occur in *Limulus* muscle). (From Squire, 1973.)

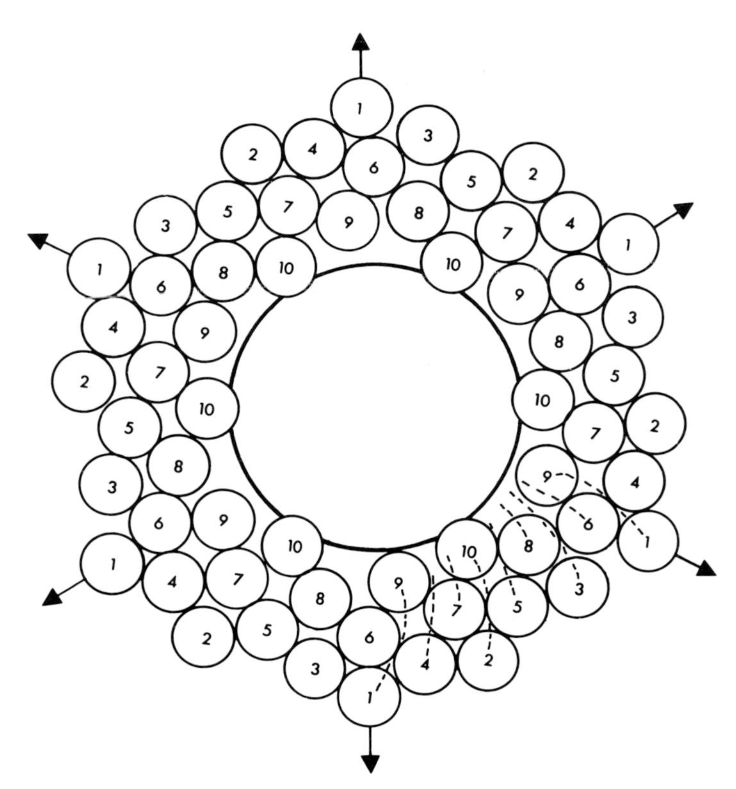

FIGURE 9.9. As for Fig. 9.7 but for a six-stranded filament of repeat 1150 Å as illustrated in Fig. 9.2d for insect muscle. (From Squire, 1973.)

quite compatible with measurements from sections of filament diameters in the range 100 to 140 Å, especially if account is taken of the possible 15% lateral shrinkage that may have occurred in these sections. The four-stranded structure in Fig. 9.8a would have an outer diameter of 170 Å with density falling off rapidly outside a diameter of 140 to 150 Å. This compares well with the measured diameters of 160 to 170 Å in *Limulus* muscle. Similarly, the six-stranded model (Fig. 9.8b) would have an outer diameter of about 225 Å with density falling off rapidly outside a diameter of about 180 to 190 Å. This compares well with measured filament diameters in insect flight muscle, *Pecten* striated muscle, and crustacean muscles, all of which fall in the range 180 to 220 Å. In addition, this six-stranded structure would have a core of diameter about 100 Å (which may or may not contain other proteins such as paramyosin), and this also agrees well with the appearance of hollow filaments in electron micrographs (Fig. 8.4).

These results show clearly that the single model derived originally for a planar myosin layer can account equally well for the structures of all of the several cylindrical filaments about which the relevant structural information is available. Since all of the arguments given in the last section about the effect of changing the angle α apply equally well to all of these cylindrical structures, it follows that the single packing scheme that seems to be the most satisfactory for the planar layers is also the most satisfactory for all of these other filaments. It is also worth noting that such a packing scheme would appear to explain directly the results of King and Young (1972), described in Chapter 6 (Fig. 6.15), in which synthetic myosin tubes were produced from slightly bent molecules tilted slightly to the tube axis. In the next section, the possible relevance of this packing scheme to the structures of the large paramyosin filaments in molluscan muscles is also considered.

9.2.3. Paramyosin Filament Structure

In Chapter 8, where paramyosin filament structure was discussed, it was mentioned that a wide variety of a spacings occur in different molluscan muscles. G. F. Elliott (1960, 1964) estimated a spacings from X-ray diffraction evidence on the assumption that the paramyosin filaments were aggregates of planar layers of paramyosin and obtained the values 250 Å for *Venus mercenaria,* 340 Å for *Crassostrea* (oyster), and 430 Å for *Mytilus* adductor muscle. These are approximately in the ratio 2 : 3 : 4. On the other hand, A. Elliott and Lowy (1970), with very similar X-ray observations, showed that if the paramyosin filaments are treated as cylinders with approximate helical symmetry, then a value of 450 Å is obtained for *Crassostrea,* and they suggested that using the same assumption, the a spacings in *Venus* and *Mytilus* adductor might be about 340 and 560 Å, respectively, giving a ratio of 3 : 4 : 5 among the a spacings in *Venus, Crassotrea,* and *Mytilus.* Since these filaments are now thought by A. Elliott (1979) to be planar structures, then the high values of 340, 450, and 560 Å for these a spacings may be further from the true values than the lower values suggested by G. F. Elliott. But in either case, it is evident that the a spacings are quantized in intervals of about 110 Å to 120 Å (e.g., 250/2 is 125 Å, 340/3 is 113 Å, and 430/4 is 108 Å).

It seems to be more than fortuitous that this spacing is close to the value of d in the myosin layers in other myosin filaments, and one is automatically led to the suggestion that there are perhaps two, three, and four myosin cross bridges, respectively, across the a spacings of these different paramyosin filaments. If so, it is conceivable that myosin packing arrangements of the kind shown in Fig. 9.3 could occur in the surface myosin layers in these filaments. As mentioned in Chapter 8, there is evidence that there is enough myosin in these different muscles to give a filament surface area of about 15,000 Å² per cross bridge just as in the cross-bridge arrangements illustrated in Fig. 9.2 (A. G. Szent-Gyorgi *et al.,* 1971). But A. Elliott (1974) has estimated that there are about two or three myosin cross bridges per *a* spacing in the oyster, which is perhaps marginally less than would be required for this model with *a* equal to about 3 × 113 Å.

The surface cross-bridge arrays that would be produced by placing two, three, four, and five myosin cross bridges at even lateral spacings about 110 Å on the nodes of Bear–Selby sets of *a* spacing 250, 340, 430, and about 550 Å, respectively, are illustrated in Fig. 9.10 (Squire, 1971). Here it has been assumed that every node in the Bear–Selby net is equivalent, and it can be seen that regular cross-bridge arrays of the kind shown in Fig. 9.2 for other myosin filaments can be produced. This equivalence of the nodes seems to be demanded by the results of A. Elliott (1974) and Sobieszek (1973) which appear to show myosin heads in clusters on each node of the Bear–Selby net (Fig. 8.37). But such an appearance might not be expected from surface myosin arrays of the kind suggested in Figs. 9.3 and 9.4 where within each myosin layer all of the myosin molecules have equivalent environments.

However, it is clear that these myosin molecules could not all be equivalent in their interaction with the underlying paramyosin core, since, in the oyster, for example, there would be myosin molecules interacting with four different positions on a single node of the Bear–Selby net. It could therefore be suggested that the intrinsic regularity of the

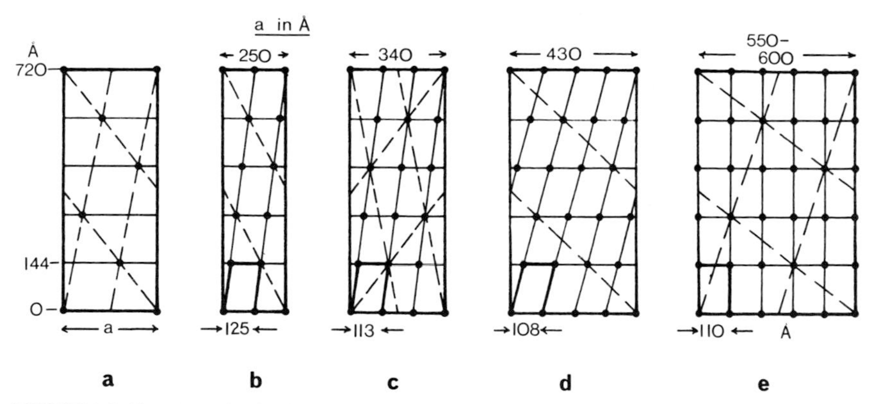

FIGURE 9.10. (a) The lattice (Bear–Selby net) on the surface of paramyosin filaments from which the myosin has been removed. (b–e) Possible subdivisions of the Bear–Selby nets on the surfaces of different paramyosin filaments in different molluscan muscles to give repeating units of area about 15,000 Å² similar to those that occur in other myosin filaments (Fig. 9.2). It has been assumed that the lattice points on the Bear–Selby nets are all equivalent. (b) *Venus* adductor muscle; (c) *Crassostrea* adductor muscle; (d) *Mytilus* adductor muscle; and (e) *Mytilus* ABRM (compare Fig. 8.38). (From Squire, 1971.)

myosin layer of equivalent molecules is perturbed by the paramyosin core to produce the observed clustering. This might be the situation *in vivo,* or it might alternatively be an effect produced by the preparative procedures for electron microscopy. The result shown in Fig. 9.38 for the paramyosin filament from *Mytilus* ABRM analyzed by Squire (1971) does seem to show clearly that at least in this case there are the predicted number (five) of myosin cross bridges per node on the Bear–Selby net, even though the surface lattice in this case is almost rectangular and slightly different in form from the other surface arrays (Fig. 9.2). This result appears to rule out the possibility that there is a myosin layer of nonequivalent myosin molecules, with, for example, three or four molecules 110 Å apart in paramyosin filaments of a spacing 5×110 Å as illustrated in Fig. 9.10e. But it would clearly be good to have direct evidence about this for all of the different types of paramyosin filament that have been discussed.

If it is assumed for the present that the surface myosin arrays on these paramyosin filaments are as drawn in Fig. 9.10 (a–d), then it is instructive to compare the myosin packing arrangements involved. If all of the nodes on the paramyosin Bear–Selby nets of all of these filaments are labeled with myosin in equivalent ways, then it is clear that the myosin arrays must each have axial repeats of about 720 to 725 Å. In the case of the structures with a spacings of 250, 340, and 430 Å, the tilt ϕ which the myosin rods would need to make to the long axis in the views normal to the filament surface (as in Fig. 9.4) would be about 9°, 9°, and 0.5°, respectively. The tilt (ϕ) in filaments with a spacings about 550 to 600 Å would be as much as 17 to 18°. These angles are clearly very much greater (except when $a = 4d \simeq 430$ Å) than those that occur in the other myosin filaments (Fig. 9.2), and one is led to wonder whether in paramyosin filaments such large angles are allowed or, alternatively, whether long myosin repeats or different myosin interactions than those in Fig. 9.10 might be involved in these filaments. Some evidence for an axial repeat longer than 720 Å in ABRM has already been obtained (Squire, unpublished observations), but it is not immediately clear to what this longer repeat relates. It must be concluded that paramyosin filaments, like other myosin-containing filaments, still need to be investigated in detail so that their structures can be defined more precisely.

9.3. Subfilament Models of Myosin Packing

9.3.1. Introduction

The underlying principle behind the general myosin filament model is that the component myosin molecules will be equivalent. In the last section a possible packing scheme for the myosin rods was introduced in which this equivalence of the myosin rods was maintained. However, that is not the only conceivable type of rod-packing arrangement that is consistent with the general model. There are, in fact, a very large number of alternative structures, each of which has the myosin rods packed into subfilaments. These subfilaments, which maintain the equivalence of the component myosin molecules, would then aggregate to form the whole filament backbone (Squire, 1979).

9.3.2. Three-Stranded Filaments

Simple Subfilament Models. As before, it is convenient to consider these subfilament models in terms of one particular myosin filament type and then to extend the conclusions to other filaments. This time, the thick filaments in vertebrate skeletal muscle will serve as the primary example. Figure 9.11 shows a radial net for the three-stranded crossbridge array on these filaments. On it are some of the different families of helices that can be drawn through the lattice points of this

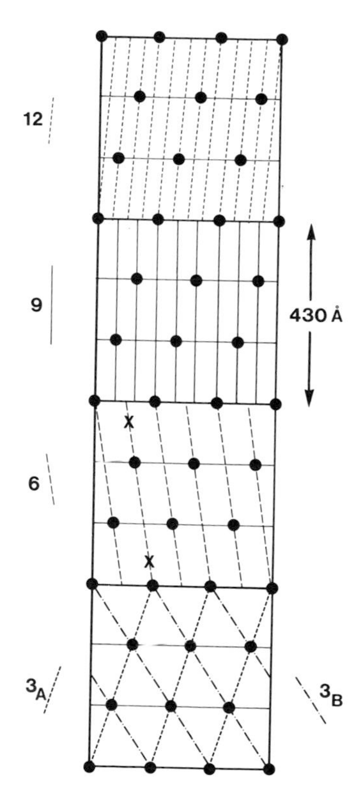

FIGURE 9.11. Radial projection of a three-stranded myosin filament of repeat 430 Å showing different sets of multistrand helices [3 (two types), 6, 9, and 12] that can be drawn through the lattice points on the filament. For discussion see text and Table 9.1. (From Squire, 1979.)

arrav. In theory, a very large number of different sets of helices could be drawn, but it is unnecessary to consider the majority of these for reasons that will become obvious later on.

In choosing a particular subfilament model, one simply needs to pick one of these families of helices such as the one marked X–X in Fig. 9.11. This particular helix is one of a family of six helices, each of which passes through one-sixth of the lattice points on the array. The helices have a pitch of 2574 Å. Along each helix, the successive cross-bridge locations are separated axially by about 288 Å, and there is an axial shift of about 144 Å between the cross bridges on adjacent helices. From the features of this particular family of helices, one could suggest that the myosin rods belonging to the cross bridges along one helix are twisted together to form a subfilament. Within this subfilament, an axial stagger of 288 Å between molecules would occur, and six such subfilaments related by an axial stagger of 144 Å would generate the whole filament. Since every molecule in one subfilament would be equivalent, as would every subfilament in the whole filament backbone, the equivalence of the myosin molecule would be maintained as required by the general model.

Another possible model would be the one generated by the family of helical strands of infinite pitch (i.e., parallel to the helix axis) indicated by the bold lines in Fig. 9.11. This time there would be nine subfilaments, each with an internal molecular stagger of 430 Å, and each staggered by 144 Å from its neighboring subfilaments. In addition, some other models are defined in Table 9.1 in terms of the pitch, intermolecular stagger, intersubfilament stagger, and subfilament diameter in each case. The filament diameters that would be produced by each model are also listed on the assumption that the myosin rods approximate to cylinders of 20 Å diameter. For example, the nine-subfilament model would have about 1450/430 or about three molecules in any subfilament, giving a subfilament diameter of about 40 Å (assuming that the myosin rods follow a helical path in each subfilament). Nine such subfilaments would generate a filament backbone in the form of a hollow cylindrical shell of outer diameter about 155 Å and core diameter about 75 Å. These are slightly larger than the measured values for this type of filament (100 to 140 Å and about 40 Å, respectively). On the other hand, a six-subfilament model would have about 1450/288 or about five molecules in any cross section, giving a subfilament diameter of about 55 Å. All six subfilaments together would therefore produce a filament of outer diameter about 165 Å and core diameter about 55 Å, also slightly different from the observed values. But here the axes of the subfilaments would make an angle of nearly 8° to the filament axis, which seems to be incompatible with the high-angle X-ray diffraction evidence unless the myosin rods in each subfilament make similar angles of ap-

TABLE 9.1. Parameters of the Simple Subfilament Models for Myosin Filaments[a]

Number of molecules per crown	Number of subfilaments per filament	Pitch of subfilaments	Repeat along subfilaments	Axial stagger between subfilaments	Approximate subfilaments diameter	Number of molecules per subfilaments	Approx. filament diameter	Approx. core diameter
3	3 (A)[b]	3×430[c]	143	0	90	9–10	170	30
3	3 (B)[b]	$\frac{3}{2} \times 430$	143	0	90	9–10	170	30
3	6	6×430	286	143	55	5	165	55
3	9	∞	430	143	40	3	155	75
3	12	12×430	572	143	35	2–3	170	100
4	4	4×430[d]	143	0	90	9–10	180–190	40
4	8	8×430	286	143	55	5	170–180	90–100
4	12	12×430	430	143	40	3	180–190	110–120
5	5	—	143	0	90	9–10	225–235	60
5	10	—	286	143	55	5	225	120
5	15	—	430	143	40	3	230	150
6	6	2×1150[e]	143	0	90	9–10	260–270	90
6	12	6×1150	286	143	55	5	265	150–160
6	18	18×1150	430	143	40	3	270	180–190

[a] Note that all dimensions are given in Å. Filament diameters are only accurate to within about 10Å. From Squire (1979).
[b] See Fig. 9.12 for the meaning of 3(A) and (B).
[c] Values are for the crossbridge lattice in Fig. 9.2b.
[d] Values are for the crossbridge lattice in Fig. 9.2c.
[e] Values are for the crossbridge lattice in Fig. 9.2d.

proximately 8° with the subfilament axis. In addition, an intermolecular interaction of 430 Å is not present in this case, and this may also argue against a six-stranded model.

Rope Models. Davey and Graafhuis (1976) have presented some interesting results on the structure of isolated thick filaments from hen pectoral muscle that had been treated with trypsin to remove the myosin heads. The resulting micrographs seemed to provide evidence for a backbone structure consisting of three subfilaments. As shown in Fig. 9.11 and in Table 9.1, such a structure could have molecules interacting with an axial stagger of 145 Å within each subfilament and a zero stagger between subfilaments. Each subfilament would follow a helical path of pitch 1296 Å and would, in fact, represent one of the strands of the three-stranded helix. In this case, each subfilament would have about 1450/144 or about nine or ten molecules in any cross section. The subfilament diameter would therefore be about 90 Å if, within each subfilament, the rods follow a simple helical path. The diameter of the whole filament would then be about 170 Å, and the subfilaments would make an angle of about 14° with the filament axis.

In this particular case, however, there is an interesting modification mentioned by Davey and Graafhuis (1976). This is that the myosin rods might first be grouped into subfilaments with an internal axial stagger of 430 Å to give a subfilament diameter of about 40 Å. Three of these subfilaments would be grouped with an axial stagger of 144 Å to form an intermediate filament, and three of these intermediate filaments would

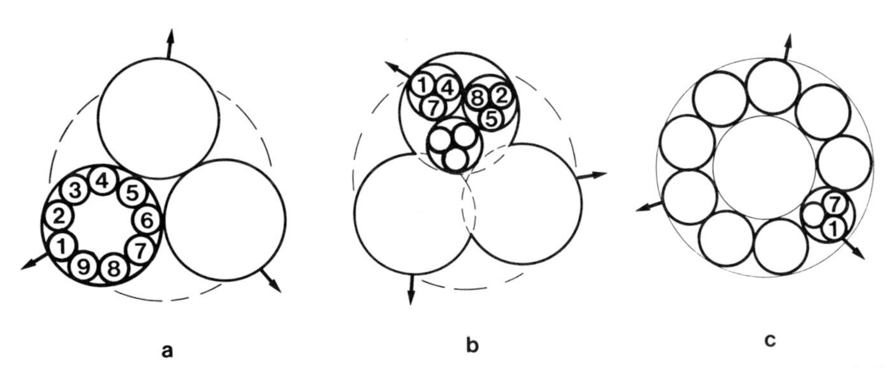

FIGURE 9.12. Cross sections of alternative subfilament models for a three-stranded myosin filament. (a) A three-subfilament model with nine molecules arranged helically in each subfilament. (b) An arrangement in which three intermediate subfilaments each comprise three smaller subfilaments about 40 Å in diameter and each containing about three molecules in any cross section. This is termed the "rope model." (c) A nine-subfilament model where each subfilament contains about three molecules per cross section as in (b). The configurations in (b) and (c) represent alternative groupings of the same type of subfilament. (From Squire, 1979.)

be grouped with no stagger to form the whole thick filament backbone. This sort of scheme is illustrated in Fig. 9.12. It is compatible with the general model, since all of the molecules would still be equivalent in this structure. It would give an intermediate filament diameter of about 80 Å and an overall filament diameter of rather less than 170 Å because of interlocking of the strands. It might be expected that in very thin transverse sections of vertebrate skeletal muscle some evidence of such a three-subunit structure might be evident right through the A-band. But appearances of that kind are rarely, if ever, observed in the bulk of the A-band, although they are sometimes observed in the M-region and in the regions where the filaments taper. Because of the twist of the intermediate filaments, evidence for a subunit structure in sections more than about 200 to 300 Å thick might not be expected. Apart from these possible objections and others given in Section 9.4.1, this model (referred to henceforth as the rope model) is extremely attractive and must be the most likely of all the subfilament structures for a three-stranded filament.

9.3.3. Multistranded Filaments

Extending the analysis of the previous section to other types of myosin filament, it is easy to see that subfilament models can be generated for any n-stranded filament in which there are Kn subfilaments (K integral). The various possibilities for four-, five-, and six-stranded filaments are summarized in Table 9.1. It can be seen that in general when K is large, then hollow filaments of very large internal diameter tend to be formed, whereas with K small, fairly large tilt angles are involved. Different axial staggers between molecules are also involved in these models. It is, however, worth noting that a structure analogous to the three-stranded rope model with two levels of subfilaments (Fig. 9.12) can be extended directly to any other filament (Fig. 9.13). Since the intermediate filaments follow the helical strands, a four-stranded filament could be composed of four such intermediate filaments, and a six-stranded structure could be formed by six intermediate filaments. The details of these models (termed rope models) are listed in Table 9.2 together with the other possibilities. From the various filament diameters produced by these models and the sizes of their hollow cores, it does appear very likely that either the uniform layer model or the rope model is the most satisfactory. More direct evidence would be needed to choose between them. As shown in the next section, in the case of crustacean muscles, direct evidence on myosin filament backbone structure is now available, and the conclusions from it are of major significance.

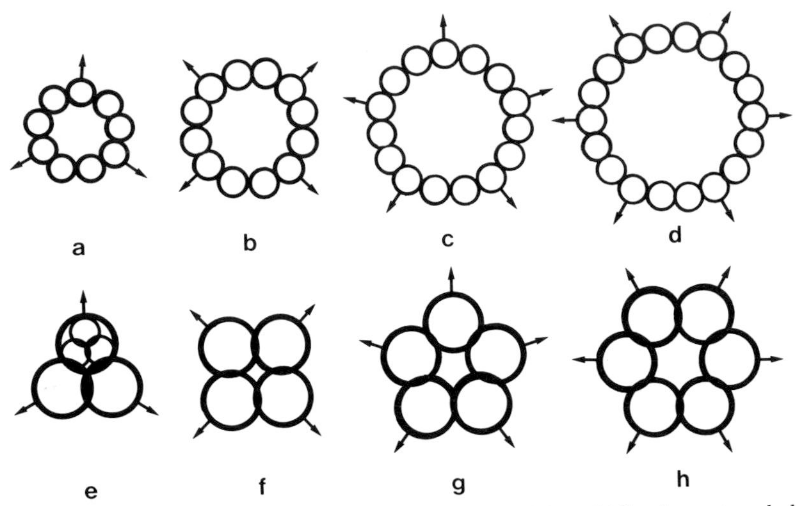

FIGURE 9.13. Alternative subfilament (a–d) and rope models (e–h) for three-stranded (a and e), four- (b and f), five- (c and g), and six-stranded myosin filaments (d and h). In all cases the models use 40 Å diameter subfilaments with about three molecules per cross section. Note that the filaments in the rope models (e–h) are smaller and more compact than the corresponding subfilament models (a–d). (From Squire, 1979.)

9.3.4. Results from Crustacean Muscles

In a recent detailed study of the low- and medium-angle X-ray diffraction patterns from a variety of slow and fast decapod crustacean muscles, Wray (1979a) found that there were reflections in the observed diffraction patterns that could only arise from the myosin filament backbones. These reflections were intermediate in position between the small-angle layer lines caused by the myosin cross-bridge array and the wide-angle diffraction from the α-helical myosin rods. But they could be interpreted as arising from the same system of layer lines as the small-angle layer lines from the cross bridges and seemed to be unrelated to the actin pattern.

The general form of these diffraction patterns is summarized in Fig. 9.14a which shows the layer line positions found in patterns from the relaxed fast abdominal flexor muscles of the lobster (Fig. 8.18a,b). Prominent in the diffraction pattern are the meridional reflection at 145 Å and the off-meridional layer lines at 308 and 274 Å. These are thought to be caused by the array of myosin cross bridges since they become weaker in patterns from rigor muscle. The axial positions of these layer lines define uniquely the screw symmetry of the cross-bridge array, and this is indicated by the black circles in Fig. 9.14b. However, the rotational symmetry of the filament which is related to the number of cross-bridge

TABLE 9.2. Parameters of the Rope Models for Myosin Filaments

Number of molecules per crown	Number of intermediate filaments per filament[a]	Pitch of intermediate filaments (Å)	Repeat along intermediate filaments (Å)	Axial stagger between intermediate filaments (Å)	Approximate diameter of intermediate filaments (Å)	Number of molecules per intermediate filaments	Approximate filament diameter[b] (Å)	Approximate core diameter (Å)
3	3	3×430	143	0	80	9–10	170	10–30
4	4	Values here depend on exact repeat along filament	143	0	80	9–10	160–180	30–40
5	5		143	0	80	9–10	210–220	50–60
6	6		143	0	80	9–10	240	80

[a] Each intermediate filament consists of three subfilaments. Each subfilament has a repeat of 430 Å and diameter 40 Å with about three molecules per cross section. Within each intermediate filament the subfilaments are staggered axially by 143 Å. From Squire (1979).

[b] Because of the uncertain amount of interlocking between the intermediate filaments, the values given are approximate and represent the largest likely diameters. In reality, they could be significantly smaller than this.

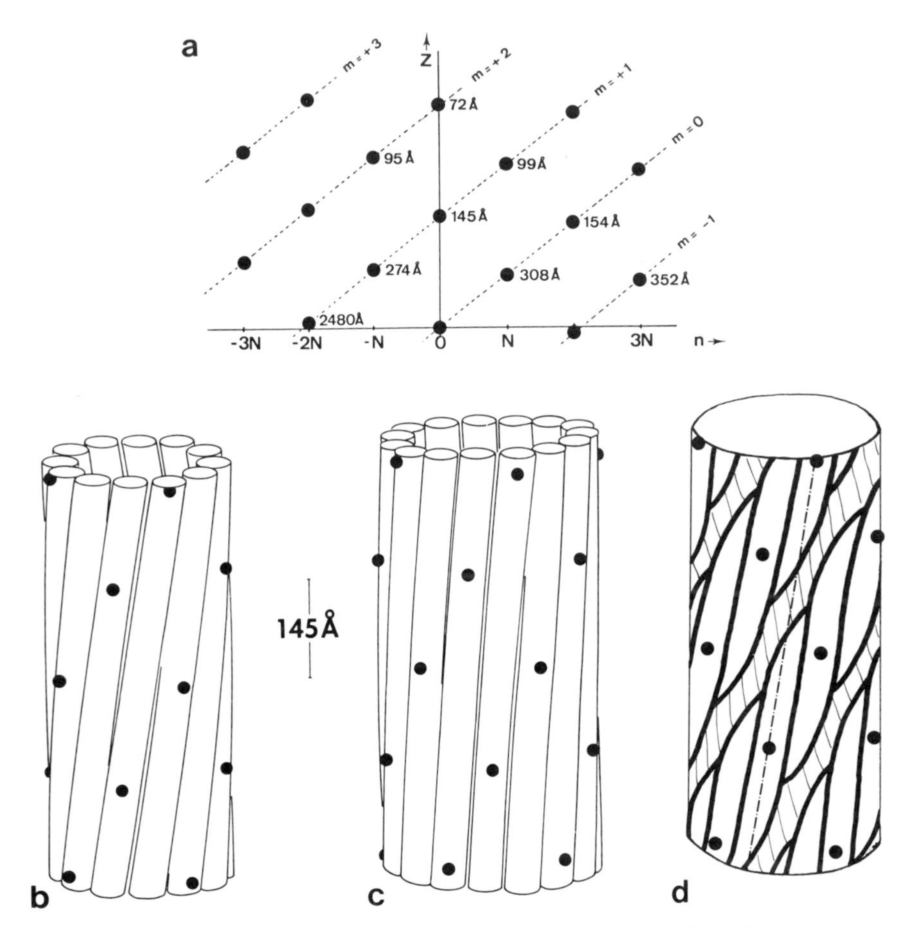

FIGURE 9.14. (a) Diagram showing the positions of the myosin layer lines in the observed X-ray diffraction patterns from lobster fast abdominal flexor muscle (see Fig. 8.18). The meridian is vertical, and 0 corresponds to the center of the pattern. The observed diffraction pattern would appear as a combination of this pattern with itself mirrored across the meridian. (b and c) Models for the myosin filaments in fast and slow crustacean muscles, respectively, as proposed by Wray (1979a). In cross section, these models would appear as Fig. 9.13b,c and are four- and five-stranded, respectively. The reflection in (a) at 352 Å and at a radius of about 40 Å from the meridian is accounted for by Wray in terms of the 40-Å subfilaments in (b) and (c) which twist slowly around the filament axis along the appropriate set of helices. (d) A very schematic diagram of a four-stranded rope structure (Fig. 9.13f) with the same screw symmetry as in (c). It can be seen that the 40-Å subfilaments in each strand of the rope lie along the same helical strands as the subfilaments in (b). Such a structure might also account for the observed crustacean diffraction patterns (a) with a prominent reflection at 40-Å radius, although the number of strands could be increased to six without making the filament diameter too large (see text). (a to c) From Wray (1979a); and (d) from Squire (1979). Note that here n and m in (a) index different reflections according to conventional helical diffraction theory (Klug *et al.*, 1958; Chapter 2, this volume).

positions per crown cannot be determined exactly for reasons that have already been discussed fully in Chapter 7.

Diffraction patterns from other fast crustacean muscles were found by Wray to be very similar in form to that summarized in Fig. 9.14a except for very slight changes in the axial positions of the layer lines. For example, the reflection at 308 Å in patterns from fast lobster muscles was at 300 Å in patterns from the fast body muscles of the crayfish, 310 Å from the fast muscles of crayfish legs, 315 Å from the fast muscles of the lobster left claw, and 318 Å in patterns from the fast muscles of both crayfish claws. Apart from these changes in spacing, the general distribution of intensity was much the same in the patterns from all of these muscles. This very clear result is a direct demonstration of the fact that, despite very slight changes in the screw symmetry of the cross-bridge arrays in these different muscles, the local geometry of the cross-bridge lattice is more or less the same in all cases (see Fig. 9.14b). As it happens, it is also very similar to the arrays shown in Fig. 9.2 for other myosin filaments. Furthermore, X-ray patterns from slow crustacean muscles, which contain thick filaments of larger diameter than those in fast muscles, also show an intensity distribution similar to that in patterns from the fast muscles. These too can be interpreted in terms of cross-bridge arrays of similar geometry (Fig. 9.14c). These beautiful new results on crustacean muscles have, therefore, provided considerable support for the fundamental ideas behind the general model of myosin filament structure.

With this conclusion in mind, we can now turn to the important new evidence that Wray (1979a) has obtained on the structure of the myosin filament backbones. This evidence consists of the observation of X-ray reflections at radial positions from the meridian of about 40 Å that form part of the system of myosin layer lines. Since these reflections remain intense in patterns from rigor muscle, they must be caused by an invariant feature of the myosin filaments, the filament backbone. In Fig. 9.14a, this reflection is the one marked 352 Å. In the particular pattern described by Wray, this layer line had peaks on it at radial positions of 43 Å, 28 Å, and 24 Å, with a broad minimum at 32 Å. Wray interpreted this feature of the pattern as being caused by subfilaments of myosin rods with a diameter of 40 Å running along the particular set of helical tracks through the cross-bridge array that would give rise to this particular layer line. This scheme, which is illustrated in Fig. 9.14b, is therefore Wray's interpretation of the backbone structure in fast crustacean muscles. Wray took the number of myosin cross bridges per crown to be four in this case, since, as shown in Fig. 9.13b, this would give a filament diameter close to the observed value of 200 Å (Jahromi and Atwood, 1969). Since the filament diameters are rather larger in the slow crustacean muscles (about 280 Å; Jahromi and Atwood, 1969), Wray suggested that the myosin filaments in this case might be five-stranded as in Fig. 9.13c and 9.14c.

Wray also extended his discussion to the structure of the thick filaments in insect flight muscles and vertebrate striated muscles. He has observed a reflection at 40-Å radius in patterns from insect flight muscles. Although the axial position (1150 Å) of this reflection is rather different from the corresponding reflections in patterns from crustaceans, the screw symmetry of the cross-bridge array is also different, and the 40-Å radius reflection can be related to the cross-bridge array in just the same way as in Fig. 9.14. Wray took this to mean that the thick filaments in insect muscle, which have a diameter of about 200 Å (see Chapter 8), might also have the structure shown in Fig. 9.14b. The problem here is that such a structure only has the heads of four molecules on each 145-Å-spaced crown, and, as discussed fully in Chapter 8, all of the other available evidence on insect thick filaments seems to suggest that this number should be six.

This discrepancy appears to lead to two alternative conclusions. One is that the number per crown of six is incorrect and that the structure of insect muscle needs to be thought through again. The other is that the interpretation of Wray's results may not be as straightforward as, at first sight, it appears. In fact, there do appear to be alternative structures that might account for Wray's results. For example, it has already been shown (Fig. 9.13h) that a rope model for a six-stranded filament is more compact than the corresponding subfilament model (Fig. 9.13d). As indicated in Fig. 9.13a, the rope model also contains 40-Å subfilaments, but these are twisted together about a common axis to give the larger intermediate filaments, six of which would form a six-stranded filament of outer diameter rather less than 240 Å, depending on the exact amount of interdigitation between the intermediate filaments (Fig. 9.13d). The structure is very like that modeled by Yagi and Matsubara (1977). One would expect in such a rope structure that the sense of twist between one level of structure and the next would change hand just as occurs in man-made ropes. Figure 9.14d illustrates very schematically the kind of result that would be obtained in such a rope structure, in this case for the four-stranded cross-bridge array as in Fig. 9.14b. Here each intermediate filament follows a right-handed helix, so the 40-Å subfilaments follow a left-handed helix. As the dotted lines indicate, even in this structure, the 40-Å subfilaments at the surface of the rope follow the same helical tracks as those in Wray's subfilament model, although in this case the density along these tracks is discontinuous. Since they follow the same tracks as in Wray's model (Fig. 9.14b), they would presumably account for the 40-Å reflection and the other features of the same layer line in the identical way. The difference in this case would be that there may be other prominent helical tracks in the structure. For example, there is a clear one in the direction of the intermediate filaments, and this would give rise to intensity on the 308-Å layer line in Fig. 9.14a. It may be possible by further analysis of Wray's patterns to choose between these alternative models. Certainly, the rope model would need to be excluded positively by features of the diffraction pattern before the subfilament model could be taken to be correct.

But, if the rope model is compatible with all of Wray's results, then the insect thick filaments could still be six-stranded, as could all of the filaments in the fast crustacean muscles and in *Pecten* striated muscle too. Slow crustacean muscles with their thick filaments of larger diameter would probably have seven- or eight-stranded cross-bridge arrays.

The main objection to Wray's model seems to be the very large size of the thick filament core as illustrated in Fig. 9.13. Although many of the observed filaments in invertebrate muscles do appear to have hollow cores, they rarely seem to be as large in comparison with their outer diamters as would be expected from the subfilament models. The corresponding rope models of the same outer diameter have smaller cores and more cross bridges than the subfilament models.

With respect to vertebrate muscles, Wray's interpretation would be be in terms of a nine-subfilament structure (Fig. 9.12 and 9.13a) in which the subfilaments are parallel to the filament axis. Wray suggested a possible modification to this structure which is described later in the chapter.

9.4. Detailed Models of Vertebrate Skeletal Muscle Myosin Filaments

9.4.1. Myosin Packing in the M-Region and Filament Tip

On the assumption that the myosin rods in the bulk of the three-stranded filament form either the uniform layer structure (Fig. 9.7) or the triple rope of intermediate filaments (Fig. 9.12), one can ask if reasonable model structures for the M-regions and tapering ends of the thick filaments in vertebrate skeletal muscle can be produced (Squire, 1973). These parts of the filament will be considered first in terms of the uniform layer structure.

Uniform Layer Model. It is now known that the bare zone (M-region) is about 1620 Å long in vertebrate skeletal muscle (measured as in Fig. 7.43). It is also known (Chapter 7) that a 145 Å repeat continues through the M-region and that the filament profile changes from triangular in one bare region to circular at the M-band and triangular again, but with a different orientation 40° away, on the other side of the M-band. In this region, the filament commonly has a less dense central region. The diameter of the circular profile at the M-band is about 100 Å.

In order to develop a bare zone packing scheme, the following assumptions have been made: (1) that the bare zone will be as tightly packed as possible, and (2) that the molecular packing arrangement will develop systematically from the arrangement in the bulk of the filament as drawn in Fig. 9.7a. On the basis of these assumptions, the packing scheme drawn in Fig. 9.15 was derived by Squire (1973). This arrangement includes the antiparallel overlaps of 430 Å and 1300 Å observed in synthetic aggregates by Cohen *et al.* (1970) and Harrison *et al.* (1971). It also includes end-to-end butting of molecules but not the antiparallel overlap of about 900 Å that has also been observed in synthetic aggregates. Single and triple cross lines join molecules interacting with overlaps of 430 and 1300 Å, respectively, in Fig. 9.15, and butting arrowheads indicate positions where end-to-end molecular butting occurs.

In this diagram the myosin rods having heads on one side of the bipolar bare zone are drawn as thin lines, and those on the other side as thick lines. Figure 9.16 shows how such an arrangement would appear in consecutive transverse sections 145 Å thick through the M-region and on one side only into the P-zone next to the M-region (see Fig. 7.46). The arrangement on level 10 is similar to Fig. 9.7, but with three rod profiles missing. Levels 6 and 6′ (see below) are located at the position of the first rows of cross bridges at the edges of the M-region. Levels 1 and 1′ lie just on each side of the center of the M-band. On the right of Fig. 9.16, views (b) to (e) indicate the probable appearance of transverse sections about 500 Å thick through various parts of this structure: (b) is a composite of levels 7 to 10, (c) of levels 3 to 6, (d) of levels 2 to 2′, and (e) of levels 3′ to 6′. It will be seen that these correspond remarkably closely to the appearances actually seen in micrographs (Figs. 7.15 and 7.16), and a 40° difference in orientation of the triangular profiles in (c) and (e) is automatically generated.

In fact, this type of structure is produced if the bulk of each half of the myosin filament is gradually built up to the structure in Fig. 9.7 using systematic 430-Å and 720-Å parallel interactions. But this inevitably means that levels 7, 8, 10, and 13 (equivalent to P2, P3, P5, and P8 in Fig. 7.46) do not have myosin cross bridges on them. Objections have therefore been raised to this model (Craig and Offer, 1976b) because antibody labeling studies with anti-S-1 (Fig. 7.45) seem to indicate the presence of cross bridges in this part of the P-zone. Although this evidence cannot yet be taken to be completely conclusive, it is instructive to con-

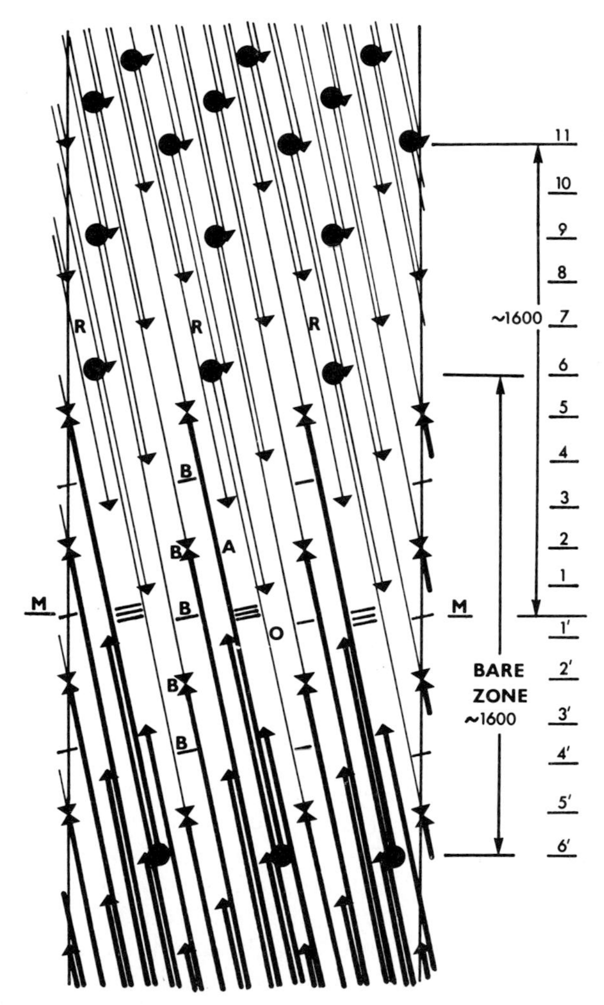

FIGURE 9.15. A possible structure (in radial projection) for the M-region of a three-stranded myosin filament in vertebrate skeletal muscle. The top half of the diagram is analogous to Fig. 9.4 but for a three-stranded filament. The lower half of the diagram corresponds to the same radial projection inverted. The filled circles represent the positions of the heads of the myosin molecules. The arrowheads indicate the other ends of the same myosin molecules. Antiparallel molecules interacting with 430- and 1300-Å overlaps are indicated by single and triple cross-lines, respectively. Positions where arrowheads meet are positions of end-to-end butting. "O" is an "up" molecule (thin lines), and "A" is a "down" molecule (thick lines). The appearances of cross sections through this structure are shown in Fig. 9.16. The position of the center of the M-band (M1) is indicated by M, and the location of the M-bridges (M1, M4, and M6) by B. It is uncertain whether or not molecules such as those marked R should be included (see text). Dimensions are in Å. (From Squire, 1973.)

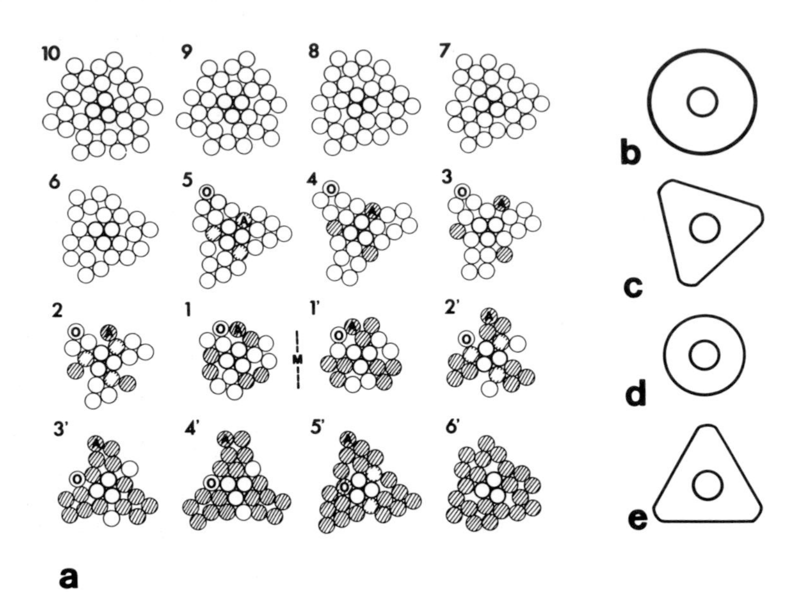

FIGURE 9.16. (a) Cross sections of different levels of the M-region packing arrangement in Fig. 9.15. "O" is an "up" molecule, and "A" a "down" molecule. Where these touch there is an antiparallel overlap of 1300 Å. (b to e) The probable appearances of thicker cross sections through the M-region: (b) composite of levels 10, 9, 8, and 7 in (a); (c) composite of levels 6, 5, 4, and 3; (d) composite of levels 2, 1, 1′, and 2′; (e) composite of levels 3′, 4′, 5′, and 6′. (From Squire, 1973.)

sider what it would imply. It would mean that according to the uniform layer packing model the complete structure as in Fig. 9.7 would occur at the edges of the bare zone (levels 6′ in Fig. 9.15). The number of molecules forming one side of the thick filament would then reduce in levels 5, 4, 3, etc. in steps of three until the last rods terminate in level 5′. Figure 9.17 shows schematically the longitudinal structure of the M-region in such a case, when compared with that in Fig. 9.15. Here it can be seen that in the former the total number of molecular profiles in any cross section remains approximately constant (at about 30) right through the M-region, whereas in the model depicted in Figs. 9.15 and 9.16, this number reduces from 30 on level 14 to 27 on level 10, to 21 on level 6, and to a minimum of about 15 on levels 2,1,1′, and 2′ which are at the M-band itself. If it is accepted that the filament profile changes from circular on about level 10 to triangular at levels 5, 4, and 3 and to circular again at levels 2 to 2′, then these two M-region structures make different predictions about the size of the molecular profiles.

These are summarized in Fig. 9.18. In brief, if the molecular number does not change through the M-region, then at the M-band, the circular profile must have about the same diameter as at level 10, whereas in the model illustrated in Fig. 9.15, the M-band profile would be about 100 to 110 Å if the profile in the bulk of the bridge region (Fig. 9.7) is about 140 Å. Similarly, in the triangular region the triangle edge would be expected to be about 188 Å long if there are 30 rods in the filaments, but only about 166 Å long if there are only about 18 rods. Although precise measurements of filament diameters in transverse sections are rather difficult, it should be possible in principle to choose between these models by an analysis of filament size through the M-region. Preliminary measurements indicate that the filament backbone does, in fact, get rather smaller through the M-region than it is in the rest of the A-band. But, once again, this evidence cannot yet be taken as conclusive.

There is another reason for believing that the model illustrated in Fig. 9.15 with gaps in the cross-bridge array may be preferable. This is that in both A-segments and cryosections of the A-band it is apparent that the filament structure changes markedly through the P-zone. This change in structure would be well explained by the model with the reducing number of myosin molecules, since, as shown in Fig. 9.15, a changing structure right through the P-zone is just what is involved in this model. In terms of the other kind of model, with all of the cross-bridge levels in the P-zone occupied (Fig. 9.17), it is not inevitable that a change in structure through the P-zone would occur. In this case, the explanation would have to be in terms of the possibility that the molecules with heads in the P-zone would still "feel" the fact that they have been involved in an antiparallel interaction in the M-region. It is also evident that the intermolecular interactions would, in this case, need

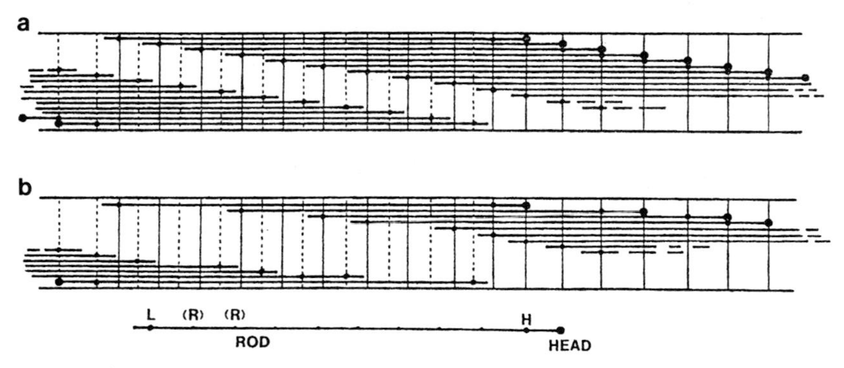

FIGURE 9.17. Schematic diagrams of the axial positions of myosin molecules in the M-region in models where (a) there are no gaps in the cross-bridge array in the P-zone and (b) there are gaps in the cross-bridge array as in Fig. 9.15 (with the R molecules missing). In (a) the effective number of myosin rods in any cross section is more or less constant right through the M-region, whereas in (b) this number reduces from the edge of the M-region to the middle.

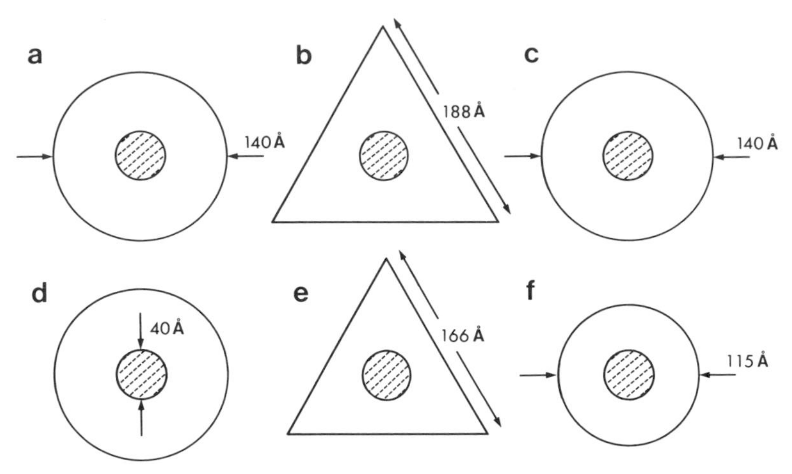

FIGURE 9.18. Cross-sectional profiles (simplified) of the myosin filament backbone in the bridge region (a and d), the bare region (b and e), and the middle of the M-band (c and f), for the models in Fig. 9.17 in which the number of molecules per cross section is constant (a to c), or reduces towards the M-band (d to f). In all cases a hollow core of 40 Å diameter is assumed and, for example, the area of the triangular profile in (b) is equal to the areas of the circular profiles in (a) and (c). Note that (b) actually appears larger than (a) or (d), whereas (e) appears slightly smaller.

to change markedly to produce the triangular bare region profiles. It is therefore possible, but not inevitable, that the changing P-zone structure could be explained by this model, whereas the other model (Fig. 9.15) would explain this appearance directly.

With regard to the tapering ends of the filaments, the uniform layer structure could explain this, if the filament taper is because of systematic removal of molecules from the full backbone structure to leave just three molecules at the tip, using only 720- and 430-Å interactions. The result is shown in Fig. 9.19b. It can be seen in this case that, once again, some of the 145-Å-spaced axial levels do not have cross bridges on them. The effect of this is to produce an arrangement in which the myosin rod packing changes right through the D-zone of the bridge region. Since such a change has been observed in electron micrographs of the D-zone, this structure is an attractive one. It would also account for the fact that, in the tapered region (but not elsewhere in the bridge region), the myosin filament backbone definitely has a three-subunit structure (Fig. 7.18) in transverse sections.

However, electron micrographs of A-segments and of cryosections only show one obvious gap in the cross-bridge array. This occurs near to the outer ends of the A-band (D19 in Fig. 7.46). Similarly, Craig and Offer (1976b) using antibody to HMM S-1 only saw this same single gap

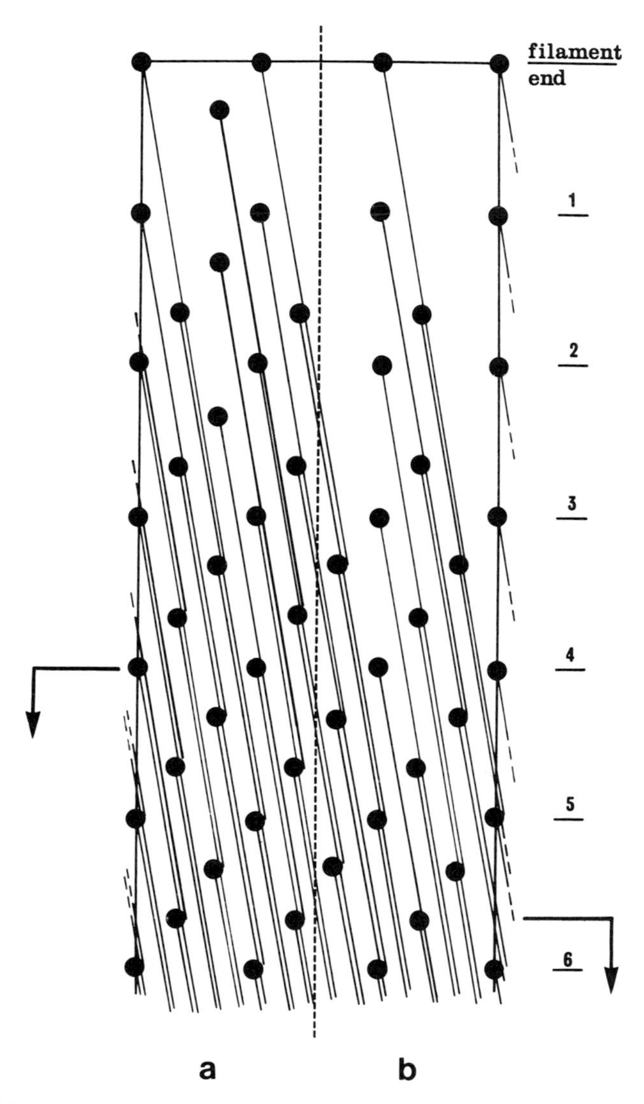

FIGURE 9.19. Radial projections (similar to that in Fig. 9.15) for the tip of a three-stranded myosin filament in vertebrate skeletal muscle. One-half of the filament has been drawn as for a structure with only one gap (at D19 in Fig. 7.46) in the cross-bridge array (a). The other half (b) has several gaps in the cross-bridge array and is produced by adhering to systematic 430- and 720-Å overlaps between myosin rods. In each case the arrowed positions indicate where the myosin packing changes (i.e., where the filament taper would start).

in the cross-bridge array. On the assumption that this is the only place where cross bridges are absent, then the structure shown in Fig. 9.19a would be generated. Here the structure changes from position D21 to position D8, but the change in structure observed in A-segments and cryosections between position D8 and D1 would not be explained directly. A possible explanation for this is given in Section 9.4.

The Rope Model. In terms of the rope model, one would expect the structure of the M-region to be produced by the gradual buildup of the various strands in the molecular rope. Since in each 40 Å diameter subfilament the basic repeat would be 430 Å, one would expect such a subfilament to build up using only the 430-Å molecular stagger. Three such subfilaments staggered initially by either 144, 288, 576, or 720 Å would then aggregate to form each intermediate (80 Å diameter) subfilament. Because of the uncertainty of the stagger in this second interaction, it is conceivable that structures analogous to those depicted in Figs. 9.15 and 9.16 for the layer model could occur with this rope model also. The models with staggers longer than 144 Å would be open to the objection that they would produce gaps in the cross-bridge array in the P-zone as before. But also, as before, they would help to explain the changing striation pattern in this region. Since the M-region would be built up using the three intermediate filaments in each half of the myosin filament, one might expect from such a model the kind of triangular and circular profiles actually observed in the M-region. This is illustrated schematically in Fig. 9.20. In a similar way, the filament taper could also be explained by such a structure. If the interactions involved

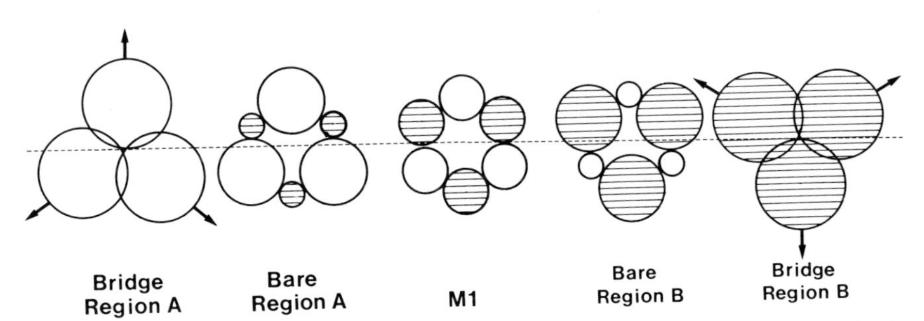

FIGURE 9.20. Possible explanation of the observed filament profiles through the M-region in vertebrate skeletal muscle for a myosin model consisting of a three-stranded rope as in Fig. 7.12b. No attempt has been made to mimic exactly the 40° rotation between bare region profiles on opposite sides of the M-band observed in sections (Fig. 7.15). The cross-hatched circles contain molecules of opposite polarity to those in the open circles (i.e., they are antiparallel). Note that it has been assumed that for some reason the helical twist of the strands in the rope that occurs eleswhere in the A-band is absent in this region; otherwise, it is very difficult to account for the M-region in terms of this kind of model. (From Squire, 1979.)

are 430 and 720 Å, then a structure similar to that depicted in Fig. 9.19b would occur inasmuch as the positions where structural changes occur in the D-zone would be the same. On the other hand, if the interactions between the 40-Å subfilaments involves a 144-Å stagger, then a structure analogous to that in Fig. 9.19a would be produced. In either case, the three-subunit profile of the filaments in this region would be explained.

The main objection to this kind of model is that the triangular filament profiles observed in the bare region and filament tip do not appear to rotate at all over a filament length of about 500 Å. These observations would only appear to be compatible with the rope model if the intermediate subfilaments, which normally twist by one-third of a revolution in 430 Å, suddenly become parallel to the filament axis in these two regions. It is evident that such a possibility seems to be rather implausible, and this result alone may be the strongest argument against the rope model for the myosin filaments in vertebrate skeletal muscle.

9.4.2. Extra Proteins and Filament Length Determination

From the analysis of the last section, it is clear that whichever filament model is taken, there are basically two kinds of structure possible for the M-regions and filament tips. In one, all of the 144-Å-spaced bridge region repeats have cross bridges on them except for D19, and in the other, gaps occur in the cross-bridge array through the P-zone and part of the D-zone in addition to D19. The two diagrams in Fig. 9.21 summarize these two possibilities in terms of the locations of cross-bridge origins. Figure 9.21a has only a single gap in the cross-bridge array at D19, and Fig. 9.21b is the structure generated using systematic 430-Å and 720-Å intermolecular staggers. Also shown in Fig. 9.21 are the ap-

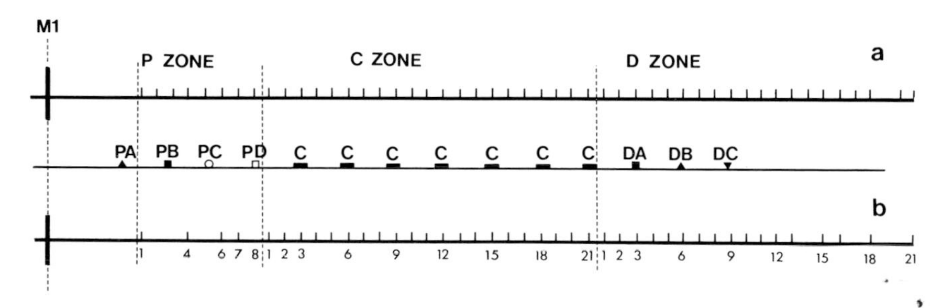

FIGURE 9.21. Axial cross-bridge positions in one-half of an A-band in vertebrate skeletal muscle for models in which (a) only level D19 is unfilled and (b) several cross-bridge levels are unfilled as in Fig. 9.15 and Fig. 9.19b. Possible positions of extra proteins (PA, PB, etc.) are also shown including the positions of C protein (C). It is not known whether in the D-zone (see Fig. 7.46) the positions DA, DB, etc. correspond to extra proteins or simply to a 430-Å repeat within the myosin filament backbone.

proximate positions of the various nonmyosin proteins (such as C protein) which are known to occur in vertebrate skeletal muscle thick filaments. It is instructive to consider why these various proteins are located in their respective positions.

Considering first model (Fig. 9.21a) with only a single gap at D19, it is perhaps somewhat surprising that four different extra proteins here labeled as PA, PB, PC, and PD interact with an apparently uniform structure through the P-zone. On the other hand, this might be explained, as mentioned earlier, if the molecules in the P-zone can still "feel" the effects of the antiparallel M-region packing. Alternatively, it might be argued that the protein PA disturbs the structure so that slightly further up the filament the structure is suitable for interaction with PB and eventually with PC and then PD. The model in Fig. 9.21b would, of course, have a systematically changing structure right through the P-zone, and the interactions of four different proteins at different positions would not be unexpected.

In either situation it would be necessary for the molecular packing to have settled down by the end of the P-zone so that a structure is reached that can interact with several C-protein molecules. However, as mentioned in Chapter 7, there is evidence from cryosections for what appear to be slightly different repeats in the C-protein stripes and the cross-bridge array. This would mean that the interaction between C protein and myosin is changing slightly through the C-zone. Assuming this to be so, then, as discussed by Squire *et al.* (1976), it seems evident that the only reason C protein would not bind to myosin with the myosin repeat would be that something structurally prevented it from doing so. One possible reason could be that C protein, which is known to be at least 350 to 400 Å long (Offer *et al.*, 1973), may actually be about 435 to 440 Å long and may run almost parallel to the filament axis. C-protein molecules linked end to end would then inevitably be related by an axial repeat of 435 to 440 Å and not by the slightly shorter myosin repeat. Since that would be just the kind of arrangement to be expected in a thick filament length-determining mechanism (H. E. Huxley and Brown, 1967), Squire *et al.* (1976) interpreted this C-zone structure in terms of such a mechanism. This was based on the following assumptions.

1. C protein will only interact with the regular part of the thick filament.
2. C-protein molecules are about 440 Å long and can link end to end by a nonconvalent interaction.
3. C-protein molecules have, along part of their length, a "sticky" region which can interact with a specific site on the myosin rod.

This region would be about 6 × (440–429) Å long (i.e., about 60 to 70 Å) and might comprise seven individual interactive sites equally spaced at axial intervals of about 11 Å.

4. C-protein molecules have a bulky region extending for 50 to 100 Å along their length, and this gives rise to the obvious unstained lines in cryosections.

The length-determining mechanism would then operate in the following way (illustrated in Fig. 9.22). The filament would first build up from the M-region outwards, with the antiparallel packing gradually changing to the polar interactions of the bridge region. The structure would have become regular by the end of the P-zone (marked A in Fig. 9.22). At point A, the structure would be such that the first three C-protein molecules can interact (at the same level) with myosin (if the filament is three-stranded). Additional C-protein molecules would then interact end to end with the first set and also with the adjacent myosin rods through the "sticky" region. This kind of assembly would continue through the C-zone, but at each C-protein position, myosin would interact with a different part of the sticky region. After seven sets of C-protein molecules had assembled in this way, a point (B) would be reached at which the sticky region of the next C-protein molecules would be out of the reach of the interaction site on the appropriate myosin rod. The binding of C protein would therefore stop at B, and this would trigger the change in packing of the myosin molecules required to terminate the filament.

The nature of this filament termination would clearly depend on whether model (a) or model (b) in Fig. 9.21 was involved. In model (b) it would be suggested that at position B the structure is such that the rod packing starts to change directly by the systematic removal of myosin molecules from the structure (Fig. 9.19b and 9.21b). In model (a) it would be necessary for the structure to be modified at position B so that protein DA in Fig. 9.21 can start to interact. This would in turn modify the structure so that DB could bind and then DC, etc. At this point the taper would start as in Fig. 9.19a.

As a final note, it is possible in model (b) that it is the extra proteins that tend to obscure the gaps in the cross-bridge array. But this explanation might be less acceptable in the D-zone than in the P-zone where many extra proteins are located. On the other hand, the experiments reported by Moos (1972) in which he tested the effect of the presence of C protein on the formation of synthetic myosin filaments may support this kind of model. Moos found that "with C-protein the filaments usually taper to a very thin tip in contrast to the rather blunt ends characteristic of pure myosin." Such a result is clearly quite compatible with the

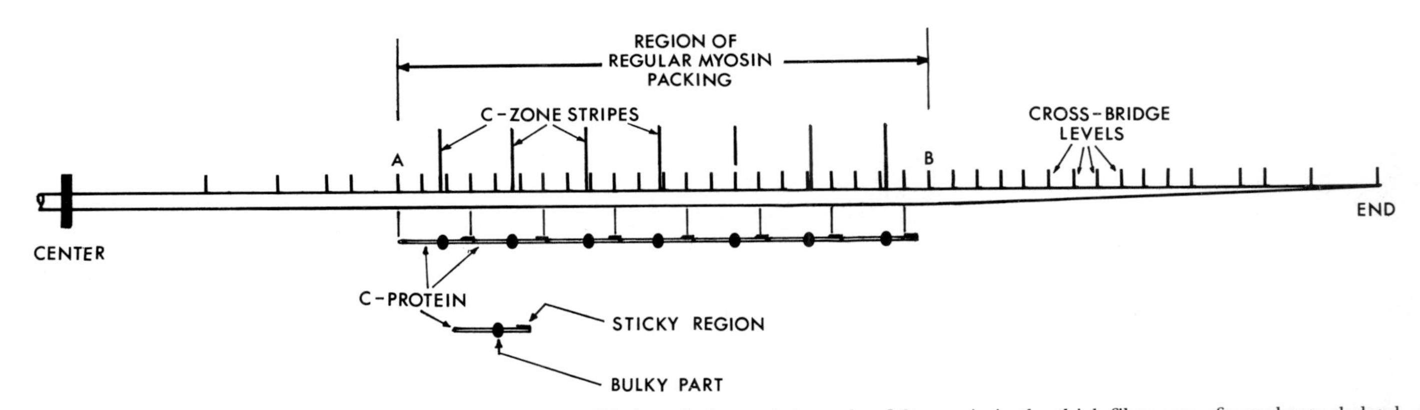

FIGURE 9.22. Diagrammatic representation of the possible length-determining role of C protein in the thick filaments of vertebrate skeletal muscle according to Squire *et al.* (1976). For details see text.

type of gap structure involved in model (b). It might also be possible that the absence of obvious triangular or three-subunit profiles in thin transverse sections through the bridge region could be because of the presence of the extra proteins in the grooves between pairs of the three intermediate subfilaments (Fig. 9.12). Another observation that might be explained by this structure is the fact that the C-protein stripes appear to be located in slightly different positions in different preparations. If the C-protein molecules lie along the filament axis, and yet axially well-localized transverse stripes are observed, then clearly specific positions along the molecules must be producing these stripes. In such a situation it might well be that the bulky region (for example) shows up in negatively stained specimens but that the interactive sticky region is strongly labeled with antibody. At the moment it is much too early to decide questions of this kind.

9.5. Models for the Myosin Filaments in Vertebrate Smooth Muscle

In Section 9.2.1 the structure of the myosin ribbons was dealt with in terms of the packing, back to back, of two planar myosin layers. On this basis the edge-view structure of the ribbons would be as indicated schematically in Fig. 9.23a. Here, end-to-end butting of the molecules must occur together with 430-Å and 720-Å antiparallel overlaps (not illustrated), since the structure could be constructed from two separate complete layers that are then put together. This structure would have the required face-polar form and would have bare edges about 3000Å long at each end (L_e).

An alternative kind of model that would demand a different kind of myosin packing would start off with a small antiparallel myosin aggregate (as suggested by Sobieszek, 1972), and these would then aggregate to form the whole ribbon. Such an arrangement could be formed into a face-polar ribbon of the kind illustrated in Fig. 9.23b,c. Here, the primary interaction is an antiparallel overlap, possibly of 430 Å or of about 700 Å (as observed in segments by Kendrick-Jones *et al.,* 1971). This structure would appear rather similar to that in Fig. 9.23a, but the bare edges would be shorter (about 3000 – (430 to 700) Å or 2300 to 2600 Å).

Small and Sobieszek (1977) and Small (1977a,b) have suggested that *in vivo* the vertebrate smooth muscle thick filaments are approximately square in profile and about 160 Å across and that they have about four cross bridges per crown. From the similarity of these filament dimensions to those in *Limulus* muscle, one might suggest that these filaments have their cross bridges arranged on some kind of four-stranded helix.

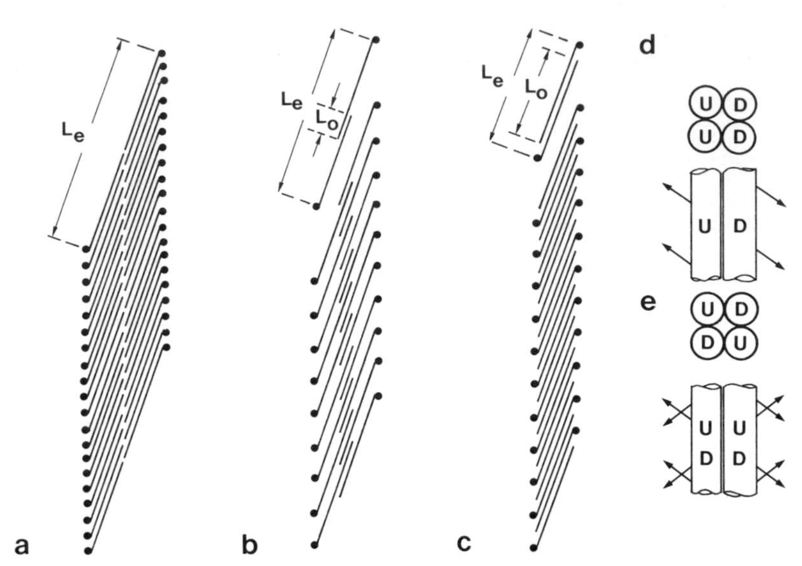

FIGURE 9.23. Schematic diagrams of various types of molecular packing models for planar layer myosin filaments in vertebrate smooth muscle seen in edge view. (a to c) Possible ways of producing face-polar assemblies in which the overlap (L_0) between antiparallel molecules varies from 0 in (a) to say 430 or 700 Å in (b) and to about 1300 Å in (c). The length (L_e) of the bare edges of these filaments vary accordingly (see text). (d and e) Two alternative ways of putting together molecular ropes as in Fig. 9.13 (e–h) to give filaments with a square cross section (upper figure in each pair) and mixed cross-bridge polarity (lower figures). Only the arrangement in (d) would have face polarity as observed in sections.

But there is no evidence either from electron microscopy or X-ray diffraction that would support a helical filament model. For example, there are no X-ray layer lines (other than the 144-Å reflection) that could be attributed to a thick filament structure, and as in *Limulus* muscle diffraction patterns, it appears that they should be there. In addition, it seems necessary that these filaments have a different kind of polarity from conventional helical filaments.

It also seems evident that the organization of this polarity must be systematic along the filaments. A filament of mixed polarity might be produced using subfilaments analogous to the intermediate filaments in the rope model if they comprised four intermediate filaments, two polarized oppositely to the other two. Figures 9.23 (d) and (e) illustrate two possible arrangements in such a scheme, one giving a face-polar filament, and the other a mixed polar filament. The bare edges formed by either of these structures would be about 1500 Å long (the length of one myosin rod), but only the face-polar structure would give bare edges on opposite sides of the filament at opposite ends. However, in this subfilament scheme it is difficult to see how a 720-Å axial repeat could be produced unless the whole structure twists slightly so that each subfilament follows a helical path. But in such a case, a change in the sense of the face polarity would occur repeatedly along the filaments, and such a change was never seen by Small and Squire (1972).

From arguments of this kind, it seems most reasonable at present to conclude that the filaments in vertebrate smooth muscle are planar layer structures and that they would have the myosin molecules packed into a scheme such as that depicted in Fig. 9.23. An intriguing feature of this scheme is that, in order to produce the 720-Å repeat observed in the myosin ribbons, the myosin rods would need to be virtually parallel to the filament axis. If the rods had needed to be at a large tilt angle to the filament axis, it might not have been reasonable to propose a planar layer structure for the ribbons, since the structure generated would have required the filament edge to be at an oblique angle (not 90°) to the transverse rows of cross bridges. Finally, it should be noted that the structures illustrated in Fig. 9.23 might possibly, but not necessarily, require some sort of core component to explain a filament backbone of width 160 Å. This kind of model would easily explain the ease of aggregation of the small ribbons into wider structures as observed by Small and Squire (1972).

9.6. Discussion

9.6.1. Summary

It has been shown in this Chapter that it now seems very likely that all of the different myosin filaments in different muscles have structures that are very closely related. All of the available evidence tends to support this idea, which is the basis of the general model of myosin filament structure.

This conclusion is interesting in itself as a study in the packing properties of two-chain α-helical coiled-coil molecules. But in addition, it has a functional significance in that it means that the cross-bridge arrays seen by the thin filaments in all of these muscles must be very similar. Indeed, it has already been noted that the lateral separation (100 to 120 Å) of adjacent cross bridges in these arrays is about the same as the lateral separation of the actin filaments in many muscles. The implications of this in terms of the geometry of the cross-bridge interactions in different muscles will be discussed in more detail in Chapter 10.

From the discussion of equivalent myosin packing in this chapter, it is evident that there are two promising classes of myosin packing model, the uniform layer model and the rope model, either of which could apply to any type of myosin filament. In addition, within each of these classes, there are a number of alternative detailed structures for the M-region and filament tips in the case of the thick filaments in vertebrate skeletal muscle. It is at present very difficult to decide among these alternatives on the available evidence, and efforts must clearly be made to produce evidence that can be interpreted unambiguously.

There are also a number of other questions that need to be answered. One of these concerns the true functions of the extra proteins in the vertebrate skeletal muscle thick filaments. The proposed length-determining mechanism would ascribe a structural role to these molecules, but it is possible that they may also have a more active role in muscular contraction, possibly in the thick filament regulation process.

Another question concerns the nature of the interaction these proteins make with myosin. The length-determining mechanism seems to demand that C-protein molecules lie more or less parallel to the filament axis. But that this is so has not yet been proven. It is conceivable that, as suggested by Offer (1972), these molecules may wrap around the circumference of the thick filament to form a kind of collar. But in such a situation, it would be hard to explain why they should interact with myosin about every 430 to 440 Å along the filament rather than at 144-Å intervals unless the myosin packing itself has a built-in 430-Å repeat. It would be even more difficult on such a model to explain the apparent difference between the myosin and C-protein repeats observed in cryosections. However, the possibility that the myosin arrangement has a built-in 430-Å repeat is worth considering, especially in view of the fact that there appear to be two types of myosin molecule (isoenzymes, Chapter 6) in single fibers. Since these isoenzymes have relative abundances in some vertebrate skeletal muscles in the ratio about 2 : 1, it is tempting to consider the possibility that every third cross-bridge level along the vertebrate skeletal muscle thick filaments corresponds to a different myosin type from the pairs of cross-bridge levels in between. The true myosin repeat would then be 430 Å and not 144 Å. In the rope model, this would require two of the three 40-Å filaments in any intermediate filament to be composed of one myosin type and the third to be composed of the other myosin. These molecules would then have only pseudoequivalence. However, labeling of isolated thick filaments with antibodies specific to the light chains of one of these two myosin types (actually specific for the alkali light chain difference peptide; see Fig. 6.3) has so far shown a more or less even distribution of antibody outside the bare regions of the myosin filaments (Lowey *et al.*, 1979). Even so, this does not disprove the possibility that there is an intrinsic 430-Å repeat of some kind in these thick filaments.

9.6.2. Models with Nonequivalent Myosin Molecules

So far in this chapter the overriding assumption behind all of the analysis has been that the myosin molecules in the bulk of the myosin filament have identical (i.e., equivalent) environments. It must be remembered that this is a first assumption, albeit a reasonable one, and for this reason it is instructive to consider what happens if the assumption breaks down. In fact, the next simplest assumption to make is that the myosin molecules group in pairs (dimers), and from the work of Harrington *et al.* (1972) (see Section 6.3.5) it would be expected that such a dimer would involve a 430-Å overlap between parallel molecules. In this case, one would need to treat the pairs of molecules as the basic structural unit, and it would be necessary to ask how such pairs could interact in equivalent ways to produce the observed cross-bridge surface lattices on different myosin filaments. It is easy to see in principle how this can be done for, say, a filament with three crossbridges per crown and an approximate repeat of 430 Å as illustrated in Fig. 9.24. For a structure with an odd number of strands, there is only one sensible mode of dimer packing and this automatically generates a new subunit axial translation of 2 × 143 Å or 286 Å. If the dimer actually has the component molecules separated axially by about, say, 411 Å rather than 430 Å, but the dimers are placed on a helix of true pitch 3 × 430 Å, then successive levels of cross bridges would be separated by 125 Å and then 161 Å and so on, as shown in Fig. 9.24b (lower part). This kind of structure is just what is observed in *Pecten* striated muscle (Millman and Bennett, 1976; see section 8.3.2 and Fig. 8.23). It could be that the basic structure in myosin filaments does involve such a dimer but that in some muscles, such as vertebrate and insect striated muscles, the difference between the two components of the dimer is hardly perceptible, whereas in others such as the *Pecten* striated muscle, the difference is slightly more marked.

Figure 9.24a illustrates the possible grouping of molecules in a transverse section of a three-stranded myosin filament composed of 430-Å dimers. It should be noted that very slight changes in orientation or axial position of one member of each pair (i.e., as little as 1° to 1 Å) will transform the structure from one with equivalent myosins to one containing myosin dimers. It is very unlikely that a difference of this sort would be detectable by the conventional electron micrographic and X-ray diffraction techniques. Note also that any such slight change would automatically produce two types of cross-bridge environments on the myosin filament surface.

Going on from the assumption that the myosin filament is composed of molecular dimers, it is clearly possible to suggest that structures

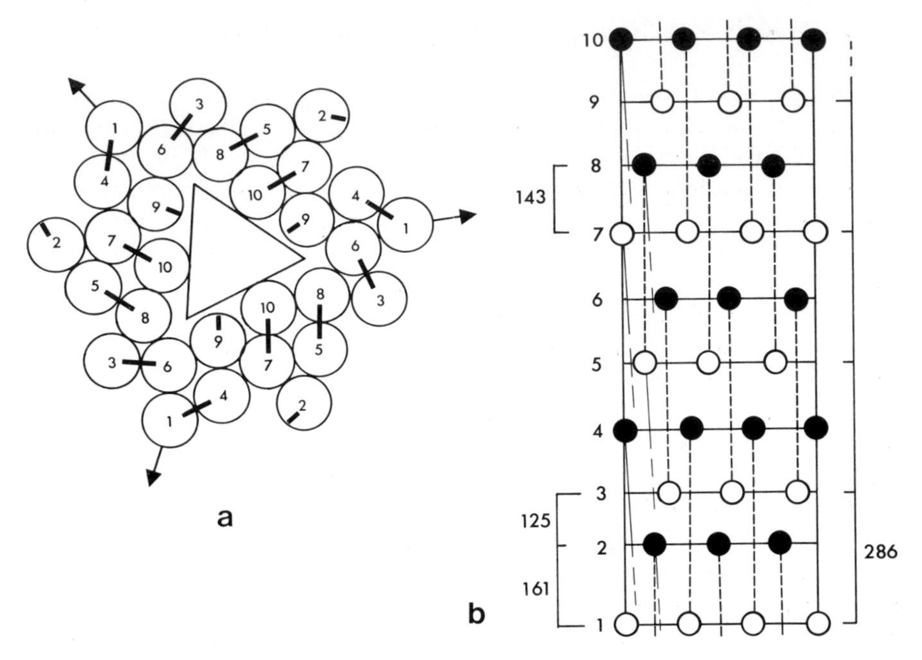

FIGURE 9.24. An example of a molecular packing model for a three-stranded myosin filament that uses a dimer of myosin molecules as the basic building unit. Paired molecules are linked by thick lines in the cross section in (a) and by dashed lines in (b). Note that in (b) a 286-Å subunit axial translation is automatically generated since there are now two classes of molecule represented by the filled and open circles. On level 2, the cross-bridge position has been displaced axially by 18 Å from the normal position 143 Å above level 1. This gives the kind of cross-bridge distribution observed in *Pecten* thick filaments (see Fig. 8.23). (From Squire, 1979.)

such as trimers, tetramers, etc. are the basic building block in the filament. But, as the complexity of the building block increases, the associated models for myosin packing become increasingly implausible. The suggestion that the filament may be composed of equivalent dimers or trimers is quite reasonable, but requiring the myosin molecules to be able to interact with their neighboring molecules in a multitude of different ways seems to be very unattractive. Even so, the myosin packing model published by Pepe (1967a) and subsequently refined (Pepe, 1971) requires a minimum of eight different molecular environments for the component myosin molecules and probably many more (Pepe and Dowben, 1977). This model is described and evaluated below.

Pepe's myosin model (Fig. 9.25) consists of myosin rods grouped in four types of structural unit arranged in a hexagonal lattice and with a variety of axial positions for adjacent structural units. In Fig. 9.25a the circles each represent one of the structural units, and the numbers in the circles represent the relative positions of units along the axis of the filament in multiples of 143 Å. The cross bridges in this structure would lie on an approximate two-stranded helix (arrows), and to account for four molecules per 143-Å repeat, it would be necessary to make each structural unit comprise two molecules. Since the molecules in each unit would be nonequivalent, and since there are four types of structural unit, then there are clearly eight different environments for the myosin molecules in the model. In addition, the model was based on Pepe's suggested structure for the vertebrate M-band, and, as described in Chapter 7, this model has now been shown to be incorrect. The model as proposed also requires the various structural units to have their axes parallel to the filament axis in the bridge region but to twist around the filament axis by 180° in a length of 220 Å in the M-band.

Apart from being highly implausible, such a scheme would hardly explain the triangularity of the filament profiles both in the bare region and at M4 (see Chapter 7) in transverse sections up to 400 Å thick through the vertebrate muscle M-region. Pepe claims the main support

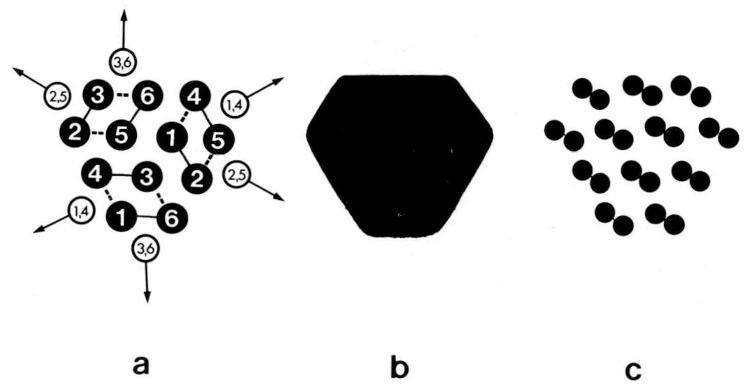

FIGURE 9.25. The model for myosin filament packing according to Pepe (1967, 1971). The circles in (a) represent structural units of parallel myosin rods that assemble in three groups of four as indicated by the joining lines. The numbers indicate the relative axial positions of the various structural units in multiples of 143 Å. Only the LMM portions of the myosin rods are included in the black circles, and these stack end to end at 860-Å axial intervals. The remainder of the myosin rod (HMM S-2) is excluded to the surface of the filament as indicated by the open circles in (a). The cross-bridge helix that is generated (arrows) does not have perfect twofold rotational symmetry but approximates a two-stranded helix as in Fig. 9.1b. (b) The filament shape that (a) might appear as in thick transverse sections according to Pepe and Dowben (1977). (c) Representation of the suggestion by Pepe and Dowben (1977) that the structural units in (a) contain two myosin molecules (20 Å in diameter) that can be successfully resolved in sections up to about 0.4 μm thick. For discussion see text. (Figures redrawn from Pepe, 1971, and Pepe and Dowben, 1977.)

for his model to be (1) the triangularity of the filament profile in the bare region—we have seen that this can readily be explained in terms of a three-stranded filament with equivalent myosin molecules—(2) The observations of triangular or blunted hexagonal filament profiles in the overlap region of the A-band in relatively thick transverse sections (1000 to 4000 Å thick; Pepe and Dowben, 1977. Pepe claims that such an appearance (Fig. 9.25b) should not be seen with a model such as that shown in Fig. 9.7 for a three-stranded filament but that his model explains the result directly. At first sight this appears to be true, but in reality the situation may be much more complex than it appears.

In the author's laboratory a comparable study of thick transverse sections of vertebrate muscle (A. Freundlich, unpublished results) has shown that blunted hexagonal filament profiles can usually be seen only when cross bridges from myosin to actin can also be seen (Fig. 9.26). The corners of the hexagon always seem to occur in the direction in which the cross bridges are pointing. What may well be happening is that a cross-bridge array on the thick filaments with ninefold rotational symmetry in projection down the filament axis is interacting with the six neighboring actin filaments to produce a structure with true threefold rotational symmetry (three being the only common factor of nine and six) and only pseudosixfold symmetry. The near-hexagonal shape of the filament profiles that is sometimes seen would then be caused by the

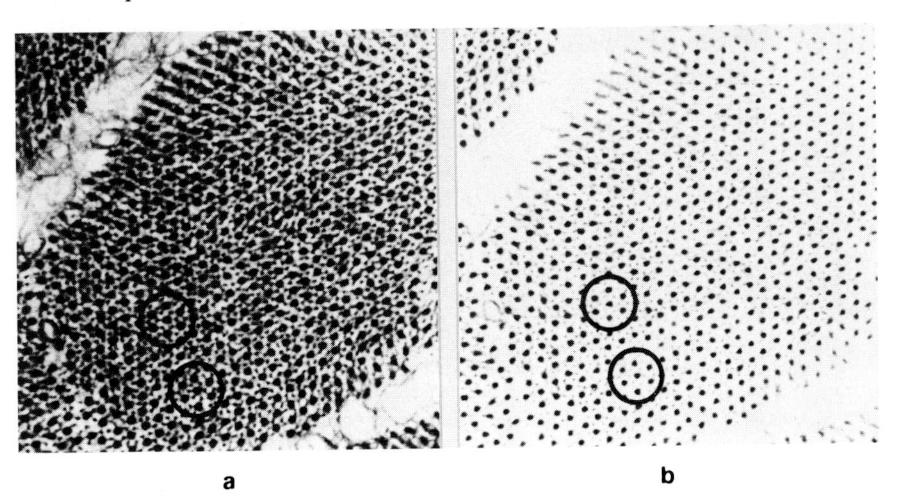

FIGURE 9.26. Electron micrograph of a thick (0.2 μm) transverse section of frog sartorius muscle showing angular thick filaments in the lightly exposed print (b) and cross bridges in the more heavily exposed print (a) that appear to emanate from the corners of the filaments in (b). The circled filaments show this quite clearly. Micrograph recorded on EM7 1-MeV electron microscope at Imperial College using a 200-kV accelerating voltage; courtesy of Dr. A. Freundlich. ×60,000 (reduced 24% for reproduction).

overlapping of the inner ends of the cross bridges radiating out towards the six surrounding actin filaments.

Pepe and Dowben (1977) and Pepe and Drucker (1972) claim to have demonstrated substructure within the backbones of myosin filaments in their electron micrographs (e.g., Fig. 9.25c). Among these claims is the suggestion by Pepe and Dowben that they can see the effects of individual myosin molecules of about 20 Å diameter in muscle sections that are about 2000 to 4000 Å thick. These are purported to show that the two molecules in each structural unit in their model twist around each other slowly. They twist so slowly, in fact, that the twist would not blur the substructure in their thick sections. In such a case, there would clearly be far more than eight molecular environments in each filament. The molecular environments even in a single structural unit would be changing continuously along the length of the filament.

In summary then, there seems to be no compelling reason for preferring Pepe's model over the much simpler and less demanding models that are available. Indeed, there are very many reasons why this model must be considered to be very implausible indeed. Apart from the implausibility of the model, there are also a number of observations described in Chapter 7 that would not be explained by such a model. Among these are the observations of M-bridges and M9-bridges arranged with threefold rotational symmetry at a single axial position along the myosin filaments. Nowhere in the Pepe model are there three equivalent interaction sites at the same axial level, so nowhere would one expect three similar molecules to interact with the thick filament at the same axial position, be they M-proteins, M9 protein, C protein or any of the other nonmyosin proteins. It should also be noted that the model does not have perfect twofold rotational symmetry either, so one would not expect two C-protein molecules to interact with myosin on each of the 14 C-protein positions along one filament. Note also that the dominant axial repeat in the Pepe model (which is not a true repeat) would be 2 × 430 Å, and no evidence for such a repeat has yet been reported.

Another possible model with nonequivalent myosin molecules is a modification of the subfilament structures described earlier in the chapter and described by Wray (1979a). Wray suggested that a nine-subfilament model (Fig. 9.12c) might collapse on itself, since the subfilaments are parallel to the filament axis, and might thus produce structures that are triangular in profile (Fig. 9.27) and that have the cross bridges functionally if not strictly equivalent. It was suggested that this might explain the existence of the "forbidden" meridional reflections in X-ray patterns from frog muscle (see Chapter 7) and that it could account for a true 430-Å repeat in the myosin filaments. Note also that the structure in Fig. 9.27b could account directly for the filament profiles

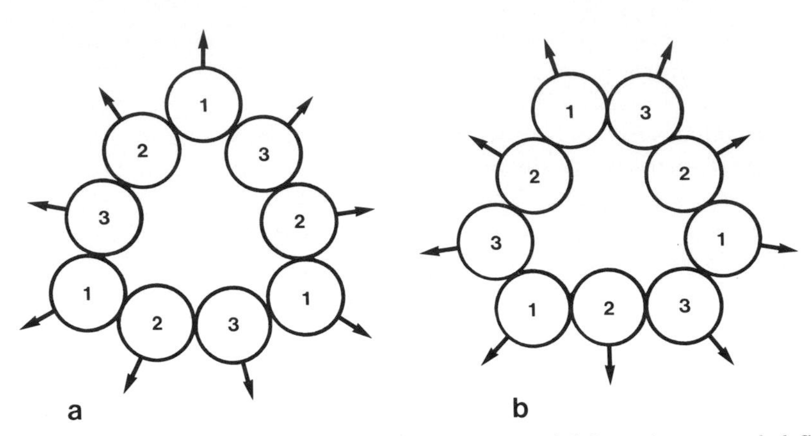

FIGURE 9.27. Modifications to the nine-subfilament model for a three-stranded filament as in Fig. 9.12c and 9.13a in which the initially cylindrical array has collapsed on itself, as proposed by Wray (1979), thus giving the myosin filament an intrinsic 430-Å repeat. See text for details.

seen by Pepe and Dowben (1977) if these are really caused by the filament backbones themselves. Evidence for a 429-Å cross-bridge repeat in human muscle has recently been obtained by Squire *et al.* (1981).

9.6.3. Structure of the Myosin Molecule

Before leaving this discussion of myosin aggregation and packing in thick filaments, it is appropriate to consider briefly what the known properties of myosin tell us about the structure of the molecules themselves. In the whole of this chapter the assumption has been made that the myosin rods approximate uniform cylinders. But of course, like paramyosin molecules, they are really two-chain α-helical coiled-coil molecules. It is commonly said that the myosin rod is similar both to paramyosin and to tropomyosin, and we have already seen in Chapter 5 that tropomyosin molecules probably have the component α-helical chains twisted into coiled coils of pitch about 137 Å. In a similar way, measurements of the near-equatorial layer lines in the high-angle diffraction patterns from a variety of muscles, although rather imprecise, all indicate that the layer line spacing is about 70 Å and, therefore, that the pitches of the two-chain coiled coils in myosin or paramyosin filaments (depending on the muscle used) are also about 140 Å. Since paramyosin and myosin both display the characteristic 144-Å axial repeat, it is tempting to conclude that this repeat may be related to the pitch of the coiled coils.

To investigate this possibility, it is necessary to consider the factors that are likely to affect the molecular packing and to relate these to the axial periodicities observed in synthetic aggregates of myosin and paramyosin. In summary, paramyosin aggregates show major repeats of 720 Å and 144 Å and a stagger of 288 Å (together with a subrepeat in paramyosin filaments of 48 Å), and myosin aggregates show major repeats of 430 Å, 144 Å, and 48 Å together with a subrepeat of 72 Å in the aggregates showing the 144-Å major repeat. A stagger of about three-eighths of 430 Å has been observed in synthetic aggregates.

As described in Chapter 4, it is possible that a favorable mode of aggregation of straight two-chain α-helical molecules might be one in which adjacent molecules are related by an axial stagger of $P/4$, where P is the pitch of one of the strands. In this way the two molecules would "fit together" to give the closest intermolecular distance as shown in Fig. 4.15. As described in Chapter 4, such a structure is thought to occur in molluscan muscle after acetone treatment (A. Elliott *et al.*, 1968a), where sampling in the equatorial and near-equatorial regions of the high-angle diffraction patterns indicates that the paramyosin molecules form a centered tetragonal arrangement. The near-equatorial layer line had a measured spacing of about 70 Å and was sampled in such a way that it seemed to represent the c axis repeat of a body-centered lattice. The axial shift between molecules would therefore be 70/2 or 35 Å, which is about one-fourth of the 144-Å paramyosin repeat. This might therefore be taken as evidence that the pitch of the paramyosin coiled coil is about 140–150 Å. It should be noted that this does not necessarily mean that equivalent points on adjacent molecules are related by a 35-Å axial stagger. An effective stagger of this amount could be produced by molecules in axial register provided that a relative 90° azimuthal rotation exists between them.

The second factor that is likely to have a major influence on the mode of aggregation of paramyosin or myosin molecules (and probably the most important factor) is clearly the amino acid sequence. From the results on the tropomyosin sequence, it seems to be almost inevitable that these other molecules will also be found to contain a periodic distribution of hydrophobic residues along the line of contact of the two chains of the coiled coil (see Trus and Elzinga, 1980). Apart from that, it is clear that the specific amino acid residues that are arranged on the outsides of the coiled-coil structure will be instrumental in defining the important interactions that can occur among molecules.

Bearing in mind these two structural features, the coiled-coil pitch and the amino acid sequence, there appear to be two obvious ways in which the known intermolecular staggers between these molecules could occur. It could be simply that the particular amino acid sequence along the myosin rods largely defines the intermolecular staggers that can occur. On the other hand, it could be that the observed intermolecular staggers are defined, in the first instance, by close packing requirements of the kind illustrated in Fig. 4.15. Specific amino acid interactions would then only serve to stabilize a particular structure, although it seems likely that the latter would be periodic with a pseudorepeat of either 144 or 430 Å.

It is convenient to consider the second of these possibilities first. If the pitch (P) is all important, then it is clear that the observed intermolecular staggers (S) must be related to P by the expression $S = NP \pm \frac{1}{4}P$, assuming that the $\frac{1}{4}P$ stagger is favorable and that adjacent molecules have identical azimuthal orientations. It seems clear from all of the available studies on two-chain α-helical coiled coils that P must be in the range 100 to 200 Å. An intermolecular stagger of say 144 Å could therefore be produced if the pitch were 4/5 × 144 = 115 Å or 4/3 × 144 = 192 Å. All other possibilities would be outside the permitted range. Taking these two possibilities for the pitch of the coiled coil, other intermolecular staggers [from $S = (N \pm \frac{1}{4})\, P$] would be as shown in Table 9.3. Significantly, in either case, the only permissible staggers that are small multiples of 144 Å are those at 430 Å and 720 Å, both of which might well be present in myosin aggregates. The observed 48-Å repeat could be explained by a pitch of 192 Å but not by one of 155 Å.

Unfortunately, the two pitch lengths discussed above do not seem to fit at all well to the observed layer line spacings from paramyosin or myosin, nor do they fit with the known tropomyosin repeat. It is therefore appropriate to ask what repeats would be generated by a pitch of 137 Å. This time, quarter staggers would give axial shifts between molecules of 445, 582, and 719 Å, all values close to multiples 3, 4, and 5 of 144 Å. A 144-Å stagger would not itself be predicted, but this need not matter, since such a repeat in paracrystals could be produced if staggers 3 and 4, 3 and 5, or 4 and 5 times 144 Å were present in the same paracrystal or filament.

Finally, there is yet another possibility if the rope model for myosin filaments is correct. Here the 40-Å subfilaments would comprise on average three myosin rods in close contact. One might then expect close packing to occur if the axial stagger is $S = (NP \pm P/3)$. In this case, a required axial stagger of exactly 430 Å could be achieved if P is 184, 161, 129, or 117 Å. On the other hand, a pitch close to that in tropomyosin and consistent with the position of the near equatorial layer line, namely one of about 133 Å, would give predicted axial staggers of about 444, 577, and 720 Å, all very close to multiples of 143 Å. A 46-Å repeat might also be observed. With all of these possibilities borne in mind, it seems to be very likely that the myosin rod has a pitch similar to that in paramyosin and tropomyosin (i.e., about 137 Å, but possibly slightly

TABLE 9.3. Possible Intermolecular Staggers (S) for Different Pitches (P) of the Myosin or Paramyosin Coiled Coil[a]

A. P = 114.7 Å	B. P = 191.1 Å	C. P = 101.2 Å	D. P = 132.3 Å	E. P = 156.4 Å
28.7 Å	47.8 Å	25.3 Å	33.1 Å	39.1 Å
86.0	143.3	75.9	99.2	117.3
143.3	334.4	126.5	165.4	195.5
200.7	430.0	177.1	231.5	273.6
258.0	525.6	227.6	297.7	351.8
315.3	621.0	278.2	363.8	430.0
372.7	716.7	328.8	430.0	508.2
430.0	812.2	379.4	496.2	586.4
487.3		430.0	562.3	664.5
544.7		480.6	628.5	742.7
602.0		531.2	694.6	
659.3		581.8	760.8	
716.7		632.4		
		682.9		
		733.5		

[a] Note that A and B are the only two values of P in the range 100 to 200 Å that are possible if the fundamental stagger (S) is assumed to be 143.3 Å and if S is given by $S = P(n \pm 1/4)$. C, D, and E are permitted values of P in addition to A and B if the fundamental stagger is taken to be 430 Å. However, C, D, and E do not generate any observed repeats except 430 Å, whereas A and B automatically generate possible staggers of 143.3, 430.0, and 716.7Å, all of which have been observed in synthetic aggregates of myosin or paramyosin.

lower, e.g., 133 Å) and that the amino acid sequence is such that shifts among molecules of 429, 576, or 720 Å (approximately) are favored, and in these situations the observed staggers are close to those that give rise to staggering of about one-third of the pitch. Only further details of the myosin sequence will allow this kind of suggestion to be tested.

9.6.4. Conclusion

Despite the title of this section, it is inevitable that this chapter must be rather inconclusive. Its purpose has been to describe some of the possibilities about molecular packing in myosin filaments rather than to assert that "myosin filaments have the following structure." Unfortunately, there is, even now, too little direct evidence about myosin packing for this kind of discussion to be more conclusive. On the other hand, one way to approach this type of problem is to define what appear to be the most promising possibilities and then to find ways in which these possibilities might be distinguished. It is hoped that this chapter will help to achieve this aim.

10

Structural Evidence on the Contractile Event

10.1. Introduction

10.1.1. Background

So far in this book the account of muscle structure has been largely concerned with the properties and static arrangements of the various myofibrillar proteins in a number of different muscle types. In this chapter and the next, we reach the heart of the matter, the nature of the contractile event itself. The aim is to describe and evaluate the available structural evidence on the contractile event and to distinguish what has actually been proven from what is commonly assumed to occur.

It is valuable to bear in mind throughout this chapter and the next a number of distinct but related questions on the contractile mechanism.

1. What evidence is there to support the idea that during muscular contraction the thick and thin myofilaments slide past each other without changing their lengths?
2. What evidence is there that leads to the idea that the contractile event is associated with the interaction of the myosin cross bridges with actin?
3. What structural evidence is there that can help to define the movements of myosin cross bridges that occur during muscular contractions?
4. What structural evidence is there that can give a clue to the number of myosin cross bridges that interact with actin at any instant of time during a contraction?

5. What steric restrictions determine the pattern of interactions of myosin cross bridges with actin?
6. What structural evidence is there that suggests how the thick and thin myofilaments might be switched on and off at the beginning and end of a contraction?
7. How are the various structural steps in the cross-bridge cycle related to the various biochemical steps involved in the hydrolysis of ATP by actomyosin?

Clearly a number of these questions have already been partially answered in previous chapters (e.g., Chapter 5 on thin filament regulation). In the present chapter some of these questions will be considered in detail, but, since they are so closely related, many of them will be considered together in the light of various pieces of structural evidence.

As a starting point three topics are briefly summarized. One is the evidence for the sliding filament model of contraction, the second deals with the nature of the force generators, and the third concerns the defined biochemical steps in the actomyosin ATPase.

10.1.2. The Sliding Filament Model, Independent Force Generators, and Cycling Cross Bridges

At the present time there appear to be five main lines of evidence that support the sliding filament model introduced in 1954 by H. E. Huxley and Hanson and A. F. Huxley and Niedergerke and that seem to discount the previous kind of contraction model which involved shortening of continuous actomyosin filaments (see A. F. Huxley, 1974, for review).

First, the proteins actin and myosin occur in separate filaments in all of the different muscle types studied so far (see Chapters 7 and 8).

Second, light and electron microscopy have indicated that A-band and thin filament lengths in all striated muscles (except possibly in *Limulus*; Chapter 8) remain constant to within 5 or 10% over a wide range of sarcomere lengths (there is no evidence for systematic changes). The amount of overlap observed between the thick and thin filaments varies linearly with the sarcomere length at which the muscle was fixed.

Third, low-angle X-ray diffraction evidence (from all muscles including *Limulus*) has shown that no major changes in axial periodicity occur in the thick and thin filaments over a wide range of sarcomere lengths (G. F. Elliott *et al.*, 1967; H. E. Huxley and Brown, 1967; Haselgrove, 1975a; Wray *et al.*, 1974), although, as described later, some very small apparent changes (~1%) in thick filament periodicity have been observed in frog sartorius muscle (Haselgrove, 1975a; H. E. Huxley *et al.*, 1980).

Fourth, the form of the X-ray diffraction pattern from an active muscle does not depend on the sarcomere length at which the muscle was stimulated but depends only on the sarcomere length of the muscle while the pattern was recorded. In other words, the pattern does not depend on the previous history of he muscle, only on its sarcomere length (Haselgrove, 1975a,b).

Finally, the form of the length–tension curve of vertebrate skeletal muscle (at least in the physiological range of sarcomere lengths) fits in well with an active tension that can increase linearly as a function of the overlap between the thick and thin filaments (Chapter 1, Fig. 1.8; Gordon *et al.,* 1966).

This last result also indicates that tension is most probably generated by a set of independent force generators distributed evenly through the A-band (except in the M-region), since the number of generators actively producing tension would then increase linearly with the amount of filament overlap. As explained in Chapter 1, Fig. 1.8, these independent force generators can reasonably be attributed to the myosin cross bridges since the shape of the length–tension curve above a sarcomere length of about 2.0 μm for vertebrate skeletal muscle follows reasonably well the number of cross bridges overlapped by the thin filaments (e.g., a plateau occurs when the thin filaments reach the M-region). One other line of evidence also supports the idea that there are independent force generators in the overlap zone. This is that the speed of shortening of a muscle under zero load ought then to be independent of the amount of overlap. That this is so has been demonstrated by A. F. Huxley and Julian (1964) and Gordon *et al.* (1966) who showed that the speed of active shortening of frog muscle under a light load from a number of different original sarcomere lengths was virtually constant. A similar conclusion can be inferred from comparative mechanical studies of muscles with different sarcomere and filament lengths. For example, in arthropod muscles (as discussed in Chapter 8) one might expect muscles with a few long sarcomeres in a given length to shorten more slowly (because of the relatively smaller number of overlap regions in series) than a similar length of a muscle containing many short sarcomeres provided, of course, that the behavior of the independent force generators in each case is identical (A. F. Huxley and Niedergerke, 1954). Such behavior has in fact been observed in arthropods (Jasper and Pezard, 1934; Atwood *et al.,* 1965). A. F. Huxley (1974) gives a very clear account of the reasons why other possible force-generating mechanisms seem less acceptable. These reasons are summarized in Chapter 11 (Section 11.2.1).

From this very brief statement of the relevant lines of evidence, it can be concluded that a reasonable working model of contraction can be taken to be one in which in most muscles the thick and thin filaments remain virtually constant in length as the sarcomere shortens and in which shortening is produced in some way by the myosin cross bridges interacting independently with the thin filaments. (Note that in this sense the term interaction does not necessarily imply physical contact.) The next step in defining the working model still further is to ask questions about the nature of this cross-bridge interaction itself.

We have seen that the LMM parts of the myosin rods are very likely buried in the backbones of myosin filaments and that, at the very most, a length equivalent to the total length of HMM (i.e., HMM S−2 + HMM S−1 $\simeq$ 600 Å) might be free to move about at the surface of these filaments (see Trinick and Elliott, 1979). It can therefore be concluded that observable changes in sarcomere length of as much as 1 μm or more could not possibly be produced by a single interaction between actin and a single cross bridge, the range of which is very likely much less than 2 × 600 Å. Similarly, more than one ATP molecule is hydrolyzed per myosin head in a contraction with a moderately high load. This kind of evidence is commonly taken to mean that myosin cross bridges must interact in a repetitive fashion with actin so that a single cross bridge may interact with several different positions along a thin filament as shortening occurs, and different cross bridges will interact with that thin filament at different times.

As described in Section 10.1.3 of this chapter, the form of the interaction is commonly assumed to be one of cross-bridge attachment to actin, cross-bridge movement on actin, thereby producing a shearing force between the thick and thin filaments, and subsequent detachment from actin, after which a cross bridge becomes ready for another attachment cycle. This conclusion is based primarily on a number of indirect structural observations that will be outlined in detail in this chapter. These are (1) that myosin cross bridges do move as a result of muscle activation, (2) that myosin cross-bridge attachment to actin filaments definitely does occur in muscles in a rigor state and in certain other defined static states, and (3) that the changes in structure that occur when a relaxed muscle is activated are in the direction of those changes that occur when a relaxed muscle goes into rigor. It will also be shown that the direct demonstration of myosin attachment to actin during muscular contraction has not been and is not likely to be an easy task. It must be said, however, that it is clearly very likely that some form of attachment occurs especially in view of the compelling result that actin activates the myosin ATPase during contraction.

In considering the type of cross-bridge cycle that might occur, it is of considerable value to bear in mind the recent fascinating evidence from studies of the biochemical kinetics of the myosin and actomyosin

ATPases (see reviews by E. W. Taylor, 1977; Trentham *et al.*, 1976; White and Thorson, 1973). This work has given an insight into the nature of the biochemical steps that occur during the hydrolysis of ATP by actomyosin and into the rates at which these steps occur. This evidence is considered in the next section.

10.1.3. Biochemical Kinetics of the Actomyosin ATPase

The science known as biochemical kinetics is itself enormous and complex, and no attempt will be made here to provide detailed justifications for the results described. A number of excellent reviews on the kinetics of the myosin and actin–myosin ATPases have been written (White and Thorson, 1973; Bagshaw *et al.*, 1974; Trentham *et al.*, 1976; Taylor, 1979; Lymn, 1979), and the reader is referred to these for details of the experiments involved.

Steady State. It is convenient to start the discussion with a summary of the results obtained by studying the steady-state ATPase of myosin and of myosin plus actin. These results can be briefly stated as follows.

1. The physiological substrate for ATP hydrolysis by myosin (in solution) is Mg-ATP (Lymn and Taylor, 1970).
2. The normal myosin ATPase rate is very low (~0.04 molecules ATP per sec per active site; Eisenberg and Moos, 1968).
3. Each myosin head will bind to both ATP and to actin, but the two binding sites involved are different (i.e., ATP can bind to the actin–myosin complex; actin can bind to myosin–ATP; Young, 1967).
4. Addition of actin to solutions containing HMM or S-1 and ATP activates the myosin ATPase to a maximum of about 10 molecules ATP per sec per active site (Eisenberg and Moos, 1970).
5. ATP has a very strong dissociating effect on the actin–myosin complex since this complex is much more dissociated in the presence of ATP than would be expected from the association rates observed in its absence (Eisenberg and Moos, 1968, 1970).
6. No change occurs in the ATPase rate at constant ionic strength when the ATP concentration changes (Eisenberg and Moos, 1968). This indicates that the dissociating effect of ATP occurs at the same site as ATP hydrolysis. On the other hand, a change in ATPase rate does occur when the ionic strength is changed, indicating that the actin–myosin interaction is partly defined by electrostatic forces.

Rapid Reaction Kinetics. The basic ATP hydrolysis reaction can be written down as

$$M + ATP + H_2O \rightarrow M \cdot ATP + H_2O \rightarrow M + ADP + P + H^+ \qquad (10.1)$$

where M is myosin (intact or HMM or S-1), ATP is adenosine triphosophate, ADP is adenosine diphosphate, and P is free inorganic phosphate. The formulae for ATP and ADP are given in Fig. 1.10.

Using techniques that stop a reaction of this kind after different times, the kinetics of the reaction can be followed. For example, Lymn and Taylor (1970) measured the formation of inorganic phosphate (either free in solution or still bound to myosin) as a function of time by using a quenched-flow apparatus of the kind shown in Fig. 10.1b. The reactants myosin and Mg-ATP were rapidly mixed, and the reaction was then stopped by an acid "quench." Their results are illustrated in Fig. 10.1c. Here it can be seen that the accumulation of inorganic phosphate is extremely rapid in the first 200 msec and that it then levels off to a steady-state rate that is similar to the steady-state ATPase rate. This early burst of phosphate accumulation showed that equation 10.1 was incomplete. It was suggested that the observations could be explained by an extended reaction

$$M + ATP \rightleftharpoons M \cdot ATP \rightleftharpoons M \cdot ADP \cdot P \rightleftharpoons M + ADP + P \qquad (10.2)$$

(Here, H_2O and H^+ have been left out for simplicity.) This reaction would explain the early burst provided that the slowest step in the reaction (the rate-limiting step) follows the formation of M·ADP·P. The consequent rapid buildup of phosphate as M·ADP·P would then be detected as the early burst in Lymn and Taylor's experiment.

Trentham *et al.* (1972) followed up this result by monitoring separately the release of free phosphate and of ADP. They found that released phosphate did not show an early burst but proceeded at the slow rate expected from the steady-state ATPase rate. Similarly, ADP was also released at a slow rate. On the other hand, it was found that the M·ADP complex (obtained by mixing ADP and M) itself dissociated at a relatively rapid rate ($\sim$2.3 sec^{-1}). As argued by White and Thorson (1973), since ADP release is rapid from its complex with myosin but slow when the initial reactants are myosin and ATP, it follows that there must be a slow step in the latter reaction that occurs before the formation of the M·ADP complex. Lymn and Taylor (1970) had already shown that an M·ADP·P intermediate must form rapidly before the rate-limiting step,

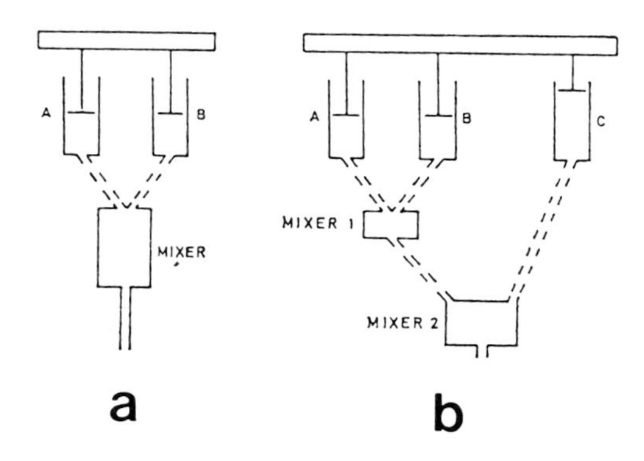

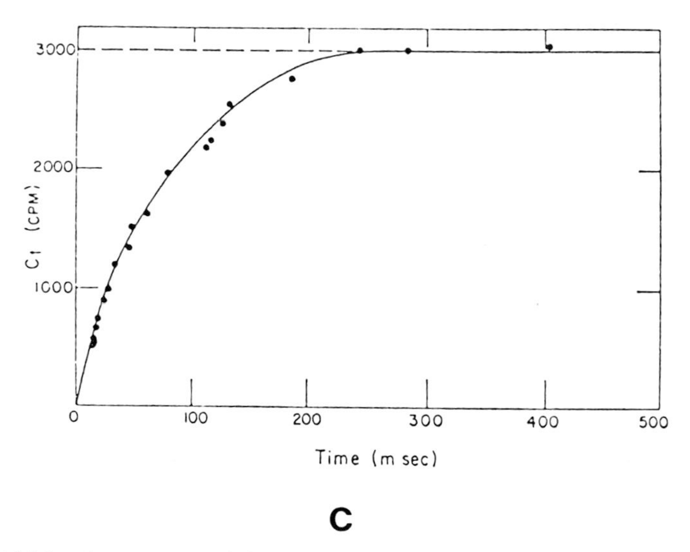

FIGURE 10.1. Apparatus used for rapid reaction kinetics: (a) stopped flow apparatus, (b) quenched flow apparatus. In (a) the reactants are held in syringes A and B and are forced into the mixing chamber within a few msec. The mixing chamber can be monitored optically to follow the progress of the reaction. In (b) the reactants in A and B as before are mixed in a small chamber (1). The flow is continuous, and the reaction proceeds in the tube between chambers 1 and 2. The reaction is quenched in chamber 2 by the addition of acid from syringe C. (From White and Thorson, 1973.) (c) Time course of phosphate accumulation from quenched flow experiments in which rabbit myosin was mixed with Mg-ATP (Lymn and Taylor, 1970). The initial ATP concentration was 32 μM. The magnitude of the burst corresponds to 1.2 moles P/mole myosin. (From White and Thorson, 1973.)

and so the whole reaction must be written

$$\mathrm{M + ATP \rightleftharpoons M^{\dagger} \cdot ATP \rightleftharpoons M^{*}ATP \rightleftharpoons M^{**}ADP \cdot P \quad M \cdot ADP \cdot P \rightleftharpoons M \cdot ADP + P \rightleftharpoons M + ADP + P} \tag{10.3}$$

assuming that phosphate release precedes ADP release. The steps with asterisks and daggers in this reaction are included because fluorescence differences have been observed (see Taylor, 1979) that indicate that some (as yet undefined) conformational difference exists in the myosin.

When actin is added to the scheme, it is clear that any of the intermediate steps in equation 10.3 could, in principle, have their counterparts with actin incorporated too. The full reaction including actin could then be written

$$\begin{array}{ccccccccccccc}
\mathrm{M} & \rightleftharpoons & \mathrm{M^{\dagger}.\,T} & \rightleftharpoons & \mathrm{M^{*}.\,T} & \rightleftharpoons & \mathrm{M^{**}.\,D.\,P} & \rightleftharpoons & \mathrm{M.\,D.\,P} & \rightleftharpoons & \mathrm{M.\,D} & \rightleftharpoons & \mathrm{M} \\
\upharpoonleft\downharpoonright & & \upharpoonleft\downharpoonright & & \upharpoonleft\downharpoonright & & \upharpoonleft\downharpoonright & & \upharpoonleft\downharpoonright & & \upharpoonleft\downharpoonright & & \upharpoonleft\downharpoonright \\
\mathrm{AM} & \rightleftharpoons & \mathrm{AM^{\dagger}.\,T} & \rightleftharpoons & \mathrm{AM^{*}.\,T} & \rightleftharpoons & \mathrm{AM^{**}.\,D.\,P} & \rightleftharpoons & \mathrm{AM.\,D.\,P} & \rightleftharpoons & \mathrm{AM.\,D} & \rightleftharpoons & \mathrm{AM}
\end{array} \tag{10.4}$$

where T = ATP and D = ADP, and unbound products are not included.

With this scheme in mind, the problem facing the kineticists was, and to some extent still is, to find the rate constants of the various alternative pathways through this reaction and, hence, to find the particular reaction pathway that is most likely to be involved in the actin–myosin ATPase. These rate constants should also make it clear which are the relatively long-lived intermediates in the reaction. One problem in doing this, however, is that the rate constants of some of these steps (e.g., that of the actin–myosin association step) are likely to be affected considerably by the local geometry of the thick and thin filament arrangements in the sarcomere. In principle, this would not matter if the kinetic experiments could be carried out on myofibrils, but this possibility is complicated by the relatively slow rates of diffusion of the reactants that would occur in such an intact system. As with the kinetic studies of the myosin ATPase, here too the necessary experiments have been carried out largely on proteins in solution. Note, however, that myofibrils are now being used for some experiments.

One of the most important observations by Eisenberg *et al.* (1972), Eisenberg and Kielley (1973), and A. B. Fraser *et al.* (1975) (who followed up the original work of Leadbetter and Perry, 1963) was that in the presence of ATP, when the ATPase rate was close to maximal (about 80%), solutions of HMM (or S-1) and actin had only about 40% (or 20%) of the HMM (or S-1) bound to actin. Their results were taken to indicate that during maximum activity only about 25% of the S-1 was bound at any instant to actin and that several S-1s can interact with a single actin monomer during the recycling time of one S-1 unit. Further experiments by Chock *et al.* (1976) have shown that the rate of recombination of actin with the myosin–products complex does not increase linearly with actin concentration but reaches a plateau at high actin concentration. This result, together with the earlier evidence on the relatively slow S-1 recycling rate, leads to the conclusion that myosin is initially in a state in which it can only bind slowly to actin and that it must rearrange to a new myosin–products complex (at a rate equal to the rate constant of the plateau) that can then bind to actin. The non-actin-binding myosin state was termed the "refractory" state by Eisenberg *et al.* (1972).

A considerable amount of information is now available about the rate constants in equation 10.3 (see White and Taylor, 1976; Lymn, 1979; Taylor, 1979), and these have led to a possible reaction sequence of the kind given in Fig. 10.2.

In the actomyosin ATPase kinetic scheme the hydrolysis step, $M^* \cdot ATP \rightarrow M^{**} \cdot ADP \cdot P$, is slow and has a rate constant similar to that of the overall ATPase rate. It is a possibility that an extra M.D.P. state in the hydrolysis step corresponds to the "refractory state." In addition, it is important to note that the association step is rather weak and is reversible to some extent so that a cross bridge could detach from actin without "using" its bound ADP · P.

With this reaction scheme in mind and with a knowledge of the appropriate rate constants and equilibrium constants, it is possible to calculate the "effective" free energy change ("basic free energy" in Simmons and Hill, 1976) associated with each step (White and Taylor, 1976; White, 1977). Clearly, the largest free energy steps are likely to be associated with the production of mechanical work. Figure 10.2 shows the positions of the largest free energy changes and indicates that the step $AM \cdot D \cdot P \rightarrow AM \cdot D + P$, having the largest effective free energy change among the 'attached steps,' is most likely to drive the work-producing part of the cross-bridge cycle.

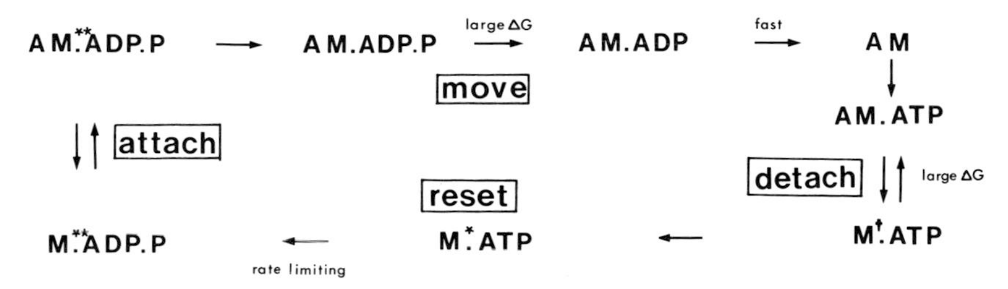

FIGURE 10.2. A possible sequence of reaction steps in the actomyosin ATPase cycle with an indication of the probable mechanical steps these correspond to. For details see text.

In terms of the structural model of cross-bridge action illustrated in Fig. 1.11 and mentioned earlier in this chapter, the various biochemical steps in the actin–myosin ATPase cycle in Fig. 10.2 can be brought together as four groups of steps:

1. The $M^{**} \cdot D \cdot P \rightarrow AM^{**} \cdot D \cdot P$ attachment step, which is weak and reversible.
2. The $AM^{**} \cdot D \cdot P \rightarrow AM \cdot D \cdot P \rightarrow A \cdot M \cdot D + P \rightarrow AM + D + P$ series of reactions, some or all of which may be associated with force generation and cross-bridge movement on the thin filaments.
3. The $AM + T \rightarrow AM \cdot T \rightarrow A + M^{\dagger} \cdot T$ detachment step (here the $^{+}$ signifies another conformational difference).
4. The $M^{\dagger} \cdot T \rightarrow M^{*} \cdot T \rightarrow M^{**} \cdot D \cdot P$ step involving some form of "refractory" state in which the myosin head is reset (but rather slowly) to be ready for a further cycle.

In the remainder of this chapter and also in Chapter 11, an account is given of the structural evidence that illuminates the nature of the cross-bridge movements that might be associated with such a cross-bridge cycle.

10.2. Structure of Defined Static States

10.2.1. Cross-Bridge Configurations in Relaxed Muscle

One of the simplest and most easily reproducible physiological states of most muscles must be the relaxed state. We have seen in a number of chapters that relaxation is associated with a low level of free calcium in the sarcoplasm. Since fixation of initially relaxed specimens for electron microscopy is itself known to induce some muscular activity and certainly produces structural changes (Reedy and Barkas, 1974; Sjöström and Squire, 1977b), it is inevitable that the only techniques that are likely to give reliable data on the relaxed state are those that are nondestructive: they must be applicable to "live" muscles in as near physiological conditions as possible. The most useful technique in this respect has so far been X-ray diffraction, and for this reason the bulk of the evidence presented in the rest of this chapter is from X-ray diffraction experiments.

Despite the apparent simplicity of the relaxed state, it is nonetheless salutory to have to acknowledge that the cross-bridge arrangement in this state is even now uncertain. In Chapter 7, a description was given of one type of approach to this problem in the case of relaxed vertebrate skeletal muscle. It will be remembered that the procedure is to account for the observed intensity distribution on the various myosin layer lines (at spacings of 429 Å, 216 Å, 143 Å etc.) by calculating the expected intensity from a model structure which is then varied to give a "best fit." The problems were that this modeling depends on a variety of factors (e.g., cross-bridge shape) that are themselves rather uncertain. Nevertheless, one or two consistent conclusions were reached in the various analyses that have been done. The most notable of these is that it now seems to be very likely that the myosin heads in relaxed muscle are situated such that when projected onto the filament axis in a radial direction their axial extent is very roughly 100 Å (Squire, 1975; Wray *et al.*, 1975; Worcester *et al.*, 1975).

What this means in terms of cross-bridge arrangement depends largely on the shape and distribution of the two heads within a single cross bridge. If the individual (S-1) heads are over 120 Å long, 30 to 40 Å in diameter, and roughly ellipsoidal in shape, then the configurations shown in Fig. 10.3a might be possible for vertebrate skeletal muscle. These comprise axial tilts of either 40° or ±20°. On the other hand, if the heads have a bulbous region at one end (about 50 to 60 Å in diameter; A. Elliott and Offer, 1978), then the tilt angle could well be much less (e.g., only 20 to 30° as in Fig. 10.3b). That the heads are more or less lined up

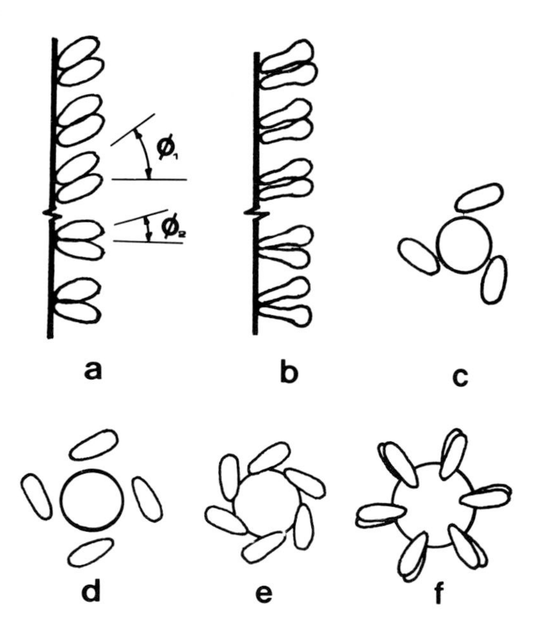

FIGURE 10.3. (a and b) In relaxed vertebrate skeletal muscle the myosin cross bridges are very probably tilted axially away from 90° to the thick filament axis (vertical lines). ϕ is probably between 20 and 40° depending on the exact shape of the myosin head. It is not yet clear whether both heads in one cross bridge tilt in the same direction (top) or in opposite directions (bottom). (c) Transverse view of a possible arrangement of the myosin heads at one axial level along the three-stranded myosin filaments in vertebrate skeletal muscle in the relaxed state. (d–f) Similar diagrams for the cross bridges from *Limulus*, *Homarus* (lobster), and *Pecten* muscles in the relaxed state according to Wray *et al.* (1975). In (f) the two heads in one cross bridge were taken to tilt axially in opposite directions. The symmetry for *Homarus* is hypothetical.

along the helices that give rise to the 429 Å layer line (see Fig. 7.34) seems to be evident from the relatively high intensity of this layer line. Also, assuming that the thick filaments in this case are three-stranded, then the center of mass of the cross bridges must be at a radius of about 120 to 130 Å as indicated in Fig. 10.3b.

This kind of model for the relaxed cross bridge can be compared with the results of Wray *et al.* (1975) on the cross-bridge locations in various invertebrate muscles. In *Limulus, Pecten,* and *Homarus* (lobster), it was found that despite some specific differences in cross-bridge location, all of these seemed to give diffraction patterns that could be accounted for most readily in terms of cross bridges that were tilted axially by between 20 and 40° and slewed aximuthally by up to 70°. The cross-bridge configurations are shown in Fig. 10.3 (d–f). In the case of *Pecten,* it was suggested by Wray *et al.* (1975) that the two heads in a single cross bridge might be tilted in opposite directions by about 20° so that one points 20° up the filament axis and one 20° down. It has also been suggested (Haselgrove, personal communication) that such a cross-bridge arrangement might also account well for the diffraction evidence from frog muscle. On the other hand, the fact that in *Pecten* (as described in Chapter 8) there may be periodic perturbations of the 146-Å cross-bridge repeat, thus giving rise to a 292-Å periodicity, adds considerable complications to the interpretation of the layer line data in terms of cross-bridge position, and one might reasonably question the conclusion of Wray *et al.* (1975) in the case of *Pecten.*

One of the features about which X-ray diffraction can give little evidence concerns the direction in which the cross bridges are tilted (i.e., are they toward or away from the M-band?). Of course, if the two heads in one cross bridge point in opposite directions, then the problem does not arise. But if both heads are tilted axially in the same direction, then it is clearly crucial to our understanding of the cross-bridge movements that take place when a muscle is activated to know which direction this is. Unfortunately, there is little direct data on this point, and it can only be suggested, for reasons that will become apparent in later parts of the chapter, that the outer ends of the bridges may tilt towards the M-band.

Finally, it should be noted that, in addition to the axial diffraction data, a considerable amount of equatorial diffraction data has been obtained from relaxed frog muscle (H. E. Huxley, 1968; Haselgrove and Huxley, 1973; Lymn and Cohen, 1975; Haselgrove *et al.,* 1976). It was mentioned in Chapter 7 that the hexagonal array of thick and thin filaments gives rise to a characteristic diffraction pattern on the equator. The most prominent reflections are the $10\bar{1}$ and $11\bar{2}$ reflections from the hexagonal lattice, but some higher order weak reflections can also be seen. It will be remembered from Chapter 7 that the effect on these equatorial reflections of shortening the sarcomere length is to increase the relative intensity of the 11$\bar{2}$ reflection relative to that of the 10$\bar{1}$, since a greater length of the thin filaments is becoming ordered at the trigonal points in the hexagonal lattice. Various estimates of the relative intensites of the 10$\bar{1}$ and 11$\bar{2}$ reflections from frog muscles at rest length (between 2 μm and 2.2 μm) have been made (H. F. Huxley, 1968; Haselgrove and Huxley, 1973; Matsubara *et al.*, 1975; Lymn and Cohen, 1975; Haselgrove *et al.*, 1976), and despite some variation, it is generally agreed that the ratio $I_{10\bar{1}}/I_{11\bar{2}}$ is about 2.5. The measured intensities of the observed higher order reflections (Lymn and Cohen, 1975) are given in Table 10.1. The significance of this equatorial evidence from relaxed muscle is more apparent when it is compared with the corresponding data from rigor and contracting muscles. A detailed discussion of it will therefore be given in Section 10.3.5, where the results from insect flight muscle will also be considered.

10.2.2. X-Ray Diffraction Evidence on Rigor Muscle

Vertebrate Skeletal Muscle. The second most well-defined static muscle state is that of rigor. It is induced by the removal of ATP from the sarcoplasm so that the dissociation step in the cycle of actin–myosin

TABLE 10.1. Intensity of the Equatorial Reflections in X-Ray Patterns from Relaxed and Rigor Frog and Insect Flight Muscles

Indices of reflection (*hkl*)	Insect muscle		Vertebrate muscle	
	Rigor[a]	Relaxed[b]	Rigor[c]	Relaxed[d]
10$\bar{1}$	100 (100)[e]	100[e]	100[e]	100[e]
11$\bar{2}$	—	—	400	44
20$\bar{2}$	123 (114)	50	—	9
21$\bar{3}$	1.2	—	—	9
30$\bar{3}$	0.8	—	—	12
22$\bar{4}$	7.9	—	—	20 (22$\bar{4}$ + 31$\bar{4}$)
31$\bar{4}$	8.0	—	—	
40$\bar{4}$	0.5	—	—	—
32$\bar{5}$	3.4	—	—	—
41$\bar{5}$	3.4	—	—	—
50$\bar{5}$	0.7	—	—	—

[a] From Holmes *et al.* (1980); peak intensities measured. Bracketed values are from Miller and Tregear (1972).
[b] From Miller and Tregear (1972).
[c] From Haselgrove and Huxley (1973); see Table 10.2.
[d] From Lymn and Cohen (1975).
[e] All intensities are normalized relative to the 10$\bar{1}$ reflection.

ATPase cannot take place. We saw in Chapter 3 that the rigor state can be induced by glycerol (which extracts the ATP), by iodoacetate (which poisons the resynthesis of ATP), or simply by exhausting the ATP supply through continued stimulation. Although rigor can hardly be said to be a normal physiological state (it occurs at death), it is nonetheless reversible in the sense that bathing a glycerinated muscle with a solution containing ATP releases the cross bridges, and then the subsequent addition of calcium allows the muscle to "work" once again. On the other hand, it has been found to be very difficult to induce a fully "relaxed" state indistinguishable from that in live relaxed muscle by addition of ATP to glycerinated vertebrate skeletal muscle, although considerable advances in that direction have been made (Rome, 1972a,b).

X-ray diffraction studies of frog sartorius and semitendinosus muscles in rigor (H. E. Huxley and Brown, 1967; H. E. Huxley, 1968; Haselgrove and Huxley, 1973; Haselgrove, 1975a) have provided clear evidence that in this state the thin filaments are heavily labeled with myosin cross bridges. This is evident from a number of results.

1. The strong myosin layer lines at orders of 429 Å in patterns from relaxed muscle (Fig. 7.5) are very much weaker, indicating that considerable movement of the myosin cross bridges away from their relaxed positions on the myosin filament helix has occurred (Fig. 10.4).

2. The actin layer lines at spacings that are orders of about 360 to 370 Å (Fig. 10.4) all become much stronger than in patterns from relaxed muscle, and occasionally the peaks on them appear to move closer to the meridian. As will be discussed later, this indicates that extra material is labeling the outsides of the thin filaments.

3. A considerable change occurs in the relative intensities of the equatorial reflections, such that the ratio $I_{10\bar{1}}/I_{11\bar{2}}$ becomes about 0.2 to 0.3 at a sarcomere length of 2 to 2.2 μm compared with 2.5 for relaxed muscle (see Table 10.2). This dramatic increase in the intensity of the $11\bar{2}$ reflection relative to the $10\bar{1}$ is a clear indication that there is more material associated with (i.e., near to) the trigonal points of the hexagonal filament lattice (or at least the $11\bar{2}$ planes) in rigor than in relaxed muscle (see Figs. 7.2 and 7.3). Electron micrographs of transverse sections of rigor muscle support this conclusion (H. E. Huxley, 1968).

4. A new set of layer lines at axial spacings of 366 Å, 241.8 Å, 186.4 Å, and 144.5 Å (Haselgrove, 1975a) appears in the diffraction patterns at a radial position much less than that of the normal actin layer lines (Fig. 10.4a). The possible origin of these layer lines will become apparent later.

In addition to the general changes in intensity that occur in the diffraction patterns from rigor muscles, some very interesting spacing changes have also been detected (Haselgrove, 1975a). Haselgrove found

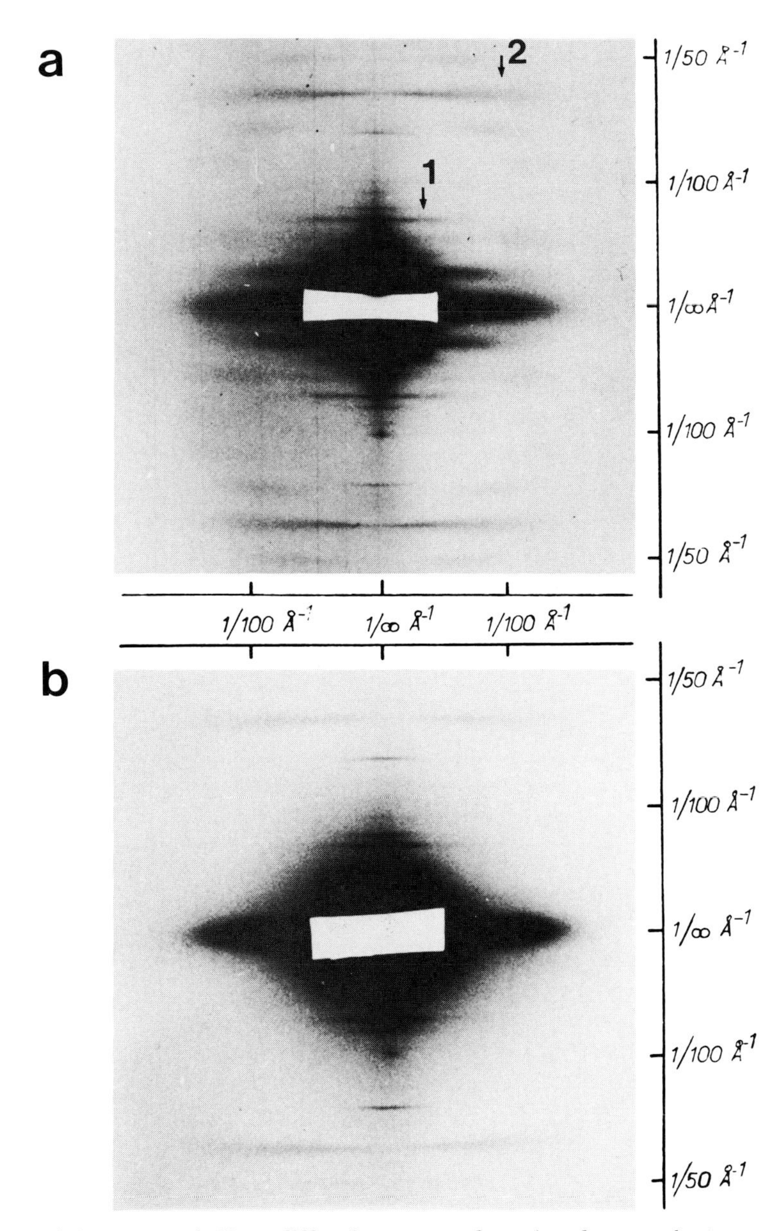

FIGURE 10.4. Low-angle X-ray diffraction patterns from rigor frog muscles (courtesy of J. C. Haselgrove). (a) Frog sartorius muscle at rest length. The vertical arrows indicate the two sets of layer lines present. Set 1 is sharp and close to the meridian, and set 2 is more diffuse and further from the meridian. See text for discussion. (b) Pattern from frog semitendinosus muscle stretched beyond overlap length. The two sets of layer lines indicated in (a) are now absent. (See Haselgrove, 1975a).

TABLE 10.2. Average Intensity Ratios $I_{10\bar{1}}/I_{11\bar{2}}$ for Frog Muscles at Rest, during Isometric Contractions, and in Rigor as a Function of Sarcomere Length[a]

Sarcomere length (μm)	Muscles at rest		Contracting muscles		
	Precontrol photographs	Postcontrol photographs	Uncorrected values	Corrected values	Muscles in rigor
1.8			0.62 ± 0.13 (2)	0.51 ± 0.06 (2)	
1.9			0.62 ± 0.14 (4)	0.50 ± 0.15 (3)	
2.0	1.74 ± 0.25 (7)	1.77 ± 0.15 (4)	0.72 ± 0.10 (4)	0.53 ± 0.15 (3)	0.28 ± 0.05 (3)
2.1	1.95 ± 0.25 (7)	1.91 ± 0.3 (6)	0.73 ± 0.09 (8)	0.56 ± 0.12 (6)	0.25 ± 0.07 (4)
2.2	2.63 ± 0.3 (5)	2.26 ± 0.3 (4)	0.99 ± 0.12 (4)	0.78 ± 0.15 (4)	0.26 ± 0.06 (7)
2.3	2.93 ± 0.5 (4)	2.87 (1)	1.56 (1)	1.16 (1)	0.31 ± 0.05 (5)
2.4	2.64 (1)	2.81 ± 0.7 (4)	1.57 ± 0.24 (4)	1.19 ± 0.25 (3)	0.47 ± 0.10 (5)
2.5	3.84 ± 0.8 (4)	3.92 ± 0.2 (3)	1.53 ± 0.12 (4)	1.28 ± 0.2 (4)	0.70 ± 0.18 (4)
2.6	5.12 ± 1.0 (2)	4.89 (1)	1.88 ± 0.20 (2)	1.50 ± 0.12 (2)	0.99 ± 0.16 (5)
2.7	6.29 ± 0.5 (4)	6.42 ± 0.3 (2)	2.12 ± 0.03 (2)	1.64 ± 0.12 (2)	1.19 ± 0.05 (2)

[a] Averaged values are given for muscles with sarcomere lengths within ±0.05 μm of the given sarcomere length and expressed as mean ± SEM; number of measurements is given in parentheses. From Haselgrove and Huxley (1973).

that the strong meridional reflection at a spacing of 143.4 Å in patterns from relaxed muscles (Table 10.3) was shifted slightly to a spacing of 144.6 Å in patterns from rigor muscles. Furthermore, this change in spacing also occurred in frog semitendinosus muscles that were stretched to a long (nonoverlap) sarcomere length (3.7 μm) before rigor treatment. It was concluded from this result that some kind of specific structural change in the myosin filaments must be taking place and that this change was not dependent on the interaction of the thick filaments with actin. Some possible explanations for this observation will be described in Section 10.3.4.

Insect Flight Muscle. Miller and Tregear (1972) have made a detailed X-ray diffraction study of the structure of glycerinated insect flight muscle (from *Lethocerus*) in the relaxed and rigor states. The situation in this case is rather different from that in vertebrate skeletal muscle in that here the actin and myosin filaments have axial repeats that are very closely related. For this reason, some of the prominent layer lines from the two filament types (see Fig. 8.8 and Section 10.4.1) have the same axial spacings (e.g., at 385 Å). The layer line spacings are listed in Table 10.4 for relaxed and rigor muscle. Despite these differences from frog muscle, very similar qualitative differences occurred in patterns from relaxed and rigor muscles. In particular, the myosin reflections were weaker, but the outer layer lines caused by the actin helix (includ-

TABLE 10.3. Layer Line Spacings in X-Ray Patterns from Relaxed and Rigor Frog Muscles

Relaxed		Rigor	
Myosin layer lines	Actin layer lines	Near meridian	Actin layer lines
(Å) 429.2 (±4)	420 (±6)	—	—
—	—	366	366
—	—	241.8	—
216.1	—	—	—
—	—	186.4	186
(143.4)	—	144.5	—
—	—	—	120.2
107.5	—	—	—
86.0	—	—	87.8
71.5	—	—	69.6
—	59.5	—	59.06
—	51.2	—	50.96

[a] These are at a radial position to be expected from a structure with the symmetry of the actin helix at a radius of about 60 to 80 Å. Reflections near the meridian would not be expected from such a helix. Layer line spacings are accurate to about 1%. Spacings are from Haselgrove (1975a).

TABLE 10.4. Layer Line Spacings in X-Ray Patterns from Insect Flight Muscle, *Limulus* Muscle and *Pecten* Cross-Striated Adductor Muscle in Both the Resting and Rigor States

Insect flight muscle[a] (Å)		*Limulus* muscle[b,c] (Å)			*Pecten* adductor muscle[c,d] (Å)			
			Rigor		Relaxed		Rigor	
Relaxed	Rigor	Relaxed	Layer line	TM/TN	Thick	Thin	Thick	Thin
					464 (S)			
		438						
389(±2)	385	(383)	383 (S)	(383 S)		367		362 (372)
230(±1)								
		219						
					200 (W)			
	195	(190)	190	(190 S)		192		191 (199)
145.1(±0.3)	145	146 (S)	146[e]		145.3 (VS)[e]		144.7	
	129		128 (S)	(128 S)				(128)
	110	109			112.4 (W)			
			96 (S)	(96 S)	96.6 (S)			
					83.9 (VW)			
			73[e]	(77 W)	72.6 (M)		72	69.3
				(64 S)	67.3 (VW)			
59.2(±0.3)	59				58.8 (M)	58.9		57.9
	55							
51	51					51.2		50.3

[a] From Miller and Tregear (1972).
[b] From Wray *et al.* (1974). Spacings are not given explicitly.
[c] S, strong; W, weak; M, medium; V, very.
[d] From Millman and Bennett (1976). Reflections from thick and thin filaments. All bracketed spacings are thought to relate to the tropomyosin/troponin (TM/TN) repeat.
[e] Here the 146-Å reflection is split across the meridian, but the second order at 73 Å is not split.

ing the 59-Å reflection) were very much stronger in patterns from rigor muscles (Fig. 10.5a). The latter could be accounted for reasonably well in terms of a helical distribution of scattering centers labeling the outsides of the thin filaments (Miller and Tregear, 1972) although the inner part of the 385-Å layer line appeared to be much too strong.

Corresponding differences were also observed in the equatorial diffraction patterns (Miller and Tregear 1970, 1971, 1972). In this case, the most prominent equatorial reflections were the 101 and the 20$\bar{2}$ reflections. Remember that in this muscle the thin filaments lie exactly between two thick ones (Fig. 8.2b), an arrangement that considerably weakens the 11$\bar{2}$ reflection so that it is masked by the strong 20$\bar{2}$ peak. The relative intensities of the 10$\bar{1}$ and 20$\bar{2}$ reflections (i.e., $I_{10\bar{1}}/I_{10\bar{2}}$) were about 2 and 0.8 to 0.9 respectively in patterns from relaxed and rigor muscles. As in the case of vertebrate skeletal muscle, such a change is consistent with there being an increase in rigor muscle of the amount of material near to the thin filament positions in the lattice.

Finally, it should be noted that the structure of rigor insect muscle is so regular that all of the low-order layer lines (e.g., at 385 Å and 190 Å) are beautifully sampled, and the whole of the low-angle diffraction pattern can be indexed in terms of a regular three-dimensional crystalline array (see Section 10.4.1).

Other Muscles in Rigor. Diffraction patterns from *Limulus* muscle in rigor (Wray *et al.*, 1974) are in some ways similar to those from insect flight muscle in that there is a very strong series of actin layer lines that index on an axial repeat of 2 × 383 Å (Table 10.4). However, an unusual feature is that the 146-Å reflection characteristic of the thick filaments is largely off-meridional, although the second order of this, at 73 Å, is apparently truly meridional. The equatorial reflections index on a hexagonal lattice of side about 650 Å.

Wray *et al.* (1974) explained the splitting of the 146-Å reflection across the meridian in terms of a systematic axial stagger (presumably by an odd multiple of 73 Å) between adjacent thick filaments. It was suggested that, since in this muscle the thick filaments appear to be free to move longitudinally, they may in rigor adopt axial positions that are favorable for cross-bridge interaction with the thin filaments.

Another notable feature of the diffraction patterns from *Limulus* muscle is the prominence of the series of meridional reflections thought to be related to the tropomyosin–troponin repeat, orders 1, 2, 3, 4, and 6 of a 385-Å axial repeat have been observed.

X-ray diffraction patterns from pecten striated adductor muscle in rigor (Millman and Bennett, 1976) are said to be similar to those from vertebrate skeletal muscle in that the actin layer lines (especially the first actin layer line at about 360 to 370 Å and the "59-" and "51-Å" reflections) are very strong (Table 10.4). The equatorial reflections, although being rather diffuse and thus indicating poor development of the filament array, were also consistent with cross-bridge attachment to the thin filaments, since the $11\bar{2}$ and $20\bar{2}$ reflections together increased in prominence relative to the $10\bar{1}$.

Summary. From this brief account it is clear that the results from many different muscle types in rigor are very consistent and indicate unambiguously that cross-bridge labeling of the thin filaments must occur. The diffraction features that are generally characteristic of rigor muscle (relative to relaxed muscle) are the following.

1. There is a change in intensity of the 59-Å actin layer line (especially close to the meridian) (Fig. 10.5a) and also of all the other low-order actin layer lines.

2. The low-order myosin layer lines apart from the orders of the 143- to 146-Å myosin subunit repeat always become much weaker unless some of the myosin and actin layer lines happen to coincide as in insect muscle.
3. The equatorial reflections change in intensity such that the ratios $I_{10\bar{1}}/I_{11\bar{2}}$ in muscles like frog sartorius muscle and $I_{10\bar{1}}/I_{20\bar{2}}$ in muscles like insect flight muscle are dramatically reduced.

But apart from these similarities among the rigor patterns from different muscles, some characteristic differences also occur and these are very significant.

1. In patterns from frog muscles, a new set of layer lines appears rather close to the meridian and with an axial repeat of about 720 Å.
2. In patterns from insect flight muscles, the meridional end of the 385-Å actin layer line is extremely intense.
3. In patterns from *Limulus* muscle, the 146-Å reflection is split across the meridian.

As mentioned above, this third effect is probably caused by a systematic axial stagger between myosin filaments of an odd multiple of 73 Å (Wray *et al.*, 1975), but little is known about the details of the structure of *Limulus* muscle, and little more can be concluded as yet. On the other hand, it is possible to analyze in detail the difference in (1) and (2) above. This is done in the next two sections.

10.2.3. Modeling of the Rigor State: Introduction

So far in this discussion of rigor muscle, the X-ray diffraction results from various muscles have been discussed in a qualitative way, and it has been asserted that some form of attachment of myosin cross bridges to the thin filaments must occur. Here we consider in more detail the form of this attachment.

The first real indication of the cross-bridge configuration in rigor muscle was from the electron microscopy of rigor insect flight muscle by Reedy *et al.* (1965). Here it was evident (Fig. 8.3) that the thin filaments were labeled by objects angled at about 45° to the filament axis. It was also apparent that these objects were pointing away from the M-band on each thin filament. After that early evidence, the structure of decorated thin filaments was worked out using the 3-D reconstruction method (Moore *et al.*, 1970; see Chapter 5), and more details were obtained of the mode of attachment of myosin heads to actin in a situation analogous to the rigor state (ATP absent). It was then clear that the characteristic arrowhead appearance of decorated thin filaments could only be produced if the myosin heads were tilted axially by about 50° and slewed aximuthally by about 60° as in Fig. 5.11.

To verify that this kind of structure actually approximated to the situation in intact rigor insect muscles, Miller and Tregear (1972) computed the diffraction pattern of an actin helix labeled with myosin heads. The model they used is as indicated in Fig. 10.5b. The myosin head was modeled as a cylinder (represented by overlapping spheres) about 110 to 120 Å long and about 50 Å in diameter, and each actin monomer by a sphere of radius 24 Å and centered at a helix radius also of 24 Å (Chapter 5, Section 5.2.2). By determining the layer line intensities as a function of the tilt (ϕ) of the myosin heads and the aximuthal slewing angle (θ), Miller and Tregear obtained a reasonable fit to parts of their data from rigor insect flight muscle with $\phi = -\theta = 35°$ (θ measured positive in the same sense as the genetic helix of pitch 59 Å; Fig. 10.5b). It is reasonable to conclude from this evidence, from the results of Moore *et al.* (1970), and also from the appearance of Reedy's (1968) transverse and longitudinal sections of rigor insect flight muscle, that the rigor interaction involves both tilting and slewing of the myosin heads on the thin filament. The actual angles involved cannot really be defined more than saying that both ϕ and $-\theta$ are roughly 45°.

One necessary feature of Miller and Tregear's computations was that in insect muscle, as in all muscles, there are insufficient myosin cross bridges to label all of the actin monomers on each of the thin filaments. For example, in a cross section through the A-band of insect muscle, there are three thin filaments to every thick filament (assumed six-stranded), so the number of actin monomers in the 1150-Å axial repeat is 126 (i.e., $(1150/27.5) \times 3$), and there are at most 48 (i.e., $1150/145 \times 6$) cross bridges to interact with them (or twice that number, 96, of myosin heads). Similarly, in a 430-Å repeat of the basic unit cell in vertebrate skeletal muscle which contains two thin filaments and one three-stranded thick filament, there would be nine cross bridges (or 18 heads) and about 32 actin monomers. These figures mean that at most only about three-fourths of the actin monomers in insect muscle and about two-thirds of those in vertebrate skeletal muscle could be labeled with myosin heads in rigor muscle. In addition, it is apparent from the flared X cross-bridge appearances seen in insect flight muscle by Reedy (1967, 1968) and the corresponding results from vertebrate skeletal muscle (Chapter 7) that there are strong indications that, of the available crossbridges, as few as two-thirds may actually attach to the thin filaments in a rigor muscle. For this reason, one would expect less than half of the actin monomers in the overlap region of the sarcomere to be labeled.

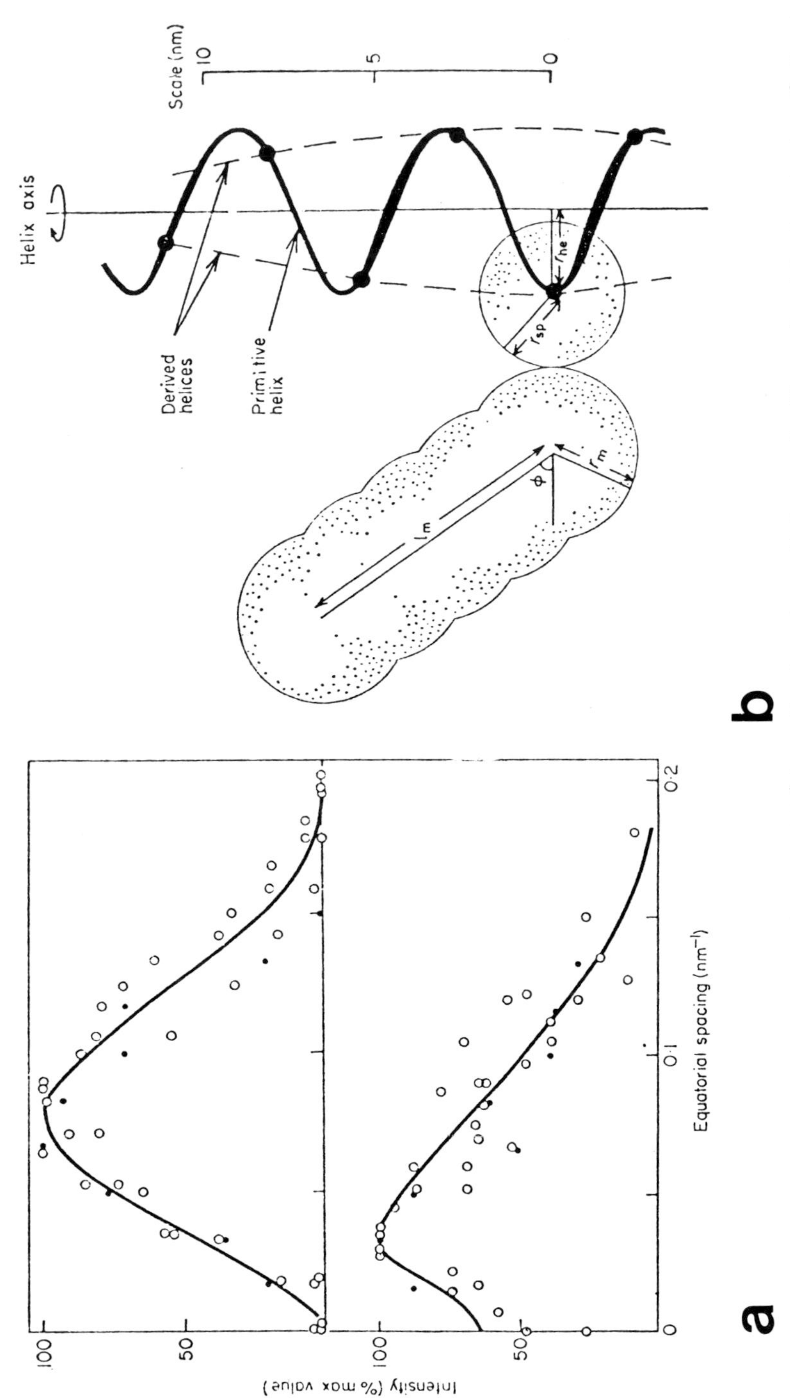

FIGURE 10.5. (a) The intensity distribution along the 59-Å layer line from insect flight muscle either relaxed (top) or in rigor (bottom) according to Miller and Tregear (1972); ○, glycerol extracted muscle; ●, live muscle. (b) The model of the cross-bridge-labeled actin filament used by Miller and Tregear (1972) to model their results from rigor insect flight muscle. A primitive helix of pitch 59 Å and subunit translation 27 Å has spherical repeating units of radius 24 Å, and these are positioned at a helix radius of 24 Å (r_{sp} and r_{he}, respectively). The myosin head is attached with an axial tilt ϕ and an azimuthal tilt θ (not shown). The best fit was obtained with $\phi = -\theta = 35°$, $r_m = 28$ Å, $l_m = 112$ Å, and myosin mass = 2 × actin mass.

In the analysis carried out by Miller and Tregear (1970), this incomplete labeling was modeled, on the assumption that the distribution of heads along the thin filaments was fairly random, by reducing the effective density of myosin heads placed on every actin monomer. As they noted, random labeling of actin monomers could be considered a form of substitution disorder (Guinier, 1963) for which the diffracted intensity is determined by the average occupant of a lattice point. In fact, the density they chose was based on the assumption that about two-thirds of the actin monomers would be labeled in rigor. Although we have seen that the true figure may actually be much less than that, their modeling accounted reasonably well for a number of features of the rigor diffraction pattern. It would also account approximately for the general distribution of intensity on the actin layer lines in other rigor muscles. On the other hand, it would not account for the specific differences that exist among the rigor patterns from different muscles, some of which were summarized at the end of the last section. The explanation for these differences seems to lie in the fact that the distributions of the cross bridges labeling the thin filaments are not random, which would make the diffraction patterns similar, but clearly follow particular groupings. Different labeling patterns in different muscles would then account for the specific differences among the various rigor diffraction patterns.

10.2.4. Modeling of the Rigor State: Insect Flight Muscle

It is probably simplest to consider this idea first in terms of the structure of rigor insect flight muscle. In the discussion given in Chapter 8 it was made clear that cross bridges do not, in fact, appear to label actin filaments randomly; they seem only to label the thin filaments at specific actin "target" areas (Reedy, 1968). In the case of the insect filament lattice, it is likely that each thin filament is approached at the same level by cross bridges from two equivalently placed thick filaments (Fig. 10.6). In addition, the actin and myosin repeats fit together systematically, and the actin filaments themselves have an approximate twofold rotation axis. This means that more or less equivalent target areas are presented to the two thick filaments at regular 385-Å axial intervals. With this structure in mind, it is very clear that labeling of the thin filaments in rigor insect muscle cannot be random but must be associated with a systematic variation every 385 Å (Squire, 1972). Barrington-Leigh *et al.* (1977) and Holmes *et al.* (1980) have suggested that the thin filament labeling might be described usefully in terms of a "probability" of myosin attachment to the thin filaments. A random distribution would then be associated with a uniform attachment proability along each thin filament, but attachment to specific target areas would be described by a

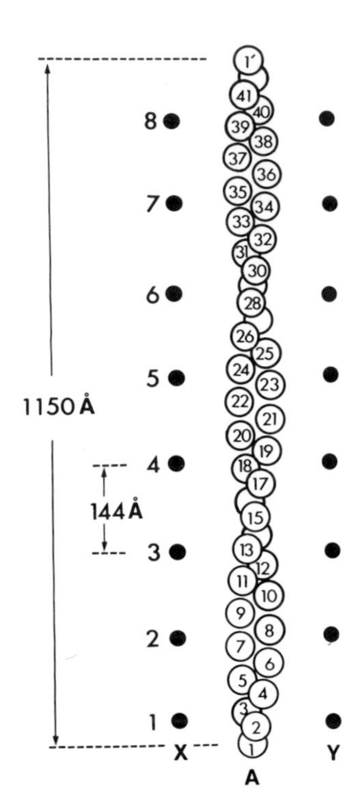

FIGURE 10.6. Representation of (A) the structure of the 28/13 helix of actin monomers as in insect flight muscle, showing the 42 actin monomers in one-half of the 2300-Å repeat of the actin–myosin assembly and the 8 axial positions of the cross-bridge levels on the two myosin filaments (X and Y) on each side of A. For details see text. (From Squire, 1979.)

periodically varying attachment probability function with a fundamental axial repeat of 385 Å. The effect of such a periodic attachment function on the diffraction pattern can be seen in Fig. 10.7c where the structure is modeled for optical diffraction. The essence of such a probability function is that one cannot say definitely that a cross bridge will have a specific attachment site, but, on average, attachment will be more likely on a group of actin monomers in the target area than on those between target areas. The model in Fig. 10.7c therefore has only target area monomers labeled, but there is a more or less random attachment within each target area to give, in this example, a total of 50% labeling of the actins. The effect of this 385-Å repeat on the diffraction patterns compared with Fig. 10.7b (random attachment) is, not surprisingly, to emphasise considerably the 385-Å reflection. The subsequent orders would not be altered very much if the probability function approximates to a sinusoidol function that would only give rise to a single diffraction order (here at 385 Å).

An additional complexity can be envisaged when it is recognized that as well as being limited by the actin target areas the cross-bridge attachment may also be restricted by the fact that the cross bridges them-

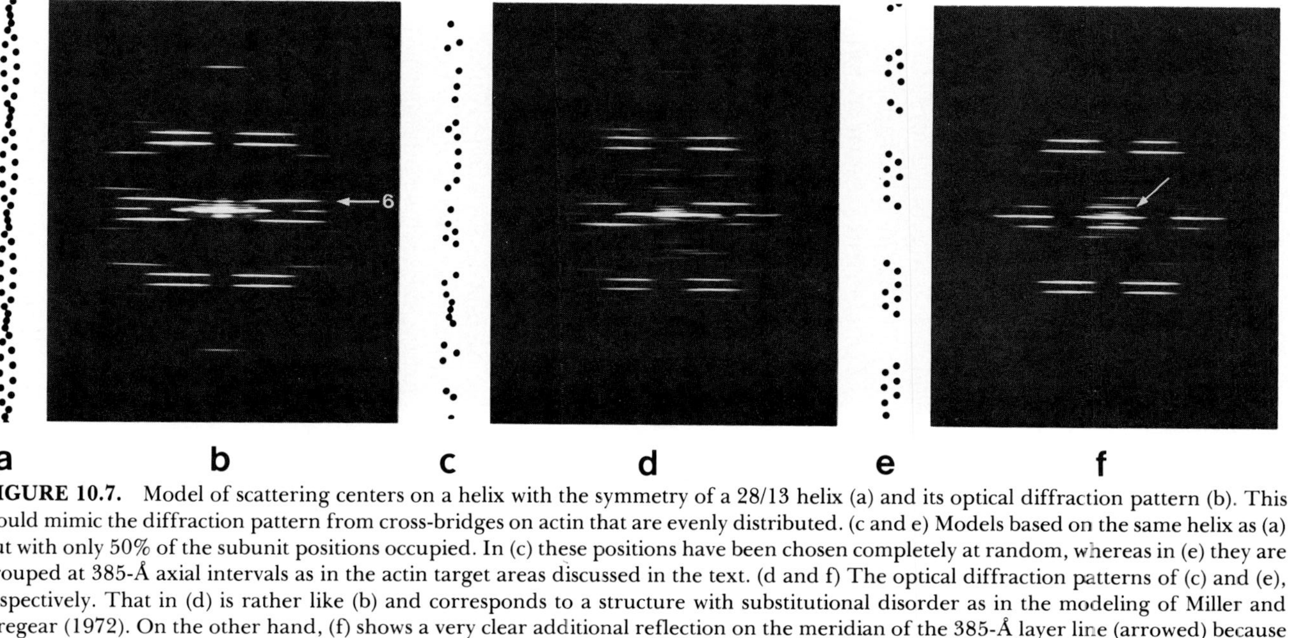

FIGURE 10.7. Model of scattering centers on a helix with the symmetry of a 28/13 helix (a) and its optical diffraction pattern (b). This would mimic the diffraction pattern from cross-bridges on actin that are evenly distributed. (c and e) Models based on the same helix as (a) but with only 50% of the subunit positions occupied. In (c) these positions have been chosen completely at random, whereas in (e) they are grouped at 385-Å axial intervals as in the actin target areas discussed in the text. (d and f) The optical diffraction patterns of (c) and (e), respectively. That in (d) is rather like (b) and corresponds to a structure with substitutional disorder as in the modeling of Miller and Tregear (1972). On the other hand, (f) shows a very clear additional reflection on the meridian of the 385-Å layer line (arrowed) because of the grouping of the scattering centers in (e) into the target areas. (From Squire, in preparation.) Layer line numbering based on 2300-Å repeat (see Figs. 10.11 and 10.12).

selves originally approach the thin filaments at regularly spaced 145-Å axial intervals. One might expect to see some evidence for this repeat in the distribution of attached cross bridges within each actin target area. But apparently, both in insect flight muscle (Tregear *et al.,* 1977; Holmes *et al.,* 1980) and in certain crustacean muscles (Wray *et al.,* 1978) where evidence for the existence of the actin modulation is strong, there is only a weak trace of the 145-Å repeat in the distribution of attached cross bridges. This situation has been interpreted to mean that the attached cross bridges in these rigor muscles have to some extent "forgotten" that they originate from the thick filaments. But this is not necessarily so for reasons given below (Squire, 1979).

As described in Chapter 8, beautiful electron micrographs of rigor insect flight muscle have been obtained by Reedy (1968). These show a characteristic flared X structure (Fig. 10.8c) in thin transverse sections (less than 200 Å thick) and a double chevron pattern in longitudinal sections (Fig. 10.8a). This particular chevron pattern was seen in myac layers—thin longitudinal sections comprising alternating thick and thin filaments as in Fig. 10.8. Such longitudinal sections of rigor muscle showed a cross-bridge-labeling pattern with a long repeat of about 1150 Å, which is eight times the myosin subunit repeat of 144 Å and three times the actin 385-Å near repeat (Fig. 10.8). As shown in Fig. 10.6, the actin filaments in insect muscle approximate a 28/13 helix of actin monomers (Miller and Tregear, 1972) with a true repeat of about 770 Å and a near repeat of 385 Å. The 1150-Å rigor repeat is therefore a beat period between the myosin and actin repeats (144 Å and 385 Å). But the true repeat of the structure is actually 2 × 1150 Å (2300 Å), since the actin true repeat of 770 Å and the myosin 1150-Å true repeat are both orders of 2300 Å. Each actin filament is located midway between two myosin filaments, and if a cross bridge labels one side of the thin filament, another will very likely label the other side at approximately the same level (e.g., one on actin 9 and one on actin 8 in Fig. 10.6; Squire, 1979). The 45° rigor cross-bridge angling gives such a pair of bridges a pointed appearance described by Reedy as a chevron. The double chevron appearance arose because often, within each 385-Å near repeat along the thin filaments, there were two chevrons separated axially by about 100 to 150 Å. This repeated double chevron pattern gave the whole myac layer its obvious 385-Å periodicity. The double chevron pattern does not, however, occur in every 385-Å repeat, and Reedy and Garrett (1977) have drawn attention to a common appearance with a pattern that approximates to double, double, and then single chevrons (2 : 2 : 1) in successive 385-Å periods. The single chevron is the least well defined of these and sometimes appears to have a faint second chevron next to it. This pattern seems to be the origin of the marked 1150-Å

FIGURE 10.8. (a) Electron micrograph of a thin longitudinal section of rigor insect flight muscle showing the typical appearance of a myac layer. (b) The optical diffraction pattern from the marked region in (a), showing the strong 385-Å and 190-Å layer lines. All of the other layer lines are relatively weak. (c) A transverse section of insect muscle in rigor showing the flared X cross-bridge configuration and the location of the myac layer. (From Haselgrove and Reedy, 1978.)

cross-bridge repeat visible in rigor insect muscle (Fig. 10.9b; Freundlich and Squire, 1980).

As described in Chapter 8, there are currently two distinct models for the flared X structures seen in transverse sections. These are illustrated schematically in Fig 10.10. Figure 10.10b illustrates the model proposed by Offer and Elliott (1978) in which the whole flared X is formed by only two myosin molecules (four heads), each of which interacts through its two heads with two different actin filaments. The other model, proposed by Squire (1972), explains the flared X in terms of four molecules each of which interacts with only a single actin filament (Fig. 10.10a). The two heads in a single molecule were imagined in this case to attach to successive actin monomers along one of the long-pitched strands of the actin helix. These models make quite distinct predictions about the cross-bridge-labeling pattern as seen in longitudinal sections. Offer and Elliott (1978) have suggested that the myosin heads could label the actin filaments at relative axial positions 0,$2P/14$, $7P/14$, $9P/14$, and P (where P is 770 Å). This does not give a 2 : 2 : 1 pattern as it stands, but could easily be modified to do so. It would correspond as it stands to labeling in Fig. 10.6 of actin monomers 5, 6, 9, and 10; 19, 20, 23, and 24; 33, 34, 37, and 38, and so on.

Haselgrove and Reedy (1978) have shown that rigor myac layers from insect have a characteristic optical diffraction pattern (Fig. 10.8b). In this, all of the layer lines can be indexed as even orders of the 2300-Å repeat. The main characteristics of the patterns are that the sixth layer line at 385 Å is the strongest feature of the pattern, followed by the 12th layer line at 192 Å. The meridional reflections on layer lines 16 (144 Å) and 18 (128 Å) are both relatively weak. Such an intensity pattern is qualitatively similar to that in X-ray diffraction patterns from rigor insect muscle (Section 10.4.1; Miller and Tregear, 1972; Barrington-Leigh *et al.*, 1977; Holmes *et al.*, 1980). But both of these patterns are in marked contrast to either X-ray patterns (Miller and Tregear, 1972; Barrington-Leigh *et al.*, 1977) or optical diffraction patterns (Reedy, 1968; Freundlich and Squire, 1980) from relaxed muscle preparations (Fig. 10.8c). In the latter, the 144-Å reflection is dominant, and the 385-Å layer line is weak. The 128-Å reflection is usually evident in all of these patterns, although in no case is it particularly strong.

Squire (1979) has carried out a preliminary analysis of the two flared X models illustrated in Fig. 10.10 by preparing masks for optical diffraction that represent actin filaments labeled with over 50 different arrangements of cross bridges and troponin. Both Offer and Elliott's rigor scheme (along with some variations) and the alternative model (Squire, 1972) with some variations, were investigated in this way. The model structures were derived with reference to Fig. 10.6. Here, X and Y are two myosin filaments, one on each side of actin A. The levels 1 to 8

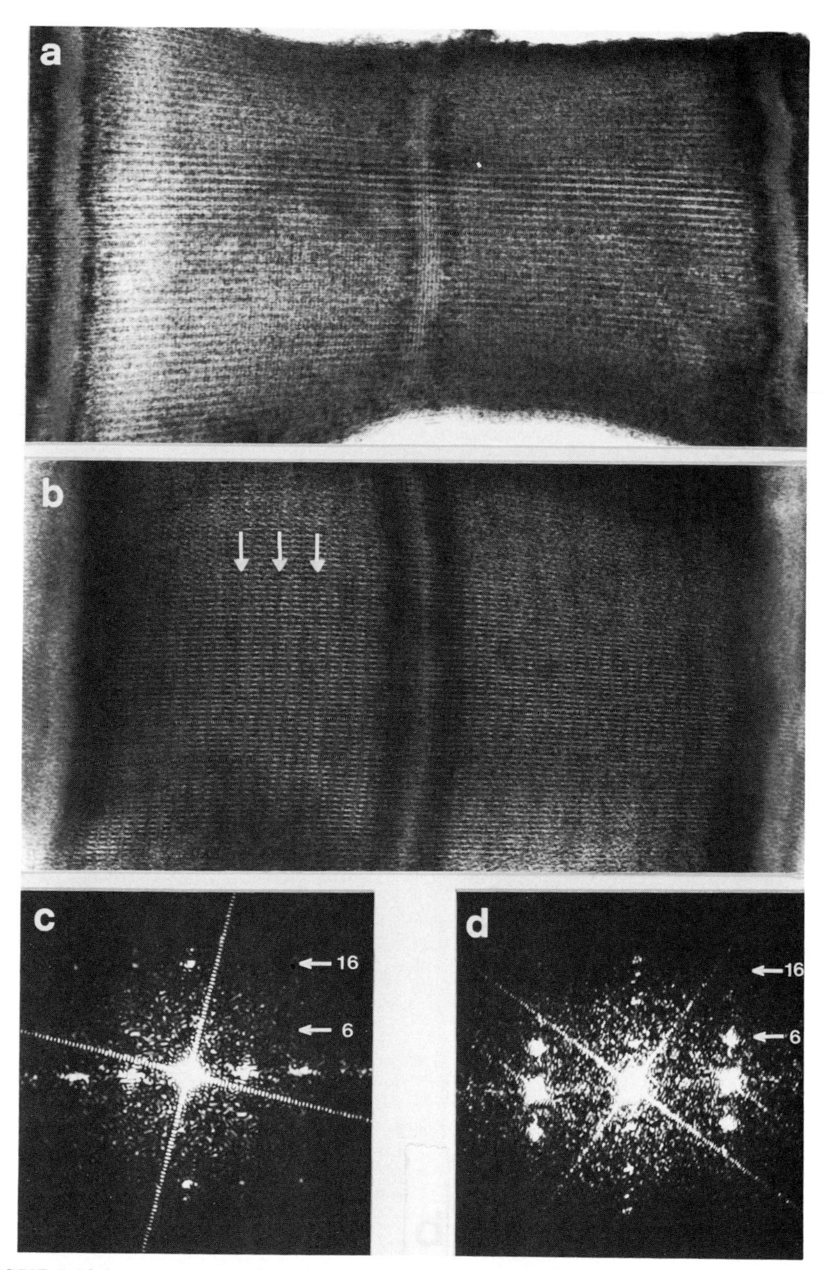

FIGURE 10.9. Longitudinal cryosections of insect flight muscle relaxed (a) or in rigor (b) and their optical diffraction patterns, (c) and (d) respectively. Note that in (c) the 145-Å layer line (16) is dominant, whereas (d), like Fig. 10.8b, has the 385-Å reflection (6) dominant. (From Freundlich and Squire, 1980.) Note also the very prominent 1150-Å period in the rigor muscle (b) (arrows).

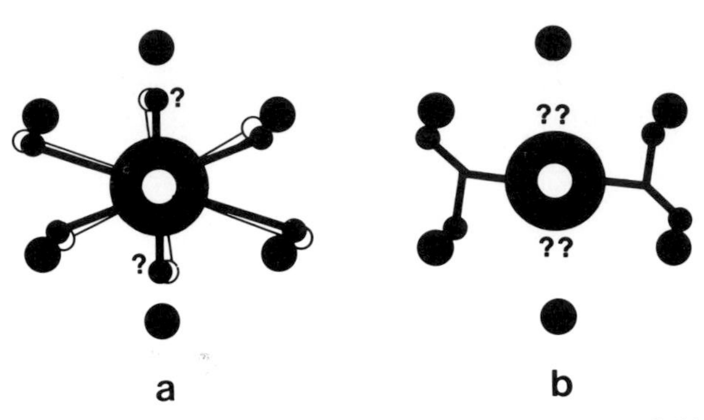

FIGURE 10.10. Two alternative explanations for the flared X cross-bridge appearance seen in transverse sections of rigor insect flight muscle. (a) The model from Squire (1972) in which the two heads in one cross-bridge (solid and hollow profiles) attach to two successive actin monomers on the same actin filament; (b) the model of Offer and Elliott (1978). In the latter, the two heads interact with two different actin filaments. (From Squire, 1979.) Note that the ? represent the possible locations of myosin heads at the same axial position on the thick filaments but not involved in the flared X.

on X and Y represent the 144-Å-spaced cross-bridge levels on these filaments. It was assumed that the azimuthal origin of a cross bridge (i.e., its position out of the plane in Fig. 10.6) was relatively unimportant in defining the probability of labeling (Squire, 1972; Haselgrove and Reedy, 1978) and therefore that at each level along X and Y there is a cross bridge ready to interact with the common actin filament.

The model structures were then generated on the assumption that the actin target areas were restricted to monomers 4 to 12, 18 to 25, 32 to 40, and so on, by postulating that a cross bridge will interact with the nearest available actin monomer to it provided that the latter is in an actin target area. Thus, cross bridge X2 will interact with actins 7 or 9, Y2 with 6 or 8, X4 with 18 or 20, Y4 with 19 or 21, X5 with 22 or 24, Y5 with 23 or 25, X7 with 33 or 35, Y7 with 34 or 36, X8 with 37 or 39, and Y8 with 38 or 40. The cross bridges on levels 1 and 6 clearly lie between actin target areas, and that on level 3 is a borderline case and may or may not interact. Thus there are 5 or 6 cross-bridge attachment levels (the chevron positions) in the 1150-Å near repeat (Reedy, 1968). According to Offer and Elliott (1978), each chevron would involve only two actin monomers, whereas Squire's model would put the two heads in one cross bridge (say X5) on successive actin monomers along one of the two actin strands (here 22 and 24); four monomers would be labeled in each chevron. In the modeling, the two troponin subunits in each 385-Å repeat were assumed to be separated axially by only 27.5 Å as suggested by Wray *et al.* (1978) and Maeda (1979). With this modeling, a number of factors can be investigated including (1) the pattern of labeling of actin by cross bridges, (2) the position of troponin relative to the cross bridges, and (3) the shape and configuration of each myosin head.

Before considering the optical diffraction, it is appropriate to consider one or two features of the very elegant X-ray diffraction analysis of rigor insect muscle recently carried out by Holmes *et al.* (1980) and mentioned earlier. These features are summarized below.

1. The high-angle actin layer lines (layer lines 27, 33, 39, and 45 on the 2300-Å repeat) are sensitive to cross-bridge shape and position and are qualitatively consistent with the modeling of Miller and Tregear (1972) and the results of Moore *et al.* (1970; Fig 5.11) on the tilt and slew of the cross bridge.
2. The low-order layer lines (6, 12, 18, etc. on the 2300-Å repeat) are much less sensitive to cross-bridge shape and attitude (i.e., the resolution is low), but they are much more sensitive to the pattern of labeling.
3. The actin attachment probability function (Q) with which they modeled their diffraction patterns treated every 385-Å actin period as being identical. No attempt was made to distinguish successive 385-Å repeats in order to generate a repeat of 1150 Å.

This last feature is of interest since it reveals the usefulness of the optical diffraction analysis of electron micrographs of rigor muscle in which the difference between successive 385-Å periods is visible. The Q function therefore provides a limitation on the average labeling pattern but gives little information on the detailed interactions within the 1150-Å near repeat.

The optical diffraction studies in Figs. 10.11 and 10.12 illustrate two types of modeling. In some cases [Fig. 10.11(a, c, g, i, and l)], point scatterers have been placed on a helix with 28/13 symmetry in order to investigate the effect of the pattern of labeling of the scattering objects on the actin filaments (the actin monomers themselves have been excluded). Figure 10.11e and the models in Fig. 10.12, on the other hand, comprise similar labeling patterns, but this time the scattering objects represent more faithfully the known cross-bridge shape (Elliott and Offer, 1978; Fig. 6.5); the effect of shape can also be monitored.

Figure 10.11a shows a labeling pattern of point scatterers with a 2 : 2 : 1 arrangement in which the chevron pair has a separation of 2.5 × 55 Å. This corresponds to labeling in Fig. 10.6 by bridges on levels 4 and 5 with actins 18, 19, 23, and 24, and those on level 2 with actins 7 and 8. The diffraction pattern from this shows a strong sixth-order layer line (385 Å) (especially near to the meridian as required), but a very much stronger 16th order at 14.4 Å. The 12th order layer line (192 Å) is very

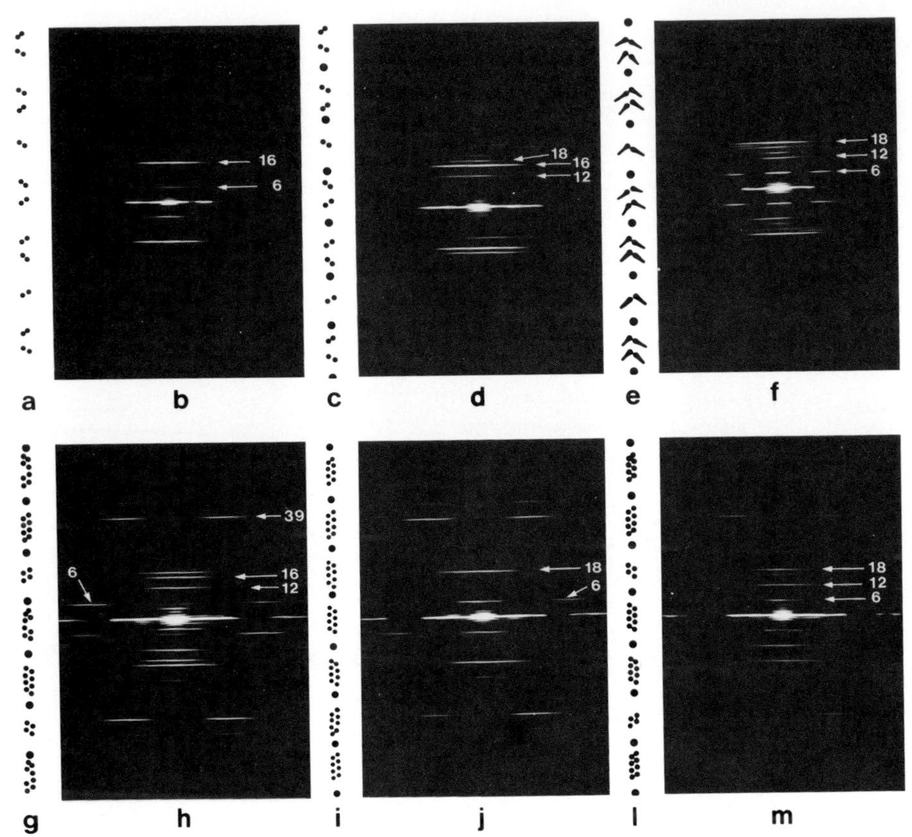

FIGURE 10.11. Various models based on the symmetry of the actin filament shown in Fig. 10.6 with different patterns of labeling on them. The actin monomers themselves are not shown. (a), (c), (e), (g), (i), and (l) Different models described in the text. (b), (d), (f), (h), (j), and (m) Their respective diffraction patterns. The layer line numbering in these patterns is based on a repeat of 2300-Å. (From Squire, 1979.)

weak. Addition of density in the position of the troponin complex (assumed to lie between actin target areas; Fig. 10.11c) causes a marked enhancement of the 192-Å meridional peak and a relative weakening of the 385-Å reflection, since it is tending to halve the 385-Å repeat. Once again, the strongest reflection by far is still the 16th order at 144 Å. Similar models with 110 Å between chevron pairs instead of 165 Å (i.e., the labeled actins are 7, 8, 18, 19, 22, 23, and so on as in Offer and Elliott's model) give a very similar result (not shown) except that now the 18th-order meridional at 128 Å is predominant rather than the 16th at 144 Å. This is because 110 Å is near to 385/3, and the third order of the 385-Å reflection is at 128 Å.

Representing the bridges more faithfully than in Fig. 10.11a gives the characteristic double-chevron pattern (Fig. 10.11e), but, once again since the 110-Å chevron separation has been included, the 128-Å (18th-order) reflection is still the strongest in the pattern. Note that this model includes the 2 : 2 : 1 chevron pattern. A 2 : 2 : 2 pattern gives a similar result except that the 12th layer line is weaker than in (f), apparently because in (e) the single chevron effectively halves that particular 385-Å repeat.

Figure 10.11g shows a model comparable to (c) in that only the pattern of labeling and not the cross-bridge shape is included, but now two actin monomers are labeled for every one in (c); this is therefore expressing the difference between the labeling patterns in Squire's model (1972) and Offer and Elliott's. It will be seen that the sixth layer line is generally much stronger relative to the 16th and 18th orders than in (d), but it is still not intense enough. The model in (g) is a 2 : 2 : 1 model; (i) shows the equivalent 2 : 2 : 2 version. It will be seen that the 12th order is much weaker in (j) than in (h) for reasons given above. Model (l) shows a distribution similar to (g) but with the separation between chevron groups consistently 2 × 55 Å, rather than the mixed 2 × 55 Å and 2.5 × 55 Å in (g). As expected, in this case the 18th-order reflection now dominates the 16th order, whereas in (h) they are of comparable intensity.

The models become much more realistic in Fig. 10.12, where objects of approximately the correct shape follow the labeling pattern in Fig. 10.11l. Figure 10.12a shows the familiar modeling of rigor invertebrate muscle thin filaments (Wray *et al.*, 1978) with attached heads bunched in the actin target areas. This is actually identical to Squire's (1972) model. For simplicity, each head has been modeled as a sphere on the end of a straight rod (Offer and Elliott, 1978; Fig. 6.5f). At the resolution being considered (no better than 128 Å), this approximation would not materially effect the conclusions reached. The diffraction patterns from model (a) in Fig. 10.12 show further enhancement of the sixth and 12th layer lines relative to the 16th and 18th orders which are now weak. The model pattern is very similar to the corresponding patterns from myac layers, and models rather similar to this (except for variations in the total number of bridges labeling within each target area) have been reported to account well for the observed X-ray diffraction patterns from insect (Barrington-Leigh *et al.*, 1977; Holmes *et al.*, 1980) and crustacean muscle (Wray *et al.*, 1978). Wray *et al.* (1978) in particular state that there is little need to consider the effect of the myosin origins of the cross bridges in modeling their rigor diffraction patterns since a structure such as that in Fig. 10.12a will, on its own, account for their observations quite well. The optical diffraction pattern in Fig. 10.12b does not conflict

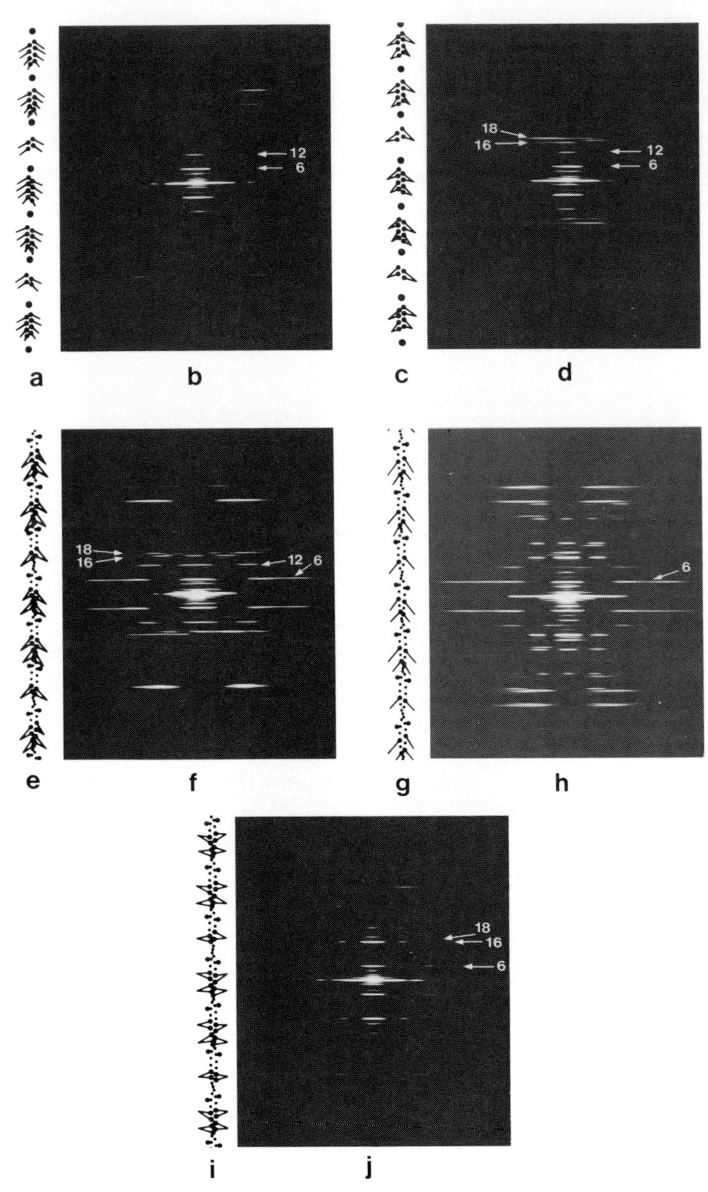

FIGURE 10.12. Further models of labeled actin filaments as in Fig. 10.11. Actin labeling according to Squire's (1972) model for rigor insect muscle is shown in (a). This is also similar to the rigor model of Wray *et al.* (1978). (c) is the same basic labeling pattern as in (a) but with pairs of attached myosin heads grouped to have a common origin in order to give the observed double-chevron appearance. (e) A more complete and realistic version of (c) in which the positions of the actin monomers are also shown. (g) A comparable model to (e) but with the labeling pattern suggested by Offer and Elliott (1978) except with a 2 : 2 : 1

with this conclusion, but this on its own is not sufficient; it is obvious that the labeling pattern in Fig. 12.12a will not account directly for Reedy's observations of double chevrons in the myac layers that are being modeled. Indeed, this is one of the criticisms of Squire's original model (1972), which Fig. 10.12a represents, given by Offer and Elliott (1978).

Fortunately, it is not difficult to suggest how this dilemma can be resolved. It must be remembered that in the type of model shown as Fig. 10.12a it is more likely than not that each row of four heads along one target area of a single actin strand would comprise two pairs of heads in which the members of each pair actually belong to the same myosin molecule. In this case, the model as drawn assumes that considerable distortion is allowed at the junction between these two heads where they join the myosin rod. But this distortion could, alternatively, be taken up where the heads attach to actin, in which case the model would appear as in Fig. 11.12c. The diffraction pattern (d) from this is still acceptable, and, in addition, an appearance very reminiscent of Reedy's double chevrons has now been produced. It is possible that such a cross-bridge arrangement, when rendered nonideal by the introduction of some disordering, will account most satisfactorily for all of Reedy's observations.

As a last check on this conclusion, two more realistic representations of the labeled thin filament have been drawn (Fig. 10.12e,g) to compare directly the models of Offer and Elliott (1978) and Squire (1979). This time the locations of the actin monomers have been included although these parts of the thin filament do not contribute significantly to the part of the diffraction pattern being considered. It is clear that the optical diffraction pattern (f) from model (e) has relatively intense sixth and 12th layer lines near the meridian relative to the 16th- and 18th-order meridional reflections. But in (h) from model (g), the 18th order at 128 Å is relatively stronger than is observed in corresponding patterns from myac layers (Haselgrove and Reedy, 1978). Although it is just possible that slight adjustment either of the position of troponin relative to the cross bridges in model (g) or of the crossbridge shape might reduce the 18th order sufficiently to make its intensity acceptably low, and only detailed computations could check this, these results, as they stand, do show that Squire's (1979) model (Fig. 10.12e) gives qualitatively the right kind of optical diffraction pattern and myac layer image, whereas the model in Fig. 10.12g may not be acceptable.

---

pattern introduced. (b, d, f, and h) The optical diffraction patterns from (a), (c), (e), and (g), respectively. For discussion see text. (i and j) A model and its diffraction pattern in which the labeling pattern is as in (e) except that the cross-bridge origins have been swung back to lie on the 145-Å repeat as in Fig. 10.6. The effect is to increase radically the intensity of the 16th order (145 Å) of the 2300-Å repeat relative to (f). In (j) the 385-Å layer line is relatively weak. (From Squire, 1979.)

One further argument supports this conclusion. The elegant X-ray diffraction analysis by Holmes *et al.* (1980) of rigor insect muscles has yielded what they believe to be a fairly reliable distribution of attachment probability within the averaged 385-Å actin repeat. This distribution is shown in Fig. 10.13a. Here the probability of attachment is above 50% over a length of actin filament of about 100 Å and is close to zero over a length of about 150 Å. A comparable distribution can be drawn for the two models illustrated as Fig. 10.12e,g. In each of these, the true repeat is 1150 Å, but it is easy to find the labeling within an average 385-Å repeat just by adding the labeling pattern in the three component repeats within each 1150 Å. Such an analysis is shown in Fig. 10.13b and c, respectively; it is readily apparent that (c) is double-peaked and markedly different from (a), whereas (b) is quite a reasonable approximation to (a). One could therefore suggest that (a) which includes only the first three components of Q (Q_1, Q_2, and Q_3) and which therefore has a

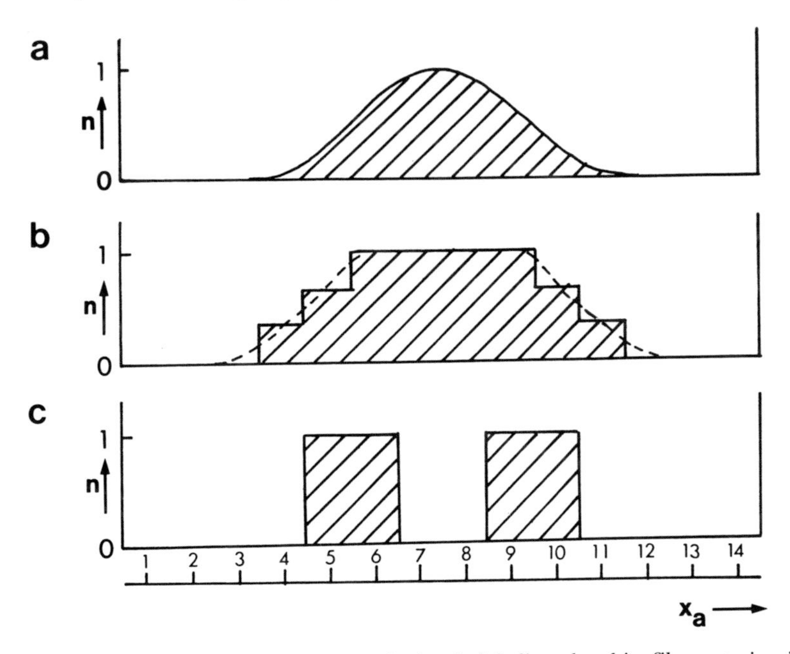

FIGURE 10.13. The distribution of myosin heads labeling the thin filaments in rigor insect flight muscle. (a) The profile of attachment probability (Q) determined by Holmes *et al.* (1980) from X-ray diffraction evidence. (b) The actual labeling distribution in the average 385-Å period according to the model of Squire (1972, 1979). (c) The actual distribution of labeling in the model shown as Fig. 5 of Offer and Elliott (1978). The parameter n is a measure of the number attached (fractional) at each actin monomer site in the three successive 385-Å repeats that form the 1150-Å period. In terms of Fig. 10.6, the monomer positions x_a go from 1 to 14, then from 14 + 1 to 14 + 14, then from 28 + 1 to 28 + 14.

resolution of only 128 Å (one-third of the repeat we are discussing) is in rather good agreement with (b) which is an exact representation of the distribution in Fig. 10.12e. The major difference is in the total area under the curves (i.e., the total number of attached bridges).

It would be difficult, I think, at this stage to find a model that agrees better with the available data than does this one (Fig. 10.12e). Of course, as it is described here, this model is incompatible with the idea that dividing cross bridges linking to two different actin filaments (Offer and Elliott, 1978) can occur. But in the less exact interactions involving some randomness of attachment which may well occur in the muscle itself, it is not inconceivable that both types of interaction can occur. Indeed, in Reedy's micrographs of myac layers (Reedy, 1968; Reedy and Garrett, 1977), it is clear that an element of disorder is present in the 2 : 2 : 1 cross-bridge-labeling pattern, and this could easily be because of the presence of a few dividing cross bridges as well as cross bridges in which both heads attach to a single actin filament.

Recently Trinick and Offer (1979) have demonstrated that binding between HMM and two actin filaments (usually antiparallel) can occur *in vitro*. However, Freundlich *et al.* (1980), from their studies of thick transverse sections of both vertebrate and insect muscle, have concluded that if both types of interaction are present, the predominant one is that in which both heads of a single myosin molecule attach to the same actin filament. Further studies like these should help to settle this question.

10.2.5. Modeling of the Rigor State: Vertebrate Muscle

It has just been shown that in insect flight muscle in rigor, there is a very characteristic distribution of attached cross bridges in the actin target areas of each actin filament such that marked 385-Å and 1150-Å periodicities are produced. The situation appears to be rather different in rigor vertebrate skeletal muscle (H. E. Huxley and Brown, 1967). There appear to be two obvious reasons for this. One is that the actin and myosin axial repeats may be very different (360 to 370 Å and 429 Å, respectively: Chapter 7), and the other is that each thin filament in the A-band lattice is approached by cross bridges from three myosin filaments 120° apart, as opposed to the two, 180° apart, in insect muscle. In this case, it seems to be self-evident that, although actin target areas very probably do occur (as described in Chapter 7; Fig. 7.64), the arrangement of cross-bridge origins around each thin filament must also have a marked effect on the distribution of attached bridges in the rigor muscle.

To investigate this possibility, we can consider the geometry of interaction in a manner similar to that given in Fig. 10.8 for insect muscle.

But before this can be done, it is necessary to know the rigor repeats of the individual thick and thin filaments, and unfortunately these are not known exactly. In the case of the myosin filaments, all of the layer lines disappear in patterns from rigor muscle except for the 144-Å meridional reflection, and this is moved slightly from its position in relaxed muscle. It is therefore evident that the thick filament repeat may have changed (Haselgrove, 1975a), but this cannot as yet be proved. If it does, what could it change to? Inspection of the radial net for the three-stranded helix (Fig. 7.9b) shows that with very little twisting, the repeat of the helical arrangement of cross-bridge origins could be altered from about 432 Å (3 × 144 Å) to about 720 Å (5 × 144 Å) as in the smooth muscles (Chapter 8) or to 1152 Å (8 × 144 Å) as in insect muscle. So these two repeats appear to be likely possibilities if the thick filament symmetry does alter (Haselgrove, 1975a,b). Although the change in spacing of the 144-Å reflection could be accounted for without invoking such a symmetry change (see Section 10.3.4), for the purposes of the present argument it will be assumed, for reasons given below, that the thick filament symmetry does change on going into rigor and that it has an axial repeat of about 720 Å. The arguments given below do not rest heavily on this assumption, but it is convenient to use it at this stage.

Regarding the structure of the thin filaments, the repeat is not even known exactly in relaxed muscle (H. E. Huxley and Brown, 1967), but it is probably about 360 to 370 Å. This would correspond to a structure intermediate between a 13/6 and a 28/13 helix. H. E. Huxley and Brown also suggested from their results that the actin helix in rigor muscle might have a slightly longer repeat than in relaxed muscle. Such a conclusion is supported by the results of Maeda (1979) and also of O'Brien *et al.* (1975) who found that addition of Ca^{2+} to paracrystals of thin filaments (*in vitro*) alters the structure from a regular 13/6 helix to a regular 28/13 helix (as in insect muscle). It is possible therefore that a structural change in the thin filament symmetry may also occur on going from the relaxed to the rigor state, but *in vivo* this might be only a slight increase in pitch of a nonintegral helix without being a distinct transition from an integral 13/6 to an integral 28/13 helix.

Another indication of the rigor structure comes from the reflections near the meridian in patterns from rigor frog muscle (Haselgrove, 1975a). The spacings of these reflections do not index exactly on any single repeat, but they approximate orders of a repeat of about 725 Å (Table 10.3). The reflections thought to arise from the cross-bridge-labeled thin filaments have similar spacings (Table 10.3) and also index on a repeat of about 725 Å. Because of this, it is tempting to take as a preliminary working structure one in which the myosin filaments and the thin filaments both have the same repeat. This repeat, for simplicity, is taken to be 5 × 144 Å or 720 Å in the first instance. The true repeat may, in fact, be a good deal longer (Haselgrove, 1975a), but the kind of argument given below would be affected little were this to be so.

We have seen in the case of insect flight muscle, as described in Chapter 8, that the six thin filaments around each thick filament are arranged on a regular helix. The thick filament "sees" equivalent positions on these thin filaments to be arranged on a two-strand 6/1 helix of pitch 2 × 385 Å. Since the thin filaments can be described as right-handed double helices, and these are arranged in a left-handed helix (Reedy, 1968), the kind of structure shown in Fig. 8.10c is generated (symmetry approximately that of space group $P6_4$). Here at the same axial level the six thin filaments are related by systematic rotations of 60°, and this fits in well with the fact that the insect Z-band contains a hexagonal arrangement of thin filaments. By having such relative rotations, all of the thin filaments in the Z-band can be in equivalent positions.

In the case of vertebrate muscle, the situation is different but in some ways analogous to the insect structure. Here the thin filaments are probably arranged both at the Z-band and in the overlap zone so that at the same level they all have identical orientations (Chapter 7). If they are each assumed to be right-handed double helices, then in this case equivalent points on the six thin filaments (with regard to the actin subunits but not troponin) as seen by the single thick filament that they surround will, as in insect, be approximately helically arranged. But this time (Fig. 10.14a) the helix of actin filaments will be right-handed (it is a pseudo 6_2 helix), whereas in insect it is left-handed. In this right-handed helix, adjacent actin filaments will, as in insect, be related by axial shifts of $P/6$ where P is the pitch of the actin helix. Taking P to be 720 Å for reasons given above gives a thin filament shift of about 120 Å (in insect, $P = 2 \times 385$ Å, and the shift is 128 Å). With this arrangement of thin filaments in mind, it is possible to investigate the kind of cross-bridge-labeling pattern that might occur in rigor vertebrate muscle. Figure 10.14b shows how this can be done. Here the six thin filaments and the radial net of the three-stranded myosin filament of repeat 720 Å have been superimposed in order to see where the cross bridges can label the actin target areas (shaded). Clearly, there are two ways that the drawing can be produced depending on the absolute hand of the cross-bridge helix. In Fig. 10.14b the three 5/1 helical strands of cross bridges (each of pitch 5 × 144 Å) are assumed to follow the same right-handed screw as the arrangement of actin filaments. This is a useful assumption since it improves the match between the cross-bridge origins and the surrounding actin target areas, but it has yet to be substantiated. In Fig. 10.14b the shaded regions on the actin filaments are taken to be those that form the main parts of the actin target areas, and on this basis, the solid arrows indicate where cross-bridge attachment is likely to occur (Squire, 1980 and in preparation).

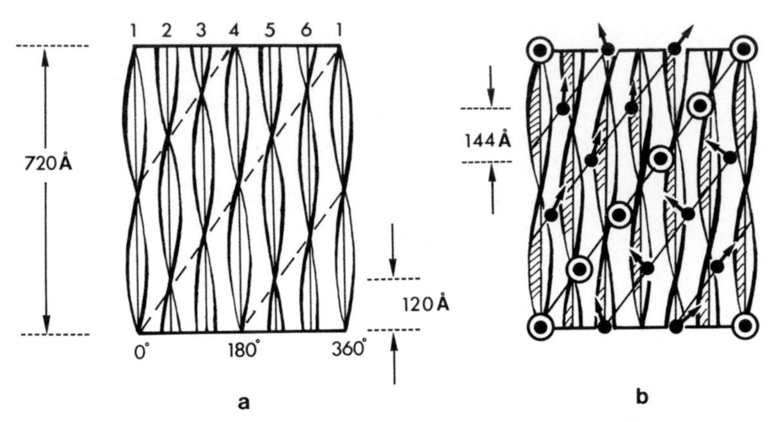

FIGURE 10.14. (a) A schematic representation of a 6_2 helix of actin filaments surrounding each thick filament as in vertebrate muscle. Each thin filament (1 to 6) is itself a right-handed helix, and these filaments are arranged in a right-handed helix of filaments. In this case the thin filament repeat is 720 Å, and the axial displacement between filaments is 120 Å (cf. Fig. 8.10). (b) The thin filament arrangement in (a) on which has been superimposed the radial net of a three-strand 5/1 helix of repeat 720 Å (a hypothetical thick filament structure; see text). The target areas on the thin filaments have been shaded, and cross bridges that can make attachments are arrowed. Circled cross-bridge positions are those for which an actin target area is not accessible. In this simplified case this clearly follows one of the 5/1 helices of crossovers in the thin filament array (see Fig. 10.15).

Two features of this arrangement are immediately obvious. First, it can be seen that, on average, cross bridges from one myosin filament label any actin filament about twice in the 720-Å repeat, a result that seems to be reasonably consistent with A. F. Huxley's (1957) observations of one such cross-link about every 400 Å. Second, it is clear that whatever the complexities of the labeling pattern, it must inevitably repeat axially every 720 Å, since both the actin and myosin repeats have been given this value. The question that remains concerns the distribution of attached bridges, and this can be considered from two points of view: the labeling on any actin filament and the cross-bridge pattern around a single thick filament.

Figure 10.15a shows a radial projection of the arrangement of the origins of the cross bridges of a single myosin filament that are attached to actin according to the scheme in Fig. 10.14b. Although there are one or two borderline cases where attachment may or may not occur, it can be seen that, on average, the attached cross bridges form a quasihelical stripe around the thick filament. On the present assumptions, this stripe obviously has a pitch of 720 Å, and the bridges are not moved far from their 145-Å axial origins. The general form of the diffraction pattern of this array can be generated (Fig. 10.15b) assuming that the structure approximates a 5/1 helix of pitch 720 Å. Layer lines at spacings of about

1/720 Å^{-1}, 1/360 Å^{-1}, 1/240 Å^{-1}, 1/180 Å^{-1}, 1/144 Å^{-1}, etc. will be produced, and it is easy to show that centers of mass at a radius from the thick filament axis of about 160 to 170 Å will give layer lines with peaks centered roughly at 1/300 Å^{-1}, just as was observed by Haselgrove (1975a) in his diffraction patterns (Fig. 10.4a). Because of the randomness in the no-three-alike lattice, there will be relatively little correlation among the labeling patterns surrounding meighboring thick filaments, so it is probably reasonable to ignore the effect of sampling on this pattern. It therefore seems that the unique feature of the rigor diffraction pattern from frog sartorius muscles (the extra layer lines) may be adequately explained in terms of the kind of cross-bridge arrangement shown in Figs. 10.14b and 10.15a.

Of course, for simplicity, the actin and myosin helices have been assumed to be integral and to repeat after 720 Å, but the measured layer lines appear to arise from a slightly longer repeat (725 to 730 Å; Table 10.3). Untwisting both of the helices slightly to increase the two repeats clearly renders the arrangement more complex, but the match between the cross-bridge helices and the array of actin filaments can be maintained. The structure in Fig. 10.15a can therefore be taken to be a first approximation to a much more complex arrangement of labeling. It is even possible, and probably very likely, that the basic thick filament

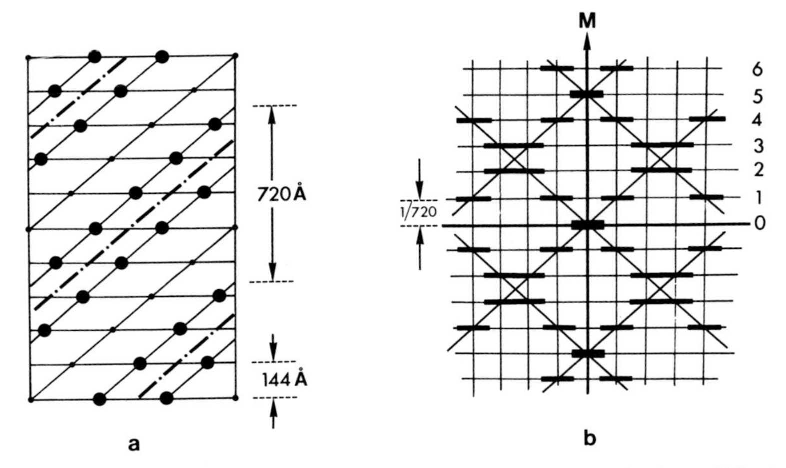

FIGURE 10.15. (a) Distribution of attached (large dots) and unattached (small dots) cross bridges in Fig. 10.14(b). The attached cross bridges clearly form a stripe of similar structures with the symmetry of a 5/1 helix of pitch 720 Å. (shown by the dot–dash line). (b) Form of the diffraction pattern to be expected from the distribution in (a). The layer lines (0 to 6 shown) are based on a repeat of 720 Å, and the first meridional is at 144 Å. This structure may account for the inner layer lines seen in Fig. 10.4a from rigor frog muscle. M, Meridian.

symmetry in rigor is the same as the symmetry (with a 430-Å axial repeat) that occurs in relaxed muscle. But, provided the helix of actin target areas repeats about every 725 Å, the pattern of cross-bridge labeling will still correspond roughly to that in Fig. 10.15a since the cross-bridge origins will still be based on an axial repeat of 144 Å and will only be displaced in an azimuthal direction. In this case, successive 725-Å repeats would not be expected to be identical, and the thick and thin filament structures would gradually get out of step (azimuthally but not axially), giving the whole arrangement a repeat of 15 × 144 Å (2160 Å). The situation with the two repeats fitting together exactly is only an approximate representation of this arrangement, but conclusions based on it will probably not be far from the truth.

Turning now to the labeling of individual thin filaments, since these are approached by cross bridges from three myosin filaments, and since there is a variation in structure every repeat (there are six repeats of 360 Å in 2160 Å), it is very likely that the labeling will be fairly uniform and will show little evidence of a marked axial repeat as in the case of insect flight muscle.

10.3. Evidence for Structural Changes during Contraction

10.3.1. Vertebrate Striated Muscles

Skeletal Muscle. We have seen that compared with diffraction patterns from relaxed vertebrate muscles, those from the rigor muscles differ in the following respects.

1. The myosin layer lines are weaker.
2. The actin layer lines are much stronger.
3. A new set of layer lines appears close to the meridian on an axial repeat of 725 Å.
4. The equatorial reflections (i.e., the $10\bar{1}$ and $11\bar{2}$) change considerably in relative intensity.
5. There are slight spacing changes of some of the reflections.

On the assumption that during contraction myosin cross bridges do attach to the thin filaments, then it might be expected that diffraction patterns recorded only when a muscle is generating tension would show changes analogous to those that occur in rigor. However, it has been found in practice that this is only true to a limited extent. G. F. Elliott *et al.* (1963,1967) H. E. Huxley and Brown (1967), H. E. Huxley (1968, 1975a,b, 1976, 1979a,b), Haselgrove and Huxley (1973), Matsubara *et al.* (1975), Haselgrove (1975a), Haselgrove *et al.* (1976), and H. E. Huxley *et al.* (1980) have carried out a systematic study of the axial and equatorial diffraction patterns from active frog sartorius and semitendinosus muscles. The diffraction patterns were usually recorded from muscles held isometrically at a known sarcomere length (monitored by laser diffraction) and stimulated electrically to give a series of tetanic responses (Chapter 1), each of duration about 1 sec. In the early work a shutter arrangement was included in the X-ray camera so that the diffraction pattern was only recorded when the muscle was generating tension above a predetermined threshold level. Nowadays, more sophisticated electronically gated counting methods are used (H. E. Huxley *et al.*, 1980).

The various observations described in the papers listed above can be summarized as follows (for muscles at rest length unless stated otherwise).

1. The spacings of the reflections in patterns from active muscle are not very different from those in patterns from the same muscle at rest, except for a 1% increase in spacing of the 143.4-Å and 71.6-Å reflections from the thick filament to values 144.8 Å and 72.3 Å (Figures from Haselgrove, 1975a; see Table 10.5.)

2. Relatively little change occurs in the intensity of the 144-Å meridional reflection, but the other myosin layer lines (at 429 Å, 216 Å, etc) become very much weaker in patterns from active muscle. H. E. Huxley and Brown (1967) and Haselgrove (1975a) estimated that the ratio I_{429}/I_{143} (peak heights) was about 0.4 in patterns from resting muscle and about 0.2 in patterns from active muscle. In absolute terms, the intensity of the 143-Å meridional reflection in patterns from active muscle was about 0.66 ± 0.10 of its value in patterns from relaxed muscle, and the corresponding value for the 429-Å reflection was 0.31 ±0.06 (Table 10.5).

TABLE 10.5. Observed Intensities of the Myosin Layer Lines in X-Ray Patterns from Frog Muscle[a]

State	Relative intensities			Spacing	
	429(L1)	215(L2)	143(L3)	"143"	"72"
Relaxed	100%	100%	100%	143.4	71.6
Contracting,[b] nonoverlap	63%	74%	74%	144.8	72.3
Contracting, rest length	31%	—	66%	144.8	72.3
Contracting, 2.9-μm sarcomere length	31%	—	61%	144.8	72.3
Rigor, all lengths	—	—	present	144.8	72.3
Relaxed (stretched before equilibration)	—	—	—	143.8	72.1

[a] From Haselgrove (1975a).
[b] Taken from Huxley (1972).

3. The whole of the diffraction pattern from the thin filaments changes little when the muscle is active except for changes in the relative intensities of the second and third layer lines. As described in detail in Chapter 5, these changes are thought to be caused by the movement of the tropomyosin molecules in the grooves of the actin helix in mediating thin filament regulation. The only other change that might be significant is a possible slight change in intensity of the 59-Å layer line. According to Haselgrove (1975a), the meridional end of this layer line increases in intensity by about 27% (±21% S.D.), whereas its outer end increases little (9% ± 15% S.D.). The possible meaning of such an increase will be discussed in Section 10.3.6. Note that similar barely significant intensity increases have been reported to occur in the 59-Å actin layer lines from active molluscan and vertebrate smooth muscles (Lowy, 1972; Vibert *et al.*, 1971, 1972).

4. Significant changes occur in the relative intensities of the $10\bar{1}$ and $11\bar{2}$ reflections on the equator of the diffraction pattern (G. F. Elliott *et al.*, 1967; Haselgrove and Huxley, 1973; Matsubara *et al.*, 1975; Haselgrove *et al.*, 1976). As summarized in Table 10.2, the relative intensity $I_{10\bar{1}}/I_{11\bar{2}}$ in patterns from muscles at rest length (2.2 μm) changes from about 2.5 for relaxed muscle to about 0.6 for active muscle to about 0.25 in rigor muscle. Note that at increasing sarcomere lengths this ratio gradually increases in patterns from muscles in all three states (Table 10.2), so that it is important to compare results from muscles at the same length. Also, recent neutron diffraction results (Worcester *et al.*, 1975) and other X-ray diffraction studies (Matsubara *et al.*, 1975) suggest that the ratios $I_{10\bar{1}}/I_{11\bar{2}}$ may be very similar in patterns from rigor and contracting muscles. The difference has been attributed to the longer relaxation time that was used between contractions in the latter studies. It was claimed that this permitted the muscle to be more fully active. The recent work by Haselgrove *et al.* (1976) and Podolsky *et al.* (1976) using position-sensitive X-ray detectors aligned along the equator showed that with maximal activation the absolute intensity of the $10\bar{1}$ reflection reduced by about 50% (±10%), whereas that of the $11\bar{2}$ reflection increased by about 100% (±25%).

5. The axial diffraction patterns of muscles stimulated after being stretched beyond overlap were recorded by Haselgrove (1975a). It was found that changes in the same direction as occurred with rest length muscles also occurred in the spacings of the 143-Å and 72-Å reflections. These were from 143.8 Å and 72.1 Å in stretched relaxed muscles to 144.8 Å and 72.3 Å in stimulated stretched muscles. Indeed, the latter spacings were found to be independent of sarcomere length (Table 10.5). This result makes it clear that, as in rigor muscle, whatever are the changes in the thick filament structure that cause this change of spacing,

they appear to be independent of the interaction of the thick filament with actin.

6. In addition to these spacing changes, it was found that the myosin layer lines also changed in intensity when a nonoverlap muscle was stimulated (H. E. Huxley, 1972a). Haselgrove (1970, 1975a) had previously found that in patterns from muscles activated at intermediate sarcomere lengths (about 2.9 μm), the first myosin layer line decreased in intensity by about the same amount as at shorter sarcomere lengths (i.e., to about 30% of the intensity from relaxed muscles). But at the nonoverlap sarcomere lengths, although an intensity drop generally did occur, it only fell to about 0.63 ± 0.11 of the resting intensity. The intensity of the second and third layer lines also fell, in both cases to 0.74 of the resting value. Note that the values quoted are for isometric contractions, but muscles contracting isotonically gave similar results (H. E. Huxley, 1972a,b). The possible meaning of these observations will be considered in detail in Section 10.3.3. But it does seem evident that some structural change must be taking place in the thick filaments even at these very long sarcomere lengths where little or no actin interaction can occur. But H. E. Huxley (1972a) was rather cautious about this and was careful to note that in a few experiments little intensity change occurred. He thought it possible that when it did occur it might have been caused by some residual overlap between the thick and thin filaments such that an interaction between these filaments could still occur. This important possibility should be kept in mind even though the majority of the available evidence seems to point the opposite way.

7. Another interesting result reported by H. E. Huxley (1972a) was that the myosin layer line pattern from rest length muscles remained weak for some time after stimulation had stopped. This suggested that after a contraction the reorganization of the myosin cross bridges back to the relaxed configuration took some time (actually about 10 sec).

8. Haselgrove and Huxley (1973) also carried out a series of contraction experiments to investigate the validity of the sliding filament model. In addition to the finding of little change of spacing in the various layer lines reported in their other papers, they studied the spacings and relative intensities of the equatorial $10\bar{1}$ and $11\bar{2}$ reflections given by muscles that were allowed to shorten actively to a predetermined length immediately prior to the X-ray exposure. It was found that the intensity ratio of the $10\bar{1}$ to the $11\bar{2}$ was just as expected for the final muscle length and did not depend on the immediate history of the muscle prior to the exposure, a result entirely consistent with the sliding filament model.

Cardiac Muscle. Apart from this comprehensive study of vertebrate skeletal muscle, a similar study of contracting vertebrate heart muscle has now been reported by Matsubara *et al.* (1977a,b). It was found that the relative intensities of the $10\bar{1}$ and $11\bar{2}$ equatorial reflections from rhythmically contracting heart muscles were different in the high-tension part of the cycle (the systolic phase) from the low-tension part (the diastolic phase) so that the ratios $I_{10\bar{1}}/I_{11\bar{2}}$ were 2.19 and 2.66, respectively. Quiescent muscles (virtually relaxed) gave a ratio $I_{10\bar{1}}/I_{11\bar{2}}$ of about 3.07, whereas rigor muscles gave a ratio of 0.37. The directions of these various changes are generally in qualitative agreement with the results from skeletal muscles except for the additional diastolic phase which has no real counterpart in voluntary muscles.

10.3.2. Insect Flight Muscle

In parallel with the study of active frog muscles, mainly by H. E. Huxley's group at Cambridge, similar studies of active insect flight muscle were in progress in J. W. S. Pringle's laboratory at Oxford. A series of papers (Tregear and Miller, 1969; Miller and Tregear, 1970, 1971, 1972; Armitage *et al.*, 1972, 1975) described the appearance of the X-ray diffraction patterns of *Lethocerus* flight muscles in various defined states and the changes that occur when a muscle is activated. The results from relaxed and rigor muscles have already been described. As noted in earlier parts of the book, insect flight muscle is unusual in that, when stimulated, its active state is oscillatory. Regulation of insect flight muscle is known to both actin-linked and myosin-linked (Lehman *et al.*, 1974; Marston and Tregear, 1974). If the level of calcium ions in the bathing medium of a relaxed glycerinated insect muscle is increased, then it is found that the thick filaments change in structure but that no tension is generated until the muscle is stretched. Miller and Tregear therefore distinguished the low-tension, Ca^{2+}-activated state from the working state. In the latter case, the muscle was mechanically oscillated by about 1 to 2% about a mean extension of 3% of the slack length of the muscle.

As mentioned earlier in the chapter, the location of the actin filaments in the insect filament lattice midway along the unit cell sides has the effect of making the $11\bar{2}$ equatorial reflections very weak. Miller and Tregear (1970) therefore studied the relative intensities of the $10\bar{1}$ and $20\bar{2}$ reflections as a function of muscle state. Table 10.1 summarizes the observed equatorial intensities relative to that of the $10\bar{1}$ reflection taken as 100% in patterns from relaxed muscle. The ratio $I_{10\bar{1}}/I_{20\bar{2}}$ changes here from about 1.9 for relaxed muscle to about 1.8 for Ca^{2+}-activated muscle, remains at about 1.8 for muscles doing oscillatory work, and changes to about 0.8 in rigor muscle.

As in the case of vertebrate muscles, these observations were interpreted in terms of a shift of mass from the myosin filaments towards the actin filaments. But whatever the interpretation, it is clear that almost all of the change that occurs in active muscle takes place when the muscle is Ca^{2+}-activated, and little further change occurs when the muscle is oscillated. This is a clear indication of Ca^{2+} activation of the thick filament assuming that the changes relate to cross-bridge movements. Apart from the $10\bar{1}$ and $20\bar{2}$ reflections, Miller and Tregear (1970) also recorded the 303 and $40\bar{4}$ reflections in patterns from relaxed muscle and found these to be less than 2% of the intensity of the $10\bar{1}$. The intensities of the higher order equatorial reflections are larger and have recently been reported by Holmes *et al.* (1980) (Table 10.1).

Studies of the layer line pattern from insect muscle were reported by Tregear and Miller (1969) and later in more detail by Armitage *et al.* (1975) who also described further experiments on the equatorial reflections. It was found that little change (actually a small increase) in intensity of the 145-Å meridional reflection occurred when the muscle was Ca^{2+} activated. But when the muscle was performing oscillatory work, the intensity of the 145-Å reflection changed in an oscillatory fashion as well.

This was determined by an ingenious experiment in which the intensities of the reflections at different phases of the contractile cycle were monitored by a counter coupled to a multichannel scaler. The results obtained by Miller and Tregear (1971) and Armitage *et al.* (1975) from measurements of the 145-Å reflection are shown in Fig. 10.16. It can be seen from this diagram that a marked periodic variation of intensity occurs at an oscillation frequency of 4.0 Hz when much work is done by the muscle; but it was also found that when the muscle is oscillated without doing useful work, either at high frequency (18Hz) or at low Ca^{2+} levels, little intensity change is observed. Since under these conditions the ATPase is also low (Tregear, 1975; Pybus and Tregear, 1972), it is likely that the oscillatory change in intensity of the 145-Å reflection is associated directly with ATP hydrolysis, and, by inference, with cross-bridge activity presumably associated with actin attachment. On the other hand, any such attachment to actin must be by only a few cross bridges at any instant since Armitage *et al.* (1975) detected little significant change in intensity during the oscillatory cycle of either the 190-Å actin layer line or the 59-Å actin layer line. Had it been observed, an increase in intensity of either of these reflections could have been interpreted with some certainty as being caused by cross-bridge labeling (see Section 10.35). Armitage *et al.* did, however, detect a slight intensity change on the outer part of the 385-Å layer line, but the intensity dropped as the tension increased, indicating that this part of the 385-Å layer

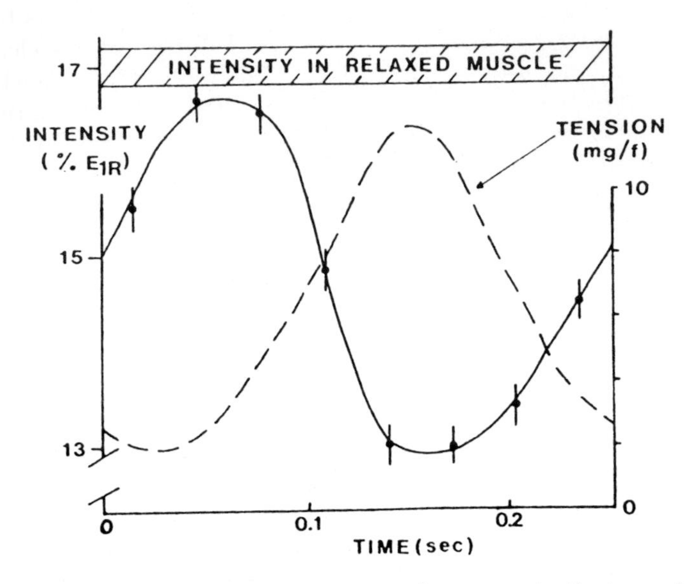

FIGURE 10.16. Variation of intensity of the 145-Å meridional reflection and the corresponding variation of tension during a 4-Hz sinusoidal oscillation of calcium-activated insect flight muscle fibers. (From Miller and Tregear, 1971.)

line was very likely caused by myosin rather than by actin labeling. Similar analysis of the $10\bar{1}$ and $20\bar{2}$ equatorial reflections showed that these too change periodically, with the $10\bar{1}$ changing more or less in antiphase to the tension and the $20\bar{2}$ lagging slightly the tension change. Little or no change was observed in any of these reflections from muscle oscillated in the absence of calcium.

The phase relationships of the various changes observed in patterns from working muscles were interpreted as being associated with movements of the myosin cross bridges, but since the average patterns recorded on film from either contracting, relaxed, or Ca^{2+}-activated muscles were substantially the same, it was concluded that those changes that did occur must be associated with the movement of relatively few cross bridges at any instant.

Cross-bridge movement as a result of Ca^{2+} activation was inferred from the result that the absolute intensities of the $10\bar{1}$ and $20\bar{2}$ equatorial and 145-Å meridional reflections increased under this treatment and that the relative intensities of the equatorials ($I_{10\bar{1}}/I_{20\bar{2}}$) changed (as described earlier) from about 1.9 in patterns from relaxed muscle to 1.8 in patterns from Ca^{2+}-activated muscle. This represents about 10% of the change in rigor muscle ($I_{10\bar{1}}/I_{20\bar{2}} = 0.82$). The interpretation of the other changes which occur will be considered in detail in the next sections.

10.3.3. Modeling: Changes in Myosin Filament Structure

Introduction. Analysis of the possible meaning of the X-ray diffraction evidence from contracting muscles is most conveniently dealt with under a number of separate headings: (1) changes in myosin filament structure, (2) the meridional pattern from vertebrate muscle and the observed spacing changes, (3) the equatorial diffraction data, and finally, (4) changes in the actin filaments. But, of course, it must be clear that any interpretation must satisfy simultaneously all of the various types of evidence that are available. This section and those that follow explore the possible structural changes that may be involved in muscular contraction.

Vertebrate Skeletal Muscle. It has been shown that when a vertebrate skeletal muscle is activated the layer lines from the myosin filaments at spacings of 429-Å and 215-Å (etc.) all get weaker. This in itself is evidence that it is largely the mobile myosin cross bridges that give rise to this layer line pattern and is consistent with these cross bridges moving from their positions in relaxed muscle. H. E. Huxley and Brown (1967) suggested that the movement might be a simple thermal "waggling" of the cross bridges during their contractile cycles. The cross bridges would therefore stabilize in a defined position in relaxed muscle but would be free to move during contraction. But it is clear from the more recent X-ray diffraction evidence that this cannot be the whole story. First, the myosin layer lines also reduce in intensity even when the muscle is stretched to a position where the thin and thick filaments do not overlap. Under the same conditions, there is a change in spacing of the strong meridional X-ray reflection from 143.4 to 144.8 Å when the muscle is activated. Taken together, these two results probably indicate (but see the caution in Section 10.3.1) that myosin filaments are "switched on" when a muscle is activated (presumably by the action of calcium, just as in the thin filaments) and that this thick filament regulation is independent of actin.

There are now a number of other results that support the idea that vertebrate thick filaments can be regulated as in many other muscles (but clearly not by the same mechanism).

1. The DTNB light chains in vertebrate myosin can bind Ca^{2+} (Morimoto and Harrington, 1974) although at nonspecific sites (Bagshaw and Kendrick-Jones, 1979).

2. The results of Harrington and his co-workers (described in detail in Chapter 6) show that there may be some structural difference between myosin filaments in the presence or absence of Ca^{2+} as detected in viscosity and sedimentation experiments (but note that the detected changes were on a relatively slow time scale).

3. The recent results of Wagner and Weeds (1977) have suggested that the DTNB light chains are located close to the junction between the myosin heads and myosin S-2. It was found that the susceptibility of this region to proteolysis can be affected by the level of Ca^{2+} present and that the DTNB light chains apparently mediate this process in some way.

4. The regulation of vertebrate muscle is much more efficient than could be explained by thin filament regulation alone (see discussion in Haselgrove, 1975a).

It is therefore very reasonable to take this Ca^{2+} activation as a likely property of the thick filament. In this case, the X-ray diffraction evidence can be interpreted to mean that in the presence of Ca^{2+} there is a structural change that gives rise both to the observed change in spacing of the 143-Å reflection and to the observed drop in layer line intensity. But having said that, several questions remain to be answered. First, what structural change could account for both of these observations? Second, why is the drop in layer line intensity less in nonoverlap muscle than in muscles at rest length? Third, why is the intensity drop the same for stretched and rest-length muscles, provided that overlap still occurs (i.e., at sarcomere lengths of 2.25 μm and 2.9 μm; Table 10.5)?

The last two of these questions can be approached in the following way. If more of a layer line intensity drop occurs in a situation where actomyosin interaction can occur than when it cannot, then interaction with actin must have additional effect in disturbing the cross-bridge regularity. It was suggested by Squire (1975) that the 30% intensity drop that occurs in diffraction patterns from activated nonoverlap muscle might well be caused by a specific change in cross-bridge configuration from a defined relaxed position to a defined Ca^{2+}-activated position. For example, the observed intensity drop on the 429-Å layer line could be accounted for by an azimuthal rotation of all of the cross bridges by, say, 30° from a relaxed position with parameters $\phi = 45–60°$, $\theta = 75–90°$, $R = 90–100$ Å (Fig. 7.33). The spacing change in the 143-Å reflection could well be associated with this change (see Section 10.3.4). The additional intensity drop that occurs in muscles with relatively short sarcomere lengths must then be associated with the actin–myosin interaction. But the "waggling" of a particular cross bridge cannot simply be a result of its interaction with a thin filament, since the same layer line intensity drop occurs even when the muscle specimen is such that only about half of the A-band overlaps these thin filaments (i.e., at a 2.9-μm sarcomere length). It must be concluded that the interaction of a few cross bridges in the overlap zone is sufficient to disrupt the regular cross-bridge arrangement and thus to allow all of the crossbridges to "waggle." Another possibility is that the thick filaments in some way "feel" the tension in the muscle (see comments in Chapter 11 on stretch activation) and that in response, the cross bridges are released to some extent from their activated positions and move about because of thermal effects. The additional layer line intensity change resulting from this could not occur in diffraction patterns from nonoverlap muscles where the tension effect would not be felt.

Diffraction patterns from rigor muscle showed no evidence for the myosin layer lines at all, but the strong meridional reflection had a measured spacing of 144.8 Å as in contracting muscle. One must conclude that in rigor the cross bridges are able to move by a considerable amount from the Ca^{2+}-activated positions in order to interact maximally with the thin filaments. A three-state model for the thick filaments can therefore be proposed.

State 1. The *relaxed* state with a defined cross-bridge position on a helix of repeat 430 Å. The strong meridional reflection is seen at a spacing of 143.4 Å.

State 2. The *activated* state with a second defined cross-bridge position (possibly induced by the binding of Ca^{2+} to the thick filaments). The cross-bridge helix repeat is about 430 Å, but the meridional reflection has moved to 144.8 Å.

State 3. The free *contracting* state that is induced in muscles at all sarcomere lengths where overlap between thick and thin filaments occurs, possibly by a propagated disruption of the cross-bridge helix from the overlap region or by a tension-sensing mechanism in the thick filaments. The myosin helix repeat may well still be about 430 Å, but the meridional spacing is 144.8 Å as in State 2.

The existence of State 3 would account for the result that the observed drop in intensity of the layer lines is the same whether the muscle is activated at a 2.25 μm or 2.9 μm sarcomere length. To account for the other observations, it is necessary to postulate that in transitions from State 1 to State 2 a 35 to 40% drop in 429-Å layer line intensity (L1) results, as do 25 to 30% drops in the intensities of the second (L2, 214-Å) and third (L3, 143-Å meridional) layer lines. The spacing of this meridional reflection also changes from 143.4 to 144.8 Å. A subsequent transition to State 3 causes a further intensity drop in the myosin layer lines so that L1 becomes about 30% of its value in patterns from resting muscles, and L3 becomes 60 to 65% of its resting value (Table 10.5). But no additional spacing change occurs.

It is possible that a fourth myosin state should be added to the list. This is the rigor state. In patterns from rigor muscles at overlap lengths, the myosin structure appears to be rather similar to that in State 3, the active state. But this would not explain why patterns from muscles stretched beyond overlap and, given the same rigor-inducing treatment, should show no trace of the myosin layer lines L1, L2, etc., that would be expected from thick filaments in State 2. It must be concluded that the rigor treatment disrupts the regularity of State 2, which by its very nature must be controlled by an indirect and subtle Ca^{2+}-sensing mechanism, so that the whole cross-bridge helix is lost. Of course, these results mean (as described in Section 10.2.4) that the thick filament repeat in rigor vertebrate muscle is not known.

However, the meridional diffraction patterns from rigor muscles at any length contain reflections that index reasonably well on a long repeat of about 2315 Å (16 × 144.8 Å; Haselgrove, 1975a). Haselgrove therefore suggested that in the rigor state the thick filament symmetry might be similar to that in insect flight muscle in having the same helix repeat (1155 Å) and axial spacing (145 Å), although having only three strands rather than the six strands in insect thick filaments. However, according to the present analysis the 1% change in spacing of the strong meridional reflection (~144 Å) occurs when the muscle is activated (State 2), and the helix repeat is still about 430 Å. Haselgrove (1975a) explained the existence of what he thought of as a mixed state—the contracting state that showed both the 430-Å layer line repeat and the 1% spacing change in the 144-Å reflection—in terms of a mixed population of rigor and relaxed thick filaments. But with the analysis given here, it would not be necessary to postulate such a mixed population, since the 430-Å periodicity and the meridional spacing of 144.8 Å would both occur in the single activated state (State 2) and consequently in contracting muscle also.

Finally, since it seems that the 1% change in the myosin periodicity can occur in filaments that still have a 430-Å repeat, one might well question Haselgrove's conclusion that the thick filament repeat in rigor muscle is not 430 Å but is 1150 Å or possibly 2315 Å (Haselgrove, 1975a). He clearly showed that the observed meridional reflections in X-ray diffraction patterns from rigor muscles at rest length or at nonoverlap were very much the same. Their spacings are given in Table 10.6. As shown here, all of these reflections will index reasonably well on a repeat of 2315 Å, but what is not known is whether these reflections come from the thick filaments or the thin filaments or from both. But even in diffraction patterns from relaxed muscle (Table 10.6), a relatively strong tropomyosin–troponin repeat of about 386 Å is evident, and this could clearly account for the observed reflections at 192 Å and 127.5 Å in patterns from rigor muscle. The reflections at 144.6 Å and 72.3 Å must, on the other hand, be related to the thick filament repeat. This leaves the very weak reflections at 227 and 212 Å and at 110.8 and 101.8 Å still to be explained. It is perhaps significant that the reflection at 101.8 Å only occurs in patterns from rigor muscles at rest length and may be an indication that it is caused by the actin–myosin interaction.

TABLE 10.6. Meridional Spacings in X-Ray Patterns from Frog Muscle in Rigor Compared with the Main Meridionals from Resting Patterns[a]

Observed spacing in rigor (Å)	Calculated order of 2315 Å	Observed spacing in relaxation	Calculated orders of	
			386 Å	144 Å
227	231.5 (10)	229.7		
212	210.45 (11)	214.3		
192	192.9 (12)	187.4	193	
144.6	144.69 (16)	143.4		144
127.5	128.61 (18)	129.0	128.7	
110.8	110.24 (21)	110.7		
101.8	100.65 (23)			
72.3	72.34 (32)	71.6		72

[a] Probable error on all measured reflections about 0.5%. From Haselgrove (1975a).

But all of these reflections are at positions where the diffraction patterns from both the M-band and the C-zone might be expected to be strong (Chapter 7). They also agree reasonably well (except for the reflection at 101.8 Å) with the spacings of reflections observed in diffraction patterns from relaxed muscle (229.7, 214.3, and 110.7 Å, respectively, Table 10.6), which might suggest that they need not be caused by a thick filament structure very different from that in relaxed muscle. There is therefore no compelling reason to postulate a change in symmetry of the thick filaments provided that the 1% spacing increase of the 143.4-Å reflection can be accounted for.

In order to account for this spacing change, it is necessary to consider the form of the diffraction pattern from relaxed muscle in some detail. This is done in Section 10.3.4.

Insect Flight Muscle. The results described in Section 10.3.2 on insect flight muscle showed conclusively that here the relaxed, Ca^{2+}-activated and contracting states are quite distinct and that Ca^{2+} activation caused a characteristic change in the equatorial and meridional diffraction pattern consistent with cross-bridge movement. On the other hand, the layer line positions and intensities in diffraction patterns from relaxed and Ca^{2+}-activated muscles show that the changes caused by calcium activation must be from one defined state to another defined state on a helix of the same pitch and subunit repeat. The result is entirely analogous to the conclusions of the last section on vertebrate skeletal muscle. Once again, it must also be postulated that when the muscle is stretched the tension generated in the thick filaments must allow a transition of the cross bridges to a third "contracting" state where force can be generated. This transition might then be part of the explanation of the phenomenon of "stretch activation" (see Chapter 11).

The beautiful electron micrographs (Fig. 8.4) published by Reedy *et al.* (1965) which showed cross connections in insect flight muscle between the thick and thin filaments in the relaxed and rigor states seemed to show a clear difference of cross-bridge angling. We have already seen that in rigor the cross bridges are attached to actin at an angle of about 45°. But it is not so clear that the 90° position of the cross bridges in preparations of fixed relaxed muscle is a true indication of the cross-bridge angling in relaxed muscle *in vivo.* It is known, for example, that the effect of fixatives such as glutaraldehyde is to "activate" a muscle either to produce tension or at least to produce some structural change. It is therefore possible that the 90° angling seen by Reedy *et al.* (1965) and by Freundlich and Squire (1980) (Fig. 10.9a) is a demonstration of the angle at which myosin cross bridges are tilted in Ca^{2+}-activated muscle and possibly the angle at which cross-bridge attachment to the thin filaments occurs prior to force generation. But, just as in vertebrate muscle, the cross bridges in relaxed insect muscle might well tilt 20 to 30° away from the 90° position in relaxed muscle, although this is by no means certain. There is as yet no clear evidence to suggest that the relaxed cross-bridge configurations in different muscles must necessarily be the same, and evidence summarized in Fig. 10.3 shows that marked differences may occur.

10.3.4. Modeling: The Meridional Pattern and the Observed Spacing Changes

The spacings of the low-angle meridional X-ray reflections in diffraction patterns from resting frog sartorius muscle according to Haselgrove (1975a) are given in Table 10.7 (see also Table 7.1). In Chapter 7, the reflections of spacing 441.8, 418.4, 394.9, and 383.0 Å were accounted for in terms of interference effects in the A-band (441.8 and 418.4 Å) and in the I-band (394.9 and 383.0 Å). The former were associated with C protein, and the latter with the tropomyosin–troponin structure in the thin filaments. Turning now to the meridional region close to the 143.4-Å reflection, there are a number of closely spaced reflections at 153.1, 150.8, 148.0, 143.4, and 141.4 Å. It is evident that these reflections are probably also caused by sampling of relatively broad intensity peaks, and from the separations of the various reflections, the interference functions involved must have periodicities of between 7,000 and 10,000 Å. Figure 10.17 shows a densitometer trace from Haselgrove (1975a) of the meridian of a diffraction pattern from a frog sartorius muscle recorded on a mirror/crystal camera (Chapter 2) with a 2-m

TABLE 10.7. The Low-Angle Meridional X-Ray Pattern from Resting Frog Muscle and Its Explanation in Terms of Thin Filaments, Myosin Cross Bridges, and C Protein

Observed spacing (Å)[a]	Orders of 7095-Å repeat (C protein)	Peaks from seven-line array of period 440 Å: peak (range)	Cross-bridge array with 8604-Å period	Thin filament reflections with 10,300-Å period
600 (±20)[b]				
529.4				
494.0				
441.8	443.45 (16)	440 (513–385)		
418.4	417.36 (17)			
394.9	394.2 (18)			395.15 (26)
383.0				381.48 (27)
365.3				367.86 (28)
352.2				355 (29)
271.3	272.9 (26)			271.05
237.7	236.5 (30)			
229.9	228.9 (31)			
222.5	221.7 (32)	220 (237–205)		
214.3	215.0 (33)		215.1 (40)	
209.8	208.7 (34)		209.86 (41)	
187.4				187.3 (55)
181.9				180.7 (57)
176.9				177.6 (58)
153.1	154.24 (46)			
150.8	150.96 (47)			
148.0	147.8 (48)		148.34 (58)	
		146.7 (154–140)	145.8 (59)	
143.4	144.8 (49)		143.4 (60)	
141.4	141.9 (50)		141.05 (61)	
110.7	110.85 (64)			
107.2	107.5 (66)			

[a] Reflections recorded using a 2-m crystal monochromator camera by Haselgrove (1975a). Probable error, unless otherwise stated, is about ±0.5%.
[b] This is a very broad diffuse reflection.

specimen-to-film distance. One feature evident from this trace is that the relatively weak reflections on the low-angle side (left) of both the 214.3 Å and the 143.4-Å reflections are considerably stronger than the reflections on the high-angle side (right).

Since evidence has been presented that C protein may have a larger repeat than myosin (about 435 to 440 Å rather than 430 Å), it is conceiv-

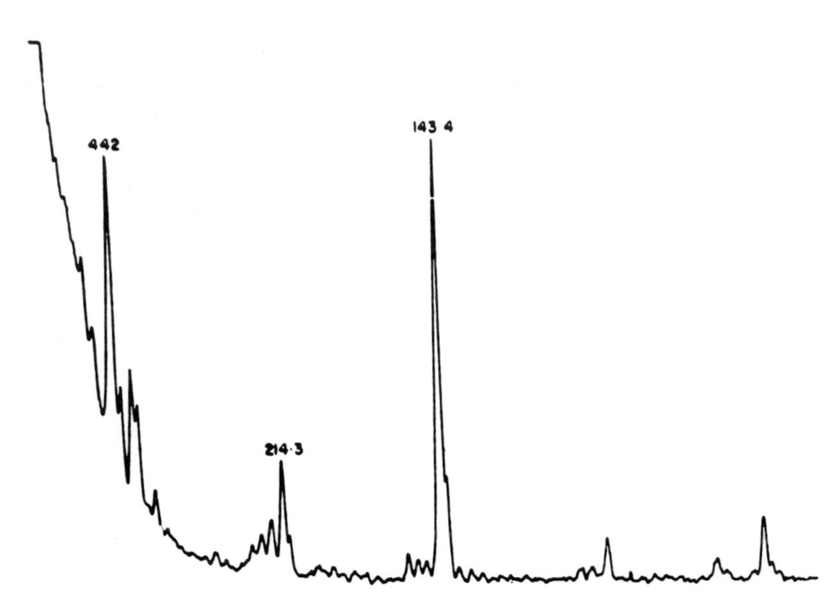

FIGURE 10.17. Densitometer trace along the meridian of the diffraction pattern of relaxed frog sartorius muscle at rest length taken on a 2-m camera by Haselgrove (1975a). The reflections at 143.4, 214.3, and 442 Å are indicated.

able that these small peaks to the longer-spacing side of the myosin peaks at 214.3, 143.4 Å etc., are also in part caused by C protein. In Chapter 7 it was suggested that the 441.8- and 418.4-Å reflections could be accounted for in terms of an interference function of repeat about 7095 Å. This same repeat will also account reasonably well for most of the other low-angle meridional reflections as indicated in Table 10.7. The table also shows that most of the reflections that do not index well on a 7095-Å repeat can be accounted for in terms of the tropomyosin–troponin periodicity of 388 Å in the I-band sampled by the I-band interference function of period about 10,300 Å (Chapter 7).

However, the all-important myosin reflection at 143.4 Å does not fit well as an order of either of these interference functions. The predicted 49th order of 7094 Å would be at 144.8 Å, a value close to the observed spacing in patterns from active and rigor muscle but not to that in patterns from relaxed muscle. The observed reflection at 143.4 Å must clearly be associated with the axial repeat of the myosin cross bridges. Its extent along the meridian is sufficiently narrow to suggest that it too is an effect of an interference function of long periodicity sampling an underlying diffraction peak from the cross bridges.

As in the case of the C-protein and the tropomyosin–troponin arrays, one could imagine that the diffraction patterns from the cross-bridge arrays in the two halves of the A-band will interfere to give the observed pattern. In this case, the interference periodicity involved must be close to a multiple of 143.4 Å. Since on any model (Chapter 9) these cross-bridge arrays would be separated by a distance of about 8600 to 8700 Å [i.e., ½(15700 − 1600) + 1600 Å], a possible repeat could be 8604 Å (143.4 × 60). As listed in Table 10.7, this repcat would have orders at spacings of 148.34, 145.8, 143.4, and 141.05 Å, all of which agree well with the spacings of the observed reflections in that region except that no peak is observed at 145.8 Å. In general, there could be two reasons for the absence of this reflection. One is that the underlying myosin diffraction pattern that is being sampled by the interference function may have zero intensity at this position. The other is that this reflection caused by the cross-bridge array might interfere destructively with a reflection of a similar spacing caused by another feature of the thick filament structure such as the filament backbone or C protein.

One important conclusion from this analysis is that it may well be the interference function and not the cross-bridge array itself that puts the observed strong reflection at a spacing of 143.4 Å (Rome, 1972). The cross-bridge array could have an axial repeat slightly different from this. It is easy to show (Table 10.7) that if the diffraction pattern from these cross bridges is required to be close to zero at 1/145.8 Å and if it is assumed that each of the two cross-bridge arrays is about 7000 Å long, then the myosin repeat would need to be about 143.5 Å as expected. But, if the 145.8-Å reflection is absent because of a destructive interference effect with a reflection from another source, then the myosin repeat could lie more or less anywhere in the range 143 to 145 Å.

Although at first sight it may seem to be more likely that the cross-bridge repeat is about 143.5 Å and that the myosin diffraction pattern is weak where the 145.8-Å reflection would be expected, it will be shown later that this may not be the truth of the matter. One thing, however, is certain. This is that the positions of the weak reflections both in the 210- to 220-Å region and the 150- to 140-Å region are consistent with a C-protein repeat different from the myosin repeat, and they would be hard to explain if these repeats were identical. In addition, the seven-line C-protein array accounts well for the number of orders seen in each group of reflections (Table 10.7). The diffraction peaks from one C-zone would be expected to extend from 513 to 385 Å, from 237 to 205 Å, and from 154 to 140 Å, respectively, and the observed reflections lie precisely within these limits. Although the M-band structure and the A-band proteins other than C protein are bound to contribute to the observed diffraction pattern (probably by giving rather diffuse peaks), it is fair to conclude that the main features of the meridional X-ray diffraction pattern from relaxed frog sartorius muscle can be accounted for quite well.

With this background, the changes in spacing that occur when a muscle is activated or goes into rigor can be discussed more easily. Haselgrove's (1975a) observations were that in diffraction patterns from muscles activated at any sarcomere length or in rigor at any sarcomere length, the 143.4-Å reflection "moves" to 144.8 Å. The only situation where an intermediate spacing was observed (143.8 Å) was in patterns from relaxed muscles stretched to a 2.9 μm sarcomere length before being allowed to equilibrate at rest length.

It seems likely that in rest-length active muscle and especially in rest-length rigor muscle, the myosin cross bridges will be sufficiently displaced from their rest positions that the reflection that remains at about 144.8 Å must be caused to a significant extent by the backbone structure of the thick filaments. Assuming that the C-protein molecules have not altered significantly in position, then part of this reflection could be caused by the 49th order of the C-protein interference function (Table 10.7). But it would be expected that the myosin rod packing in the filament backbone would also contribute in this region. This suggests that in both rigor and activated thick filaments (State 2), the backbone structure has a repeat of about 145 Å, but since the observed meridional X-ray reflection is still very narrow along the meridian, it is clear that the backbone structure is reasonably regular over a large part of the filament; as before it is likely that the diffraction patterns from the two halves of the thick filament interfere with each other. Now, in the modeling of the thick filament described in Chapter 9, it was suggested that the myosin rods in the thick filament backbone may only be equivalently packed in the C-zone. Thus, the diffraction pattern from these rods might well be sampled by the same interference function (of period 7095 Å) as the C-protein reflections. A strong peak at a spacing of 144.8 Å would therefore be expected, but it might be flanked by weaker reflections at spacings of 147.82 and 141.9 Å. Alternatively, if the backbone structure is more or less regular in sections 4 to 13 inclusive of the bridge region (Chapter 9), the interference function would have a period of about 8540 Å and would give only a single strong peak at 144.8 Å (59th order).

Assuming then that the reflection at 144.8 Å is partly caused by the filament backbone and by C protein, it is necessary to account for the fact that such a reflection is not observed in diffraction patterns from relaxed muscle. In fact, there are two obvious reasons for not seeing this reflection. One is that the backbone periodicity may itself change from 143.4 Å in relaxed muscle to 144.8 Å in activated or rigor muscles as suggested by Haselgrove (1975a), and the other is that the periodicity might remain constant but that in patterns from relaxed muscle the backbone reflection at 144.8 Å interferes destructively with another reflection, presumably from the cross bridges (Rome, 1972), and the reflection is not seen. In fact, it can be shown (Squire, unpublished) that the 145.8-Å reflection expected from the cross-bridge array (Table 10.7) probably has the right amplitude and phase and is sufficiently close to the backbone reflection at 144.8 Å that destructive interference can occur.

Two models can therefore be suggested to account for the "shift" in spacing of the 143.4-Å reflection. First, the whole myosin filament may undergo a slight, specific 1% extension (from State 1 to State 2) so that its axial repeat changes from 143.4 to 144.8 Å, and the helix pitch changes from 3 × 143.4 to 3 × 144.8 Å (i.e., 434.4 Å). This change must inevitably be associated with a change in azimuthal position of the cross bridges, but little change in axial tilt of the cross bridges would be possible. This would alter the effective separation of the two halves of the A-band by either more or less than the observed 1%. The second model is one in which the thick filament repeat remains virtually the same (144.8 and 434 Å) the whole time. But in diffraction patterns from relaxed muscle, the cross bridges in the two halves of the A-band are separated by say 8604 Å, giving a strong myosin peak at 143.4 Å, whereas on activation by calcium (State 2) the cross bridges move both azimuthally and axially so that the periodicity of the interference function increases by 1% (to 8690 Å), and the myosin reflection is now sampled at 144.8 Å. In this case, each cross bridge would need to move axially by about 40 to 45 Å away from the M-band. If, in relaxed muscle, the cross bridges are axially tilted by 20 to 30° from a direction normal to the filament axis, one could suggest that this tilt might be towards the M-band in relaxed muscle (Fig. 10.18a) and that on activation the tilt might reduce to about 0 to 10° (Fig.

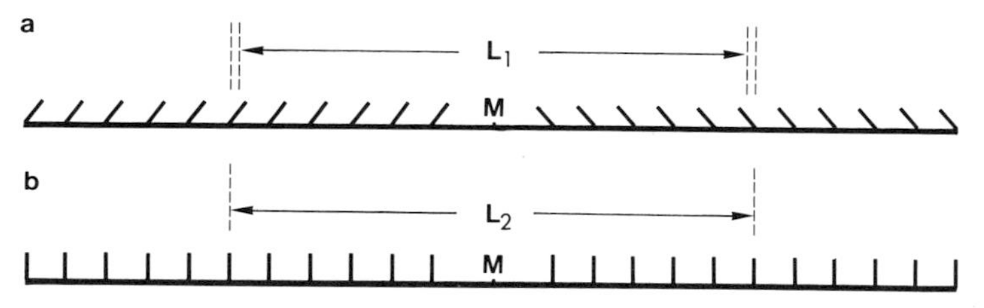

FIGURE 10.18. Simulation of a possible scheme in which cross bridges are tilted towards the M-band (M) in relaxed vertebrate muscle and swing to a tilt closer to 90° to the filament axis when the muscle is activated. This would increase the spacing of the separation of the two halves of the cross-bridge array from L_1 in (a) to L_2 in (b), and this could explain the 1% change in spacing of the 144-Å reflection (see text).

10.18b) so that the cross bridges are pointing more or less straight out from the filament surface in the activated state. Until more evidence is available, it will be hard to choose between these alternatives. But for the purposes of the present discussion, the important point is that in vertebrate skeletal muscle there is a structural difference between relaxed and activated myosin filaments that is very probably related to a systematic alteration of the cross-bridge configuration.

10.3.5. Modeling: The Equatorial Diffraction Data

Vertebrate Skeletal Muscle. The differences in the equatorial diffraction patterns from vertebrate skeletal muscles in different states (Haselgrove and Huxley, 1973) are that the ratio $I_{10\bar{1}}/I_{11\bar{2}}$ changes from about 2.5 for relaxed muscle to 0.6 for active muscle to about 0.25 for rigor muscle (although the figure for contracting muscle is in dispute—Matsubara *et al.,* 1975; Worcester *et al.,* 1975—and may be closer to the rigor value). For reasons described in detail in Chapter 7, the relative increase in intensity of the $11\bar{2}$ reflection is consistent with the presence of more material at the trigonal points of the unit cell (Table 10.1). This is shown again in Fig. 10.19, where it can be seen that these trigonal points lie in the $11\bar{2}$ planes of the lattice and between the $10\bar{1}$ planes. But it is important to know more about the specific mass movements that cause the observed changes.

In Chapter 2 it was shown that a particular diffracted beam from a specimen can be described in terms of an amplitude (e.g., of the alternating electric or magnetic field associated with the electromagnetic X radiation) and a particular phase angle relative to some standard (such as the phase of the undiffracted beam). When the diffraction pattern is recorded, the blackening observed on the film is a function of the square

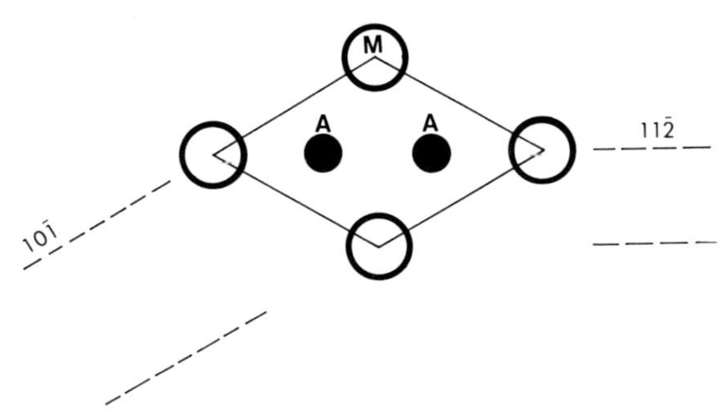

FIGURE 10.19. Diagram to show the actin filaments (A) in the unit cell of the A-band in vertebrate striated muscle lie between the $10\bar{1}$ planes through the myosin filaments (M) but along the $11\bar{2}$ planes.

of the amplitudes of the different diffracted beams but is independent of their phase. For this reason phase information is lost.

The "phase problem" is crucial, however, since in order to reconstruct an image of the diffracting object (as is done in a microscope), it is necessary to know both the amplitudes and phases of the diffracted beams. Since the phases do not come directly from a recorded diffraction pattern, it is necessary to think of ways of finding out what they might be. Fortunately, in some special cases the phase problem is simplified. One such case is when the diffracting object is centrosymmetric. This means that if there is a particular electron density at point x, y, z in space, then an identical electron density will also be found at $-x, -y, -z$. The effect of this is to restrict the possible values of the phases of the diffracted beams so that instead of being anywhere in the range 0 to 360° (see Fig. 2.2) they can only have the values 0 or 180°. In general, if a diffraction pattern from any object is recorded, and the amplitudes $F(hkl)$ of the reflections of indices (hkl) are measured, it is possible to calculate the electron density $\rho(xyz)$ at a particular point x, y, z in the object. This general Fourier synthesis expression can be written

$$\rho(xyz) \propto \sum_{hkl} F(hkl)\exp[-2\pi i(hx + ky + lz)]$$

(i.e., electron density is the sum over all reflections of the product of the structure factor and an exponential term in hkl and xyz). The general problem is that $F(hkl)$ (the structure factor) is defined both by an amplitude $|F(hkl)|$ and by a phase angle that is anywhere from 0 to 360°. However, if the structure is centrosymmetric, the expression for $\rho(xyz)$ simplifies to

$$\rho(xyz) \propto \sum_{hkl} (\pm)|F(hkl)|\cos 2\pi\,(hx + ky + lz)$$

where the only ambiguity is the sign of each term in the series. This sign depends on whether the relative phase of any term is 0 or 180°. In the present discussion we are concerned with the diffraction pattern on the equator where, by definition, $l = 0$. So the expression given above simplifies to

$$\rho(xy) \propto \sum_{\mathrm{hk}} (\pm)|F(hk0)|\cos 2\pi(hx + ky)$$

This now means that the electron density, $\rho(xy)$ of the diffracting object when viewed down the z axis can be determined if the relevant structure amplitudes, $|F(hko)| = \sqrt{I(hko)}$, are known and if the sign of each term can be determined. However, the amount of detail (resolution) that can be seen in the final electron density image will depend on how many terms are included in the summation with which $\rho(xy)$ is calculated. The more terms that are included, the better will be the resolution. In practical terms, all of the relevant reflections actually observed in a diffraction pattern are normally included in such a Fourier synthesis, but usually the reflections with high indices (*hkl*) that occur to the outer edges of a diffraction pattern are very weak for a variety of reasons and cannot be included. As explained in Chapter 2, such high-angle reflections are associated with relatively small interplanar spacings (d_{hkl}) in the diffracting object, and since these are not included in the Fourier synthesis, the diffracted information about fine details in the object is not included in the image. In fact, the resolution in the image is usually defined in terms of the spacing (d_{hkl}) of the reflection (*hkl*) that, of those used in the Fourier synthesis, is furthest out from the center of the diffraction patterns.

In the case of vertebrate skeletal muscle, the only reflections that have been recorded satisfactorily from relaxed, rigor, and active muscles are the $10\bar{1}$ and $11\bar{2}$ reflections. The latter corresponds to a *d* spacing of about 200 Å. At a resolution of only 200 Å, little detail of the internal structure of the actin and myosin filaments would be visible in an image of the unit cell. But, since in the unit cell the filaments themselves are arranged in a centrosymmetric fashion, it is reasonable to assume that the phases for these two reflections are either 0 or 180°. The phase factors (or signs) in the Fourier synthesis expression will be ± 1, but since these are relative, one can assume that the sign for the $10\bar{1}$ reflection term is +, which leaves doubt only about the sign of the $11\bar{2}$ term.

H. E. Huxley (1968) and Haselgrove and Huxley (1973) obtained Fourier synthesis images for the unit cell in vertebrate skeletal muscle in the relaxed, active, and rigor states using their observed values for the relative intensities of the $10\bar{1}$ and $11\bar{2}$ reflections. They assumed in each case that the phases for both of these reflections were the same. The results they obtained are shown in Fig. 10.20. It is clear that the myosin filaments (M) are very massive in relaxed muscle, but the actin positions (A) are not; whereas in active muscle the actin positions are considerably denser, and in rigor they are very strong indeed, in both cases at the expense of the myosin filaments which reduce slightly. These images therefore reinforce the conclusions described earlier that in rigor the myosin cross bridges, which are the only structures thought to be mobile enough, move from the relaxed position close to the myosin filaments to positions where they bind to the thin filaments. This conclusion was also

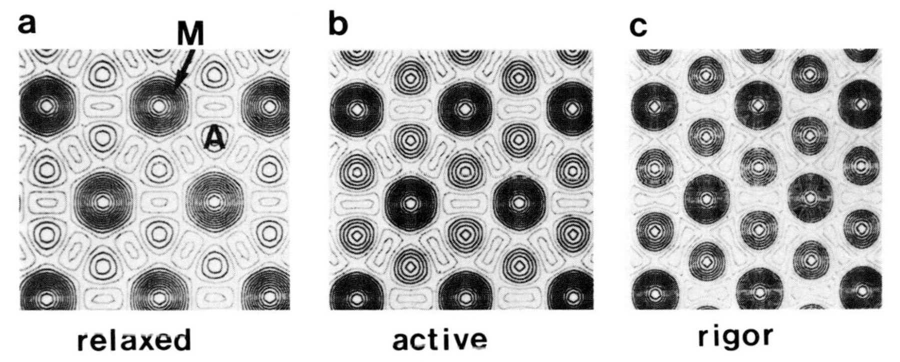

FIGURE 10.20. Fourier projections along the muscle axis of the electron density distribution in frog sartorius muscles at a sarcomere length of 2.2 μm. The projections were calculated using only the $10\bar{1}$ and $11\bar{2}$ equatorial reflections. The strongest peaks correspond to the positions of the myosin filament. (a) Live resting muscle; (b) contracting muscle; (c) rigor muscle. (From Haselgrove and Huxley, 1973.)

supported by electron micrographs of transverse sections of relaxed and rigor muscles (H. E. Huxley, 1968).

For various reasons given in Chapters 7 and 8 and in Sections 10.2.3 and 10.2.4, it now seems likely that not all of the available cross bridges can, in fact, attach to actin in rigor muscle; possibly only about 50 to 75% do so. But the rigor structure (Fig. 10.20c) provides a yardstick against which the results from contracting muscles can be gauged. It gives an indication of how many cross bridges can attach to actin in favorable circumstances.

However, there is a problem in interpreting the Fourier synthesis images in Fig. 10.20. This is that strictly the summation should have included the $F(000)$ term. This would have added a uniform background to the electron density image and would therefore have shown where the background electron density level should be. Interpretation of the computed images therefore depends on what is assumed about where the background ought to be. Haselgrove and Huxley (1973) considered two possibilities, (1) one in which it was assumed that the background density level corresponded to the lowest density value seen in the whole of the calculated density map, and (2) one in which the background level was taken to be the lowest density seen on the line joining the actin and myosin filaments (see Fig. 10.20). Using these assumptions, the relative masses of the myosin and actin filaments were estimated from the electron density maps to give a ratio A/M. The filaments were taken to extend to the minimum of the observed filament profile. On this basis, assumption (1) above always gave slightly higher A/M values than assumption (2). For resting, contracting, and rigor muscles, the high values were 0.23, 0.36, and 0.61, respectively, and the low values were 0.12, 0.27, and 0.52, in all cases for rest-length muscle. Note that these values for contracting muscle took no account of the gradual drop in the tension level that occurred during the X-ray diffraction experiment. Haselgrove and Huxley therefore compensated for this to produce a "corrected" value for $I_{10\bar{1}}/I_{11\bar{2}}$ of 0.78 and corrected values for A/M of 0.39 (high) and 0.30 (low). Note also that the results of Matsubara *et al.* (1975) gave slightly lower values for $I_{10\bar{1}}/I_{11\bar{2}}$ in contracting muscles and this would lead to slightly higher values for A/M, values rather close to the rigor values.

Since these calculations can only be expected to give an approximate indication of the mass changes that occur (see discussion in Chapter 11), it will be assumed for the purposes of the discussion that follows that A/M is about 0.2 in relaxed muscle, about 0.35 to 0.5 in contracting muscle, and about 0.55 to 0.6 in rigor muscle. Assuming that this change in relative mass corresponds to a reduction of mass (ΔM) on each thick filament and an increase of the same total mass in the thin filaments (actually + $\Delta M/2$ per thin filament) then the observed ratio (A/M_{contr} must be given by

$$\left(\frac{A}{M}\right)_{\text{contr}} = \frac{A_{\text{rel}} + \Delta M/2}{M_{\text{rel}} - \Delta M}$$

(A_{rel} and M_{rel} are the relative masses of the actin and myosin filaments, respectively, in relaxed muscle.)

Since $A_{\text{rel}}/M_{\text{rel}}$ is 0.2, then

$$\left(\frac{A}{M}\right)_{\text{contr}} = \frac{0.2M_{\text{rel}} + \Delta M/2}{M_{\text{rel}} - \Delta M}$$

From this expression it can easily be shown that if $(A/M)_{\text{contr}}$ is 0.4, then $\Delta M/M$ is 0.22, if it is 0.5, $\Delta M/M$ is 0.3, and if it becomes 0.55 or 0.6, the $\Delta M/M$ becomes 0.33 to 0.36 (rigor). This was interpreted by Haselgrove and Huxley to mean that about 33 to 36% of the initial mass of the myosin filament is transferred to the vicinity of the actin filaments in a rigor muscle. The amount transferred in active muscle depends on which of the various published figures for $I_{10\bar{1}}/I_{11\bar{2}}$ is used but is probably about 20 to 30% of the initial myosin filament mass. Put another way, this is about half of the mass transferred in rigor using Haselgrove and Huxley's values and about 80 to 90% of the rigor transfer using the figures given by Matsubara *et al.* (1975) and Worcester *et al.* (1975).

The equatorial diffraction data should, in principle, be able to tell us how many cross bridges are close to the thin filaments at any instant in contracting muscle. But, unfortunately, the resolution in the Fourier images obtained using just the $10\bar{1}$ and $11\bar{2}$ reflections is so poor that few details of the locations of the cross bridges are revealed. However, as noted by Haselgrove and Huxley (1973), the figures of 50% (or 80%) of the rigor transfer in active muscle can also be taken to be a maximum value for the number of bridges, relative to those attached in rigor, that are actually attached to the thin filaments at any instant in active muscle. The number could clearly be much less than this, and it may well be that all of the cross bridges move on activation but that their mean position in active muscle is somewhere between the actin and myosin filaments, with only a small proportion attached at any instant.

It is instructive to consider the observed *A*/*M* ratios for relaxed and rigor muscle in terms of the alternative models for the rotational symmetry of the thick filaments (Squire, 1972). A possible approach to this is illustrated in Fig. 10.21). Here the structure of a four-stranded filament is illustrated in projection down the filament axis. The inner dotted circle shows the approximate position within which H. E. Huxley (1968) and Haselgrove and Huxley (1973) measured the myosin filament mass in their Fourier synthesis maps. It is at a radius of about 140 to 150 Å from the myosin filament axis. In this case, the center of mass of the myosin

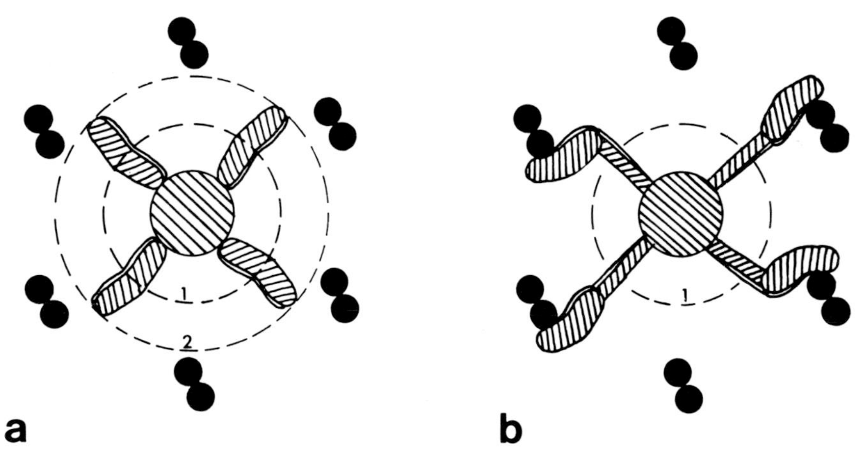

FIGURE 10.21. Diagrams of transverse sections of vertebrate striated muscle, in this case taken to contain four-stranded myosin filaments, as an example of the kind of mass transfer that might occur in going from the relaxed to the rigor state. The filled circles represent the actin filaments, and the shaded circles the thick filament backbone. The two heads in one myosin cross bridge (shaded) are assumed to lie on top of each other in this axial view of the structure. The broken circle is an arbitrary division between the myosin and actin peaks (see text) at a radius of about 140 to 150 Å from the thick filament axis. (a) Relaxed state; (b) rigor state.

cross bridges in relaxed muscle must be at a radius of about 155 Å (see Chapter 7), and less than half (say one-third) of the cross-bridge mass will therefore be included in the 140- to 150-Å radius. If it is assumed that the part of the cross-bridge mass outside this radius contributes to the general background in the electron density map and not specifically to the thin filament peaks, then the expected ratio of the myosin to actin peaks can be calculated approximately using the known molecular weights of myosin, actin, tropomyosin, and troponin (and ignoring other minor protein components) and is about 0.1. For the two-stranded model the mean cross-bridge radius would be only about 95 Å from the filament axis, and virtually the whole of each cross bridge might then be included in the myosin peak within a 140- to 150-Å radius. The A/M ratio in this case would be about 0.17. In the case of the three-stranded filament with cross bridges centered at a radius of about 125 to 130 Å, about two-thirds of each cross bridge would be included in the myosin peak, and the A/M ratio would be about 0.2.

The corresponding ratio for a rigor muscle can also be computed assuming that all the cross bridges attached to actin are included in the actin peak and lost completely from the myosin peak and that those not attached to actin remain in their relaxed positions. Table 10.8 shows the results for the situations where 50%, 75%, or 100% of the available cross bridges attach to actin. If, alternatively, it is assumed (probably more realistically) that those cross bridges not attached to actin occupy an intermediate position so that each contributes only about one-third to one-half of its mass to the myosin peak, then the figures for the rigor two- and three-stranded filaments increase slightly to the figures given in brackets in Table 10.8. From these figures, which should not be taken too literally but which probably give a feel for the kind of values that

TABLE 10.8. Calculated Values for the Ratio A/M in Relaxed and Rigor Muscles Containing Two-, Three-, or Four-Stranded Elements

Number of strands	Radius of center of gravity of cross bridges (Å)	Fraction of bridge inside 140 to 150 Å radius of myosin	Relaxed A/M ratio	Rigor ratios[a] (% attachment)		
				100%	75%	50%
2	95	1	0.17	1.2	0.7 (0.8)	0.45 (0.55)
3	125	2/3	0.2	0.85	0.6 (0.65)	0.45[b] (0.5)
4	155	1/3	0.1	0.50	0.4	0.3
Observed ratios			0.2		0.6	

[a] All figures rounded off to nearest 0.05. Bracketed values assume that unattached bridges contribute about 40% of their mass to the thick filament peak.
[b] Note that with 25% attachment with a 3-stranded filament, the ratio A/M is 0.38 (0.30).

might be expected, it is clear that the two- and three-stranded models give reasonable ratios for relaxed muscle but that the four-stranded model gives a value that is rather low. Similarly the two- and three-stranded filaments give reasonable rigor ratios (close to 0.6) if, respectively, about 60% and about 70% of the available cross bridges attach to actin in rigor muscle. But the four-stranded model can barely account for the observed rigor A/M ratio with 100% of the cross bridges attached to actin.

Looking at these calculations another way and saying that the thick filaments are very likely three-stranded shows that the observed results can be accounted for very well if only about 70% of the cross bridges attach to actin in rigor muscle, and this figure is entirely consistent with the analysis given at the end of Chapter 7 and earlier in this chapter where it was shown that the probability of rigor attachment for about one-third of the cross bridges was likely to be very low since they would be situated far from an actin target area. The equatorial data therefore supports (but does not prove) the earlier conclusions. Furthermore, the observed A/M ratio for contracting muscles (say 0.35 to 0.5) would be consistent with about 25 to 60% of the available cross bridges actually being attached to (or very close to) actin at any instant.

Of course, it could reasonably be said that this kind of analysis is extracting a great deal of information from rather low-resolution data, and it is clearly of crucial importance to obtain more details of the nature of the mass transfer that occurs from the relaxed state during contraction and in rigor, especially since it is not only the number of bridges attached but also their configuration (Lymn, 1978) that affects the observed intensities (see Section 11.3.2). As far as the equatorial data are concerned, this demands the recording of reflections other than the 101 and $11\bar{2}$ reflections from muscles in all three states. In fact, with the advent of a new kind of X-ray detector—the linear position-sensitive X-ray counter (Chapter 2)—it is now possible to record equatorial information at higher angles of diffraction (therefore corresponding to a better resolution) with much greater facility. These detectors give a lower instrumental background, and weak reflections can be observed and measured more easily. They also give a considerable gain in speed over conventional X-ray film.

Using this kind of recording equipment, Lymn and Cohen (1975) obtained accurate intensity values for the first seven equatorial diffraction peaks in the patterns from relaxed frog sartorius muscle at rest length (Table 10.9). Unfortunately, attempting to account for these intensities is rather difficult. As described earlier, with a knowledge of the phases of the reflections, an electron density map could be calculated directly using Fourier synthesis. But the phases are not known. In addition, as the resolution of the Fourier synthesis improves, it becomes less and less clear that it is valid to assume that the structure is centrosymmetric. Of course, if the myosin filaments have two- or four-fold rotational symmetry, then the structure probably is centrosymmetric to quite a high resolution. But if it is three-stranded, then it will not generally be centrosymmetric at high resolution, but it may appear so at low resolution.

There is another problem as well. This is that each of the recorded diffraction peaks is in fact the sum of a number of separate components. For example, in a simple hexagonal lattice (Fig. 10.19) there are three different sets of planes like the $10\bar{1}$ planes that have exactly the same interplanar spacing. All of these planes therefore contribute to the same diffraction peak—the recorded $10\bar{1}$ peak. If the contents of the unit cell have three-fold rotational symmetry, then each set of diffracting planes will be associated with an identical view of the unit cell contacts, and the resulting diffracting peaks will therefore be equal. But if the unit cell contents do not have this symmetry, then the density profiles on the

TABLE 10.9. Intensities of the Equatorial Reflections in X-Ray Patterns from Resting Frog Muscle According to Lymn and Cohen (1975) and from the Models Shown in Fig. 10.24 According to Haselgrove *et al.* (1976)[a]

Model length of cross bridges	1 Long	2 Short	3 Short	4 Long	5 Long	6 Short	Experimental
Reflection							
1,0	100	100	100	100	100	100	100
1,1	190 (324)	47	33	30	26	39	44
2,0	47	7	3	36	4	7	9
2,1 1,2	62	2	11	12	20 (36)	9	9
3,0	16	0	18	155	35	4	12
2,2	34	21 (2)	1	11	19	10	20
3,1 1,3	30	0	1	23	85	19	(20, bracketed with 2,2)
$I_{1,0}/I_{0,0}$	0.112	0.515	0.525	0.124	0.081	0.434	—

[a] To avoid sampling errors intensity was calculated directly as

$$I_{hk} = (F_{hk})^2/(h^2 + k^2 + hk)^{1/2}$$

where

$$F_{hk} = \Sigma_x \Sigma_y \rho_{xy} \exp[-2\pi i(hx + ky)]$$

and where h and k are the indices of the reflection and x and y are the fractional cell coordinates of each density point. I_{hk} was calculated for all points in the reciprocal lattice, and the values for those points that superimpose on the pattern were then added. Values where Lymn's quoted intensities are significantly different are underlined with Lymn's values given in parentheses. The intensity of the 0,0 reflection was calculated directly as

$$I_{0,0} = (\Sigma_x \Sigma_y \rho_{xy})^2$$

different sets of planes will be different, the reflections will therefore be unequal, and the observed diffraction peak will contain a greater intensity contribution from some diffraction planes than from others.

When carrying out a Fourier synthesis it is necessary to define the contribution from each of the component peaks. This means that it normally has to be assumed that the diffracting structure in a hexagonal lattice has threefold symmetry,* otherwise one needs to know the structure of the object beforehand, in which case the calculation of an electron density map by Fourier synthesis is rather less valuable.

With this in mind, it is clear that in order to proceed with an equatorial Fourier synthesis for muscle, one needs to be satisfied that (1) the structure is centrosymmetric and (2) the structure has threefold rotational symmetry. These assumptions were implicit in the calculations described earlier (Fig. 10.20) of electron density maps from only the $10\bar{1}$ and $11\bar{2}$ reflections (Haselgrove and Huxley, 1973). Using the recorded X-ray reflections out to the $31\bar{4}$ reflection (which corresponds to a *d* spacing of about 100 Å) means that the unit cell is being considered at 100-Å resolution. The two- and four-stranded models for the myosin filament would have respectively six and 12 cross bridges evenly spaced around the filament backbone in a view down the filament axis. They would therefore automatically have both centrosymmetry and threefold rotational symmetry. On the other hand, the three-stranded filament would have nine cross bridges spaced evenly around the filament backbone. Since nine is not divisible by two, the structure is not centrosymmetric, but since it is divisible by three it does have threefold symmetry. But even in this case, at a resolution of 100 Å, it seems likely that the three-stranded filament might appear to be centrosymmetric. This can be seen in Fig. 10.22a where the filament backbone and cross-bridge arrangement are drawn approximately to scale. It can be seen that even on the assumption that the two heads in one cross bridge are arranged on top of each other in an axial direction and if the cross bridges stick straight out (radially) from the filament axis, neighboring cross bridges will be sufficiently close together that it is unlikely that they will appear as individual entities at a resolution of only 100 Å. If, as seems likely, the cross bridges are slewed azimuthally as in Fig. 10.22b, then the smearing out of the cross-bridge density will be even more marked.

Two conclusions follow from this. One is that it is probably valid to assume centrosymmetry in carrying out an analysis of the equatorial diffraction pattern. The other is that it may well be unreasonable to

*For some reflections where the multiplicity is more than six, an additional assumption that the structure contains three mirror planes needs to be made.

also seen in the computed Patterson function. A filament that appears to be circular and to have uniform electron density in projection down its axis would give rise to a peak in the Patterson synthesis that is circularly symmetrical and has a bell-shaped profile. But in the observed Patterson synthesis, the peaks at the origin, although having more or less circular symmetry, have shoulders around them. This feature is still present to some extent if the sixth and seventh equatorial peaks are left out of the synthesis. This is important, since if these reflections are included in the synthesis, it is necessary to assume (as yet without a great deal of confidence) that the $22\bar{4}$ and $13\bar{4}$ reflections, which were estimated as a single peak by Lymn and Cohen (1975), have equal intensities.

Despite this, it is not unreasonable to conclude (Squire and Child, unpublished data) that the observed shoulders in the Patterson synthesis are providing information either about the myosin cross bridges or about the internal structure of the thick filaments (such as their hollow centers?). But the detailed analysis by this method requires much further work to be done. Suffice it to say that the circular symmetry of this shoulder supports the kind of model shown in Fig. 10.22 in which the cross bridges provide only a featureless shelf of density around the thick filament backbone. Patterson syntheses from the model structures of Lymn and Cohen (1975) and Haselgrove *et al.* (1976) have also been computed and are shown in Fig. 10.27. From these syntheses it is fair to conclude that, as would be expected, the rigor model (I) is nothing like the relaxed structure and that, of the various relaxed structures, models 5 and 6 seem at first sight to be the most satisfactory.

Vertebrate Cardiac Muscle. Before leaving this discussion of vertebrate muscle, it is important to make mention of conclusions that can be reached from the recent results of Matsubara *et al.* (1977a,b) on the equatorial diffraction patterns from active heart muscle. It will be remembered that it was found that the ratio $I_{10\bar{1}}/I_{11\bar{2}}$ in this case changed from 3.07 in the quiescent phase, to 2.66 in the diastolic phase to 2.19 during the systolic phase and to 0.37 in rigor. Interpreting these results in the same way as Haselgrove and Huxley (1973), Matsubara *et al.* computed Fourier electron density maps for the four muscle states and then determined the mass transfer as a function of that which occurred in rigor. They found that in the systolic phase about 70 to 71% of the rigor mass transfer occurred and that about 51 to 52% transfer occurred in the diastolic phase. The systolic value is similar to that obtained for contracting skeletal muscle (50 to 80% transfer). The lower value reported by Matsubara *et al.* (1977a) was attributed by Matsubara *et al.* (1977b) to the submaximal tension generated by the cardiac muscle preparations during the X-ray exposures in their earlier experiments.

However, an interesting feature of these results is that the diastolic phase is not equivalent to relaxed muscle but has the cross bridges further away from the thick filaments than in relaxed muscle. It is conceivable that this is a demonstration of an intermediate cross-bridge state, such as the Ca^{2+}-activated state (State 2) discussed earlier, that in this case is inhibited either from actin attachment or from going through a contraction cycle possibly by a refinement of the thin filament regulation mechanism.

Insect Flight Muscle. The equatorial diffraction results obtained for relaxed, Ca^{2+}-activated, contracting, and rigor insect flight muscles by Miller and Tregear (1970) and by Armitage *et al.* (1975) were described briefly earlier in the chapter. The measured reflections were the $10\bar{1}$ and $20\bar{2}$ reflections, and the intensity ratios $I_{10\bar{1}}/I_{20\bar{2}}$ for relaxed, Ca^{2+}-activated, contracting (at the peak of the tension cycle), and rigor muscle were about 1.94, 1.85, 1.81, and 0.86, respectively. These results were analyzed by Miller and Tregear (1970) in a manner later used by Haselgrove and Huxley (1973) (except that here only one-dimensional data were obtained), and it was concluded (Fig. 10.28) that about 10% to 20% of the number of bridges attached in rigor were attached to (or very close to) actin in contracting muscle. It was also concluded that the number attached (or close to) actin in the Ca^{2+}-activated state was also about 10 to 20%.

In this muscle the actin filaments are on a line joining the centers of adjacent thick filaments, and anything that tends to orient the cross bridges toward the thin filament positions instead of pointing in one of the 48 different cross-bridge positions expected for the relaxed six-strand 16/1 helical structure would significantly alter the observed diffraction pattern. It therefore seems quite conceivable that Ca^{2+} activation causes a general azimuthal swing of the cross bridges in the same direction towards the thin filaments from whatever position they have in relaxed muscle. This swing might well be associated with a small outward radial movement, and this would also affect the observed diffracted intensities. This might also be similar to what happens in vertebrate muscle where Ca^{2+} activation would cause the cross bridges to swing azimuthally from their relaxed positions towards the trigonal points in the unit cell (Fig. 10.25). But the whole problem needs to be studied in much more detail before this can be certain.

10.3.6. Changes in the Thin Filaments

The most significant of all of the changes in the X-ray diffraction patterns from contracting muscles would be the detection of changes in the reflections from the thin filaments. These could provide conclusive evidence of cross-bridge attachment to actin during contraction. But,

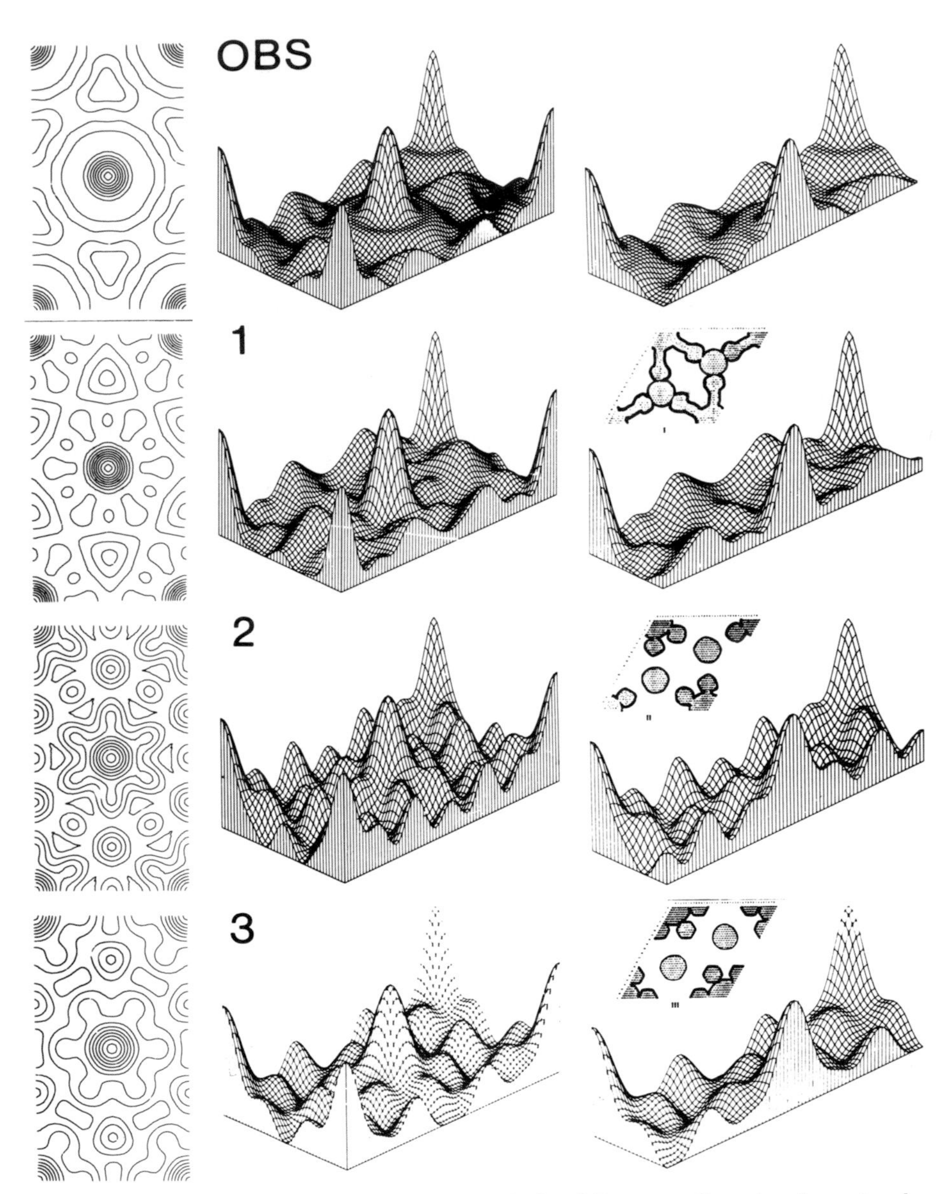

FIGURE 10.27. Comparison between the calculated Patterson function from the observed equatorial intensities (Fig. 10.26) and those from the six models shown in Fig. 10.24. The equatorial intensities used to calculate these syntheses are given in Table 10.8. On the left are axially viewed contour maps of density, in the center are perspective views of the same distribution, and to the right are the same views sectioned down the central $11\bar{2}$ planes. In these perspective views the vertical axis measures density. Note that most syntheses are very unlike the observed synthesis except possibly those from models 5 and 6. (From Squire and Child, in preparation.)

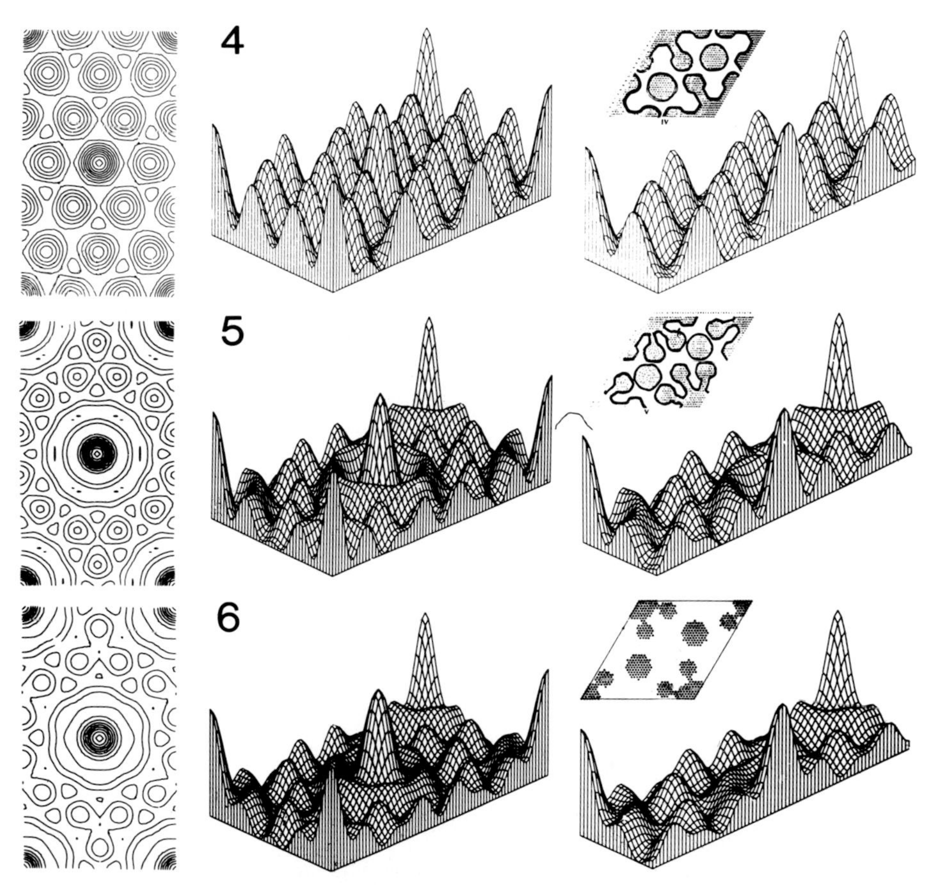

FIGURE 10.27 *Continued*

apart from the changes in the second and third actin layer lines caused by the movement of tropomyosin involved in thin filament regulation (Chapter 5), little change in the diffraction pattern from actin has been observed. In studies of contracting insect muscles, no significant change at all was observed by Armitage *et al.* (1975) on the 190-Å and 59-Å layer lines. Active vertebrate and molluscan muscles did, however, give a barely significant increase of intensity on the 59-Å layer line (Haselgrove, 1975a; Vibert *et al.*, 1972).

Since these slight changes provide the only available data on actin labeling, it is instructive to consider how any increase in the intensity of the 59-Å layer line could be explained. The necessary analysis was carried out in detail by Parry and Squire (1973). It was shown that of the various possible causes of the intensity change, (1) changes in structure of the actin monomers, (2) changes in structure of the tropomyosin

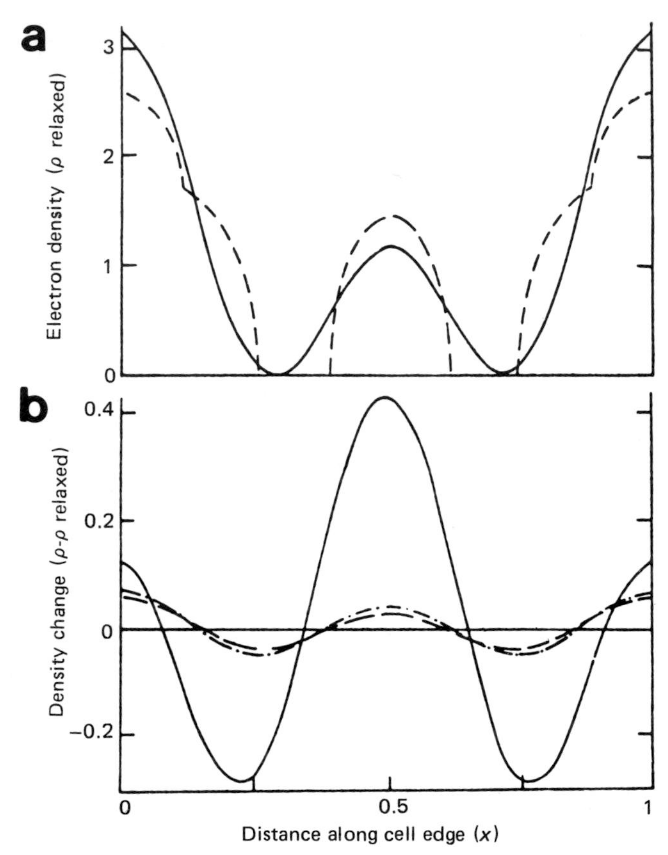

FIGURE 10.28. (a) Electron density map of the projection onto one edge of the unit cell of the A-band filament lattice in relaxed insect flight muscle. The cell is mapped from the center of one myosin filament to the center of the next. One-third of the thin filaments appear at the cell edge, and two-thirds in the center. The solid line is derived from the observed equatorial intensities. The dashed line is derived from the probable values of the positions and relative weights of A-filaments, thin filaments, and cross bridges (0.33, 0.22, and 0.45, respectively). (b) Electron density shift relative to the relaxed distribution on going into rigor (solid line), during calcium activation (dashed line), and during mechanical oscillation of active muscle (dot–dash line) at the phase of peak tension. Note the magnified ordinate relative to (a). (From Miller and Tregear, 1970.)

molecules, and (3) addition of extra material to the thin filaments, only the latter could give an intensity increase of as much as the observed 20%. This latter possibility clearly must relate to the attachment of cross bridges to the thin filaments. But, since it seems likely that the cross bridges will actually change their orientation while attached to actin in order to cause the necessary relative motion of the thick and thin filaments during contraction (see Chapter 11), it is likely (Fig. 10.29) that only the actin end of each cross bridge will be static enough to cause significant labeling of the thin filament. The density in the remainder of the cross bridge will tend to be smeared out over a large volume and will, therefore, contribute little.

Parry and Squire, therefore, modeled the myosin attachment in active muscle by adding spherical scattering units to each actin monomer and then computing the resulting thin filament diffraction pattern. It was concluded that a sphere of radius 15 Å and molecular weight about 3000 attached on the outside of each actin monomer would produce intensity increases of about 25% and 20% respectively on the inner ends of the 59-Å and 51-Å layer lines. Combining this effect with the tropomyosin movement from the "off" to the "on" position during activation gave computed intensity distributions on the first seven actin layer lines in the diffraction patterns of active and relaxed muscles as shown in Fig. 10.30. Here the required changes on the second and third layer lines and on the 59-Å layer line are all obtained. Note in addition that Haselgrove (1975a) detected less intensity change on the 59-Å layer line at a radius of 1/100 $Å^{-1}$ than at 1/180 $Å^{-1}$, and this is also reproduced to some extent by the modeling.

This observation also indicates that the measured increase at 1/180 $Å^{-1}$ could not be because of improved thin filament ordering since in that case it would be expected that the outer end of the layer line would increase in intensity relatively more than the inner end, not less as observed.

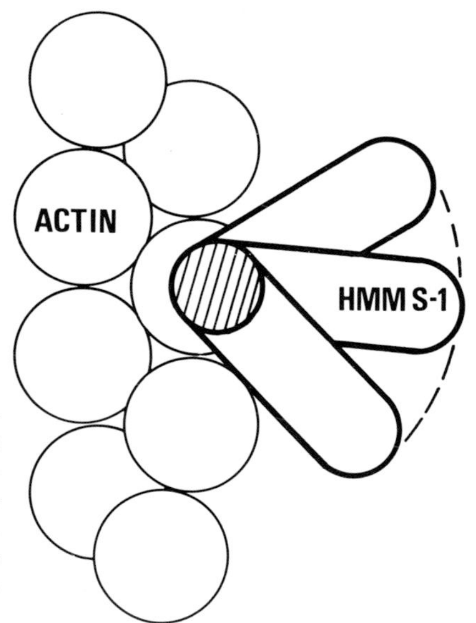

FIGURE 10.29. Diagrammatic representation of the possible movement of a cross bridge during contraction, showing that there is always likely to be a region of relatively high electron density close to the actin attachment site. In the calculations given in Fig. 10.30, this region was represented as a sphere of radius 15 Å (shaded). (From Parry and Squire, 1973.)

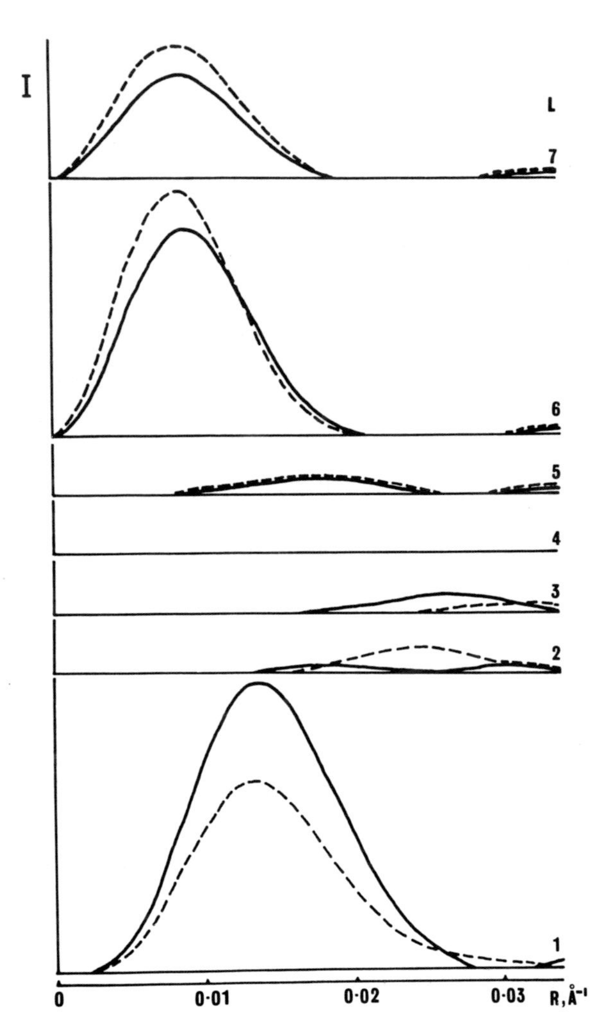

FIGURE 10.30. The calculated intensity distribution (in arbitrary units) on layer lines 1 to 7 of the 13/6 actin filaments discussed in Chapter 5 (Fig. 5.25), both relaxed (solid line) and labeled as in Fig. 10.29 on every monomer (dashed line), thus mimicking contracting muscle. (From Parry and Squire, 1973.)

The question then arises: What significance can be attached to the apparent molecular weight of 3000 of the object labeling each actin monomer? Parry and Squire interpreted this as follows. If a fraction F of each cross bridge attached to the thin filaments is sufficiently well ordered to contribute significantly to the 59-Å layer line, if a fraction N of the available actin monomers have these cross bridges attached at any instant, and if the molecular weight of a cross bridge is M, then the object labeling the thin filaments, when averaged over each actin monomer, would have an effective molecular weight M' of MFN. Taking M to be 115,000, F to be about 0.1 (since the sphere used had radius 15 Å, and the the cross-bridge size is roughly 150 Å × 45 Å × 30 Å; Moore *et al.*, 1970), and taking M' to be the estimated value of 3000, then a value of N of about 25% is obtained.

In vertebrate muscle there are about four times as many actin monomers as myosin molecules (Tregear and Squire, 1973), and in a rigor muscle containing three-stranded myosin filaments, only about two-thirds of the available cross bridges would be attached to actin for reasons given earlier. Assuming that the two myosin heads in each cross bridge attach to two different actin monomers in rigor muscle, then about 30 to 35% of the available actin monomers would be labeled by a myosin head. A 25% attachment in active muscle (although inevitably a very approximate figure) would therefore correspond to about 70 to 80% of the rigor attachment. This may possibly be slightly high, but it is in approximate agreement with the results on mass transfer obtained from the equatorial diffraction data where the comparable figure is 50% to 80% attachment. But whether or not these figures are absolutely right, the fact remains that the observed marginal increase in intensity on the 59-Å layer line, provided it is genuine (Haselgrove, 1975a), could only be satisfactorily explained if the myosin cross bridges actually do attach to actin in a contracting muscle.

The uncertainty that remains is therefore in the data and not in the interpretation of the data. At its face value it supports the idea that actin attachment does occur, and from the likely figures from other sources on the proportion of cross bridges likely to be attached during contraction in vertebrate muscle, an increase in intensity on the 59-Å layer line of more than 20 to 25% would not be expected. Experimentally, the need is not therefore to look for a greater increase on the 59-Å layer line but is to define the increase that does occur with much more precision—a task that is by no means easy.

10.4. Artificially Modified Muscle Structures

10.4.1. Introduction

The static muscle states that have been discussed so far, the rigor and resting states, provide useful information about the structures of the myosin and actin filaments and about the constraints on actin labeling that can occur. But they throw relatively little light on the different structural changes involved in the contractile cycle itself. In order to probe this particular question, use has recently been made of a variety of analogues of ATP which, it was hoped, would stop the contractile event at particular points and thus produce synthetic states in which a particular part of the contractile cycle has been frozen in. The particular ligands that have been used are α,β-methylene-ATP [ATP(α,-CH_2)], adenosine 5′-0-(3-thiotriphosphate) [ATP(γ-S)], and β,γ-imido-ATP (AMP · PNP). The effects of using ADP and pyrophosphate have also been studied.

It has been shown by Mannherz *et al.* (1973) that ATP (α,β-CH_2) is a substrate for rabbit myosin and that it dissociates actomyosin with a cleavage rate about 1000 times slower than that of ATP. ATP (γ-S) is also cleaved more slowly than ATP (Goody and Eckstein, 1971), although its turnover rate is higher. It is expected that the major component of the steady-state complex with myosin under saturating conditions with either of these analogues will be M·ATP as in Fig. 10.2. The third analogue, AMP·PNP, is different in that it is not cleaved at all (Yount *et al.,* 1971a,b), and the rate of release of the uncleaved analogue is very temperature dependent (Goody *et al.,* 1975). Goody *et al.* (1976), Marston *et al.* (1976, 1979), Barrington-Leigh *et al.* (1977), and Beinbrech (1977) have described the effects of these various analogues on the structure of glycerinated insect flight muscle. Similarly, Lymn and Huxley (1972) and Lymn (1975) have described the effect of AMP·PNP on rabbit psoas muscle. The conclusions from these various studies are given in the next sections.

10.4.2. The Effects of Different ATP Analogues

Insect Flight Muscle. The studies so far of muscles treated with either ADP or with the analogues ATP(α,β-CH_2) and ATP(γ-S) have shown that the states induced are relatively simple and well-defined. ADP treatment leaves a structure not much different from the normal rigor state (Rodger and Tregear, 1974), although Beinbrech (1977) and Marston *et al.* (1979) suggest there may be slight changes, and treatment with either ATP(α,β-CH_2) or ATP(γ-S) leaves structures little different from relaxed muscle (Barrington-Leigh *et al.,* 1972; Goody *et al.,* 1975).

Because of the different biochemical properties of AMP·PNP, it might be expected that a different state would result, as indeed it does. But the new state that is produced is both complex and interesting. Figures 10.31 and 10.32 compare the diffraction patterns of insect muscle in various states as published by Barrington-Leigh *et al.* (1977).

Figures 10.31a,b are the low-angle and medium-angle patterns, respectively, from a normal rigor muscle. As described earlier in this chapter, it includes the very strong sixth layer line (385 Å) especially near to the meridian ($10\bar{1}$ row line), a significant 12th layer line and relatively

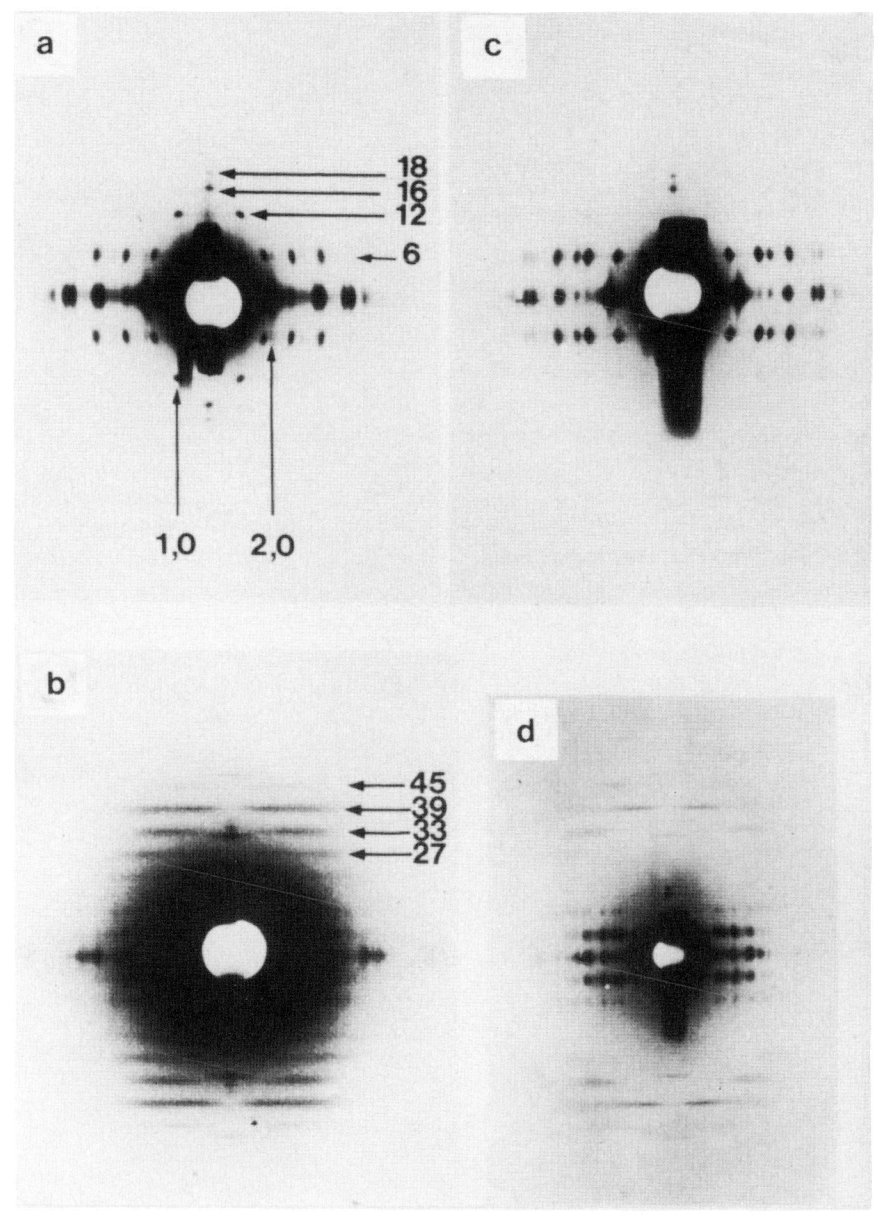

FIGURE 10.31. (a and b) X-ray diffraction pattern obtained from glycerinated *Lethocerus* flight muscle in the absence of nucleotides. The scale in (b) is twice that in (a). Horizontal arrows indicate layer lines (orders of 2300 Å), and vertical arrows indicate row lines. (c and d) Comparable patterns to (a) and (b) but from glycerinated muscle incubated in 20 mg/ml myosin subfragment 1 at 4°C. (All patterns from Barrington-Leigh *et al.*, 1977.)

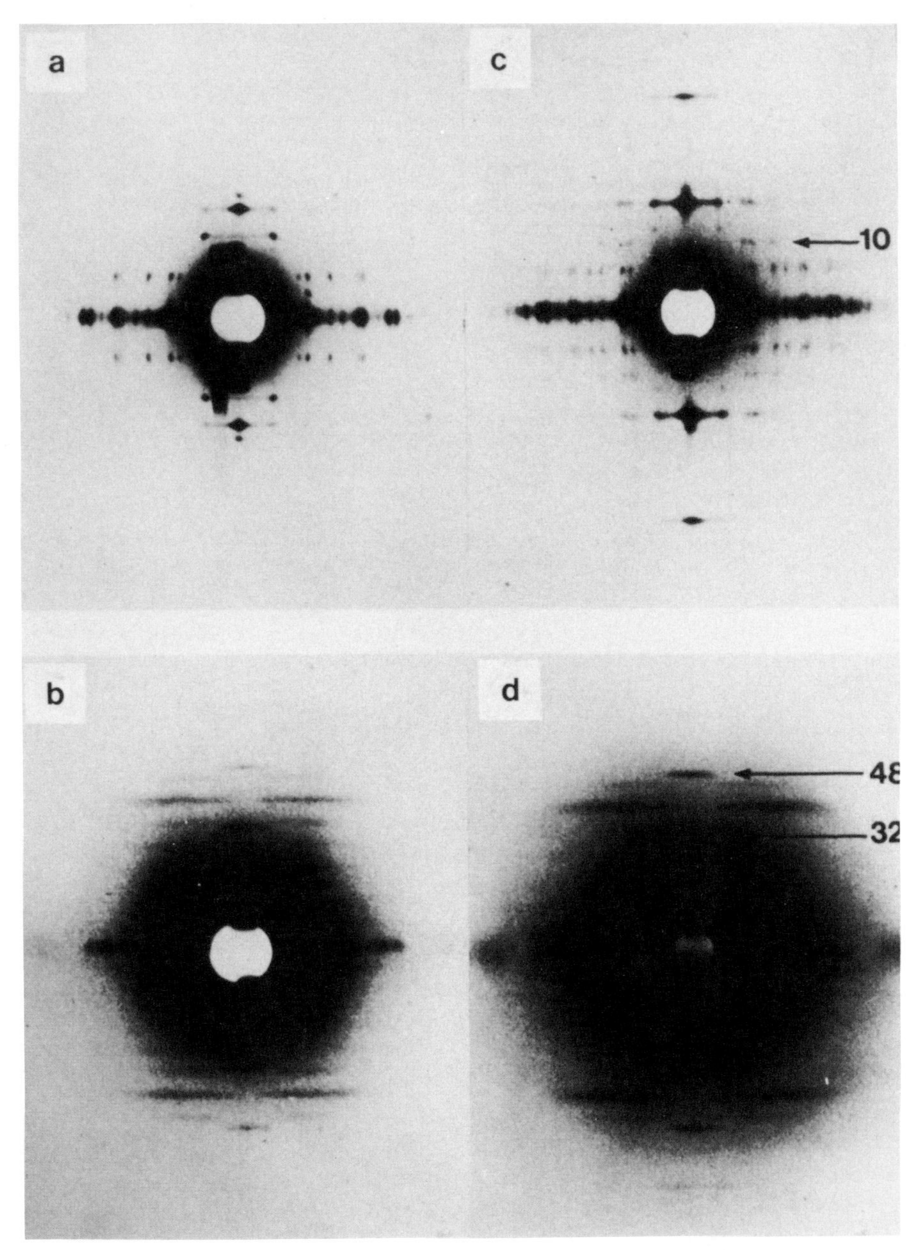

FIGURE 10.32. Similar patterns to those in Fig. 10.31 but in this case from muscles bathed in either 1 mM AMP·PNP (a and b) or 15 mM Mg-ATP (c and d). Horizontal arrows identify layer lines on the 2300-Å repeat. (From Barrington-Leigh *et al.,* 1977.)

weak 16th- and 18th-order meridional reflections. Figures 10.32c,d show equivalent patterns for resting muscle (ATP added). The sixth layer line is now relatively weak, but the 16th-order meridional reflection at 145 Å has become very strong.

Figures 10.32a,b show what happens when AMP·PNP is used. In this case, the sixth-order layer line at 385 Å is rather like the rigor layer line but with a weaker peak on the $10\bar{1}$ row line (Fig. 10.31a), but, unlike rigor, the 16th-order peak is also strong. The ratio of the intensities of the equatorial reflections also changes so that $I_{10\bar{1}}/I_{20\bar{2}}$ is about 2.2 in patterns from relaxed muscle, about 0.8 in patterns from rigor muscle, and about 1.6 for AMP·PNP-treated muscle. These various changes appear at first sight to indicate that the AMP·PNP state is a simple sum of separate relaxed and rigor structures, but this is clearly not so. For example, AMP · PNP-treated muscle gives a strong $20\bar{2}6$ reflection and a weak $11\bar{2}6$* peak, whereas in rigor patterns the reverse is true. We definitely are dealing with a new state, and it is of great importance to define its structure.

Two independent lines of evidence help to do this. First, longitudinal sections of insect muscles irrigated with AMP·PNP have been studied by Marston *et al.* (1976) and by Beinbrech (1977), and these clearly seem to show a structure rather like that of relaxed muscle in that the cross bridges appear to be oriented roughly at 90° to the filament axis. Figure 10.33 compares micrographs of rigor muscle and AMP·PNP-treated muscle together with their optical diffraction patterns. The differences between the optical diffraction patterns are comparable to the changes in the X-ray diffraction patterns described earlier. Second, mechanical experiments on AMP·PNP-treated muscle (Marston *et al.*, 1976, 1979) have shown that relative to the tension–length and 5-Hz stiffness–length diagrams of rigor muscle, those from the same muscle treated with 500 μM AMP·PNP were displaced to longer sarcomere lengths without altering their shapes (Fig. 10.34). Relaxation with ATP greatly reduced the slope of both plots, and ADP treatment left both plots relatively unaltered. The displacement caused by AMP·PNP corresponds to a maximum relative sliding of the actin and myosin filaments of about 20 to 30 Å.

The stiffness measurements clearly show that the AMP·PNP state is one where cross bridges are still bound to actin. But the electron microscopy and X-ray diffraction evidence indicates that the predominant angle of attachment of these bridges is not the same as in rigor; it is much more nearly at 90° to the thin filament axis rather than the 45° that occurs in rigor. A model such as that in Fig. 10.12i might explain the

*This is the four-index notation h, k, i, l for a hexagonal structure where $h + k + i = 0$.

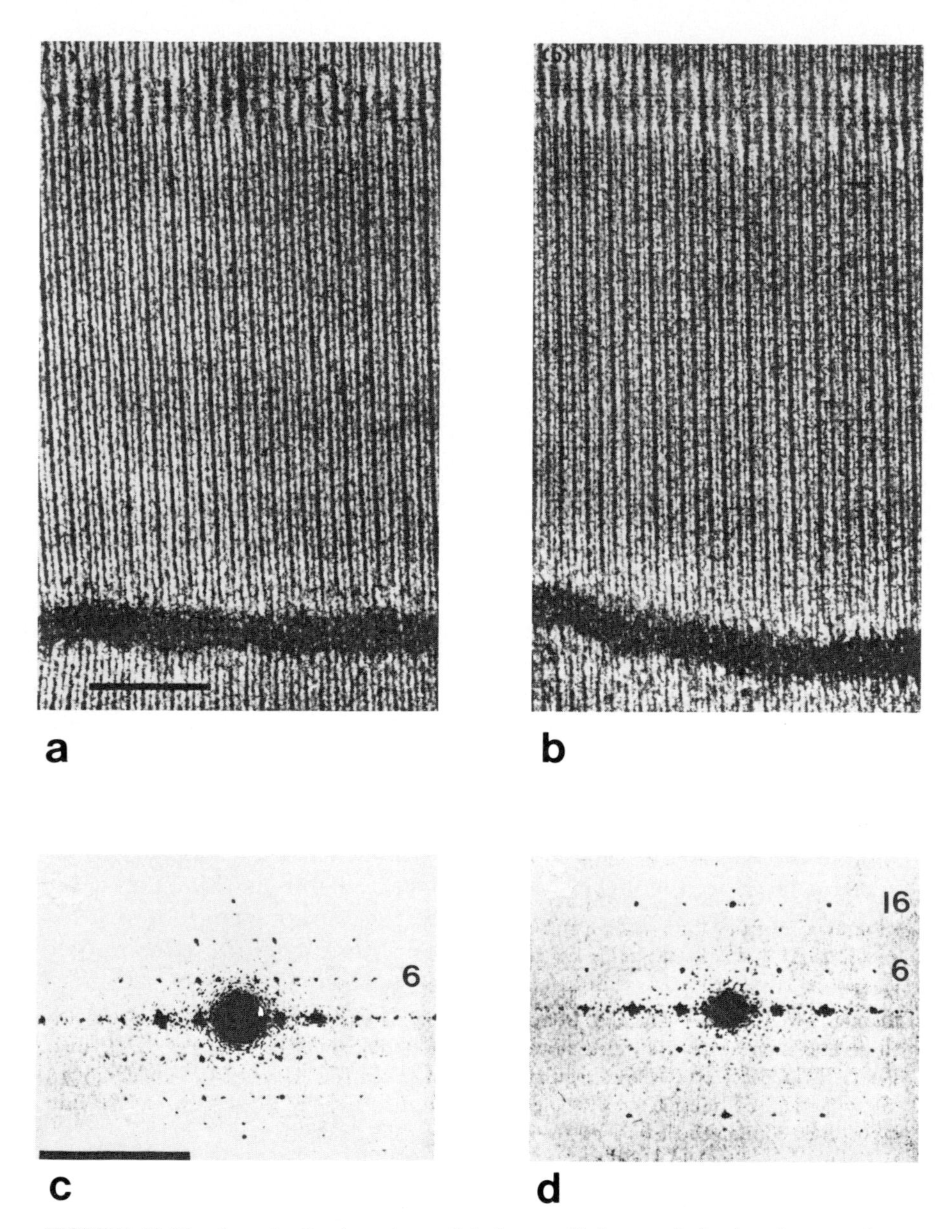

FIGURE 10.33. Longitudinal sections of *Lethocerus* flight muscle in the absence of nucleotide (a) and in the presence of 200 μM AMP·PNP (b). Scale bar, 5000 Å. (c and d) The optical diffraction patterns of whole sarcomeres of the specimens illustrated in (a) and (b), respectively. Scale bar, 1 $Å^{-1}$. Note the relatively strong 145-Å layer line ($l = 16$) in (d). (From Marston *et al.*, 1976.)

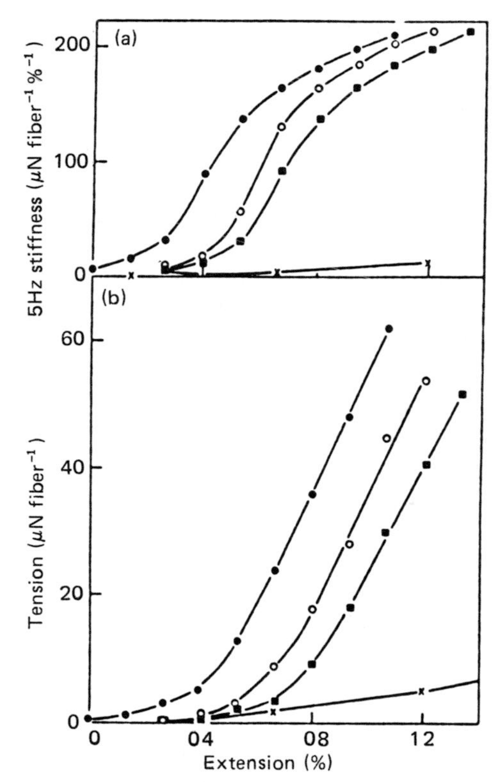

FIGURE 10.34. The relation between (a) stiffness and (b) tension in *Lethocerus* flight muscle to its extension. Glycerinated muscle before (●) and after (○) immersion in 500 μM AMP·PNP. The effects of ADP (■) and ATP (x) are also shown. Length changed by 0.12% increments at 10s intervals; stiffness calculated from oscillatory tension due to a 0.18% peak to peak amplitude, 5 Hz oscillation. (From Marston *et al.,* 1976.)

observations. The displacement of the tension and stiffness curves to longer sarcomere lengths would also be consistent with rotations of the 45°-angled rigor bridges towards the 90° position, but unfortunately it is possible that some redistribution of the attached cross bridges may also occur, and this would complicate the problem. Another possibility is that some of the apparent effects in electron micrographs are caused by the unattached bridges in the rigor muscle taking up a regular structure with a 90° angle. This might conceivably be produced by the action of the fixative rather than AMP · PNP. It is therefore clear that the AMP · PNP state is as yet ill-defined and may, like rigor, involve two different populations of cross bridges. Solving its structure should help considerably in defining the structural events involved in the contractile cycle.

Vertebrate Muscle. Lymn and Huxley (1972) and Lymn (1975) have reported X-ray diffraction studies of vertebrate muscles (rabbit psoas) treated with AMP·PNP and inorganic pyrophosphate. Their results are in many ways comparable to those for insect flight muscle. For example, the X-ray patterns from AMP·PNP-treated muscle are intermediate between those of rigor and relaxed muscle without being a simple mixture of the two patterns. The main changes that occur relative to the vertebrate rigor patterns are the following.

1. The ratio of the intensities of the $10\bar{1}$ and $11\bar{2}$ equatorial reflections changes dramatically from about ¼ in patterns from rigor muscle to about 1/2.5 for AMP·PNP-treated muscle and about 1/1.2 for ATP-treated muscle.

2. The normal outer rigor layer lines (Fig. 10.4a) virtually disappear, but the meridional reflections near 144 Å and 72 Å increase in intensity while maintaining their rigor spacings (e.g., about 72.8 Å) rather than reverting to the values for relaxed muscle (e.g., about 71.6 Å; Table 10.5)

3. Although the rigor layer lines disappear, there is no sign of the usual relaxed layer lines that are orders of a 429-Å repeat.

4. The X-ray diffraction results depend markedly on temperature for AMP·PNP-treated muscles. The results described above are for muscles at 0 to 4°C. At a temperature of 22°C, the whole diffraction pattern becomes much more like that from a rigor muscle. Intensity measurements indicated that the 22°C AMP·PNP treatment produced a state where head attachment to actin occurred, but the structure was not strictly identical to rigor (Lymn, 1975).

5. Dos Remedios *et al.* (1972) showed that AMP·PNP at pH 8.0 (as used in the work described above) causes a reduction in tension and stiffness (5 Hz) of glycerinated psoas fiber bundles. Fluorescence measurements indicated a similar myosin S-1 configuration in muscle treated with either ATP or AMP·PNP.

6. Lymn (1975) found that AMP·PNP-treated muscles could be slowly stretched to long sarcomere lengths without an apparent diminution in the quality of their X-ray diffraction patterns, a result taken to indicate that the attached cross bridges must be capable of breaking and reforming within seconds.

To summarize, these results are quite comparable to those from insect flight muscle, and they probably reflect a similar kind of structure. The continued presence of the longer axial repeats in the AMP·PNP patterns (e.g., 72.8 Å) is quite consistent with the model suggested earlier (Section 10.3.3) in which the cross bridges need to be both ordered on the myosin helix and tilted away from 90° in order to generate the smaller periodicity (see Fig. 10.21). It is not therefore necessary to suggest that the basic thick filament repeat in vertebrate muscles changes between the relaxed and any other state (Lymn, 1975).

A further discussion of the various stages in the contractile cycle will be given in Chapter 11.

10.4.3. S-1 Labeling Studies

Insect Flight Muscles. The structure of insect flight muscle in rigor, as described in Section 10.2.3, is one in which attachment of myosin heads to the thin filaments follows a well-defined pattern: it occurs in the target areas. The effects of this particular labeling pattern are to modify the diffraction pattern from that of a completely labeled thin filament by introducing the effect of a marked 385-Å periodicity in the labeling probability (Barrington-Leigh *et al.,* 1977; Holmes *et al.,* 1980). This gives rise to additional peaks in the basic thin filament diffraction pattern (Fig. 10.7). It has already been shown that the strong $10\bar{1}6$ reflection in patterns from rigor muscle (Fig. 10.31a) is caused by this labeling pattern.

Another feature of the observed pattern that is explained by this is the medium-angle layer line pattern that can be seen in Fig. 10.31b. If the thin filament were uniformly labeled with myosin heads, then the layer lines of number 27, 33, 39, 45, and so on should show evidence of the helic cross pattern typical of a simple helix (as in Fig. 5.6 or 5.8c). But Fig. 10.31b does not show such an intensity distribution; on the contrary, each layer line shows rather a similar distribution with peaks at a similar radius from the meridian. The periodic labeling pattern in rigor explains this feature (Barrington-Leigh *et al.,* 1977) since the complete structure can be thought of as a product between a completely labeled thin filament and a periodic function representing attachment probability (the Q function; see Section 10.2.3). Since the rigor structure can be represented by such a product, from the convolution theorem (Chapter 2), we know that the rigor diffraction pattern should be produced by a convolution of the pattern of a completely labeled thin filament with that of the attachment probability function. The latter repeats every 385 Å so its transform will be a series of meridional orders of this 385-Å repeat.* Then on each such order will be placed the helix cross patterns from a fully labeled filament to generate the required convolution. The effect of this is that layer lines 27, 33, 39, and 45, which are all successive orders of the 385-Å layer line repeat, will contain intensity contributions from a number of adjacent layer lines in the patterns from a simple helix, and

*Note that the absence of meridional reflections on layer lines 6 and 12 in rigor is not inconsistent with this, since the space group in insect muscle would make these reflections systematically absent.

their final character will be altered considerably from a simple helix cross.

In order to test this kind of conclusion from their analysis of rigor insect flight muscle, Barrington-Leigh *et al.* (1977) studied diffraction patterns from insect muscles permeated by a solution of myosin heads (HMM S-1). It was thought that these would fill in any gaps in the labeling of the rigor thin filament and so make uniform the attachment function *Q*. In so doing their theory would predict that the resulting diffraction pattern should be more like that of a simple helix cross than the rigor pattern. Figures 10.31c,d show X-ray patterns from an S-1-labeled muscle, and clearly the prediction is fulfilled; a very clear helix cross pattern is now visible on the outer layer lines. In addition, the $10\bar{1}6$ peak in Fig. 10.31c is much weaker as would be expected. This result clearly confirms the analysis of rigor insect muscle given in Section 10.2.4.

Vertebrate Skeletal Muscle. The kind of S-1 labeling experiment carried out by Barrington-Leigh *et al.* (1977) was carried out originally on rabbit psoas muscle by Rome (1972). Although the X-ray diffraction patterns were not of comparable resolution to those in Fig. 10.31 from insect, they did show changes in the same direction. The layer lines caused by the actin helix were enhanced, indicating good labeling of the thin filaments. But of crucial interest here is the effect on the "extra" layer lines, orders of a 725-Å repeat, that are close to the meridian in diffraction patterns from rigor frog muscle (Haselgrove, 1975a; Table 10.3). Earlier in this chapter these were explained in terms of a specific labeling pattern (a helical stripe of pitch 725 Å) on the rigor thin filaments. As in the case of insect muscle, filling in the gaps between the native rigor bridges with additional free S-1 moieties might be expected to modify these extra peaks in the diffraction pattern. Rome (1972) commented, on the basis of her fairly low-resolution pictures, that there was "no enhancement of these layer lines in the S-1-labeled patterns." She did not, however, say that they were present or that they had disappeared. This aspect of her results is therefore inconclusive, and further work at higher resolution still needs to be done to test the model.

10.4.4. Scallop Muscle

The structure of scallop muscle (*Pecten* striated muscle) has been described in Chapter 8. It is one of the muscles that have direct myosin regulation by calcium through the regulatory (EDTA) myosin light chain (Kendrick-Jones *et al.*, 1976; Simmons and Szent-Gyorgyi, 1978; Chapter 6, Section 6.2.3). Vibert *et al.* (1978) have studied X-ray diffraction patterns from rigor scallop muscle (produced by glycerol extraction or by detergent skinning) and from similarly prepared muscles from which the EDTA light chain had been removed. The thick filaments in the latter preparation were therefore rendered insensitive to calcium. The rigor diffraction pattern showed the characteristically strengthened 385-Å and 59-Å actin layer lines. The desensitized muscles showed a similar pattern but with the inner end of the 385-Å reflection considerably enhanced by a factor of about three. The 193-Å layer line changed in a similar way. As discussed in Chapter 6, it is quite possible that regulation of scallop myosin involves the cooprative behavior of the two myosin heads in each molecule, although this has yet to be proven. In a similar way, Vibert *et al.* (1978) suggested that a change in structure of the rigor attached bridges when the muscle is treated with EDTA to remove the regulatory light chain could also be caused by cooperativity between the heads. The exact structural change has yet to be determined, but it should eventually help to define the relationship between the two heads.

10.5. Summary

In this chapter we have seen that in certain well-defined static muscle states there are probably different arrangements of the myosin cross bridges. In relaxed muscle they are well-ordered on the myosin filament helix. In rigor muscle they are mainly bound to the thin filaments with a well-defined 45° tilt, although some heads remain unattached; the pattern of labeling in rigor varies among different muscles; the structure of AMP·PNP-treated muscles is rigorlike in that the cross bridges label the thin filaments but has cross bridges with a predominant tilt nearer to 90° than in rigor, and the unattached bridges in rigor may have adopted a new configuration with AMP·PNP treatment. In insect flight muscle, there is a Ca^{2+}-activated state that may not involve actin attachment of cross bridges, and a similar state, as yet not so well documented, may occur in vertebrate muscle. Finally, all of the available evidence from contracting muscles suggests that actin attachment probably does occur but that relatively few cross bridges are attached to actin at any instant and that even when attached, a variety of angles of attachment probably occurs. In Chapter 11, these various results are put together to provide a coherent story about the nature of the contractile event, and the story is correlated with a brief account of the available mechanical and biochemical data on cross-bridge activity.

11

Discussion

Modeling the Contractile Event

11.1. Introduction

It is now well established that the myosin cross bridges are implicated in the contractile event and that it is likely that their attachment angle on the actin filaments varies during the cross-bridge cycle, thus producing tension. In this final chapter the attachment–detachment cycle of the cross bridges is considered in more detail, both in terms of the available mechanical data, of which only a brief summary will be given, and in terms of the various theories that have been proposed to account for it. Preliminary X-ray diffraction tests of these theories are then described and evaluated, as is the recent evidence on the time course of cross-bridge movement during contraction. The chapter ends with a brief statement of the probable sequence of events that is involved in producing tension following a stimulus and also of the structural experiments that in the future are likely to be of most value in defining with more certainty the true nature of the contractile event.

11.2. Evidence from Mechanical Experiments

11.2.1. Early Experiments

Concurrent with the detailed structural and biochemical studies on muscle that have been carried out in recent years, considerable advances have also been made on muscle mechanics, and the degree of experimental sophistication involved has become very high. Fortunately, two really excellent review articles have been written on this topic, one by White and Thorson (1973) and one by A. F. Huxley (1974). Both reviews provide a historical perspective on the progress of mechanical experiments, and the reader is referred to these for details of the many experiments that can only be considered briefly in this chapter.

The logic of the presentation given here will follow fairly closely that of A. F. Huxley (1974), in whose laboratory many of the relevant experiments have been carried out and who has contributed more than any other to the improvements in experiment design that have enabled the latest detailed results on the mechanics of the cross-bridge cycle to be obtained.

It was shown in Chapter 1 that following nervous stimulation, an action potential is set up in the surface membranes of muscle cells and is propagated towards the ends of the cell. In vertebrate skeletal muscle, this action potential also spreads inwards through the cell along the T-tubules and results in the release of calcium from the terminal cisternae of the neighboring sarcoplasmic reticulum (Fig. 1.25). But in all muscles the calcium released as a result of stimulation binds to either the thick or thin filaments or both (as described in Chapters 5 and 6) and, in so doing, the filaments are activated, and contraction can take place. Relaxation is induced by the active transport of calcium into the longitudinal elements of the sarcoplasmic reticulum. We have seen that this calcium regulation, according to current theories, controls the interaction of the myosin heads with the actin filaments and that contraction may be associated with cross-bridge movement (rotation) while attached to actin.

Throughout this book this kind of contraction theory has been given prominence, but it is as well that the reasons for the success of this theory should be clear, as should be the reasons for rejecting other possibilities. These reasons are given below.

Before the sliding filament theory was introduced in 1954 (H. E. Huxley and Hanson, 1954; A. F. Huxley and Niedergerke, 1954), a considerable amount of mechanical data had already been obtained, largely in A. V. Hill's laboratory. The early theories based on the available data mainly used the idea that muscle, when activated, behaved like a stretched spring (Weber, 1846; Fig. 11.1a). This basic idea was refined by the introduction of damping of the spring by viscous forces (Fig. 11.1b) (analogous to a spring in heavy oil) and by adding a series elastic component (Fig. 11.1c). All of these models possess the common property that for a given degree of shortening "the amount of energy available should be independent of the manner of contraction provided that

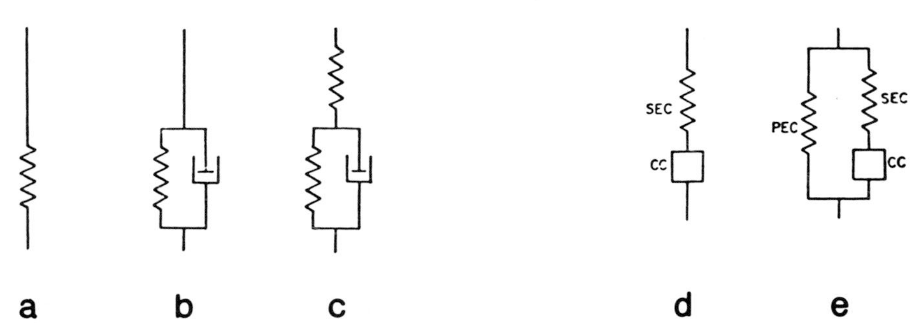

FIGURE 11.1. Various mechanical models for the behavior of muscle. A, B, and C are examples of the viscoelastic or cocked spring models of muscle. A. Simple spring. B. Damped spring. C. Damped spring plus series elastic element. D and E are two forms of A. V. Hill's model. D is a two-element model comprising a contractile component (CC) in series with a series elastic component (SEC). E is like D but with a parallel elastic component (PEC) included. This accounts for the elasticity of resting muscle at high degrees of stretch; at short sarcomere lengths the two-element model (D) is sufficient.

all methods of release of energy are taken into account" (White and Thorson, 1973).

It was the observation by Fenn (1924), now known as the Fenn effect, that the total energy released in the form of heat plus mechanical work in frog sartorius muscle was greater in twitches when shortening was allowed than in isometric twitches, that eventually proved the downfall of the cocked spring models. However, it is as well to note that the requirement that mechanical work should not be reconverted to chemical energy during contraction, a requirement necessary to this deduction from the Fenn effect, was not shown to be reasonable until 1973 (Curtin and Davies, 1973).

Further improvements in both mechanical and thermal measurements of muscle by A. V. Hill (1938) led to a new kind of model. Hill found that under a wide variety of conditions a muscle developed more heat while shortening than it did in isometric contractions, and the amount of extra heat produced was solely dependent on the amount of shortening (x). The rate of heat production during shortening is therefore $a(\mathrm{d}x/\mathrm{d}t) = aV$, where a is the relevant constant of proportionality between heat and shortening and V is the shortening velocity. Hill further showed experimentally that the rate of heat production plus the rate of mechanical working (PV) depended linearly on the load on the muscle, being zero when $P = P_0$, the isometric force. From these two relationships we obtain $(P + a)V = b(P_0 - P)$ where b is another constant which can be determined by measuring the extra energy liberated as a function of the load.

This relation is normally termed the Hill equation. Hill visualized these results in terms of a new three-element model (Fig. 11.1e) which contained a contractile component (CC) in series with a series elastic component (SEC) and both in parallel with a parallel elastic component (PEC). The SEC was introduced because it had been shown by Levin and Wyman (1927) that instantaneous length changes were also associated with instantaneous tension changes. Similarly, the PEC was introduced since relaxed muscles will exert tension when stretched (see Fig. 1.14). The contractile component (CC), which was thought to generate tension actively, was assumed to follow the behavior described by the Hill equation. Of course, this equation makes predictions about the velocity (V) for a given load P. Hill found that the values of a and b obtained by this means agreed well with the corresponding values obtained from the energy measurements for frog and toad muscle. The Hill equation therefore seemed to provide a good simulation of the constant-velocity behavior of contracting muscles.

But it was not tested critically until Jewell and Wilkie (1958) measured the characteristic properties of the SEC and CC. The PEC was ignored (Fig. 11.1d), since their experiments were carried out as sarcomere lengths at which the PEC tension was negligible. Jewell and Wilkie measured the initial development of tension in an isometric tetanus and then the length changes resulting from a rapid step drop in tension. Figure 11.2a–c illustrates the form of the results obtained in this way. The step tension drop results in an instantaneous shortening of the muscle followed by a slower steady shortening. Oscillations were observed between the two phases of shortening, but at the time these were largely caused by mechanical defects in the experimental apparatus.

If Hill's model (Fig. 11.1e) is correct, then the initial length change should result from elastic shortening of the SEC, and subsequent shortening should follow the properties of the CC as defined by the Hill equation. Assuming that Hill's equation is followed immediately after the step, extrapolation back from one of the constant velocity lines (xy in Fig. 11.2b) to the time of the step (z) should yield one point on the velocity–force curve of the CC and one point on the length–tension curve of the SEC.

With the characteristics of the SEC and the CC defined in this way, one can calculate the rate of redevelopment of tension during a tetanus following a step length change of magnitude just sufficient to reduce the tension to zero. This is shown in Fig. 11.2d for the results from Jewell and Wilkie (1958). Also shown in this figure are Jewell and Wilkie's measurements of the increase in tension with time during the rising phase of a tetanus (⦶) and following a quick release (⊖) sufficient to reduce the tension to zero. It is clear that there is very poor agreement between the observations and the predictions from Hill's model and his equation. The reason for this now appears to be that events of considerable interest take place immediately after the tension or length steps. Therefore, Hill's equation is not obeyed instantaneously but only applies to the steady state. The conclusions from this must be that models that predict the events that occur within the first few milliseconds following a length or tension step are needed to account for Jewell and Wilkie's results, and that apparatus of very fast time resolution is needed to explore experimentally the initial changes that occur following the step.

Clearly in this process of theory development, the advent of the sliding filament model was crucial, and in Chapters 1 and 10 reasons have already been given for favoring models in which the relative sliding of the thick and thin filaments is caused by independent force generators, probably the myosin cross bridges. But the sliding filament model is compatible with other types of contraction hypotheses that do not involve cycling cross bridges. A. F. Huxley in his 1974 review summarized these other models in a very helpful and stimulating way and gave good reasons for discounting most of them. The essence of his argument is given below.

1. Podolsky (1959) suggested that the ends of the thin filament might attach to a myosin filament, and the overlapping part of the thin filament shorten. This idea is untenable since the ends of the thin filament move towards the M-band into the region devoid of cross bridges when the sarcomere shortens.

2. Several models have not yet been discounted, but there is no direct evidence for them. Among these are models involving a vernier action between sites on the thick and thin filaments (Spencer and Worthington, 1960; H. E. Huxley, 1964b), models in which one of the types of filament undergoes cyclic changes in length and is temporarily attached to the other filament type for an instant during each cycle (Hanson and Huxley, 1955; H. E. Huxley, 1960; Asakura *et al.*, 1963), and models in which the shortening is produced by thin longitudinal filaments that run between the actin and myosin filaments (McNeill and Hoyle, 1967; Hoyle, 1968; see Section 7.4.3.).

3. Models have been proposed that involve shortening of sarcomeres that exhibit constant-volume behavior when lateral expansion occurs. The model of G. F. Elliott *et al.* (1967) involved electrostatic repulsive forces between filaments, and that of Ullrick (1967) involved expansion of the Z-line. These ideas seem to have been discounted by results of Matsubara and Elliott (1972) that indicate that in some muscle fibers constant-volume behavior is lost when the fibers are skinned, yet they will still contract.

4. Several theories involve the maximization of interaction sites between thick and thin filaments. As A. F. Huxley (1974) has shown, these

theories (H. E. Huxley and Hanson, 1954; Yu *et al.,* 1970) fail to account either for the observed length–tension curve (Fig. 1.8), for the fact that shortening can occur beyond the length where overlap is complete, or for the decrease of energy liberation per unit change of length as the speed of shortening is increased (Fig. 11.3). According to such models, the total energy released as heat and work will just be proportional to the change in the amount of overlap, which is contrary to the observations of A. V. Hill (1938).

5. The final class of theories, of which there are now many examples, all involve the action of independent force generators linearly distributed along the overlap region of the A-band. As described earlier, these account satisfactorily for the length–tension curve (Fig. 1.8) and for the observations that the speed of shortening per sarcomere under zero load is largely independent of overlap or of sarcomere length in different muscles whose filaments vary in length (Jasper and Pezard, 1934; Atwood *et al.,* 1965; A. F. Huxley and Julian, 1964; Gordon *et al.* 1966).

Thus the idea that there are independent force generators seems

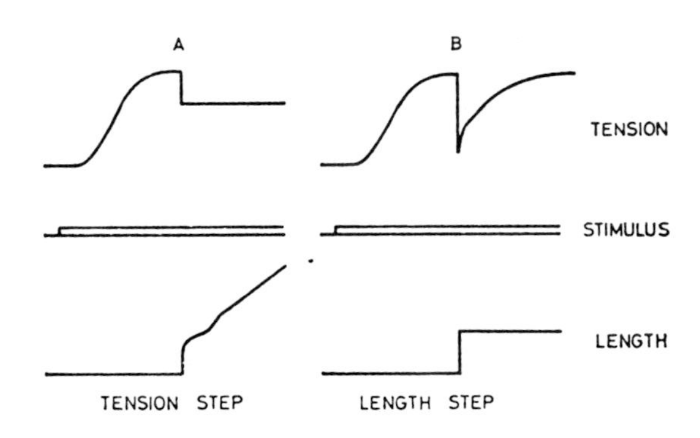

FIGURE 11.2. (a) Characteristics of the two most useful types of mechanical experiment performed on active muscle during tetanus. In both cases, the full tetanic tension is first developed under isometric (constant length) conditions. In tension-step experiments, the tension is then changed rapidly to a new level, and the resulting length changes of the muscle recorded. In length-step experiments, the length of the muscle is forced to a new value very rapidly, and the resulting tension changes are observed. (Decrease of length upwards; from White and Thorson, 1973.) (b) Length change resulting from a step reduction of tension. Decrease of length upwards. (From Jewell and Wilkie, 1958). (c) Series of experimental curves as in (b) for length changes resulting from different changes in tension (to the values in grams shown alongside each trace). The origins of the traces have been shifted for clarity, and the dots on the horizontal lines denote time intervals of 1 msec. (From Jewell and Wilkie, 1958.) (d) The measured development of tension (⦶) during the rising phase of a tetanus and (⊖) following a quick release just sufficient to bring the tetanic tension to zero. The closed circles show the predicted time course of development of tension on Hill's two-element model. (From Jewell and Wilkie, 1958.)

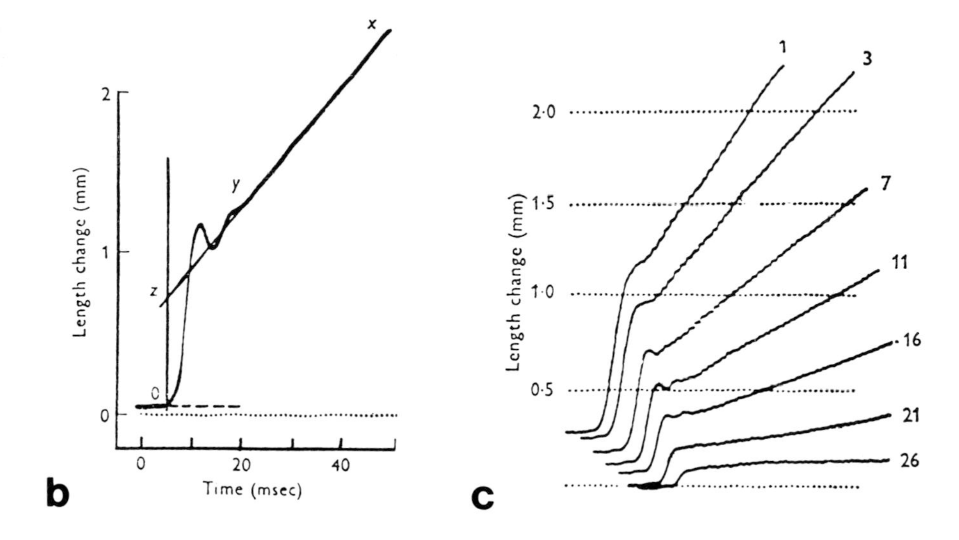

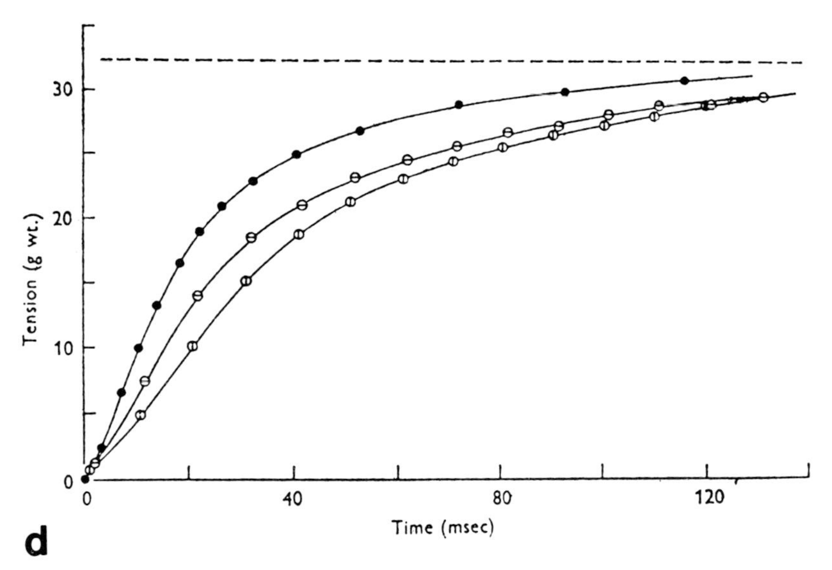

FIGURE 11.2 *Continued*

to be reasonable, but several alternative theories based on this idea have been put forward (Weber 1956, 1958; A. F. Huxley, 1957; Tonomura *et al.*, 1961; Davies, 1963; H. E. Huxley, 1969; A. F. Huxley and Simmons 1971b, 1972; McClare, 1972). Some of these alternatives are considered in the next sections.

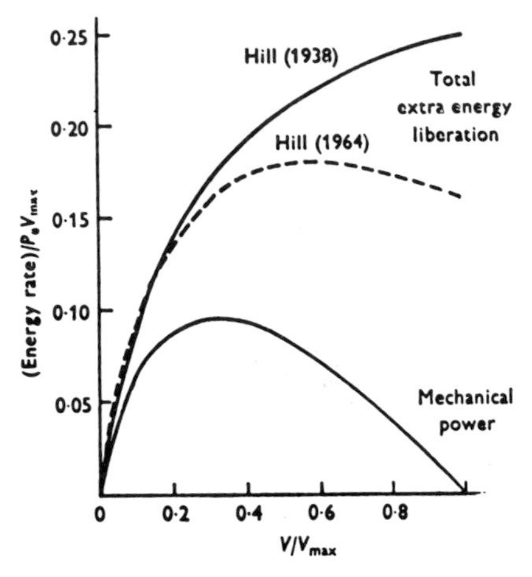

FIGURE 11.3. The rate of energy liberation as a function of shortening speed during steady shortening in a tetanic contraction. The lower curve represents the rate of doing external work calculated from the Hill equation, and the upper curves show the rate of appearance of heat plus work in excess of the heat produced in an isometric contraction. The extra energy liberated per unit change of length is equal to the slope of the line joining the origin to the appropriate point on one of the upper curves. (From A. F. Huxley, 1974.)

11.2.2. A. F. Huxley's 1957 Model

A great deal of evidence has been presented throughout this book that supports the idea that it is the myosin heads or cross bridges that are the independent force generators in muscle. It is important, then, to see how such cross bridges can account for the observed experiments on muscle mechanics. One of the most significant approaches to this problem was that of A. F. Huxley (1957) who devised a model in which, in his own words, "The structural and chemical assumptions were the simplest I could think of that would cause contraction to occur" (from A. F. Huxley, 1974).

His model was based on the diagram shown here as Fig. 11.4a. It included "side pieces" (cross bridges in current terminology) that are elastically linked to the myosin filament shaft and are capable of binding to sites A (only one shown) on the actin filaments. Attachment occurs spontaneously but may be reversible, and its probability $f(x)$ is taken to be a function of displacement x of the A site from the mean cross-bridge position. Similarly, $g(x)$ is the probability of detachment per unit time, and this process involves ATP binding. The values of $f(x)$ and $g(x)$ as a function of x (Fig. 11.4b) were chosen to make attachment likely for

FIGURE 11.4. Mechanism of contraction (a) proposed by A. F. Huxley (1957). The side piece (M) elastically connected to the thick filament is assumed capable of binding to sites A (only one shown) on the thin filament. Attachment occurs spontaneously but may be reversible; detachment occurs principally by a process that involves the hydrolysis of an ATP molecule. Attachment is accelerated by an enzyme attached to the thick filament placed to the right of O and therefore effective only when M is displaced in this direction by Brownian motion. As soon as an attachment is made, the elastic connection exerts a force

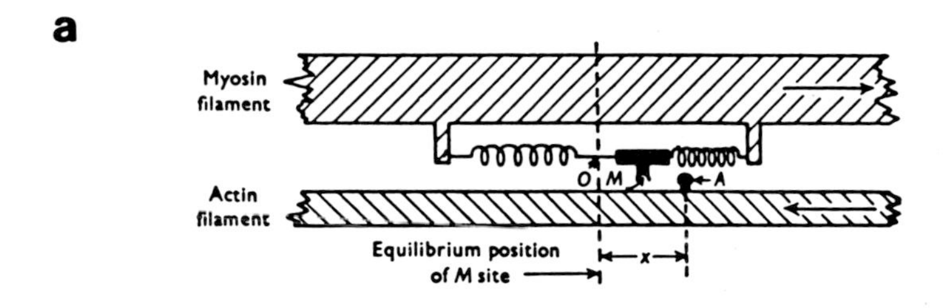

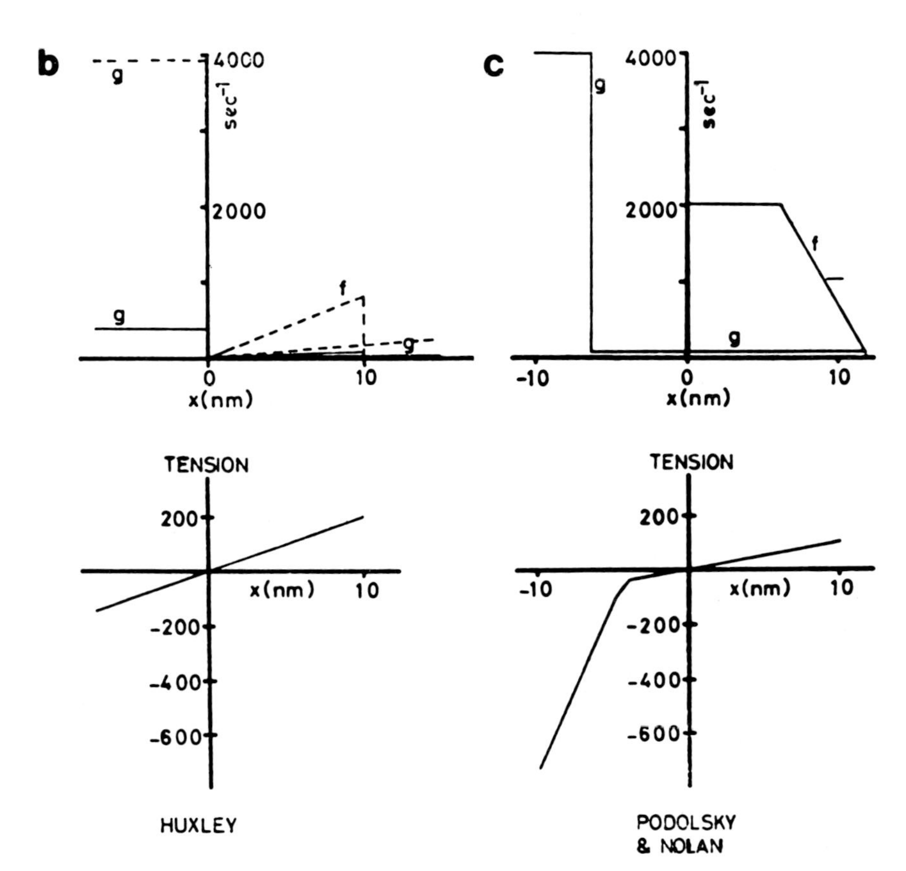

in the direction tending to shorten the muscle. The detachment process (utilizing ATP) is accelerated by another enzyme attached to the thick filament and placed to the left of O. Detachment therefore occurs rapidly after the sliding movement has brought the AM complex to positions where the elastic connection exerts a force that resists shortening. (From A. F. Huxley, 1974.) (b and c) The rate constants of attachment (f) and detachment (g), and tension versus distortion (x) in A. F. Huxley's (1957) model (a), as determined by Julian (1969) (b), and as adapted by Podolsky and Nolan (1972) (c). (From White and Thorson, 1973.)

positive values of x and detachment likely for negative values of x. By assuming that detachment is slow unless shortening occurs and the cross bridge is allowed to complete the working part of its stroke, the heat of shortening can be explained.

A result of A. V. Hill (1938) was that the energy liberated per unit change of length becomes less as the speed of shortening increases (Fig. 11.3). This can be explained if at high speeds of shortening the filaments are moving so fast that a cross bridge cannot cycle fast enough to allow it to interact with every potentially available actin site that rushes past (Needham, 1950). The rate of attachment $f(x)$ must therefore not be too high. A further observation that needs to be accounted for (Fenn, 1924; Hill, 1938; Aubert, 1944a,b, 1948; Abbott *et al.*, 1951; Abbott and Aubert, 1951; Hill and Howarth, 1959; Wilkie, 1968; Curtin and Davies, 1973) is that if the muscle is stretched, the rate of energy liberation (heat produced minus work done on the muscle) is much less than during an isometric contraction. This can be incorporated into A. F. Huxley's model by assuming that the attachment process is reversible so that a cross bridge can detach without going through its working stroke or using ATP (an additional rate constant then needs to be included).

Using the rate constants $f(x)$ and $g(x)$ shown in Fig. 11.4b (values calculated by Julian, 1969), A. F. Huxley's 1957 theory could account satisfactorily for both the force–velocity curve from A. V. Hill's results (Hill, 1938; Fig. 11.5) and the relationship between load and energy liberation. It also explains in a natural way the difference between the effects of shortening and lengthening in the force–velocity curve (Fig. 11.5), where the speed of lengthening under a load greater than the isometric tension (P_0) is much slower than would be expected from extrapolation from shortening speeds for loads less than P_0 (Hill, 1938; Katz, 1939). Note, however, that only the kinetic properties of the model lead to this agreement (Deshcherevski, 1968), and other models with different chemical and structural assumptions but the same kinetics would serve as well in accounting for the observations.

11.2.3. Podolsky's Model

Evidence obtained since the formulation of A. F. Huxley's model has shown it to be unable to explain two classes of result. One is illustrated in Fig. 11.3, where it is seen that Hill's more accurate estimates of the rate of liberation of energy as a function of speed (Hill, 1964) showed that the "shortening heat" drops at high shortening speeds and goes through a maximum, in this case at V/V_0 equal to 0.5 to 0.6. As noted by A. F. Huxley (1973), this could be explained by a slight modification of his original model if attachment is assumed to be a

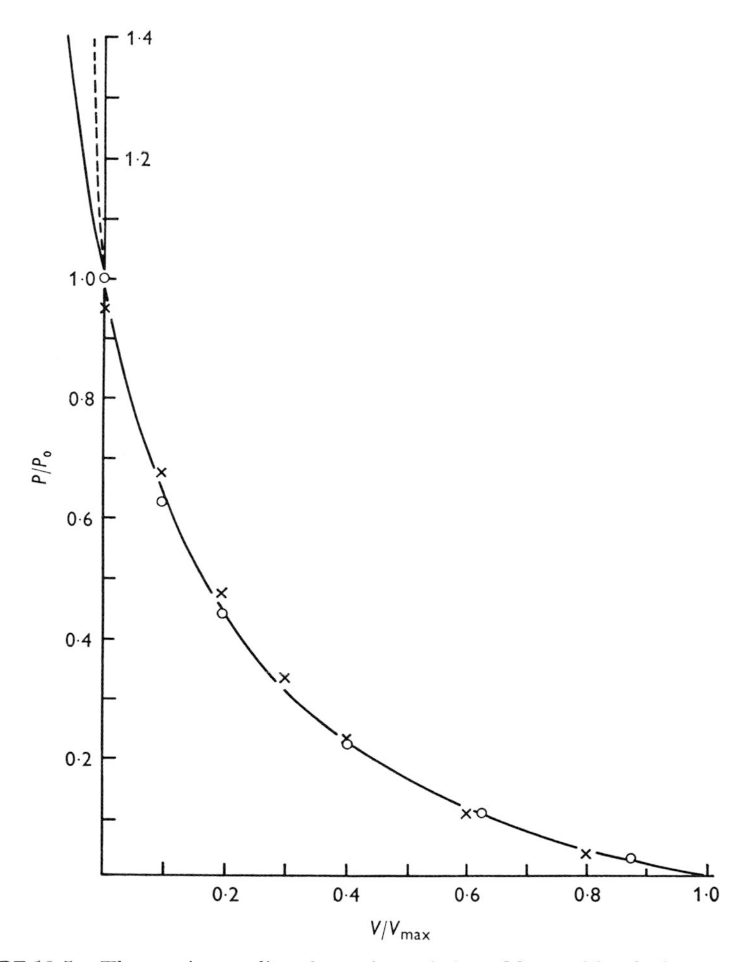

FIGURE 11.5. The continuous line shows the variation of force with velocity as predicted by the Hill equation (see text) with $P_0/4 = 4$, appropriate for frog muscle at 0°C. This equation is a good fit to the observations at loads below P_0 (shortening). In lengthening, Katz (1939) found the relation shown as the dotted line. The crosses are from Fenn and Marsh's relation $P = W_0e^{-aV} - kV$, with $W_0 = 0.95\ P_0$, $a = 3.4/V_{max}$, and $k = 0.03P_0/V_{max}$, values chosen to fit the line as well as possible. In their curve fitting, Fenn and Marsh did not make use of a measured P_0. The circles are values calculated from A. F. Huxley's (1957) model. (From A. F. Huxley, 1974.)

two-state process so that detachment can easily take place without net chemical change before the second step has occurred.

The second class of observations is a refinement of the length step and tension step experiments of Jewell and Wilkie (1958) in which the recording and control equipment has been made sufficiently fast that details of the transient changes that result from rather small step changes can be observed (Podolsky, 1960; Civan and Podolsky, 1966). Figure 11.6 illustrates the kinds of responses that are obtained in such experiments. It can be seen that following either a sudden change in load which gives a velocity transient (upper diagram in Fig. 11.6) or a sudden change in length which gives a tension transient (lower diagram), there is an oscillatory response before the steady state is reached.

Figure 11.7A shows a series of predicted transients in tension step experiments calculated by Civan and Podolsky (1966) on the basis of A. F. Huxley's 1957 model for different step magnitudes. It can be seen that oscillatory transients would not be predicted by the model. For this reason, Podolsky and Nolan (1972) modified the rate constants $f(x)$ and $g(x)$ in Huxley's model to the values shown in Fig. 11.4c. In this way, an oscillatory response was obtained as shown in Fig. 11.7B. The origin of this oscillatory behavior, to quote White and Thorson (1973), is "the gap between zero displacement and the value of x at which the detachment probability becomes very high, aided by the steep increase in stiffness of

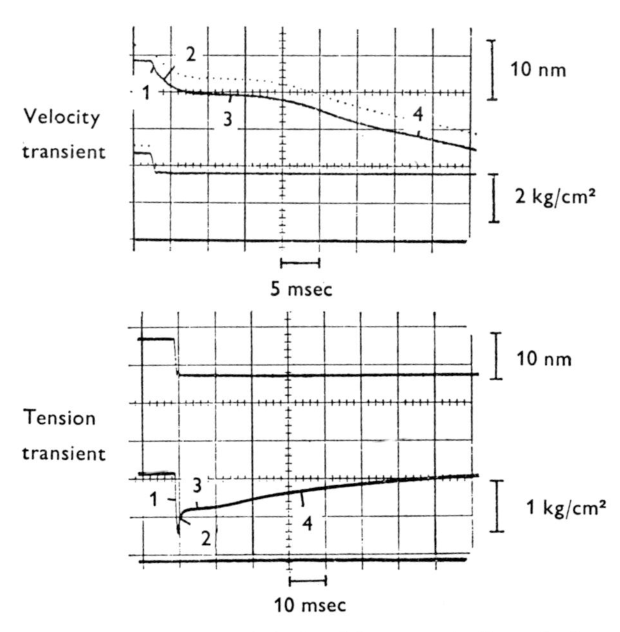

FIGURE 11.6. Mechanical transients in isolated frog muscle fibers during tetanic stimulation. The velocity transient is the time course of the length change (upper trace) resulting from a sudden alteration in load (middle trace). The tension transient is the time course of tension change (middle trace) resulting from a step change in length (upper trace). In both cases the bottom trace is the tension base line. The numbers represent corresponding phases in the two types of transient as detailed in the text. (From A. F. Huxley, 1974.)

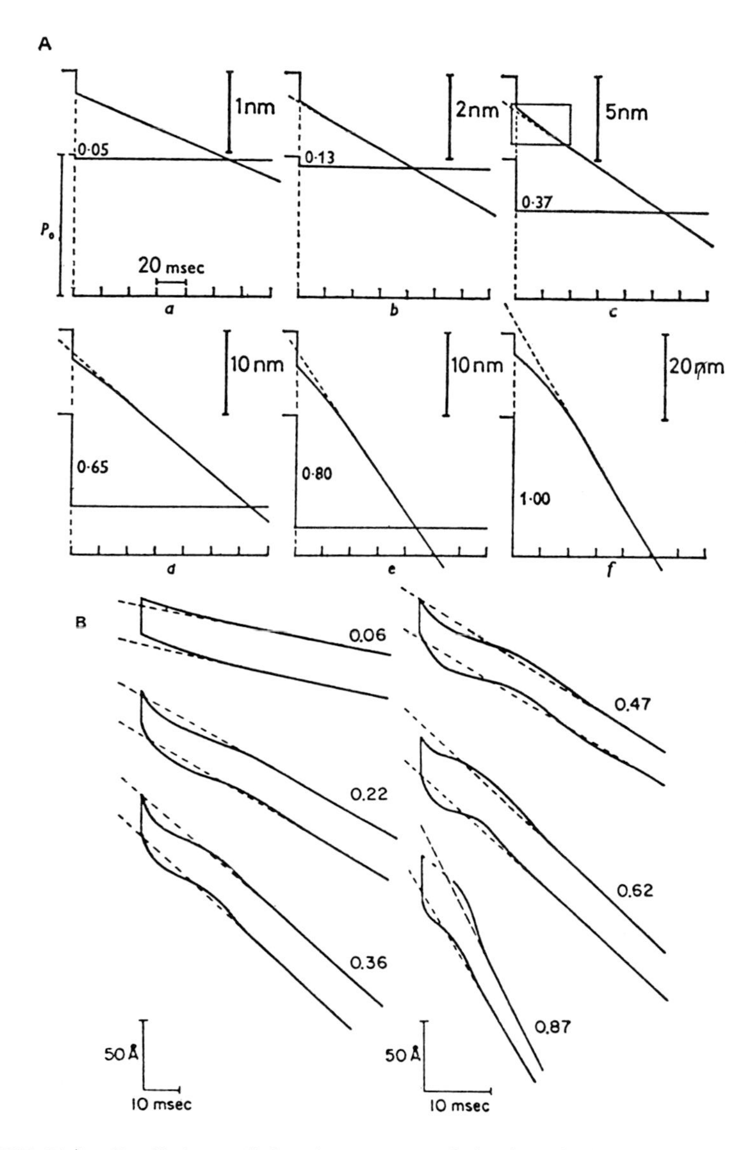

FIGURE 11.7. Predictions of the time course of the length changes obtained in a tension-step experiment (velocity transient) on the Huxley (1957) model with (A) Huxley's (1957) values for f and g (Fig. 11.4b) and (B) Podolsky and Nolan's distributions of f and g (Fig. 11.4c). A is taken from Civan and Podolsky (1966) but with the scales modified to account for the magnitudes of the rate constants in Fig. 11.4b. The upper trace is of length, and the lower trace is tension. In B, from Podolsky and Nolan (1972), the upper trace is the experimental length record, and the lower trace is computed from the distributions of Fig. 11.4c. Note that the initial instantaneous length changes have been omitted. The magnitudes of these depend, in their model, on the amount of extra elasticity in series with the cross bridge. In both parts the figures alongside the records denote the fraction of the full isometric force to which the tension has been reduced. (From White and Thorson, 1973.)

attached bridges in the region of negative x. The effect of the gap is to introduce a delay between the effect of bridge attachment (which occurs at a very high rate in regions of positive x) and bridge detachment."

Let us now turn to the type of experiment carried out by Jewell and Wilkie (1958) and illustrated in Fig. 11.2. It seems that the formulations of A. F. Huxley (1957) and Descherevski (1968) will account well for the predicted time course of the redevelopment of tension following a length step just sufficient to cause the tension during an isometric tetanus to fall to zero. It also appears that Podolsky and Nolan's model will probably do just as well (White and Thorson, 1973). However there are significant differences between the two classes of model, apart from their ability to account for the length transients shown in Fig. 11.7. In particular, the number of cross bridges attached under different steady-state conditions are clearly different in the two cases. If the number of bridges attached at displacement x at time t is $n(x,t)$ then the rate of change of $n(x,t)$ with time is a function of $f(x)$ and $g(x)$ such that

$$\frac{dn(x,t)}{dt} = f(x)\,[1 - n(x,t)] - g(x)n(x,t)$$

By making use of the appropriate values for $f(x)$ and $g(x)$, the distribution of attached bridges during steady-state shortening conditions can be calculated (A. F. Huxley, 1957; Podolsky and Nolan, 1972). The results from this are illustrated in Fig. 11.8, and it is clear that A. F. Huxley's model predicts a decrease in attachment number when speed increases, whereas Podolsky and Nolan's predictions show an increased attachment under the same conditions. We shall return to this difference later in the chapter since, in theory, it can be investigated by X-ray diffraction methods.

Another feature of Podolsky and Nolan's model was that they needed to include a series elastic element in series with the single cross bridge to account for the observed response to step tension reductions; the observed instantaneous length changes were attributed equally to the cross-bridge elasticity and to the series elasticity.

Further inspection of the transients shown in Figs. 11.6 and 11.7 shows that they can be considered in terms of four phases (A. F. Huxley, 1974). Phase 1 is the instantaneous phase resulting from an instantaneous elasticity in the muscle, which a priori might be caused by both the cross bridge and a series elastic element (e.g., in the Z-band). Phase 2 (next 1–2 msec) is associated with either a rapid tension recovery in a tension transient or a rapid early shortening during a velocity transient. Phase 3 (next 5–20 msec) shows an extreme reduction or reversal of tension recovery or of shortening speed, and Phase 4 shows either a

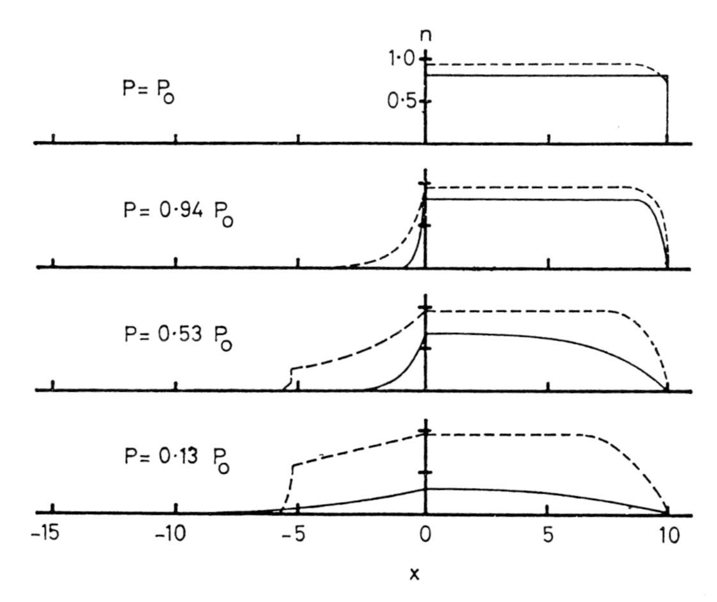

FIGURE 11.8. Prediction from A. F. Huxley's (1957) model of the distribution of fraction of attached crossbridges (n) as a fraction of distortion (x in Fig. 11.4a) at various steady tensions P (indicated as fractions of the full isometric tension P_0). Solid lines, distribution calculated using f and g values in Fig. 11.4b (A. F. Huxley, 1957); dashed lines, distribution calculated using Podolsky and Nolan's values (Fig. 11.4c). (From White and Thorson, 1973.)

gradual recovery of tension with asymptotic approach to isometric tension or a steady shortening sometimes with a slight damped oscillation. Phases 3 and 4 can be attributed to detachment and reattachment of cross bridges with kinetics similar to those in Huxley's 1957 model. However, for detachment to give rise to the tendency for tension to fall in Phase 3, it is necessary to suggest that accelerated detachment begins when the cross bridge is still producing tension. This is a requirement not included in Huxley's model but one that will account well for Phases 3 and 4 if it is included (Julian *et al.,* 1973). Phase 2 is the most critical of all the phases studied and is the most difficult to account for. Podolsky and Nolan (1973) proposed that the instantaneous recovery of tension could be the result of bridges not attached to actin in the isometric case that were suddenly moved by the length step to positions where actin attachment was sterically possible. However, since at least some (probably most) of the instantaneous elasticity of the system resides in the cross bridges, it would be expected that the extra attached bridges in this case would make the muscle stiffer. Ford *et al.* (1974) have tested this possibility by applying a second length step towards the end of the early recovery phase (phase 2), and they found a stiffness no greater than in the isometric condition.

A. F. Huxley and Simmons (1971a,b; 1972) have treated Phase 2 in quite a different but illuminating way, the details of which are given in the next section.

11.2.4. A. F. Huxley and Simmons' Model

Figure 11.9 illustrates the detailed form of Phases 1 and 2 on a very fast time base in the tension transients caused by different length steps (from A. F. Huxley, 1974). A. F. Huxley and Simmons (1971a,b, 1973) noted the similarity between this kind of response and that of a Voigt element, a spring (Hooke's constant, k_1; Fig. 11.10A) in series with a parallel combination of spring (k_2) and dashpot (viscous component). In analyzing the response of such an element to step length changes, it is convenient to consider the extreme tension (T_1; Fig. 11.10B) reached instantaneously after the length step and the equilibrium tension (T_2) which the tension trace approaches after some time. T_1 is a measure of the properties of k_1, since the dashpot cannot move instantaneously. After this initial change, the dashpot will gradually lengthen under the influence of the increased tension in k_1 until an equilibrium is reached when the tension in k_1 and k_2 is equalized. The time course of the tension change from T_1 to T_2 will be exponential with a time constant that depends on k_2 and the properties of the dashpot.

If different length steps are applied, then the values of T_1 and T_2 should vary linearly, provided the springs are linear, as illustrated in Fig. 11.10D. The T_1 value gives the stiffness of k_1, whereas that of T_2 gives the combined stiffness of k_1 and k_2 in series. So k_2 can be determined from Fig. 11.10c by subtracting S_1 (the length step that gives tension T on the T_1 curve) from S_2 (the length step that gives tension T on the T_2 curve); k_2 is then $T/(S_2 - S_1)$ (White and Thorson, 1973).

Figure 11.11 shows an experimental plot of T_1 and T_2 against length step magnitude obtained by A. F. Huxley and Simmons (1971b). The T_1 curve is very similar to that for a Voigt element except for becoming slightly less steep at lower tensions, a common characteristic of biological elastomers (A. F. Huxley, 1974). However the T_2 curve is significantly different from that of a Voigt element. It is not linear, being almost horizontal for small steps and then dropping considerably for longer step decreases in length. This suggests that the components responsible for the tension recovery are not just passive visco-elastic elements as in the Voigt model but are active tension generators. Also, the displacement of the T_2 curve about 60 Å (possibly 80 Å; Ford *et al.*, 1977) to the left of the T_1 curve "suggests that the active element in the cross bridge is

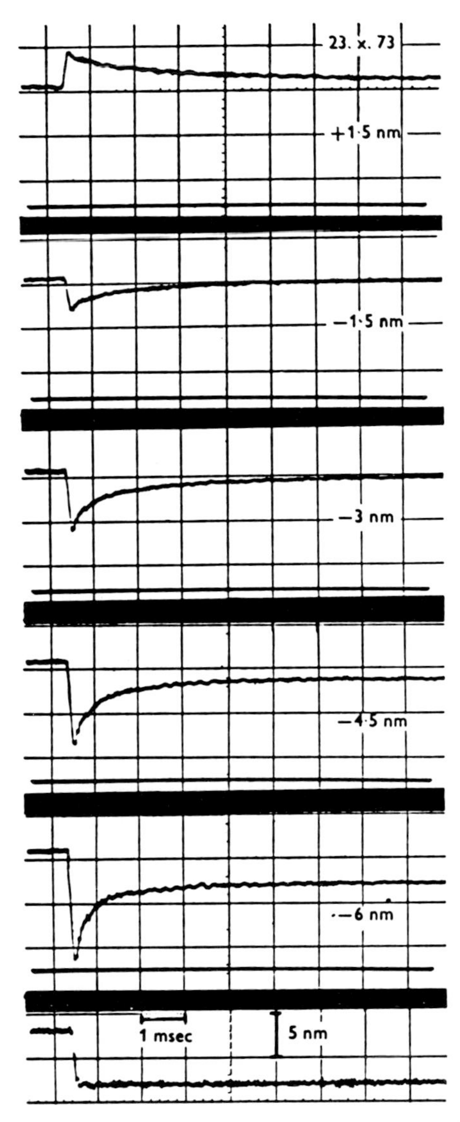

FIGURE 11.9. Family of tension transients as in Fig. 11.6 but on a time scale ten times as fast. The step change of length is complete in 0.2 msec, whereas in Fig. 11.6 it took about 1 msec. The distance specified in each of the upper panels is the amount of the imposed length change in each half sarcomere. The lower panel is the length record. The transients show phases 1 and 2 of Fig. 11.6. (From A. F. Huxley, 1974.)

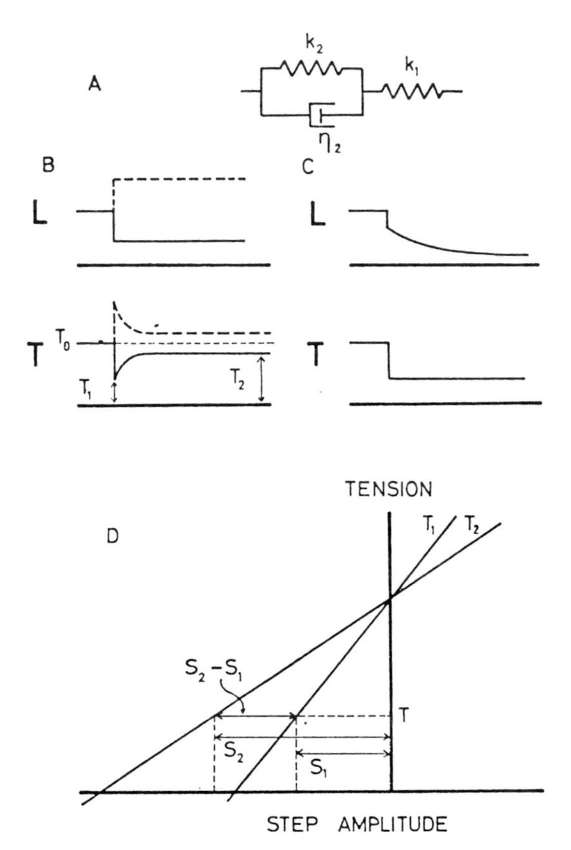

FIGURE 11.10. Linear model used to illustrate Huxley and Simmons' analysis of their length-step experiments (Fig. 11.9). A is a Voigt element, a damped spring in series with an undamped spring. B and C are the responses of the model to length-step and tension-step experiments, respectively. D represents the analysis of these length step experiments described in the text. (From White and Thorson, 1973.)

capable of taking up approximately 60 Å of shortening while maintaining a tension not much less than that which it exerts in an isometric contraction" (A. F. Huxley, 1974). A. F. Huxley and Simmons (1971a) further showed that the T_1 and T_2 curves plotted for muscles initially at two different sarcomere lengths (2.2 μm and 3.2 μm) scaled down directly in proportion to the amount of overlap between the thick and thin filaments (Fig. 11.11). This clearly implies that much of the instantaneous elasticity (k_1 in Fig. 11.10) actually resides in the cross bridge itself, as well as in any truly external series elastic component such as the Z-band.

The cross bridge therefore seems to contain an instantaneous elastic element in series with an active component which can maintain tension while taking up about 60 to 80 Å of length change (A. F. Huxley 1974;

Ford *et al.*, 1977) The elastic element is also such that the tension in it is reduced to zero by a length reduction of about 40–60 Å (Ford *et al.*, 1977; Shoenberg *et al.*, 1974) per half sarcomere (Fig. 11.11). The whole assembly could therefore operate over a range of about 120 Å, a value that agrees reasonably well with those calculated by H. E. Huxley (1960) and Davies (1963) on the basis of the tension exerted per cross bridge and the work that can be done per mole of ATP.

A. F. Huxley and Simmons (1971b; 1972) visualized their results in terms of the swinging cross bridge model of H. E. Huxley (1969) illustrated in Fig. 1.11, with the modification that the S-2 link between the myosin head and rod was given the properties of the instantaneous elastic element. They were, however, very careful to point out that this is only one of several possible locations of this series elasticity, and others will be discussed later.

The form of the initial response to a step decrease in length (Fig. 11.9) can then be considered in the following purely mechanistic terms. In a muscle held isometrically so that no relative filament movement can occur at all, the tension is generated by the cross bridge swinging from orientation A (zero tension) to orientation B (tension T) in Fig. 11.12a, thus stretching the S-2 spring by up to about 40–60 Å. Thus, if the muscle is instantaneously allowed to shorten by 40–60 Å, the spring

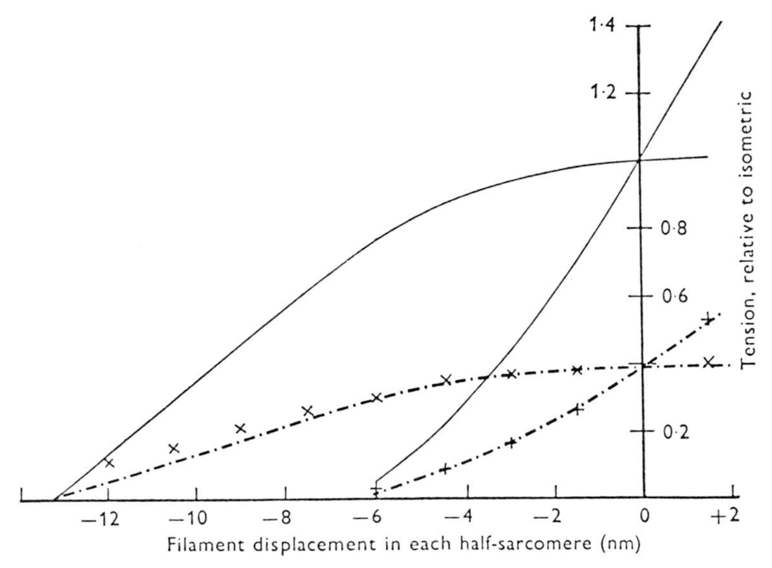

FIGURE 11.11. The solid lines show the observed T_1 and T_2 curves (Fig. 11.10) from a frog muscle fiber with a sarcomere length of 2.2 μm, and the broken lines are the same curves scaled down by a factor of 0.39. They fit well with the crosses which are the T_1 and T_2 curves of the same fiber at a sarcomere length of 3.1 μm, where the overlap is reduced to 39% of that at 2.2 μm. (From A. F. Huxley, 1974.)

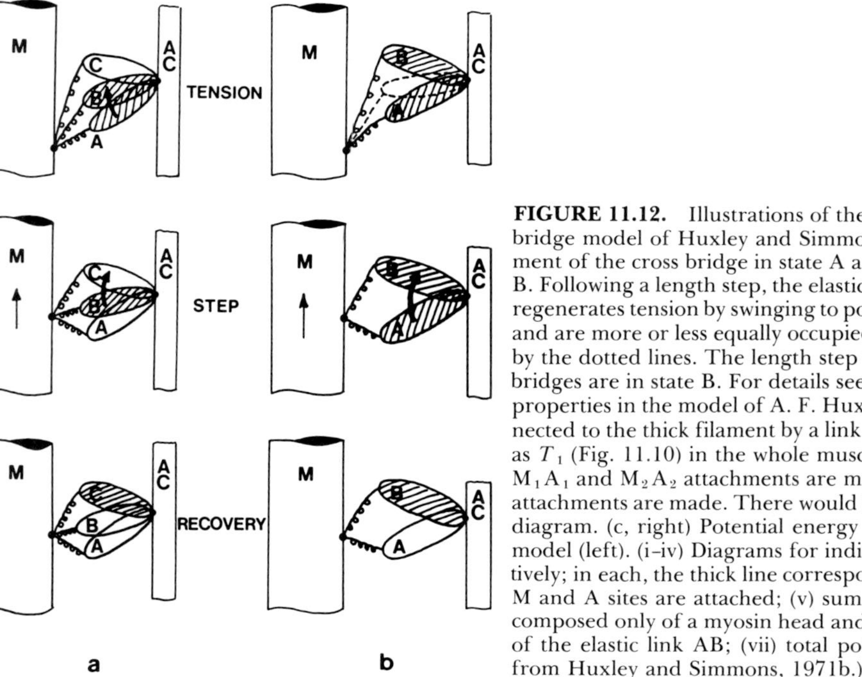

FIGURE 11.12. Illustrations of the three-state (a) and two-state (b) versions of the cross-bridge model of Huxley and Simmons (1971b). In (a) the top figure shows initial attachment of the cross bridge in state A and the rapid transition to the tension-producing state B. Following a length step, the elastic element on A is shortened, and the cross bridge then regenerates tension by swinging to position C. In (b) states A and B are both tension-bearing and are more or less equally occupied so that the average cross-bridge position is as shown by the dotted lines. The length step then tends to alter the distribution so that most cross bridges are in state B. For details see text. (c, left) Diagram showing assumed cross-bridge properties in the model of A. F. Huxley and Simmons (1971b). The myosin head H is connected to the thick filament by a link AB containing the undamped elasticity that shows up as T_1 (Fig. 11.10) in the whole muscle fiber. The full line shows the head position where M_1A_1 and M_2A_2 attachments are made, and the dotted lines are where M_2A_2 and M_3A_3 attachments are made. There would be three stable positions for the head according to this diagram. (c, right) Potential energy diagrams for the Huxley and Simmons cross-bridge model (left). (i–iv) Diagrams for individual attachments A_1, M_2A_2, M_3A_3, M_4A_4, respectively; in each, the thick line corresponds to the range of θ within which the corresponding M and A sites are attached; (v) sum of (i) to (iv), giving the potential energy of a system composed only of a myosin head and a thin filament; (vi) potential energy from stretching of the elastic link AB; (vii) total potential energy of the complete system. (Reproduced from Huxley and Simmons, 1971b.)

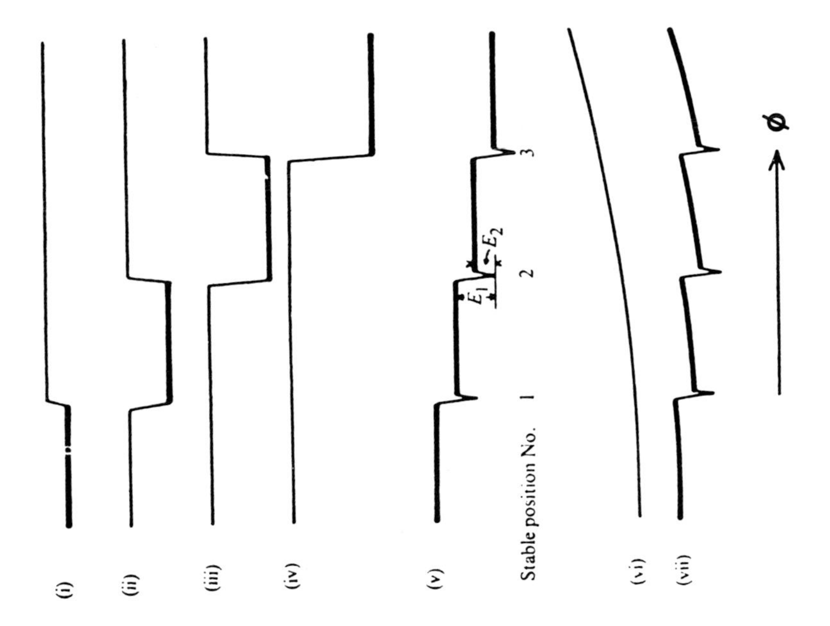
(i)
(ii)
(iii)
(iv)
(v)
Stable position No. 1 2 3
E_1
E_2
(vi)
(vii)
φ

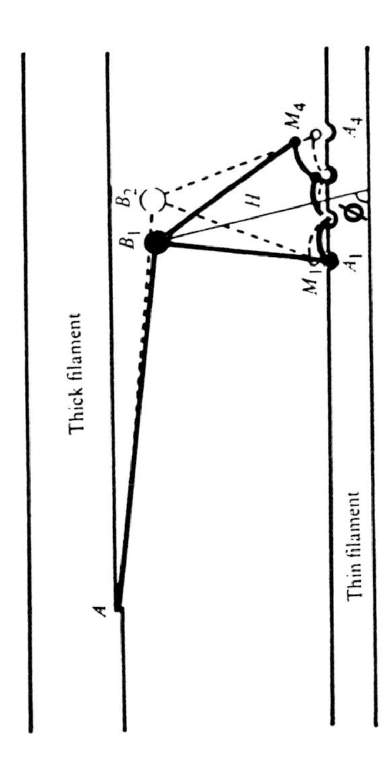
Thick filament
Thin filament
A
B_1
B_2
H
M_1
M_4
A_1
A_4
φ
c

tension will instantaneously fall to zero as will the total tension in the muscle (point *a* on Fig. 11.11). However, the T_2 curve shows that some tension can now be recovered, and this implies that the cross bridge, which is now restricted in movement by the stretched spring, can rotate to a third orientation C, thus stretching the S-2 spring again by about 60 to 80 Å (point *b* on Fig. 11.11). If, however, the length step is initially 110 to 120 Å, then the S-2 spring first shortens by 40 to 60 Å (to zero tension) and then becomes slack (or possibly exerts negative tension; A. F. Huxley, 1974; Ford *et al.*, 1977) during a further 60 to 80 Å of shortening. This slack is then rapidly taken up by a transition of the cross bridge from orientation B to C where it returns the spring to its original unstretched length at which it exerts no tension (point *c* on Fig. 11.11). Intermediate steps produce some recovery of tension (between *b* and *c* on the T_2 curve; Fig. 11.11) since some net stretching of the S-2 spring is still possible.

Huxley and Simmons (1971b) actually looked at the results in a slightly different but analogous way. They assumed for their initial calculations that there are two attached states (A and B in Fig. 11.12b) and that in isometric contractions both states are equally populated so that the "average" cross bridge position can be thought of as being halfway between the two states. Shortening of the muscle by 60 Å then releases the tension in the S-2 linkage of the hypothetical "average" cross bridge, but the cross bridges are then more likely to swing to the higher tension state (State B) where the spring is stretched once again and the tension recovers. The essential difference between the two descriptions is that in Fig. 11.12a it is assumed that cross bridges spend a relatively long time in the single orientation B during isometric contraction, whereas in Fig. 11.12b the two attached states are assumed to be roughly equally populated and at least one is tension bearing. The two-state attachment model (Fig. 11.12b) has been shown by A. F. Huxley and Simmons (1971b) to account reasonably well for the observed variation in the rate constant of the recovery phase in Fig. 11.9 (where the recovery is faster for longer length steps), and they claim that such a model will probably account equally well for the observed relations between load, speed, and heat production as does A. F. Huxley's 1957 model. Podolsky's variation on this model (Podolsky *et al.*, 1969) in which the recovery is explained in terms of sudden increased cross-bridge attachment, is here represented by a sudden transition between different attached states of cross bridges, and thus both models can account for the observed transients. However, the two-state model leads to low values for the force and work per cross bridge (A. F. Huxley and Simmons, 1971b) since transitions between the states depends on Brownian motion and is slow. So it seems likely that at least a three-state model (Fig. 11.12a) may be necessary unless the results can be explained by a completely different mechanism (Eisenberg and Hill, 1978; see later discussion).

The advantage of the new Huxley and Simmons model is that the initial step (see Fig. 11.12a) is an attachment step that is not tension producing. The A. F. Huxley (1957) and Podolsky theories both assumed that cross bridges already bear tension before they attach to the thin filament at large displacements. But, to quote A. F. Huxley and Simmons (1971b), "thermal motion would so seldom bring the cross bridge to such a large deflection that it might be impossible to account in that way for the rapidity of contraction that some real muscles achieve."

To summarize, the new model of A. F. Huxley and Simmons suggests that cross bridges can be attached to actins in a number of configurations (metastable states) with different energies. The distribution of bridges among these states depends on the tension in the series elastic component (visualized here as in the S-2 linkage) and on the energy available to the cross bridge which can cause a jump between attached states. One might presume that the energy necessary to allow a cross bridge to move from one position to stabilize in a state at which it exerts higher tension on the S-2 linkage is derived from ATP hydrolysis; each step may relate to one or more of the biochemical steps in the acto myosin ATPase. Figure 11.12c from Huxley and Simmons (1971b) summarizes their ideas about the potential energy diagram for the different attached states.

Of course, the mechanisms illustrated in Fig. 11.12 are considerable oversimplifications, since the degree of stretch of S-2 will clearly depend on the initial relative axial positions of the actin and myosin sites, and these are not uniquely defined even in isometric contractions. When length changes are permitted as in isotonic contractions, it is clear that the actin and myosin sites must be continuously changing in position, and both the configurations of head attachment and the degree of stretch in the series elastic element will be varying rapidly.

11.2.5. Insect Flight Muscle

As explained in Chapters 1 and 8, one of the most fascinating properties of insect (fibrillar) flight muscle is that when it is stretched at the frequency of the wing beat in the insect its response is to produce delayed changes in tension (Machin and Pringle, 1960). A similar observation by Jewell and Ruegg (1966) on glycerinated insect muscle showed clearly that this response was caused by the components of the myofibril and not by any part of the nervous control system.

One of the most significant mechanical properties of this muscle is that when relaxed it is very much stiffer than is vertebrate muscle

(Machin and Pringle, 1959). This has been attributed to the C-filaments (Chapter 8, Section 8.2.3) which are thought to link the myosin filaments to the Z-band and therefore to provide mechanical continuity through the sarcomere (White, 1967; Ashhurst 1977; Pringle, 1977). The only available alternative explanation, that there are some long-lived fixed cross bridges on actin, possibly giving a response mediated through the paramyosin in the thick filaments as in molluscan muscle (Bullard *et al.*, 1972), seems to be disproved by the observation (White, 1967) that the high resting stiffness is maintained up to extensions of 9%, probably much too far for single cross-bridge interactions.

The effect of the mechanical continuity through the C-filaments on an active muscle is thought to be that stretching the muscle will thereby transmit tension to the thick filaments which in some way then promote cross-bridge attachment and tension generation. But whatever the effect, it clearly must be very rapidly reversible since oscillations at as great a frequency as 1000 Hz or more (Pringle, 1949) have been observed.

An alternative mechanism for stretch activation (Pringle, 1977) might be one involving regulation of the thin filament, possibly connected with movement of the tropomyosin molecules. However, it is hard to imagine how stretch could be sensed directly by the thin filaments unless they are mechanically connected, for example, to the M-band. So far no such connections have been observed in electron micrographs. But if it is the C-filaments that provide the necessary continuity, then two classes of mechanism for stretch activation seem possible. In one, the thick filament is modified in some way so that the attachment rate of myosin to actin is altered. This could be achieved either by mechanically blocking attachment or by slowing down one or more of the recovery steps in the actomyosin ATPase cycle (Fig. 10.2) when stress is reduced, either process possibly being mediated by the paramyosin in the thick filaments. The alternative mechanism assumes that some cross bridges in a calcium-activated muscle are attached to actin (with a slow turnover rate) and that when stretch occurs, these cross bridges exert a mechanical (or cooperative) effect on the tropomyosin and actin in the thin filaments, so that cross-bridge attachment is made easier.

The complex mechanical properties of insect flight muscle, which have been reviewed by White and Thorson (1973), Pringle (1974), and Tregear (1975, 1979), have been modeled in a variety of ways in purely mathematical terms with fair, but not total, success. The models have involved either two steps (attached/detached; Thorson and White, 1969); three steps (1 attached, 2 detached; White, 1972) or four steps (2 detached, 2 attached; R. H. Abbott, 1977), in each case with one or more of the steps taking place at a rate that is a function of muscle length. In this way the stretch activation behavior of insect muscle can be reproduced with a fair degree of success. However, it seems clear that the models involving two or more attached states as in the theory of A. F. Huxley and Simmons (1973) will be needed to account satisfactorily for the available data (Herzig, 1977).

Insect muscle is clearly a highly specialized muscle, but it is as well to note that under suitable mechanical conditions many muscles other than insect fibrillar muscles will display stretch activation although to a much smaller degree (see Steiger, 1977 for survey). It is likely, therefore, that the property of the myofibril responsible for this phenomenon is present in all muscles but is highly developed in insect muscle, possibly because of the direct C-filament link from the myosin filaments to the Z-band. In other muscles with less direct mechanical continuity and that are not in any case mechanically loaded by a suitable resonant system, the stretch activation property, although present, is relatively poorly developed.

Finally the recent mechanical experiments of R. H. Abbott and Cage (1979) deserve some consideration. Here it was found that the tension output of insect muscles mechanically oscillated by a very small amount at gradually increasing degrees of initial stretch varied periodically with stretch at about 380-Å intervals. This was taken to relate to 385-Å repeats in the thin and thick filaments (Wray, 1979b), so that cross-bridge matching with the actin target areas might vary periodically with the amount of filament overlap. However, in electron micrographs of rigor insect muscle it can often be seen that the 385-Å repeat associated with cross-bridge labeling of the thin filaments is in the same axial position across the sarcomere even when axial displacements of 1 × 145 Å or 2 × 145 Å can be seen in the M-band (see Chapter 8; Fig. 8.12). It might reasonably be questioned whether a strong 385-Å repeat along the myosin filaments is necessary to explain these observations; certainly the 385-Å repeat in a six-stranded helix (Fig. 8.7) is not particularly marked. Wray (1979b) has pointed out that a four-stranded filament would have a much more marked 385-Å repeat. Interpretation of this fascinating observation will clearly require more detailed knowledge of the structure of the insect A-band and must await further evidence.

11.3. Equatorial X-Ray Diffraction Evidence on Cross-Bridge Kinetics

11.3.1. Evidence on the Number of Attached Bridges

We have seen in Chapter 10 that in the past the relative intensities of the $10\bar{1}$ and $11\bar{2}$ equatorial reflections from vertebrate muscles have been taken as an indication of the location of the myosin cross bridges in a particular muscle state (H. E. Huxley, 1968; Haselgrove and Huxley,

(1976) illustrate the problems involved in carrying out this kind of experiment.

11.3.2. Interpretation of Equatorial Diffraction Data

A detailed discussion of the equatorial diffraction data from vertebrate and insect muscles and the published interpretations of these data were given in Chapter 10 (Section 10.3.5). But we have just seen that such data are now being used as a direct probe of cross-bridge behavior during contraction, and in the next section it will be shown that these equatorial reflections are also being used to monitor the time course of cross-bridge movement as tension is developed. It is as well, therefore, to

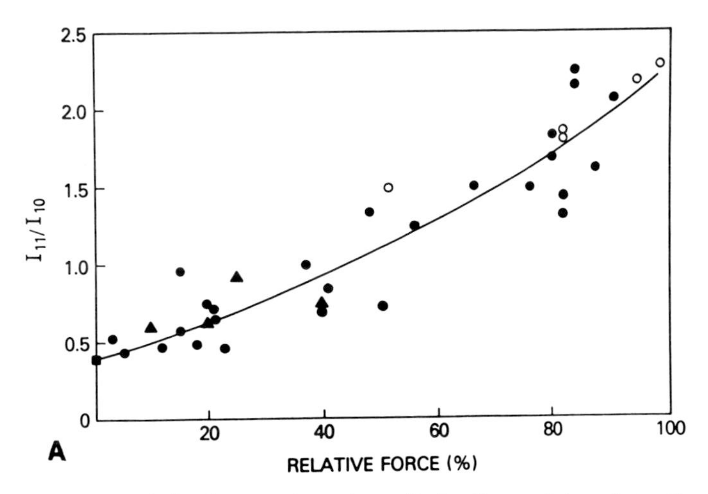

FIGURE 11.13. (a) Relation between intensity radio $I_{11\bar{2}}/I_{10\bar{1}}$ and the relative force (P/P_0). The intensity ratio is the average of the ratios from two sides of the patterns obtained by Yu *et al.* (1979). Contracture force P is the average of the range of force recorded during the exposure period. Filled circles, preparations activated by a constant flow of cold caffeine-Ringer's solution. Each data point was obtained by a single activation cycle. Triangles, preparations cooled to 4–6°C by flow of cold caffeine-Ringer's solution. Each data point was the average (in force and in intensity ratio) of several exposures. Open circles, values of contracture forces were corrected for shortening against the tendon. The increase in the $10\bar{1}$ spacing was between 2 and 3%. The correction was based on the regression line ($I_{10\bar{1}}/I_{11\bar{2}}$ vs sarcomere length) reported by Podolsky *et al.* (1976). Square, mean value in the patterns of the resting state obtained immediately before each activation. The solid line is the least squares fit to all the data points; there is an upward curvature. (From Yu *et al.* 1979.) (b and c) The intensities of the $10\bar{1}$ and $11\bar{2}$ reflections normalized with respect to the undiffracted beam and the values in the resting state. $R_{10\bar{1}}$ is $(I_{10\bar{1}}/I^*)_{contr}/(I_{10\bar{1}}/I^*)_{rel}$, where I* is the attenuated intensity of the direct beam. The definition of $R_{11\bar{2}}$ is analogous. The lines represent the best theoretical fit assuming cross bridges to be in one of only two states (attached, detached; see text), and the maximum fraction of attached heads is 1 when the force is P_0. (From Yu *et al.*, 1979.)

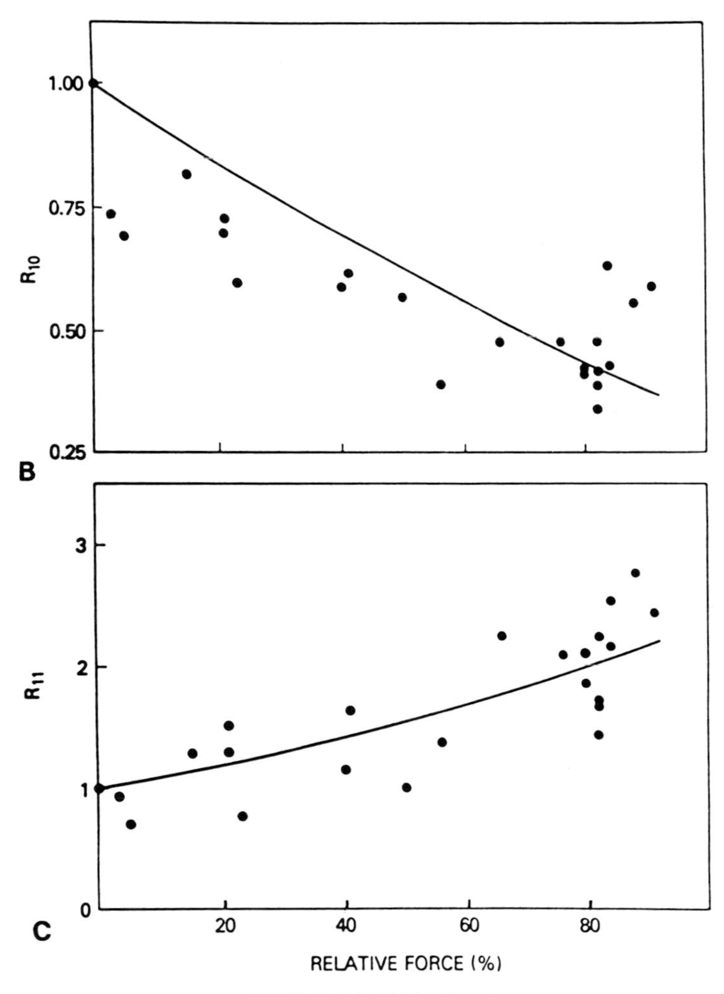

FIGURE 11.13 *Continued*

summarize here what kind of information these reflections are likely to provide.

On the assumption that in higher vertebrate muscles the thick filaments are three-stranded and in a no-three-alike superlattice, the average thick filament structure will appear as in Fig. 10.25. On the equator of the diffraction pattern there is no evidence for the superlattice. This means that the observed density at the thick filament positions of the lattice is an average of two filaments, each with ninefold rotational symmetry in projection, but with a 180° rotation between them (giving 18-fold symmetry). To fit the observed layer line profiles the cross bridges need to extend out to a radius of 180 to 200 Å from the thick filament axis. The total effect is that the thick filament backbone is surrounded by a broad flat shelf of electron density which is featureless at a resolution of say 100 Å. It has already been suggested in Chapter 10 that when a muscle is activated the first effect is for the actin and myosin filaments to be switched on by Ca^{2+} binding. As explained, the thick filament switch could be one that locks the cross bridges onto the thick filament backbone in the absence of Ca^{2+} but rapidly releases the bridges when Ca^{2+} is bound. The cross bridges then swing azimuthally to account for the change in the 429-Å and 215-Å layer line intensities (Haselgrove, 1975a; H. E. Huxley *et al.*, 1980), and they may move very slightly (~20°) in an axial direction to account for the "apparent" change in spacing of the meridional 143-Å peak (see Chapter 10 and Rome, 1972). This activation of the thick filament would not, therefore, be associated with a large change in radial position of the cross bridges, a change that would be detected on the equator of the diffraction pattern. The azimuthal swing of the cross bridges would not affect the equator at all in the region we are considering since the rotational symmetry and azimuthal position of the cross bridges would not become apparent within a radius of about 1/85 $Å^{-1}$ if the thick filament has nine-fold symmetry or about 1/40 $Å^{-1}$ if, as is more likely, the thick filament symmetry appears to be 18-fold (Fig. 10.25). As predicted from this, H. E. Huxley (1979b) has demonstrated that when frog semitendinosus muscle is stimulated at sarcomere lengths beyond overlap (i.e., when according to Haselgrove, 1975a, the myosin layer lines still drop in intensity because of activation of the muscle), no significant change is seen in the intensity of the $10\bar{1}$ equatorial reflection. Activation, assuming it occurs, must therefore be associated with an azimuthal or slight axial tilt of the cross bridges rather than a radial movement. Note however that H. E. Huxley *et al.* (1980) have questioned whether the layer line pattern changes significantly under these conditions.

The next point is that if contraction merely involved a radial movement of all of the cross bridges while they remained myosin-centered, then the effect would be to reduce the ratio $I_{10\bar{1}}/I_{11\bar{2}}$, but the ratio would very probably remain greater than one. It would not go down from about 2.0 for relaxed muscle to the observed value of about 0.5 during maximal tension generation if only radial motion were involved. In order for changes of this magnitude to occur with three-stranded filaments, it is necessary for cross bridges to become actin-centered and to become localized at a small mean radius from the thin filament axis. The smooth shelf of cross-bridge density in relaxed muscle would therefore, on initial activation, become a density distribution modulated by the physical presence of the thin filaments to have marked sixfold rotational symmetry in which cross bridges tend to point slightly tangentially to the six neighboring actin filaments. The cross bridges would then bind and change their state on actin, and their center of gravity would swing away from myosin. We have seen that anything that tends to increase the value of F_A in the structure factor, as would such a redistribution of cross bridges, would have the effect both of reducing the absolute intensity of the $10\bar{1}$ reflection and of increasing the value of $I_{11\bar{2}}$. The various experimental observations of changes in the ratio $I_{10\bar{1}}/I_{11\bar{2}}$ and in the absolute magnitudes of each reflection must therefore be indicative of cross bridges becoming localized around actin rather than being myosin-centered, and the most obvious cause of this must clearly be that actin attachment occurs.

With that important point established then the question remains whether the intensity ratio $I_{10\bar{1}}/I_{11\bar{2}}$ is sensitive to the number of attached bridges. The answer to this is clearly yes, since in general F_A will increase with the number of bridges attached, and so the ratio will change. But this is not the only factor determining the ratio $I_{10\bar{1}}/I_{11\bar{2}}$. It has been shown by Lymn (1978) that the configuration of the attached cross bridges also markedly modifies this intensity ratio. Once again, the azimuthal position of the bridge around the actin filament is relatively unimportant, but the radius of the center of gravity of the bridge from the axis of the thin filament is crucial.

Lymn considered two attached configurations, one (A) at roughly 90° to the filament axis (Fig. 11.14b), and one (B) at about 45° (Fig. 11.14c) as in the reconstructions of Moore *et al.* (1970; Fig. 5.11). Figure 11.14a shows the variation of $I_{11\bar{2}}/I_{10\bar{1}}$ calculated by Lymn for these two configurations as a function of the number of attached cross bridges. (Note that those cross bridges not attached to actin were assumed to remain in their relaxed positions in the myosin helix, an assumption that may be suspect since the unattached bridges may, for example, move continuously under Brownian motion. However the effect on the observed intensities would not be large) Figure 11.14a clearly shows that the ratio $I_{11\bar{2}}/I_{10\bar{1}}$ increases rapidly with increased attachment but that the configuration (A or B) also has a marked effect. For example, the given ratio $I_{11\bar{2}}/I_{10\bar{1}}$ of say 2.0 could be caused, according to this model, by 40% attachment in configuration B or by 20% attachment in configuration A. If there are several attached states, then the effect of the various configurations on the ration $I_{11\bar{2}}/I_{10\bar{1}}$ will, by implication, depend on the distribution of cross bridges among these various states. Only when this distribution is reasonably constant will the ratio $I_{11\bar{2}}/I_{10\bar{1}}$ be influenced largely by attachment number. The results of Yu *et al.* (1979) on muscles contracting isometrically (Fig. 11.13) and presumably containing a more or less constant distribution of cross-bridge configurations, are therefore

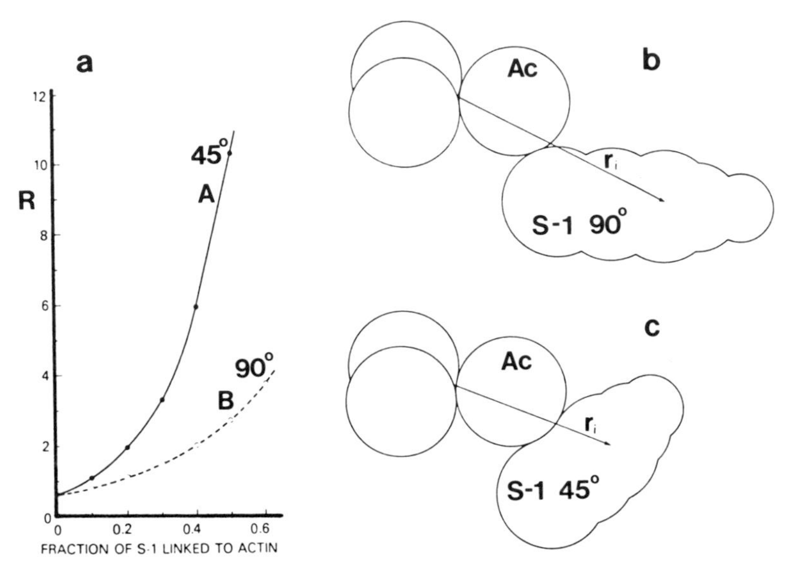

FIGURE 11.14. Theoretical variation of $I_{10\bar{1}}/I_{11\bar{2}}$ (= R) with attachment number (a) calculated by Lymn (1978). Configuration A (45°) is similar to the rigor complex (c) which in (c) is viewed down the thin filament axis. r_j is a measure of the radial separation of the cross bridge from this axis. Configuration B in (a) is a structure with the axial tilt (ϕ) of the cross bridge about 90° from the thin filament axis (b). Here the S-1 is also slewed from the position in (c), and the value of r_j is considerably greater than in (c). This has a marked effect on the diffraction pattern as seen in (a). (From Lymn, 1978.)

reasonable. But, on any model, the effect of shortening will inevitably lead to a change in the distribution of attached states, and so the interpretation of the results of Podolsky *et al.* (1976) and H. E. Huxley (1979b) on isotonically contracting muscles is likely to be extremely complex. It will be necessary to know (1) the structure of the different attached configurations, (2) the distributions of cross bridges between these states, and (3) the total number of attached bridges in order to account for the data. If all that were known, we should know a great deal about the mechanism of force generation.

11.3.3. The Time Course of Cross-Bridge Movement

We have seen that in the case of isometric contractions under steady-state conditions, the individual intensities $I_{10\bar{1}}$ and $I_{11\bar{2}}$ and the ratio $I_{11\bar{2}}/I_{10\bar{1}}$ are all likely to be sensitive measures of the number of attached cross bridges at any instant. It is not necessarily true, however, that this number can be related directly to the number attached in rigor, as was assumed by H. E. Huxley (1968) and Haselgrove and Huxley

(1973) and described in the last chapter. However, the equatorial intensities can reasonably be taken as a first indication of cross-bridge movement and attachment during the course of an isometric twitch. In the last few years the advent of the linear position-sensitive detector coupled with high-power X-ray generators (Chapter 2) has allowed the intensities of the $10\bar{1}$ and $11\bar{2}$ reflections to be monitored with a time resolution of at least 2–3 msec (H. E. Huxley, 1975a,b, 1976, 1979a,b), sufficient to be able to follow the changes during a twitch. Other similar studies at a slightly slower time resolution have also been carried out (Matsubara and Yagi, 1978).

The main conclusions from the work of H. E. Huxley (1975a,b, 1979a,b) are illustrated in Fig. 11.15. In this particular case, the intensity of the $10\bar{1}$ reflection and the change in tension are plotted against time. It was found that study of the $10\bar{1}$ was better than study of the $11\bar{2}$ reflection since one of the effects of tension generation could be internal shortening of the muscle which was found to increase the total intensity in the $10\bar{1}$ peak. But the effect that could be followed (Fig. 11.15) was a decrease in $10\bar{1}$ intensity at a rate faster than the development of tension. In this case any complications from internal shortening would tend to underestimate the changes in the $10\bar{1}$ reflection. Experiments of this kind on a large number of specimens have led to the conclusion that

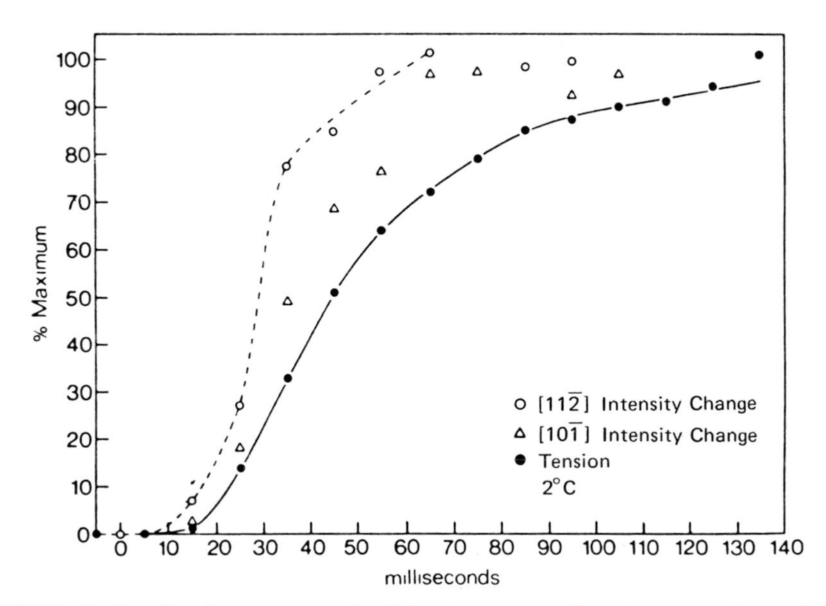

FIGURE 11.15. Simultaneous records of the time course of tension generation and the variation of $I_{10\bar{1}}$ published by H. E. Huxley (1979a) and showing the marked lag between the changes in equatorial intensity and the development of tension.

about 50% of the X-ray change has occurred at the point where only about 30% of peak tension has been developed. It is perhaps not altogether surprising that the cross bridges have to move and attach to and rotate on actin before tension is generated, but these new results should eventually help to show how fast these events occur.

At the other end of a contraction it is also of interest to follow the time course of relaxation. H. E. Huxley (1972a) reported that the myosin layer lines in patterns built up over a long time do not return to precontraction form until several seconds after a tetanus; the layer lines become weak while tension is generated, and they remain weak for several seconds after the tension relaxation. However, using better equipment and a short total exposure H. E. Huxley *et al.* (1980) have concluded that the layer line changes occur in step with the tension changes, and Podolsky *et al.* (1976) had earlier found, using the equatorial reflections, that the return of the myosin heads to their relaxed configuration after a contraction is complete within 100 msec after tension relaxation. But Yagi *et al.* (1977) have studied this further (Fig. 11.16) and have concluded that the experiments of Podolsky *et al.* (1976) may not have been repeated often enough to provide reliable statistics on this particular problem. In their own studies they concluded that the changes in the relative intensity $I_{10\bar{1}}/I_{11\bar{2}}$ occurs in two stages, a fast decrease concurrent with the fast tension relaxation and a slower decrease lasting a further few seconds (Fig. 11.1b). It was reported that the difference between the intensity ratio before and after the contraction did not become insignificant until 7 sec after the end of the tetanus. Yagi *et al.* (1977) interpreted these two

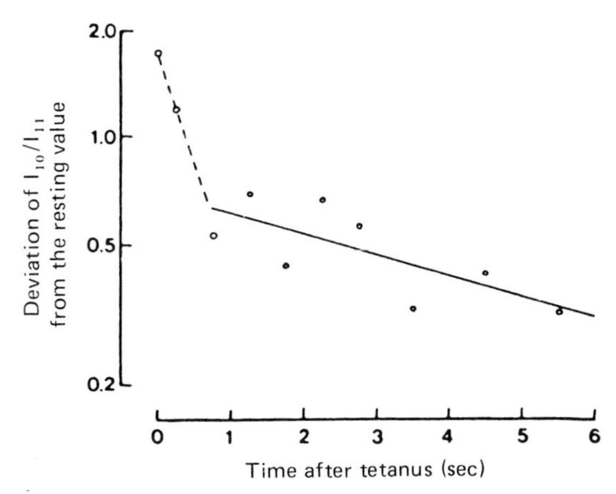

FIGURE 11.16. Deviation of $I_{10\bar{1}}/I_{11\bar{2}}$ from the resting value during relaxation following a tetanic response in frog sartorius muscle. The time is measured from the onset of relaxation. It is clear that the observed intensity ratio does not return to close to the resting value for several seconds after relaxation. (From Yagi *et al.*, 1977.)

phases solely in terms of radial cross-bridge movement by an analysis similar to that of Haselgrove and Huxley (1973), but for reasons given earlier this may well be too simple an approach when considering cross-bridge movements associated with contraction.

As detailed in the next section, all of these observations can be related to a possible sequence of changes involved in activation and tension generation.

11.4. Scenarios for the Cross-Bridge Cycle

11.4.1. Location of the Instantaneous Cross-Bridge Elasticity

We have seen in Section 11.2.4 that one of the conclusions of A. F. Huxley's research group (e.g., A. F. Huxley, 1974; Ford *et al.,* 1977) is that there is instantaneous elasticity somewhere in the cross-bridge–actin assembly in addition to any elasticity in the Z-band. In isometric contractions this carries tension that can be reduced to zero by a shortening of frog muscle by about 40 to 60 Å per half sarcomere. In Fig. 11.12 this elasticity was visualized, as in the diagrams of A. F. Huxley and Simmons (1971b), as residing in the myosin S-2 linkage between the myosin head and the thick filament backbone. But in principle this elasticity could reside anywhere in the cross-bridge–actin assembly. A. F. Huxley (1974) has illustrated some of the alternatives (Fig. 11.17), but other possibilities exist. In addition, there are a number of alternative models for the configurational differences among the different attached myosin states. In Fig. 11.17b the elasticity is shown in the S-2 link, and three attachment states are illustrated in which the structural changes occur within the myosin head rather than at the actin attachment site (cf. Fig. 11.7a). In Fig. 11.17c the elasticity is shown as being a property of the actin molecules; but as in (b), the three attached states occur with a single orientation of cross-bridge attachment to the actin filament as a whole. In some ways this alternative is preferable to those in Figs. 11.12 or 11.17a,b, since it seems very unlikely that the two-chain α-helical coiled-coil structure of myosin S-2 could allow it to have the necessary springlike properties. Indeed, it is not even necessary to visualize the elasticity in terms of a spring at all, although as in Figs. 11.12 and 11.17 this is a convenient way of drawing it. Any property of the attached cross bridge that would vary the tension in the bridge more or less linearly with attachment angle (so that it effectively runs up and down the sides of a potential energy well) could provide the appropriate elasticity (Eisenberg and Hill, 1978). This can be visualized as in Fig. 11.18. Here a cross bridge in stable position 1 will undergo the next step in the ATPase

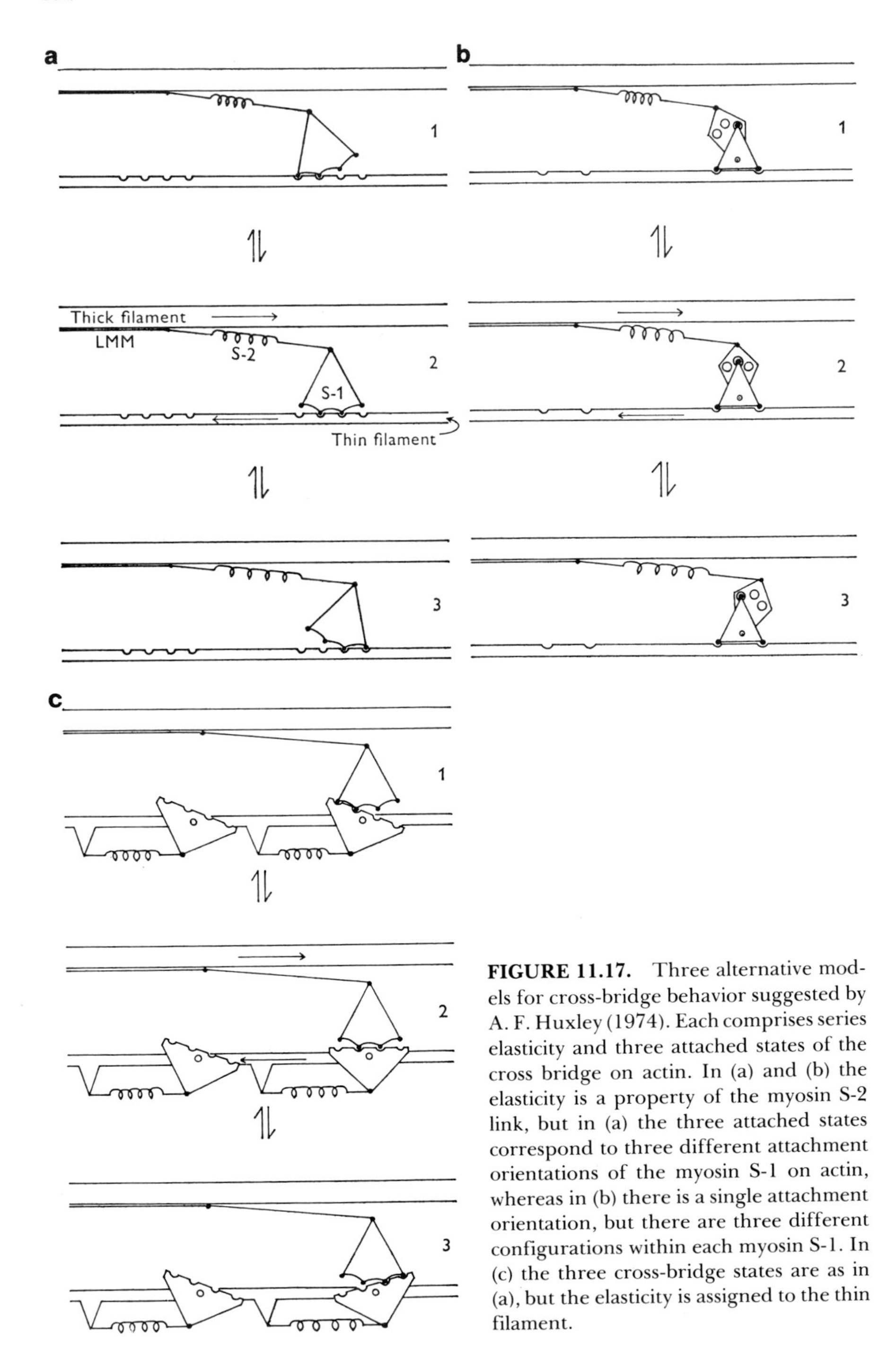

FIGURE 11.17. Three alternative models for cross-bridge behavior suggested by A. F. Huxley (1974). Each comprises series elasticity and three attached states of the cross bridge on actin. In (a) and (b) the elasticity is a property of the myosin S-2 link, but in (a) the three attached states correspond to three different attachment orientations of the myosin S-1 on actin, whereas in (b) there is a single attachment orientation, but there are three different configurations within each myosin S-1. In (c) the three cross-bridge states are as in (a), but the elasticity is assigned to the thin filament.

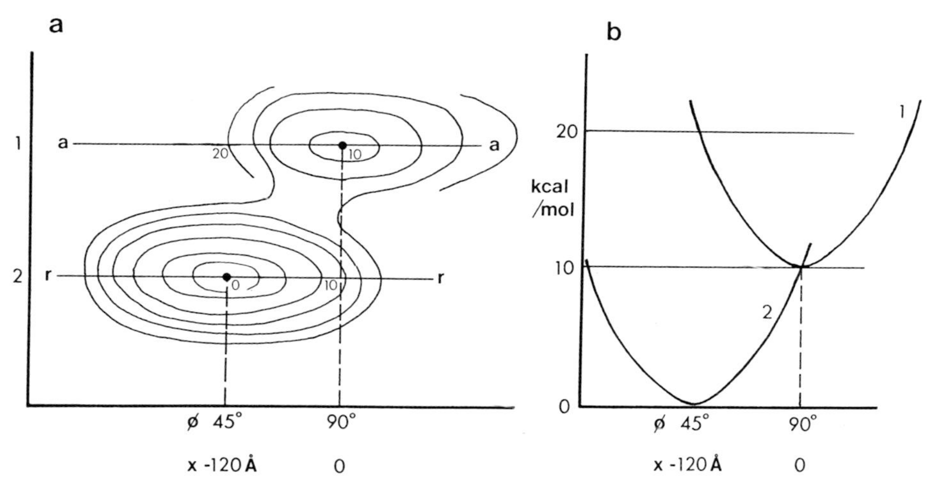

FIGURE 11.18. An alternative model to explain the instantaneous elasticity of the attached cross bridge (Eisenberg and Hill, 1978). Here there are two attached states (1 and 2), each of which is defined by a rather broad energy well in the ϕ direction (axial tilt of the cross bridge from the thin filament axis). Variation of angular tilt within each state occurs along lines a–a or r–r, and this appears as a series elasticity. Transition from state 1 to state 2 (parallel to the vertical "change of state" axis) can occur preferentially at $\phi = 90°$ where no change in angular potential energy is involved. The numbers in (a) are in kcal/mol; (b) is an alternative way of visualizing the situation in (a) (the parabolic energy wells are obvious here). (After Eisenberg and Hill, 1978.)

cycle such that the energy barrier to the next state is crossed, and the cross bridge will then tend to move in the angular energy well of state 2. Each state is associated with an angularly broad potential energy well, but depending on the conditions, the cross bridge may be prevented from moving within this well. Only a change in length of the muscle would move the cross bridge up and down the sides of the well. Movement towards $\phi = 45°$ in state 2 through any intermediate attachment states that may occur will then only take place if the cross bridge is in the correct biochemical state and if movement is physically possible. Note also that with such a system only two attached states are needed (not three as in A. F. Huxley and Simmons model (see Section 11.2.4) since transitions between states can be very fast in this case; no movement need be associated with them.

Because of these various possibilities, it is clear that there is great uncertainty about the true nature and location of the instantaneous elasticity, and only further evidence will allow its location to be defined. For the purposes of the remaining discussion, it will be presumed, as in earlier diagrams (e.g., Fig. 11.12), to reside in a springlike S-2, but this is only for ease of diagrammatic representation.

11.4.2. Structural Steps in the Cross-Bridge Cycle

From the various pieces of evidence already described in previous chapters it is now possible to put together a likely sequence of events involved in activation and force generation in muscle. For simplicity, the discussion will be confined to data from vertebrate skeletal muscle.

As described in Chapter 10 the biochemical data on the actomyosin ATPase shows that the principal pathway can probably be written tentatively as:

$$
\begin{array}{ccccc}
\mathrm{AM\cdot ADP\cdot P} & \xrightarrow{3} & \mathrm{AM\cdot ADP + P} & \xrightarrow{4} & \mathrm{AM} \\
{\scriptstyle 2}\uparrow & & & & \uparrow\downarrow{\scriptstyle 5} \\
\mathrm{AM^{**}\cdot ADP\cdot P} & & & & \mathrm{AM\cdot ATP} \\
{\scriptstyle 1}\uparrow\downarrow & & & & \downarrow{\scriptstyle 6} \\
\mathrm{A + M^{**}\cdot ADP\cdot P} & \xleftarrow{8} & \mathrm{A + M^{*}\cdot ATP} & \xleftarrow{7} & \mathrm{A + M^{\dagger}\cdot ATP}
\end{array}
$$

where steps 1 and 5 are probably rapidly reversible and the other steps are predominantly in the forward direction. Since *in vivo* the concentrations of ATP are saturating (Marston and Taylor, 1977), this cycle can be thought of in terms of four main steps

$$
\begin{array}{ccc}
\mathrm{AM\cdot ADP\cdot P} & \xrightarrow{b} & \mathrm{AM} \\
{\scriptstyle a}\uparrow\downarrow & & \uparrow\downarrow{\scriptstyle c} \\
\mathrm{A + M\cdot ADP\cdot P} & \xleftarrow[d]{} & \mathrm{A + M\cdot ATP}
\end{array}
$$

Step (a) is the attachment step which is a rapid equilibrium; step (b) is the slowest step (rate limiting) and probably corresponds to force generation (see Fig. 10.2; H. D. White 1977a,b); step (c) involves detachment (another rapid equilibrium); and step (d) involves the resetting of the cross bridge ready for another cycle.

In addition to this basic cycle, it is also necessary to consider the effects of calcium. Here step (a) cannot proceed in the absence of calcium, and in addition there may be a structural difference between M · ADP · P in the presence and absence of calcium as required by the activation step discussed in Chapter 10 (Section 10.).

Taking all of these factors into account, together with the suggestions of A. F. Huxley and Simmons (1971b) and the available knowledge of the structures of relaxed, rigor, and AMP · PNP-treated muscles, the structural scheme illustrated in Fig. 11.19 can be proposed.

Here state 0 is relaxed muscle in which the cross bridges tilt slightly towards the M-band in a configuration that is locked onto the filament backbone in the absence of calcium. Activation of the muscle by calcium

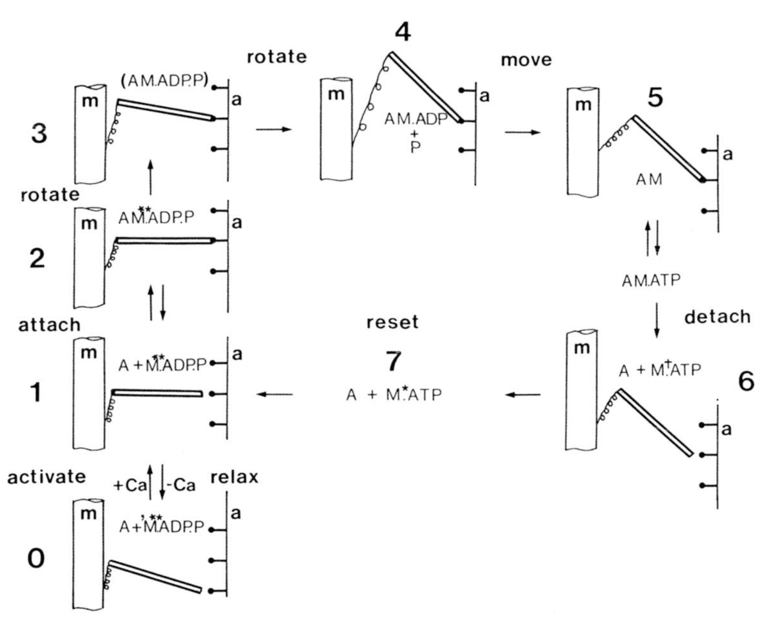

FIGURE 11.19. A possible sequence of structural events during the cross-bridge cycle together with the likely biochemical state of the myosin shown in each case. State 0 is the relaxed state which differs from State 1 in the axial and azimuthal tilt of the cross bridge (m, myosin; a, actin). Transition between states 0 and 1 is mediated by calcium. State 2 is the first rather weak attached state and is rapidly reversible. Binding is stronger in state 3 which may be at the same tilt as state 2 or at a slightly different angle as drawn. State 4 is the "rigor" cross-bridge configuration which may be reached by myosin in state AM:ADP before AM is reached. State 5 shows movement between the actin and myosin filaments which may of course occur simultaneously with changes in cross-bridge tilt (states 3 and 4). State 6 is the initial detached state following which the cross bridge resets biochemically through several intermediate states and also swings back to the equilibrium detached tilt at about $\phi = 90°$ that exists in the presence of calcium (state 1). On relaxation, cross bridges initially revert to state 1, and the subsequent transition to state 0 is relatively slow (see Fig. 11.16). The elasticity of the cross bridge is shown here in the myosin S-2, but this is merely for ease of pictorial representation.

released from the sarcoplasmic reticulum switches on the thin filaments via the tropomyosin–troponin system and the thick filaments by a mechanism yet to be defined, so that state 1 is reached. State 1 differs from state 0 in that the cross bridges have swung axially away from the M-band to account for the apparent movement of the 143-Å reflection (Chapter 10, Section 10.3.4) and have swung azimuthally to enable the cross bridges to reach the actin filaments. Whether or not such a swing would affect the layer line intensities (Haselgrove, 1975a; H. E. Huxley *et al.,* 1980) is unclear. But it would not affect the equatorial intensities unless different cross bridges behaved differently (see later comments). In overlap muscles, state 1 is rapidly followed by attachment (state 2; AM · ** ADP · P) which at first is weak and rapidly reversible. But a transition to state 3 (AM · ADP · P) strengthens the binding and possibly also begins to stretch the instantaneous elasticity (here shown in myosin S-2). Several attached states may now follow until the 45°-angled state (akin to rigor) is obtained after the release of the ADP and P_i products (state 5) and, of course, a simultaneous shortening of the muscle (state 6) may occur. On reaching this 45° angling, the cross bridge can detach once again and reset itself, ready for a further cycle, through the steps AM · ATP → A + M† · ATP → A + M* · ATP → A + M** · ADP · P which is where the cycle started. Note that between states 6 and 1 the detached cross bridge needs to swing back through 45° to an orientation perpendicular to the filament axis. This could presumably be because the most stable configuration of the detached cross bridge is at 90° and only the strength of the actomyosin interaction through states 3 to 5 can displace the head from this position. Once detached it will inevitably revert to the 90° position. It seems unlikely that the biochemical state of the detached cross bridge would affect its position.

Similarly the absence of calcium alters the equilibrium of state 1 so that the optimum position is now as in relaxed muscle. If this transition is to an energy well that is relatively shallow, the transition from state 1 to state 0 may be slow as observed by Yagi *et al.* (1977). Relaxation may therefore be associated with rapid switching off of the thin filaments with a rapid tendency for cross bridges to move to state 1, followed by what may be a relatively slow subsequent transition to state 0. Such a transition to state 0, being slow, is probably also dependent on a very fine balance of forces between the thick filament backbones and the cross bridges. It may well explain why it has been found difficult to relax glycerinated muscles to reproduce exactly the diffraction patterns presumed to come from state 0 (Rome, 1972). The synthetic balance of ions may be slightly wrong, or the glycerol may have disrupted the mechanism in some subtle way while leaving other aspects of the contractile mechanism virtually intact. The marked effect of ionic balance on *Limulus* muscle structure when relaxed, has already been noted by Wray *et al.* (1974) (see Chapter 8, Section 8.2.4).

Turning now to the rapid equilibrium between states 1 and 2, this seems to imply that attachment in state 2 is rather weak (it is defined by a shallow energy well), and it could well be that it is not highly directional. In particular, it seems to be likely that attachment of a cross bridge to actin can probably occur in a range of azimuths, since the need for a cross bridge to attach at one specific configuration relative to the actin attachment site would inevitably make the interaction relatively slow. Further, the helical structure of the thin filaments causes actin subunits to be presented to the cross bridges at different azimuths even within a single actin target area. The cross-bridge attachment can therefore be visualized in terms of the energy diagram in Fig. 11.20. Here the energy well for attachment at 90° to the thin filament axis ($\phi = 90°$) is very shallow (i.e., it is a weak, reversible interaction) and broad in the azimuthal direction measured by θ. Transitions from state 2 to other attached states could either be initially to a deep well in the middle of the shallow well at $\phi = 90°$ or to a more specific well at $\phi \neq 90°$. Further transitions to well-defined energy troughs at different axial tilts could then occur until the highly specific rigor attachment state (AM: state 5) is reached. Here both ϕ and θ have well-defined values.

One interesting conclusion from the azimuthal breadth of the $\phi = 90°$ energy well would be that attached states at $\phi = 90°$, such as that induced by AMP · PNP (see Section 10.4), could be relatively well defined in ϕ and very poorly defined in θ. The bridges would most likely point directly back to the points of origin on the adjacent thick filaments. Modeling such states in terms of a concerted swing of attached bridges from rigor to a 90° position well defined in θ would, therefore, not account for the observations (Barrington-Leigh *et al.*, 1977). It is also

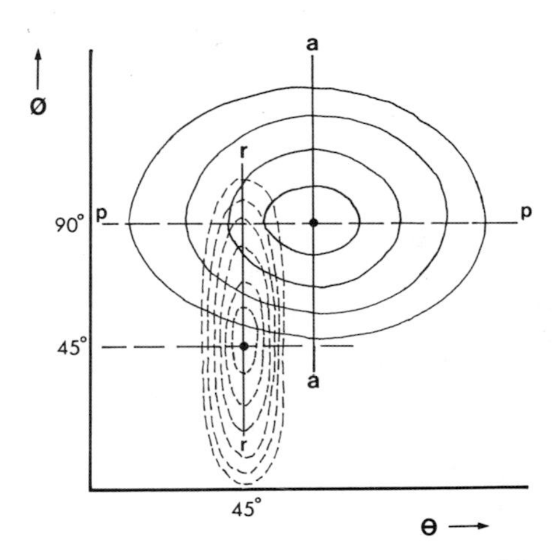

FIGURE 11.20. A highly schematic illustration of the potential energy of an attached cross bridge as a function of the axial tilt (ϕ) and the azimuthal tilt (θ) of the cross-bridge axis. The 90° attached state is shown to be relatively ill-defined in θ and better defined in ϕ, so that in muscle states involving this configuration the attached cross bridges may well be constrained by the S-2 link to point directly back towards their point of origin on the thick filament whatever the azimuth of the actin attachment site relative to that thick filament. The 45° attached state may be much better defined in θ (dotted lines). It may also be broad in the ϕ direction (but not necessarily) as explained in Fig. 11.18. Instantaneous cross-bridge elasticity will be produced by movement parallel to a–a for cross bridges stable in the 90° state and will be parallel to r–r for rigor cross bridges. There would be other potential energy wells for cross bridges in different metastable states.

possible that the structure of fixed relaxed muscle in which cross-bridge attachment to actin seems to occur can also be described by attachments that are specific in ϕ but variable in θ (Freundlich *et al.*, 1980). Computer modeling of the effects of this nonspecific attachment clearly needs to be carried out to test this possibility.

The range of attachment azimuths on actin would also mean that attachment of cross bridges would occur without their needing to move markedly in a radial direction from state 1. In this case it could be that the azimuthal swing of cross bridges from state 0 is a swing that moves the cross bridges until they approach the six actin filaments; they are already at a large enough radius for this. As mentioned earlier, the effect of this is that the initial ninefold symmetry of the relaxed cross-bridge array may be modified towards a structure with pseudosixfold symmetry as the bridges swing by different amounts to positions adjacent to the six neighboring thin filaments. It is conceivable that this could affect the intensity of the $10\bar{1}$ equatorial reflection at the start of activation (Yu *et al.*, 1979; Fig. 11.13) without markedly affecting the $11\bar{2}$, since the cross bridges have not yet become localized (radially) around the actin filaments. Such an effect would not, of course, be apparent in nonoverlap muscles (H. E. Huxley, 1979b) where there are no thin filaments to induce the tendency towards sixfoldness in the cross-bridge array. This and the other possibilities summarized in Fig. 11.19 await further experimental data, but I believe they provide a useful working hypothesis about the structural events involved in activation and force generation.

11.5. Conclusion: Future Prospects

It must be clear from the length and detail in this book that an enormous amount of data on muscle is available; and even a book like this cannot hope to do justice to all of the different possible structural approaches to the problem of muscular contraction. But despite this wealth of information, several crucial features of the actomyosin interaction remain unclear, and here I wish to list several of these unknowns together with possible ways in which more information on some of them might be obtained in future studies.

It is now very clear that the sliding filament model is substantially correct, even if it turns out that variations on the theme can occur as in *Limulus* muscle (Chapter 8). It also seems beyond doubt that cross-bridge attachment to actin does occur and that the attached cross bridges can change their angle of attachment to the thin filaments as part of the force-producing process. But the original concept of H. E. Huxley (1969), in which the two heads of one myosin molecule were thought to act cooperatively to produce the observed change in angle, seems to have been superseded by models in which the heads can act independently and in which they may even attach to different actin filaments (Freundlich *et al.,* 1980). Similarly, the H. E. Huxley (1969) model with the simple cross-bridge swing from say a 90° attachment position to a 45° attachment now seems to have been replaced by a mechanism, based on the same idea, with several different stable attachment configurations of the heads among which the heads move very rapidly; they spend relatively little time between these stable positions (see Fig. 11.12c; A. F. Huxley and Simmons, 1971b). The Eisenberg and Hill, (1978) model is similar except that the biochemical changes of state need not occur simultaneously with the changes in attachment angle. In addition to these models of contraction, the outline mechanisms of calcium regulation in both thick and thin filaments are known (even though the details are not), and the three-dimensional ultrastructures of a variety of muscle types are becoming increasingly well documented.

Despite this wealth of knowledge on muscle the following things, each of great importance, are at best only partially understood (the list is an expanded version of one given by A. F. Huxley, 1974).

1. What is the structure of the myosin head?
2. Is the structure of the myosin head variable or constant between different muscle states?
3. What is the structure of the actin monomer?
4. Does the cross bridge really work by attaching and tilting on actin?
5. Which structure is the elastic element in the cross bridge?
6. What is the nature of this elasticity?
7. What structure undergoes the stepwise changes during cross-bridge attachment?
8. Does attachment occur between a single myosin head and a single actin subunit or with two actin subunits?
9. Do the two heads in one cross bridge tend to attach to actin subunits in the same thin filament or in two different filaments?
10. Do both heads produce tension at the same time?
11. Are two heads needed within a single cross bridge for efficient tension production? If not why are there two heads?
12. What kind of interaction is there between the myosin head and the thin filaments when attachment occurs?
13. What are the structures of the intermediate states of attachment of cross bridges to actin (if they occur), and how many states are there?
14. What is the locking mechanism of the cross bridges on the thick filaments in relaxed muscle if it exists?

15. What defines the lengths of the thick and thin filaments?
16. What is the role of C protein and the other nonmyosin thick filament proteins?
17. What defines the differences among the thick filaments found in different muscles?
18. What are the detailed structures of relaxed, rigor, and AMP · PNP muscles?
19. What is the mechanism of stretch activation in insect flight muscles?
20. What is the mechanism of catch in molluscan muscles?

This long list is by no means exhaustive, but it probably includes most of the problems that are being considered by muscle biophysicists at the present time. Some of the problems have been discussed in this book, and others are completely unknown. But it is appropriate to consider briefly how some of them might be answered, how progress is to be made.

Of course, one of the most desirable objectives would be to produce good single crystals of actin and of myosin heads and also of the other important muscle proteins and to determine their structures by means of X-ray crystallography. Subsequently, if actin and myosin could be cocrystallized under different conditions, then the nature of the actin–myosin interaction could be defined in some detail. But so far only crystals of actin have been prepared (see Chapter 5, Section 5.2.1), and even these have not yet yielded a structure. Unfortunately, crystallization of the myosin head has not so far been achieved with sufficient success to enable X-ray crystallography to be carried out.

In the absence of such a direct approach we need to consider what else can be done. One of the most pressing structural problems seems to be to define in detail the structures of the different static muscle states such as relaxation, rigor, AMP · PNP, and so on. The technology is already available for carrying out X-ray experiments on active muscle with a very fast time resolution (H. E. Huxley, 1979b; H. E. Huxley *et al.,* 1980); synchrotron X-ray sources and fast X-ray detectors have permitted that. But until the structures of the associated static states are known, there is little hope of being able to interpret the data from active muscle in a realistic way.

How then should one attempt to find the structures of these static states? Undoubtedly they will be defined by the complementary use of information from X-ray diffraction and electron microscopy, and the major improvements are likely to come on the electron microscopy side. In addition, the associated mechanical data on these states will have to fit any structural model that is produced. Possible advances in electron microscopy are likely to come in two ways: one is in the preparation of tissue prior to its study in the microscope, and the other is to improve the data acquisition and specimen quality in the microscope. One possible method of improving the tissue preparation is to use sections of frozen muscle (Sjöström and Squire, 1977b). Not only do such sections provide different (often much clearer) data than conventional plastic sections, they also allow the possibility of "freezing in" the instantaneous structure (on a millisecond time scale) of muscles at different stages of the contractile cycle. But plastic sections can also yield exceptionally good data if suitable care is taken in the preparations (Reedy, 1968; Luther and Squire, 1978, 1980) and if image-analysis techniques, including computer methods, are used to the full. A combination of such sections with comparable X-ray diffraction results could help and is helping to define these static states. In the microscope itself, improvements can be expected if the specimen is cooled to very low temperatures so that radiation damage is reduced, if low dose imaging methods are used, and if advantage is taken of the scanning transmission electron microscopes with which the image can be manipulated and the contrast artificially enhanced.

Apart from these techniques, there are other structural probes that are beginning to find their application in muscle studies, although in some cases the techniques are not yet well enough defined to provide unambiguous results. Among these tools are nuclear magnetic resonance, fluorescence from dye-labeled molecules (e.g., myosin heads) (Mendelson *et al.*, 1973), light scattering (Carlson *et al.*, 1972; Newman and Carlson, 1980), and electron paramagnetic resonance (Thomas *et al.*, 1980; Thomas and Cooke, 1980).

Muscular contraction remains a fascinating topic for study; muscle provides a unique combination of structural regularity and biological function that is amenable to a wide range of experimental approaches. But, above all, muscular contraction is vital to us all in our everyday lives. Many of us take our muscles for granted, but to those with muscular dystrophy this is surely not the case. It remains one of the hopes of the author that through understanding the workings of normal healthy muscle, an insight will be obtained into the structural effects of this pernicious disease. The study of muscle therefore has intellectual satisfaction and direct practical application (in the future), both of which provide the stimulus for further research. The next few years promise to be of great interest to muscle biophysicists, and it is very likely that at least some of the problems listed above will have answers to them soon. If this book helps to further this study by stimulating others it will have served its purpose.

References

Abbott, B. C., and Aubert, X. M. (1951) *Proc. R. Soc. Lond.*[*Biol.*] **139,** 104–117.
Abbott, B. C., Aubert, X. M., and Hill, A. V. (1951) *Proc. R. Soc. Lond.* [*Biol.*] **139,** 86–104.
Abbott, B. C., and Lowy, J. (1958) *J. Physiol.* (*Lond.*) **141,** 398–407.
Abbott, R. H. (1977) In *Insect Flight Muscle* (Tregear, R. T., ed), pp. 269–273, North-Holland, Amsterdam.
Abbott, R. H., and Cage, P. E. (1979) *J. Physiol.* (*Lond.*) **289,** 32P–33P.
Aberg, A. K., and Axelsson, J. (1965) *Acta Physiol. Scand.* **64,** 15–27.
Adelstein, R. S., and Conti, M. A. (1975) *Nature* **256,** 597–598.
Adelstein, R. S., and Conti, M. A. (1976) *Cold Spring Harbor Conference on Cell Proliferation,* Vol. 3, Book B, pp. 725–738.
Adelstein, R. S., Chacko, S., Barylko, B., Scordilis, S. P., and Conti, M. A. (1976) In *Contractile Systems in Non-Muscle Tissues* (Perry, S. V., Margreth, A., and Adelstein, R. S., eds.), pp. 153–163, North-Holland, Amsterdam.
Afifi, A. K., Smith, J. W., and Zellweger, H. (1965) *Neurology* **15,** 371–381.
Aidley, D. J. (1971) *The Physiology of Excitable Cells,* Cambridge University Press, London.
Amphlett, G. N., Syska, S. A., Brown, M. D., and Vrbova, G. (1975) *Nature* **257,** 602–604.
Apathy, S. (1894) *Arch. Mikrosk. Anat.* **46,** 886–911.
April, E. W. (1975) *Nature* **257,** 139–141.
April, E. W., and Wong, D. (1976) *J. Mol. Biol.* **101,** 107–114.
April, E. W., and Brandt, P. W., and Elliott, G. F. (1971) *J. Cell Biol.* **51,** 72–82.
April, E. W., Brandt, P. W., and Elliott, G. F. (1972) *J. Cell Biol.* **53,** 53–65.
Armitage, P. M., Miller, A., Rodger, C. D., and Tregear, R. T. (1972) *Cold Spring Harbor Symp. Quant. Biol.* **37,** 379–387.
Armitage, P. M., Tregear, R. T., and Miller, A. (1975) *J. Mol. Biol.* **92,** 39–53.
Arndt, U. W., Gilmore, D. J., and Boutle, S. H. (1972) *Adv. Electron. Electron Phys.* **33B,** 1069–1075.
Arnott, S., and Wonacott, A. J. (1966) *J. Mol. Biol.* **21,** 371–383.
Arnott, S., Dover, S. D., and Elliott, A. (1967) *J. Mol. Biol.* **30,** 201–208.
Asakura, S., Taniguchi, M., and Oosawa, F. (1963) *J. Mol. Biol.* **7,** 55–69.
Ashhurst, D. E. (1967a) *J. Cell. Sci.* **2,** 435–444.
Ashhurst, D. E. (1967b) *J. Mol. Biol.* **27,** 385–389.
Ashhurst, D. E. (1977) In *Insect Flight Muscle* (Tregear, R. T., ed), pp. 57–73, North-Holland, Amsterdam.

Ashhurst, D. E., and Cullen, M. J. (1977) In *Insect Flight Muscle* (Tregear, R. T., ed) pp. 9-14, North-Holland, Amsterdam.
Ashley, C. C., and Ridgeway, E. B. (1968) *Nature* **219,** 1168-1169.
Ashley, C. C., and Ridgeway, E. B. (1970) *J. Physiol.* (*Lond.*) **209,** 105-130.
Ashton, F. A., Somlyo, A. V., and Somlyo, A. P. (1975) *J. Mol. Biol.* **98,** 17-29.
Astbury, W. T. (1947) *Proc. R. Soc. Lond.* [*Biol.*] **134,** 303-328.
Astbury, W. T. (1949) *Exp. Cell. Res. Suppl.* **I,** 234-247.
Astbury, W. T., and Spark, L. C. (1947) *Biochim. Biophys. Acta* **1,** 388-392.
Atwood, H. L. (1972) In *The Structure and Function of Muscle,* Second Edition (Bourne, S. H., ed.), pp. 421-489, Academic Press, New York.
Atwood, H. L., Hoyle, G., and Smyth, T., Jr. (1965) *J. Physiol.* (*Lond.*) **180,** 449-482.
Auber, J. (1967) *Am. Zool.* **7,** 451-456.
Auber, J. (1969) *J. Microsc.* (*Paris*) **8,** 197-232.
Auber, J., and Couteaux, R. (1963) *J. Microsc.* (*Paris*) **2,** 309-324.
Aubert, X. (1944a) *C.R. Soc. Biol.* (*Paris*) **138,** 1011-1012.
Aubert, X. (1944b) *C.R. Soc. Biol.* (*Paris*) **138,** 1048-1050.
Aubert, X. (1948) *Arch. Int. Physiol.* **55,** 348-361.
Aubert, X., Roguet, M. L., and Van Der Elst, J. (1951) *Arch. Int. Physiol.* **59,** 239-241.
Bagshaw, C. R. (1977) *Biochemistry* **16,** 59-67.
Bagshaw, C. R., and Kendrick-Jones, J. (1979) *J. Mol. Biol.* **130,** 317-336.
Bagshaw, C. R., and Reed, G. H. (1977) *FEBS Lett.* **81,** 386-390.
Bagshaw, C. R., Eccleston, J. F., Eckstein, F., Goody, R. S., Gutfreund, H., and Trentham, D. R. (1974) *Biochem. J.* **141,** 351-364.
Baguet, F., and Gillis, J. M. (1968) *J. Physiol.* (*Lond.*) **198,** 127-143.
Bailey, K. (1946) *Nature* **157,** 368-369.
Bailey, K. (1948) *Biochem. J.* **43,** 271-279.
Bailey, K. (1954) In *The Proteins,* Vol. IIB (Neurath, H., and Bailey, K., eds.), pp. 951-1055, Academic Press, New York.
Bailey, K., Gutfreund, H., and Ogston, A. G. (1948) *Biochem. J.* **43,** 279-281.
Barany, M., and Close, R. J. (1971) *J. Physiol.* (*Lond.*) **213,** 455-474.
Barany, M., Barany, K., Gaetjens, E., and Bailin, G. (1966) *Arch. Biochem. Biophys.* **113,** 205-221.
Barrington-Leigh, J., and Rosenbaum, G. (1976) *Annu. Rev. Biophys. Bioeng.* **5,** 239-270.
Barrington-Leigh, J., Holmes, K. C., Mannherz, H. G., Rosenbaum, G., Eckstein, F., and Goody, R. (1972) *Cold Spring Harbor Symp. Quant. Biol.* **37,** 443-447.
Barrington-Leigh, J., Goody, R. S., Hofman, W., Holmes, K., Mannherz, H. F., Rosenbaum, G., and Tregear, R. T. (1977). In *Insect Flight Muscle* (Tregear, R. T., ed.), pp. 137-146, North-Holland, Amsterdam.
Bear, R. S. (1945) *J. Am. Chem. Soc.* **67,** 1625-1626.
Bear, R. S., and Selby, C. C. (1956) *J. Biophys. Biochem. Cytol.* **2,** 55-85.
Beinbrech, G. (1977) In *Insect Flight Muscle* (Tregear, R. T., ed.), pp. 147-160, North-Holland, Amsterdam.
Bennett, P. M. (1976) *Proceedings of the Sixth European Congress on Electron Mïcroscopy, Vol. II,* Ben Shaal, Y., ed.), pp. 517-519, Tal International, Israel.
Blundell, T. L., and Johnson, L. N. (1976) *Protein Crystallography,* Academic Press, New York.
Bradbury, E. M., Brown, L., Downie, A. R., Elliott, A., Fraser, R. D. B., and Hanby, W. E. (1962) *J. Mol. Biol.* **5,** 230-247.
Bragg, W. L. (1933) *The Crystalline State,* Vol. 1, Bell and Sons, London.
Bragg, W. L. (1939) *Nature* **143,** 678.

Bragg, W. L. Kendrew, J. C., and Perutz, M. F. (1950) *Proc. R. Soc. Lond.* [*A*] **203,** 321-357.
Brann, L., Dewey, M. M., Baldwin, E. A., Brink, P., and Walcott, B. (1979) *Nature* **279,** 256-257.
Bray, D. (1972) *Cold Spring Harbor Symp. Quant. Biol.* **37,** 567-571.
Bremel, R. D., and Weber, A. (1975), *Biochim. Biophys. Acta* **376,** 366-374.
Bremel, R. D., Murray, J. M., and Weber, A. (1972) *Cold Spring Harbor Symp. Quant. Biol.* **37,** 267-275.
Bridgen, J. (1971) *Biochem. J.* **123,** 591-600.
Brooke, M. H., and Kaiser, K. K. (1970) *Arch. Neurol.* **23,** 369-379.
Brown, L., and Trotter, I. F. (1956) *Trans. Faraday Soc.* **52,** 537-548.
Bulbring, E., and Golenhofen, K. (1967) *J. Physiol.* (*Lond.*) **193,** 213-224.
Bulbring, E., Lin, N. C. Y., and Schofield, G. (1958) *Q. J. Exp. Physiol.* **43,** 26-37.
Bullard, B., and Reedy, M. K. (1972) *Cold Spring Harbor Symp. Quant. Biol.* **37,** 423-428.
Bullard, B., and Sainsbury, S. M. (1977) *Biochem. J.* **161,** 399-403.
Bullard, B., Luke, B., and Winkelman, L. (1973) *J. Mol. Biol.* **75,** 359-367.
Bullard, B., Bell, J., and Luke, B. M. (1977) In *Insect Flight Muscle* (Tregear, R. T., ed.), pp. 41-52, North-Holland, Amsterdam.
Buller, A. J., Eccles, J. C., and Eccles, R. M. (1960) *J. Physiol.* (*Lond.*) **150,** 417-439.
Buller, A. J., Mommaerts, W. F. H. M., and Seraydarian, K. (1969) *J. Physiol.* (*Lond.*) **205,** 581-597.
Burke, M., and Harrington, W. F. (1971) *Nature* **233,** 140-142.
Burke, M., and Harrington, W. F. (1972) *Biochemistry* **11,** 1456-1462.
Burnstock, G. (1970) In *Smooth Muscle* (Bulbring, E., Brading, A., Jones, A., and Tomita, T. eds.), pp. 1-69, Edward Arnold, London.
Cannan, C. M. M. (1950) Ph.D Thesis, M.I.T., Cambridge, Massachusetts.
Carlson, F. D., Bonner, B., and Fraser, A. (1972) *Cold Spring Harbor Symp. Quant. Biol.* **37,** 389-396.
Carlsson, L., Nystrom, L. E., Lindberg, U., Kannan, K. K., Cid-Dresdner, H., Lovgren, S., and Jornvall, H. (1976) *J. Mol. Biol.* **105,** 353-366.
Carsten, M. E., and Katz, A. M. (1964) *Biochim. Biophys. Acta* **90,** 534-541.
Caspar, D. L. D., Cohen, C., and Longley, W. (1969) *J. Mol. Biol.* **41,** 87-107.
Chacko, S., Conti, M. A., and Adelstein, R. S. (1977) *Proc. Natl. Acad. Sci. USA* **74,** 129-133.
Chaplain, R. A., and Tregear, R. T. (1966) *J. Mol. Biol.* **21,** 275-280.
Chi, J., Rubinstein, N., Strahs, K., and Holtzer, H. (1975) *J. Cell Biol.* **67,** 523-537.
Chock, S. P., Chock, P. B., and Eisenberg, E. (1976) *Biochemistry* **15,** 3244-3253.
Choi, J. K. (1962) *Fifth International Congress on Electron Microscopy, Philadelphia,* Vol. 2 (Breese, S. S., Jr., ed.), Abstract M-9, Academic Press, New York.
Chou, P. Y., and Fasman, G. D. (1974) *Biochemistry* **13,** 222-245.
Chowrashi, P. K., and Pepe, F. A. (1977) *J. Cell Biol.* **74,** 136-152.
Civan, M. M., and Podolsky, R. J. (1966) *J. Physiol.* (*Lond.*) **184,** 511-534.
Clarke, M., and Spudich, J. A. (1974) *J. Mol. Biol.* **86,** 209-222.
Cochran, W., Crick, F. H. C., and Vand, V. (1952) *Acta. Crystalogr.* **5,** 581-586.
Cohen, C. (1975) *Sci. Am.* **233**(5), 35-46.
Cohen, C., and Hanson, J. (1956) *Biochim. Biophys. Acta* **21,** 177.
Cohen, C., and Holmes, K. C. (1963) *J. Mol. Biol.* **6,** 423-432.
Cohen, C., and Longley, N. (1966) *Science* **152,** 794-796.
Cohen, C., and Szent-Gyorgyi, A. G. (1957) *J. Am. Chem. Soc.* **79,** 248.
Cohen, C., and Szent-Gyorgyi, A. G. (1960) *Fourth International Congress of Biochemistry* (*Vienna, 1958*) Vol. 8 (Neurath, H., and Tuppy, H., eds.), pp. 108-118, Pergamon Press, Oxford.

Cohen, C., Lowey, S., and Harrison, R. G., Kendrick-Jones, J., and Szent-Gyorgyi, A. G. (1970) *J. Mol. Biol.* **47,** 605–609.
Cohen, C., Szent-Gyorgyi, A. G., and Kendrick-Jones, J. (1971a) *J. Mol. Biol.* **56,** 223–237.
Cohen, C., Caspar, D. L. D., Parry, D. A. D., and Lucas, R. R. (1971b) *Cold Spring Harbor Symp. Quant. Biol.* **36,** 205–211.
Cohen, C., Caspar, D. L. D., Johnson, J. P., Nauss, K., Margossian, S. S., and Parry, D. A. D. (1972) *Cold Spring Harbor Symp. Quant. Biol.* **37,** 287–297.
Collins, J. H. (1974) *Biochem. Biophys. Res. Commun.* **58,** 301–308.
Collins, J. H., and Elzinga, M. (1975) *J. Biol. Chem.* **250,** 5915–5920.
Collins, J. H., Potter, J. D., Horn, M. J. Wilshire, G., and Jackman, N. (1973) *FEBS Lett.* **36,** 268–272.
Cooke, P. (1975) *J. Cell Biol.* **68,** 539–556.
Cooke, P. H., and Chase, R. H. (1971) *Exp. Cell Res.* **66,** 417–425.
Cooke, R., and Franks, K. E. (1978) *J. Mol. Biol.* **120,** 361–373.
Cork, C., Fehr, D., Hamlin, R., Vernon, W., Xuong, N. H., and Perez-Mendez, V. (1973) *J. Appl. Crystallogr.* **7,** 319–323.
Coulson, C. A. (1961) *Valence, Second Edition,* Oxford University Press, London.
Couteaux, R. (1960) In *The Structure and Function of Muscle,* Vol. 1 (G. H. Bourne, ed.), pp. 337–380, Academic Press, New York.
Craig, R. W. (1975) Ph.D Thesis, University of London, London.
Craig, R. W. (1977) *J. Mol. Biol.* **109,** 69–81.
Craig, R., and Megerman, J. (1977) *Cell Biol.* **75,** 990–996.
Craig, R., and Offer, G. (1976a) *Proc. R. Soc. Lond.* [*Biol.*] **192,** 451–461.
Craig, R., and Offer, G. (1976b) *J. Mol. Biol.* **102,** 325–332.
Crick, F. H. C. (1952) *Nature* **170,** 882–883.
Crick, F. H. C. (1953a) *Acta Crystallogr.* **6,** 685–688.
Crick, F. H. C. (1953b) *Acta. Crystallogr.* **6,** 689–697.
Crowther, R. A., and Amos, L. A. (1971) *J. Mol. Biol.* **60,** 123–130.
Cullen, M. J. (1977) In *Insect Flight Muscle* (Tregear, R. T., ed.), pp. 357–366, North-Holland, Amsterdam.
Cummins, P., and Perry, S. V. (1972) *Biochem. J.* **128,** 106P.
Cummins, P., and Perry, S. V. (1973) *Biochem. J.* **133,** 765–777.
Cummins, P., and Perry, S. V. (1974) *Biochem. J.* **141,** 43–49.
Curtin, N. A., and Davies, R. E. (1973) *Cold Spring Harbor Symp. Quant. Biol.* **37,** 619–626.
Davey, C. L., and Graafhuis, A. E. (1976) *Experientia* **32,** 32–34.
Davey, D. F. (1976) *Aust. J. Exp. Biol. Med. Sci.* **54,** 441–447.
Davies, R. E. (1963) *Nature* **199,** 1068–1074.
Depue, R. H., and Rice, R. V. (1965) *J. Mol. Biol.* **12,** 302–303.
DeRosier, D. J., and Klug, A. (1968) *Nature* **217,** 130–134.
DeRosier, D. J., and Moore, P. B. (1970) *J. Mol. Biol.* **52,** 355–369.
Deshcherevski, V. I. (1968) *Biofizika* **13,** 928–935.
de Villafranca, G. W. (1961) *J. Ultrastruct. Res.* **5,** 109–115.
Devine, C. E., and Somlyo, A. (1971) *J. Cell Biol.* **49,** 636–649.
Dewey, M. M., Walcott, B., Colflesh, D. E., Terry, H., and Levine, R. J. C. (1977) *J. Cell Biol.* **75,** 366–375.
Dos Remedios, C. R., Millikan, R. G. C., and Morales, M. F. (1972) *J. Gen. Physiol.* **59,** 103–120.
Draper, M. H., and Hodge, A. J. (1949) *Aust. J. Exp. Biol. Med. Sci.* **27,** 465–503.
Dubowitz, V., and Brooke, M. (1973) *Muscle Biopsy: A Modern Approach,* Saunders, Philadelphia.
Eaton, B. L., and Pepe, F. A. (1972) *J. Cell Biol.* **55,** 681–695.

Eaton, B. L., and Pepe, F. A. (1974) *J. Mol. Biol.* **82,** 421–423.
Ebashi, S. (1960) *J. Biochem.* **48,** 150–151.
Ebashi, S. (1976) *Annu. Rev. Physiol.* **38,** 293–313.
Ebashi, S. (1980) *Proc. R. Soc. Lond.* [*Biol.*] **207,** 259–286.
Ebashi, S., and Ebashi, F. (1964) *J. Biochem.* **55,** 604–613.
Ebashi, S., and Kodama, A. (1966) *J. Biochem.* **59,** 425–426.
Ebashi, S., and Endo, M. (1968) *Prog. Biophys. Mol. Biol.* **18,** 123–185.
Ebashi, S., Kodama, A., and Ebashi, F. (1968) *J. Biochem.* **64,** 465–467.
Ebashi, S., Endo, M., and Ohtsuki, I. (1969) *Q. Rev. Biophys.* **2,** 384–385.
Ebashi, S., Ohtsuki, I., and Mihashi, K. (1972) *Cold Spring Harbor Symp. Quant. Biol.* **37,** 215–223.
Eccles, J. C., Eccles, R. M., and Lundberg, A. (1968) *J. Physiol.* (*Lond.*) **142,** 275–291.
Edgerton, V. R., and Simpson, D. R. (1971) *Exp. Neurol.* **30,** 374–378.
Edsall, J. T. (1930) *J. Biol. Chem.* **89,** 289–313.
Edström, L., and Kugelberg, E. (1968) *J. Neurol. Neurosurg. Psychiatry* **31,** 424–433.
Eisenberg, E., and Hill, T. L. (1978) *Prog. Biophys. Mol. Biol.* **33,** 55–82.
Eisenberg, E., and Kielley, W. W. (1973) *Cold Spring Harbor Symp. Quant. Biol.* **37,** 145–152.
Eisenberg, E., and Moos, C. (1968) *Biochemistry* **7,** 1486–1489.
Eisenberg, E., and Moos, C. (1970) *J. Biol. Chem.* **245,** 2451–2456.
Eisenberg, E., Dobkin, L., and Kielley, W. W. (1972) *Proc. Natl. Acad. Sci. USA* **69,** 667–671.
Elder, H. Y. (1975) In *Insect Muscle* (Usherwood, P. N. R., ed.), pp. 1–74. Academic Press, New York.
Elliott, A. (1965) *J. Sci. Instrum.* **42,** 312–316.
Elliott, A. (1968) In *Symposium on Fibrous Proteins, Australia* (Crewther, W. G., ed.), pp. 115–123, Butterworths, London.
Elliott, A. (1971) *Phil. Trans. R. Soc. Lond.* **261,** 197–199.
Elliott, A. (1974) *Proc. R. Soc. Lond.* [*Biol.*] **186,** 53–66.
Elliott, A. (1979) *J. Mol. Biol.* **132,** 323–341.
Elliott, A., and Lowy, J. (1970) *J. Mol. Biol.* **53,** 181–203.
Elliott, A., and Malcolm, B. R. (1958) *Proc. R. Soc. Lond.* [*Biol.*] **249,** 30–41.
Elliott, A., and Offer, G. (1978) *J. Mol. Biol.* **123,** 505–519.
Elliott, A., Lowy, J., Parry, D. A. D., and Vibert, P. J. (1968a) *Nature* **218,** 656–659.
Elliott, A., Lowy, J., and Squire, J. M. (1968b) *Nature* **219,** 1224–1226.
Elliott, A., Offer, G., and Burridge, K. (1976) *Proc. R. Soc. Lond.* [*Biol.*] **193,** 43–53.
Elliott, G. F. (1960) Ph.D Thesis, University of London, London.
Elliott, G. F. (1963) *J. Ultrastruct. Res.* **9,** 171–176.
Elliott, G. F. (1964) *Proc. R. Soc. Lond.* [*Biol.*] **160,** 467–472.
Elliott, G. F. (1967) *J. Gen. Physiol.* **50,** 171–184.
Elliott, G. F., and Lowy, J. (1968) *Nature* **219,** 156–157.
Elliott, G. F., Lowy, J., and Worthington, C. R. (1963) *J. Mol. Biol.* **6,** 295–305.
Elliott, G. F., Lowy, J., and Millman, B. M. (1967) *J. Mol. Biol.* **25,** 31–45.
Ellis, R. A. (1965) *J. Cell Biol.* **27,** 551–563.
Elzinga, M., and Collins, J. H. (1972) *Cold Spring Harbor Symp. Quant. Biol.* **37,** 1–7.
Elzinga, M., and Collins, J. H. (1977) *Proc. Natl. Acad. Sci. USA* **74,** 4281–4284.
Englemann, T. W. (1881) *Arch. Ges. Physiol. Mensch. Tiere* **25,** 538.
Fahrenbach, W. H. (1964) *J. Cell Biol.* **22,** 477–481.
Faruqi, A. R. (1975) *J. Phys. Eng.* **8,** 633–635.
Faruqi, A. R., and Huxley, H. E. (1978) *J. Appl. Crystalogr.* **11,** 449–454.
Fawcett, D. W. (1966) *An Atlas of Fine Structure,* Saunders, Philadelphia.
Fawcett, D. W. (1968) *J. Cell Biol.* **36,** 266–270.
Fenn, W. O. (1924) *J. Physiol.* (*Lond.*) **58,** 373–395.

Ford, L. E., Huxley, A. F., and Simmons, R. M. (1974) *J. Physiol. (Lond.)* **240,** 42–43.
Ford, L. E., Huxley, A. F., and Simmons, R. M. (1977) *J. Physiol. (Lond.)* **269,** 441–515.
Foster, G. A. R., Jr. (1930) *J. Text. Inst.* **21,** T18.
Frank, G., and Weeds, A. G. (1974) *Eur. J. Biochem.* **44,** 317–334.
Franks, A. (1955) *Proc. Physiol. Soc. B* **68,** 1054–1064.
Franks, A. (1958) *Br. J. Appl. Physiol.* **9,** 349–352.
Franzini-Armstrong, C. (1970) *J. Cell. Sci.* **6,** 559–592.
Franzini-Armstrong, C. (1973) *J. Cell Biol.* **58,** 630–642.
Franzini-Armstrong, C., and Porter, K. R. (1964) *J. Cell Biol.* **22,** 675–696.
Frasca, J. M., and Parks, V. R. (1965) *J. Cell Biol.* **25,** 157–161.
Fraser, A. B., Eisenberg, E., Keilley, W. W., and Carlson, F. C. (1975) *Biochemistry* **14,** 2207–2214.
Fraser, R. D. B., and MacRae, T. P. (1961) *J. Mol. Biol.* **3,** 640–647.
Fraser, R. D. B., and MacRae, T. P. (1973) *Conformation in Fibrous Proteins and Related Synthetic Polypeptides,* Academic Press, London.
Fraser, R. D. B., and Millward, G. R. (1970) *J. Ultrastruct. Res.* **31,** 203–211.
Fraser, R. D. B., MacRae, T. P., and Miller, A. (1965) *J. Mol. Biol.* **14,** 432–442.
Frearson, N., and Perry, S. V. (1975) *Biochem. J.* **151,** 99–107.
Frearson, N., Focant, B. W. W., and Perry, S. V. (1976a) *FEBS Lett.* **63,** 27–32.
Frearson, N., Solaro, R. J., and Perry, S. V. (1976b) *Nature* **264,** 801–802.
Freundlich, A., and Squire, J. M. (1980) In *Fibrous Proteins: Scientific, Industrial and Medical Aspects,* Vol. II (Parry, D. A. D. and Creamer, L. K., eds.), pp. 11–21, Academic Press, London.
Freundlich, A., Luther, P. K., and Squire, J. M. (1980) *J. Muscle Res. Cell Motil.* **1,** 321–343.
Friedman, M. H. (1970) *J. Ultrastruct. Res.* **32,** 226–236.
Fujime-Higashi, S., and Ooi, T. (1969) *J. Microsc.* **8,** 535.
Greaser, M. L., Yamaguchi, M., Brekke, C., Potter, J., and Gergely, J. (1972) *Cold Spring Harbor Symp. Quant. Biol.* **37,** 235–244.
Greenfeld, N., and Fasman, G. D. (1969) *Biochemistry* **8,** 4108–4116.
Guinier, A. (1963) *X-ray Diffraction,* Freeman, London.
Gauthier, G. F., and Lowey, S. (1977) *J. Cell Biol.* **74,** 760–779.
Gauthier, G. F., Lowey, S., and Hobbs, A. W. (1978) *Nature* **274,** 25–29.
Gazith, J., Himmelfarb, S., and Harrington, W. F. (1970) *J. Biol. Chem.* **245,** 15–22.
Gillis, J. M., and O'Brien, E. J. (1975) *J. Mol. Biol.* **99,** 445–459.
Glauert, A. M. (1975) *Practical Methods in Electron Microscopy,* Vol. 3, Part 1, *Fixation, Dehydration and Embedding of Biological Specimens,* Elsevier/North-Holland, Amsterdam.
Godfrey, J. E., and Harrington, W. F. (1970) *Biochemistry* **9,** 894–908.
Goldstein, M. A., Schroeter, J. P., and Sass, R. L. (1974) *J. Cell Biol.* **63,** 14a.
Goldstein, M. A., Schroeter, J. P., and Sass, R. L. (1977) *J. Cell Biol.* **75,** 818–836.
Gonatas, N. K. (1966) *J. Neuropathol. Exp. Neurol.* **25,** 409–421.
Goody, R. S., and Eckstein, F. (1971) *J. Am. Chem. Soc.* **93,** 6252.
Goody, R. S., Holmes, K. C., Mannherz, H. G., Barrington-Leigh, J., and Rosenbaum, G. (1975) *Biophys. J.* **15,** 687.
Goody, R. S., Barrington-Leigh, J., Mannherz, M. G., Tregear, R. T., and Rosenbaum, G. (1976) *Nature* **262,** 613–614.
Gordon, A. M., Huxley, A. F., and Julian, F. J. (1966) *J. Physiol. (Lond.)* **184,** 170–192.
Greaser, M. L., and Gergely, J. (1971) *J. Biol. Chem.* **246,** 4226–4233.
Greaser, M. L., Yamaguchi, M., Brekke, C., and Gergely, J. (1972) *Cold Spring Harbor Symp. Quant. Biol.* **37,** 235–244.
Hall, C. E. (1956) *Proc. Natl. Acad. Sci. USA* **42,** 801–806.
Hall, C. E., Jakus, M. A., and Schmitt, F. O. (1945) *J. Appl. Physiol.* **16,** 459–465.
Hammond, K. S., and Goll, D. E. (1975) *Biochem. J.* **151,** 189–192.

Hamoir, G. (1969) *Angiologica* **6,** 190–227.
Hamoir, G. (1973) *Phil. Trans. R. Soc. Lond.* [*B*] **265,** 169–181.
Hamoir, G. and Laszt, L. (1962) *Nature* **193,** 682–684.
Hanson, J. (1967) *Nature* **213,** 353–356.
Hanson, J. (1968) In *Muscle* (Ernst, E., and Straub, F. B., eds.) p. 93, Akademiai Kiado, Budapest.
Hanson, J. (1973) *Proc. R. Soc. Lond.* [*Biol.*] **183,** 39–58.
Hanson, J., and Huxley, H. E. (1953) *Nature* **172,** 530–532.
Hanson, J., and Huxley, H. E. (1955) *Symp. Soc. Exp. Biol.* **9,** 228–264.
Hanson, J., and Huxley, H. E. (1957) *Biochim. Biophys. Acta* **23,** 250–260.
Hanson, J., and Lowy, J. (1959) *Proc. Physiol. Soc. Lond.* **149,** 31–32P.
Hanson, J., and Lowy, J. (1961) *Proc. R. Soc. Lond.* [*Biol.*] **154,** 173–196.
Hanson, J., and Lowy, J. (1962) *Proceedings of the Fifth International Congress for Electron Microscopy,* Vol. 2 (Breese, S. S., Jr.), Abstract 09, Academic Press, New York.
Hanson, J., and Lowy, J. (1963) *J. Mol. Biol.* **6,** 46–60.
Hanson, J., and Lowy, J. (1964) *Proc. R. Soc. Lond.* [*Biol.*] **160,** 523–524.
Hanson, J., Lowy, J., Huxley, H. E., Bailey, K., Kay, C. M., and Ruegg, J. C. (1957) *Nature* **180,** 1134–1135.
Hanson, J., O'Brien, E. J., and Bennett, P. M. (1971) *J. Mol. Biol.* **58,** 865–871.
Hanson, J., Lednev, V., O'Brien, E. J., and Bennett, P. M. (1972) *Cold Spring Harbor Symp. Quant. Biol.* **37,** 311–318.
Hardwicke, P. M. D., and Hanson, J. (1971) *J. Mol. Biol.* **59,** 509–516.
Harrington, W. F., and Burke, M. (1972) *Biochemistry* **11,** 1448–1455.
Harrington, W. F., Burke, M., and Barton, J. S. (1972) *Cold Spring Harbor Symp. Quant. Biol.* **37,** 77–85.
Harrison, R. G., Lowey, S., and Cohen, C. (1971) *J. Mol. Biol.* **59,** 531–535.
Hartshorne, D. J., and Dreizen, P. (1972) *Cold Spring Harbor Symp. Quant. Biol.* **37,** 225–234.
Hartshorne, D. J., and Mueller, H. (1967) *J. Biol. Chem.* **242,** 3089–3092.
Hartshorne, D. J., and Mueller, H. (1968) *Biochem. Biophys. Res. Commun.* **31,** 647–653.
Hartshorne, D. J., and Pyun, H. Y. (1971) *Biochim. Biophys. Acta* **229,** 698–711.
Hartt, J. E., Yu, L. C., and Podolsky, R. J. (1977) *Biophys. J.* **17,** 171a.
Hartt, J. E., Yu, L. C., and Podolsky, R. J. (1978) *Biophys. J.* **21,** 87a.
Haselgrove, J. C. (1970) Ph.D Thesis, University of Cambridge, Cambridge.
Haselgrove, J. C. (1972) *Cold Spring Harbor Symp. Quant. Biol.* **37,** 341–352.
Haselgrove, J. C. (1975a) *J. Mol. Biol.* **92,** 113–143.
Haselgrove, J. C. (1975b) In *Comparative Physiology—Functional Aspects of Structural Materials* (Bolis, L., Maddrell, H. P., and Schmidt-Nielsen, K., eds.), pp. 127–138, North-Holland, Amsterdam.
Haselgrove, J. C., and Huxley, H. E. (1973) *J. Mol. Biol.* **77,** 549–568.
Haselgrove, J. C., and Reedy, M. K. (1978) *Biophys. J.* **24,** 713–728.
Haselgrove, J. C., Stewart, M., and Huxley, H. E. (1976) *Nature* **261,** 606–608.
Hasselbach, W. (1953) *Z. Naturforsch.* **8b,** 449–454.
Hasselbach, W. (1964) *Prog. Biophys. Biophys. Chem.* **14,** 169–222.
Hatano, S., and Takahashi, K. (1971) *J. Mechanochem. Cell. Motil.* **1,** 7–14.
Hayat, M. A. (1972) *Basic Electron Microscopy Techniques,* Van Nostrand Reinhold, New York.
Head, J. F., and Perr , S. V. (1974) *Biochem. J.* **137,** 145–154.
Heide, G. (1958) *Proc. 4th Int. Conf. on Electron Microscopy, Berlin,* pp. 87–90.
Heide, G. (1960) *Naturwissenschaften* **14,** 313–317.
Heide, G. (1963) *Z. Angew. Phys.* **15,** 116–128.
Heide, G. (1964) *Z. Angew. Phys.* **17,** 73–75.

Heizmann, C. W., and Eppenberger, H. M. (1978) *J. Biol. Chem.* **253,** 270–277.
Hellam, D. C., and Podolsky, R. J. (1969) *J. Physiol.* (*Lond.*) **200,** 807–819.
Henderson, R., and Unwin, P. N. T. (1975) *Nature* **257,** 28–32.
Herzig, J. W. (1977) In *Insect Flight Muscle* (Tregear, R. T., ed.), pp. 209–220, North-Holland, Amsterdam.
Heumann, H. G. (1970) *Experientia* **26,** 1131–1132.
Hill, A. V. (1938) *Proc. R. Soc. Lond.* [*Biol.*] **126,** 136–195.
Hill, A. V. (1964) *Proc. R. Soc. Lond.* [*Biol.*] **159,** 297–318.
Hill, A. V., and Howarth, J. V. (1959) *Proc. R. Soc. Lond.* [*Biol.*] **151,** 161–193.
Hitchcock, S. E. (1973) *Biochemistry* **12,** 2509–2515.
Hitchcock, S. E. (1975a) *Biochemistry* **14,** 5162–5167.
Hitchcock, S. E. (1975b) *Eur. J. Biochem.* **52,** 255–263.
Hitchcock, S. E., Huxley, H. E., and Szent-Gyorgyi, A. G. (1973) *J. Mol. Biol.* **80,** 825–836.
Hodge, A. J. (1952) *Proc. Natl. Acad. Sci. USA* **38,** 850–855.
Hodge, A. J. (1959) *Rev. Mod. Phys.* **31,** 409–425.
Hodges, R. S., and Smillie, L. B. (1970) *Biochem. Biophys. Res. Commun.* **41,** 987–994.
Hodges, R. S., and Smillie, L. B. (1972a) *Can. J. Biochem.* **50,** 312.
Hodges, R. S., and Smillie, L. B. (1972b) *Can. J. Biochem.* **50,** 330.
Hodges, R. S., Sodek, J., Smillie, L. B., and Jurasek, L. (1972) *Cold Spring Harbor Symp. Quant. Biol.* **37,** 299–310.
Hoh, J. F. Y. (1975) *Biochemistry* **14,** 742–746.
Hoh, J. F. Y., and Yeoh, G. P. S. (1979) *Nature* **280,** 321–322.
Holmes, K. C., Tregear, R. T., and Barrington-Leigh, J. (1980) *Proc. R. Soc. Lond.* [*Biol.*] **207,** 13–33.
Holt, J. C., and Lowey, S. (1977) *Biochemistry* **16,** 4398–4402.
Holtzer, A., Lowey, S., and Schuster, T. M. (1961) In *Molecular Basis of Neoplasia, Symposium on Fundamental Cancer Research, 15th, Houston, Texas,* pp. 259–280, University of Texas Press, Austin, Texas.
Holtzer, A., Clark, R., and Lowey, S. (1965) *Biochemistry* **4,** 2401–2411.
Horowitz, J., Bullard, B., and Mercola, D. (1979) *J. Biol. Chem.* **254,** 350–355.
Hosemann, R., and Bagchi, S. N. (1962) *Direct Analysis of Diffraction by Matter,* North-Holland, Amsterdam.
Hoyle, G. (1964) In *The Contractile Process,* p. 71, Little, Brown, Boston.
Hoyle, G. (1968) *J. Exp. Zool.* **167,** 471–486.
Hoyle, G., McAlear, J. H., and Selverston, A. (1965) *J. Cell Biol.* **26,** 621.
Hoyle, G., and McNeill, P. A. (1968) *J. Exp. Zool.* **167,** 487–522.
Huddart, H. (1975) *The Comparative Structure and Function of Muscle,* Pergamon Press, Oxford.
Huggins, M. L. (1943) *Chem. Rev.* **32,** 195–218.
Huggins, M. L. (1952) *J. Am. Chem. Soc.* **74,** 3963–3964.
Huriaux, F., Pechere, J. F., and Hamoir, G. (1965) *Angiologica* **2,** 15–43.
Huxley, A. F. (1957) *Prog. Biophys.* **7,** 255–313.
Huxley, A. F. (1973) *Proc. Roy. Soc. B* **183,** 83–86.
Huxley, A. F. (1974) *J. Physiol.* (*Lond.*) **243,** 1–43.
Huxley, A. F., and Julian, F. J. (1964) *J. Physiol.* (*Lond.*) **177,** 60–61.
Huxley, A. F., and Niedergerke, R. (1954) *Nature* **173,** 971–972.
Huxley, A. F., and Simmons, R. M. (1971a) *J. Physiol.* (*Lond.*) **218,** 59–60.
Huxley, A. F., and Simmons, R. M. (1971b) *Nature* **233,** 533–538.
Huxley, A. F., and Simmons, R. M. (1972) *Cold Spring Harbor Symp. Quant. Biol.* **37,** 669–680.
Huxley, A. F., and Taylor, R. E. (1955) *Nature* **176,** 106–108.

Huxley, A. F., and Taylor, R. E. (1958) *J. Physiol. (Lond.)* **144,** 426–441.
Huxley, H. E. (1951) *Discuss. Faraday Soc.* **11,** 148–149.
Huxley, H. E. (1952) Ph.D. Thesis, University of Cambridge, Cambridge.
Huxley, H. E. (1953a) *Biochim. Biophys. Acta* **12,** 387–394.
Huxley, H. E. (1953b) *Proc. R. Soc. Lond.* [*Biol.*] **141,** 59–62.
Huxley, H. E. (1956) In *Proceedings of the Stockholm Congress on Electron Microscopy,* pp. 260–261, Almgrist and Wiksell, Stockholm.
Huxley, H. E. (1957) *J. Biophys. Biochem. Cytol.* **3,** 631–648.
Huxley, H. E. (1960) *The Cell,* Vol 4 (Brachet, J., and Mirsky, E. A., eds.), pp. 365–481, Academic Press, New York.
Huxley, H. E. (1961) *Circulation* **24,** 328–335.
Huxley, H. E. (1963) *J. Mol. Biol.* **7,** 281–308.
Huxley, H. E. (1964a) *Nature* **202,** 1067–1071.
Huxley, H. E. (1964b) *Proc. Roy. Soc. Lond. B* **160,** 442–448.
Huxley, H. E. (1967) *J. Gen. Physiol.* **50,** 71–83.
Huxley, H. E. (1968) *J. Mol. Biol.* **37,** 507–520.
Huxley, H. E. (1969) *Science* **164,** 1356–1366.
Huxley, H. E. (1971a) *Proc. R. Soc. Lond.* [*Biol.*] **178,** 131–149.
Huxley, H. E. (1971b) *Biochem. J.* **125,** 85P.
Huxley, H. E. (1972a) *Cold Spring Harbor Symp. Quant. Biol.* **37,** 361–376.
Huxley, H. E. (1972b) In *The Structure and Function of Muscle,* Second Edition, Vol. I, *Structure,* Part I (Bourne, G. H., ed.), pp. 301–387, Academic Press, New York.
Huxley, H. E. (1975a) *5th Int. Biophys. Cong. Copenhagen,* 553.
Huxley, H. E. (1975b) *Acta. Anat. Nippon.* **50,** 310–325.
Huxley, H. E. (1976) In *Cell Motility, Cold Spring Harbor Conf. Cell Prolif.* **3,** 115–126.
Huxley, H. E. (1979a) In *The Molecular Basis of Force Development in Muscle* (Ingels, N. B., ed.), pp. 1–13. Palo Alto Medical Research Foundation, Palo Alto, California.
Huxley, H. E. (1979b) In *Fibrous Proteins: Scientific, Industrial and Medical Aspects,* Vol I (Parry, D. A. D., and Creamer, L. K., eds.), pp. 71–95, Academic Press, London.
Huxley, H. E., and Brown, W. (1967) *J. Mol. Biol.* **30,** 383–434.
Huxley, H. E., and Hanson, J. (1954) *Nature* **173,** 973–976.
Huxley, H. E., and Hanson, J. (1957) *Biochim. Biophys. Acta* **23,** 229–249.
Huxley, H. E., and Zubay, G. (1960) *J. Mol. Biol.* **2,** 10–18.
Huxley, H. E., Faruqi, A. R., Bordas, J., Koch, M. H. J., and Milch, J. R. (1980) *Nature* **284,** 140–143.
Jahromi, S. S., and Atwood, H. L. (1969) *J. Exp. Zool.* **171,** 25–38.
Jasper, H. H., and Pezard, A. (1934) *C. R. Hebd. Seanc. Acad. Sci. Paris* **198,** 449–501.
Jewell, B. R. (1969) *J. Physiol. (Lond.)* **149,** 154–177.
Jewell, B. R., and Ruegg, C. (1966) *Proc. R. Soc. Lond.* [*Biol.*] **164,** 428–459.
Jewell, B. R., and Wilkie, D. R. (1958) *J. Physiol. (Lond.)* **143,** 515–540.
Johnson, P., and Perry, S. V. (1968) *Biochem. J.* **110,** 207–216.
Johnson, P., and Smillie, L. B. (1975) *Biochem. Biophys. Res. Commun.* **64,** 1316–1322.
Johnson, W. H. (1962) *Physiol. Rev. (Suppl.)* **5,** 113–143.
Johnson, W. H., Kahn, J., and Szent-Gyorgyi, A. G. (1959) *Science* **130,** 160–161.
Julian, F. J. (1969) *Biophys. J.* **9,** 547–570.
Julian, F. J., Sollins, K. R., and Sollins, M. R. (1973) *Cold Spring Harb. Symp. Quant. Biol.* **37,** 685–688.
Kabsch, W., Mannherz, H. G., and Suck, D. (1980) *J. Muscle Res. Cell Motil.* **4,** 451.
Kaminer, B., and Bell, A. L. (1966) *J. Mol. Biol.* **20,** 391–401.
Kaminer, B., Szonye, E., and Belcher, C. D. (1976) *J. Mol. Biol.* **100,** 379–386.
Kaminer, M. J. (1969) *J. Mol. Biol.* **39,** 257–264.

Katchburian, A., Burgess, M. C., and Johnson, F. R. (1973) *Experientia* **29,** 1020-1022.
Katsura, I., and Noda, H. (1973) *J. Biochem.* **73,** 245-256.
Katz, A. M., and Hall, E. J. (1963) *Circ. Res.* **13,** 187-198.
Katz, B. (1939) *J. Physiol. (Lond.)* **96,** 45-64.
Katz, B. (1966) *Nerve, Muscle and Synapse,* McGraw-Hill, New York.
Kawamura, M., and Maruyama, K. (1970) *J. Biochem.* **67,** 437-457.
Kelly, D. E. (1967) *J. Cell Biol.* **34,** 827-840.
Kelly, D. E., and Cahill, M. A. (1972) *Anat. Rec.* **172,** 623-642.
Kelly, R. E., and Rice, R. V. (1968) *J. Cell Biol.* **37,** 105-116.
Kendrick-Jones, J. (1974) *Nature* **249,** 631-634.
Kendrick-Jones, J. (1975) *26th Mosbach Colloquium on Molecular Basis of Motility* (Heilmeyer, L., Rüegg, J. C., and Wieland, T., eds.), pp. 122-136, Springer-Verlag, New York.
Kendrick-Jones, J., Lehman, W., and Szent-Gyorgyi, A. G. (1970) *J. Mol. Biol.* **54,** 313-326.
Kendrick-Jones, J., Szent-Gyorgyi, A. G., and Cohen, C. (1971) *J. Mol. Biol.* **59,** 527-529.
Kendrick-Jones, J., Szent-Kiralyi, E. M., and Szent-Gyorgyi, A. G. (1976) *J. Mol. Biol.* **104,** 747-775.
Kielley, W. W., and Harrington, W. F. (1960) *Biochim. Biophys. Acta* **41,** 401-421.
King, M. V., and Young, M. (1972) *J. Mol. Biol.* **65,** 519-523.
Klug, A., and Berger, J. E. (1964) *J. Mol. Biol.* **10,** 565-569.
Klug, A., and DeRosier, D. J. (1966) *Nature* **212,** 29-32.
Klug, A., Crick, F. H. C., and Wyckoff, H. W. (1958) *Acta Crystallogr.* **11,** 199-213.
Knappeis, G. G., and Carlsen, F. (1962) *J. Cell Biol.* **13,** 323-335.
Knappeis, G. G., and Carlsen, F. (1968) *J. Cell Biol.* **38,** 202-211.
Kominz, D. R., Carroll, W. R., Smith, E. N., and Mitchell, E. R. (1959) *Arch. Biochem. Biophys.* **79,** 191-199.
Kretzschmar, K. M., Mendelson, R. A., and Morales, M. F. (1976) *Biophys. J.* **16,** 126a.
Kretzschmar, K. M., Mendelson, R. A., and Morales, M. F. (1978) *Biochemistry* **17,** 2314-2318.
Kugelberg, E. (1973) In *New Developments in Electromyography and Clinical Neurophysiology,* Vol. 1 (Desmedt, J. E., ed.), pp. 2-13, Karger, Basel.
Kühne, W. (1864) *Untersuchungen über das Protoplasma und die Contractilität,* Engelmann, Leipzig.
Lamvik, M. K. (1978) *J. Mol. Biol.* **122,** 55-68.
Landon, M. F., and Oriol, C. (1975) *Biochem. Biophys. Res. Commun.* **62,** 241-245.
Lane, B. P. (1965) *J. Cell Biol.* **27,** 199-213.
Lazarides, E. (1980) *Nature* **283,** 249-255.
Leadbetter, L., and Perry, S. V. (1963) *Biochem. J.* **87,** 233-238.
Lednev, V. V. (1974) *Biofizika* **19,** 116-121.
Lehman, W. (1977a) In *Insect Flight Muscle* (Tregear, R. T., ed.), pp. 277-284, North-Holland, Amsterdam.
Lehman, W. (1977b) *Biochem. J.* **163,** 291-296.
Lehman, W., and Szent-Gyorgyi, A. G. (1975) *J. Gen. Physiol.* **59,** 375-387.
Lehman, W., Kendrick-Jones, J., and Szent-Gyorgyi, A. G. (1972) *Cold Spring Harbor Symp Quant. Biol.* **37,** 319-330.
Lehman, W., Bullard, B., and Hammond, K. (1974) *J. Gen. Physiol.* **63,** 553-563.
Lehninger, A. L. (1975) *Biochemistry,* Second Edition, Worth Publishers.
Levin, A., and Wyman, J. (1927) *Proc. R. Soc. Lond. B* **101,** 218-243.
Levine, R. J. C., Elfvin, M., Dewey, M. M., and Walcott, B. (1976) *J. Cell Biol.* **71,** 273-279.
Lipson, H. S., and Taylor, C. A. (1958) *Fourier Transforms and X-Ray Diffraction,* Bell, London.

Lipson, H. S. (1972) *Optical Transforms,* Academic Press, London.
Locker, R. H., and Daines, G. J. (1979) In *Fibrous Proteins: Scientific, Industrial and Medical Aspects,* Vol. II (Parry, D. A. D., and Creamer, L. K., eds.), pp. 43-55, Academic Press, New York.
Locker, R. H. and Leet, N. G. (1975) *J. Ultrastruct. Res.* **52,** 64-75.
Locker, R. H. and Leet, N. G. (1976) *J. Ultrastruct. Res.* **55,** 157-172.
Locker, R. H., and Schmitt, F. O. (1957) *J. Biophys. Biochem. Cytol.* **3,** 889-896.
Locker, R. H., Daines, G. J., and Leet, N. G. (1976) *J. Ultrastruct. Res.* **55,** 173-181.
Longley, W. (1975) *J. Mol. Biol.* **93,** 111-115.
Low, B. W., and Edsall, J. T. (1968) In *Currents in Biochemical Research* (D. E. Green, ed.), p. 398, Wiley-Interscience, New York.
Lowey, S. (1971) In *Subunits in Biological Systems, Part A* (Timasheff, S. N., and Fasman, G. D. eds.), pp. 201-259, Marcek Dekker, New York.
Lowey, S. (1979) In *Fibrous Proteins: Scientific, Industrial and Medical Aspects* (Parry, D. A. D., and Creamer, L. K., eds.), pp. 1-25. Academic Press, New York.
Lowey, S., and Cohen, C. (1962) *J. Mol. Biol.* **4,** 293-308.
Lowey, S., and Margossian, S. S. (1974) *J. Mechanochem. Cell. Motil.* **2,** 241-248.
Lowey, S., and Risby, D. (1971) *Nature* **234,** 81-85.
Lowey, S., Kucera, J., and Holzer, A. (1963) *J. Mol. Biol.* **7,** 234-244.
Lowey, S., Goldstein, L., and Luck, S. (1966) *Biochemistry* **345,** 248-254.
Lowey, S., Goldstein, L., Cohen, C., and Luck, M. (1967) *J. Mol. Biol.* **23,** 287-304.
Lowey, S., Slayter, H. S., Weeds, A., and Baker, H. (1969) *J. Mol. Biol.* **42,** 1-29.
Lowey, S., Silberstein, L., Gauthier, G. F., and Holt, J. C. (1979) In *Motility in Cell Function,* Academic Press, New York.
Lowy, J. (1972) *Bull. Zool.* **39,** 119-138.
Lowy, J., and Hanson, J. (1962) *Physiol. Rev.* (*Suppl.*) **5,** 34-42.
Lowy, J., and Millman, B. M. (1959) *J. Physiol.* (*Lond.*) **149,** 68p.
Lowy, J., and Millman, B. M. (1963) *Phil. Trans. R. Soc. Lond.* [*Biol.*] **246,** 105-148.
Lowy, J., and Mulvany, M. J. (1973) *Acta Physiol. Scand.* **88,** 123-136.
Lowy, J., and Small, J. V. (1970) *Nature* **227,** 46-51.
Lowy, J., and Vibert, P. J. (1967) *Nature* **215,** 1254-1255.
Lowy, J., Millman, B. M., and Hanson, J. (1964) *Proc. R. Soc. Lond.* [*Biol.*] **160,** 525-536.
Lowy, J., Poulsen, F. R., and Vibert, P. J. (1970) *Nature* **225,** 1053-1054.
Lowy, J., Vibert, P. J., Haselgrove, J. C., and Poulsen, F. R. (1973) *Phil. Trans. R. Soc. Lond.* [*Biol.*] **B265,** 191-196.
Luft, J. M. (1961) *J. Biophys. Biochem. Cytol.* **9,** 409-414.
Luft, J. H., and Wood, R. L. (1963) *J. Cell Biol.* **19,** 46A.
Luther, P. K. (1978) Ph.D. Thesis. University of London, London.
Luther, P. K., and Squire, J. M. (1978) *J. Mol. Biol.* **125,** 313-324.
Luther, P. K., and Squire, J. M. (1980) *J. Mol. Biol.* **141,** 409-439.
Luther, P. K., Munro, P. M. G., and Squire, J. M. (1981) *J. Mol. Biol.* (submitted).
Lymn, R. W. (1978) *Biophys. J.* **21,** 93-98.
Lymn, R. W. (1979) *Annu. Rev. Biophys. Bioeng.* **8,** 145-163.
Lymn, R. W., and Cohen, G. H. (1975) *Nature* **258,** 770-772.
Lymn, R. W., and Huxley, H. E. (1972) *Cold Spring Harbor Symp. Quant. Biol.* **37,** 449-453.
Lymn, R. W., and Taylor, E. W. (1970) *Biochemistry* **9,** 2975-2983.
Lymn, R. W., and Taylor, E. W. (1971) *Biochemistry* **10,** 4617-4624.
MacArthur, I. (1943) *Nature* **152,** 38-41.
MacDonald, R. D., and Engel, A. G. (1971) *Cell. Biol.* **48,** 431-437.
Machin, K. E., and Pringle, J. W. S. (1959) *Proc. R. Soc. Lond.* [*Biol.*] **151,** 204-225.

Machin, K. E., and Pringle, J. W. S. (1960) *Proc. R. Soc. Lond.* [*Biol.*] **152,** 311-330.
Mackean, D. G. (1962) *Introduction to Biology,* John Murray, London.
Maeda, Y. (1978) Ph.D. Thesis, Nagoya University, Nagoya.
Maeda, Y. (1979) *Nature* **277,** 670-672.
Mannherz, H. G., Barrington-Leigh, J., Holmes, K. C., and Rosenbaum, G. (1973) *Nature* [*New Biol.*] **241,** 226-229.
Margossian, S. S., and Cohen, C. (1973) *J. Mol. Biol.* **81,** 409-413.
Margossian, S. S., and Lowey, S. (1973) *J. Mol. Biol.* **74,** 301-311.
Markham, R., Frey, S., and Hills, G. J. (1963) *Virology* **20,** 88-102.
Markham, R., Hitchborn, J. H., Hills, G. J., and Frey, S. (1964) *Virology* **22,** 342-359.
Marston, S. B. (1973) *Biochim. Biophys. Acta* **305,** 397-412.
Marston, S. B. (1977) In *Insect Flight Muscle* (Tregear, R. T., ed.), pp. 293-305, North-Holland, Amsterdam.
Marston, S. B. and Tregear, R. T. (1972) *Nature* **235,** 23-24.
Marston, S. B., and Tregear, R. T. (1974) *Biochim. Biophys. Acta* **333,** 581-584.
Marston, S. B., Rodger, C. D., and Tregear, R. T. (1976) *J. Mol. Biol.* **104,** 263-276.
Marston, S. B., Tregear, R. T., Rodger, C. D., and Clark, M. L. (1979) *J. Mol. Biol.* **128,** 111-126.
Martonosi, A. (1962) *J. Biol. Chem.* **237,** 2795-2803.
Martonosi, A. (1968) *J. Biol. Chem.* **243,** 71-81.
Masaki, T., and Takaiti, O. (1972) *J. Biochem.* **71,** 355-380.
Masaki, T., and Takaiti, O. (1974) *J. Biochem.* **75,** 367-380.
Mashima, H., and Yoshida, T. (1965) *Jpn. J. Physiol.* **15,** 463-477.
Matsubara, I., and Elliott, G. F. (1972) *J. Mol. Biol.* **72,** 657-669.
Matsubara, I., and Yagi, N. (1978) *J. Physiol.* (*Lond.*) **278,** 297-307.
Matsubara, I., Yagi, N., and Hashizume, H. (1975) *Nature* **255,** 728-729.
Matsubara, I., Suga, H., and Yagi, N. (1977a) *J. Physiol.* (*Lond.*) **270,** 311-320.
Matsubara, I., Kamiyama, A., and Suga, H. (1977b) *J. Mol. Biol.* **110,** 1-7.
Maunder, C. A., Dubowitz, V., Hall, T. A., and Yarom, R. (1976) In *Electron Microscopy 1976. Proceedings of the 6th European Congress on Electron Microscopy, Jerusalem, Israel, 1976,* Vol. II, *Biological Sciences* pp. 229-231, Tal International, Israel.
McClare, C. W. F. (1972) *J. Theor. Biol.* **35,** 569-595.
McLachlan, A. D., and Stewart, M. (1975) *J. Mol. Biol.* **98,** 293-304.
McLachlan, A. D., and Stewart, M. (1976) *J. Mol. Biol.* **103,** 271-298.
McLachlan, A. D., Stewart, M., and Smillie, L. B. (1975) *J. Mol. Biol.* **98,** 281-291.
McNeill, P. A., and Hoyle, G. (1967) *Ann. Zool.* **7,** 483-493.
Mendelson, R. A., and Cheung, P. (1976) *Biophys.* **16,** 126a.
Meldelson, R. A., and Kretzschmar, K. M. (1979) *Biophys. J.* **25,** 20a.
Mendelson, R. A., Morales, M. F., and Botts, J. (1973) *Biochemistry* **12,** 2250-2255.
Mercola, D., Bullard, B., and Priest, J. (1975) *Nature* **254,** 634-635.
Mihalyi, E., and Harrington, W. F. (1959) *Biochim. Biophys. Acta* **36,** 447-466.
Mikawa, T., Nonomura, Y., and Ebashi, S. (1977) *J. Biochem.* **82,** 1789-1791.
Mikawa, T., Hirata, M., Saida, K., Nonomura, Y., Ebashi, S., and Kakicuhi, S. (1978) Abstracts 6th International Congress of Biophysics, Kyoto, p. 233.
Miller, A. (1975) In *Muscle Contraction, M.T.P. Int. Rev. Sci.: Biochemistry,* Ser. 1, Vol. 12 (Blashkco, H., ed.), pp. 137-168, Butterworths, London.
Miller, A., and Tregear, R. T. (1970) *Nature* **226,** 1060-1061.
Miller, A., and Tregear, R. T. (1971) In *Symposium on Contractility* (Podolsky, R. J., ed.), pp. 205-228, Prentice Hall, Englewood Cliffs.
Miller, A., and Tregear, R. T. (1972) *J. Mol. Biol.* **70,** 85-104.
Millman, B. M. (1963) Ph.D. Thesis, London University, London.

Millman, B. M., and Bennett, P. M. (1976) *J. Mol. Biol.* **103,** 439–467.
Mitsui, T. (1978) *Adv. Biophys.* **10,** 97–135.
Moffitt, W., and Yang, J. T. (1956) *Proc. Natl. Acad. Sci. USA* **42,** 596–603.
Moore, P. B., and DeRosier, D. J. (1970) *J. Mol. Biol.* **50,** 293–295.
Moore, P. B., Huxley, H. E., and DeRosier, D. J. (1970) *J. Mol. Biol.* **50,** 279–292.
Moos, C. (1972) *Cold Spring Harbor Symp. Quant. Biol.* **37,** 93–95.
Moos, C., Offer, G., Starr, R., and Bennett, P. (1975) *J. Mol. Biol.* **97,** 1–9.
Morgan, M., Perry, S. V., and Ottaway, J. (1976) *Biochem. J.* **157,** 687–697.
Morimoto, K., and Harrington, W. F. (1972) *J. Biol. Chem.* **247,** 3052–3061.
Morimoto, K., and Harrington, W. F. (1973) *J. Mol. Biol.* **77,** 165–175.
Morimoto, K., and Harrington, W. F. (1974) *J. Mol. Biol.* **83,** 83–97.
Mueller, H. (1966) *Biochem. Z.* **345,** 300–321.
Mühlrad, A., Corsi, A., and Granata, A. L. (1968) *Biochim. Biophys. Acta* **162,** 435–443.
Murphy, A. J. (1971) *Biochemistry* **11,** 2622–2627.
Murray, A. C., and Kay, C. M. (1972) *Biochemistry* **11,** 2622–2627.
Nagy, B. (1966) *Biochim. Biophys. Acta* **115,** 498–500.
Nagy, B., and Jencks, W. P. (1962) *Biochemistry* **1,** 987–996.
Nakamura, A., Sreter, F., and Gergely, J. (1971) *J. Cell Biol.* **49,** 883–898.
Namba, K., Wakabayashi, K., and Mitsui, T. (1978) In *Diffraction Studies of Biomembranes and Muscles and Synchrotron Radiation* (Mitsui, T., ed.), pp. 177–216, The Taniguchi Foundation, Taniguchi, Japan.
Needham, D. M. (1950) *Biochim. Biophys. Acta* **4,** 42–49.
Needham, D. M., and Shoenberg, C. (1964) *Proc. R. Soc. Lond.* [*Biol.*] **160,** pp. 517–522.
Needham, D. M., and Shoenberg, C. F. (1968) In *Handbook of Physiology,* Section B, Vol IV (Code, C. F., and Heidel, W., eds.), pp. 1793–1810, American Physiological Society, Washington.
Needham, D. M., and Williams, J. M. (1963a) *Biochem. J.* **89,** 534–545.
Needham, D. M., and Williams, J. M. (1963b) *Biochem. J.* **89,** 546–551.
Needham, D. M., and Williams, J. M. (1963c) *Biochem. J.* **89,** 552–561.
Newman, J., and Carlson, F. D. (1980) *Biophys. J.* **29,** 37–48.
Noelken, M. E. (1962) Ph.D. Thesis, Washington University.
Nonomura, Y. (1968) *J. Cell Biol.* **39,** 741–745.
Nonomura, Y., Drabikowski, W., and Ebashi, S. (1968) *J. Biochem.* (*Tokyo*) **64,** 419–422.
O'Brien, E. J., and Couch, J. (1976) *Proceedings 6th European Congress on Electron Microscopy, Israel,* Vol. II, pp. 153–155, Tal International, Israel.
O'Brien, E. J., Bennett, P. M., and Hanson, J. (1971) *Phil. Trans. R. Soc. Lond.* [*Biol.*] **261,** 201–208.
O'Brien, E. J., Gillis, J. M., and Couch, J. (1975) *J. Mol. Biol.* **99,** 461–475.
Offer, G. (1972) *Cold Spring Harbor Symp. Quant. Biol.* **37,** 87–96.
Offer, G. (1974) In *Companion to Biochemistry* (Bull, A. T., Lagnado, J. R., Thomas, J. O., and Tipton K. F., eds.), pp. 623–671, Longmans, London.
Offer, G. (1976) *Proc. R. Soc. Lond.* [*Biol.*] **74,** 653–676.
Offer, G., and Eliott, A. (1978) *Nature* **271,** 325–329.
Offer, G., Moos, C., and Starr, R. (1973) *J. Mol. Biol.* **74,** 653–676.
Ohtsuki, I. (1974) *J. Biochem.* (*Tokyo*) **75,** 753–765.
Olander, J. (1971) *Biochemistry* **10,** 601–609.
Olander, J., Emerson, M. F., and Holtzer, A. (1967) *J. Am. Chem. Soc.* **89,** 3058–3059.
Ooi, T., and Fujime-Higashi, S. (1971) *Adv. Biophys.* **2,** 113–153.
Osborne, M. P. (1967) *J. Insect Physiol.* **13,** 1471–1482.
Page, S. (1964) *Proc. R. Soc. Lond.* [*Biol.*] **160,** 460–466.
Page, S. (1968) *J. Physiol.* (*Lond.*) **197,** 709–715.

Page, S., and Huxley, H. E. (1963) *J. Cell Biol.* **19,** 369–390.
Panner, B. J., and Honig, C. R. (1970) *J. Cell Biol.* **44,** 52–61.
Parry, D. A. D. (1969) *J. Theor. Biol.* **24,** 73–84.
Parry, D. A. D. (1970) *J. Theor. Biol.* **26,** 429–435.
Parry, D. A. D. (1974) *Biochem. Biophys. Res. Commun.* **57,** 216–224.
Parry, D. A. D. (1975a) *J. Mol. Biol.* **98,** 519–535.
Parry, D. A. D. (1975b) *Nature* **256,** 346–347.
Parry, D. A. D. (1976) *Biochem. Biophys. Res. Commun.* **68,** 323–328.
Parry, D. A. D., and Elliott, A. (1967) *J. Mol. Biol.* **25,** 1–13.
Parry, D. A. D., and Squire, J. M. (1973) *J. Mol. Biol.* **75,** 33–55.
Parry, D. A. D., and Suzuki, E. (1969) *Biopolymers* **7,** 189–197.
Pauling, L., and Corey, R. B. (1953) *Nature* **171,** 59–61.
Pauling, L., Corey, R. B., and Branson, H. R. (1951) *Proc. Natl. Acad. Sci. USA* **37,** 205–211.
Peachey, L. D. (1958) *Biophys. Biochem. Cytol.* **4,** 233–242.
Peachey, L. D. (1965) *J. Cell Biol.* **25** (pt. 2), 209–231.
Pearlstone, J. R., Carpenter, M. R., Johnson, P., and Smillie, L. B. (1976) *Proc. Natl. Acad. Sci. USA* **73,** 1902–1906.
Pepe, F. A. (1966) *J. Cell Biol.* **28,** 505–525.
Pepe, F. A. (1967a) *J. Mol. Biol.* **27,** 203–225.
Pepe, F. A. (1967b) *J. Mol. Biol.* **27,** 227–236.
Pepe, F. A. (1971) *Prog. Biophys.* **19,** 75–96.
Pepe, F. A. (1972) *Cold Spring Harbor Symp. Quant. Biol.* **37,** 97–108.
Pepe, F. A. (1975) *J. Histochem. Cytochem.* **23,** 543–562.
Pepe, F. A., and Dowben, P. (1977) *J. Mol. Biol.* **113,** 119–218.
Pepe, F. A., and Drucker, B. (1972) *J. Cell Biol.* **52,** 255–260.
Pepe, F. A., and Drucker, B. (1975) *J. Mol. Biol.* **99,** 609–617.
Pepe, F. A., and Drucker, B. (1979) *J. Mol. Biol.* **130,** 379–393.
Pepe, F. A., Chowrashi, P. K., and Wachsberger, P. (1975) *Comparative Physiology—Functional Aspects of Structural Materials* (Bolis, L., Madchell, H. P., and Schmidt-Nielsen, K., eds.), pp. 105–120, North-Holland, Amsterdam.
Perrie, W. T., and Perry, J. V. (1970) *Biochem. J.* **119,** 31–38.
Perrie, W. T., Smillie, L. B., and Perry, J. V. (1973) *Biochem. J.* **135,** 151–164.
Perry, S. V. (1967) *Prog. Biophys.* **17,** 325–381.
Perry, S. V. (1955) In *Methods in Enzymology,* Vol. 2 (Colowick, S. P., and Kaplan, N. O., eds.), pp. 582–588, Academic Press, New York.
Perry, S. V., and Corsi, A. (1958) *Biochem. J.* **68,** 5–12.
Perry, S. V., and Grey, T. C. (1956) *Biochem. J.* **64,** 184–192.
Perry, S. V., Cole, H. A., Head, J. F., and Wilson, F. J. (1972) *Cold Spring Harbor Symp. Quant. Biol.* **37,** 251–262.
Perry, S. V., Cole, H. A., Morgan, M., Moir, A. J. G., and Pires, E. (1975) *FEBS Lett.* **31,** 163–176.
Perutz, M. F. (1951) *Nature* **167,** 1053–1054.
Philips, G. N., Jr., Lattman, E. E., Cummins, P., Lee, K. Y., and Cohen, C. (1979) *Nature* **278,** 413–417.
Philpott, D. E., and Szent-Gyorgyi, A. G. (1954) *Biochim. Biophys. Acta* **15,** 165–173.
Pires, E., Thomas, S. V., and Thomas, M. A. (1974) *FEBS Lett.* **41,** 292–296.
Podlubnaya, Z. A., Kalamkarova, M. B., and Nankina, V. P. (1969) *J. Mol. Biol.* **46,** 591–592.
Podolsky, R. J. (1959) *Ann. N.Y. Acad. Sci.* **72,** 522–537.
Podolsky, R. J. (1960) *Nature* **188,** 666–668.

Podolsky, R. J., and Nolan, A. C. (1972) In *Contractility of Muscle Cells and Related Processes* (Podolsky, R. J., ed.) pp. 247-260, Prentice Hall, Englewood Cliffs, N.J.
Podolsky, R. J., Nolan, A. C., and Zaveller, S. A. (1969) *Proc. Natl. Acad. Sci. USA* **64,** 504-511.
Podolsky, R. J., St. Onge, R., Yu, L., and Lymn, R. W. (1976) *Proc. Natl. Acad. Sci. USA* **73,** 813-817.
Pollard, T. D., and Weihing, R. R. (1974) *CRC Crit. Rev. Biochem.* **2,** 1-65.
Porter, K. R., and Palade, G. E. (1957) *J. Biophys. Biochem. Cytol.* **3,** 269-300.
Potter, J. D. (1974) *Arch. Biochem. Biophys.* **162,** 436-44.
Potter, J. D., and Gergely, J. (1974) *Biochemistry* **13,** 2697-270.
Potter, J. D., and Gergely, J. (1975) *J. Biol. Chem.* **250,** 4628-4633.
Pringle, J. W. S. (1949) *J. Physiol. (Lond.)* **108,** 226-232.
Pringle, J. W. S. (1967) *Prog. Biophys. Mol. Biol.* **17,** 1-60.
Pringle, J. W. S. (1968) In *Aspects of Cell Motility* (Miller, P. L., ed.), pp. 67-86, Cambridge University Press, Cambridge.
Pringle, J. W. S. (1972) In *Structure and Function of Muscle* (Bourne G. H., ed.), pp. 491-541, Academic Press, New York.
Pringle, J. W. S. (1974) *Symp. Biol. Hung.* **17,** 67-78.
Pringle, J. W. S. (1977a) In *Insect Flight Muscle* (Tregear, R. T., ed.), pp. 177-196, North-Holland, Amsterdam.
Pringle, J. W. S. (1977b) In *Insect Flight Muscle* (Tregear, R. T., ed.), pp. 337-344, North-Holland, Amsterdam.
Pringle, J. W. S. (1978) *Proc. R. Soc. Lond.* [*Biol.*] **201,** 107-130.
Pybus, J. H., and Tregear, R. T. (1972) *Cold Spring Harbor Symp. Quant. Biol.* **37,** 655-660.
Ramsey, R. W., and Street, S. F. (1940) *J. Cell Comp. Physiol.* **15,** 11-34.
Rayns, D. G. (1972) *J. Ultrastruct. Res.* **40,** 103-121.
Rayns, D. G., DeVine, C. E., and Sutherland, C. L. (1975) *J. Ultrastruct. Res.* **50,** 306-321.
Reedy, M. K. (1964) *Proc. R. Soc. Lond.* [*Biol.*] **160,** 433-542.
Reedy, M. K. (1965) *J. Cell Biol.* **26,** 309-311.
Reedy, M. K. (1967) *Am. Zool.* **7,** 465-481.
Reedy, M. K. (1968) *J. Mol. Biol.* **31,** 155-176.
Reedy, M. K. (1971) In *Symposium on Contractility* (Podolsky, R. J., ed.) pp. 229-246, Prentice Hall, Englewood Cliffs.
Reedy, M. K. (1976) *Biophys. J.* **16,** 126a.
Reedy, M. K., and Barkas, A. E. (1974) *J. Cell Biol.* **63,** 282a.
Reedy, M. K., and Garrett, W. E., Jr. (1977) In *Insect Flight Muscle* (Tregear, R. T., ed.), pp. 115-135, North-Holland, Amsterdam.
Reedy, M., Holmes, K. C., and Tregear, R. T. (1965) *Nature* **207,** 1276-1280.
Reedy, M. K., Bahr, F. G., and Fischmann, D. A. (1972) *Cold Spring Harbor Symp. Quant. Biol.* **37,** 397-422.
Rees, M. K., and Young, M. (1967) *J. Biol. Chem.* **242,** 4449-4458.
Reger, J. (1964) *J. Ultrastruct. Res.* **10,** 48-57.
Reger, J. (1967) *J. Ultrastruct. Res.* **20,** 72-82.
Reid, N. (1975) *Practical Methods in Electron Microscopy,* Vol. 3, Part II: *Ultramicrotomy* (Glauert, A. M., ed.), North-Holland, Amsterdam.
Reisler, E., Burke, M., Josephs, R., and Harrington, W. F. (1973) *J. Mechanochem. Cell. Motil.* **2,** 163-179.
Reynolds, E. W. (1963) *J. Cell Biol.* **17,** 208-212.
Rice, R. V. (1961a) *Biochim. Biophys. Acta* **52,** 602-604.
Rice, R. V. (1961b) *Biochim. Biophys. Acta* **53,** 29-43.

Rice, R. V. (1964) In *Biochemistry of Muscle Contraction* (Gergely, J., ed.), pp. 41-51 Little, Brown, Boston.
Rice, R. V., Brady, A. C., DePue, R. H., and Kelly, R. E. (1966) *Biochemistry* **345,** 370-394.
Rice, R. V., Moses, J. A., McManus, G. M., Brady, A. D., and Blasik, L. M. (1970) *J. Cell Biol.* **47,** 183-196.
Rice, R. V., McManus, G. M., DeVine, C. E., and Somlyo, A. P. (1971) *Nature* **231,** 242-243.
Robertson, J. D., Bodenheimer, T. S., and Stage, D. E. (1963) *J. Cell Biol.* **19,** 159-199.
Robison, W. G., and Lipton, B. H. (1969) *J. Cell Biol.* **43,** 117a.
Rodger, C. (1973) D. Phil. Thesis, Oxford University, Oxford.
Rodger, C. D., and Tregear, R. (1974) *J. Mol. Biol.* **86,** 495-497.
Rome, E. (1972a) *Cold Spring Harbor Symp. Quant. Biol.* **37,** 331-339.
Rome, E. (1972b) *J. Mol. Biol.* **65,** 331-345.
Rome, E., Offer, G., and Pepe, F. A. (1973a) *Nature* [*New Biol.*] **244,** 152-154.
Rome, E., Hirabayashi, T., and Perry, S. V. (1973b) *Nature* [*New Biol.*] **244,** 154-155.
Rosenbluth, J. (1965a) *J. Cell Biol.* **25,** 495-515.
Rosenbluth, J. (1965b) *J. Cell Biol.* **26,** 579-591.
Rosenbluth, J. (1967) *J. Cell Biol.* **34,** 15-33.
Rosenbluth, J. (1968) *J. Cell Biol.* **36,** 245-259.
Rosenbluth, J. (1972) *J. Cell Biol.* **54,** 566-579.
Rowe, A. J. (1964) *Proc. R. Soc. Lond.* [*Biol.*] **160,** 437-441.
Rowe, R. W. D. (1973) *J. Cell Biol.* **57,** 261-277.
Rudall, K. M. (1956) In *Lectures on the Scientific Basis of Medicine,* Vol. 5, pp. 217-230, British Postgraduate Medical Federation, University of London, London.
Rüegg, J. C. (1958) *Biochem. J.* **69,** 46P.
Rüegg, J. C. (1961) *Proc. R. Soc. Lond.* [*Biol.*] **154,** 224-249.
Saide, J. D., and Ullrick, W. C. (1973) *J. Mol. Biol.* **79,** 329-337.
Salmons, S., and Sreter, F. (1976) *Nature* **263,** 30-34.
Sanger, J. W. (1971) *Cytobiology* **4,** 450-466.
Sanger, J. W., and Szent-Gyorgyi, A. G. (1964) *Biol. Bull.* **127,** 391.
Sarkar, S. (1972) *Cold Spring Harbor Symp. Quant. Biol.* **37,** 14-17.
Schiaffino, S., Hanzlikova, V., and Pierobon, S. (1970) *J. Cell Biol.* **47,** 107-119.
Schollmeyer, J. E., Goll, D. E., Robson, R. M., and Stromer, M. H. (1973) *J. Cell Biol.* **59,** 306.
Segal, D. M., Himmelfarb, S., and Harrington, W. F. (1967) *J. Biol. Chem.* **242,** 1241-1252.
Selby, C. C., and Bear, R. S. (1956) *J. Biophys. Biochem. Cytol.* **2,** 71-85.
Sender, P. M. (1971) *FEBS Lett.* **17,** 106-110.
Seymour, J., and O'Brien, E. J. (1980) *Nature* **283,** 680-682.
Shafiq, S. A., Dubowitz, V., Peterson, H. De L., and Milhorat, A. T. (1967) *Brain* **90,** 817-828,
Shoenberg, C. F., and Haselgrove, J. C. (1974) *Nature* **249,** 152-154.
Shoenberg, C. F., Ruegg, J. C., Needham, D. M., Schirmer, R. H., and Nemetchek-Gansle, H. (1966) *Biochem. Z.* **345,** 255-266.
Shoenberg, M. (1979) *Biophys. J.* **25,** 114a.
Shoenberg, M., Wells, J. B., and Podolsky, R. J. (1974) *J. Gen. Physiol.* **64,** 623-642.
Shy, G. M., Engel, W. K., Somers, J. E., and Wanko, T. (1963) *Brain* **86,** 793-810.
Simmons, R. M., and Hill, T. L. (1976) *Nature* **263,** 615-618.
Simmons, R. M., and Szent-Gyorgyi, A. G. (1978) *Nature* **273,** 62-64.
Sjöström, M., and Squire, J. M. (1977a) *J. Mol. Biol.* **109,** 49-68.
Sjöström, M., and Squire, J. M. (1977b) *J. Microsc.* (*Oxf.*) **111,** 239-278.
Sjöström, M., and Thornell, L. E. (1975) *J. Microsc.* (*Oxf.*) **103,** 101-112.

Sjöström, M., Edman, A. C., Thornell, L. E., Schiaffino, S., and Squire, J. M. (1981) (in preparation).
Slayter, H. S., and Lowey, S. (1967) *Proc. Natl. Acad. Sci. USA* **58,** 1611-1618.
Small, J. V. (1968) In *Proceedings of the 4th European Regional Conference on Electron Microscopy* (Bocciarelli, D. S., ed.), pp. 609-610, Tipografia Poliglotta Vaticana, Rome.
Small, J. V. (1974) *Nature* **249,** 324-327.
Small, J. V. (1977a) *J. Cell Sci.* **24,** 327-349.
Small, J. V. (1977b) In *Biochemistry of Smooth Muscle* (Stephens, N. L., ed.), pp. 379-411, University Park Press, Baltimore.
Small, J. V., and Sobieszek, A. (1972) *Cold Spring Harbor Symp. Quant. Biol.* **37,** 439-442.
Small, J. V., and Sobieszek, A. (1977) *J. Cell Sci.* **23,** 243-268.
Small, J. V., and Squire, J. M. (1972) *J. Mol. Biol.* **67,** 117 149.
Small, J. V., Lowy, J., and Squire, J. M. (1971) *Proceedings of the 1st European Biophysics Congress, Vienna,* Vol. 5 (Broda, E., Locker, A., and Springer-Lederer, H.), pp. 419-423, Verlag der Wiener Medizinischen Akadamie, Vienna.
Sobieszek, A. (1972) *J. Mol. Biol.* **70,** 741-744.
Sobieszek, A. (1973) *J. Ultrastruct. Res.* **43,** 313-343.
Sobieszek, A. (1977) In *The Biochemistry of Smooth Muscle* (Stephens, N. L., eds), pp. 413-443, University Park Press, Baltimore.
Sobieszek, A., and Bremel, R. D. (1975) *Eur. J. Biochem.* **55,** 49-60.
Sobieszek, A., and Small, J. V. (1972) *Cold Spring Harbor Symp. Quant. Biol.* **37,** 109-111.
Sobieszek, A., and Small, J. V. (1973) *Phil. Trans. R. Soc. Lond.* [*Biol.*] **265,** 203-212.
Sobieszek, A., and Small, J. V. (1976) *J. Mol. Biol.* **102,** 75-92.
Sodek, J., Hodges, R. S., Smillie, L. B., and Jurasek, L. (1972) *Proc. Natl. Acad. Sci. USA* **69,** 3800-3804.
Somlyo, A. P., and Somlyo, A. V. (1970) *Pharmacol. Rev.* **22,** 249-353.
Somlyo, A. P., Somlyo, A. V., Devine, C. E., and Rice, R. V. (1971) *Nature* **231,** 243-246.
Somlyo, A. P., Devine, C. E., Somlyo, A. V., and Rice, R. V. (1973) *Phil. Trans. R. Soc. Lond.* [*Biol.*] **265,** 223-229.
Somlyo, A. V. (1979) *Cell Biol.* **80,** 743-750.
Spencer, M. (1969) *Nature* **223,** 1361-1362.
Spencer, M., and Worthington, C. R. (1960) *Nature* **187,** 388-391.
Spudich, J. A., and Watt, S. (1971) *J. Biochem. Chem.* **246,** 4866-4871.
Spudich, J. A., Huxley, H. E., and Finch, J. T. (1972) *J. Mol. Biol.* **72,** 619-632.
Squire, J. M. (1969) Ph.D. Thesis, University of London, London.
Squire, J. M. (1971) *Nature* **233,** 457-462.
Squire, J. M. (1972) *J. Mol. Biol.* **72,** 125-138.
Squire, J. M. (1973) *J. Mol. Biol.* **77,** 291-323.
Squire, J. M. (1974) *J. Mol. Biol.* **90,** 153-160.
Squire, J. M. (1975) *Annu. Rev. Biophys. Bioeng.* **4,** 137-163.
Squire, J. M. (1977) In *Insect Flight Muscle* (Tregear, R. T., ed.) pp. 91-114, North-Holland, Amsterdam.
Squire, J. M. (1979) In *Fibrous Proteins Scientific Industrial and Medical Aspects,* Vol. I (Parry, D. A. D., and Creamer, L. K., eds.), pp. 27-70, Academic Press, New York.
Squire, J. M. (1980) *J. Muscle Res. Cell Motil.* **1,** 450.
Squire, J. M., and Elliott, A. (1969) *Mol. Cryst. Liquid Cryst.* **7,** 457-468.
Squire, J. M., and Elliott, A. (1972) *J. Mol. Biol.* **65,** 291-321.
Squire, J. M., Sjöström, M., and Luther, P. K. (1976) *Proceedings of the Sixth European Congress on Electron Microscopy, Jerusalem,* pp. 91-95, Tal International, Israel.
Squire, J. M., Harford, J. J., and Sjöström, M. (1981) *J. Mol. Biol.* (in press).

Sreter, F. A., Romanul, F. C. A., Salmons, S., and Gergely, J. (1974) In *Exploratory Concepts in Muscular Dystrophy* (Milhaat, A. T., ed.), International Congress Series No. 333, pp. 338–343, Experta Medica, Amsterdam.

Starling, E. H., and Lovatt-Evans, C. (1962) In *Principles of Human Physiology,* 13th Edition (Davson, H., and Eggleton, M. G., eds.), Churchill, London.

Starr, R., and Offer, G. (1971) *FEBS Lett.* **15,** 40–44.

Starr, R., and Offer, G. (1978) *Biochem. J.* **171,** 813–816.

Steiger, G. J. (1977) In *Insect Flight Muscle* (Tregear, R. T., ed.), pp. 221–268, North-Holland, Amsterdam.

Stewart, M. (1975) *Proc. R. Soc. Lond.* [*Biol.*] **190,** 257–266.

Stewart, M., and McLachlan, A. D. (1975) *Nature* **257,** 331–333.

Stewart, M., and McLachlan, A. D. (1976) *J. Mol. Biol.* **103,** 251–269.

Stone, D., Sodek, J., Johnson, P. and Smillie, L. B. (1974) *FEBS Lett.* **31,** 125–136.

Straub, F. B. (1943) *Stud. Inst. Med. Chem. Univ. Szeged* **2,** 3–15.

Stromer, M. H., Tabatabai, L. B., Robson, R. M., Goll, D. E., and Zeece, M. G. (1976) *Exp. Neurol.* **50,** 402–421.

Sutoh, K., and Harrington, W. F. (1977) *Biochemistry* **16,** 2441–2449.

Suzuki, A., Goll, D. E., Singh, I., Allen, R., and Stromer, M. H. (1976) *J. Biol. Chem.* **251,** 6860–6870.

Szent-Gyorgyi, A. (1942) *Stud. Inst. Med. Chem. Univ. Szeged* **1,** 17.

Szent-Gyorgyi, A. (1943) *Stud. Inst. Med. Chem. Univ. Szeged* **3,** 76.

Szent-Gyorgyi, A. (1949) *Biol. Bull.* **96,** 140–161.

Szent-Gyorgyi, A. (1951) *Chemistry of Muscular Contraction,* 2nd revised edition, p. 10, Academic Press, New York.

Szent-Gyorgyi, A. (1953a) *Chemical Physiology of Contraction in Body and Heart Muscle,* Academic Press, New York.

Szent-Gyorgyi, A. (1953b) *Arch. Biochem. Biophys.* **42,** 305–320.

Szent-Gyorgyi, A. G. (1951) *J. Biol. Chem.* **192,** 361–369.

Szent-Gyorgyi, A. G. (1975) *Biophys. J.* **15,** 707–723.

Szent-Gyorgyi, A. G., and Cohen, C. (1957) *Science* **126,** 697–698.

Szent-Gyorgyi, A. G., Mazia, D., and Szent-Gyorgyi, A. (1955) *Biochim. Biophys. Acta* **16,** 339–342.

Szent-Gyorgyi, A. G., Cohen, C., and Philpott, D. E. (1960) *J. Mol. Biol.* **2,** 133–142.

Szent-Gyorgyi, A. G., Cohen, C., and Kendrick-Jones, J. (1971) *J. Mol. Biol.* **56,** 239–258.

Szent-Gyorgyi, A. G., Szentkiralyi, E. M., and Kendrick-Jones, J. (1973) *J. Mol. Biol.* **74,** 179–203.

Taylor, C. A., and Lipson, H. (1964) *Optical Transforms,* Bell, London.

Taylor, E. W. (1977) *Biochemistry* **16,** 732–740.

Taylor, E. W. (1979) *CRC Crit. Rev. Biochem.* **6,** 103–164.

Thomas, D. D., and Cooke, R. (1980) *Biophys. J.* **32,** 891–906.

Thomas, D. D., Ishiwata, S., Seidel, J. C., and Gergely, J. (1980) *Biophys. J.* **32,** 873–889.

Thornell, L. E. (1973) *J. Mol. Cell. Cardiol.* **5,** 409–417.

Thorson, J. W., and White, D. C. S. (1969) *Biophys. J. 9,* 360–390.

Tonomura, Y., Yagi, K., Kubo, S., and Kitagawa, S. (1961) *J. Res. Inst. Catalysis* (*Hokkaido*) **9,** 256–286.

Toselli, P. A., and Pepe, F. A. (1968) *J. Cell Biol.* **37,** 445–461.

Totsuka, T., and Hatano, S. (1970) *Biochim. Biophys. Acta* **223,** 189–197.

Tregear, R. T. (1973) In *Insect Muscle* (Usherwood, P. N. R., ed.), pp. 357–400, Academic Press, New York.

Tregear, R. T. (1977) In *Insect Flight Muscle* (Tregear, R. T., ed.). pp. 137–146, North-Holland, Amsterdam.

Tregear, R. T., and Mendelson, R. A. (1975) *Biophys. J.* **15,** 455–467.
Tregear, R. T., and Miller, A. (1969) *Nature* **222,** 1184–1185.
Tregear, R. T., and Squire, J. M. (1973) *J. Mol. Biol.* **77,** 279–290.
Trentham, D. R., Bardsley, R. G., Eccleston, J. F., and Weeds, A. G. (1972) *Biochem. J.* **126,** 635–644.
Trentham, D. R., Eccleston, J. F., and Bagshaw, C. R. (1976) *Q. Rev. Biophys.* **9,** 217–281.
Trinick, J. A. (1973) Ph.D. Thesis, University of Leicester, Leicester.
Trinick, J. A., and Elliott, A. (1979) *J. Mol. Biol.* **131,** 133–136.
Trinick, J. A., and Lowey, S. (1977) *J. Mol. Biol.* **113,** 343–368.
Trinick, J. A., and Offer, G. (1979) *J. Mol. Biol.* **133,** 549–556.
Trombitas, K., and Tigyi-Sebes, A. (1979) *Nature* **281,** 319–320.
Trus, B. L., and Elzinga, M. (1980) *Structural Aspects of Recognition and Assembly in Biological Macromolecules, 7th Annual Katzir-Katchalsky Conference* (in press).
Tsao, T. C., Bailey, K., and Adair, G. S. (1951) *Biochem. J.* **49,** 27–36.
Tsao, T-C., Kung, T-H., Peng, C-M., Chang, Y-S., and Tsou, Y-S. (1965) *Sci. Sin.* **14,** 91–105.
Tsuboi, K. K. (1968) *Biochim. Biophys. Acta* **160,** 420–434.
Turner, D. C., Wallimann, T., and Eppenberger, H. M. (1973) *Proc. Natl. Acad. Sci. USA* **70,** 702–705.
Ullrick, W. C. (1967) *J. Theor. Biol.* **15,** 53–60.
Ullrick, W. C., Toselli, P. A., Saide, J. D., and Phear, W. P. C. (1977) *J. Mol. Biol.* **115,** 61–74.
Usherwood, P. N. R. (1975) *Insect Muscle,* Academic Press, New York.
Vibert, P. J. (1968) Ph.D. Thesis, London University, London.
Vibert, P. J., Lowy, J., Haselgrove, J. C., and Poulsen, F. R. (1971) *First European Biophysics Congress, Vienna,* Vol. 5 (Broda, E., Locker, A., and Springer-Lederer, H., eds.), pp. 409–413, Verlag der Wiener Medizinischen Akademie, Vienna.
Vibert, P. J., Haselgrove, J. C., Lowy, J., and Poulsen, F. R. (1972) *J. Mol. Biol.* **71,** 757–767.
Vibert, P. J., Szent-Gyorgyi, A. G., Craig, R., Wray, J., and Cohen, C. (1978) *Nature* **273,** 64–66.
Wachsberger, P., and Pepe, F. A. (1974) *J. Mol. Biol.* **88,** 385–391.
Wagner, P. D., and Weeds, A. G. (1977) *J. Mol. Biol.* **109,** 455–473.
Walker, I. D., and Stewart, M. (1975) *FEBS Lett.* **58,** 16–18.
Wakabayashi, T., Huxley, H. E., Amos, L. A., and Klug, A. (1975) *J. Mol. Biol.* **93,** 477–497.
Walliman, T., Turner, D. C., and Eppenberger, H. M. (1977) *J. Cell Biol.* **75,** 297–317.
Walliman, T., Pelloni, G., Turner, D. C., and Eppenberger, H. M. (1978) *Proc. Nat. Acad. Sci. U.S.A.* **75,** 4296–4300.
Warren, R. C., and Hicks, R. M. (1971) *J. Ultrastruct. Res.* **36,** 861–874.
Weber, A. (1959) *J. Biol. Chem.* **234,** 2764–2769.
Weber, A. (1975) In *Functional Linkage in Biomolecular Systems* (Schmitt, O., Schneider, D. M., and Crothers, D. M., eds.), pp. 312–318, Raven Press, New York.
Weber, A., and Herz, R. (1963) *J. Biol. Chem.* **238,** 599–605.
Weber, A., and Murray, J. (1973) *Physiol. Rev.* **53,** 612–673.
Weber, A., and Winicur, S. (1961) *J. Biol. Chem.* **236,** 3198–3202.
Weber, E. (1846) *Muskelbewegung, Handworterbuch der Physiologie,* p. 1.
Weber, H. H. (1956) In *Verhandlungen der Deutschen orthopädischen Gessellschaft, 44th Congress,* pp. 13–27, Enke, Stuttgart.
Weber, H. H. (1958) In *The Motility of Muscle and Cells,* pp. 33–36, Harvard University Press, Cambridge, Massachusetts.
Weeds, A. G. (1969) *Nature* **223,** 1362–1364.

Weeds, A. G., and Lowey, S. (1971) *J. Mol. Biol.* **61,** 701-725.
Weeds, A. G., and Pope, B. (1971) *Nature* **234,** 85-88.
Weeds, A. G., and Pope, B. (1977) *J. Mol. Biol.* **111,** 129-157.
Weeds, A. G., and Taylor, R. S. (1975) *Nature* **257,** 54-56.
Weeds, A. G., Trentham, D. R., Kean, C. J. C., and Buller, A. J. (1974) *Nature* **247,** 135-139.
Weeds, A. G., Hall, R., and Spurway, N. C. (1975) *FEBS Lett.* **49,** 320-324.
Weihing, R. R., and Korn, E. D. (1972) *Biochemistry* **11,** 1538-1543.
Weisel, J. W. (1975) *J. Mol. Biol.* **98,** 675-681.
Weisel, J. W., and Szent-Gyorgyi, A. G. (1975) *J. Mol. Biol.* **98,** 665-673.
Werber, M. M., Gaffin, S. L., and Oplatka, A. (1972) *J. Mechanochem. Cell. Motil.* **1,** 91-95.
White, D. C. S. (1967) D. Phil. Thesis, University of Oxford, Oxford.
White, D. C. S. (1972) *Cold Spring Harbor Symp. Quant. Biol.* **37,** 201-213.
White, D. C. S., and Thorson, J. (1973) *Prog. Biophys. Mol. Biol.* **27,** 175-255.
White, H. D. (1977) *Nature* **267,** 754-755.
White, H. D., and Taylor, E. W. (1976) *Biochemistry* **15,** 5818-5826.
Wilkie, D. R. (1968) *J. Physiol.* (*Lond.*) **195,** 157-183.
Wilkie, D. R. (1976) In *Molecular Basis of Motility* (Heilmeyer, L. M. C., Jr., Ruegg, J. C., and Wicland, T. L., eds.), pp. 69-78, Springer-Verlag, New York.
Wilkinson, J. M., and Grand, R. J. A. (1975) *Nature* **271,** 31-35.
Wilkinson, J. M., Perry, S. V., Cole, H. A., and Trayer, I. P. (1972) *Biochem. J.* **127,** 215-228.
Winkelman, L., and Bullard, B. (1977) In *Insect Flight Muscle* (Tregear, R. T., ed.), pp. 285-290, North-Holland, Amsterdam.
Woods, E. F. (1967) *J. Biol. Chem.* **242,** 2859-2871.
Woods, E. F. (1969) *Biochemistry* **8,** 4336-4344.
Woolfson, M. M. (1970) In *An Introduction to X-ray Crystallography* Cambridge University Press, Cambridge.
Worcester, D. L., Gillis, J. M., O'Brien E. J., and Ibel, K. (1975) *Brookhaven Symp. Biol.* **27,** 101-114.
Worthington, C. R. (1959) *J. Mol. Biol.* **1,** 398-401.
Wray, J. S. (1979a) *Nature* **277,** 37-40.
Wray, J. S. (1979b) *Nature* **280,** 325-326.
Wray, J. S., Vibert, P. J., and Cohen, C. (1974) *J. Mol. Biol.* **88,** 343-348.
Wray, J. S., Vibert, P. J., and Cohen, C. (1975) *Nature* **257,** 561-564.
Wray, J. S., Vibert, P. J., and Cohen, C. (1978) *J. Mol. Biol.* **124,** 501-521.
Wu, J-Y., and Yang, J. T. (1970) *J. Biol. Chem.* **245,** 212-218.
Yagi, N., and Matsubara, I. (1977) *J. Mol. Biol.* **117,** 797-803.
Yagi, N., Ito, M. H., Nakajima, H., Izumi, T., and Matsubara, I. (1977) *Science* **197,** 685-687.
Yamaguchi, M., Greaser, M. L., and Cassens, R. (1974) *J. Ultrastruct. Res.* **48,** 33-58.
Yamamoto, K., Yanagida, M., Kawamura, M., Maruyama, K., and Noda, H. (1975) *J. Mol. Biol.* **91,** 463-469.
Yang, J. T., and Wu, C-S.C. (1977) *Biochemistry* **16,** 5785-5789.
Young, M. (1967) *Proc. Natl. Acad. Sci. USA* **58,** 2393-2400.
Young, M., Blanchard, M. H., and Brown, D. (1968) *Proc. Natl. Acad. Sci. USA* **61,** 1087-1094.
Young, M., King, M. V., O'Hara, D. S., and Molberg, P. J. (1972) *Cold Spring Harbor Symp. Quant. Biol.* **37,** 65-76.
Yount, R. G., Babcock, D., Ballantyne, W., and Ojala, D. (1971a) *Biochemistry* **10,** 2484-2489.

Yount, R. G., Ojala, D., and Babcock, D. (1971b) *Biochemistry* **10,** 2490–2496.
Yu, L. C., Douben, R. M., and Kornacher, K. (1970) *Proc. Natl. Acad. Sci. USA* **66,** 1199–1205.
Yu, L. C., Lymn, R. W., and Podolsky, R. J. (1977) *J. Mol. Biol.* **115,** 455–464.
Yu, L. C., Hartt, J. E., and Podolsky, R. J. (1979) *J. Mol. Biol.* **132,** 53–68.
Zobel, C. R., and Carlson, F. R. (1963) *J. Mol. Biol.* **7,** 78–89.

Suggested Further Readings

Techniques

Diffraction and Image Analysis

Beeston, B. E. P., Horne, R. W., and Markham, R. (1973) *Practical Methods in Electron Microscopy,* Vol. 1, Part II, *Electron Diffraction and Optical Diffraction Techniques,* North-Holland, Amsterdam.

Blundell, T. L., and Johnson, L. N. (1976) *Protein Crystallography,* Academic Press, New York.

Fraser, R. D. B., and MacRae, T. P. (1973) *Conformation in Fibrous Proteins,* Academic Press, New York.

Holmes, K. C., and Blow, D. (1965) *The Use of X-Ray Diffraction in the Study of Protein and Nucleic Acid Structure,* Interscience, New York.

Lipson, H. S. (1970) *Crystals and X Rays,* Wykeham Science Publications Ltd., London.

Lipson, H. S., and Taylor, C. A. (1958) *Fourier Transforms and X-Ray Diffraction,* Bell and Sons Ltd., London.

Misell, D. L. (1978) *Practical Methods in Electron Microscopy,* Vol. 7, *Image Analysis, Enhancement and Interpretation,* North-Holland, Amsterdam.

Taylor, C. A., and Lisbon, H. S. (1964) *Optical Transforms,* Bell and Sons Ltd., London.

Electron Microscopy

Agar, A. W., Alderson, R. H., and Chescoe, D. (1974) *Practical Methods in Electron Microscopy,* Vol. 2, *Principles and Practice of Electron Microscope Operation,* North-Holland, Amsterdam.

Glauert, A. M. (1975) *Practical Methods in Electron Microscopy,* Vol. 2, Part I, *Fixation, Dehydration and Embedding of Biological Specimens,* North-Holland, Amsterdam.

Hayat, M. A. (1972) *Basic Electron Microscopy Techniques,* Van Nostrand Reinhold, New York.

Lewis, P. R., and Knight, D. P. (1977) *Practical Methods in Electron Microscopy,* Vol. 5, Part I, *Staining Methods of Sectioned Material,* North-Holland, Amsterdam.

Reid, N. (1975) *Practical Methods in Electron Microscopy,* Vol. 3, Part II, *Ultramicrotomy,* North-Holland, Amsterdam.
Weakley, B. S. (1972) *A Beginners Handbook in Electron Microscopy,* Churchill Livingstone, London.

Texts on Muscle

Introductory Books and Reviews

Aidley, D. J. (1971) *The Physiology of Excitable Cells,* Cambridge University Press, London.
Bendall, J. R. (1969) *Muscles, Molecules and Movement,* American Elsevier, New York.
Buller, A. J. (1975) *The Contractile Behaviour of Mammalian Skeletal Muscle,* Oxford University Press, London.
Carlson, F. D., and Wilkie, D. R. F. (1975) *Muscle Physiology,* Prentice Hall, Englewood Cliffs.
Cohen, C. (1975) *Sci. Am.* **233** (November), 36–45.
Hill, A. V. (1965) *Trails and Trials in Physiology,* Edward Arnold, London.
Huxley, H. E. (1969) *Science* **164,** 1356-1366.
Huxley, H. E. (1973) *Nature* **243,** 445–449.
Katz, B. (1966) *Nerve, Muscle and Synapse,* McGraw-Hill, New York.
Murray, J. M., and Weber, A. (1974) *Sci. Am.* **230** (February), 69-71.
Offer, G. (1974) In *Companion to Biochemistry* (Bull, A. T., Lagnado, J. R., Thomas, J. O., and Tipton, K. F., eds.), pp. 623–671, Longmans, London.
Wilkie, D. R. (1976) *Muscle,* Second Edition, Edward Arnold, London.

Advanced Review Articles and Texts

Alexander, R. M. (1968) *Animal Mechanics,* Sidgewick and Jackson, London.
Bourne, G. H. (1972) *The Structure and Function of Muscle, Second Edition,* Vols. 1–4, Academic Press, New York.
Cold Spring Harbor Symposium on Quantitative Biology (1973) *The Mechanism of Muscle Contraction,* Cold Spring Harbor Laboratory, New York.
Ebashi, S., and Endo, M. (1968) *Prog. Biophys. Mol. Biol.* **18,** 123-183.
Fraser, R. D. B., and MacRae, T. P. (1973) *Conformation in Fibrous Proteins,* Academic Press, New York.
Goldman, R., Pollard, T., and Rosenbaum, J. (1976) *Cell Motility,* Cold Spring Harbor Laboratory, New York.
Hanson, E. J. (1968) *Q. Rev. Biophys.* **1,** 177-216.
Hill, T. L. (1974) *Prog. Biophys. Mol. Biol.* **28,** 267-340.
Hill, T. L. (1975) *Prog. Biophys. Mol. Biol.* **29,** 105-160.
Huddart, H. (1975) *The Comparative Structure and Function of Muscle,* Pergamon Press, New York.
Huxley, A. F. (1974) *J. Physiol. (Lond.)* **243,** 1–43.
Ingels, N. B. (1979) *The Molecular Basis of Force Development in Muscle,* Palo Alto Medical Research Foundation, Palo Alto, California.
Needham, D. (1971) *Machina Carnis,* Cambridge University Press, London.
Parry, D. A. D., and Creamer, L. K. (1980) *Fibrous Proteins: Scientific, Industrial and Medical Aspects,* Academic Press, New York.

Pepe, F. A., Sanger, J. W., and Nachmias, V. T. (1979) *Motility in Cell Function: Proceedings of the First John Marshall Symposium on Cell Biology,* Academic Press, New York.

Squire, J. M. (1975) *Annu. Rev. Biophys. Bioeng.* **4,** 137–163.

Taylor, E. W. (1972) *Annu. Rev. Biochem.* **41,** 577–616.

Taylor, E. W. (1973) In *Current Topics in Bioenergetics,* Vol. 5, pp. 201–231, Academic Press, New York.

Taylor, E. W. (1979) *CRC Crit. Rev. Biochem.* **6,** 103–164.

Tregear, R. T. (1977) *Insect Flight Muscle,* Elsevier/North-Holland, Amsterdam.

Uehara, Y., Campbell, C. R., and Burnstock, G. (1976) *Muscle and Its Innervation, An Atlas of Fine Structure,* Edward Arnold, London.

Usherwood, P. N. R. (1975) *Insect Muscle,* Academic Press, New York.

Weber, A., and Murray, J. (1973) *Physiol. Rev.* **53,** 612–673.

White, D. C. S., and Thorson, J. (1975) *The Kinetics of Muscle Contraction,* Pergamon Press, New York.

Woledge, R. (1971) *Prog. Biophys.* **22,** 37–74.

Index